Atlas of the Breeding Birds of Humboldt County, California

John E. Hunter
David Fix
Gregory A. Schmidt
Jude Claire Power

A Publication of Redwood Region Audubon Society

ISBN 0-9760380-0-5

Library of Congress Control Number: 2004116154

Published by Redwood Region Audubon Society
P. O. Box 1054
Eureka, CA 95502-1054 U.S.A

Printed in the United States of America by Reischling Press, Inc., Seattle, Washington

Printed on recycled, acid-free paper.

Cover design and illustration by Gary A. Bloomfield

TABLE OF CONTENTS

FOREWORD

What you hold in your hands represents the first systematic effort to map the distribution of the birds that nest in Humboldt County. Fieldwork by both birders and professional biologists began in spring 1995 and concluded in fall 1999. The Atlas Steering Committee then tackled the monumental task of weaving together historical records, crates of data forms, distribution maps, published and "gray" literature, original illustrations, and comments from scientific reviewers and copy editors to create this finished product. On behalf of the birding community, I congratulate the many participants who contributed to this decade-long citizen science effort.

The local artists whose work graces these pages and the community members who commissioned drawings of their favorite breeding birds deserve a special thank you. These skillful renderings of more than 60 species add much to the publication's appeal.

Humboldt County's geographic isolation has kept it relatively unspoiled. Development here will never match the scale common in many other portions of the Golden State. Yet, the pace of life behind the Redwood Curtain has accelerated as well. Development here is inevitable. Establishing a baseline of breeding bird distribution is critical to understanding how these populations change in response to alteration or loss of habitat. Many people have chosen to live in Humboldt in reaction to having witnessed the over development occurring elsewhere in California. As a result, I think it's fair to characterize the "Humboldt ethos" as one that acknowledges the natural beauty of this unique area and gives voice to a fervent desire to maintain open space and a high quality of life. I hope that this document will be widely used to aid in directing prudent, well-planned growth in sync with this ethos.

As naturalist Joseph Wood Krutch said, "There is all the difference in the world between looking at something and living with it." For the past 10 years, those associated with the Atlas have truly lived with Humboldt. Atlasers slogged through wet meadows beneath its highest peaks to delineate the breeding grounds of Lincoln's Sparrows. They hiked the idle railroad tracks through the Eel River Canyon, the point of discovery for Rufous-crowned Sparrows. They continued westward until the Pacific Ocean blocked their path, searching the sand dunes of Clam Beach for Snowy Plovers. They boarded a boat, bound for waters about the offshore rocks, trying to unlock the nesting secrets of seabirds such as the Rhinoceros Auklet.

I'd like to say several words about the value of the descriptive writing style that permeates the species accounts. Look around you: the natural world is being diminished daily. It needs all the allies it can get. Many of us are keenly aware of the graying of our birding constituency and how critical it is to expose young minds to the vibrant and fascinating world of birds. Today, many distractions and temptations serve as barriers to experiencing the natural world. The written word *can* compete with cell phones, the internet, television, and video games *if* it is honest, takes chances, and – in this case – successfully evokes the sounds, sights, and smells of a mid-summer morning atlasing expedition. In my opinion, the Humboldt Atlas succeeds on all counts.

Those of us who consider ourselves defenders of wild places and the creatures that dwell there have reached a strategy crossroads. It seems that, as a culture, we are uncomfortable with scientists acting as advocates. We value the input of these experts, but feel that to be unbiased, their statements should be presented dispassionately, as disembodied third-person "tech-speak." The problem with this "just the facts, ma'am" delivery is that it fails to convey *why* we should care about diminished habitats or their denizens. This is where the humanities come in. Writers, artists, photographers, and poets are adept at conveying the wonder and mystery of birds. I'm encouraged by the burgeoning body of literature that explores the cross-pollination between art and science. I consider the *Atlas of the Breeding Birds of Humboldt County, California* to be a significant addition to this genre.

Tom W. Leskiw
Atlaser and former Steering Committee member
Eureka, California
February 2005

To my parents, James and Cynthia Hunter, for providing me with the opportunity to indulge myself in my own personal interests (including birds), and to my daughter Liberty Hunter, for helping me gain perspective on those indulgences.
John E. Hunter

To my father, Bob Fix, who showed me my first wild bird nest, and to Steve Herman, who showed me that the collection of data and the appreciation of beauty can coexist.
David Fix

To my wife Kristin, one of the original Atlas steering committee members, who was instrumental in launching the Atlas project and in obtaining grants that helped us complete our fieldwork, and to our sons Sam and Carlo–the reason she left the Atlas project. And to my brother Scott, who spurred my early interest in nature by allowing me to tag along on his outdoor excursions, then later guided me toward a formal education in wildlife biology.
Gregory A. Schmidt

To the birds that so gracefully adorn my life, and to my father, Jack Power, who encouraged my interest in the natural world and gave me my first binoculars.
Jude Claire Power

ACKNOWLEDGMENTS

This project was conducted in part under the auspices of the Redwood Region Audubon Society (RRAS), which greatly benefited the project by providing seed money, field volunteers and insurance for them, administrative support, and funding for various aspects of manuscript preparation and printing. Most notably, they provided funds to one of the coauthors, David Fix, to expedite the writing of draft species accounts. We appreciate the support of numerous members of the RRAS Board of Directors including Jan Andersen, Frank Anderson, Louise Bacon-Ogden, Gary Bloomfield, Jim Clark, Hal Genger, Luke George, Pablo Herrera, Robert Hewitt, Tom W. Leskiw, Chet Ogan, Jude Claire Power, C. J. Ralph, Gregory Schmidt, Kristin Schmidt, and Shana Stearn. Elaine David and Frances Madrone each provided administrative support as Treasurers of RRAS. Six Rivers National Forest was the primary partner to the Audubon Society, providing maps, supplies, administrative and computer support, and person-hours for Atlas fieldwork. Without their support, fieldwork would not have been completed within the 5-year study period. Specific Six Rivers National Forest projects from which data were drawn for the Atlas included the Marbled Murrelet Range and Distribution Study (Hunter *et al.* 1998, Schmidt *et al.* 2000) and administrative surveys for Northern Goshawk, Spotted Owl, and other species.

A few individuals who served as Atlas Steering Committee members contributed in a substantial way to the completion of the Atlas and deserve special recognition. Robert Hewitt first suggested a Humboldt atlas, and enthusiastically assisted in fieldwork, fund raising, and other aspects of the Atlas. Rob's business, LBJ Enterprises, also contributed data tidbits from many of their far-flung projects and assisted in fund raising. Kristin Schmidt participated in early planning efforts as well as fieldwork and provided strong support, most importantly as our liaison to the Six Rivers National Forest, where she was instrumental in obtaining grants from the National Fish and Wildlife Foundation (NFWF). A real field workhorse, Tom W. Leskiw not only went about the routine work of systematically gathering data in blocks and recording casual observations wherever they were encountered, but he also organized and conducted numerous special trips to assess rare and local breeders, and was a stalwart on "block-buster" outings and training sessions. Tom also helped coordinate access and assisted with a variety of other Atlas matters.

Atlas planning was greatly facilitated by input from atlasing veterans from other California counties including Betty Burridge (Sonoma), Barbara Carlson (Riverside and San Bernardino), Robin Leong (Napa), Tim Manolis (Sacramento), Don Roberson and Chris Tenney (Monterey), and David Shuford (Marin). Don Roberson also provided training to Humboldt atlasers early in the project—he and Rita Carratello even completed a few blocks while in town!

In the spirit of openness and mutual benefit, two local timber companies welcomed and facilitated our survey efforts: Pacific Lumber Company and Simpson Timber Company. We greatly appreciate land access and data provided by Sal Chinnici and Dan Dill (representing Pacific Lumber Company) and Lowell Diller, Catherine Hibbard and Thomas MacDonald (representing Simpson Timber Company). Both companies rolled out the red carpet despite our pesky calls about gate keys, maps, *etc.*, and often sent an employee along to show us that "special spot" or to ensure our safety in a working forest landscape. Further, both companies shared freely large amounts of data collected during their scientific and administrative studies, and in the process contributed to a collaborative atmosphere among local biologists. Our coverage of Humboldt County would have been greatly deficient without the earnest participation of these two companies. Nancy Karracker at Louisiana Pacific also provided assistance to atlasers.

Since Humboldt County is largely (66%) privately owned, a field effort such as this would not have been possible without the cooperation of numerous small private landowners. Due to fear and loathing associated with local environmental issues (mostly timber and salmon), the sanctity of private property rights, the reality of marijuana cultivation, and concerns for legal liability, gaining access to private lands was especially difficult. An AmeriCorps volunteer with the Six Rivers National Forest, Holly Taylor, conducted a huge mail campaign to elicit landowner permission for blocks with no known access. George Frey at the Six Rivers National Forest and Misha Schwarz assisted with land ownership research. Many of these landowners were strident and colorful with their negative replies, but the following individuals and families graciously allowed us to look for breeding birds on their land. For the record, none of the following persons who gave us access were robbed, sued, or lost their property rights as a result of environmentalist property takings following their participation in the Atlas project:

John W. Annable
Ronald & Mimi Aronson
Virginia Becker
Gary & Patricia Belli
Richard Bolin
Larry Bowermaster
Ernest D. Boynton
Joy Call
Linda Candiotti
Ken Cordry

Susan & Earl Corp
Lawrence & Eileen Crain
Wendy R. Crisp
Richard & Lori Cutler
Jan & Linda Derksen
Pat Dorn
Levia Downing
Marion Drew
Thomas K. Evans
Elizabeth M. Filmer
Jonell Friedkin
Cindy Glover
Joe & Blanche Grinsell
Michael A. Guerriero
Wayne E. Henley
George Hindley
Thomas & Helene Keter
Robert E. Lemieux
Albert Lewis
Charlie Liphart
Eugene B. Lucas
Scott MacDougall
John Mahony
Jon & Suzanne McBride
Andy McBride
Mark & Grace Megazzi
Frank & Naomi Melewicz
Lindsay Merryman
Michael Meyer
John Mitchell
Jack & Jonelle Monschke
Harold Mulvey
Michael Myers
Stephen Nesvold
Dezh Pagen
John & Cheryl Palmer
Jerold & Bettie Phelps
Todd Phelps
Joseph & Wileeta Philbrick
Robert & Diane Propster
Douglas & Jean Quittenton
Allan & Helen Ridley
Lauri Rose
Gerald D. Rush
Terry E. Russ
Ron & Orinda Samuelson
Norm Saterly
Richard Scheinman
Roy & Barb Shumard
Martin Simpson
Kathleen Smith
Randall Speck
James L. Tauscher
Daniel & Vonne Thomas
Robert Tiller
Carl Tissol
Edward Traverse
Dan & Claire Trower
Robert Urmann
Rosemary Valentine
Herminigildo Vallee
Martin Watson
Stephen Weissbluth
Hartwell Welsh
Whale Gulch Jimmy
Leslie & Nancy Whatley
Jessie Wheeler
Keith & Margaret Wheeler
Lawrence & Vonda Wilson
Robert Young
Joseph E. Zanone

The U.S. Coast Guard and Heartwood Institute also provided access to some sites. A heartfelt thanks to all the others who said "go ahead".

The following individuals and organizations donated items for Christmas Bird Fairs that netted the Atlas project some much needed funds:

Abruzzi
Arcata Books
Arcata Community Pool
Arcata Co-op
Arcata Exchange
Gary Bloomfield
Curley's Restaurant
Joan Dunning
Brendon Farley
Ferndale Repertory Theater
Finnish Country Sauna and Tubs
Garden Gate
Keith Hansen
HealthSport
Holly Yashi
Hotel Carter
Humboldt State University Center Activities
Humboldt State University Natural History Museum
Humboldt Brewing Company
Humboldt's Finest
Norm Ingle
LBJ Enterprises
Tom W. Leskiw
Los Bagels
Mad River Biologists
Mad River Gardens
M/V Madaket
Moon's Play 'n Learn
Stephanie Morrissette
National Wildlife Foundation
Natural Selection
Nilsen Feed & Grain
North Coast Repertory Theater
Northern Mountain Supply

Northtown Books
Plaza Design
Pottery Farm
Ramone's Bakery & Cafe
Bill Rodstrom
Samurai
Strictly For The Birds
Talisman
Thanksgiving Coffee Company
The Blacksmith Shop
The New Outdoor Store
Tin Can Mailman
Tomo Japanese Restaurant
Catherine Walling
Wildberries Marketplace
Withy-windle
Yakima Products

Many of the principal atlasers were professional field biologists, which was both a blessing (they provided many casual observations), and a curse (everyone was too busy during the field season to focus on difficult blocks). In addition, the unexpected non- or unreliable participation by many local birders and biologists, combined with Humboldt County's large size and extensive private land ownership, put us way behind our anticipated field schedule by the end of Year Three. Thanks to the generosity of the NFWF, we were able to obtain grants to hire a field ornithologist, David Fix, to conduct Atlas surveys; this rapidly got our data collection back on schedule. NFWF grant monies also helped fund some boat-based Atlas surveys to offshore seabird colonies. Jeff Jacobsen provided his boat, *R/V Tsitika Blue*, for that work. Mad River Biologists and Six Rivers National Forest also donated time and person-hours as partners in NFWF Challenge Cost-Share grants. The Humboldt Area Foundation also provided much appreciated financial support. Adobe Systems Inc. kindly donated desktop publishing and image editing software, and the Environmental Systems Research Institute, Inc. (ESRI) donated ARC/INFO Geographic Information System (GIS) software. Local residents Preston & Nancy Spruance, Jay Sooter of Jay Sooter's Pure Water Spas, and Joshua Kinch and Lez Walker of Sweethome Realty donated funds to the Atlas project and their gestures were greatly appreciated.

Humboldt County—the home of Humboldt State University, and virtually every state and federal natural resource management agency, as well as numerous private wildlife consulting firms and other non-governmental organizations—is truly a hotbed of ornithological activity. The resulting bonanza of incidental observations helped fill out maps of some of the more at-risk, special, or otherwise high-profile bird species. Most organizations were eager to help. Our gratitude goes to the following organizations and individuals for contributing material, brainpower, or data. Journalists Barbara Henry and the late David R. Anderson helped spread the word in local newspapers. Ken Anderson of California State Parks and Recreation provided atlasers with fee waivers to conduct surveys on State Park lands. Louise Bacon-Ogden and David Ogden provided atlasers with discounts on field guides and other gear at their Eureka store, Strictly For The Birds. Jane Bardolf and Jamie Bettaso provided data from their studies in the Pilot Creek watershed. Esther Burkett of the California Department of Fish and Game in Sacramento provided Marbled Murrelet records. James Bland provided information on Blue Grouse. Amedee Brickey at the Arcata Office of the U.S. Fish and Wildlife Service provided support. Phil Capitolo provided Double-crested Cormorant nest count data. Carla Cicero of the Museum of Vertebrate Zoology, Berkeley (MVZ), René Corado of the Western Foundation of Vertebrate Zoology, Camarillo (WFVZ), Tamar Danufsky of the Humboldt State University Wildlife Department Museum (HSU), and Douglas Long of the California Academy of Sciences, San Francisco (CAS) assisted with specimen identification and historical data. Anthony Desch shared Hooded Oriole information. Andrew Forde of the U.S. Geological Survey provided data on Band-tailed Pigeons. Kevin Foerster, Kim Forest, and AmeriCorps volunteer Carey Goss conducted surveys on and gathered data from the Humboldt Bay National Wildlife Refuge. Alan B. Franklin and Peter Carlson of the Spotted Owl Comparative Demography project at Humboldt State University provided records for Spotted Owls and other owl species. Luke George of Humboldt State University kindly shared his banding and other bird data and encouraged his students to do the same. Gina Gillich of the Automobile Association of Southern California provided unfolded AAA maps. Gordon Gould of California Department of Fish and Game in Sacramento provided Spotted Owl records. Stanley W. "Doc" Harris helped out by providing training to fledgling atlasers early in the project; we greatly appreciate his constant willingness to share both his extensive personal knowledge and his unique historical database. Each of us had a copy of Doc's *Northwestern California Birds* (1991, 1996) in hand during every phase of the project. Gjon C. Hazard provided data from extensive point counts in redwood forests. Gary S. Lester, Paul F. Springer, Richard A. Erickson, and Brian Woodbridge also provided historical records. Gary Lester also provided a draft compilation of early breeding records for Humboldt County; these data helped focus some of our early efforts. Steve Hawks of the Bureau of Land Management provided topographic maps for much of the county, facilitated access to some very difficult areas, and provided Spotted Owl survey data. J. Mark Higley, Teresa Godfrey, and Heather Brown of Hoopa Valley Tribal Forestry provided data collected during wildlife surveys on the Hoopa Valley Indian Reservation. Immediately recognizing the value of an Atlas, J. Mark Higley directed his staff to work on specific problem blocks whenever possible. Karen Kovacs of California Department of Fish and Game in Eureka provided heron monitoring data and Osprey nest locations. Ron LeValley of Mad River Biologists provided data from Snowy Plover monitoring efforts as well as other projects, and helped train atlasers. Pat Lorenson of the California Department of Fish and Game provided data from the department's Hunter Survey. Mike Martischang from Six Rivers National Forest and Rusty Scalf provided GIS support. Sherri Miller of the U.S. Forest Service's Redwood Sciences Laboratory in Arcata provided Marbled Murrelet records. Chet Ogan helped train atlasers and provided other support. Bruce Oppenheim, Chris Flann, Jeremy Kwolek, and Bryan Yost of the California Waterfowl Association supplied information on nesting Hooded Mergansers and other species that used their Wood Duck nest boxes. Kimberly A. Rodriguez of UC Cooperative Extension helped coordinate access. Bill Rodstrom and Elaine David wrote grant proposals. Keith Slauson and Kathleen Dickson helped organize members of Humboldt State University's Conservation Unlimited into a Breeding Bird-a-thon. Dick Spight gave a few lucky atlasers a chance to survey for birds up to 100 miles off the Humboldt coast aboard the *M/V Blitzen* in early August 1999. Craig Strong of Crescent Coastal Research in Crescent City provided records collected during offshore Marbled Murrelet surveys. Jim Tietz and Kris Baker helped compile training materials. Rick Wallen of Redwood National Park provided breeding records. Catherine Walling helped with newsletters and other administrative tasks. Tim Young at Humboldt Redwoods State Park facilitated access. Personnel at the Roads Division of the Humboldt County Department of Public Works provided detailed road maps. Sadly, a few tax-dollar-supported biologists declined to contribute their Humboldt County bird data.

Other data sources tapped included U.S. Geological Survey's Breeding Bird Surveys (BBS) including the Alderpoint, Friday Ridge, Holmes, Honeydew, Korbel, Lonesome Ridge, Pilot Creek, and Rio Dell routes and, when appropriate, National Audubon Society's Christmas Bird Count (CBC) data including the Arcata, Centerville Beach to King Salmon, and Willow Creek counts. We also gleaned some information from the journal variously known as American Birds (AB), National Audubon Society Field Notes (NASFN), Field Notes (FN), and North American Birds (NAB).

A special thanks goes to the relatively small group of birders who collected a large percentage of the total data used to create this Atlas. The following individuals each contributed at least 300 records to the final database and completed at least eight blocks: R. J. Adams (RJA), David Fix (DFx), John Gaffin (JG), Gjon C. Hazard (GCH), Robert Hewitt (RHw), J. Mark Higley (JMH), John E. Hunter (JEH), Tom W. Leskiw (TWL), Chet Ogan (COg), Jude Claire Power (JCP), Gregory A. Schmidt (GAS), Keith Slauson (KMS), and Paul F. Springer (PFS). Their efforts alone accounted for 60% of all records and 70% of all blocks completed, and thus constituted the "backbone" of the Atlas project. Our hats are also off to all the folks who said they would complete a block—and then did it.

The following observers either submitted records directly to us, or their observations were otherwise obtained:

R. J. Adams
David R. Anderson
M. Anderson
L. Angel
C. Arguello
Susan Armstrong
Pam Arneson
Tim Ash
Louise Bacon-Ogden
Mark & Melinda Bailey
Bill Baker
Kris Baker
Jane Bardolf
Maurie Beck
Michael Beeve
S. Bennitt
Kathleen & Eric Bentler
Jamie Bettaso
John Biggin
Gary Bloomfield
Kory Bonner
Tim Bowden
Brian Boyd
Sam Brennan
G. Brooks
Heather Brown
Bruce & Lynne Bryan
Susie Bulger
Seth Bunnell
Esther Burkett
Bryon Cariss
Peter Carlson
Don Carpenter
M. Carpenter
Rita Carratello
Kevin Charkins
Sal Chinnici
Billy Colegrove
Gary Colegrove
Joshua Colegrove
L. Coleman
Mark Colwell
Maya Conrad
Cindy Cool
Don Copeland
James Cotton
Brent Critch
Bob Damschroeder
S. Daniel
Jim Danzenbaker
Shawna Dark
Elaine David
Scott Davies
Anthony Desch

Brenda & Pete Devlin-Craig
Tad Diaz
Duke Diehl
Dan Dill
Lowell Diller
Jan Derksen
Linda Doerflinger
S. Downing
K. Drake
Jeff Dunk
Joan Dunning
Kevin Dyer
Bill Eastwood
Elias Elias
Eric Elliott
B. Evasic
T. Filliater
Mark Fisher
Sam Fitton
David Fix
Kevin Foerster
Lee Folliard
Alan Franklin
Brad Freeman
Mark Freitas
Gary & Jan Friedrichsen
John Gaffin
Rudy Galindo
Dan Garrison
Luke George
Esther Gilchrist
Teresa Godfrey
B. Golden
Armand Gonzales
Carey Goss
Matt Gostin
C. Grable
C. Grant
George Green
Colinda Gutiérrez
Tony Hacking
Judy Haggard
Patricia Halpin
Robin Hamlin
Keith Hamm
A. Hammond
Nina Hapner
Camille Haroldsen
Heidi Harris
Jay Harris
Stanley W. Harris
Bryan Hawes
Steve Hawks
Gjon C. Hazard
Pablo A. Herrera
Robert Hewitt
Catherine Hibbard
J. Mark Higley
Matt Hinton
Tom Hinz
Joe Hiss
Gretchen Hoiness
Deb Hom
C. Hovnanian
John E. Hunter
Kim Hunter
Liberty A. Hunter
Ken Irwin
Jeff Jacobsen
Joe James
Colin Jewett
Leilani Jones
Larry Karsteadt
Christine Keil
Gail Kenny
Mark Kenney
C. Kent
K. Kent
Jeff Kidd
Tom Kirk
Rich Klug
S. Koller
Karen Kovacs
Diana Kriger
Don Kudrna
Mary Kuehner
Pat Larson
Carol Lawrence
Tom W. Leskiw
Gary Lester
Kelsey Lester
Lauren Lester
Ron LeValley
Sharon Levy
Harmon Lewis
Paul Lohse
J. Brett Lovelace
Vanessa Loverti
A. Maehler
Patty Marsh
Marla Marshall
Matthew Marshall
Sherri Marubayashi
M. Mattax
Brad Mauney
Merry Maxwell
Sean McAllister
Juliene McCovey
R. McCoy
Lynn McCulloch
Michelle McKenzie
Pat McNeil
S. McVicor
Carl Meyers
Brian Michaels
Curtis Miller
Sherri Miller
Bill Minehan
J. Moore
Karly Moore
Paul Morey
Michael Morris
Mark Morrissette
Stephanie Morrissette
Kristie Nelson
Laura Nelson
E. Nielsen
Brad Norman
Ros Novick
Wally Obie
W. O'Brien
Sue O'Connell
Brian O'Donnell
Chet Ogan
David Ogden
Bruce Oppenheim
J. Osborn
Jason Osburn
Antonio Padilla
Dezh Pagen
Steve Pagliughi
John Palmer
N. Pappani
Linda Parkinson
L. Pascoe
J. Pearson
Merritt Perry
Susan Petersen
M. Peterson
Aaron Pole
Kathy & Steve Pollard
Chuck Powell
Marna Powell
Jude Claire Power
John Price
N. Price
Carla Ragatz
Richard Ridenhour
Bruce Riley
Don Roberson
Lynn Roberts
Scott Robinson
Bill Rodstrom
Terry Roelofs
Paul Roush
Steve Rovell
Ben Russell
Kevin Ryan
T. Salberg
Phil Sanders
J. Schimke
Gregory A. Schmidt
Kristin Schmidt
Pete Schmidt
Samuel Schmidt
Scott Schmidt
Misha Schwarz
Cecelia & Ed Secor
Jay Seeger
D. Shaw
Ed Sher
K. Shimizu
Jim Shreeve
Mernie Shreeve
J. Sibley
M. Sieben
Steve Sillett
Dottie Simmons
Keith Slauson
Denise Smith
M. Smith
Sue Sniado
Jay Sooter
Lorentz Sorrell
Paul F. Springer
Virginia Springer
Pete Spruance
S. Ssutu
Peggy Stebbins
Denise Steenblock-Smith
John C. Sterling
T. Stokes
Craig Strong
Kris Sundeen
Robert Sutherland
Holly Taylor
Margaret Taylor
Susan Taylor
Stan Thiesen
Joel Thompson
Jim Tietz
Clyde Trimble
Steve Tucker
Susie Van Kirk
Margaret Varga
Sandra Von Arb
Mary Wachs
Matt Wachs
Cindy Walker
Tom Wallace
Catherine Walling
Justin Ward
R. Ward
Leon Warmuth
Jim Watkins
W. Weber
D. Weinberg
Kevin Westfall

Deb Whitaker
Margaret Widdowson
William "Bud" Widdowson
Melanie Wilbourn
Lan Sing Wu
Scott Yeager
Bryan Yost

The decision was made early in the planning process to solicit only Humboldt County artists for illustrations and cover design, to impart a decidedly local flavor to the final product. Further, individual sponsors were sought to fund the illustrations for bird species of their choice, thereby expanding community involvement in the project. Their financial contributions were divided between the artists and the Atlas publication fund, benefiting everyone in the process. Our sincerest thanks to the art sponsors and the fine artists that distinguish our county (Table 1).

The Atlas greatly benefited from reviews of particular sections or species accounts by Mark Colwell, Alan B. Franklin, Ralph J. Gutiérrez, Robert Hewitt, Sue Leskiw, Ron LeValley, Kristin Schmidt, and Paul F. Springer. We were quite fortunate that Richard A. Erickson, Stanley W. Harris, Tom W. Leskiw, Gary S. Lester, and Don Roberson agreed to review either large sections or complete drafts of the Atlas. Gary S. Lester was the botanical reviewer and compiled our plant list. James and Virginia Waters did yeoman's work on final copy editing, and completed a huge amount of detailed work on a breakneck schedule to help us meet our publication deadline. The Atlas was greatly improved by their skilled efforts. Given the large number of people and the extended time period involved, some important individuals may have been inadvertently omitted here. We apologize to reviewers if any of their hard work was not incorporated into the Atlas; we did so at our own peril. We offer a heartfelt thanks to those who helped the project in any way.

TABLE 1 : ARTISTS AND ART SPONSORS

ARTIST	SPECIES	SPONSOR
Carol Andersen	Western Screech-Owl	Helen McKenna & Allan Ridley
	Belted Kingfisher	Harvey & Jeanette Carroll
	Common Raven	Lan Sing Wu
	Spotted Towhee	Helen McKenna & Allan Ridley
David R. Anderson	Gadwall	Contributed by artist's family
	Hutton's Vireo	Contributed by artist's family
	Bewick's Wren	Contributed by artist's family
	MacGillivray's Warbler	Contributed by artist's family
Gary Bloomfield	Leach's Storm-Petrel	C. J. & Carol Ralph
	American Bittern	Elyse M. Kelly
	Great Blue Heron	Barbara Adams
	Snowy Plover	Robert Hewitt
	Black Oystercatcher	Robert Frey
	Spotted Sandpiper	Contributed by the artist
	Pigeon Guillemot	Jim & Virginia Waters
	Common Nighthawk	Contributed by the artist
	Vaux's Swift	Elias Elias
	Anna's Hummingbird	Betsy & Bryant Harrell
	White-headed Woodpecker	Gregory & Kristin Schmidt
	Western Scrub-Jay	Jay Sooter & Margaret Taylor
	Barn Swallow	C. J. & Carol Ralph
	Winter Wren	Jim & Virginia Waters
	Marsh Wren	C. J. & Carol Ralph
	Western Bluebird	Sally Christensen
	Townsend's Solitaire	Michael Morris
	Hermit Thrush	Jude Claire Power
	California Thrasher	C. J. & Carol Ralph
	Wilson's Warbler	C. J. & Carol Ralph
	Yellow-breasted Chat	Carl Chatfield & Susan Anderson
	Green-tailed Towhee	Gregory & Kristin Schmidt

TABLE 1 (CONTINUED): ARTISTS AND ART SPONSORS

ARTIST	SPECIES	SPONSOR
Gary Bloomfield (cont.)	Fox Sparrow	C. J. & Carol Ralph
	Black-headed Grosbeak	C. J. & Carol Ralph
	Brown-headed Cowbird	C. J. & Carol Ralph
	Purple Finch	C. J. & Carol Ralph
	American Goldfinch	Louise Bacon-Ogden
Joan Dunning	Ruffed Grouse	Robert Hewitt
	Violet-green Swallow	Tom Cockle & Carol Lawrence
	Black-capped Chickadee	Jim & Donna Clark
	Chestnut-backed Chickadee	John & Linda Gaffin
	Swainson's Thrush	Luke George
	Varied Thrush	Robert Hewitt
	Western Tanager	Elaine David
	Rufous-crowned Sparrow	John E. Hunter
Susan Fox	Great Egret	Hal Genger family
	Black-crowned Night-Heron	Marie Granshaw
	Northern Goshawk	Doug Wallace
	Golden Eagle	Michael Meyer
	Flammulated Owl	Janice Andersen
	Grasshopper Sparrow	Jude Claire Power
Chris Frolking	White-tailed Kite	Margaret Varga
	Bald Eagle	Chet Ogan
	Peregrine Falcon	Phyllis Tomich
	California Quail	Frances Madrone
	Pileated Woodpecker	Mark & Melinda Bailey
Paula Golightly	Hooded Merganser	Tim McKay
	Red-shouldered Hawk	Matt Wachs
	Gray Jay	Bojan & Norm Ingle
Rick Kruse	Wood Duck	Catherine Walling
	Common Merganser	Jude Claire Power
	Barn Owl	Judith Louise
	American Dipper	Gary Falxa & Gayle Garman
Linda Parkinson	American Kestrel	Duke Diehl
	Red-breasted Nuthatch	John & Barbara Schmidt
	Cedar Waxwing	Gretchen Ziegler
Andrea Pickart	Douglas-fir, madrone, tanoak, and redwood sprigs	Used by permission of the artist
Todd Telander	American Avocet	Paul & Virginia Springer
	Marbled Murrelet	David Fix
	Spotted Owl	Tom & Sue Leskiw
	Black Phoebe	David Fix

INTRODUCTION

Despite all of the technological gadgetry, high-powered statistical analyses, and detailed research associated with modern field ornithology, basic information such as where particular birds breed is far less documented and understood than commonly thought. This is especially true at the finer scales, where spatial and temporal variation in the ever-shifting patterns of bird distribution are most pronounced. Accordingly, our primary goal was to map the distribution of each bird species found to breed within Humboldt County, California during a 5-year survey period (1995–1999). Our distribution maps establish a county-wide baseline that biologists and others can use in the future to detect trends in breeding bird distribution. No such baseline existed prior to this systematic effort. This "snapshot in time" of county-wide breeding bird distribution will hopefully become a valuable resource for a wide variety of individuals ranging from backyard birders to research scientists.

We are certain that bird abundance and distribution will continue to ebb and flow for reasons both natural and anthropogenic. In addition to helping detect any future changes in the local breeding avifauna, we hope our efforts will also shed some light on the changes that have occurred already. Consequently, we have included much information on Humboldt County breeding birds that predates our survey period. Any new discoveries regarding breeding birds that occurred after completion of the 1999 breeding season were painstakingly ignored herein; those data await the compilers of some future Humboldt Breeding Bird Atlas.

We also intended to contribute to the general knowledge of regional and statewide breeding bird distribution. A number of other California county breeding bird atlases already have been published. Some of these, such as the Marin (Shuford 1993), Monterey (Roberson and Tenney 1993), Orange (Gallagher 1997), and Sonoma (Burridge 1995) county atlases, were available during the fieldwork phase of this project. Other California breeding bird atlases are in various states of completion. We hope that our contribution to this larger effort will provide a more comprehensive understanding of regional and statewide breeding bird distribution and aid land-use planners, conservationists, policy makers, and others interested in the bigger picture.

Still another use of this Atlas will be for the more site-specific planning of agriculture, construction, mining, timber harvest, and other projects that may affect breeding birds. While at a scale far more detailed than that available in bird identification guides, the resolution of maps in this Atlas admittedly are too coarse for project-planning efforts that require site-specific information such as the location of nest sites or home ranges. While our Atlas falls short of meeting this particular need, it does provide biologists and land managers with relative indications of which species are likely to be affected, and when and where additional fieldwork may be necessary. Whether or not birds are included on the various lists of rare, threatened, endangered, sensitive, and special concern species is influenced by political, economic, biological, and other factors. Some future biologists may spend their entire careers on species not currently on anyone's radar screen. So, while detailed studies of high-visibility species necessarily take place at some point after the conservation alarm is sounded, specific and local information on most other bird species is limited or nonexistent. We believe our work may fill part of this information gap. To help meet our goal of providing information useful for project planning, we also provide information we gathered on nesting chronology. Some knowledge of the dates within which critical breeding activity takes place is essential if disturbance of nesting birds is to be minimized or prevented during project implementation.

The means to completing this project were an end unto themselves. We sought to marshal the ever-growing number of local birders into a collective endeavor that would contribute to the conservation and understanding of birds and their habitats. In the process we hoped that, as participants, we would learn about local birds while improving our own abilities. In short, the Atlas project provided a great excuse to go birding! We also hope this effort will give something back to those in the birding community, and the ecotourism economy it helps support, by assisting both local and visiting birders in observing particular species of interest. In this regard, the Atlas is a recreational tool, intended to improve the high level of enjoyment and appreciation that most birders already possess.

Finally, this Atlas will be an educational resource. Teachers, students, or anyone interested in learning more about local birds will find this work useful. The more serious field ornithologists, such as graduate students and others wanting to launch a more detailed study, can use this work to direct them to geographic areas where their efforts will be most fruitful. The questions raised by this Atlas, or any inadequacies of our broad-brush approach, also may serve to stimulate further, more detailed surveys or research.

Clearly, the graphical presentation of the breeding distribution of Humboldt County's birds is the single most important product of this project. While some explanatory text is necessary, we sought to restrict our interpretations and ancillary goals to a few basic areas so that this book would be finished during our lifetimes. Along those lines, we chose not to summarize the extensive life history information readily available elsewhere (*e. g.*, Ehrlich *et al.* 1988, Shuford 1993, and *The Birds of North America* species accounts published by the AOU and The Academy of Natural Sciences). Yet, because so much formal bird research has been conducted within

Humboldt County, we attempted to incorporate as much of that local work as possible. This should make that information more accessible, provide some insights into regional variation in habitat use, and highlight other aspects of breeding biology. Focusing on the local literature also was in keeping with the "Humboldt-centric" emphasis of this project. Since this Atlas attempts to establish a baseline, we did not include any studies in which all fieldwork was conducted after 1999.

While our primary goal was to provide a relatively detailed assessment of county-wide bird distribution over a short time frame (5 years), it was necessary to incorporate some historical information for context. As a result of that effort, this Atlas also can function as an annotated list of Humboldt breeding birds. Thus, we also have included information on species for which no possible, probable or confirmed breeding evidence was reported during the Atlas period (Appendix A). To provide a more complete picture of Humboldt's breeding avifauna, we have included much data (both published and unpublished) that was collected prior to the Atlas period. Our attempt to establish a comprehensive list of the breeding birds will not replace the excellent compilations of Yocom and Harris (1975) and Harris (1991, 1996), which deal with the entire northwestern California area, and include non-breeders. For completeness, duplication of information already published in *Northwestern California Birds* by Harris (1991, 1996) and others could not be avoided. In fact, rather than reinventing the wheel, we typically borrowed directly from Harris' (1996) descriptive terminology for regional status and abundance. It should be noted that, while the breeding evidence codes we used worked quite well for mapping distributions at the county scale, they may be too coarse for evaluating historical or infrequent records of rarer species. Thus, in the case of a few species for which only limited breeding evidence exists, we evaluated all available information and exercised our best judgment as to the "official" breeding status of the species in Humboldt County.

We collected data on only where and when birds were breeding, and did not collect specific data on breeding bird habitats. We are well aware of the hazards of extrapolation beyond the data collected, as well as conjecture regarding data *not* collected. However, we consider the basic, albeit unstructured, natural history knowledge residing in the heads of atlasers to be potentially useful to others. Therefore, we took great liberties in reporting our qualitative impressions of bird distribution and habitat associations. Although imprecise, we believe that these gestalt-like descriptions can provide useful depictions of where birds breed in Humboldt County.

As mentioned above, we sought to provide information on breeding chronology. Since we did collect information on *when* birds were conducting specific breeding activities, we capitalized on this rich source of information. To describe nesting chronology, we typically provide dates of meaningful breeding behaviors that bracket or otherwise delineate the breeding season. No rules for selecting these records were established; all available data for each species were reviewed carefully with an eye for dates and behaviors that we believe best illustrated the local breeding season. Obviously, our bird observation database was used extensively for this. However, these data are biased toward later (higher) breeding evidence, so we made use of any and all bird data from the Atlas period, including Atlas records that had been superseded by higher-level breeding evidence and thus not preserved in the final bird observation database, which was used to build the distribution maps. It is important to note that breeding code usage statistics pertain only to the subset of records in the final bird observation database. Whenever "earliest" or "latest" records are discussed, this pertains to seasonal timing (one may disregard the year); earliest *historical* records are so labeled.

To describe chronology, or to support discussions on other aspects of breeding bird biology, we have included as many of the actual and complete observations (species, date, breeding evidence, location, and observer) as possible. While some readers may initially find these data cumbersome to read, including them allows individuals with divergent or unanticipated interests to garner additional or different information than would have been available were only our interpretations of the data provided. Block designations are provided for all Atlas records, but not for any pre-Atlas records. Because of space limitations, such details are often omitted from published works, but a county atlas such as this is not so constrained. If all those records are not enough for the hard-core bird dataphiles out there, we will provide our final bird observation database in its entirety to any interested party upon request. With these data, others can conduct their own analyses on a variety of subjects including atlasing methodology and large-scale habitat associations.

Another area for which we gathered no specific data pertained to conservation concerns. Nevertheless, we opted to take this opportunity to relay to the reader some of our thoughts regarding local bird conservation. While our concerns about the continuing declines in bird populations emboldened us to discuss conservation issues, it was our extensive local field experience that convinced us that doing so would add value to the Atlas. Perhaps the issues discussed herein will stimulate some productive thought and discussion by others. Any opinions expressed in the Atlas are those of the authors only, and not those of any other Atlas contributor, RRAS, or any of our past or present employers. It is our sincerest hope that the information and ideas expressed in this work will enhance the reader's appreciation of Humboldt's rich avifauna, and will increase understanding of the role birds play in our environment.

STUDY AREA

OVERVIEW

Humboldt County is located in northwestern California, and is bordered on the west by the Pacific Ocean (Figures 1 & 2). The 14th largest of California's 58 counties, Humboldt comprises about 2,287,000 acres (3573 mi^2) of land and 307,000 acres (480 mi^2) of water (McGinnis *et al.* 1996), and is larger than the states of Delaware and Rhode Island. The topography is dominated by a series of mostly northwest-to-southeast trending mountain ridges, contrasting with a few coastal lowland areas formed where large drainages meet the ocean (Figure 3).

Standing in the northern and interior portions of the county, the Klamath Mountains feed the Klamath and Trinity rivers and countless smaller rivers and streams (Figure 4). This system drains into the ocean north of Humboldt in neighboring Del Norte County. Elevation ranges from sea level to 6956 feet at Salmon Mountain in the Klamath Mountains, which border Siskiyou and Trinity counties. Most of this mountainous area is in the 60% or greater slope class.

The California Coast Ranges, found in the coastal and southern portions of the county, generally are composed of low to moderate relief linear ridge systems. A good example is the 40-mile long South Fork Mountain, the longest continuous ridge in North America. These ranges provide the sources for the Mad, Van Duzen, Eel, Bear, and Mattole rivers, and numerous tributaries. For reference, the species distribution maps contained herein display the Klamath, Trinity, Mad, Van Duzen, Eel, South Fork Eel, and Mattole rivers, as well as Redwood Creek. The southeast and southern borders of Humboldt are adjacent to Trinity and Mendocino counties, respectively.

Despite an abundance of water, the county's rugged topography has precluded the natural formation of large lakes. Humboldt's "lakes" are best characterized as ponds; the largest, Fish Lake, is a mere 28 acres (USDA 1995). A few other small lakes, such as Water Dog and Mill Creek lakes, are found in the Trinity Alps east of the Hoopa Valley Indian Reservation. There are also many small eutrophic slump ponds in the Six Rivers National Forest (Marcot 1978). Numerous old mill ponds also dot the landscape. Many of these are now mostly dry and overgrown with willows, although some, such as Aldergrove and Essex ponds, still provide open-water habitat with emergent vegetation.

Humboldt County has roughly 120 miles of coastline, about 70 miles of which are sandy beaches. Some of the more well known include Clam, Mad River, and Centerville beaches. The remaining 50 miles of coastline are dominated by rocky intertidal habitats with small offshore rocks and islands. Also prominent in Humboldt are numerous coastal lagoons and bays. The Humboldt Lagoons, comprised of Big, Dry, Stone and Freshwater lagoons, are located along the northern coast and provide varying amounts of sheltered fresh and brackish water habitats. Devlin-Craig (1992) provided more information on the habitats and birds of the Humboldt Lagoons.

Figure 1. Humboldt County, California.

Small bays are found at Trinidad and Shelter Cove, but Humboldt Bay, California's second largest enclosed bay after San Francisco Bay, is relatively large (223 mi^2). Not only is Humboldt Bay the most important commercial harbor in northwestern California, but this estuary is also of great biological importance to the region (Barnhart *et al.* 1992). More than half of the North and South bays of Humboldt Bay are exposed mud flats at low tide, and there are about 4600 acres (7.2 mi^2) of eelgrass beds within the bay. In addition to most of the aforementioned rivers, many other waterways, including Redwood Creek, Little River, Jacoby Creek, Elk River, Salmon Creek, and the Mattole River, also form estuaries of varying sizes where they meet the marine environment.

CLIMATE

The climate of Humboldt County is classified as maritime Mediterranean, characterized by moderate temperatures and distinct wet (winter) and dry (summer) seasons. However, coastal and inland areas typically experience quite different conditions. In the summer, low-lying coastal areas are influenced by an often persistent "marine layer", creating cool, humid weather. During the summer and fall, Eureka has some form of fog during 21% of days, with a mean maximum temperature in August of 61 °F (Elford and McDonough 1964). The inland extent of these summer maritime influences is limited by the mountainous topography, which restricts much of the summer fog and overcast to coastal areas. Notable exceptions are almost daily incursions of cool, moist air up major drainages. For example, on a typical summer morning, coastal fog may extend to Weitchpec on the Klamath, to Dyerville on the Eel, and to Grizzly Creek on the Van Duzen River. In these more inland areas under a marine influence, the fog and overcast often dissipate each day, revealing sunny skies by early afternoon. Temperatures on the coast are quite stable, often varying less than 25 °F from summer to winter. In Eureka the average minimum temperature for December and January is about 42 °F.

While it is generally hot during the summer at inland locations, temperature gradients within the county can be extreme. For example, temperatures in Willow Creek can be over 100 °F, while not many miles toward the coast, temperatures can hover in the 60s beneath overcast skies. Average relative humidity during the summer at inland sites is 40–50%. The hot and dry summer conditions in the interior of the county can create conditions favorable for vegetation fires. Outside the marine influence zone, the Coast Ranges tend to be drier than the Klamath Mountains, resulting in higher fire frequencies and correspondingly higher natural seral stage diversity.

Figure 2. Selected place names of Humboldt County, California.

During the late summer of 1999, the lightning-caused Megram Fire burned about 125,000 acres (195 mi^2), about half of which was in Humboldt County on the Six Rivers National Forest and Hoopa Valley Indian Reservation. This was one of the largest fires in the history of northwestern California and will undoubtedly affect breeding bird distribution in that portion of the Klamath Mountains for years to come.

Winter precipitation in both coastal and inland areas is high, relative to the rest of the state, with 90% of the precipitation falling from October through April. Fall and winter storms from the northern Pacific Ocean drop much of their moisture at higher elevations, with many areas above 3000 feet receiving 70+ inches of precipitation annually. Summer thunderstorms are rare on the coast, but occur over inland ridges, although resulting precipitation usually is insignificant. Average annual precipitation at established weather stations ranges from a low of about 40 inches at Eureka to a high of about 69 inches at Richardson Grove State Park. Sites with more coastal influences such as Orick, Scotia, and Shelter Cove average 68, 49, and 62 inches per year, respectively. Annual averages at the low elevation inland sites of Orleans, Hoopa, Willow Creek, and Alderpoint are 53, 58, 52, and 51 inches, respectively. January mean minimum temperatures for inland areas are generally in the mid-30s, with inland areas experiencing wider temperature fluctuations than coastal areas throughout the year. Winter snow often persists at elevations above 4000 feet, with some high elevation sites under snow for all but the summer and early fall months in some years. Coastal areas experience freezing temperatures only a few times a year and snow rarely. Indeed, no measurable snowfall occurred in coastal areas during the 5-year Atlas period.

Floods are a part of the history and a powerful source of landscape dynamics in Humboldt County. Many of the

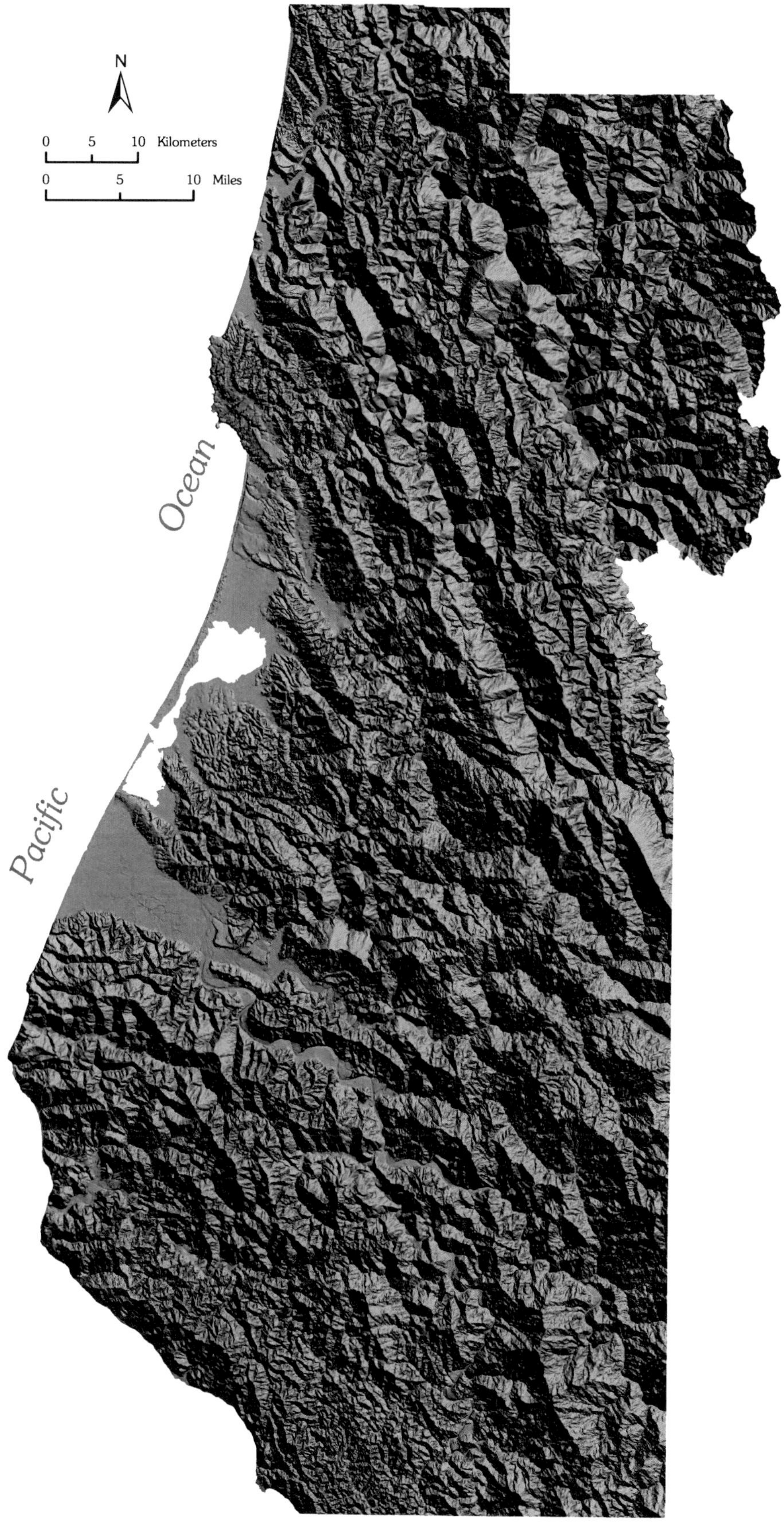

Figure 3. Topography of Humboldt County, California.

larger floods occur during "rain-on-snow" events (CDWR 1965), and are exacerbated by watershed degradation from logging and other impacts. The December 1964 flood was the greatest in north coast history, causing extensive flooding and great economic loss in the Mad and Eel river lowlands. While this event prompted flood control projects such as levee construction along lower Redwood Creek near Orick, it probably also reinvigorated riparian habitat in some lowland areas. Farther up the drainages, however, riparian soils and vegetation have not recovered from the extensive scouring that occurred. Some towns, such as Pepperwood and Weott, were nearly completely swept away: in Weott the water level during the 1964 flood was 35 feet above the sidewalk! While the 1964 flood was an extremely rare event, smaller flood events can and do occur irregularly; the New Year's Day Flood of 1997 was deemed a "50-year" event and caused significant damage in some areas.

The most noteworthy weather related event during the Atlas period was one of the most intense El Niño-Southern Oscillation events ever recorded in northern California. Unusually warm and unproductive ocean waters were present from the spring of 1997 through the spring of 1998. As often happens, this event was followed in 1999 by a La Niña event, during which cool, productive waters occurred because of the resumption of upwelling caused by strong northwest winds. Unfortunately, human-made events also impacted the marine environment during the Atlas period. In November 1997 the *M/V Kure* spilled about 5000 gallons of bunker oil into Humboldt Bay, and in September 1999 the *M/V Stuyvesant* spilled another 2000 gallons of oil near the entrance of the bay. After these spills, 951 and 1272 oiled birds, respectively, were collected; these numbers may represent only the tip of the iceberg with respect to the number of bird casualties.

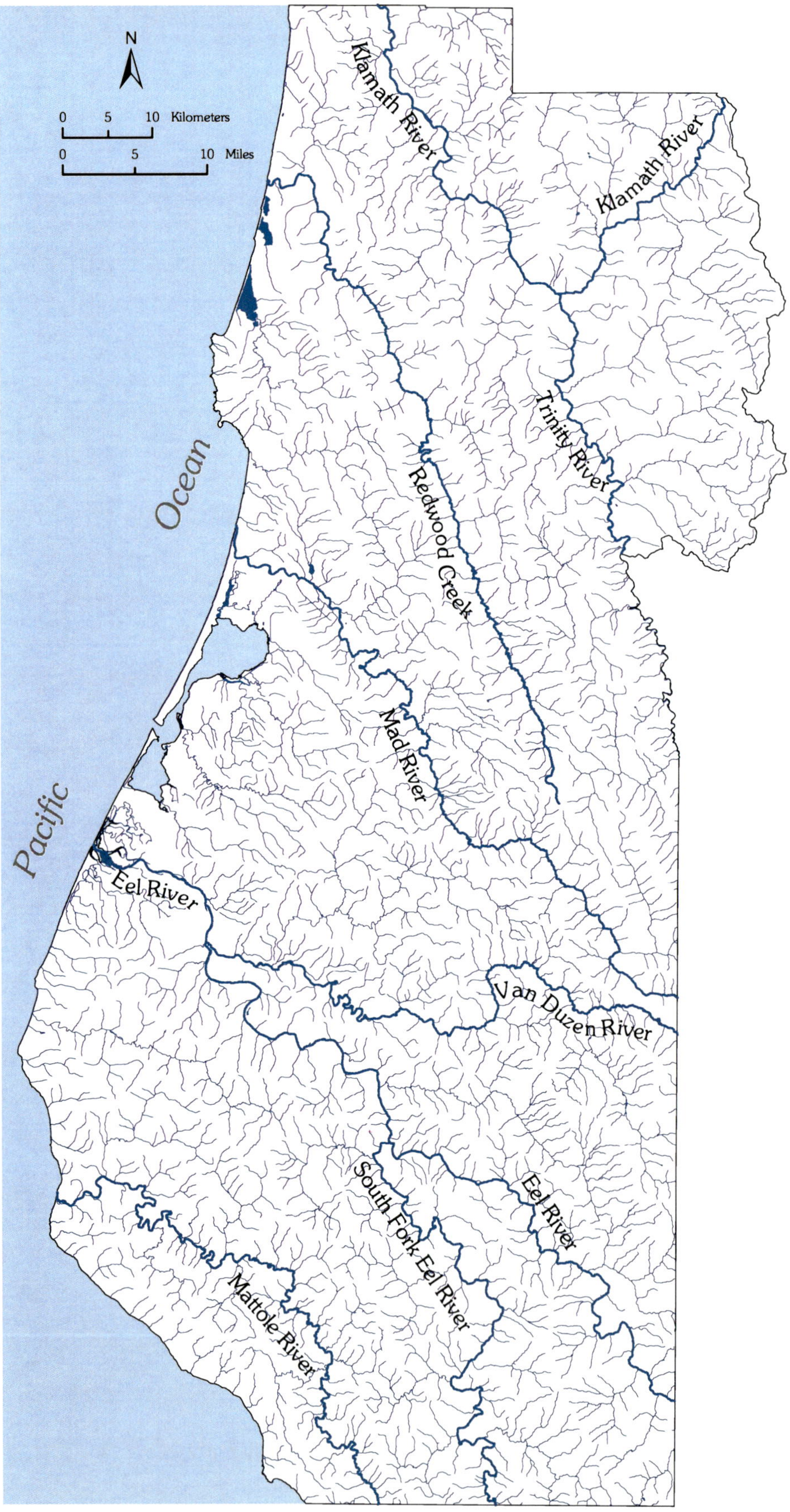

Figure 4. Rivers and perennial streams of Humboldt County, California.

LAND USE AND OWNERSHIP

Humboldt County has probably been inhabited by humans for at least 5000 years. Between 1000 and 1500 years ago, complex cultures developed that were marked by an increased population and greater emphasis on processing and storing food for the winter. Any effects these early inhabitants may have had on the local flora and fauna as a result of burning, hunting, and gathering are unknown. Aboriginal groups inhabiting the Humboldt area at the time of European contact included the Yurok, Karuk, Wiyot, Hupa, Chilula, Whilkut, Mattole, Nongatl, Lassik, and the Sinkyone (Kroeber 1925).

The entrance to Humboldt Bay was "discovered" by Europeans in 1806. By the mid-1800s, pack trains originating at Humboldt Bay sought routes into the Trinity County area to supply gold miners. Gold mining, both placer and lode, played an important role in the early history and economy of Humboldt. Old pack trails and the impacts of unrestrained mining are still apparent in some areas of interior county. Following on the heels of the explorers, trappers, and miners, the modern settlement of Humboldt was well in progress by the late 1800s. Sheep and cattle grazing became increasingly important and was most widespread in the Coast Ranges. Along with livestock came the inevitable predator removal, fence building, introduction of non-native grasses, burning to convert to rangeland, and other associated activities. Many of the early efforts to produce livestock were directly related to the need to provide fresh meat to an ever growing population of loggers.

The initial logging in Humboldt occurred primarily in lowland coastal redwood forests. The first lumber mill in Humboldt (situated on the Bay) was constructed in 1850, and by 1854 there were nine (Noss 2000). Around 1850, the towns of Eureka and Arcata were formed. Although logging methods were labor intensive and slow, many areas in the coastal lowlands were already logged over by the 1880s. The post-World War II building boom created a huge demand for lumber, further increasing harvest rates. The timber boom peaked in 1959. Fortunately, Humboldt Redwoods and Prairie Creek Redwoods state parks had been set aside in the 1920s and 1930s, and Redwood National Park was created in 1968. Today, some intensively managed coastal redwood forests are on their third or fourth harvest rotation.

It was during the post-World War II boom period that the inland Douglas-fir forests also began to experience widespread intensive timber harvest. Maximization of wood-fiber production was the primary management goal on these lands for many decades. The listing of the Northern Spotted Owl as "threatened" under the Endangered Species Act in 1990, followed by the listing of numerous anadromous fishes, along with other increased environmental regulations, greatly slowed the rate of clearcutting of Humboldt's mostly federally administered Douglas-fir forests.

The widespread use of fire by Indians, then ranchers, and finally by loggers to clear brush and logging slash generally came to an end in the mid-1940s. Changes in forest policy and post-war improvements in firefighting capability prompted this shift (Veirs 1982). While large wildfires have occurred in Humboldt, the attempt to exclude fire was largely successful. As a result, many of the areas previously converted to rangeland have now grown in with redwood. In addition, many timber stands that formerly had a relatively open understory can now be characterized as "doghair" thickets, with the density of trees and other vegetation probably much greater now than before the mid-1900s. Today, the purposeful use of fire to manipulate vegetation occurs on a relatively small scale in Humboldt County, and is most often used to prepare clearcut areas for replanting.

Overall, land use in Humboldt County is 74% forestry, 10% agriculture, 6% public use, 4% residential, 3% water resources, 2% industrial, and 1% commercial. With roughly three-quarters of the total area in timberlands, it is no surprise that forestry and timber harvesting are still important components of the local economy. In 1996 there were more than 20 sawmills operating in Humboldt County. Humboldt usually leads California counties in annual timber harvest, with wood and paper products its top export. About 5% of the total Humboldt workforce is employed directly in forestry, fishing, or farming. Crab, salmon, and bottom fish are important fisheries, and dairy farming thrives in the coastal lowlands. Services, retail sales, and government continue to be the fastest growing sectors of the economy. Recreation and tourism are of growing economic importance in Humboldt County. Currently there is little or no metallic mineral extraction. However, numerous stone and rock quarries are active, and sand and gravel mining occurs annually on river bars in the Mad, Van Duzen, Trinity, and Eel rivers. Also, about 30 natural gas wells operate in the county.

Land ownership in Humboldt is represented by a variety of entities including federal and state agencies (28%), tribal groups (5%), the County of Humboldt and incorporated cities (1%), and large (including corporate) and small private landowners (66%) (Figure 5).

As of January 1999, about 59% of Humboldt's total human population of 128,100 lived in the area surrounding Humboldt Bay. About 19% of the population resides north of the bay, while 22% live to the south. Approximately 67,000 Humboldt residents live outside the seven incorporated cities of Arcata, Eureka, Blue Lake, Ferndale, Fortuna, Rio Dell, and Trinidad (Figure 2).

LAND COVER AND BIRD HABITAT

While there are a wide variety of habitats within Humboldt County, at the county scale there are five broad cover types that influence breeding bird distribution (Figure 6; USDA 1997). Many bird habitats, such as riparian corridors, influence local bird distribution, but occur in patches or slivers too small to display at the county scale. Harris (1996) provided additional information on habitats and their associated birds in northwestern California.

The Urban–Agricultural land cover type contains the most diverse habitats in Humboldt but accounts for only about 4% of the county. This cover type predominates in the coastal lowlands, especially around Humboldt Bay, and includes riparia, wetlands, dune systems, coastal scrub, urban, agricultural, and numerous other types. In addition to high habitat diversity, these readily accessible areas also receive a disproportionate amount of coverage by birders, causing an unavoidable bias in the mapped breeding distribution of some species. This phenomenon is addressed in the species accounts where appropriate.

Riparian habitat occurs within each of the five cover types, but is more apparent and distinct in the lowlands, as a result of deeper soils and wider meander zones, than in the mountains. Most low elevation riparian areas are dominated by a mix of red alder and willow. A few contain significant amounts of black cottonwood. Red elderberry and cascara are common riparian shrubs. Most of the remaining lowland riparian areas occur in the alluvial floodplains near Ferndale and Loleta (the Eel River Bottoms), and where the Mad River long ago entered Humboldt Bay (the Arcata Bottoms). Smaller bottomland areas are found along the Mad River near Blue Lake and the mouth of Redwood Creek in Orick. Although some lowland cottonwood riparian areas

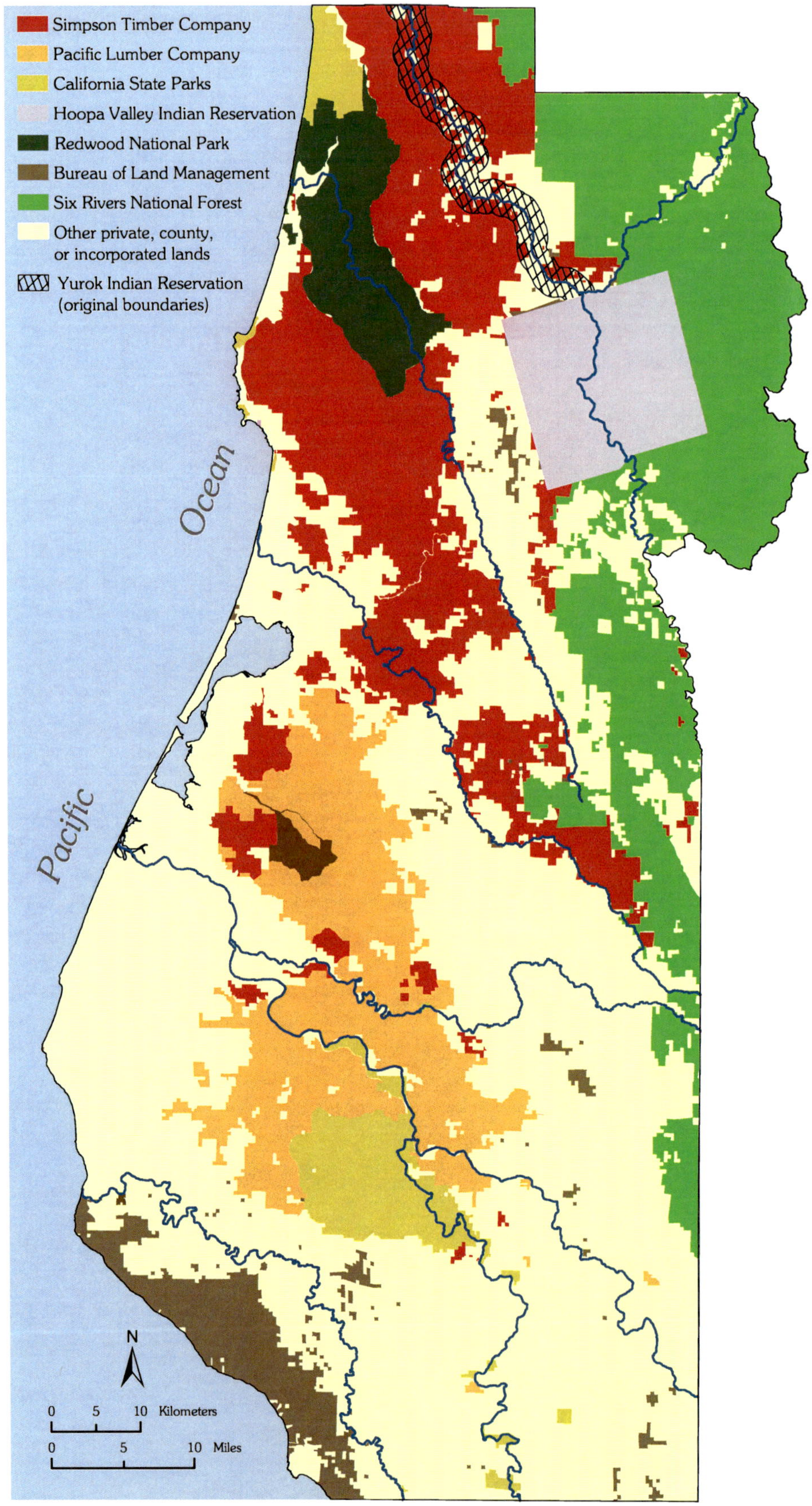

Figure 5. Land Ownership in Humboldt County, California.

and wetlands still exist in these bottomlands, most have been converted to pastures or have been built upon. Less than 5% of an estimated 10,000 acres (15.6 mi^2) of rich bottomland riparian habitat still exists along the lower Eel River (Kelly 1987).

Since Humboldt Bay is only two-thirds of its original size as a result of diking, dredging, draining, filling, and sedimentation, only a fraction of the once extensive wetlands and marshes remain. Less than 10% (about 900 acres) of the original saltmarsh still persists around the bay, and about 700 acres of other brackish and freshwater marsh types can still be found. Burton (1972), Base (1982), and Jacobson (1986) studied various aspects of habitat and avian use of saltmarsh around Humboldt Bay. Gerstenberg (1972) also described saltmarsh, as well as numerous other habitats around the Bay. Common saltmarsh species include dense-flowered cordgrass and pickleweed.

Probably the most noteworthy freshwater pond and marsh habitat in Humboldt is the Arcata Marsh and Wildlife Sanctuary (hereafter Arcata Marsh). Located at the northeast corner of North Humboldt Bay, this 174-acre site was constructed in 1981 to provide both wastewater treatment and wildlife habitat. See Tousley (1982), Spitler (1985), Higley (1989), Grosz (1992), and Schlafmann (1992) for details on the birds and habitats of the Arcata Marsh. Freshwater marshes at Ocean Ranch, Eel River Wildlife Area also provide some of the highest quality coastal freshwater habitat in northwestern California.

Many of Humboldt's sandy beaches adjoin dune systems supporting a characteristic suite of habitats including beach strand, coastal brushfields, dune hollows, riparian/wetland, and coastal conifer forest. Willow, yellow bush lupine, coyote brush, blue blossom, silk tassel, wax myrtle, and shore pine are common species. Hill (1977) and Sferra (1988) provided details on these habitats and their associated

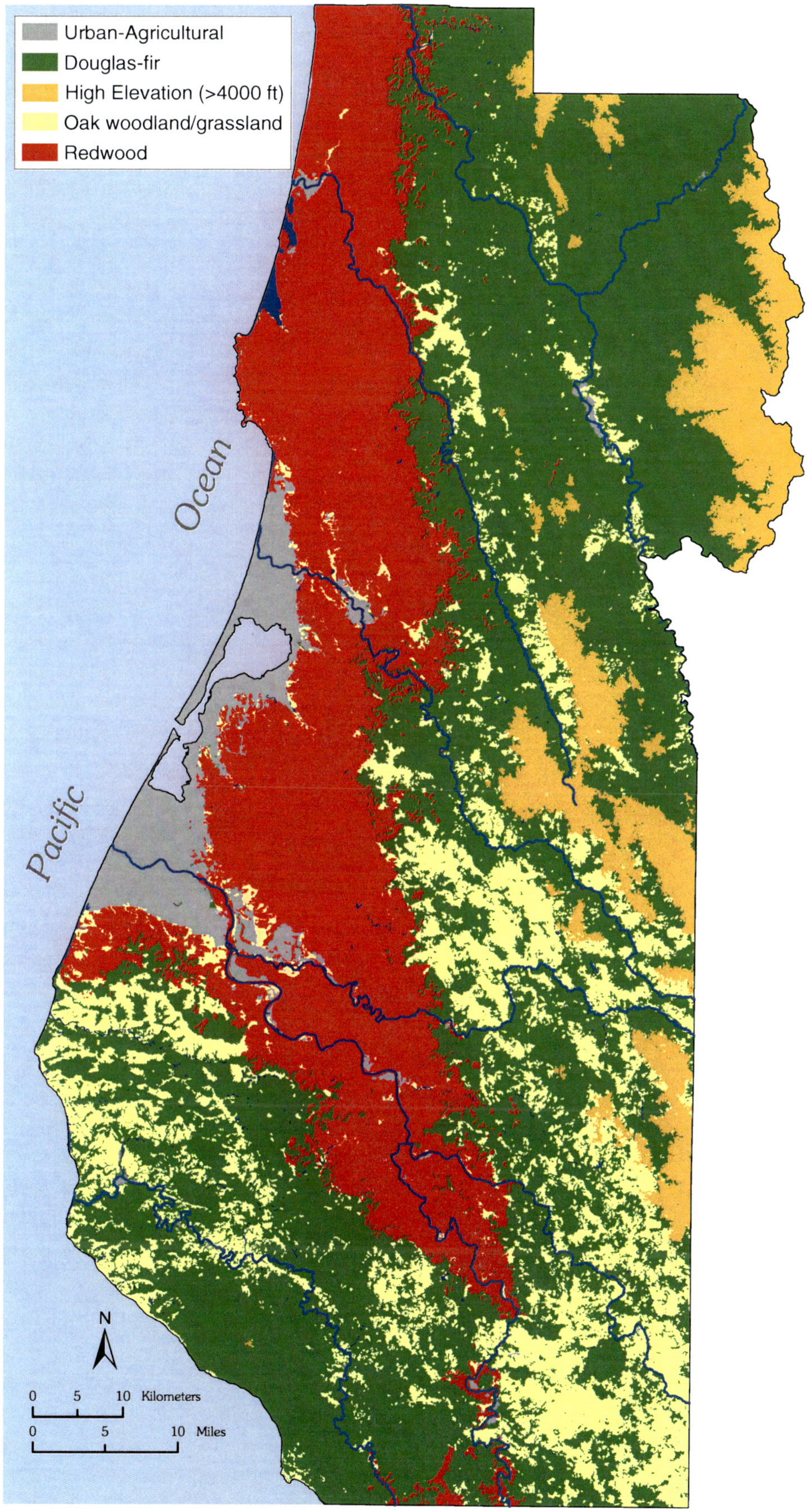

Figure 6. Primary land cover types, Humboldt County, California.

birds. Among the buildings, vacant lots, and ornamental plantings of Eureka, the Hoopa Valley, and the many other small towns and settled areas, urban and suburban habitats provide unique nesting and foraging opportunities for some of Humboldt's breeders. The exotic plant species, year-round water sources, bird feeders, and more moderate microclimates found in these urban and suburban areas have undoubtedly resulted in some relatively recent changes in breeding bird distributions. Perennially green lowland pastures, both well grazed and rank, occupy most of the Redwood Creek, Mad River (Arcata), Elk River, and Eel River bottoms, as well as some smaller alluvial plains.

When one thinks of Humboldt, fog-shrouded stands of ancient redwoods might come to mind. Indeed, redwood forests comprise about 24% of the county. The distribution of redwood forest is largely coincident with the inland influence of cool, wet marine air. While redwoods predominate here, other conifers such as Sitka spruce, grand fir, western hemlock, and Douglas-fir are also found, as are a number of hardwood species including red alder and bigleaf maple. Various technological and economic forces resulted in early harvest practices in the redwood region that were less "clean" than would be seen later in an era of industrial even-age silviculture (O'Dell 1996). These early "messy" forestry practices, combined with the phenomenal resprouting and growth rates of redwoods, helped in many cases to produce structurally diverse second-growth forests. However, as technology has improved, forestry has become "tidier", and redwood forests are becoming progressively less structurally diverse.

Today, Humboldt's redwood forests exist in a patchwork or complicated mix of seral stages including freshly cut/herbaceous, brush stage, small trees, mature trees, and old growth. Large snags are rare or absent in most areas. Scattered residual ancient redwood trees still persist in some intensively managed areas. The earliest seral stages of redwood forest often contain sword fern, poison-oak, red huckleberry, salmonberry, thimbleberry, salal, red alder, and other species. Less than 5% of the original old-growth redwood stands remain protected in parks, reserves, and other conservation areas. See Noss (2000) for more on the history and vegetation of the redwood zone, and Fisher (1902a, 1902b), Hazard and George (1999), and Hazard (2002) for additional information on the birds of redwood forests.

The redwood cover type certainly enjoys a far higher public profile, but forests dominated by Douglas-fir are the more widespread cover type in Humboldt, covering approximately 51% of the county. These forests extend east from the coastal redwood zone upslope to the high elevation (> 4000 ft) true fir zone, and become less contiguous southward in the county. South of Cape Mendocino, the King Range and Mattole Valley are dominated by Douglas-fir forest, although toward the coast beyond the Mendocino County line, redwood again dominates. Numerous hardwoods (especially tanoak, but also California black oak, Pacific madrone, and others) and other conifer species also occur in the Douglas-fir forests, but many of these stands, especially in the Coast Ranges, have a minimal hardwood midstory. Pockets dominated by hardwoods are not uncommon, such as large stands of canyon live oak that occur on lower elevation sites with rocky soils and steep, southern exposures. River gorges provide rock and cliff habitats such as those found along the Weitchpec Bluffs.

In the Douglas-fir forests, the practice of clearcutting followed by intensive site preparation and replanting has been used almost exclusively. As a result, in contrast to Humboldt's redwood forests, the inland Douglas-fir forests exist in a more clearly defined patchwork of generally distinct seral stages. Selective logging and other less impactive harvest methods have been more recently introduced in these forests. Brushy areas in this zone contain numerous species including tanoak, Oregon-grape, buck brush, deer brush, salal, rhododendron, Sierra gooseberry, and evergreen huckleberry. Some estimates put the amount of mature and old-growth Douglas-fir remaining in the Six Rivers National Forest at about 30%, well above similar areas in southern Oregon. Jimerson *et al.* (1996) provided details on Humboldt's Douglas-fir and Douglas-fir/tanoak forests. Humboldt's Douglas-fir forests are the most studied of the major land cover types; information on their bird communities was provided by Raphael (1984, 1987), Raphael *et al.* (1988), Ralph *et al.* (1991), Sniado (1999), and Bardolf (2001).

Given the limited spatial extent of grasslands and oak woodland/grasslands and the strong association between these two types, grassland and oak were lumped into a single cover type. Oak woodland/grasslands occupy about 13% of Humboldt and are generally found in the drier southern portions of the county. This cover type also includes areas such as the Bald Hills, prairies and grasslands near Kneeland, Bear River Ridge, and Cape Mendocino. Coastal balds also occur on marine terraces such as those found near Table Bluff and McKinleyville, although many of these are rapidly disappearing due to urbanization. Others, such as at Elk Head, are being encroached upon by conifers due to fire exclusion and lack of grazing.

Oregon white oak and, to a lesser extent, California black oak are the dominant hardwood species of this cover type, and most grasslands are dominated by introduced annual grasses such as hedgehog dogtail, sweet vernal grass, and California oatgrass. Humboldt's oak woodland/grasslands are often mixed with, or adjacent to, stands of Douglas-fir. Shrubs of this zone include California buckeye, redbud, toyon, and several *Ceanothus* species. Because many of the oak woodland/grasslands are privately owned, restricted access has resulted in this cover type being one of the most poorly studied or birded habitats in Humboldt County.

Fisher (1904) briefly discussed some of the birds of the oak woodland/grassland type in the Hoopa Valley.

The "high country" of Humboldt starts at about 4000 feet in elevation, where white fir begins to replace Douglas-fir as the predominant conifer species. The high elevation cover type occupies only 7% of Humboldt County. White fir is the dominant species in this zone, although other conifer species such as incense-cedar are common. Hardwoods are much less common in high elevation forests than in redwood or Douglas-fir forests, although some, such as giant chinquapin, can be found. Shasta red fir is the predominant conifer species at Humboldt's highest elevations (*e. g.*, Blake Mountain, Salmon Mountain, Lonesome Ridge).

High elevation montane brushfields occur as a result of timber harvest or may persist as climax vegetation. Typical brush species include snowbrush, *Arctostaphylos* spp., *Ribes* spp., bitter cherry, and silk tassel. While wet meadows and similar habitats are found scattered about the Humboldt landscape at all elevations, some of the more interesting meadows, such as Groves Prairie, are found within this cover type. In some high elevation areas such as Horse Mountain or the Lassics, significant stands of Jeffrey pine occur, but these areas are relatively small. These open pine forests are typically associated with serpentine soils. They have an understory of low shrubs such as huckleberry oak and bush chinquapin, or bare soil. The highest portions of the eastern peaks often are dominated by talus, cliffs, or other rocky substrates. Most of the high elevation areas in Humboldt are publicly owned, yet because of their remoteness receive only occasional summer visits by birders or others with an awareness of birds. While no local studies have focused on this cover type, some of the Douglas-fir studies (*e. g.*, Sniado 1999) have provided information on the birds of the high country.

METHODS

Our methods generally followed those of Roberson and Tenney (1993), our intent being to maintain compatibility with other California county atlases so that individual atlases could be compared and potentially synthesized into a larger scale meta-analysis. For more information on the history, biases, and limitations of grid-based distribution mapping see Udvardy (1981), Laughlin *et al.* (1982), Roberson and Tenney (1993), and Shuford (1993).

We used a GIS (ArcInfo v.7.1, Environmental Systems Research Institute, Redlands, California) to generate a grid of 25-km^2 (5-by-5-km) cells over Humboldt County. The coordinates (UTM projection; UTM Zone 10; Datum NAD27 CONUS) for the northwest corner of the grid were 380,000 Easting, 4,595,000 Northing and of the southeast corner were 470,000 Easting, 4,425,000 Northing. The grid was overlaid on a Humboldt County boundary map in GIS and all grid cells that intersected the boundary were retained. Each grid cell, or "block", was assigned a unique alpha-numeric designator with letters A through R assigned along the horizontal axis and numbers 1 through 34 along the vertical axis (Figure 7). Blocks were also assigned unique names based on prominent geographic features or locations of special ornithological interest (Appendix B).

Blocks were dropped from the survey if they were composed almost entirely of ocean waters. Two exceptions to this rule were blocks e04 (Redding Rock) and e11 (Pilot Rock) since they contained islands used by nesting seabirds. We also dropped blocks if <1% of their total area was composed of land within Humboldt County. An exception to this was block o14 (Salyer), which was composed entirely of Trinity County lands, but was completely surrounded by blocks containing Humboldt County lands. A total of 425 blocks were retained for survey.

Given the county's large area and the limited availability of observers, we used a system of priority and non-priority blocks to create a more even distribution of our effort across the county. Different survey requirements were used for priority and non-priority blocks (see below). Each block in the northernmost row contained <25% Humboldt County land and were *a priori* assigned as non-priority blocks (Figure 7). Starting at block g02, each southeast-corner block within each contiguous set of four blocks (in this case blocks g02, h02, g03, and H03) was designated as a priority block. Each non-priority block was indicated by a lowercase letter (*e. g.*, c21), and priority blocks were indicated by uppercase letters (*e. g.*, J27). Any priority block with less than 50% of its land in Humboldt County was downgraded to non-priority status. This resulted in a total of 91 priority and 334 non-priority blocks (Figure 7). Most peripheral blocks contained portions of Del Norte, Siskiyou, Trinity, or Mendocino counties. Breeding observations were retained if they occurred within any of the 425 blocks, regardless of county.

Within each block we sought to determine the breeding status of as many species as possible. In each block, for each species encountered and suspected as a breeder, we assigned one of the following three hierarchical categories of breeding status: "confirmed", "probable", or "possible". The assignment of breeding status was based on the breeding evidence observed (Table 2). The one- and two-letter codes used for these categories of breeding evidence were used in all aspects of data collection and management. These categories were also hierarchical, from the strongest evidence of breeding (NY: nest with young seen or heard), down to the weakest evidence of breeding (X: species present in suitable nesting habitat during its breeding season). Lower-level breeding evidence for any species in any block was always superseded by higher-level breeding evidence when available. However, any confirmed breeding evidence was considered acceptable. We did not attempt to increase the level of breeding evidence by searching for nests, because such an effort could have affected nest success. Evidence of breeding was collected primarily from 1 Feb–30 Sep each year during the 5-year period 1995–1999.

The "completion" of a priority or a non-priority block was subjectively established with the intent of maximizing the information gathered in each block, while taking into account the limited number of observers available. Two criteria were used to determine when priority blocks were considered completed. A priority block was considered complete when at least 45% of all possible breeding species encountered were confirmed breeding *or* when at least 60% of all possible breeding species were recorded as either confirmed or probable, as long as at least 30% of all possible breeding species were confirmed. In addition, we established the guideline that at least 40 possible breeding species be recorded in each priority block. A non-priority block was considered completed when at least 20 species were recorded as possible, probable, or confirmed. This criterion was abandoned for some blocks (a27, a28, b19, c30, e04, and e11), which were composed of greater than 75% ocean waters. In all blocks, an attempt was made to survey all of the habitats present, but this goal was not met in many blocks because access was limited. Inadequacies or other limitations in survey coverage are discussed in the individual species accounts.

Three primary methods were used to collect breeding evidence data. First, individual observers (hereafter referred to as "atlasers"), or groups of atlasers, would survey specific blocks until completion criteria were met. In addition, to ensure that the entire Humboldt County breeding avifauna was represented in our data, some areas were specifically surveyed *a priori* by atlasers for known specialty species or communities. Second, we solicited and collected any incidental or "casual" observations of breeding evidence that occurred during the study period. Sources for these casual observations included direct input from birders, wildlife biologists, and others, as well as data gleaned from the Arcata Birdbox, the Redwood Region Audubon Society's newsletter *The Sandpiper*, and unpublished sources. Third, we obtained sets of observations of breeding evidence from individuals and organizations involved in organized research and management-oriented projects (see the Acknowledgments section for a list of these projects).

Data recorded for each block consisted of species, breeding evidence observed, date of observation, and observer. In addition, observers submitted other pertinent information at their discretion. Birds considered migrants, breeders outside of breeding habitat, or birds otherwise determined to be non-breeders were deleted from the final bird observation database used to create the final distribution maps. Records of lower-level breeding codes that had been superseded by higher-level codes were either not reported by atlasers or were deleted from the final bird observation database. However, all entered records collected during the 5-year survey period were saved in a separate archive data table. Estimates of abundance, habitat amounts in blocks, party-hours spent collecting data, or other variables were not recorded because of difficulty in obtaining these estimates from atlasers and the presumed high variability among individual atlasers' estimates.

Each record submitted was reviewed for proper code usage and reasonableness, and supporting details were sought for any unexpected, rare, or otherwise unusual reports. Data were

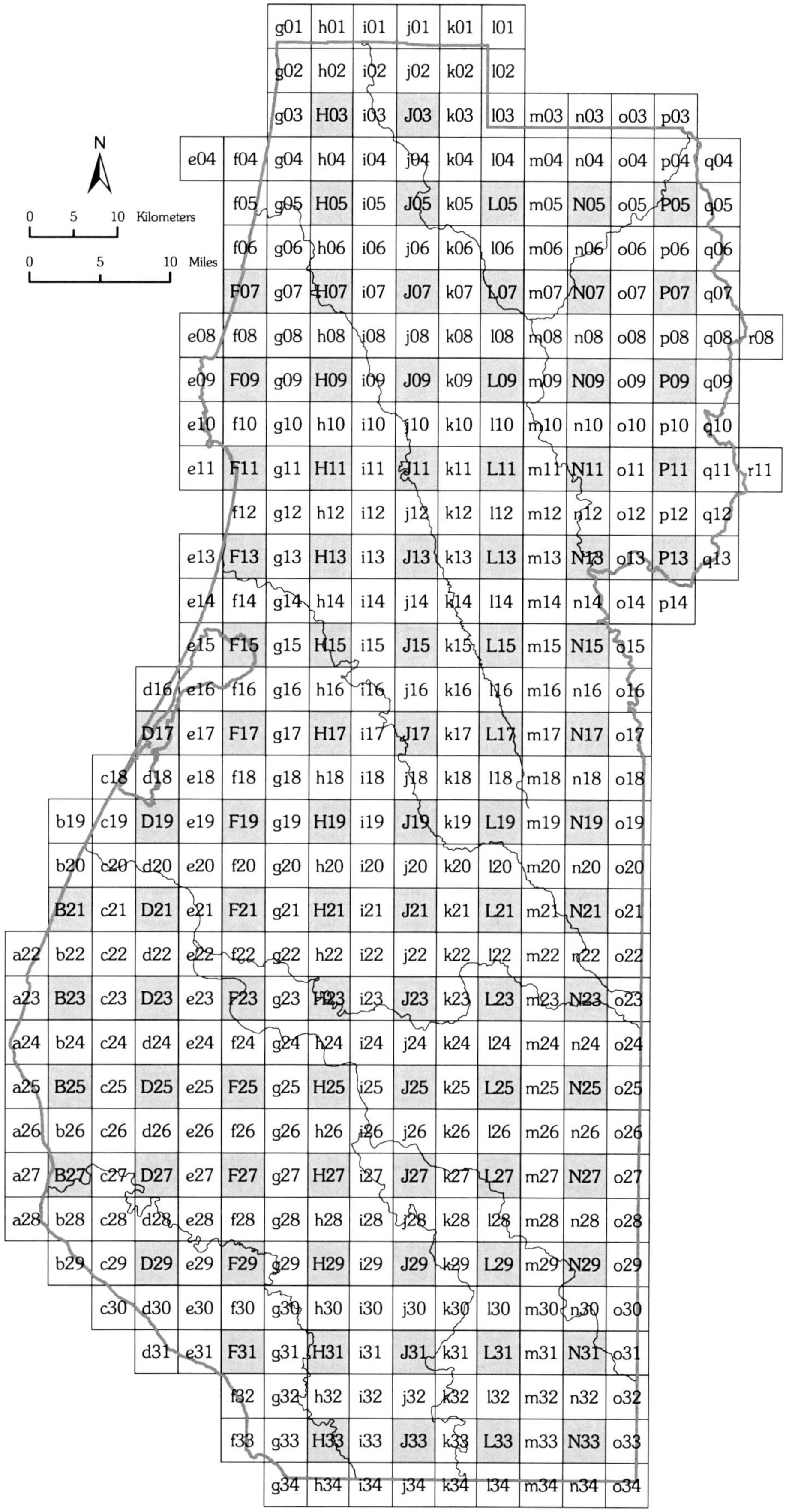

Figure 7. Survey block system. Shaded grid cells are priority blocks, unshaded cells are non-priority blocks. Block dimensions are 5x5 km.

managed using PARADOX, a relational database management program, and species distribution maps were created using ArcMap 8.3 (ArcGIS, Environmental Systems Research Institute, Redlands, California). Species-level bird taxonomy followed the seventh edition of the *Check-list of North American Birds* (AOU 1998). Scientific names of higher plants followed Hickman (1993); a list of common and scientific names of all plants mentioned in text can be found in Appendix C.

TABLE 2 : BREEDING EVIDENCE, RANKED STRONGEST TO WEAKEST, USED TO DETERMINE BREEDING STATUS. Eggs or young of Brown-headed Cowbirds confirmed both the cowbird and the host species.

CODE	EVIDENCE OBSERVED
	CONFIRMED BREEDING
NY	**N**est with **Y**oung. An adult seen or heard feeding young at the nest was considered NY, not FY.
NE	**N**est with **E**ggs. Included a bird on a nest that was apparently incubating eggs.
FS	**F**ecal **S**ac carried by an adult.
ON	**O**ccupied **N**est. Adults entering or leaving nest sites in circumstances that indicated an occupied nest was present, such as a nest hole or other structure that was too high to see the contents.
FY	**F**eeding fledged **Y**oung. Dependent young either seen or heard being fed by adults. Not used in circumstances where adults with highly mobile dependent young may have originated in another block.
CF	**C**arrying **F**ood for young. Did not include adults of some species (*e. g.*, Western Gulls, Caspian Terns, raptors) when food was thought to be part of courtship displays or was potentially destined for another block.
PY	**P**recocial **Y**oung. Used for waterfowl, shorebirds, rails, and gallinaceous birds that hatch precocial young.
FL	Recently ***FL***edged young of altricial species incapable of sustained flight. Used for dependent young of resident species if within natal territory, even if the young could fly. This code was not supposed to be used in circumstances where highly mobile fledglings may have originated in another block.
UN	***U***sed **N**est or eggshells found. This was the only code that was used for observations outside of 1 Feb–30 Sep.
PE	**P**hysiological **E**vidence of breeding. Used on a live or dead bird in the hand, when the presence of a highly vascularized brood patch, or a cloacal protuberance in a male, was found.
NB	***N***est **B**uilding. Not used for wrens (other than Bewick's, Rock, and Canyon) and woodpeckers.
DD	**D**istraction **D**isplay or injury-feigning. This included broken wing displays of waterfowl, shorebirds, rails, grouse, and quail, and dive-bombing behavior by raptors or other species in nesting habitat.
CN	**C**arrying **N**est material. Not used for some species (*e. g.*, raptors) when nest material was thought to be part of a courtship display.
	PROBABLE BREEDING
B	**B**uilding of nests by House, Winter, and Marsh Wrens, and hole excavation by woodpeckers.
N	Visiting probable ***N***est site, such as when swallows or swifts are observed investigating potential nest holes, or suspicious entry or exit from a bush or tree.
C	**C**ourtship behavior or **C**opulation. Not used for hummingbirds.
A	***A***gitated behavior or ***A***nxiety calls by adults in circumstances that suggested a nest or young was nearby.
T	***T***erritorial behavior, such as chasing other birds. This included mobbing of hawks or corvids by blackbirds, *etc.*, and counter-singing behavior. Not used for hummingbirds.

TABLE 2 (CONTINUED): BREEDING EVIDENCE, RANKED STRONGEST TO WEAKEST, USED TO DETERMINE BREEDING STATUS. Eggs or young of Brown-headed Cowbirds confirmed both the cowbird and the host species.

CODE	EVIDENCE OBSERVED
	PROBABLE BREEDING (CONTINUED)
M	***M***ultiple males. Used when seven or more singing males of the same species were detected singing on territory within a block on a single day. Used only for resident species that have distinctive territorial songs, or for migratory species with territorial songs generally between the "safe dates" of 1 Jun–10 Jul.
S	***S***inging male present in the same territory on two dates that were at least a week apart. Used only when male was within breeding habitat, and generally between the safe dates of 1 Jun–10 Jul for migratory species. Included grouse booming, woodpecker drumming, owl hooting, and other territorial noise-making.
P	***P***air observed interacting in suitable habitat during the breeding season. For species that breed in group situations like Acorn Woodpecker, at least one of each sex within an apparent territorial group. Not used for hummingbirds. This criterion was also used for species that were not sexually dimorphic, if behavioral interactions indicated a pair bond or intention to mate.
	POSSIBLE BREEDING
X	Bird or birds were present within or near suitable breeding habitat, within their breeding season, and within proper geographic range. Not used for suspected transients, migrants, or foraging birds not in breeding habitat.

RESULTS

We collected more than 32,000 bird observation records during the 5-year Atlas period (1995–1999). However, approximately 45% of these records were omitted from our final bird observation database because of: 1) duplication, 2) higher-ranking records superseding lower-ranking breeding records, 3) low reliability, 4) species not considered a likely breeder, or for other reasons. We retained 17,957 records in the final bird observation database, which included only the single highest-ranking breeding observation for each species for each block. These records were used to produce the breeding distribution maps contained in this Atlas. For all Atlas blocks combined, possible breeding accounted for 52% of all final bird observation database records, probable for 22%, and confirmed for 26%. In contrast, for the 91 priority blocks only, possible breeding accounted for 35% of the 5405 records, probable 24%, and confirmed 41%.

A total of 181 species were recorded as possible, probable, or confirmed breeders during the Atlas period: 170 confirmed; seven probable, and four possible. Distribution maps were produced for all 181 species except the Peregrine Falcon. We discuss the status of an additional 16 species that had past evidence of possible, probable, or confirmed breeding in Humboldt County, but for which breeding evidence was not recorded during the Atlas period (Appendix A). Distribution maps were not created for these 16 species.

As a result of restricted access, low habitat or bird diversity, records omitted during data checking, or simple counting errors in the field, the block completion criteria of total species observed was not met in some non-priority (b22, l26, and m22) and priority (H05, H27, and J03) blocks. We were unable to obtain access to the privately owned Bloody Run block (l22), and thus did not survey that block. Considering that access restrictions were so formidable in the rural and privately-owned portions of Humboldt, we were quite pleased that only a single block of the 425 remained completely off limits to atlasers for the duration of the project. For all blocks combined, the average number of species detected was 42 (Appendix B; n = 424 blocks, range = 3–97 species, SD = 16.8 species). Priority blocks had an average of 59 species (n = 91 blocks, range = 34–97 species, SD = 13.3 species). Non-priority blocks had an average of 38 species (n = 333 blocks, range = 3–82 species, SD = 14.5 species). The Crannell block (F11) had the highest number of species with 97 (Table 3). The high number of species in the Crannell block was probably due as much to the close proximity of that block to the "block owner's" (JEH) residence as to the high habitat diversity there.

The most widely distributed breeding species (those with any breeding evidence in the most blocks) were Steller's Jay (380 blocks), Common Raven (352), Pacific-slope Flycatcher (351), Chestnut-backed Chickadee (335), and Dark-eyed Junco (331; Table 4). For 10 species, breeding evidence was recorded in only one block. Appendix D lists each species and the total number of blocks in which it was found, highest-ranking breeding evidence attained, and a breakdown of the breeding codes used.

TABLE 3: BLOCKS WITH 70 OR MORE SPECIES DETECTED

BLOCK NUMBER	BLOCK NAME	BLOCK OWNER	NUMBER OF SPECIES
F11	Crannell	Hunter, J.	97
H15	Mad River Hatchery	Springer, P.	93
N13	Willow Creek	Leskiw, T.	90
H23	Van Duzen County Park	Powell, C.	86
F15	North Arcata Bay	Meyers, C.	83
h14	Blue Lake	Brown, H.	82
F23	Scotia	Bacon-Ogden, L.&D.	81
D21	Waddington	Karsteadt, L.	80
F13	McKinleyville	Schmidt, G.	80
o05	Orleans West	Hewitt, R.	79
g14	Essex Pond	Hunter, J.	78
i08	Childs Hill Prairie	Ogan, C.	78
o16	Todd Ranch	Hunter, J.	77
F17	Freshwater West	Springer, P.&V.	76
J29	Lower Salmon Creek	Gaffin, J.	76
L25	Brightman Ranch	Morris, M.	76
N33	Island Mountain Road	Secor, C.&E.	76
N23	Dinsmore	Shreeve, M.&J.	75
D19	Hookton Slough	Foerster, K.	74
N11	Tish Tang a Tang Creek	Power, J.	74
m11	Hoopa Airport	Higley, M.	74
B23	Bear River Ridge West	Bacon-Ogden, L.&D.	72
L09	Hog Ranch Ridge	Higley, M.	71
P05	Orleans East	Slauson, K.	71
g05	Orick	Devlin-Craig, B.&P.	71
B21	Centerville	Ogan, C.	70

Species confirmed breeding in the most blocks were Spotted Owl (167 blocks; 39% of all blocks), Chestnut-backed Chickadee (154; 36%), American Robin (143; 34%), Dark-eyed Junco (139; 33%), and Steller's Jay (132; 31%; Appendix D). However, these confirmation rates are low when compared to priority blocks alone, where survey effort between blocks was fairly equal. Indeed, confirmation rates for species detected in at least 45 of the 91 priority blocks ranged from a high of 86% (European Starling) to a low of only 2% (Northern Pygmy-Owl; Table 5), and the confirmation rates for the aforementioned species ranged from 82% (Chestnut-backed Chickadee) to 72% (Spotted Owl). Breeding confirmation rates calculated using only priority block data provide a more accurate representation of the relative ease or difficulty of confirming a species using standard atlasing methods. An obvious exception is the Spotted Owl, which was intensively monitored by government agencies and private timber companies in the county during the Atlas period, inflating its breeding confirmation rate well beyond what would be expected for an owl species using our methods (Table 5). A total of 10 owl species were detected in the county during the Atlas period and six species, including the Spotted Owl, were detected in one or more priority blocks. Breeding confirmation rates for the five other owl species found in priority blocks ranged from a high of 33% for the Barn Owl to a low of 2% for Northern Pygmy-Owl (Table 5). Flammulated Owl, Great Horned Owl, and Northern Saw-whet Owl, were 17%, 9%, and 6%, respectively.

Of the 181 species that were recorded as possible, probable, or confirmed breeders during the Atlas period, eight had previously been confirmed in Humboldt, but were not confirmed during the Atlas period (Fork-tailed Storm-Petrel, Leach's Storm-Petrel, Sora, Cassin's Auklet, Long-eared Owl, Short-eared Owl, Canyon Wren, and Northern Parula). Prior to the Atlas project, 181 species had been confirmed as breeders in Humboldt County. Of these 181 known historical breeders, nine were not recorded as possible, probable, or confirmed breeders during the Atlas surveys (Green-winged Teal, Lesser Scaup, Ruddy Duck, Ring-necked Pheasant, Black-necked Stilt, Caspian Tern, Forster's Tern, Rufous Hummingbird, and Bank Swallow; Appendix A). We upgraded five new species to confirmed status during the Atlas period (American Bittern, Rhinoceros Auklet, Barred Owl, White-throated Swift, and Rufous-crowned Sparrow), for a post-Atlas confirmed breeding bird list for Humboldt County of 186 species. Some of these "new" breeders probably nested in Humboldt prior to the Atlas study, but Atlas evidence, either alone or in conjunction with historical data, prompted us to upgrade them to confirmed status.

In addition to confirmed Humboldt breeders, there are 11 suspected Humboldt nesters for which confirmation is lacking. Of these, we recorded some level of breeding evidence for four species (Common Poorwill, Calliope Hummingbird, Mountain Bluebird, and Indigo Bunting). We did not record any breeding evidence for the remaining seven suspected breeders (Northern Pintail, Blue-winged Teal, Ring-necked Duck, Clapper Rail, Yellow-billed Cuckoo, Rose-breasted Grosbeak, and Yellow-throated Warbler; Appendix A). Using both historical and Atlas data, we prepared species accounts for each of Humboldt's 197 confirmed, possible, or probable breeding bird species.

TABLE 4: SPECIES DETECTED IN THE MOST BLOCKS

RANK	/SPECIES	# OF BLOCKS	% OF BLOCKS
1	Steller's Jay	380	89
2	Common Raven	352	83
3	Pacific-slope Flycatcher	351	82
4	Chestnut-backed Chickadee	335	79
5	Dark-eyed Junco	331	78
6	Northern Flicker	325	76
7	American Robin	324	76
8	Wilson's Warbler	312	73
9	Hutton's Vireo	295	69
10	Black-headed Grosbeak	290	68
11	Warbling Vireo	287	67
12	Western Tanager	284	66
13	Wrentit	282	66
14	Orange-crowned Warbler	278	65
15	Turkey Vulture	275	64
16	Song Sparrow	273	64
17	Spotted Owl	272	64
18	California Quail	266	62
19	Hermit Thrush	266	62
20	Hermit Warbler	259	61
21	Cassin's Vireo	256	60
22	Spotted Towhee	248	58
23	Winter Wren	248	58
24	Red-breasted Nuthatch	247	58
25	Red-tailed Hawk	246	58
26	Golden-crowned Kinglet	241	56
27	Allen's Hummingbird	240	56
28	Hairy Woodpecker	229	54
29	Brown Creeper	225	53
30	Black-throated Gray Warbler	222	52
31	Lazuli Bunting	208	49
32	Pileated Woodpecker	206	48
33	Swainson's Thrush	206	48
34	Western Wood-Pewee	202	47
35	Purple Finch	193	45
36	Band-tailed Pigeon	191	45
37	Olive-sided Flycatcher	184	43
38	Western Scrub-Jay	184	43
39	Mountain Quail	178	41
40	Northern Pygmy-Owl	176	41
41	Violet-green Swallow	166	39
42	Mourning Dove	165	38
43	Barn Swallow	163	38
44	MacGillivray's Warbler	158	37
45	Acorn Woodpecker	156	36
46	Bushtit	150	35
47	Anna's Hummingbird	148	34
48	American Goldfinch	145	34
49	Lesser Goldfinch	142	33
50	Western Screech-Owl	141	33

There were a number of miscellaneous observations from the Atlas period that, under a strict interpretation of Atlas breeding criteria, would indicate possible or probable breeding status. However, these records probably do not represent actual breeders. Several vagrant warbler species were detected singing on territory for at least a week. A singing male Chestnut-sided Warbler (*Dendroica pensylvanica*) was at Bayside (g15) 20 Jun–9 Jul 1995 (DFx, JCP). Magnolia Warblers (*Dendroica magnolia*) sang 29 May–10 Jun 1997 at Elk Head (e10; TWL), 1–9 Jun 1997 at Big Lagoon (f08; K. Irwin, DFx), 13–18 Jun 1998 at College Cove (e10; G. Lester, TWL), and 18–25 Jun 1996 at Elk Head (e10; G. Bloomfield, J. Tietz). A Northern Waterthrush (*Seiurus noveboracensis*) did the same 24 Jun–5 Jul 1997 near Trinidad (F11; K. Irwin, TWL). Hooded Warblers (*Wilsonia citrina*) sang 30 Jun–6 Jul 1996 at the Arcata Marsh (F15; J. Harding) and 1–31 Jul 1998 at McKinleyville (F13; G. & L. Lester). Two species never before recorded in Humboldt County were detected for the first time during Atlas surveys: a singing male Black-chinned Sparrow (*Spizella atrogularis*) was observed near Branstetter Ridge (B25) 3–10 Jul 1999 (DFx, M. Wachs, JCP, G. Lester), and a female Great-tailed Grackle (*Quiscalus mexicanus*) was first seen on the Arcata Plaza (F15) 5 Jun 1999 (J. Luther *et al.*).

TABLE 5: SPECIES CONFIRMATION RATES IN PRIORITY BLOCKS

Includes all species detected in at least 45 of the 91 priority blocks.

RANK	/SPECIES	# BLOCKS BREEDING CONFIRMED	TOTAL BLOCKS IN WHICH DETECTED	CONFIRMED %
1	European Starling	39	45	86
2	Chestnut-backed Chickadee	73	89	82
3	Barn Swallow	39	48	81
4	Steller's Jay	69	89	77
5	Dark-eyed Junco	63	82	76
6	American Robin	68	90	75
7	Spotted Owl	49	68	72
8	Song Sparrow	55	77	71
9	Black Phoebe	29	45	64
10	Violet-green Swallow	33	53	62
11	Wilson's Warbler	51	82	62
12	Bushtit	36	60	60
13	Common Raven	48	89	53
14	Western Scrub-Jay	33	62	53
15	Black-throated Gray Warbler	29	59	49
16	Spotted Towhee	34	69	49
17	Orange-crowned Warbler	35	72	48
18	Red-breasted Sapsucker	23	47	48
19	Hairy Woodpecker	35	74	47
20	Red-tailed Hawk	37	78	47
21	Western Tanager	35	74	47
22	Lazuli Bunting	30	64	47
23	Hutton's Vireo	39	84	46
24	California Quail	35	79	44
25	Hermit Warbler	32	72	44
26	Mountain Quail	22	50	44
27	Black-headed Grosbeak	33	80	41
28	American Goldfinch	18	45	40
29	Cassin's Vireo	29	71	40
30	Pacific-slope Flycatcher	37	91	40
31	Northern Flicker	34	87	39
32	Lesser Goldfinch	23	59	38
33	Warbling Vireo	30	78	38
34	Wrentit	28	75	37
35	Downy Woodpecker	17	48	35
36	Swainson's Thrush	19	55	34
37	Acorn Woodpecker	15	47	31
38	Allen's Hummingbird	24	76	31
39	Brown Creeper	20	69	28
40	Red-breasted Nuthatch	20	69	28
41	Hermit Thrush	19	72	26
42	Purple Finch	16	64	25
43	Anna's Hummingbird	12	53	22
44	Winter Wren	15	71	21
45	Western Wood-Pewee	13	65	20
46	Golden-crowned Kinglet	14	72	19
47	Band-tailed Pigeon	11	68	16
48	Olive-sided Flycatcher	9	61	14
49	Mourning Dove	7	56	12
50	Pileated Woodpecker	6	62	9
51	Turkey Vulture	4	77	5
52	Northern Pygmy-Owl	1	48	2

SPECIES ACCOUNTS

PIED-BILLED GREBE *Podilymbus podiceps*

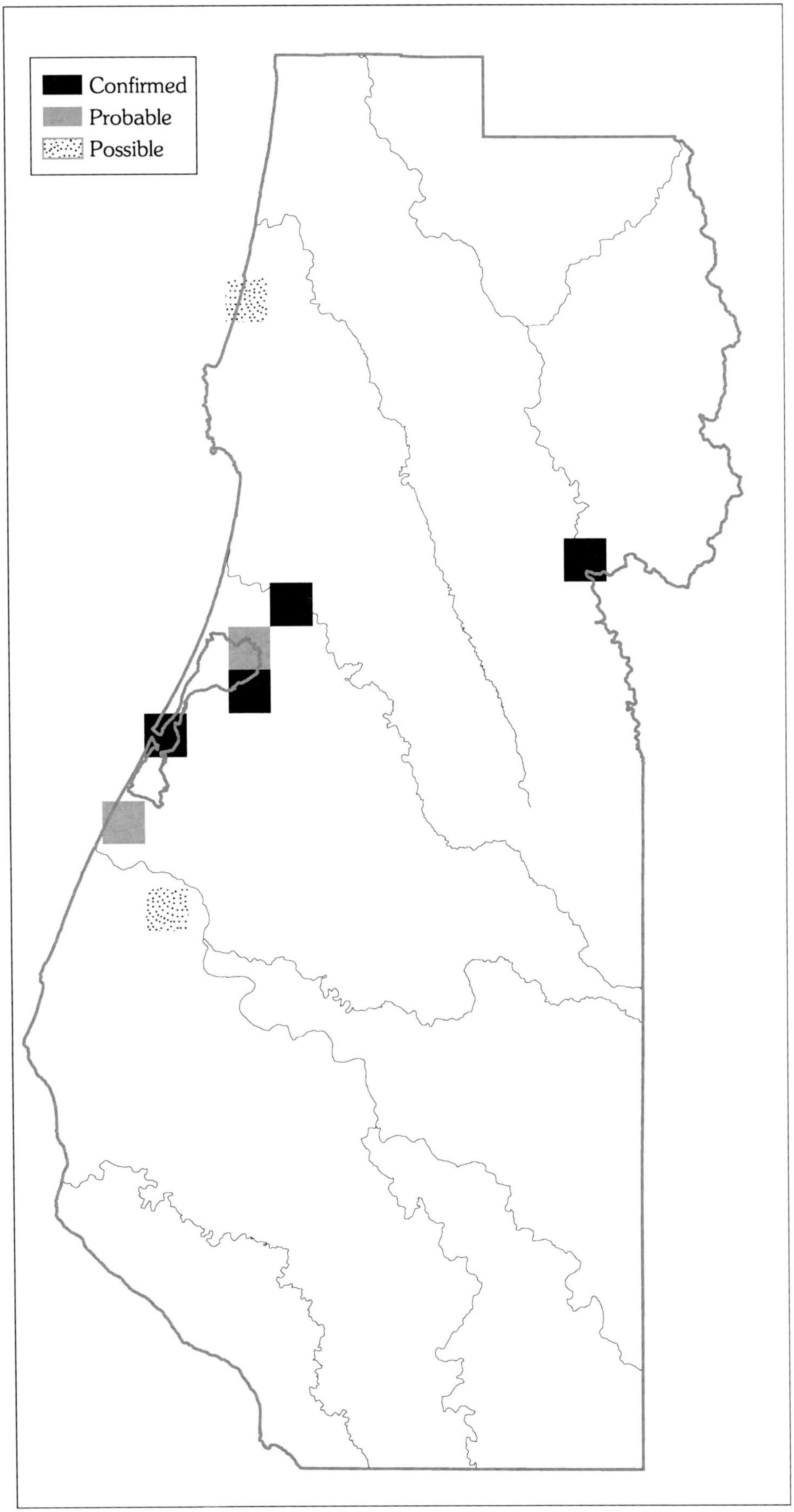

Pied-billed Grebes are small brown waterbirds of freshwater marshes and other shallow-water settings. Common, widespread, and highly migratory, they are given to occasional appearances at even the smallest ponds and marshes with open water. Because they frequently use ponds at such venues as city parks and golf courses, they are recognized by many persons, including those only slightly acquainted with wild birds. They are highly adapted to life at and below the surface. Pied-bills sometimes express air from their bodies, steadily submerging so that only their head and bill remain above water. In an earlier era, this notable behavior earned the species the colorful names "water witch", "sink-duck", and "hell-diver". The lobed toes common to all grebes give them speed and maneuverability beneath the surface—yet their legs are set so far astern that these birds are nearly helpless on land.

As Humboldt County offers comparatively little freshwater marsh habitat, Pied-billed Grebe has been known chiefly as a spring and fall migrant and winter resident. Summarizing this species' status in northwestern California, Yocom and Harris (1975) described Pied-billed Grebe as an uncommon resident and breeder that is more numerous in winter, and mentioned Humboldt's first historical breeding record, of a brood of two young seen 11 Jun and 12 Jul 1960 in the Arcata Bottoms (C. Yocom, S. Harris). Harris (1996) modified this treatment, calling the bird an "uncommon winter visitor and migrant; locally rare to uncommon breeder and summer resident", and noted that a small summer population, consisting of isolated pairs and their offspring, was augmented by the arrival of migrants and wintering birds. While Pied-billed Grebes are apparently restricted to freshwater habitats as breeders in Humboldt, autumn migration brings many to estuarine waters, including the lower Mad and Eel rivers and nearly the entire Humboldt Bay system. South

Humboldt Bay supports dozens of Pied-bills during the colder months, particularly over beds of eelgrass southwest of Fields Landing.

During the Atlas period, considerable numbers of Pied-billed Grebes occurred from October to April on ponds, marshes, lagoons, estuaries, and along slow reaches of the major rivers within the county. These migrants or wintering birds disappeared by May, with the exception of confirmed breeding reports in four blocks, probable breeding in two, and mere possible breeding in two others.

Pied-billed Grebes construct floating or semi-floating mats of dead marsh vegetation that serve as nests. These are loosely anchored to emergent stems or to the bottom of the marsh. In marshes with scant emergent vegetation, nests may be readily observed; in stands of bulrush or cattail, they may be well hidden. As the larger freshwater marshes in Humboldt County are typically grown to cattails, yellow pond-lily, or marsh pennywort, it is not surprising that no nests were detected during the Atlas period. Each of the Atlas confirmations was based on broods of precocial young. Three of the four reports were from freshwater wetlands in the coastal lowlands, and one was well inland. Near the coast, precocial young were encountered at Fay Slough Wildlife Management Area (f16) 31 May 1996 (P. Schmidt); at Essex Pond (g14) 15 Jul 1995 (JEH); and at the "tallow works" south Eureka sewage treatment marshes (D17) 2 Jul 1995 (JEH) and 18 Jul 1998 (DFx). The only confirmed report from an inland site was of a brood at the water feature on the Bigfoot Golf Course just northeast of Willow Creek (N13) 4 May 1996 (DFx, JCP, E. Elias, S. McAllister). A pair was also seen there 13 May 1995 (JEH). There is also an unusual pre-Atlas record from the Bigfoot Golf Course of an adult with three one-third-grown young on 26 Dec 1993 (S. Harris *et al.*; Harris 1996).

Probable breeding status was supported by courtship behavior at the Arcata Marsh (F15) 5 May 1998 (DFx) and by the presence of a pair well within the expected nesting season at Eel River Wildlife Area (c19) 4 Jul 1996 (S. McAllister). In addition, birds were reported near Ferndale (D21) 7 May 1995 (L. Karsteadt) and at Stone Lagoon (F07) 18 May 1996 (JEH). Some other locations of confirmed breeding in Humboldt include Big Lagoon (Devlin-Craig 1992) and the Arcata Marsh (Harris 1996). Historically, breeding conditions for these grebes have probably improved in Humboldt County as freshwater ponds and wetlands were created in areas where there formerly were few or none (Harris 1996).

FORK-TAILED STORM-PETREL *Oceanodroma furcata*

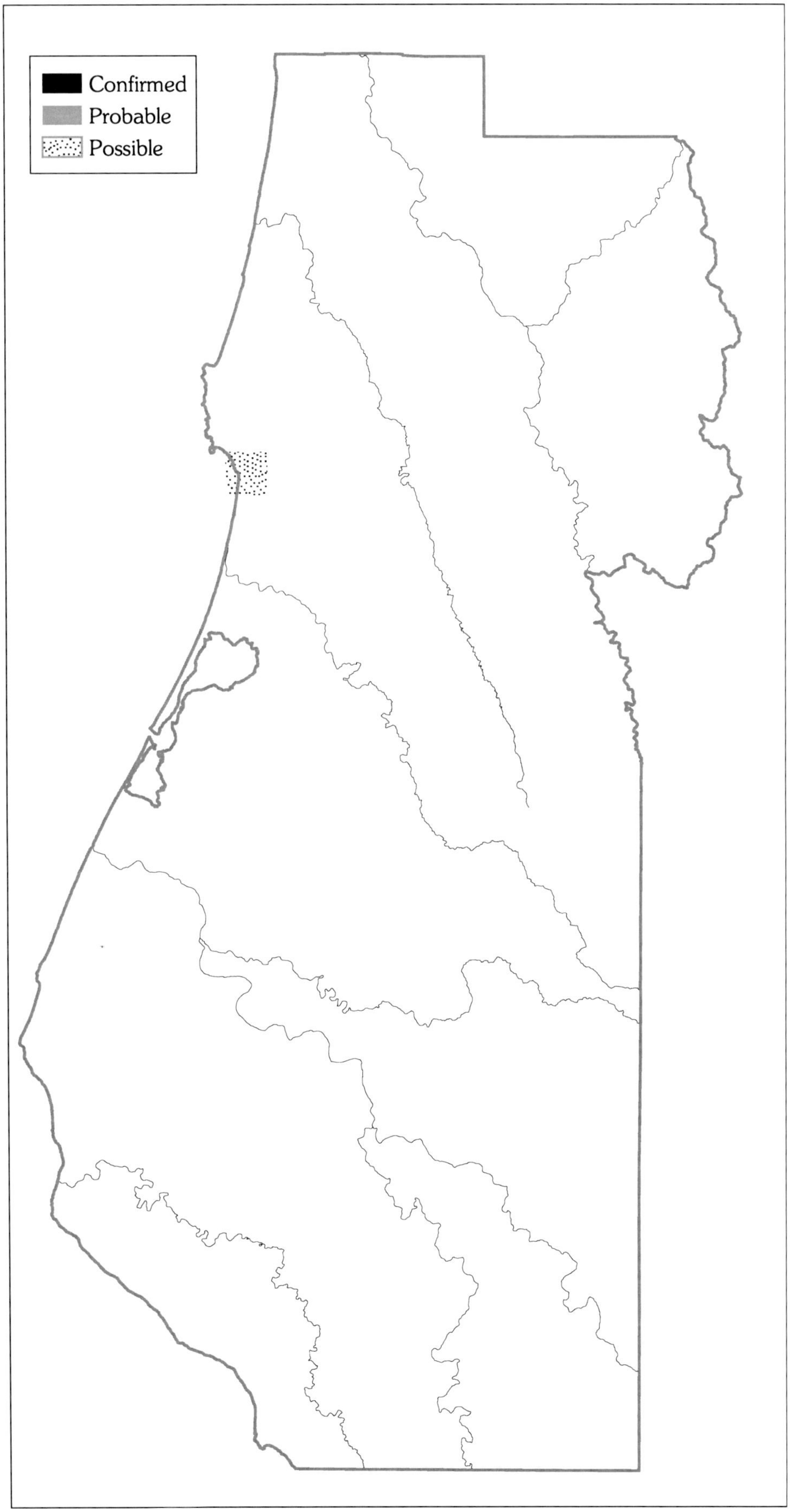

The two species of storm-petrels that breed in Humboldt County exhibit different patterns of seasonal activity and, thus, detectability. Fork-tailed Storm-Petrel is much the rarer of the two. It is a bird of colder North Pacific waters, with the great bulk of the population nesting north of California. They appear in spring in some years right along the coast, seen just within or shortly beyond the entrance to Humboldt Bay and around the harbor in Trinidad. These incursions may be correlated with warmer-water years, in which general upwelling across the continental shelf is reduced, perhaps compelling the birds to forage close inshore. During the day in the nesting season they are seldom seen from shore, as they forage miles offshore over the outer continental shelf and continental slope. However, a keen-eared birder listening to the nocturnal sounds of inshore sea-stacks south of Trinidad might discern their high-pitched scraping calls against the background ruckus-and-hullabaloo of the much more abundant Leach's Storm-Petrels.

Fork-tailed Storm-Petrels are very rare and local breeders on a few small islands and offshore rocks along the coast of Humboldt County. Clay (1916) found eggs and downy young in burrows in the sandy ground of Blank (Sugarloaf) Rock near Trinidad 18 Jun 1916. Nesting in sandy soil is somewhat unusual for these petrels, which typically segregate themselves from Leach's by nesting in rock crevices (Clay 1916, Harris 1974). Harris (1974, 1991) listed Pewetole Island and Green, Flatiron (Off-Trinidad), Blank, Prisoner, Split, Button, Tepona, and Little River (Camel) rocks as either potential or known breeding sites in Humboldt County. Breeding has been confirmed on Blank (Clay 1916, Dawson 1923), Tepona, Green, Prisoner, Little River (Osborne 1972, Harris 1974), and Flatiron (Harris 1991) rocks. However, the actual current status and distribution of this largely nocturnal species is unknown because of the lack of any recent survey information for this species. Apparently, some systematic surveys had been made in recent years, but those data were not available. Harris (1974) provided considerable detail on storm-petrel nesting in Humboldt.

While not confirmed during the Atlas period, a single possible breeder was seen near Tepona Rock (F11) 8 May 1996 (S. Harris, R. Ridenhour). As Tepona, Button, and Prisoner rocks all have been confirmed as storm-petrel breeding stations in the past, it would appear that this sighting may have pertained to a locally nesting bird. On 3 Aug 1999 only two Fork-tailed Storm-Petrels were counted in northern Humboldt waters 75–100 miles offshore (D. Spight, S. Harris, DFx, GCH, JEH, TWL, JCP *et al.*).

Soil erosion and vegetation destruction caused by nesting cormorants may have resulted in local extirpation on Flatiron and Blank rocks and may have reduced numbers at Little River Rock (Harris 1991). The fewer than 100 nesting pairs estimated by Harris (1974) at Little River Rock may represent this species' southernmost colony, as well as California's largest (Sowls *et al.* 1980). Other threats to this species include marauding mammalian predators, oil spills, and human visits to nest rocks.

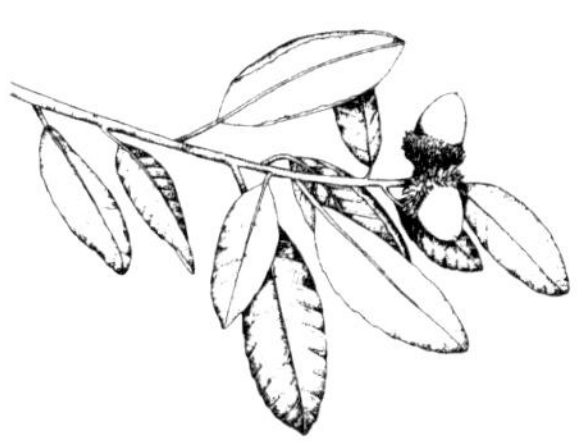

LEACH'S STORM-PETREL *Oceanodroma leucorhoa*

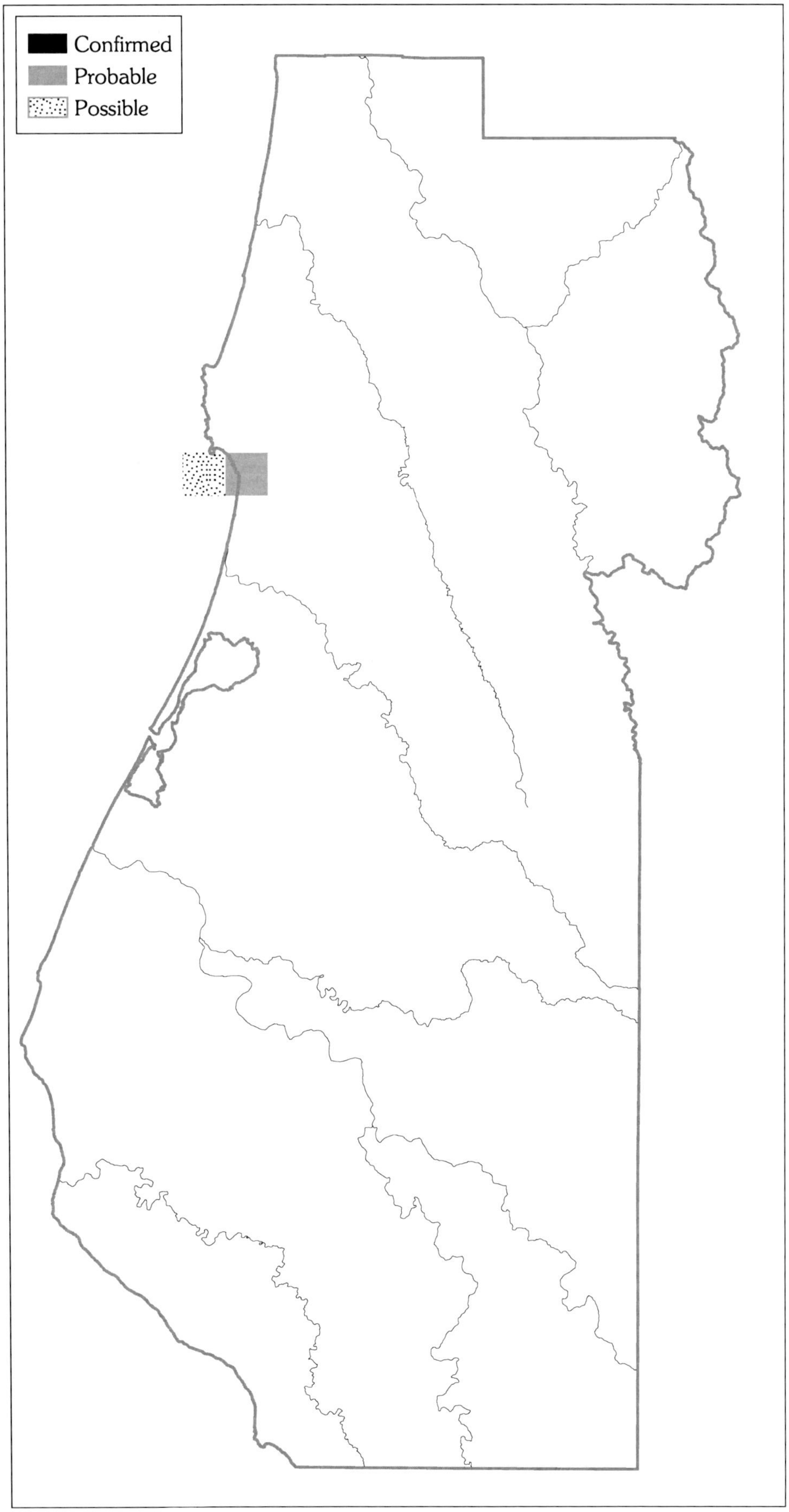

Where no land can be seen, Leach's Storm-Petrels are at ease. They spend most of their lives at sea, returning to the coast only to breed on islands. Other than commutes to and from nest sites, nearly all of their flight activity takes place well beyond the familiar western horizon one sees from coastal points. Many forage 100 miles or more offshore during the breeding season, and during daylight hours, are rarely seen from shore or on the ocean over the continental shelf. Their nocturnal behavior probably evolved in response to gulls and other diurnal inshore predators (Ainley and Boekelheide 1990). They seem generally to be present in waters well out over the continental slope and seaward. Because of this, few persons other than storm-petrel researchers, marine biologists, fishermen, and birders willing to pay for a special trip offshore have gotten to know these birds.

The range of Leach's Storm-Petrel takes in the North Pacific and North Atlantic Ocean basins. In the Pacific, the species breeds on islands from Japan and Russia through the Aleutian Islands and southward to Baja California. This is the most numerous of the four species of storm-petrel breeding in California, with a population estimated at 18,300 in 1980 (Sowls *et al.* 1980). The birds are known or suspected to nest at 13 sites in the state, with 91% of the estimated total occurring on islands of Humboldt and Del Norte counties (Sowls *et al.* 1980). Little River (Camel) Rock is believed to have been the site of the largest colony known in California, supporting an estimated 5000 pairs (Harris 1974).

During the Atlas period, Leach's Storm-Petrels were detected in two blocks in the Trinidad area. Each site is recognized as having hosted nesting birds of this species. At midnight 4 May 1996, Leach's Storm-Petrel became Species Number One for a "Big Day" team when numerous birds were heard calling, and seen with the aid of a powerful spotlight, flying about and landing on Tepona Rock opposite Luffenholtz Point (F11; DFx, JCP, E. Elias, S. McAllister). Although many birds were observed visiting probable nest sites at night, this behavior did not meet the minimum criterion for confirmed breeding. Additional reports of activity at the Tepona Rock colony at night were made 4 May 1999 (COg, J. Sooter), and 10–15 birds were there 8 May 1996 (S. Harris, R. Ridenhour). Two possible breeders were also seen in the vicinity of Prisoner Rock (e11) 19 May 1995 (JEH, TWL). More than 1000 Leach's Storm-Petrels were seen 60–100 miles west of the Humboldt County coast 3 Aug 1999 (D. Spight, S. Harris, DFx, GCH, JEH, TWL, JCP *et al.*). Toward dusk, most, but not all, were seen flying steadily eastward, as if en route to their nesting islands. Some of these far-offshore birds, however, probably nest at sites outside of Humboldt County.

Leach's Storm-Petrels have long been known to breed on small islands and sea-stacks along Humboldt's northern coast. The discovery of breeding storm-petrels on Flatiron (Off-Trinidad) Rock in 1910 by oologist C. Clay spurred the beginning of an era in which he and other workers collected many specimens and eggs almost annually from Trinidad and at Crescent City. This collecting continued into the 1920s (Harris 1996) and provided evidence of these storm-petrels breeding in the county. Since that time, only a few surveys of nesting seabirds in Humboldt have been undertaken. Sowls *et al.* (1980) estimated 1640 birds present on several sea-stacks collectively referred to as the "Trinidad Bay rocks" and 160 on nearby Prisoner Rock. While no nesting took place at Button Rock in 1970 (Osborne 1972), there were about 1000 Leach's there in 1980, possibly birds that had been displaced from Little River Rock. Harris (1974) estimated there were 6590 burrow openings on Little River Rock in 1966, and also provided details on the nesting chronology, habitat, activity patterns, distribution, predation, and other aspects of the biology of the species at this and other sites in Humboldt County.

Harris (1996) considered Leach's Storm-Petrel an abundant local breeder in northwestern California. However, the actual current status and distribution of this largely nocturnal species is unknown because of the lack of any recent survey effort targeting this species. Harris (1974, 1996) listed Pewetole Island and Green, Flatiron (Off-Trinidad), Blank (Sugarloaf), Prisoner, Split, Button, and Tepona rocks and Little River Rock as either potential or known breeding sites in Humboldt County. Following Clay's discovery of breeding Leach's on Flatiron Rock in 1910, breeding has been confirmed on Blank Rock (Clay 1916, Dawson 1923), "Trinidad Bay Rocks" (probably Split, Button and Tepona rocks; Osborne 1972), Little River Rock (Harris 1974), Prisoner Rock (Sowls *et al.* 1980), and Green and Flatiron rocks (Harris 1991).

Harris (1991) noted that colonies known from Blank Rock as late as 1916 and at Flatiron Rock in 1934 apparently either no longer existed or persisted in much reduced numbers a half-century later (Osborne 1972), with none detected at either site. Loss of soil was postulated as a reason for loss of these colonies. Sowls *et al.* (1980) thought that overcollecting and accelerated soil erosion caused by early egg collectors may have contributed to declines on these particular rocks. While Leach's Storm-Petrels had been recorded at Green Rock as late as the 1930s, only one dead bird was found there in 1970 (Osborne 1972). Harris (1996) also stated that Little River Rock had been impacted by soil loss subsequent to the establishment of a Double-crested Cormorant colony at that site. In addition to apparently deteriorating conditions on nest rocks, these birds are at risk from oil spills, natural predators, and disturbance by human visitors.

BRANDT'S CORMORANT *Phalacrocorax penicillatus*

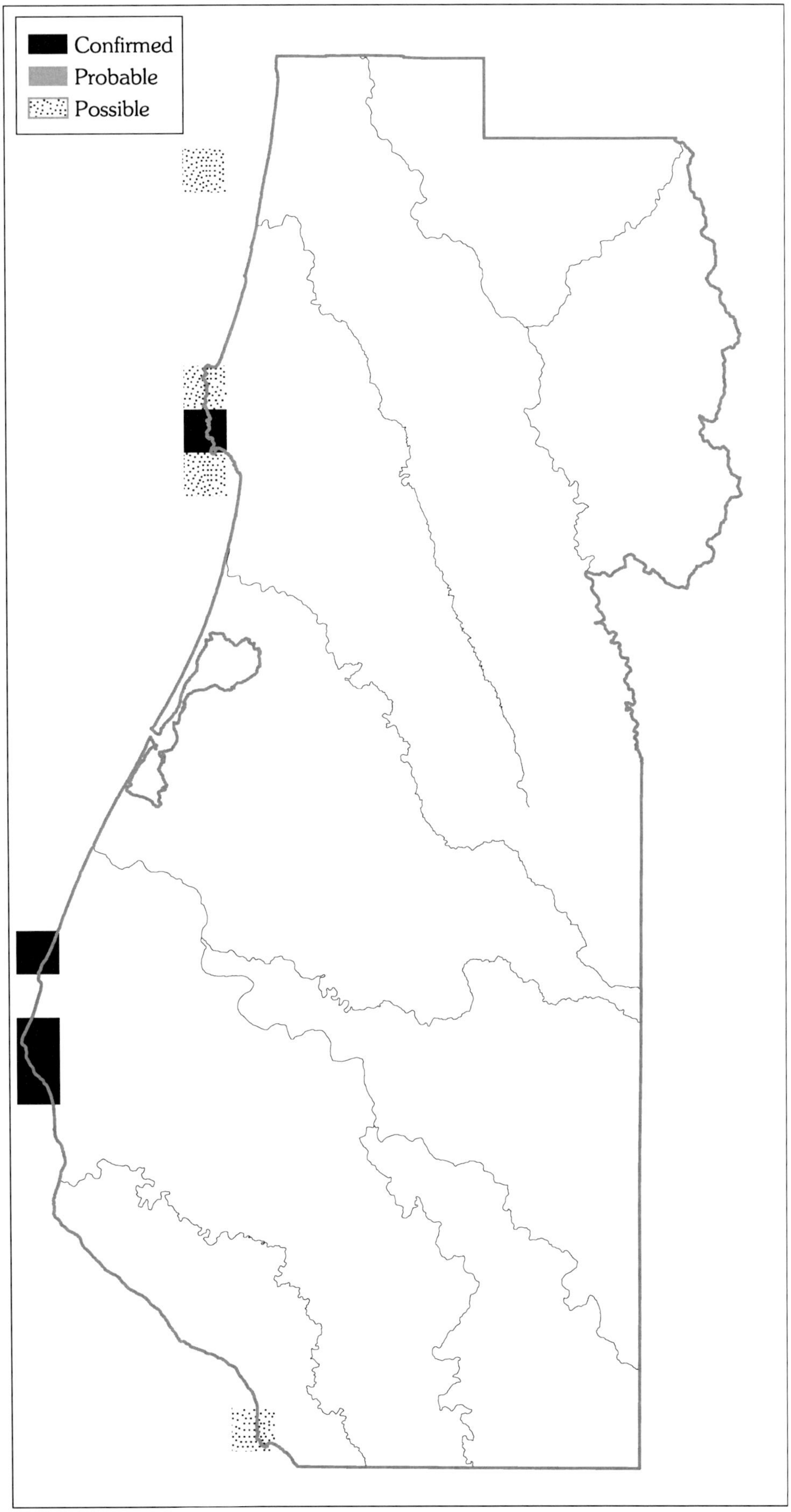

Nearshore ocean habitats from extreme southeastern Alaska to the Pacific coast of Mexico are home to Brandt's Cormorants. These are large, heavy-billed cormorants, appearing all dark at any distance. Adults wear a blue throat pouch bordered below by a pale brown band, and have wispy white plumes on the face and upper back during the breeding season. Seventy-five percent of the world population occurs along the northern and central California coast (Ainley and Boekelheide 1990), with an estimated 930 breeding pairs in Humboldt County (Ainley and Whitt 1973). Most nesting colonies occupy the crests or slopes of offshore sea-stacks.

Brandt's Cormorant was first formally reported from Humboldt County when three birds were taken from a fishing net 12 Jun 1910 at Trinidad (Clay 1911). Since that time the species has been widely noted in northwestern California. It has been considered a common summer resident and breeder, rare in winter (Harris 1996). That characterization held true for Humboldt County during the Atlas period, with the qualifications that it is a fairly common migrant, is locally common from April through late summer, and occurs in lower numbers widely along the coast away from breeding colonies.

While foraging, these cormorants are most numerous in fairly shallow water over the inner continental shelf, seldom venturing more than a few miles offshore opposite Humboldt shores. A few use lower South Humboldt Bay, the Eureka ship channel, and the lowermost Eel River estuary from July to November and infrequently at other times of year. Brandt's Cormorants appear to use the Humboldt Lagoons irregularly in spring and summer. Adults with nestlings to feed commute many miles north and south of colony sites, almost always flying individually but often foraging in the company of Common Murres, gulls, and similar birds concentrated above productive feeding places within several miles of shore. Evidence of widespread foraging activity is indicated by the fact that these cormorants were detected in 13 Atlas blocks that contained no potential nesting habitat (DFx *et al.*). Talmadge (1953b) reported single Brandt's Cormorants foraging with Double-crested Cormorants well inland at the confluence of the Trinity and Klamath Rivers during late August to November salmon runs. There have been no other such reports.

Brandt's Cormorants create a spectacle at colonies. Arriving and departing members of pairs are continually evident. Nesting birds gather billfuls of marine algae or other vegetation from at or near the ocean's surface, then transport this material to the colony where it is incorporated into the messy structures that serve as nests. Breeding evidence was recorded in eight Atlas blocks, with confirmations in half of those. Occupied nests in colonies at Green Rock and Flatiron Rock (both e10) 29 Jun 1995 (DFx) were at Trinidad-area sites known for long use. A portion of Humboldt County coast near Cape Mendocino contains several offshore rocks of both historical and recent use. Colonies with occupied nests were seen at Sugarloaf Island (a24) 4 Jun 1995 (D. Ogden), Steamboat Rock (a25) 10 Jun 1998 (DFx), and at False Cape Rock (a22) 23 Jun 1995 (DFx). The number of pairs nesting at these sites was not estimated. Additionally, Brandt's Cormorants were noted as possible breeders at Point Delgada (f33) 10 Apr 1999 (KMS), near Patricks Point (e09) 29 Jun 1995 (DFx), at Redding Rock (e04) 31 Jul 1998 (DFx), and at Pilot Rock (e11) 31 Jul 1998 (J. Jacobsen). While Redding Rock would seem to offer the extensive sloping ground favored by these cormorants for nesting, it was heavily used as a haul-out by sea lions during the Atlas period and appeared unsuitable for a colony. Brandt's Cormorant colonies are known to shift location from year to year (Sowls *et al.* 1980). While this probably occurs in Humboldt County, it was not specifically noted by atlasers.

Annual northward post-breeding dispersal by Brandt's Cormorants is known through recoveries of cormorants banded on the Farallones. Most recoveries have come from Humboldt Bay and along the Oregon coast (Ainley and Boekelheide 1990). Apparent fall migration of flocks of this species along the Humboldt County coast is occasionally remarked upon by shore-based birders. Brandt's Cormorants are distinctly less numerous in late fall, winter, and early spring; local CBC data indicate that only a very few are to be expected in late December and early January. Numbers swell in April with the return of prospective breeders and subadults.

As is true for other seabirds, Brandt's Cormorants face human-caused challenges ranging from physical disturbance of nesting birds to entanglement in fishing line and nets. Perhaps foremost among these threats is the possibility of an oil spill near a large colony. These cormorants are highly social, not only when nesting but at many foraging venues, and this increases the risk of sudden loss of large numbers of birds. However, mortality of Brandt's Cormorants during oil spills is only infrequently reported, possibly because of a greater tendency for these cormorants to sink (Sowls 1980, Wahl 2001). One long-time Humboldt observer (C. Clay) thought that the size and number of nesting colonies had increased over time (Osborne 1972, Harris 1996).

DOUBLE-CRESTED CORMORANT *Phalacrocorax auritus*

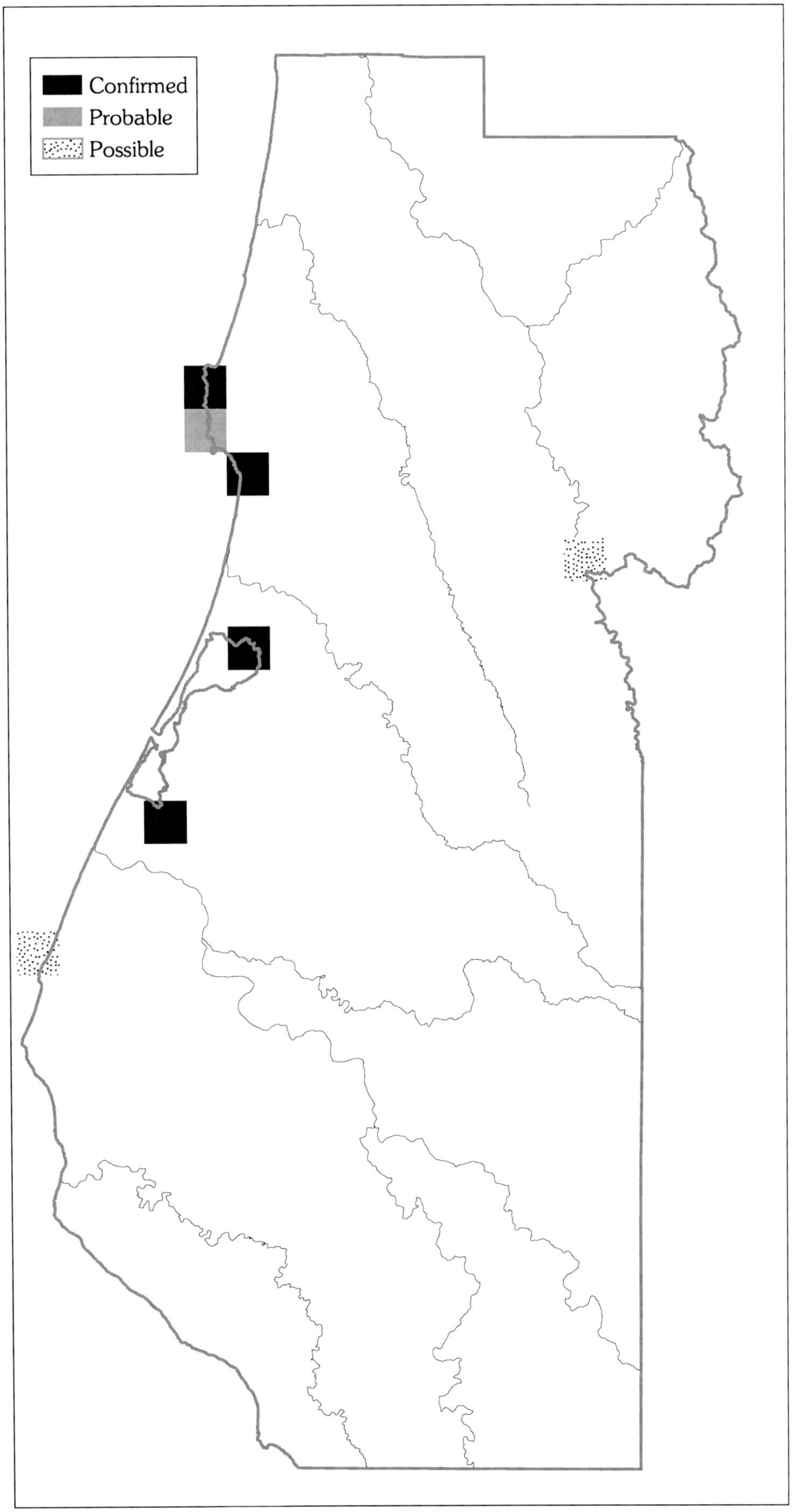

Sitting atop wharves with spread wings or flying in shifting formation low over the water, Double-crested Cormorants are among the most conspicuous waterbirds wherever they are found. Their social nature is exhibited by nesting, fishing, roosting, and flying in groups, although a lone bird, pair, or threesome also are familiar sights. Following the domestic ban on the use of DDT and similar pesticides in 1972, the fortunes of Double-crested Cormorants improved across their North American range. These widespread cormorants have prospered, apparently owing not only to "cleaner" food chains but to reduced persecution, diverse introduced fish populations, and a continent-wide increase in maturing reservoirs. However, the population in Humboldt County today is apparently smaller than in the late nineteenth century, when Townsend (1887) reported a crowded Mad River tree-roost several hundred yards long where "the place was a Babel from their commotion"; the likes of which have not been seen since.

Double-crested Cormorant has long been known as a common resident and breeder along the north coast (Harris 1996). Breeders from interior California and North America probably winter along the Humboldt coast. A few, mostly subadult, birds may be found along inland portions of the Klamath, Trinity, and Eel rivers at nearly any time of year, though unlike their conspecifics in the interior U.S., they apparently do not breed in such settings here. Tree nesting, a strategy used with great success at several Oregon estuaries, has also yet to be documented in Humboldt County. At some difficult sites to observe along Big, Stone, and Freshwater lagoons, these cormorants have been seen roosting in large numbers in Sitka spruce trees, but nesting has not been confirmed there (J. & V. Waters). Inland observations of these birds offer the possibility of the establishment of future breeding colonies, especially since the Humboldt population is apparently expanding. Continued high reproduction may exert population pressure sufficient to cause Double-crested Cormorants to adopt inland or tree-nesting, should all currently available colony sites reach maximum capacity. The semi-regular breeding season presence of birds along the Trinity River at the confluence of Willow Creek (*e. g.*, 12–28 Apr 1996; TWL, JEH) prompted inclusion of that block (N13) as a possible breeding location.

The bulk of Humboldt's population, however, is coastal. Locally, Double-crested Cormorants favor freshwater and estuarine situations, often becoming abundant at river entrances. Birds over the open ocean are an uncommon sight. They have been known to nest at a handful of coastal locations: Sea Gull Rock, Sea Lion Rock, White Rock, Little River Rock, Old Arcata Wharf, Sugarloaf Island (Sowls *et al.* 1980), and False Cape Rocks. In 1997, there were an estimated 451 nests at these locations (P. Capitolo *et al.*, unpubl. data).

The Atlas survey period coincided with a well observed increase and geographic shift in the population of Double-crested Cormorants breeding in Humboldt County. First reported during the summer of 1993 (Harris 1996), a colony at Teal Island in southeast South Humboldt Bay (D19) increased in size in subsequent years. Teal Island is an extensive semicircular levee, long abandoned and now discontinuous in places. It is separated from the adjacent mainland portion of the Salmon Creek Unit of Humboldt Bay National Wildlife Refuge by the sinuous Hookton Slough and Salmon Creek outflows. The Double-crested Cormorant colony is distributed along the island's crest. Five hundred seven nests were counted at this site in 1995, 768 in 1996 (C. Goss), and 1189 in 1997 (P. Capitolo *et al.*, unpubl. data). The formal confirmed breeding record for Teal Island (D19) during the study period was a nest with eggs 28 May 1996 (K. Foerster). Cormorants may have moved to this site as the Arcata Wharf and the duck-hunting platforms in the bay either fell apart or were dismantled.

At least during the first few years of Atlas surveys, Double-crested Cormorants also nested on the Old Arcata Wharf, a massive, burned dock nearly in the center of North Humboldt Bay. For example, occupied nests were reported there (F15) 19 Apr 1995 (S. Harris). Isolation, freedom from disturbance, and a setting surrounded by fish-bearing waters have caused this site to support large assemblages of Double-crested Cormorants throughout the year for decades. Estimates for breeding birds here ranged from 322 to 424 over four censuses during 1971–1980 (Sowls *et al.* 1980), and 19 nests were counted there in 1997 (P. Capitolo *et al.*, unpubl. data.). Eggshell thickness, examined during a study of the Arcata wharf colony by Ayers (1975), revealed that shells of unhatched eggs collected on the wharves were 17–18% thinner than shells of eggs from museums collected prior to 1947, and shells found crushed on the wharves averaged 24% thinner than museum eggshells. Judging by band returns and pesticide levels known in Humboldt Bay, Ayers (1975) suggested that the Double-crested Cormorants nesting on the bay were contaminated elsewhere.

Nesting Double-crested Cormorants were also recorded during Atlas surveys on nearshore sea-stacks. Little River Rock was colonized by two pairs in 1972, increasing to 200+ pairs by the mid-1980s (Harris 1996). An occupied nest was seen on Little River Rock (F11) 26 Mar 1995 (JEH). An occupied nest was also reported 17 May 1998 on a sea-stack just south of Palmers Point, between Patricks Point and Elk Head (e09; JEH). By the mid-1990s, it became apparent that a sizable portion of the delicate grassy turf on the crown of Little River Rock had been defoliated by cormorant excrement. The situation appeared worse at Teal Island, where breeding cormorants had broadly denuded that site. First nesting in and subsequently killing thickets of tall coyote brush, by the mid-1990s they could nest only on the ground; fragments of dead coyote brush were then used in nest construction. By 1999, the island supported live coyote brush only on its southern and eastern portions, where cormorants had not yet nested.

Although not known to nest anywhere around the north-coastal Humboldt Lagoons, Double-crested Cormorants did frequent Big, Stone, and Freshwater lagoons in considerable numbers in spring and summer during the Atlas period. Such foraging venues would seem to be within easy flight distance for cormorants breeding at sites in the Trinidad area. Cormorants with nestlings at the Old Arcata Wharf routinely commute 6 miles across intervening pastureland to and from the Mad River estuary, while Harris (1996) alluded to birds foraging on Humboldt Bay but roosting at night at Trinidad.

PELAGIC CORMORANT *Phalacrocorax pelagicus*

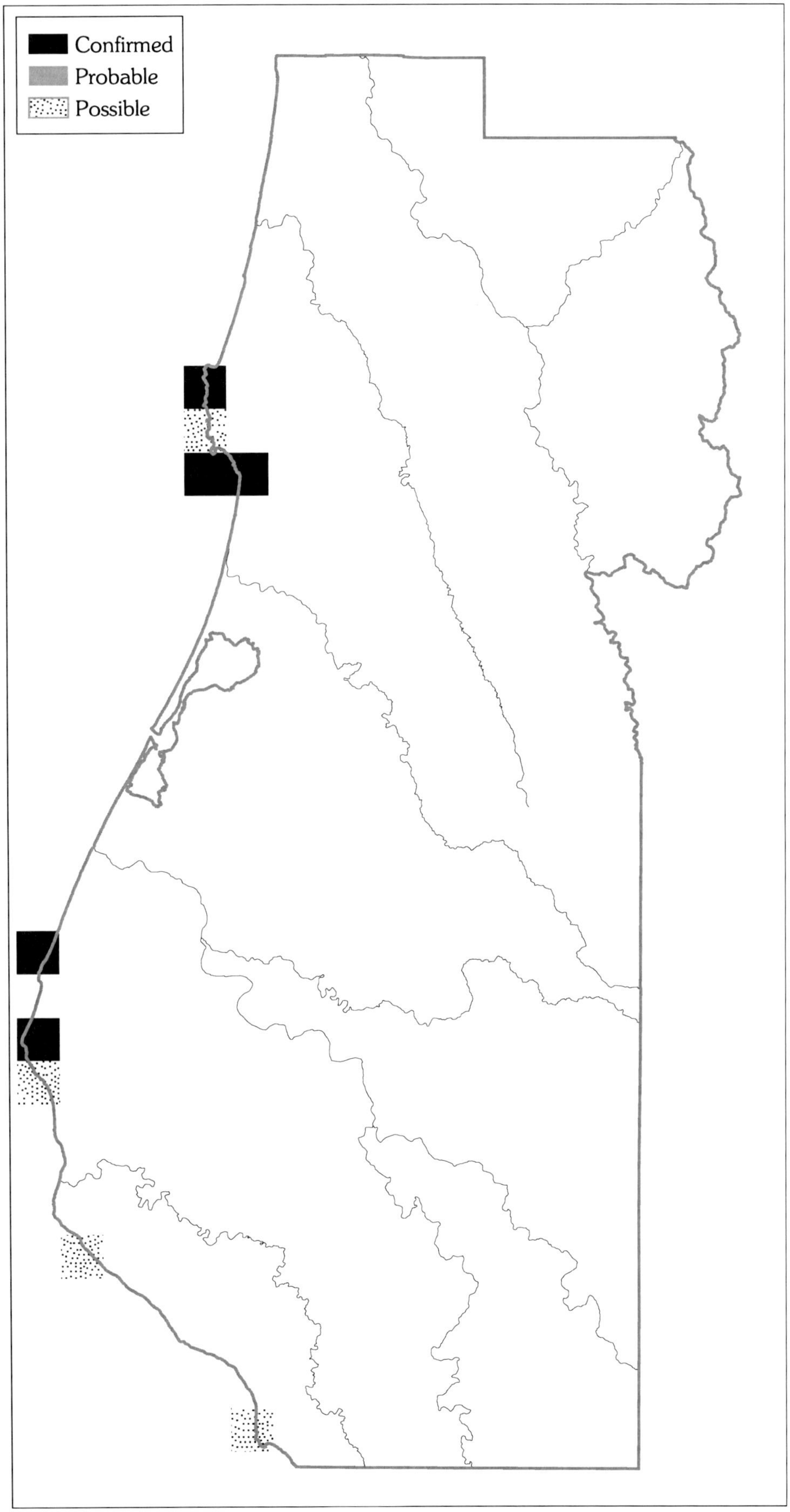

Pelagic Cormorants are slender black waterbirds, resident along cooler North Pacific coasts and well represented in northern California. They are year-round residents in Humboldt County, although the overall distribution of birds varies somewhat by season. They occur only in nearshore ocean waters, with regular use by a few birds of the Humboldt Bay entrance and the Eel River mouth. Pelagic Cormorants are quite common year-round, but it is during the breeding season that they are most appreciated by persons other than birders and naturalists. From April into August, pairs pay frequent visits to nests on sea cliffs, some in proximity to sites of human recreation. They are, therefore, a common element in the spray-misted backdrop of color and activity experienced by beachcombers, surfers, and fishermen during the warmer months in the Trinidad area.

As is true for other common and highly visible waterbirds, this species was first mentioned in the annals of ornithological record in Humboldt County long ago. Clay (1911) provided one of the earliest Humboldt reports, from Trinidad, and he termed the species common there. With reference to northwestern California, Harris (1996) characterized Pelagic Cormorant as a "common local resident and breeder, more conspicuous in summer than winter." This status persisted through the Atlas period. While no recent population estimates have been published, Ainley and Whitt (1973) stated that there were 252 breeding pairs in Humboldt County.

In the manner of Brandt's Cormorants and Common Murres, Pelagic Cormorants forage widely north and south along the coast, often well away from nest sites. In marked contrast to Brandt's and Double-crested cormorants, Pelagics are quite solitary. At least in Humboldt County, they neither assemble in any numbers on the water nor fly in lines. Incidental detections during boat surveys for Marbled Murrelets in June and July have indicated that individual Pelagic Cormorants are scattered sparingly along much of Humboldt's coastline (DFx *et al.*). This suggests either that adults commute miles from the nest or that a nonbreeding population occurs without respect to shore type. Birds known as adults by their white flank patches are sometimes seen flying north or south well away from possible breeding sites. This apparent commuting is displayed far less than it is by Brandt's Cormorants. Whether this is owing to greater numbers of Brandt's Cormorants within the region or to an inherent difference in the foraging strategy employed by each species is unclear.

The phenomenon described above was considered during data-checking and map preparation. The mapping of detections in about one-half of all coastal blocks during the course of Atlas fieldwork would have misrepresented the status of the species as a breeder. With an understanding that many reports fell in blocks with no suitable nesting habitat, such records (pertaining to nine blocks) were dismissed prior to map creation. This species was ultimately reported as either a confirmed or possible breeder in nine of the 425 Atlas blocks. There were no probable breeding reports received from atlasers.

While nesting, Pelagic Cormorants occupy two distinct and separate stretches of coastline within Humboldt County. This distribution is broadly shared by many seabirds breeding locally. These areas extend from Patricks Point (e09) south to Little River Rock (F11) and, 30+ miles south of there, from False Cape Rock (a22) to Sugarloaf Island (a24). Confirmed breeding was recorded in three blocks in the northern portion of the local range and in two to the south. At Wedding Rock (e09) nests with young were seen 31 Jul 1998 (JCP). Occupied nests were noted near Trinidad (e11) 19 May 1995 (JEH), and on rocks offshore from Luffenholtz Point (F11) 22 Jun 1995 (JEH, D. Roberson). Just north of these two blocks, suitable nest sites are abundant to the immediate north of Trinidad Head. In this area, Harris (1996) lists the known nesting locations of Green, Puffin, and Flatiron (Off-Trinidad) rocks, so it is somewhat surprising that this cormorant was not confirmed breeding there. An Atlas-period sighting of Pelagic Cormorants 18 Apr 1998 in this block (e10; JEH) was granted possible breeding status.

The southern segment of the range in Humboldt County takes in the outer Cape Mendocino area, where several places offer appropriate sea cliffs. Occupied nests were seen 23 Jun 1995 on False Cape Rocks (a22) and on Sugarloaf Island (a24; both DFx). Birds seen in blocks with plausible nesting habitat 24 May 1997 south of Punta Gorda (b29; H. Taylor), 10 Jun 1998 at Steamboat Rock (a25; DFx), and 21 Jun 1995 at Shelter Cove (f33; D. Roberson) merited possible breeding status.

Pelagic Cormorants use sea cliffs above the splash zone with narrow ledges or cupped hollows for nesting. Cliffs typically support one to many pairs of cormorants. Pairs of Western Gulls may occupy similar niches on cliffs used by cormorants, and Pigeon Guillemots exploit crannies within or below the elevational limit of cormorant nest placement. Both offshore and mainland nesting takes place in Humboldt County.

Although direct human effects on nest sites are probably uncommon, Pelagic Cormorants face a spectrum of challenges created by human activities. Oil spills and entanglement in fishing nets or line are ever-present threats. Since these poorly named birds spend nearly all of their lives over shallow waters of the inner continental shelf, they may be exposed to chemical contaminants at river and estuarine outfalls. Contrasting with Brandt's and Double-crested Cormorants and Brown Pelican (*Pelecanus occidentalis*), no documentation of significant eggshell thinning attributable to pesticide loading had been noted in Pelagic Cormorants in California as of 1980 (Sowls *et al.* 1980).

Competition with humans for food may also pose a problem for Pelagic Cormorants. A relictual "groundfish" trawl fleet, once much larger, operates from Eureka, targeting rockfish and similar bottom-dwelling species. The general North Pacific trawl industry has been blamed for declines in rockfish that have been described both as insubstantial and critical. It is worth noting that the health of populations of rockfish, a group with low reproductive potential, may bear specifically on the welfare of Pelagic Cormorants. Ainley and Boekelheide (1990) pointed out that size fluctuation in the Farallon Island population of Pelagic Cormorants has been closely tied to changes in the abundance of juvenile rockfish.

AMERICAN BITTERN *Botaurus lentiginosus*

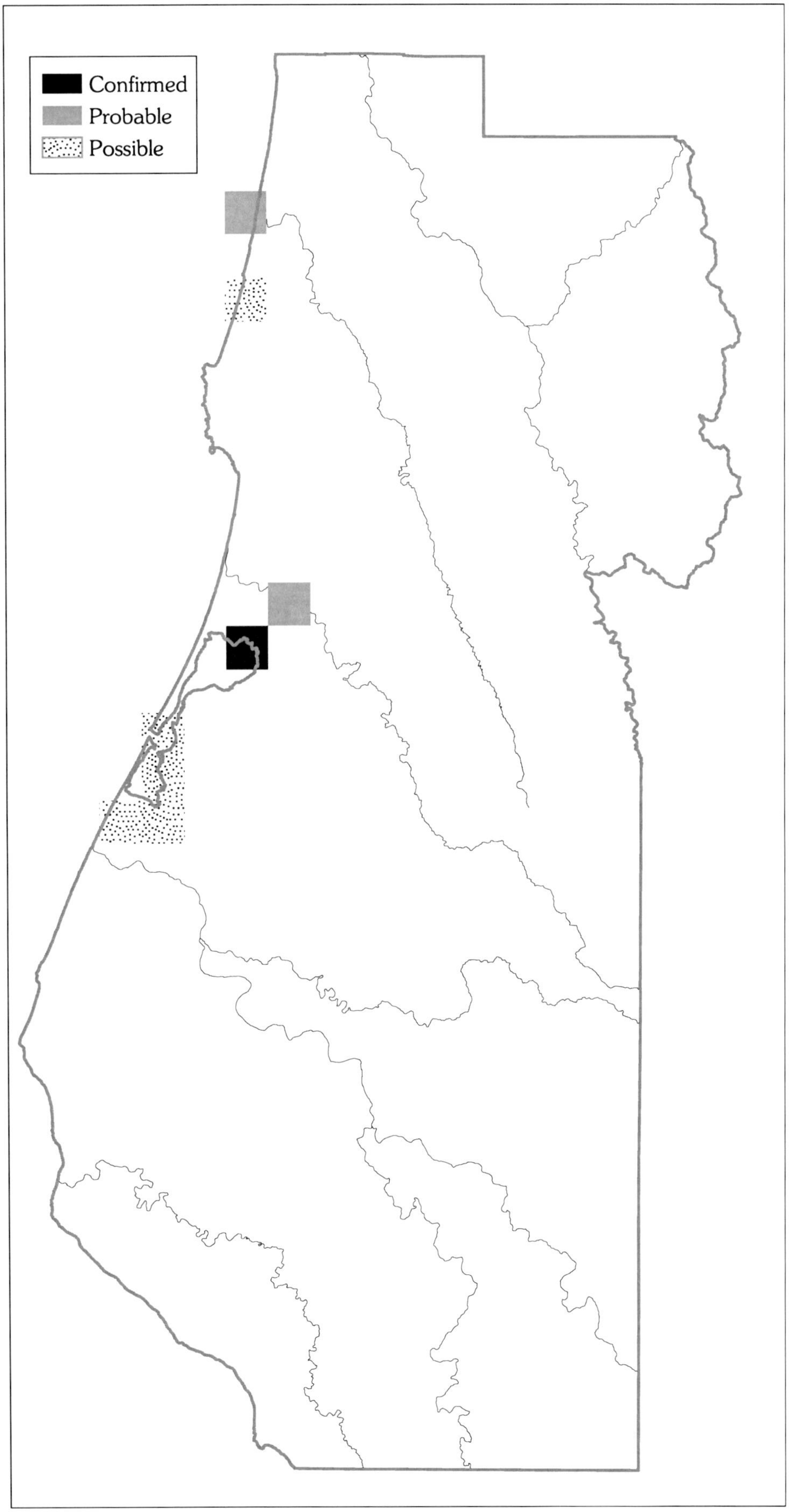

Big brown herons given to protracted moments of unease before human eyes, American Bitterns require the quiet seclusion of freshwater marshes. As such settings are uncommon and localized in Humboldt County, few of these birds are able to breed here. While also present in some brackish and saltwater marshes, breeding status in these habitats in Humboldt and elsewhere in California (Shuford 1993) is unclear. C. Fiebig considered the bittern of irregular occurrence at Humboldt Bay (Townsend 1887). More recently considered a rare to uncommon resident and breeder in northwestern California (Harris 1996), the bittern is most typically encountered from October into mid-April. Confirmed breeding records are few and, prior to the Atlas period, nonexistent for Humboldt County proper. While the validity of such an old report is unknown, Yocom and Harris (1975) mentioned that "M.P. Hansen recalls mowing over nests of bitterns in lowland hayfields near Loleta in the 1920s."

During the 5-year survey period, only a very few detections of American Bitterns were made during the breeding season. Each of these occurred at wetlands along or near the coast from the Redwood Creek estuary to the south end of Humboldt Bay. While long suspected to be a breeder here, the county's first confirmed breeding—and the only confirmation during the Atlas period—was not recorded until 29 Jul–3 Aug 1995, during which time two still-downy fledglings were seen and photographed at Butcher Slough logpond at the Arcata Marsh (F15; G. Green). This pond is a managed freshwater impoundment with broad-leaved cattail, marsh pennywort, and similar emergent aquatic vegetation. A bittern had been heard earlier in the season giving characteristic "pumping" territorial calls from within the marsh. Attention to the outcome of the suspected nesting provided a stamp of achievement for marsh-bird enthusiasts involved in the Atlas project.

American Bitterns were also recorded in seven other blocks during the Atlas period. During late May and early June 1995, a pumping bittern was heard in a rushy pasture edge swale near the end of Hufford Road in the lower Redwood Creek estuary (f05; DFx, JCP), indicating that a territorial male was in residence. In subsequent years, cattle were allowed access to this isolated wetland and trampled it to the extent that the vegetative cover was no longer suitable. The other probable breeding report also referred to a male pumping 29 Apr–15 Jul 1995 at Essex Pond (g14), a large old millpond northeast of Arcata that is heavily grown to yellow pond-lily (JEH). Because Essex Pond is such fine habitat, and is so extensive, it is possible that American Bitterns regularly summer there (S. Harris, pers. comm.). As Essex Pond is privately owned and somewhat secluded, it was not intensively surveyed by atlasers.

The five possible breeding observations involved single birds encountered only once: 18 Mar 1996 on the Salmon Creek Unit of Humboldt Bay National Wildlife Refuge (d18; M. Hinton); 21 Apr 1996 in an unmanaged marsh just southwest of the junction of Hookton Slough Road with Eel River Drive (D19; L. Angel); 28 Apr 1996 at Eel River Wildlife Area (c19; DFx); 14 May 1995 at the south Eureka sewage ponds north of the end of Pound Road (D17; GCH); and 18 May 1996 at Dry Lagoon (F07; JEH).

Most potential American Bittern habitat in the county occurs in publicly administered wetlands. Thus, the future of this species as a breeding bird here hinges on proper management by government agencies overseeing these strategic sites.

GREAT BLUE HERON *Ardea herodias*

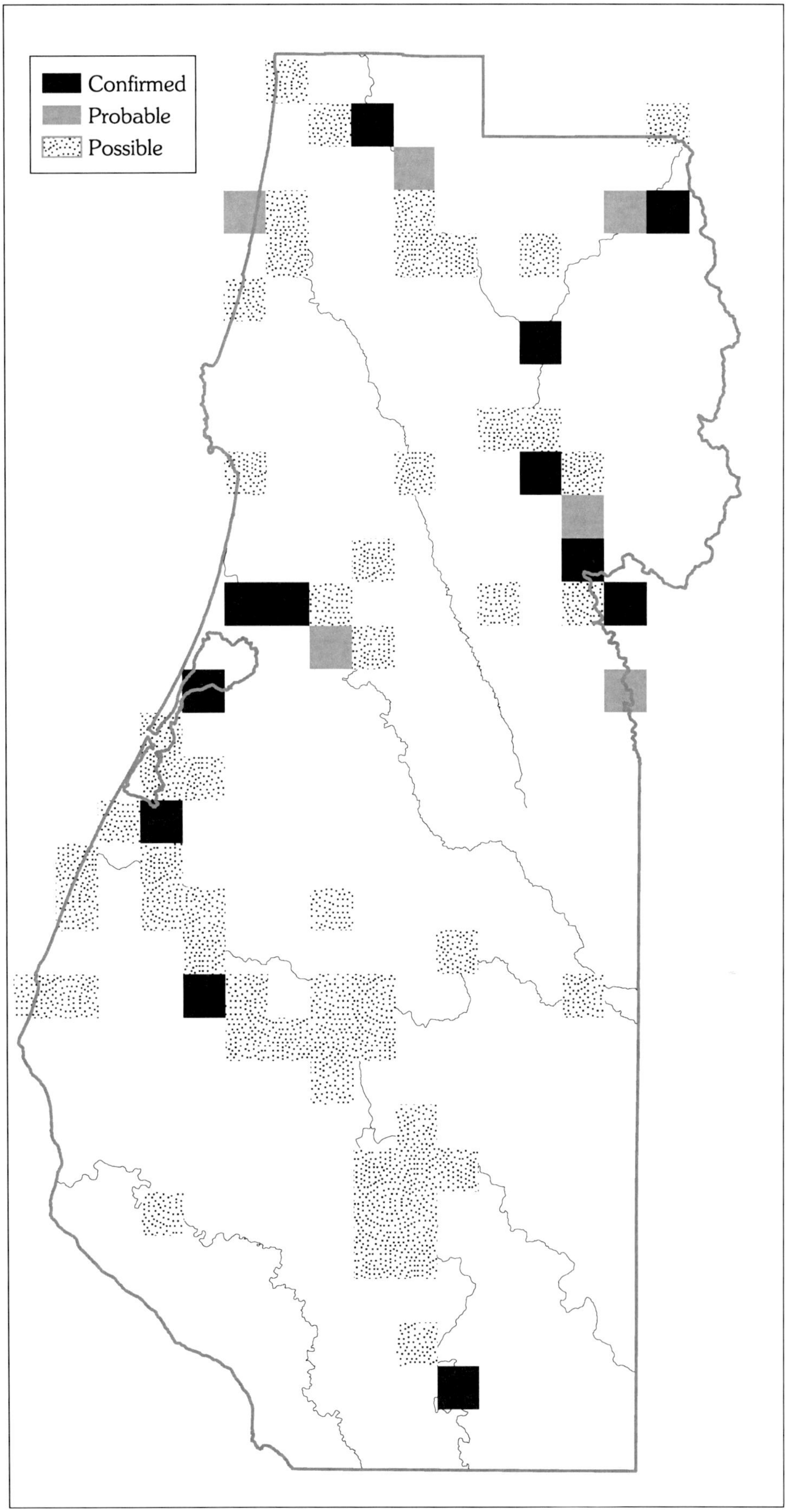

Nearly every body of water within the Great Blue Heron's extensive range in temperate North America is visited by these birds at some point in the year. Humboldt County is no exception. In northwestern California, the Great Blue is the most widely distributed heron and is a common resident and breeder (Harris 1996). These herons can be seen foraging along ponds, rivers, marshes, and lagoons, as well as in outer-coastal habitats such as beaches and rocky shorelines. While by far the largest numbers of birds occur about Humboldt Bay and the Eel River delta, a few may be found at nearly any time of year at any sizable wet area. Upland habitats such as cattle pastures are also frequented, but unlike Great Egrets they appear to prefer to feed in the absence of cattle. These birds are also occasionally noted flying over seemingly unsuitable habitat, such as coniferous forests, perhaps en route to or from some favored feeding site hidden among the trees.

Although Great Blue Herons are resident here in a strict sense, there may exist migratory, wintering, or post-breeding components of the population. These remain cryptic owing to the solitary nature of these birds, and to absence from places where they would be truly unexpected. While nesting, Great Blue Herons are widespread but localized. Unlike egrets, these herons nest at many sites throughout the county and do so either as single pairs or in groups. During the Atlas period, confirmed breeding was noted in 12 blocks. Generally, nesting seems to be restricted to the coastal slope and inland along major waterways.

A distinct pattern of reports of possible breeding from the Eel River and Van Duzen River corridors suggests that undetected colonies of herons may exist in these areas, although a few breeding confirmations were reported from these locales. For example, three nests with young were found 24 May

1995 in a 43-inch-diameter Sitka spruce in a stand of 70-year-old coast redwoods about 1500 feet from the Eel River (e23; C. Hibbard).

The Klamath, Trinity, and Mad rivers also host their share of nesting Great Blue Herons. An entire rookery, composed of about eight Douglas-fir trees containing nests with eggs and young, was illegally cut down 4 Apr 1996 (the earliest occupied nest reported) along Warren Creek Road (g14) near the lower Mad River (*fide* J. Schimke). Only slightly less gruesome was the account of a young Great Blue Heron forcibly taken and eaten by an adult Bald Eagle, from a lone nest north of Orleans along the Klamath River (P05) 2 Jul 1996 (the latest nest with young reported; T. Kirk). Other waypoints in the chronology of this particular nest were an adult incubating eggs 13 Jun 1996, and one small nestling seen 27 Jun 1996 (T. Kirk).

These herons also nested in the Hookton Road (Table Bluff) rookery (D19), as evidenced by an occupied nest there 26 Apr 1996 (J. Kidd). The multi-species heron rookery on Indian Island (e16) contained nesting Great Blue Herons in each year of the Atlas period. Observations from this site included adults carrying nest material 29 Apr 1995 (JEH), nests with eggs 5 May 1996 (KMS), and nests with young 27 Jun 1996 (*fide* K. Kovacs). This species has nested at Indian Island since at least 1943, with the highest count there of 87 nests in 1972 (Beals 1994). In 1996, nine adults, nine nests, and eight young were counted there (*fide* K. Kovacs). Beals (1994) also provided other details on nesting of Great Blue Heron at Indian Island, and Kelly (1987) discussed a colony in the Eel River delta.

GREAT EGRET *Ardea alba*

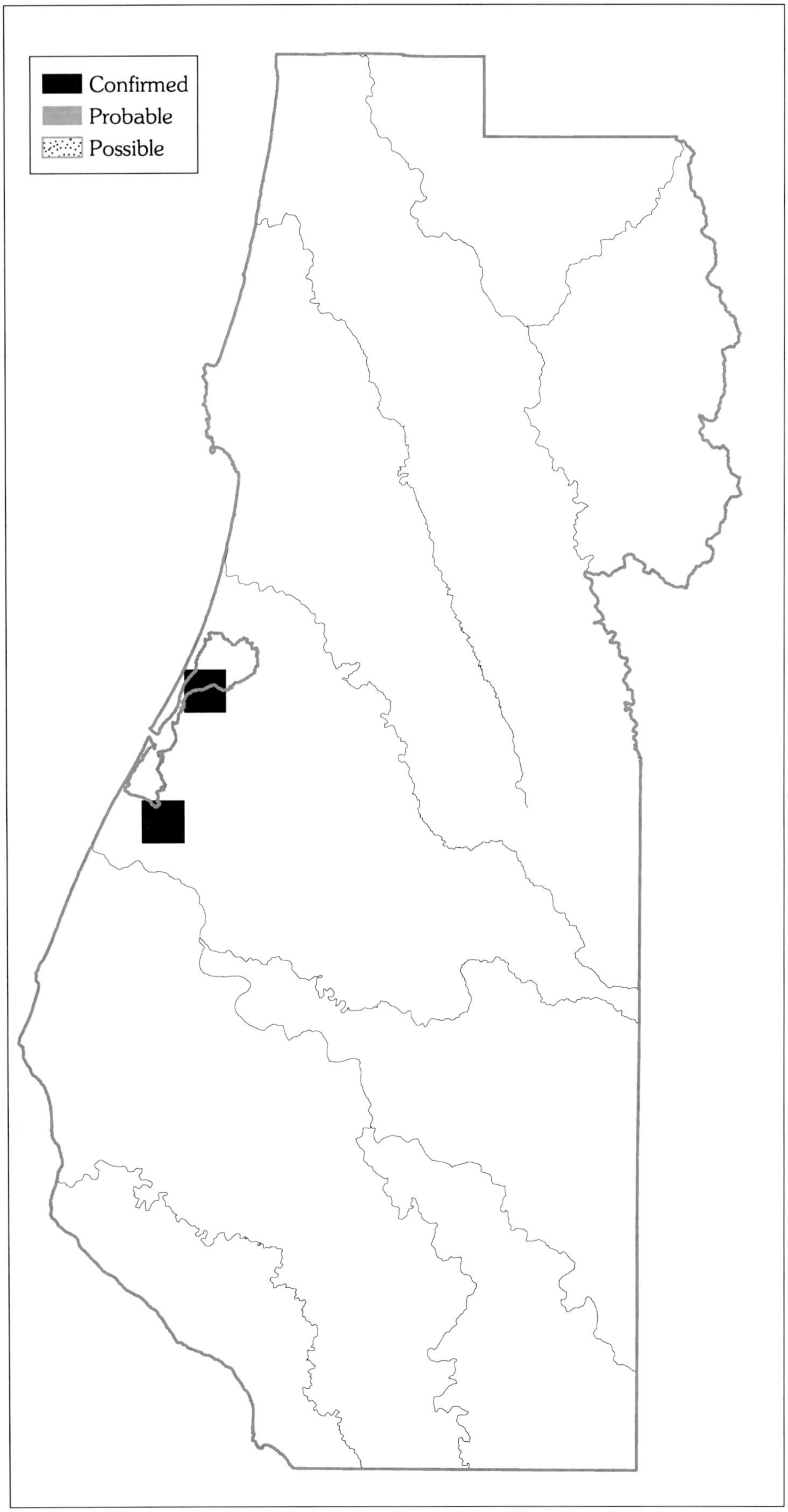

Tall, white, and fond of wide open spaces, few birds are as eye-catching as the Great Egret. This large heron is among the world's most widespread birds and occurs throughout much of temperate North America. California's abundant shores and wetlands offer them a bounty of foraging, roosting, nesting, and dispersal habitat. Nearly extirpated during a millinery craze a century ago, Great Egrets have bounced back. Although a great deal of the historical habitat in California has been lost, some has been restored and, in some cases, created where none previously existed. Habitat changes around Humboldt Bay may have favored this species locally. It is a common resident and locally common breeder in and about the coastal lowlands of Humboldt County, breeding at two sites on Humboldt Bay (Harris 1996).

The Great Egret has a well-documented history of increasing occurrence, growth in numbers, and eventually the establishment of a regular nesting population in the county. While Townsend (1887) "occasionally noted" this species prior to the extensive diking and draining that created pasturelands around Humboldt Bay, limited winter foraging habitat may have precluded establishment of a resident population. Great Egrets forage only in the daytime, and because in winter both daily low tides may occur during the same long night, winter foraging habitat may have been unavailable temporally. The increased availability of pastures, drainage ditches, and other impoundments may have provided extensive foraging habitats that are now available throughout the year (Harris 1996). Harris (1996) summarized the phenomenon of increasing numbers of Great Egrets as a gradual buildup beginning about 1930, with substantial numbers present in late summer and early fall by the mid-1940s.

By the late 1950s the birds had nested in a Sitka spruce grove near Table

Bluff Lighthouse (Yocom and Harris 1975), and by 1960 they were known to be nesting in Monterey cypresses on Indian (Gunther) Island (Harris 1996). Since Indian Island had supported a large colony of Great Blue Herons and Black-crowned Night-Herons since at least the mid-1940s, it is possible that Great Egrets had begun nesting there at some earlier time. There were 60–65 nests on Indian Island in 1967–1968 (Yull 1972) and 251 nests by 1972 (Ives 1973a, 1973b). While information is incomplete, it appears that the Table Bluff rookery was not occupied continuously through this time period.

While CBC data indicate steadily increasing numbers in northwestern California into the 1970s, since then the population may have stabilized (Harris 1996). A count of Great Egrets visible from several viewpoints around North Humboldt Bay 28 Jul 1999 revealed a minimum of 230 birds in sight simultaneously, while a similar count done from the eastern shores of South Humboldt Bay during the same tide resulted in at least 260 in sight at once. As 100–125 birds were also at the Indian Island rookery, it would appear that roughly 600 Great Egrets were present on Humboldt Bay that day (DFx). The proportion of these birds that were breeding was unclear for various reasons, including the fact that post-breeding Great Egrets are known to disperse northward.

During the Atlas period, the rookery at Indian Island (e16) was occupied each year. However, since annual monitoring of nesting areas was not conducted, only two actual records of breeding from that site were received: adults carrying nest material 29 Apr 1995 (JEH) and nests with eggs 5 May 1996 (KMS). Nineteen young were counted at Indian Island in 1996 (*fide* K. Kovacs). Certainly, the sight of scores of adult and nestling Great Egrets in and around the colony on the island continues to be among the most impressive colonial waterbird spectacles to be seen in Humboldt County. The other rookery observed during the Atlas period was in a stand of maturing exotic pines on the north slope of Table Bluff, just above Hookton Road (D19). As is the case on Indian Island, Great Egrets nesting there do so in the crowded company of Great Blue Herons and Black-crowned Night-Herons. This site is several miles east of the historical Table Bluff site mentioned above. Whether the current Table Bluff rookery (D19) was occupied each year during the Atlas was unclear. Reports were received of occupied nests there 15 May 1997 (M. Wachs) and 22 May 1996 (K. Foerster).

While reports were received of Great Egrets in 27 other blocks, it was apparent that all of these records represented foraging birds away from nesting areas. For example, egrets often lingered around the Redwood Creek estuary well into spring during the Atlas period and the behavior of these birds did not suggest breeding. Thorough scanning of nearby forested hillsides also did not yield any evidence of nesting. Other summer foraging areas where these birds were commonly seen included the Arcata Bottoms; the Eel River delta and its backwaters; the lower Mad River floodplain; the shoreline of Humboldt Bay; and even along the edge of Highway 101 between McKinleyville and Fortuna. Shortly before sunset, these feeding birds head for the colony, either individually or in loose groups, some flying several miles across intervening bayland and cityscape. By deep dusk, most birds appear to have returned to the colony. In winter, the rookeries appear to be little used as roosts. A large Great Egret winter roost located in a thick second-growth stand of conifers on the Humboldt State University campus has been known for years. Schlorff (1978) studied foraging ecology of Great Egrets around Humboldt Bay, and Beals (1994) studied nest sites at Indian Island.

While this bird appears to be doing well in Humboldt County, loss or degradation of nest groves and human disturbance around active nests are potential threats. Independence Day fireworks displays over Humboldt Bay are known to cause havoc among the nesting herons and egrets on Indian Island, although the actual impacts on reproduction have not been quantified.

SNOWY EGRET *Egretta thula*

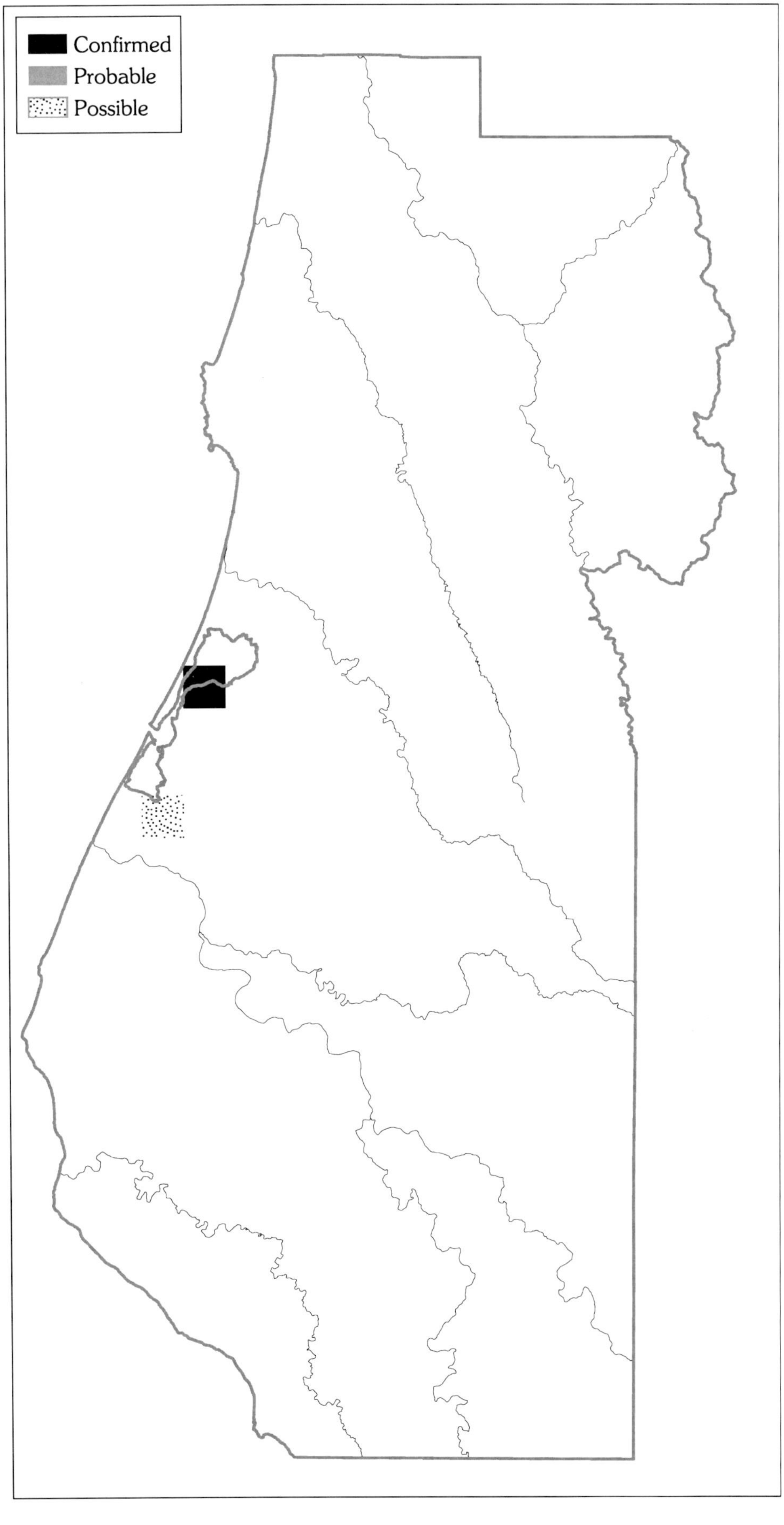

Gauzy white plumes worn by breeding adult Snowy Egrets brought disaster to the species more than a century ago. As fashionable ornaments to women's hats, the plumes of this and other species fueled a savage millinery market. Attractive colonial-nesting waterbirds such as egrets, herons, and grebes bore the brunt of the late spring assault, when hunters shot as many adults as possible, leaving clutches of eggs abandoned and dependent nestlings to starve. An organized public outcry against the practices of the millinery trade secured protection for the birds, and was the phenomenon that spawned the National Audubon Society.

Response to the new protections afforded this bird became apparent after the close of the 1930s Dust Bowl era, and today the Snowy Egret has reclaimed nearly all of its historical range. In addition, the northern frontier of its range has become increasingly dynamic, with new outposts gained in some regions. As is the case for herons and other waterbirds as a group, reduced persecution and the establishment of federal, state, and private wetland reserves have each served to enable this comeback. Snowy Egrets are again widespread in California. Indeed, they are among but a handful of birds that nest nearly to the four corners of the state. In northwestern California this egret is presently a common resident and uncommon breeder (Harris 1996).

Snowy Egrets are present throughout the year around Humboldt Bay and, to a lesser extent, at a few favored sites in the Eel River delta. They are numerous but are decidedly less common than Great Egrets. With respect to habitat preferences and foraging range, Snowies are generally less visible than Great Egrets, although concentrations of some size frequently assemble at preferred foraging venues. During the Atlas period, their abundance and distribution in Humboldt County appeared to change in only minor ways. As was anticipated, confirmed breeding occurred in the mixed-species heronry on Indian Island (e16) at the south end of North Humboldt Bay. The formal report was of nests with young 27 Jun 1996 (*fide* K. Kovacs). During the course of Atlas fieldwork, these birds were observed during the breeding season in five other blocks around Humboldt Bay, as well as along the Eel River at Fortuna (e21) 9 Jul 1996 (R. LeValley). As it became apparent that nesting took place at only one location (e16), and was reasonably possible in only one other (D19), these five outlying blocks were not assigned possible breeding status. A Snowy Egret reported 4 Jul 1996 from the vicinity of the Hookton Road heron rookery (D19; S. McAllister) prompted a designation of possible breeding in that block.

The history of the Snowy Egret as a breeding bird in Humboldt County is detailed by Harris (1996). None of the early observers mentioned them. Harris (1996) recognized the report of a single bird on Humboldt Bay in March 1954 as the first record from the county. Following the first sighting, only three more single Snowy Egrets were seen through 1964. By 1968, seven to 10 birds were regular from October through December (S. Harris, R. Gerstenberg). The first indication of breeding was a report of adults carrying nest material at Indian Island 14 Apr 1968 (P. Yull; Harris 1991). By the early 1970s, actual nesting had been observed among the other herons in the big Indian Island heronry, with three nests and fledglings photographed there in 1972 (Ives 1973b, Harris 1991). Subsequent nest counts at Indian Island were seven in 1977, 18 in 1978, 23 in 1979, 10 in 1980, and none in 1983 (Harris 1991). Beals (1994) described Snowy Egret nest sites at Indian Island. A June 1996 census at Indian Island counted four nests, seven adults, and three young (*fide* K. Kovacs). At least 70 birds were present around Humboldt Bay 28 Jul 1999 (DFx).

Breeding biology of Snowy Egrets in Humboldt County has been known rather circumspectly, with observers first noticing an initial loose assemblage about Indian and Woodley islands. The presence of adults in the Indian Island cypresses was then noted beginning in April. Nestlings may be discerned with a spotting scope in early summer. Juveniles are seen on mudflats and in tidal sloughs throughout the Humboldt Bay system in late summer. This pattern was noted during each of the five Atlas survey years.

Adult Snowy Egrets are somewhat restricted in their foraging range in the Humboldt Bay bottomlands. Most feeding is done in association with water (Posternak 1987, Harris 1996). Notably little feeding takes place in upland sites such as pastures and rank unmanaged grassland. Unlike Great Egrets, Snowies do not commonly forage on median strips along Highway 101 or in roadside ditches, and they are rarely seen in the company of cattle. Increased use of flooded pastures in winter was noticed beginning in the mid-1990s.

CATTLE EGRET *Bubulcus ibis*

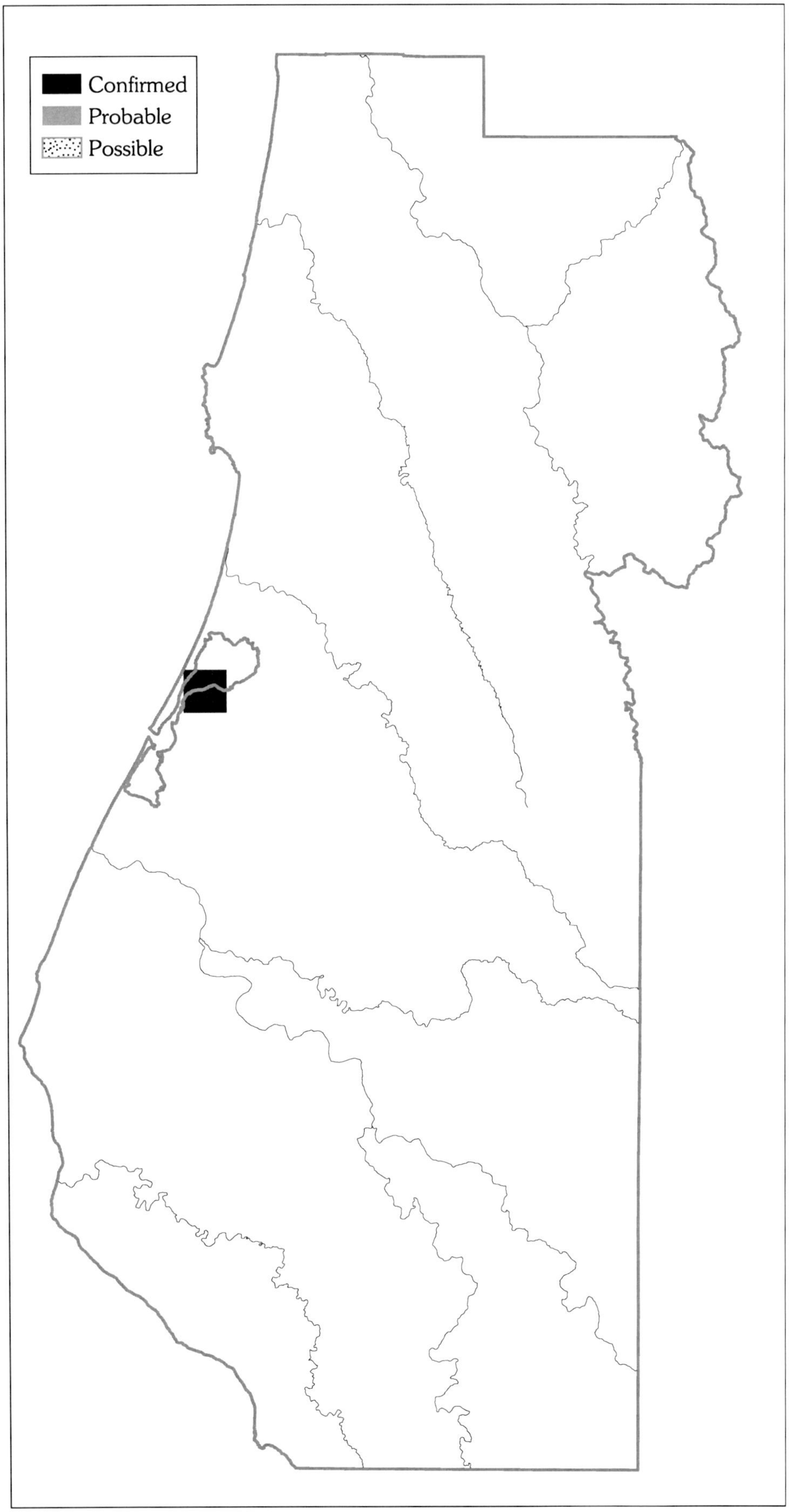

Cattle Egrets are white herons of upland habitats, peculiar for their habit of associating closely with herds of grazing ungulates. They are native to the Old World. Apparently unassisted by humans, they emigrated to northeastern South America in the 1930s where they consorted with domestic cattle. During the 1940s, Cattle Egrets appeared in Florida and began breeding in mixed heron and egret colonies there. In subsequent decades, the species has expanded its range in North America and has now been recorded in all 50 states and most Canadian provinces and territories. Expansion in the West was concentrated roughly in the period 1960–1980, during which time these egrets established themselves as breeding birds in southern and, less commonly, central California. Some 25,000 have nested in recent years in the Imperial Valley.

The earliest historical northern California (and Humboldt County) record of Cattle Egret was of two birds in the Arcata Bottoms 13 Dec 1966 (Harris and Yocom 1967). (A published report of a bird in McKinleyville 6 Jul 1966 [Grosz and Dole 1967] was apparently an error; subsequent examination of photographs showed it to have been an albino Black-crowned Night-Heron [Harris 1991].) The egret has occurred each winter since then (Harris 1996). Yocom and Harris (1975) described it as a "rare winter visitant, becoming more common." Humboldt County's first breeding confirmation was of three nests on Indian Island in 1978, and nine adults with three young known to have fledged there recorded 9 Jul 1978; this was followed by two nests there in the summer of 1980 (S. Kelso; Harris 1996). There was no verified breeding reported from 1981–1995. However, more than 100 birds summered in the Arcata Bottoms in 1985 (G. Lester; AB 39:957). By the beginning of the Atlas period, Cattle Egret was considered a common winter visitor, sporadically casual to rare in summer, and a casual breeder in northwestern California (Harris 1996), although there are no records suggestive of breeding in northwestern California outside of Humboldt County.

During the Atlas period, birders reported only five observations of Cattle Egrets and no breeding; most of these egrets were not near Indian Island. One or two were in the Arcata Bottoms 18 Apr–8 May 1995 (GCH, RJA); one in breeding plumage was at Woodley Island Pond 8 May 1996 (S. Harris, R. Ridenhour); three were at the Arcata Marsh 22 Jun 1998 (S. Harris) and one was there 17 Jul 1998 (P. Lohse); four were in the Arcata Bottoms 1 May 1999 (K. Irwin); and one bird in breeding plumage was on Vance Road, Samoa 10 Jul 1999 (K. Irwin). Thus, one can imagine the surprise of local birders when California Department of Fish and Game provided results of heron and egret monitoring conducted during the period (*fide* K. Kovacs). While the agency did not monitor Indian Island (e16) from 1981–1995, five adults, three nests, and one young were reported there 27 Jun 1996 (L. Coleman); three adults and no nests were observed there 10 Jun 1997 (V. Loverti); five adults, five nests, and two young were there 25 May 1998 (V. Loverti); none were seen in 1999. Also reported were three birds on Woodley Island 29 May 1998 (V. Loverti) and 14 adults roosting there 10 Jul 1996 (V. Loverti). No Cattle Egrets were observed by Fish and Game personnel at any other heron nesting area monitored by them.

The general peak in numbers in Humboldt from the late 1970s through the late 1980s (Harris 1996) was similar to a peak noted during the same period in Santa Cruz County (D. Suddjian, unpubl. data), suggesting a recent declining trend for this species in northern California. While a northward dispersal through California in late autumn has become anticipated each year, most local observers agree that the number of birds reaching the north coast by the mid-1990s was far smaller than in the several preceding decades. Very few of the dispersants that do arrive here in the fall linger beyond January. Between 1995 and 1999, the Arcata and Centerville Beach to King Salmon CBCs reported an average of 10 and three Cattle Egrets per year, respectively. In contrast, the Centerville CBC averaged 34 birds per year from 1975 to 1985.

GREEN HERON *Butorides virescens*

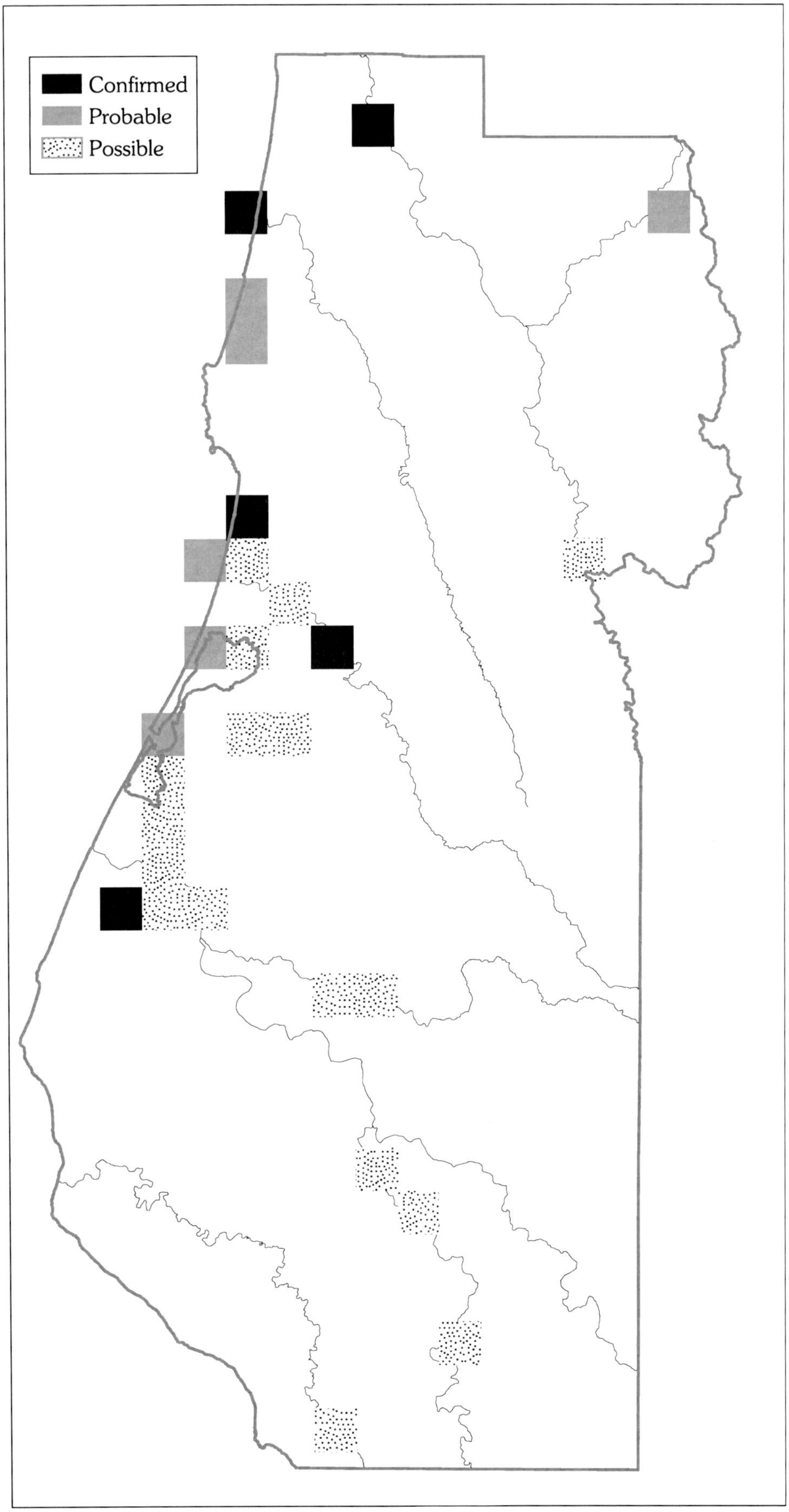

Pond shores, quiet streamsides, and estuarine backwaters are favored hunting grounds for Green Herons. These are small dark herons, with none of the flash of egrets or the handsome cut of night-herons. If this were a deficit, it would be balanced by the simple charm experienced by those who happen to encounter them. Green Herons are most frequent in bottomland countryside, where floodplains or stream corridors offer foraging sites featuring still water in shade. Sluggish rivers, wooded creek meanders, protected tidal channels, and willow-dotted marshes appeal to them. Where the pace of human activity is constant but restrained, Green Herons exploit settings such as small-craft basins, golf course water features, and ornamental ponds. This species is representative of a distinctive genus with far-flung members; the general Green Heron "theme" has proven successful in many parts of the world.

The western portion of this species' somewhat disjunct North American range includes coastal and interior California at lower elevations. Green Herons were infrequently noted during the Atlas period; they were reported from only 28 (7%) of the 425 total blocks. Confirmed breeding was reported from five blocks. With respect to the total number of birds detected, this species, along with American Bittern, would seem to be much in the minority among heron species occurring in the county. Green Herons are found more often in coastal or near-coastal areas of Humboldt County than inland. This is suggested by the cluster of some 16 blocks that stretch from near Westhaven (f12) south to the south side of the Eel River delta (e21). This loose cluster comprised more than half of all blocks with detections. Whether this pattern is due to higher abundance or more frequent birding at popular wetland sites in these blocks cannot be made clear from Atlas data.

Atlas results mirror preexisting perceptions of the status of the species in Humboldt. With indirect reference to the county, Harris (1996) summarized Green Heron in northwestern California generally as a "rare to uncommon summer resident and breeder; rare in winter. Thinly distributed throughout the region." The first report of this species from Humboldt County was one at Myers Flat along the Eel River 4 Jun 1921 (C. Clay; Harris 1996). Recalling this minor historical note was a Green Heron seen near Myers Flat (j28) 25 May 1995 (JG), one of three blocks (i27, j28, k31) in which these birds were detected along the Eel River beyond the coastal influence.

Other inland reports included an adult visiting a probable nest site along the Klamath River near Orleans (P05) 18 Jun 1996 (KMS) as well as one of five confirmed breeding reports, an occupied nest found 15 Jul 1999 in a coast redwood about one-half mile from the Klamath River (i03; A. Gonzales). Possible breeding was recorded at the confluence of Willow Creek with the Trinity River at Willow Creek (N13) 27 Apr 1995 (TWL); along the Mattole River northwest of Whitethorn (H33) 18 Jun 1995 (K. Dyer); and along the Van Duzen River 15 May 1995 (H23; C. Powell) and 2 Jun 1998 (i23; JEH). The sparse detections of Green Herons in the interior of Humboldt County suggest that this species is uncommon there. However, the cryptic plumage and slow movements of these birds, combined with their reclusive habits, cause them to be easily overlooked. Add to this the incomplete coverage of inaccessible streamsides during the course of the Atlas surveys, and it seems likely that the distribution of Green Heron in Humboldt is wider than indicated by the map.

Coastally, Green Herons were seen in both freshwater and estuarine settings. Quiet streamsides and lagoon shores are well recognized habitats. Old log ponds in the settled bottomlands also attract these birds, as do the managed wetlands of the Arcata Marsh and other public wildlife areas that provide suitable cover. The lower portions of streams and sloughs draining into Humboldt Bay and the Eel River estuary have also been known to be inhabited by these herons. Around Humboldt Bay proper, mudflats under old docks and wharves have been used to a limited extent.

Along the coast, there were five probable breeding observations, which spanned the dates of 16 May 1999 (f08; G. Bloomfield) to 24 Jun 1995 (e15; GCH). Confirmations in coastal areas were obtained in four blocks. Adults carrying nest material were seen near Clam Beach (f12) 14 Jun 1996 (TWL) and at the Ferndale sewage ponds (c21) 17 Jun 1995 (JEH). Fledged young were seen in the Redwood Creek estuary (F5) 11 Jul 1998 (B. Freeman) and near Blue Lake (H15) 19 Aug 1995 (PFS).

Although they are surely susceptible to practices that degrade or destroy habitat, it seems that Green Herons have benefited from some effects of human activity. The persistence of neglected log ponds, the establishment of managed wetland parcels, and the addition of human-altered shoreline habitats such as boat basins and breakwaters has afforded this species a selection of foraging sites unavailable in the ages before modern settlement. Given this species' capacity for coexistence at the muddy edges of the human footprint, it would seem that no serious issues attend the future of the Green Heron in Humboldt County.

BLACK-CROWNED NIGHT-HERON *Nycticorax nycticorax*

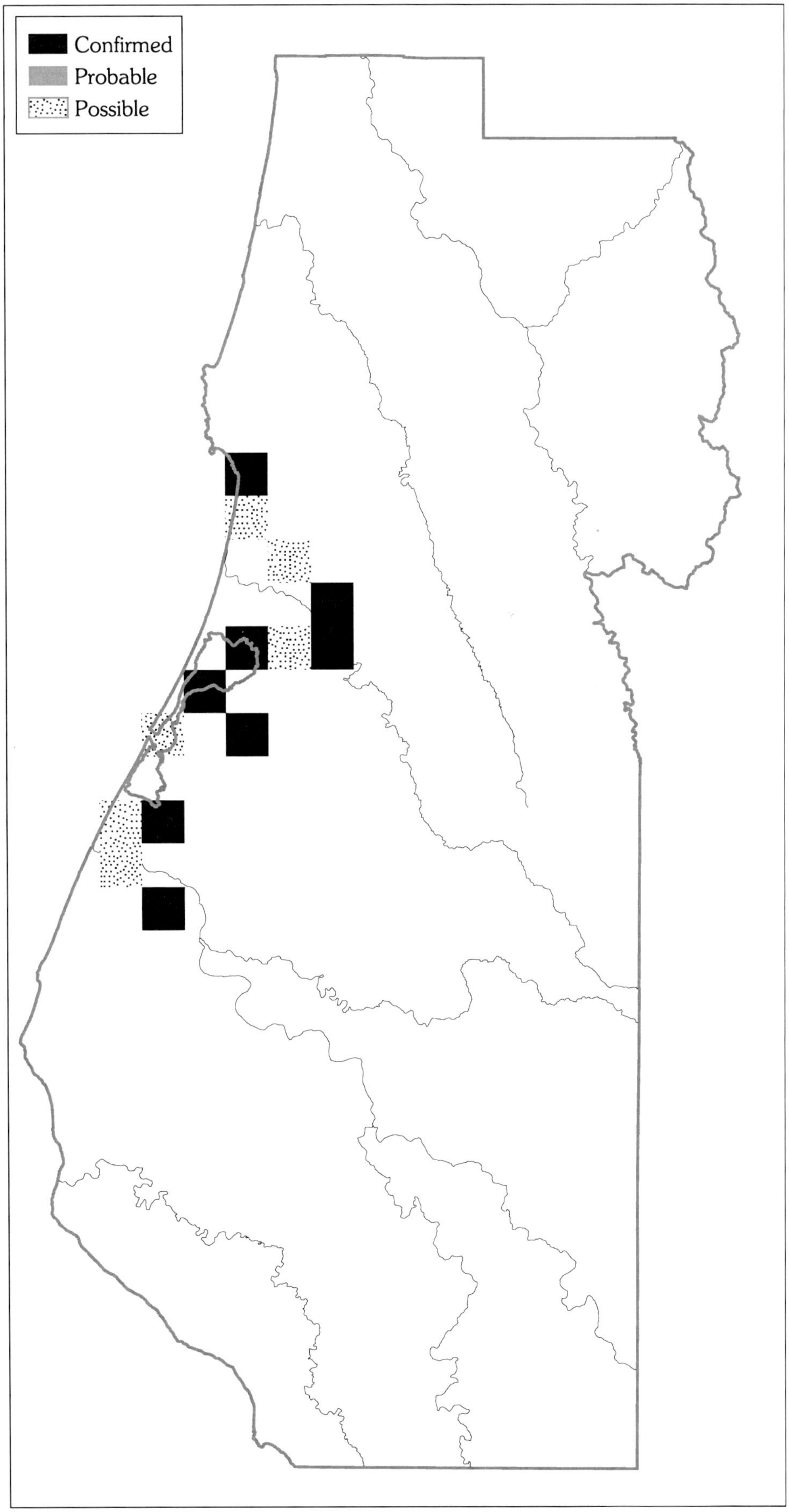

As dusk overtakes the auburn blush of sunset, Black-crowned Night-Herons stir from daytime roosts and fly to outlying foraging sites. The squawk of a night-heron moving unseen through the gloom overhead is a sound of marshlands, riversides, and harbors, but also of heavily managed urban landscapes often adjacent to such places. This species is the most northerly and most widespread of the two night-herons in North America. They are common birds at many of the places in which they are found in California. Forested and mountainous regions of the state lack sufficient habitat to support more than the occasional visitor, but scattered individuals or small groups may appear almost anywhere.

Black-crowned Night-Herons have long been known as permanent residents in Humboldt County. Townsend (1887) reported them as abundant at roosts on the Eel River in November and December 1885, and C. Clay witnessed 1000 birds in the air over the north spit of Humboldt Bay 22 Jan 1911 (Harris 1996). Since then, night-herons have not been estimated in four figures. The earliest historical record of confirmed breeding in Humboldt was that of eggs collected in Eureka 13 Apr 1924 (F. Smith; MVZ 1995). More recently, Harris (1996) characterized the Black-crowned Night-Heron in northwestern California as a common resident and breeder that seems to be confined to coastal lowlands.

Their status in Humboldt County was believed to be well understood at the beginning of the Atlas period. Historically, night-herons have bred at a number of sites in the lowlands of the Mad River, around Humboldt Bay, and in the Eel River delta. Non-breeding roosts, generally vacant or nearly so in late spring and summer, assemble in or around the edges of towns as well as in shrubby wetland settings. Few Black-crowned Night-herons have ever been detected in the interior of the county. None of three reports from well inland during the Atlas survey period carried a likelihood of local breeding and each was deleted from the final distribution map. Reports of Black-crowned Night-Herons during the course of Atlas surveys provided no great surprises, basically reiterating what was known of their distribution in Humboldt prior to the inception of the project.

Black-crowned Night-Herons were confirmed breeding in eight blocks, spread from Little River south to Ferndale. All sites can be considered coastal. Nest building was observed at Blue Lake (h14) 5 Jun 1999 (KMS). A dead chick was found beneath the cypresses above Moonstone Beach near Crannell (F11) 8 Jun 1997 (*fide* TWL), and an occupied nest was reported at Butcher Slough logpond at the Arcata Marsh (F15) 10 Jul 1996 (*fide* K. Kovacs). Night-herons carrying nest material were recorded at the large multi-species heronry on Indian Island in Humboldt Bay (e16) 29 Apr 1995 (JEH). They nested there throughout the study period. The large heron colony on this island also supports Great Blue Herons, Great Egrets, Snowy Egrets, and Cattle Egrets. In 1996, 45 adults, 24 nests, and seven young night-herons were counted at Indian Island (*fide* K. Kovacs). Daytime flights to and from the colony site were seen frequently, with most night-herons appearing to use the canopy of the Monterey cypress stand for nesting. At evening twilight, an irregular procession of night-herons leaving Indian Island and flying in all directions might be observed by anyone driving across the Samoa Bridge or observing from nearby Woodley Island. At least one juvenile bird was reported from the heronry above Hookton Road on the north side of Table Bluff (D19) 4 Jul 1996 (S. McAllister). This colony was previously located elsewhere along the slopes of Table Bluff, but became established in its Atlas-period location in the early 1990s.

While foraging during the breeding season, Black-crowned Night-Herons in Humboldt County use a variety of aquatic and estuarine settings. These include Humboldt Bay tidal channels, sloughs, and saltmarsh edge; estuaries of the Little, Mad, and Eel rivers; managed waterfowl impoundments of Arcata Marsh; Humboldt Bay National Wildlife Refuge; Eel River, Fay Slough, and Mad River Slough Wildlife Areas; and Indian, Woodley, and Daby islands west of Eureka. Dairy pastures appear generally to be avoided, in contrast to the frequent and perennial use of such habitat by Great Egrets and Great Blue Herons. Many night-herons, chiefly subadults, roost during the day in summer on piling clusters or ramshackle wharves at many sites scattered along the Humboldt Bay waterfront from Eureka to Fields Landing. By early fall, increased attendance becomes apparent at winter roost sites such as Mad River County Park, lower Beith Creek in Sunny Brae, the Butcher Slough logpond at the Arcata Marsh, the Carson House Mansion and Hilfiker Avenue in Eureka, and in the towns of Loleta, Ferndale, and Fortuna.

TURKEY VULTURE *Cathartes aura*

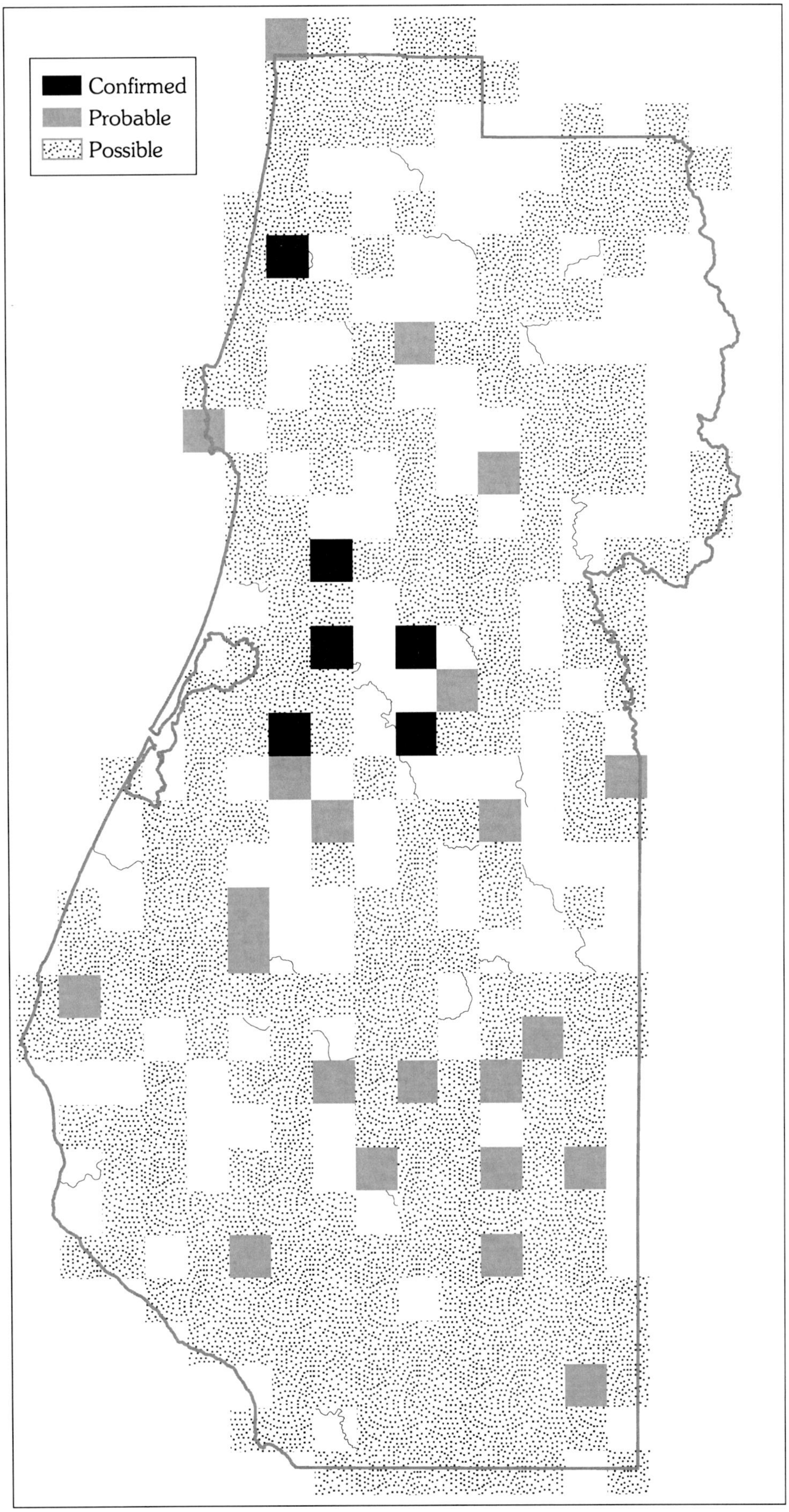

Few birds advertise their presence while foraging as conspicuously as Turkey Vultures. Though not abundant, they were recorded in 65% of all blocks during the Atlas period, and in more than 90% of the priority blocks. And yet few large birds nest in such secrecy. The map of Turkey Vulture in Humboldt County during the Atlas period looks as if the ink might have run. Like the broad sweep of dark wings against the sky, the signature of abundant blocks in which no more than possible breeding evidence was obtained produces a portrait of widespread occurrence—but with only a few breeding confirmations.

While roughly half of the breeding Turkey Vultures in Humboldt are in the air during the early stages of their nesting efforts, the other half must sit in the shadows of secluded nest sites. Nests are not often discovered by humans. Between 1995 and 1999, there were only six breeding confirmations. Set against the total of 275 blocks with detections of Turkey Vultures, the proportion of blocks in which breeding was confirmed was only slightly more than 2%—the lowest percentage for any bird found routinely during Atlas fieldwork. This phenomenon was anticipated at the beginning of the project, and it has manifested itself in the pages of other California county breeding bird atlases (Roberson and Tenney 1993, Shuford 1993, Burridge 1995).

Throughout northwestern California, this vulture is a common migrant and summer visitor (Harris 1996). From the start of the historical record, Turkey Vultures have been common migrants and summer visitors in Humboldt County. Breeding records from the early era of Humboldt ornithology are few. Harris (1996) reported two old records of nests in southern Humboldt, one "100 feet above river at base of rock 30 feet high, slight overhang, no nest" near Island Mountain 18 May 1928 (G. Atwell), the other "from an old root in a marsh" along the Mattole River in June 1930 (G. Howatt).

Because they find carrion chiefly by smelling it, these vultures are able to forage efficiently over nearly all terrestrial and shoreline habitats, including over Humboldt Bay. The distribution map indicates that Turkey Vultures may be expected nearly anywhere in the county during the breeding season. This would appear to be especially true across the southern one-third of Humboldt from Cape Mendocino east through the Van Duzen River corridor, a region in which there were no conspicuous gaps between blocks with detections. In line with the more thorough representation there, the preponderance of blocks with probable breeding also occurred in southern Humboldt. All but one of the probable breeding records were of pairs of vultures. In the county's northern two-thirds, Turkey Vultures were ultimately reported from only about six out of every 10 blocks. Two small but discernible gaps in overall detection occurred there, one taking in blocks draining to the Klamath River canyon or Redwood Creek (centered in the vicinity of k05) and another in mountains east of the Hoopa Valley Indian Reservation (p08). Interior central Humboldt County also was an area where quite a few no-detection blocks fell.

Given that this species was detected essentially throughout the county, it is interesting to note that breeding confirmations were somewhat clustered. It is unknown if this is an artifact of Atlas sampling or if some other factor, such as the presence of potential nest sites in large basal hollows or "goose pens" in old coast redwoods, is at work. An adult on two eggs was found in a burned-out redwood stump 15 Mar 1995 on Simpson Timber Company lands (H15), and on a subsequent visit, the observer (L. Diller) was greeted by two adults and two hissing chicks. Two other chicks were seen in a rocky area on Simpson land (J15) but no other details were available (J. Thompson). Few details were available for any of the other four breeding confirmations, but the earliest report of a fledgling was 31 May 1998 (H13; GAS). Reports of pairs spanned the dates of 3 Apr 1996 (n32; M. Bailey) to 29 Jul 1996 (F21; JEH). The only non-pair probable breeding record submitted was a report of a vulture going into a cave near the confluence of Grouse Creek and the South Fork Trinity River (o18) 25 Jun 1995 (C. Kent).

Humboldt County's expansive sweep of lightly peopled countryside offers food for Turkey Vultures wherever life leaves off. During the warmer months, the shadows of soaring Turkey Vultures pass daily over forested slopes, rangeland, highway corridors, dunes and beaches, and the margins of the urban environment. Having scavenged the results of natural animal mortality for ages, these vultures now benefit from roadkill along rural and suburban roads. Diseased or foundered cattle and sheep constitute an additional food source in rangeland or pasturage where dead stock is not recovered. Prehistorically, marine mammal and elk (*Cervus canadensis*) carcasses undoubtedly constituted a greater proportion of their diet than they do nowadays. Turkey Vultures probably fed alongside California Condors (*Gymnogyps californianus*) in Humboldt (Smith 1916, Grinnell 1933, Harris 1996).

Although they are widespread during spring, summer, and fall, few Turkey Vultures are seen outside the coastal and near-coastal lowlands from November through February. Talmadge (1954) considered observations of a bird in the winter of 1953–1954 in Hoopa to be quite unusual, and this species has subsequently been reported only once (26 Dec 1989; E. Gilchrist) on the Willow Creek CBC. In marked contrast to the situation inland, the coastal lowlands around Humboldt Bay and the Eel River delta currently support a significant wintering population, which appears to be larger than it was a few decades ago. The Centerville Beach to King Salmon CBC recorded an average of 128 vultures per count 1995–1999. Together with birds present in the Sacramento Valley, these birds represent the northernmost Turkey Vultures regularly wintering in North America.

CANADA GOOSE *Branta canadensis*

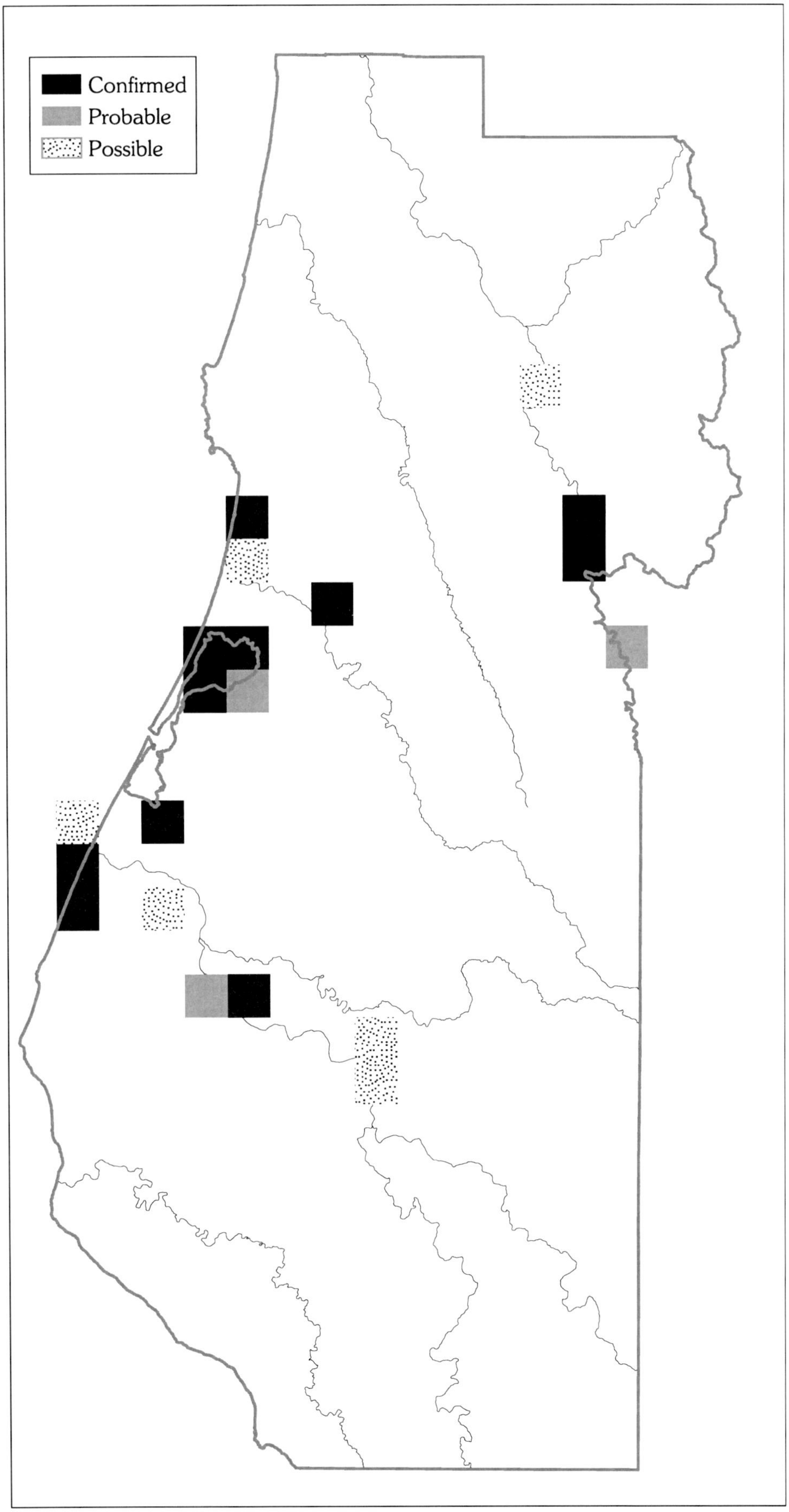

The story of the Canada Goose as a breeding species in Humboldt County is similar to that in many other places in North America where, historically, the birds were not known to nest. Present only since the 1980s, Canada Geese have become well established and numerous around Humboldt Bay, in the Eel River delta, and in outlying pasturelands. During the Atlas period they were encountered in 20 (5%) of the 425 total blocks.

Although representatives of other races have long been known to occur in the county, nothing points to there ever having been a breeding population of "honkers" in Humboldt. Harris (1996) related that none of the early collectors had specimens of the introduced and currently resident form, *B. c. moffitti*, the Western Canada Goose. Beginning in 1987 and ending in 1992, hundreds of these geese were introduced into the county's coastal bottomlands through a cooperative propagation plan proposed by the Northcoast Waterfowler's Association and implemented by the California Department of Fish and Game. Justification for the yearly translocation of pre-flight goslings, as well as some adults, included giving the taxpayers some visible benefit from habitat acquisitions around the bay, the economic benefits of enhanced waterfowl hunting, and facilitating public education about waterfowl (Head 1983).

In cooperation with the Pacific Coast Waterfowl Association, many of the translocated geese were trapped near Reno, Nevada and transported to Humboldt County. Yearly releases beginning in 1987 were 102, 117, 136, 99, 80, and 111 birds, for a total transplant of 645 (*fide* K. Kovacs). They were released on public wildlife areas around Humboldt Bay and in the Eel River delta. The birds promptly established a resident population consisting of transplanted geese and their progeny. Prior to this concerted effort, a few pairs had been released at Willow Creek. These releases account for the earliest historical record of confirmed breeding in the county, of precocial young along the Trinity River near Willow Creek 27 May 1987 (P. Paton).

Censuses of *moffitti* Canada Geese following the summer molt in Humboldt County were coordinated and conducted by the California Waterfowl Association during at least 4 years in the mid-1990s. Areas surveyed were the Arcata Bottoms, Humboldt Bay National Wildlife Refuge, and the Eel River delta. While the thoroughness of effort increased somewhat with each survey, combined totals for Canada Geese at the three general locations were 530, 1060, 1215, and 1737 during the 4 years 1993–1996, indicating strong reproduction and a growing population. It was the impression of some local observers that this increase leveled off at some point in the mid- to late 1990s.

Canada Geese nesting in Humboldt County concentrate notably within the managed waterfowl habitat and estuarine sloughs of the Salmon Creek Unit of Humboldt Bay National Wildlife Refuge, where in spring multiple pairs of geese or vigilant ganders can be seen scattered across the wet expanse of bottomlands. The pastures and wetland edge of the Arcata Bottoms, the Arcata Marsh, and the upper saltmarsh edge of Humboldt Bay are also well inhabited during spring. Higher ground provided by dikes, piles of dredge spoil from ditches, and other minor earthworks attendant to the dairy industry serve as favored lookouts for territorial ganders and, possibly, as nest sites. Mad River (e15), McDaniel (F15), Eureka (e16), and Hookton (D19) sloughs each supported nesting Canada Geese during the study period. Geese were also confirmed breeding in the lower Eel River estuary (b20, B21); along the Eel River at Rio Dell (F23); at Bigfoot Golf Course just north of Willow Creek (N13); in McKinleyville (f12); along the Mad River at Blue Lake (h14); and along the Trinity River in the block containing the confluence of Horse Linto Creek (n12). It will be interesting to see whether the current breeding population continues to expand and becomes well established along the Eel and Trinity rivers.

The most-reported evidence noted in the 11 blocks with confirmed breeding was precocial young (eight blocks), with the earliest record 11 Apr 1997 (e15; GCH). Two occupied nests were reported, one at Arcata (F15) 24 Apr 1998 (P. Lohse) and one near Centerville (B21) 25 May 1998 (COg). A nest with eggs at Humboldt Bay National Wildlife Refuge 3 May 1998 (KMS) was fitting with respect to location, as numerous pairs were thought to nest on the Salmon Creek and Hookton Slough units of the refuge (D19) during the Atlas period. Indeed, the 63 nests counted in 1996 on Teal Island (C. Goss) indicate a semi-colonial density there. Adults with maturing broods often form multi-family creches in summer. These assemblages spend much time along and near tidal sloughs, whose upper reaches offer protective cover during the adult molt. It can be noted that, during the Atlas period, Canada Geese with numbered neck bands continued to be widely observed at all seasons.

The wisdom of purposefully introducing yet another alien species, carrying with it the potential for unforeseen or adverse environmental consequences, has been questioned and is of concern to some. Efforts to convert the precious few rank pastures persisting in coastal lowland Humboldt County to managed goose grazing fields should be strongly discouraged. These places typically support thriving rodent populations, and thus are of great value to raptors. Such habitat "improvement" degrades or eliminates these *Microtus* factories, which feed native White-tailed Kites, Northern Harriers, American Kestrels, Barn and Short-eared owls, shrikes (*Lanius* spp.), and other birds of prey.

WOOD DUCK *Aix sponsa*

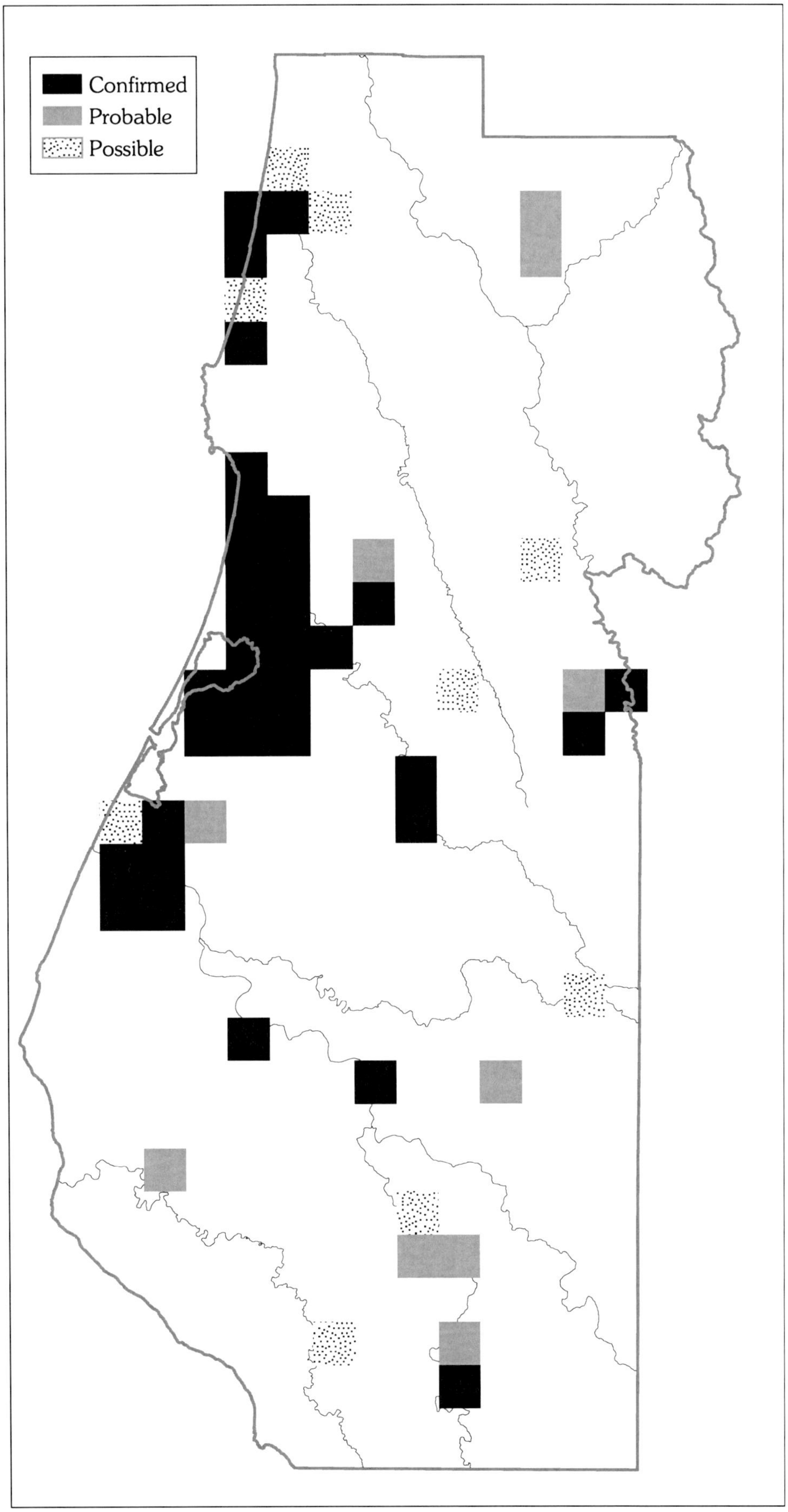

Wood Ducks are among the most beautiful waterfowl. Strongly contrasting colors, some with subtle iridescence and presented in amazing patterns, dress the drake during most of the year. The hen, while not striking, appeals with her mousy tones, expressive voice, and her guardianship of dependent ducklings. As well, Wood Ducks are typically encountered in pretty places. Ponds, sloughs, shaded lake shores, stream meanders and backwaters, and freshwater marshes with some open water but plenty of cover attract these birds. Once hunted into scarcity, over the past century these birds have recovered in numbers. Wood Ducks are now a familiar sight over much of temperate North America. On the Pacific Slope they occur commonly from British Columbia southward through California. Within this range they are most widespread and most common during the warmer months.

In some sections of their range, Wood Ducks are quite common and are widely encountered. Flocks of dozens assemble where habitat is extensive and of high quality, particularly in regions where Wood Duck nest box programs have been implemented. However, their habitat is somewhat restricted in Humboldt County. Freshwater ponds and the floodplains and lower canyons of rivers and streams supported most of the Wood Ducks reported during the Atlas period. Atlasers encountered these birds at old millponds, sewage ponds, persistently flooded pasture edges, and streams heavily shaded by deciduous hardwoods. Naturally occurring freshwater ponds are rare in Humboldt, and so the few remaining, neglected millponds of a bygone era are an attraction favored by many Wood Ducks in the county. This species was detected in 52 (12%) of all blocks.

Wood Ducks are present throughout the year in Humboldt County, but numbers in winter appear to be far smaller than during the breeding sea-

son, with the birds more localized. During the warmer months, pockets of suitable habitat across much of the county support nesting birds. The map resulting from Atlas surveys indicates that most Wood Ducks in the county in spring and summer occurred along the coast and in the coastal lowlands. Illustrating this was the loose cluster of varied-status blocks extending from Prairie Creek Redwoods State Park (g04) southward to Big Lagoon (f08); a larger cluster in the lower Mad River corridor and around the Humboldt Bay floodplain from McKinleyville (F13) to Eureka (D17); and a strong presence in the general Eel River delta. Wood Ducks were also encountered widely, though more sparingly, inland. Reports came from along or near the stream courses of all of the principal drainages in the county with the exception of the Mattole River. Reports away from the coast were best represented along the combined Eel-South Fork Eel rivers from the Salt River (c21) upstream to Garberville (k32). In both the Eel and the South Fork Trinity River watersheds, birds were encountered at isolated ponds as well as in riverine settings.

Originally restricted to nesting in large woodpecker cavities and in other hollows in trees, Wood Ducks readily accept nest boxes, provided that they enjoy some freedom from disturbance, adequate open water, and a food source. Starting on the north coast in 1991, a Wood Duck nest box program coordinated by the California Waterfowl Association was active throughout the Atlas period. Nearly all of the Wood Duck nest reports (and many for other species as well) received during the Atlas period were the result of periodic checks of nest boxes. In 1999 alone, 152 nest boxes were maintained in Humboldt County, and these produced an amazing 707 ducklings (*fide* B. Oppenheim).

The degree to which the breeding population of Wood Ducks in Humboldt County relies upon artificial nest sites is unknown, but it is apparently substantial. When the long-term and continuing decline in numbers of large dead and dying trees with suitable nest cavities is considered, it would appear likely that nest boxes have become important to this (and other) species locally. It would be difficult for even the most cautious or skeptical biologist to conjure up something bad about the Wood Duck nest box program.

Among the 33 blocks with breeding confirmations, the 21 occupied nests spanned the dates of 5 Apr 1997 (g14; *fide* RHw) to 15 Aug 1995 (F11; B. Oppenheim). Of all the nest records, only two specified that eggs were present: 23 Apr 1996 (F17; RJA) and 22 May 1996 (D19; K. Foerster). The earliest of nine reports of precocial young was 10 May 1995 (g05; C. Arguello).

Yocom and Harris (1975) termed the Wood Duck an uncommon resident and breeder in northwestern California, commenting that the breeding population had apparently decreased at least in local areas since 1960, and that it "responds to artificial nests." Redefining the species' status within approximately the same region, Harris (1996) called it a "locally common resident and breeder" that "readily uses nest boxes". He also noted this bird's frequent use of old, acidic millponds grown to yellow pond-lily, a situation that persisted during the Atlas period.

GADWALL *Anas strepera*

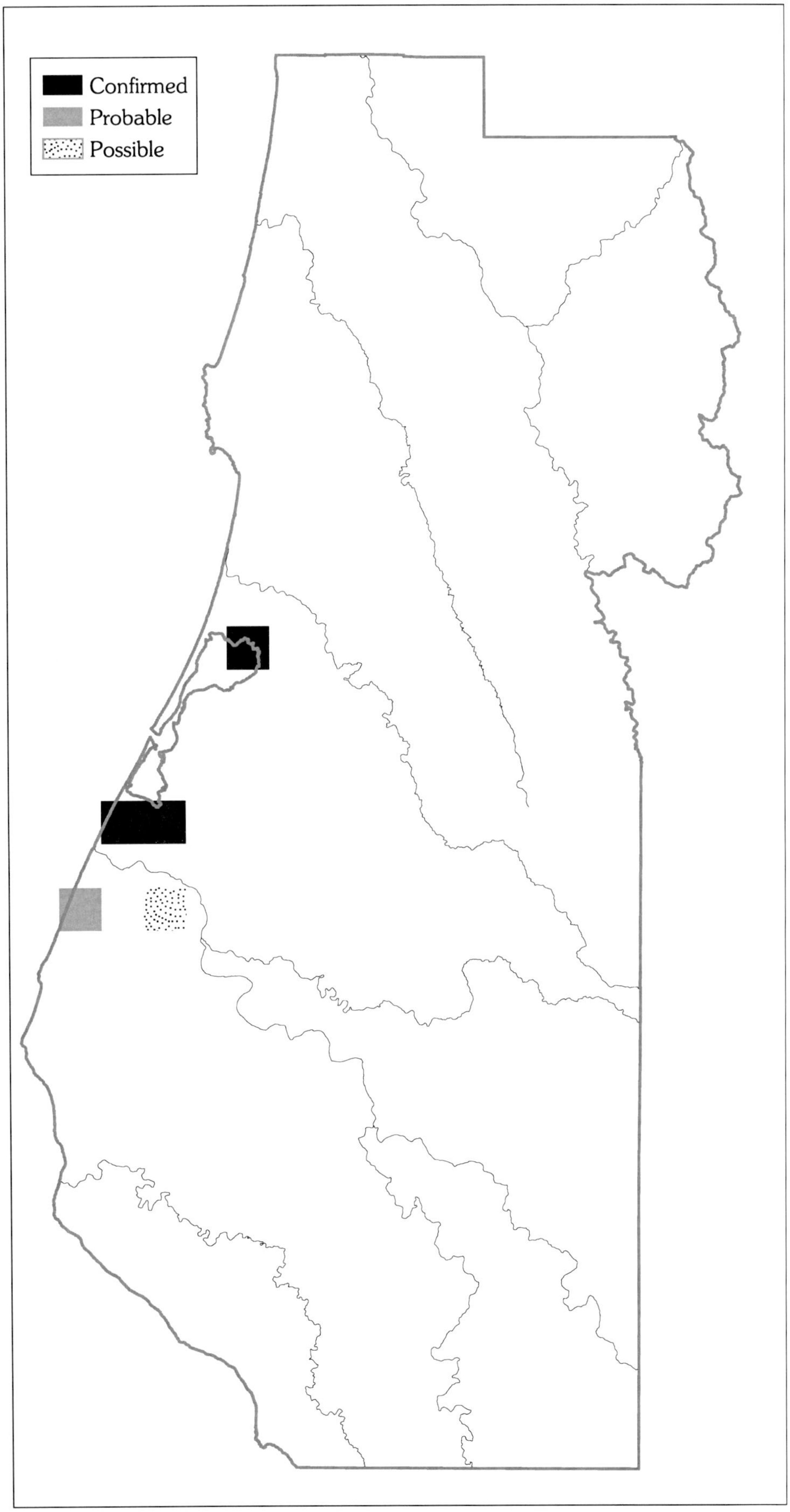

Deceptively plain-looking at a distance, the stolid Gadwall charms all who obtain good views with its appealing, vermiculated feather patterns. Gadwalls are large dabbling ducks, found over much of the world and quite common in western North America. Considered the most common breeding duck at Lake Earl in Del Norte County, over most of northwestern California this species is an uncommon to rare migrant and winter visitor and a rare breeder (Harris 1996). Although they are not especially common in Humboldt County, the token pair or small sub-flock of Gadwalls may frequently be distinguished among mixed-species assemblages of surface-feeding ducks in migration and in winter. This is less often the case in late spring and summer, when the small Humboldt breeding population appears to be confined to a few local freshwater ponds and impoundments. As similar-looking hen Mallards raise their broods throughout spring and summer, broods of Gadwall accompanied by their nondescript mothers may be overlooked. Humboldt's first historical record of breeding was of a female with ducklings 20 Aug 1990 at the Arcata Marsh (S. Harris, R. Grosz). Seven additional broods were recorded in the coastal lowlands in 1992 and 1993, with none reported in 1991 or 1994 (*fide* S. Harris).

During the Atlas period, breeding was confirmed in three blocks. Females with ducklings were recorded at the Arcata Marsh (F15) from 6 Jun 1995 (E. Elias) through 2 Aug 1998 (DFx); a nest with eggs was found 6 Jun 1996 at the Salmon Creek Unit of Humboldt Bay National Wildlife Refuge (D19; *fide* K. Foerster); and females with broods were seen between 2 Jul 1995 (JEH) and 25 Jul 1996 (B. Deuel) at Eel River Wildlife Area (c19). In addition, a pair of Gadwalls seen 25 May 1998 east of Centerville Beach (B21; COg) was suspected of summering in the area. It is possible that a few pairs of this species bred elsewhere in the Humboldt

Bay lowlands or in the Eel River delta during the study period. It seems likely that some nesting Gadwalls were overlooked owing to the reclusive behavior and hidden nests of breeding waterfowl, the similarity of Gadwall and Mallard hens, and to modest atlasing coverage of the more extensive wetlands. In addition, the increased height of marsh vegetation during the rather late nesting season of Gadwalls may hinder observation of broods.

Gadwalls in Humboldt County are typically found in wetland and bayshore habitats of the coastal bottomlands. While nesting, they are partial to larger, marsh-edged ponds. Nearly all such habitats occurring locally have been human-created or are otherwise managed for waterfowl. After fledging, broods of Gadwall begin to associate loosely with growing late summer concentrations of ducks and other waterfowl, both on nesting ponds and in adjacent wetlands. Still-water habitats such as flooded fields, rush-dotted pastureland, sewage ponds, and the muddy margins of sloughs draining through estuarine mudflats are favored.

MALLARD *Anas platyrhynchos*

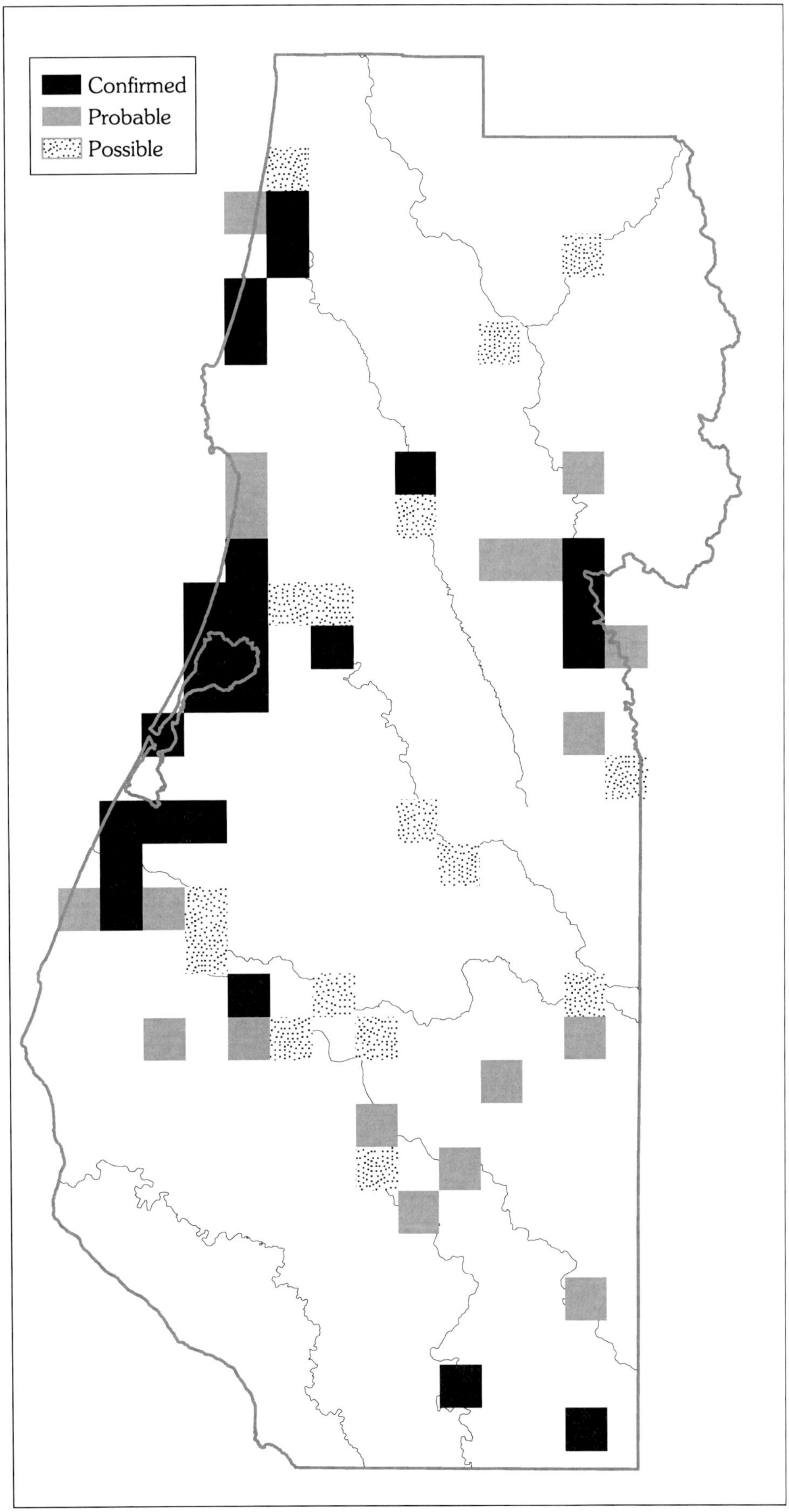

From the time most persons are very young, the image of a duck held in one's mind is of a stout, low-slung bird with a rounded green head, spatulate yellow bill, a raucous *quack!*, and an uneasy manner when more than a moment's waddle from water. Throughout much of the Northern Hemisphere, the Mallard is the origin and embodiment of this image. It is widespread, abundant, and inhabits a great many different wetland habitats. From this species came many of the domesticated Mallard-like duck varieties seen on park ponds, strains in existence now for ages. Mallards have long been among the most-hunted waterfowl. They are resilient birds, able to persist in marginal, highly fragmented, and ephemeral wetlands. They quickly investigate and breed in newly created habitats as they become available.

Characterizing the Mallard in northwestern California, Harris (1996) described it as a "common resident and breeder. The most common nesting waterfowl in the region." Mallards are certainly the most conspicuous breeding ducks in Humboldt County, where they are resident in appropriate habitat throughout the year. Although occurrence in numbers is restricted to the larger wetlands in coastal bottomlands, (*e. g.*, Eel River Wildlife Area, Humboldt Bay National Wildlife Refuge, and the Arcata Marsh), this duck may be anticipated wherever freshwater bodies exist. No small pond tucked among trees, no backwater overarched by alders is overlooked by these birds. Sloughs, unmanaged wetlands, persistently ponded corners of pastures, drainage ditches, and undisturbed streamsides also provide habitat. Unlike most waterfowl that breed in Humboldt County, they are also encountered in wetlands and along watercourses in rugged terrain well into the hills.

Some sites host these ducks almost continuously, although the population undergoes great turnover during north- or southbound migration or other movements. Many more are present within the county in the colder months than during the breeding season; impoundments around Humboldt Bay, outlying wetlands, and, to a lesser degree, dairy pastures attract flocks of from dozens to hundreds. Yet each year this species remains numerous and widespread at the close of the spring waterfowl migration, remaining so through the summer.

While Mallards were quite widespread—they were found in 59 (14%) of the 425 total blocks—Atlas results indicate that Common Merganser had a somewhat broader breeding distribution (77 blocks), with Wood Duck ranking third (51 blocks). Mallard distribution during the study period closely paralleled that of Wood Duck; indeed, each species was noted in loose association with the other at many locations. As were Wood Ducks, Mallards were found chiefly along the coast, from the Redwood Creek estuary through the Humboldt Lagoons, about the Mad River and Humboldt Bay lowlands, and in suitable habitat at many locations inland. Distribution of Mallards along and near the Eel River above its floodplain was apparently more uniform than was the case for Wood Duck. Large areas of mostly mountainous, forested landscape drained by large streams and small rivers appeared to host fewer breeding Mallards; these are areas where Common Mergansers populate many blocks unsuitable for Mallards.

No doubt abetted by their comparatively broad habitat requirements and a tolerance of human activity, breeding evidence was reported in many blocks. From February into May, pairs or courting-party threesomes of Mallards were seen scattered throughout the near-coastal lowlands and in some riverine situations well into the interior of the county. All 18 reports of probable breeding were of pairs, 10 of which were in May. Broods of ducklings became evident beginning in April. Later, mixed assemblages of molting adults and young were noted at or near breeding sites.

Mallard ducklings trundling along in Mama's close escort furnished the most conspicuous evidence of breeding. No surprise was that, among the 25 blocks with confirmations, this status was based on observation of precocial young in 23, ranging from 14 Apr 1996 (f16; S. Harris) to 4 Aug 1996 at the McKinleyville sewage ponds (F13; JEH). Reports of broods were well scattered from April through early August. It should be noted that, while first broods were ordinarily seen during April and May at regular breeding sites, "new" broods appeared well into midsummer, suggesting second nesting attempts. Harris (1996) noted that in "most recent years the Arcata Marsh and Oxidation Pond has held 20 to 40 broods." Wheeler and Harris (1970) estimated that in 1964 about 900 ducklings from 200 broods were produced around Humboldt Bay, and they provided additional information on nest sites, habitat, and nesting chronology.

A nest with eggs 12 May 1996 was the highest evidence reported from Humboldt Bay National Wildlife Refuge (D19; KMS), an important center of breeding activity. Similarly, an occupied nest at the Arcata Marsh (F15) 10 Jul 1996 (*fide* K. Kovacs) also ranked as the best code noted at this well known "Mallard factory."

Mallards assemble in flocks of varying size in wetlands of the coastal lowlands following nesting. Freshwater and upper-estuarine settings may receive daily use by scores of these birds in late summer and early fall. Because these flocks probably are composed of many summering or locally produced birds, it is difficult for the average birder to perceive at what point the annual influx of birds originating from points to the northward of Humboldt County takes place. Mallards are preceded in their southward migration by Northern Pintails, Northern Shovelers, and Green-winged Teals, each of which become conspicuous at the same places used by Mallards by August. By October, Mallards become common at most freshwater wetlands, though they are typically outnumbered by the species mentioned above. Large pure flocks are uncommon, occurring most often in "sheet water" habitat of flooded pastures and in crop stubble well puddled by winter storms. Such concentrations are observed most frequently in December and January; it is possible that these may refer to early northbound migrant flocks on the move.

CINNAMON TEAL *Anas cyanoptera*

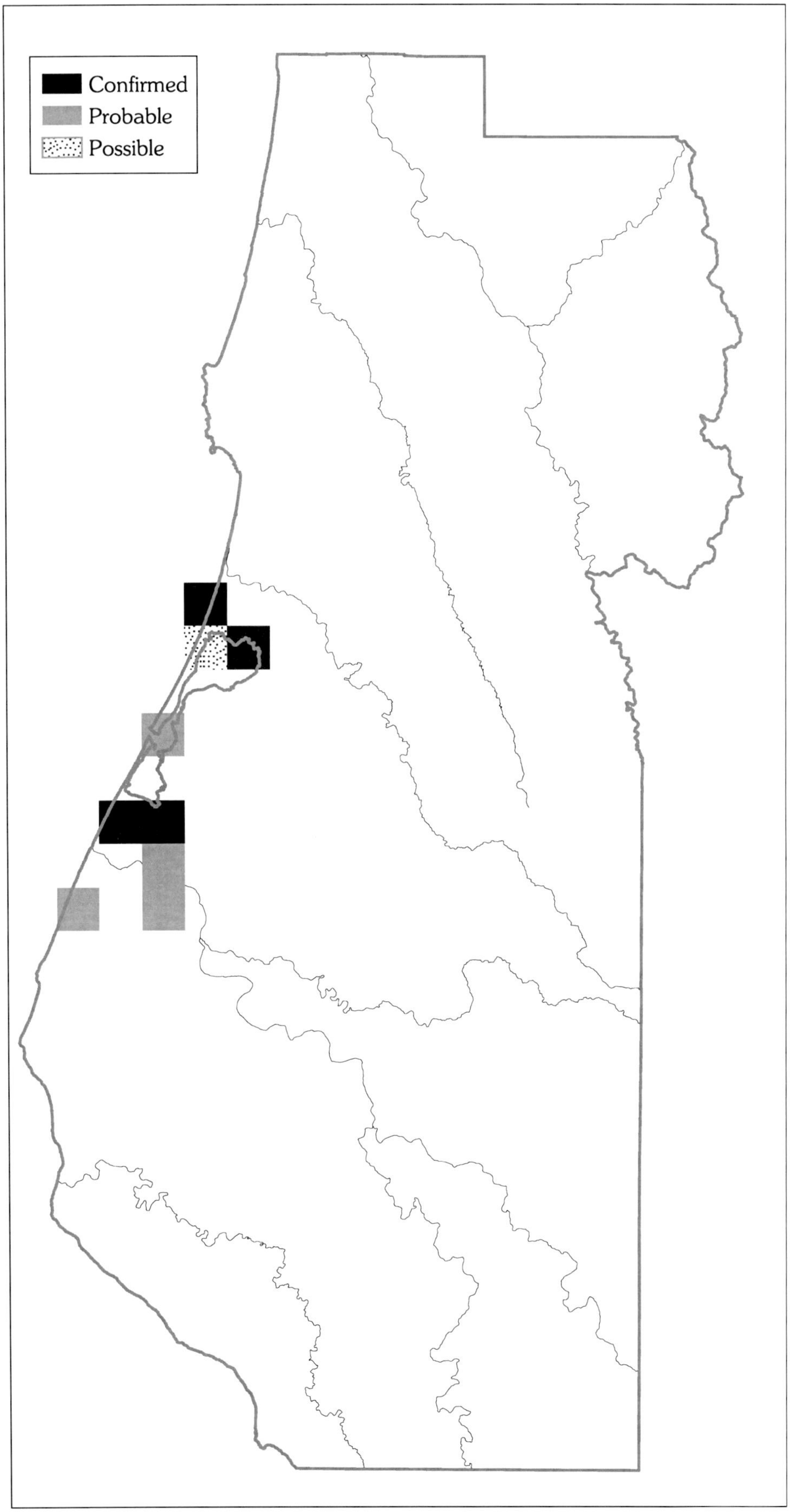

The Cinnamon Teal is among the "signature species" of freshwater and alkaline marshes of the American West. Most Cinnamon Teal breeding in California occurs in large managed wetlands of the intermountain basin country east of the Cascade-Sierra and in the Central Valley. Where size and quality are sufficient, marshes of the coastal lowlands are also used for nesting.

The freshwater ponds and wetlands clustered irregularly in Humboldt County's central coastal lowlands have long been recognized as supporting breeding Cinnamon Teal. Harris (1996) related a changing historical dynamic, from absence or scarcity early in the 20th century to the status of a common breeding bird by the 1970s. Late spring and summer occurrences in the Arcata Bottoms increased during the 1950s and 60s. The earliest historical breeding record was of a female on eggs 25 May 1951 at an unspecified location in the county (Dollahite and Anderson 1952). Harris and Wheeler (1965) report on a male hybrid Blue-winged × Cinnamon Teal attending a female teal sp. and brood near Arcata 10 May 1964. During 1964–1966, random searches in the Arcata Bottoms located 15 Cinnamon Teal nests (Wheeler and Harris 1970). Wheeler and Harris (1970) estimated that, throughout the bottomlands around Humboldt Bay, there were 491 and 221 young produced in 1964 and 1965, respectively. Cinnamon Teal began breeding at the Arcata Marsh at some point after its establishment in 1981. These teal have been seen in summer on the Salmon Creek Unit of Humboldt Bay National Wildlife Refuge, and breeding was noted on ponds at Eel River Wildlife Area in the early 1990s. They were considered a "common migrant and breeder; uncommon or local in winter" by Harris (1996).

In such a rugged and well forested region as Humboldt County, the presence of managed wetlands can be imagined to make a great deal of difference for birds favoring ponds and marshes. This is particularly likely for waterbirds that avoid saltwater or estuarine settings. The conversion of many sections of original Humboldt Bay saltmarsh to ditched pastureland foretold a better future for Cinnamon Teal long ago. With the development of wetland projects such as the Arcata Marsh, Humboldt Bay National Wildlife Refuge, and Eel River Wildlife Area, freshwater habitat was made available for these ducks where little had existed previously.

Atlasers found Cinnamon Teal in nine (2%) of the 425 blocks. All blocks with detections were around Humboldt Bay and in the Eel River delta. Breeding was confirmed in four blocks, was judged probable (all reports were of pairs) in another four, and was possible in the remaining block. All confirmations involved the observation of broods of precocial young accompanying females. At the Arcata Marsh (F15), broods were reported 7–25 May 1996 (C. Meyers *et al.*) and 5 Jun 1998 (S. Harris). During the project, Cinnamon Teal appeared to nest on every one of the ponds and marshes within the combined Arcata Marsh and sewage pond complex, with the exception of Klopp Lake. The early May brood at the Arcata Marsh was unusual since most do not appear until midsummer—when the season's earlier-hatched Mallard ducklings on the same pond may be nearly fully grown. At Ocean Ranch, Eel River Wildlife Area (c19), broods were reported 2 Jul 1995 (JEH) and 4 Jul 1996 (S. McAllister). Broods were also reported at Hookton Slough (D19) 26 Jul 1997 and at Mad River Slough (e14) 1 Aug 1995 (both C. Bartolotta).

NORTHERN SHOVELER *Anas clypeata*

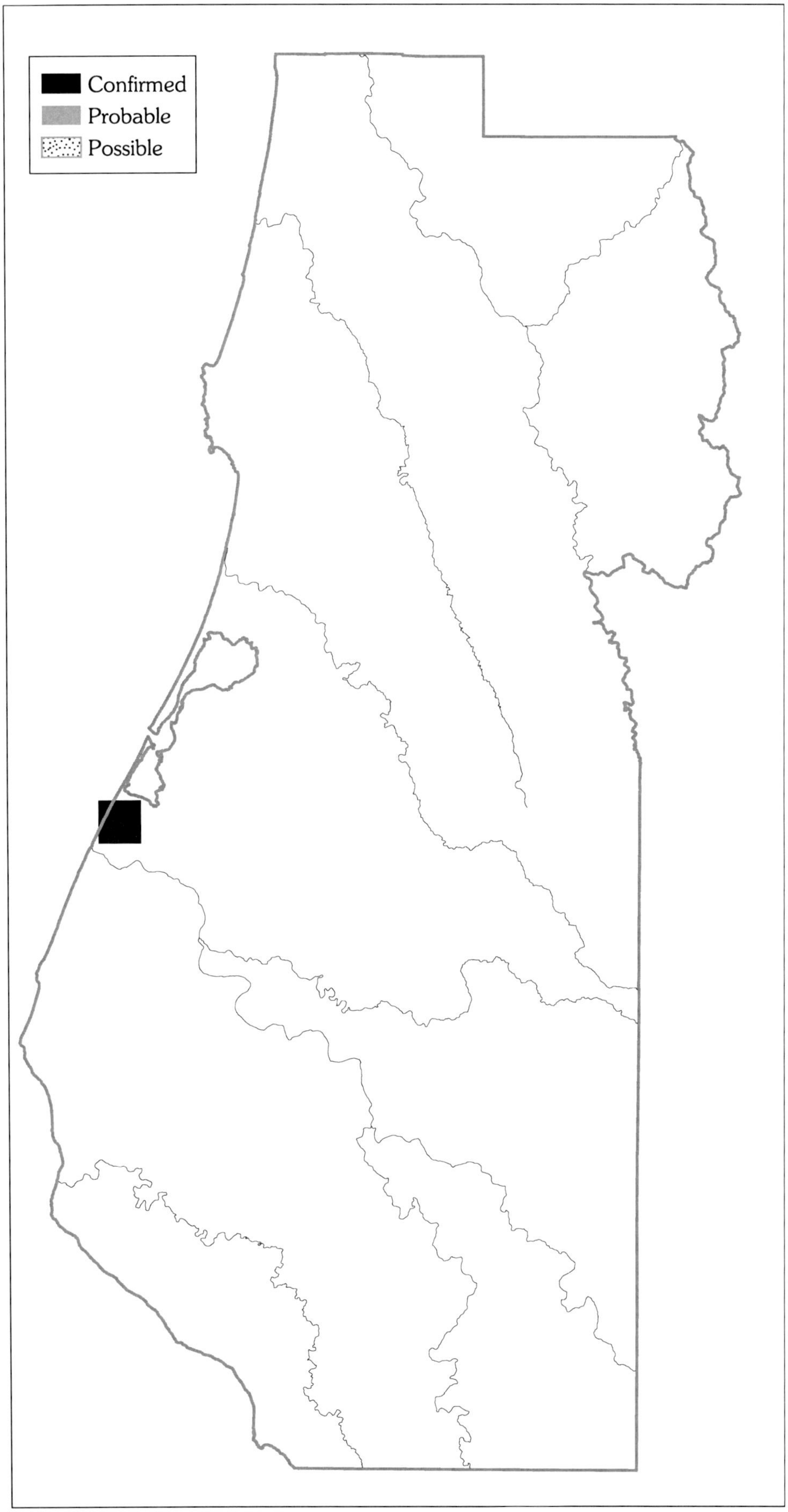

An outsize bill draws one's eye to Northern Shovelers, appealing teal-like ducks of ponds and marshes. Delicate lamellae along their bill margins strain bits of vegetation and other organic material as the birds feed along with heads held low, guppering at the water's surface. As well, these birds are graceful on the wing, dropping quickly onto favored open wetlands and "sheet water" in flooded pastures with deft and synchronized movements. Shovelers are widespread in the Northern Hemisphere, and are found at some point in the year at nearly all significant wetlands in California. The species is a locally common migrant and winter visitor in northwestern California (Harris 1996).

The seasonal status of Northern Shoveler in Humboldt has been one of local abundance during spring and fall migration and to a lesser extent in winter, as well as a casual breeder. Harris (1996) had records in his files of from one to three broods in nine summers from 1960–1995 at five sites in northwestern California, three of which were in Humboldt County: Arcata Bottoms, Arcata Marsh, and the Arcata sewage ponds. During the Atlas period, shovelers were reported at these and similar locations. For example, they were reported at the Arcata Marsh (F15) 18 May 1997 (PFS), the McKinleyville sewage ponds (F13) 22 May 1998 (JEH), and at Hookton Slough (D19) 22 May 1996 (K. Foerster). However, such occurrences, while of interest owing to date, were probably northbound birds. Spring migration of shovelers continues well into May, involving mostly discrete flocks, with few pairs lingering later. Yocom (1961) provided details on the chronology of Humboldt's first documented breeding in the Arcata Bottoms near the old Mad River channel. He found a pair acting suspiciously 20 May 1960, a female doing a distraction display 11 Jun 1960, and the female with two ducklings there 21 Jun 1960.

A hen Northern Shoveler with a brood of precocial young was seen on the ponds at Ocean Ranch, Eel River Wildlife Area (c19) 4 Jul 1996 (S. McAllister), providing the Atlas project with its only breeding confirmation. A pair of shovelers was also seen there 2 Jul 1995 (JEH). Mallards, Gadwall, and Cinnamon Teal also were observed with broods of ducklings at this favored nesting site at various times during Atlas fieldwork. At the close of the study, it appeared that Northern Shoveler could be considered only a peripheral or marginal breeding bird in Humboldt County. Given that the Atlas surveys managed but a stab at censusing waterfowl production on the county's marshes, it is conceivable that this species nests more frequently in Humboldt than the 5-year "snapshot" has indicated.

HOODED MERGANSER *Lophodytes cucullatus*

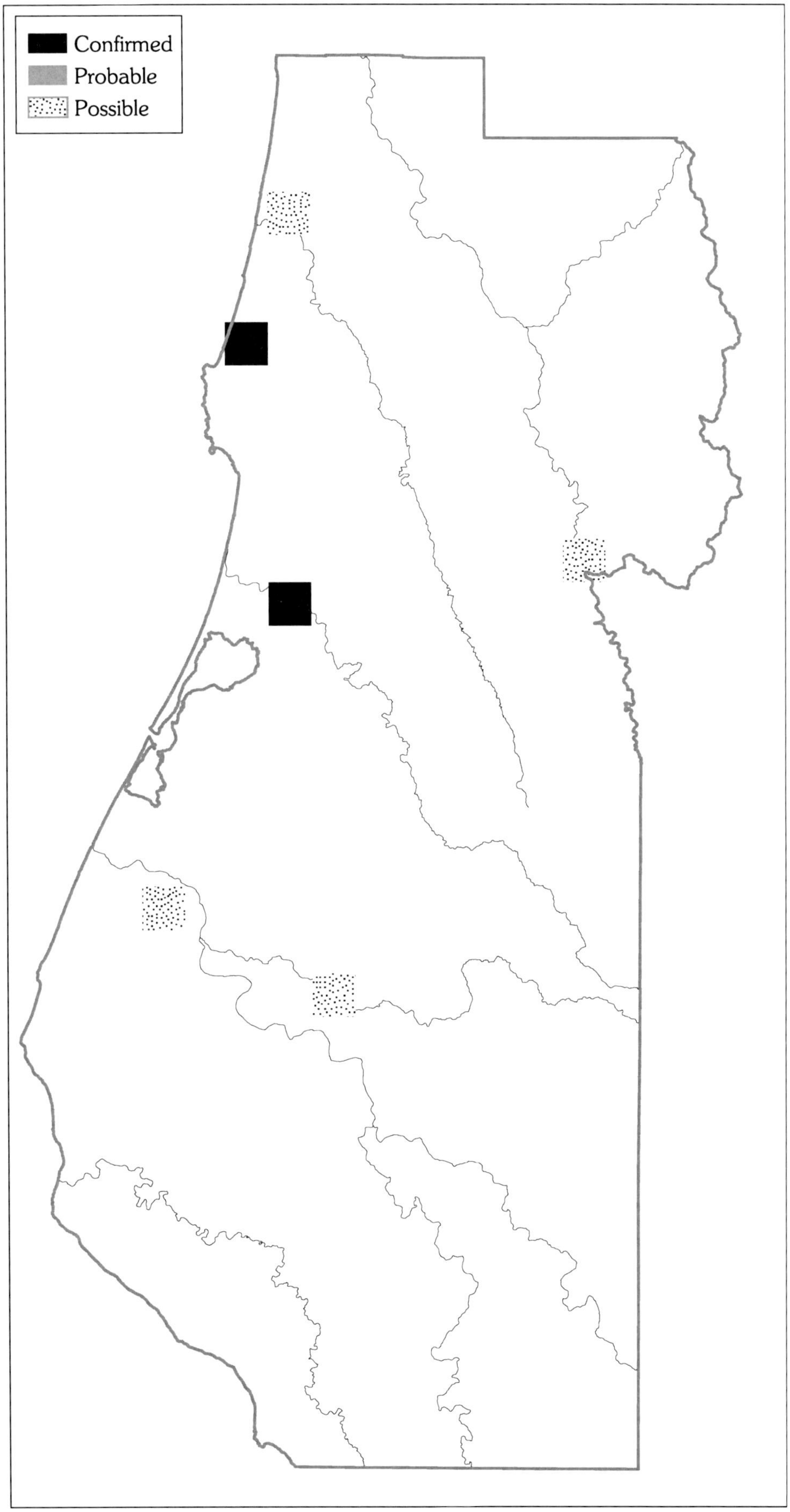

Hooded Mergansers are small fish-eating ducks with lovely and expressive crown feathering. Fond of shadows and stillness, they avoid estuarine and "big water" situations, typically resorting to secluded lakes, ponds, and river eddies. These mergansers generally breed in southern boreal Canada and in parts of the northern United States. As ice covers northern lakes and beaver ponds, the birds move south, with the bulk of the population wintering in the southeastern states. Smaller numbers occur along the temperate Pacific Coast and sparingly through the inland West. This species is of regular occurrence in Humboldt County from late fall to early spring. Numbers are never great; sightings typically involve singles or pairs. Flocks of as many as a dozen birds are occasionally encountered. Few are noted from April through September.

This duck has long been recognized as a rare breeding species in California. Hot summers, parched surroundings, and unfavorable water conditions at most of the state's lakes and ponds may not suit the species well, leaving it with but limited opportunities for breeding. These birds have, however, nested at several locations in northern California, including at least once in Del Norte County (Harris 1996), and at sites in the southern Cascade Range and in the Central Valley. Breeding by this species is somewhat more expected in lowland western Oregon (Gilligan *et al.* 1994, Marshall *et al.* 2003). Harris (1996) characterized the Hooded Merganser as a "locally uncommon to rare winter visitor and migrant; casual breeder" in northwestern California and mentioned a single report of nesting in Humboldt County: a female and brood on Maple Creek at Big Lagoon 4 May 1981 (P. McLaughlin).

It became clear during the course of the Atlas project that Hooded Mergansers must still be considered at best a rare or occasional nesting bird in the county. Two confirmed breeding

records were submitted. A female Wood Duck incubating five Hooded Merganser eggs was discovered in a box erected for Wood Ducks at Essex Pond (g14) 12 Jun 1998 (P. McNeil). Three of the five merganser eggs hatched, along with 13 Wood Duck eggs. Thirteen of her own eggs were being incubated 8 May 1999 by a Hooded Merganser hen in another Wood Duck box at Big Lagoon Mill Pond (f08; B. Yost). All 13 eggs subsequently hatched.

Evidence of possible breeding was obtained in four other blocks. A female was on a flooded meadow adjacent to Prairie Creek along Davison Road north of Orick (g05) 14 Apr 1998 (DFx). A pair was on the favored pond at Bigfoot Golf Course near Willow Creek (N13) 22 Apr 1995 (TWL), yet the birds were not seen again and may have been late northbound spring migrants. In light of this possibility, the report is treated conservatively, rather than granting it probable breeding status. Single birds of unstated sex were in a pond near Riverside Park along the Van Duzen River (H23) 21 May 1996 (C. Powell), and on a slough near Ferndale (D21) 29 May 1995 (L. Karsteadt).

COMMON MERGANSER *Mergus merganser*

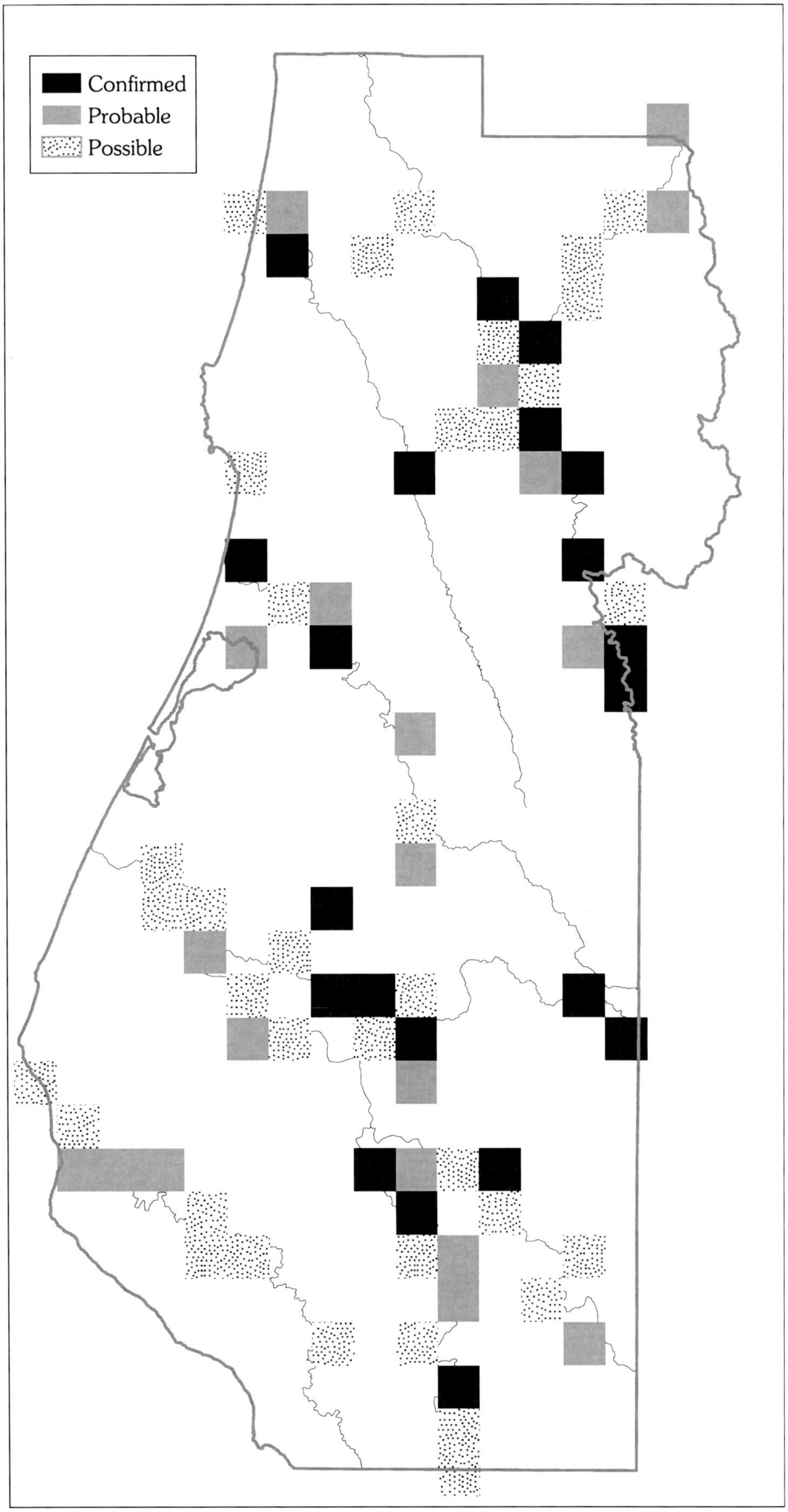

Common Mergansers are large ducks with serrated bills adapted to grasping fish. These waterfowl enjoy a great range in the Northern Hemisphere. The distribution of Common Merganser in the breeding season includes all of the major rivers and many of the larger creeks in Humboldt County. During the Atlas period, confirmed breeding was reported from 21 of the 77 blocks with mergansers reported. Paralleling the situation exhibited by other aquatic and streamside birds, Reports of Common Mergansers fell largely in blocks bisected by the Klamath, Trinity, Mad, Van Duzen, Eel, and Mattole rivers. That this was no coincidence is attested by a pattern of use of the county's main rivers summarized by Harris (1996), extending well back in the historical record (Wilder 1916a).

Common Merganser has been considered a common resident and breeding species in northwestern California (Harris 1996). While preferred reaches of the larger rivers may have a few Common Mergansers on them every day of the year, their distribution and abundance vary seasonally. In March, Common Mergansers appear either to increase in numbers or to otherwise become more visible. In their largely white, pink-flushed alternate plumage, males are especially conspicuous. Pairs are routinely seen from routes of travel fronting rivers and larger creeks until May. All but two of 20 reports of probable breeding pertained to pairs, which spanned the dates of 15 Mar 1996 (k30; JEH) to 26 May 1996 (p03; KMS). Females visiting likely nest sites accounted for the balance of probable breeding reports.

Amount and type of forest cover on adjacent slopes varies greatly, but since these birds typically nest in tree cavities, there must be a strong association with mature or old-growth conifer forest. However, many Common Mergansers nest in more oak-dominated surroundings. An occupied nest, or possibly multiple nests, attended by up to three females was found 29 May 1995 in a large oak along Redwood Creek (J11; D. Kriger).

After incubation starts, males retire for the summer molt while females are left to tend broods of ducklings. Evidence in the remaining 20 of 21 blocks with confirmed breeding was precocial young. Observations of these broods spanned the dates of 20 Jun 1996 (k32; JEH) to 10 Aug 1995 (N13; TWL). These family groups were conspicuous at gravel bars and eddies on the big rivers and the lower reaches of their tributaries. Such reports were most frequent along the Trinity, Eel, and Van Duzen rivers. Despite consistent presence along the length of the Mattole River, no breeding confirmations were made there.

Foreman (1975) estimated that there was one merganser brood for every 2.3 miles of Humboldt and Del Norte County river he studied, and that there were 1318 mergansers in the two counties. Judging from the number of confirmed and probable merganser breeding reports (41 blocks) during the Atlas period, this species rivals Mallard and Wood Duck (both with 43 blocks with such reports) as the most numerous breeding duck in Humboldt County. During late July and August, creches of ducklings and, later, apparent combined family groups assemble at favored sites. These flocks may contain dozens of birds or more. Extensive bar-and-eddy configurations of the Trinity, lower Mad, Eel, and Van Duzen rivers are well visited by these ducks after nesting. This suggests a general downstream shift of the summer population. Gravel bars offer preening and roosting sites, sheltered backwaters, and quick access to escape afforded by the river current. These post-breeding concentrations diminish at some point during fall.

Modest numbers of mergansers winter along Humboldt's major rivers. Because these ducks avoid saltwater, and as Humboldt County offers no large freshwater lakes or reservoirs (with the exception of Freshwater Lagoon), no great concentrations ordinarily develop anywhere in the county in winter. See Foreman (1975, 1976a, 1976b) for more information on the abundance, habitat, distribution, activity patterns, and reproductive behavior of this species in Humboldt County.

OSPREY *Pandion haliaetus*

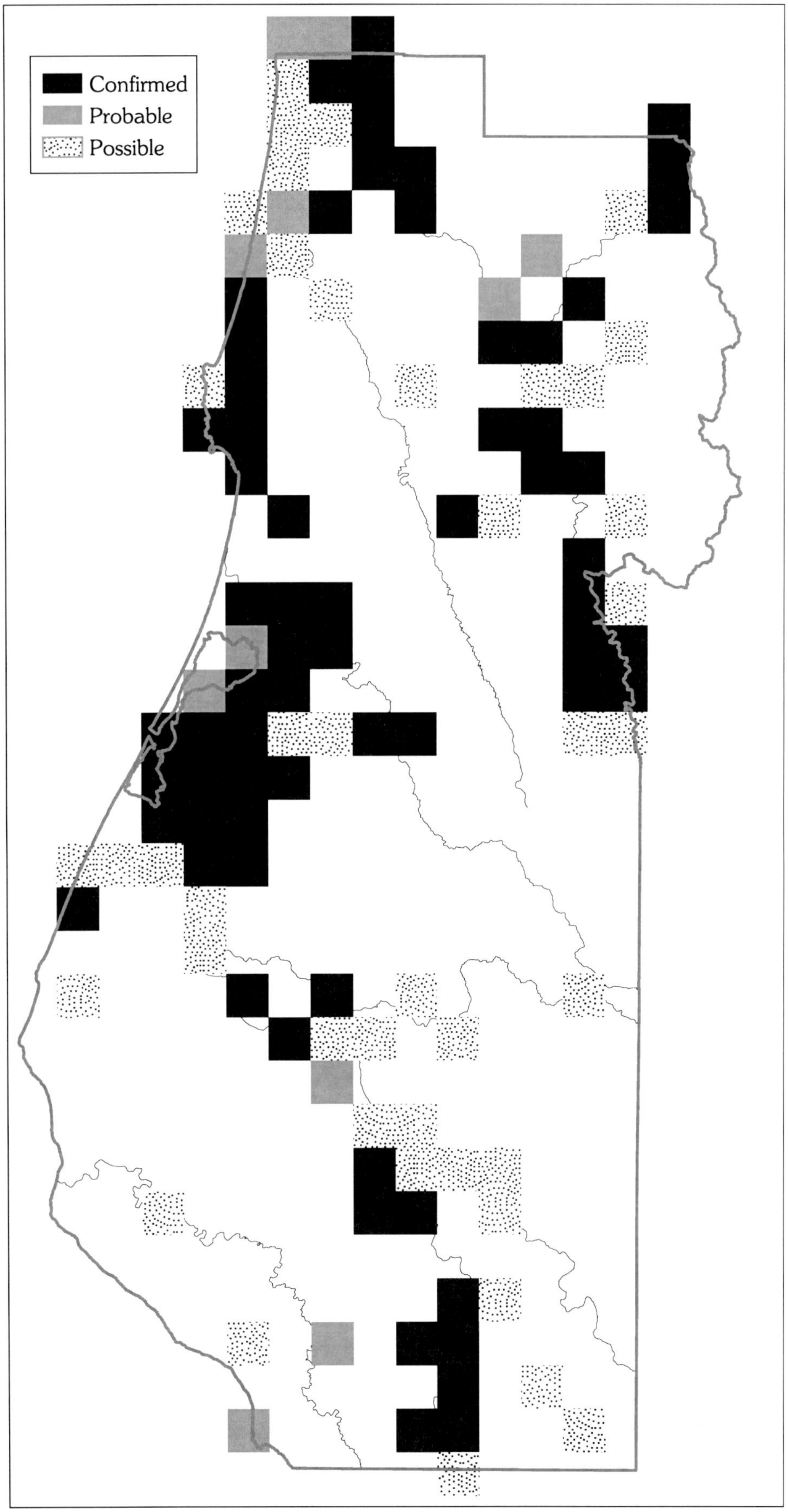

Married to water, Ospreys are seen far from it only during the necessity of overland migration. These raptors know a life-long daily routine of flying over the surface and plunging feet-first to capture small and medium-sized fishes. This skill has been honed to excellence by natural selection. Their success is attested by their presence at lakes, rivers, and coasts throughout much of the world. Known to be augmented by but few other items, fish constitute nearly the entire diet of the Osprey. Subsisting largely on mid- and higher trophic level fishes, Ospreys suffered concentrations of pesticide residues and consequent eggshell thinning and clutch loss during the 1950s and 1960s. The ban on domestic use of DDT and other organochlorine pesticides in 1972 resulted in a population rebound, aided in some areas by the erection of nesting platforms for the birds. Diverse fisheries in maturing reservoirs in the continental interior have provided additional opportunities. Today few clean rivers and lakes are without their nesting Ospreys.

It is a sure thing that the Osprey has been familiar not only to modern bird observers, but to anyone who has kept an eye on the sky during the period of human habitation in Humboldt County. These are conspicuous birds of prey, vocal at times for prolonged periods, and their bulky treetop nests are difficult to overlook when built adjacent to open country. They live in and about places where people like to live, and along routes of travel in river corridors. This phenomenon results in great visibility of the breeding population in the county. It would seem that the exhibition of such a large proportion of individuals present is shared with few other species. Ospreys were commonly encountered during all 5 years of the Atlas period and were ultimately reported from 120 (28%) of all blocks.

In a pattern recalling the distribution of other streamside birds, the Osprey map reflects this species' need for open water. Detections were numerous along nearly the entire coast or inner coastline from the Del Norte County line south through the Eel River lowlands, but were entirely lacking from Cape Mendocino through the shores at the base of the King Range. A pair at Shelter Cove (f33) 21 Jun 1995 (D. Roberson, R. Carratello) constituted an isolated report of probable breeding from the extreme southern Humboldt County coast. The coastward section of the range also takes in the lower Klamath River, Redwood Creek, each of the Humboldt Lagoons, the lower Mad River, and tributaries of the lower Eel River.

Away from the coast, Ospreys were widely reported along the Klamath, Trinity, Van Duzen, Eel, and South Fork Eel rivers and the lower reaches of their larger tributaries. Oddly, presence in the Mattole River was represented only by a single bird between Honeydew and Petrolia (d28) 3 August 1996 (C. Keil). Large areas of interior Humboldt County offer no suitable habitat for Ospreys. Therefore it is not surprising that many middle- and higher-elevation blocks well removed from sizable streams had no reports.

Owing to the commanding situations these birds select as nest sites, occupied nests were the most-reported evidence of confirmed breeding. These were noted in 62% of all blocks reporting Ospreys. Nests in Humboldt County are almost always built either in flat- or broken-topped live conifers or in conifer snags. In contrast to what has become routine at many locations elsewhere in North America, the use of channel markers and power poles as nest sites in the county is nearly unknown. While many nests are situated at water's edge or shortly upslope, some are well upslope. Within the coast redwood belt, nests are often built on the tops of broken or deformed trees or snags. In the interior of the county, Douglas-fir is, by a wide margin, the tree species most often used as a nest support. No use of the widely scattered exotic eucalyptus, Monterey cypress, or Monterey pines planted in lowland western Humboldt County was specifically reported.

By mid-May, Ospreys are seen most often in association with nest sites and adjacent waters. Males engage in frenzied, highly vocal display flights over their territories and over nearby forested hillsides. Harris (1996) stated that nesting was underway by late May. Suggesting that at least some eggs were laid in April were reports of adults seen carrying food, presumably for nestlings, 3 May 1996 (g12; GAS) and 8 May 1995 (D17; B. Riley). This behavior was reported as late as 31 Jul 1996 near Hoopa (m11; *fide* JMH). Of the 41 blocks in which one or more occupied nests furnished the highest confirmed breeding evidence, more than 80% were spread evenly from April through June; the earliest occupied nest was 21 Mar 1995 (N13; JEH) and the latest was 10 Aug 1996 (016; R. McCoy). It is likely that many of these records involved nestlings too small to observe. There were seven blocks in which young were actually seen in the nest. The earliest of these was 6 Jun 1995 (f16; L. George) and the latest was 22 Jul 1996 (g16; GCH).

Apparent family groups of Ospreys in loose association are a fairly common sight in midsummer, so it is somewhat surprising that fledged young served as highest confirmed breeding evidence in only one block, near Little Lost Man Creek (H05) 3 Jul 1999 (JEH). This may be a function of the greater detectability of an occupied nest, which is a higher-ranking breeding code. Together, the pattern of reports suggests that juveniles produced in Humboldt generally fledge in June and early July.

Ospreys are strongly migratory. This is made apparent by the near absence of this species in Humboldt County from November through February. These birds typically appear in March and April. In many years a perceptible influx of through-migrants or summer residents has been noted in April, with numbers of Ospreys assembling daily at river mouths and around the Humboldt Bay entrance when shad (*Alosa sapidissima*) or other anadromous fish are running. Exceptionally, 19 were in sight at once 26 Apr 1995 at the Klamath River mouth in adjacent southern Del Norte County (JEH, DFx).

In northwestern California, Harris (1996) characterized Osprey as a "common summer resident and breeder; rare in winter. Most birds leave by early October with a few stragglers on major foraging areas in winter." There were 59 known breeding pairs in Humboldt, Del Norte, northern Mendocino, and western Trinity counties in 1972 (French 1972, French and Koplin 1977). That there were more Atlas blocks with breeding confirmations (66) in Humboldt County alone from 1995–1999 surely reveals a recovery of abundance in recent decades. With many Atlas blocks boasting multiple nests, it would appear that the population in the county probably numbered several hundred pairs during the study. But while these birds have recovered well, land managers should attempt to minimize human disturbance near active nests. Monitoring of some Humboldt and Mendocino County nests indicated that disturbance can negatively affect Osprey reproduction (Levenson and Koplin 1984). Levenson (1976) also conducted detailed observations of an active Osprey nest near Jacoby Creek. Ueoka (1974) studied foraging, prey, nest sites, and productivity around Humboldt Bay.

WHITE-TAILED KITE *Elanus leucurus*

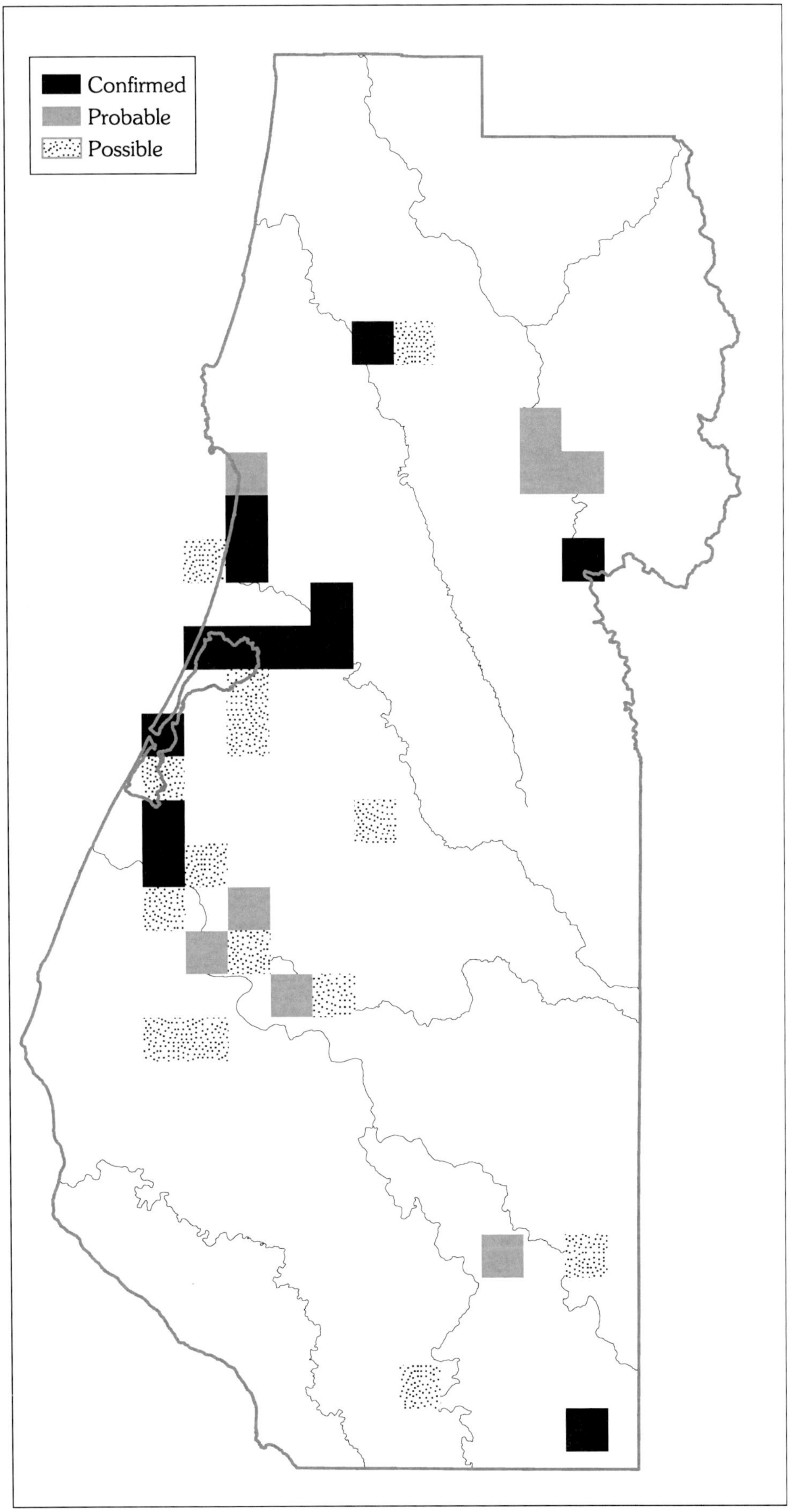

White-tailed Kites in Humboldt County punctuate an expanse of varied greens with bold points of animated whiteness. They attract the eye of all who enjoy ephemeral beauty, either as a temporary ornament to treetops or as white's sudden shadow, passing above and beyond. Although not as highly aerial as some other kites, they have a willowy, windblown grace while in commuting flight and are, at all times, at seeming ease on the wing. White-tailed Kites spend much time hovering as they hunt small mammals, their slender bodies held diagonally, wings pumping and tail flaring to maintain a chosen position in the sky. If prey is spotted, a kite will let its wings splay loosely above its body to control a slow-motion, foot-first dive intended to capture the item. More often than not, the object of such keen attention is a California vole (*Microtus californicus*), known to be the chief element of the kite's diet in coastal northwestern California (Dunk and Cooper 1994). Because kites hunt in exposed grassy places at the cusp of town and countryside—such as in valleys, streamside groves, around farms, and along roadsides and freeway median strips—they are seen often, and are admired by many.

During the Atlas period, White-tailed Kites were reported in 35 (8%) of the 425 total blocks, with confirmation of breeding in 13 blocks. They were encountered chiefly in sections of the county well known for supporting them. However, the block-by-block effort revealed birds in some localities at which there had been few, or no, previous reports.

White-tailed Kites in Humboldt County live in places well grown to grass, weed patches, and wetland vegetation. As is true for nearly any bird one might name, the quality, extent, and contiguity of habitat appear to dictate the occurrence and abundance of kites in the region. Sections of the county offering the greatest available habitat include the coastal lowlands from Little River through the Mad River floodplain and the Humboldt Bay coastal plain to the uppermost Eel River floodplain (F11 to e22); in several blocks of the Eel River and lower Van Duzen River (f22, g23, H23); along Bear River Ridge (d24, e24); in the lower Trinity River canyon at Willow Creek (N13) and around Hoopa (m10, m11, N11); and in four blocks in southeastern Humboldt. There were also reports in two adjacent blocks in the Bald Hills (i08, j08), in one of which breeding was confirmed.

Extensive stretches of rolling, bucolic foothill grassland occur in portions of interior Humboldt County, in which no kites were reported. An exception was a report of one southeast of Kneeland (i19) 14 Jun 1998 (K. Moore). With so much suitable-looking habitat, is it possible that some birds were simply overlooked? Surely, few raptors advertise their presence as effectively as White-tailed Kites. Perhaps some critical element is missing from these landscapes. Is the prey base insufficient? Is the degree or timing of grazing throughout the county's interior rangelands not to their liking?

White-tailed Kites appear strongly partial to less-grazed or fallow grasslands. This propensity is exhibited by the tendency of kites to hunt even small, somewhat isolated patches of such habitat, and to hunt less in pasturage stocked with beef cattle or dairy cows. This may be because California voles are virtually absent in grazed areas with vegetation less than 4 inches tall (Dunk and Cooper 1994). Although a great deal of the central Humboldt County coastal lowland terrain does not support ungrazed or neglected grassland, where it does occur White-tailed Kites are regularly seen and can be numerous. This is nowhere better illustrated than in the south Arcata Bottoms. Use by kites of the unmanaged grassland just north of the Humboldt Bay levee and along the Mad River Slough was typically high during the mid-1980s through the Atlas period, whereas use of outlying active pasture was comparatively low.

Although exceptions may be noted, sites favored by kites are characterized by grasses from 1 to 4 feet in height, with a matted layer of "decadent thatch" beneath. Scattered annual forbs such as radish, mustard, wild carrot, and thistle are generally present. Mounds of blackberry occur commonly. Stringers or clumps of saplings and smaller trees, either deciduous hardwoods or conifers, are present at many sites. A freshwater wetland component is sometimes present, but often is not. A timber wall created by a stand of taller trees nearby at valley edge or on slopes may serve as an attraction for prospective nest or roost sites.

Indicating that breeding activity began at least as early as March during the Atlas period were records of courtship 13 Mar 1997 (g23; TWL) and of an adult carrying nest material 20 Mar 1996 (f12; GAS). There were no breeding confirmations in April. Occupied nests were reported from four blocks, spanning the dates of 2 May 1996 (e15; J. Kidd) to 11 Jun 1998 (i08; COg). The earliest fledged young were noted at Hookton Slough (D19) 30 May 1996 (JCP). A fledgling near Fernbridge (d20) 8 Jul 1998 was carried off and eaten by a young Peregrine Falcon (S. McAllister)!

The status of the White-tailed Kite in Humboldt County has been one of scarcity during the early period of historical record, followed by decades of increase in population and range within the county. The earliest historical record in Humboldt was of an apparent pair near Miranda in July 1924 (Clay 1926). The local range expansion was further documented by Kimsey (1955), who described the next round of eight observations 1951–1953 from Dows Prairie south to Loleta, but noted that no nests had yet been found. By 1975, Yocom and Harris considered this species to be an uncommon resident and breeder. They also listed what appears to be the first documented Humboldt breeding record, of a pair that nested in a Sitka spruce and fledged two young near Indianola in May 1967. The coastal lowlands of Del Norte County were colonized soon thereafter.

Continued on page 389

BALD EAGLE *Haliaeetus leucocephalus*

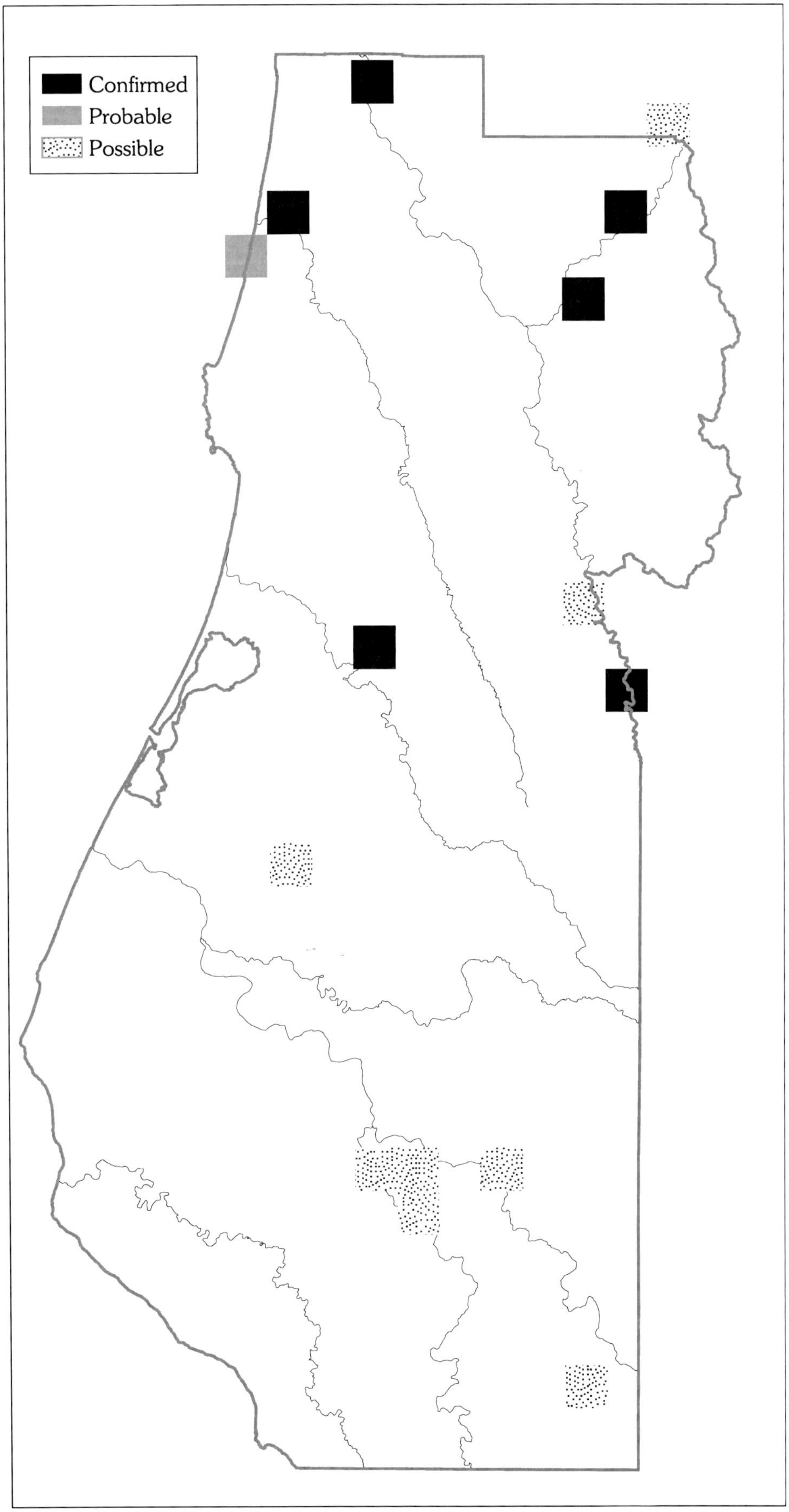

The Bald Eagle is one of eight large raptors of coasts and waterways collectively referred to as sea eagles. It is the only member of the genus typically occurring in continental North America, and is one of the most widespread birds of prey. Bald Eagles are common in some regions, uncommon but of regular occurrence over most of their range, and are nowhere wholly unexpected during migration. Until recently, Bald Eagles have endured an uneasy cohabitation with humans. From the beginning of European settlement, they were killed indiscriminately. Federal protection came only in 1940.

While persecution by direct means abated, however, a new threat came to the fore. Because they are apex predators and scavengers of large animals, particularly fish, Bald Eagles are prone to the harsh effects of concentrations of environmental contaminants in the food they eat. For several decades following World War II, these birds suffered dramatic loss of reproductive vitality from DDT and other organochlorine pesticides. As the national emblem of the U.S. and as a symbol of wild lands, this bird was among those whose unfolding plight was documented by Rachel Carson and other concerned biologists. Wholesale loss of breeding Bald Eagles at most historically occupied sites in the lower 48 states by the late 1960s sparked public outcry, providing impetus for the domestic ban on DDT enacted in 1972. Cooperative action begun before it was too late has resulted in a remarkable comeback for Bald Eagles, which have since increased tenfold as nesting birds in the U.S. outside Alaska. As a result, by the end of the Atlas period, this species was under consideration for removal from the federal list of endangered species (USFWS 1999a).

Bald Eagles have been uncommon in Humboldt County during the period of historical record. Referring to northwestern California without special reference to the county, Harris (1996) termed this species a "rare to uncommon resident and local breeder," further noting that populations increase in winter. Harris pointed to a sketchy historical record as a source of difficulty in determining the comparative abundance of eagles in Humboldt in modern times. Among few nest reports in his extensive files is the record of a pair that nested on an offshore pinnacle at False Cape in 1939 (*fide* P. Detrich; S. Harris, pers. comm.). Between 1970 and 1981 there were no known active nests in the county. By 1990 there were two (Jurek 1990). The first "modern" breeding record was a nest with young 30 Jul 1984 on the Humboldt side of the South Fork Trinity River near Todd Ranch (J. Mattison, T. Burton).

During the Atlas period, breeding was confirmed at six sites in as many blocks. Additionally, sightings of single eagles during the breeding season were made in three adjacent blocks along the South Fork Eel River (i27, J27, j28), in two blocks along the main Eel River (L27, n32), in upper Yager Creek near Carlotta (g20), Willow Creek (n14), and near Somes Bar (p03). The only probable breeding record was of a pair 30 May 1998 hanging around Freshwater and Stone lagoons (f06; M. Morris). As eagle numbers recover, breeding should be looked for in these and other areas in the county.

Nesting along the upper Klamath River near Orleans was known in the late 1980s (S. Harris, pers. comm.). During the Atlas period there were two breeding confirmations in this area. The Waakar site near the mouth of Camp Creek (o05), first discovered in 1995, was active through at least 1998. Some waypoints in the chronology of this nest include a bird seen apparently incubating eggs as early as 12 Mar 1996; the earliest chick seen 7 May 1998; an approximately 7-week-old nestling 28 May 1996; two approximately 8-week-old chicks in the nest 19 Jun 1997; an adult in an empty nest 7 Jul 1997; and a fledgling still very near the nest 20 Jul 1998 (all S. Sniado). The second site was near the confluence of Allen Creek with the Klamath River (N07). During the Atlas period, eagles nested there from 1995 through at least 1997. The earliest seasonal date that birds were seen on this nest was 3 Apr 1995, the earliest nestling was observed 16 May 1995, and by 22 May 1995 two nestlings were present (all S. Sniado). Adults were observed feeding young at this site as late as 1 Jul 1996 (T. Kirk). This site was first brought to light in 1994.

Farther down the Klamath River, an occupied Bald Eagle nest was observed near the confluence of Blue Creek (i02) 23 Jul 1996 (R. LeValley). On 15 Mar 1995, an adult incubating eggs was observed near the Mad River in the vicinity of Canon Creek (i15). Later in the season, a single eaglet was seen in the nest, which was located in an 80-year-old coast redwood surrounded by clearcuts (L. Diller). Two young were fledged there in 1996. This site was first discovered in 1990 and was known to be active through 1997. Near Orick (g05), a nest was suspected in 1998 and an active nest was confirmed in 1999 (no details available).

Along the South Fork Trinity River in the vicinity of Todd Ranch, nesting was suspected in 1981 through 1983. From 1984 through 1990, eagles had varying levels of nesting success there (Jurek 1990). During the Atlas study, a nest with young was recorded along the South Fork Trinity River (o16) 4 Jun 1996 (*fide* D. Hom).

Bald Eagles in Humboldt County are strongly tied to open water and undisturbed shorelines. Tall exposed perches, freedom from harassment at nest sites, and a long line of sight are repeated elements in settings favored by these birds. River corridors and estuaries attract scattered individuals thought to be migrants, or otherwise nonresident, from October to March, especially when anadromous fish are on the move. Open bottomlands well populated with waterfowl—such as Humboldt Bay National Wildlife Refuge and the lower Eel River delta—received use by very small numbers of adult or immature Bald Eagles in each of the intervening winters of the Atlas period. Although occasional birds are seen well away from water, for example, over foothill rangeland or along ridgelines, it does not appear that extensive upland foraging takes place. Bald Eagles thought to be actively migrating north or south are rarely reported in the county.

NORTHERN HARRIER *Circus cyaneus*

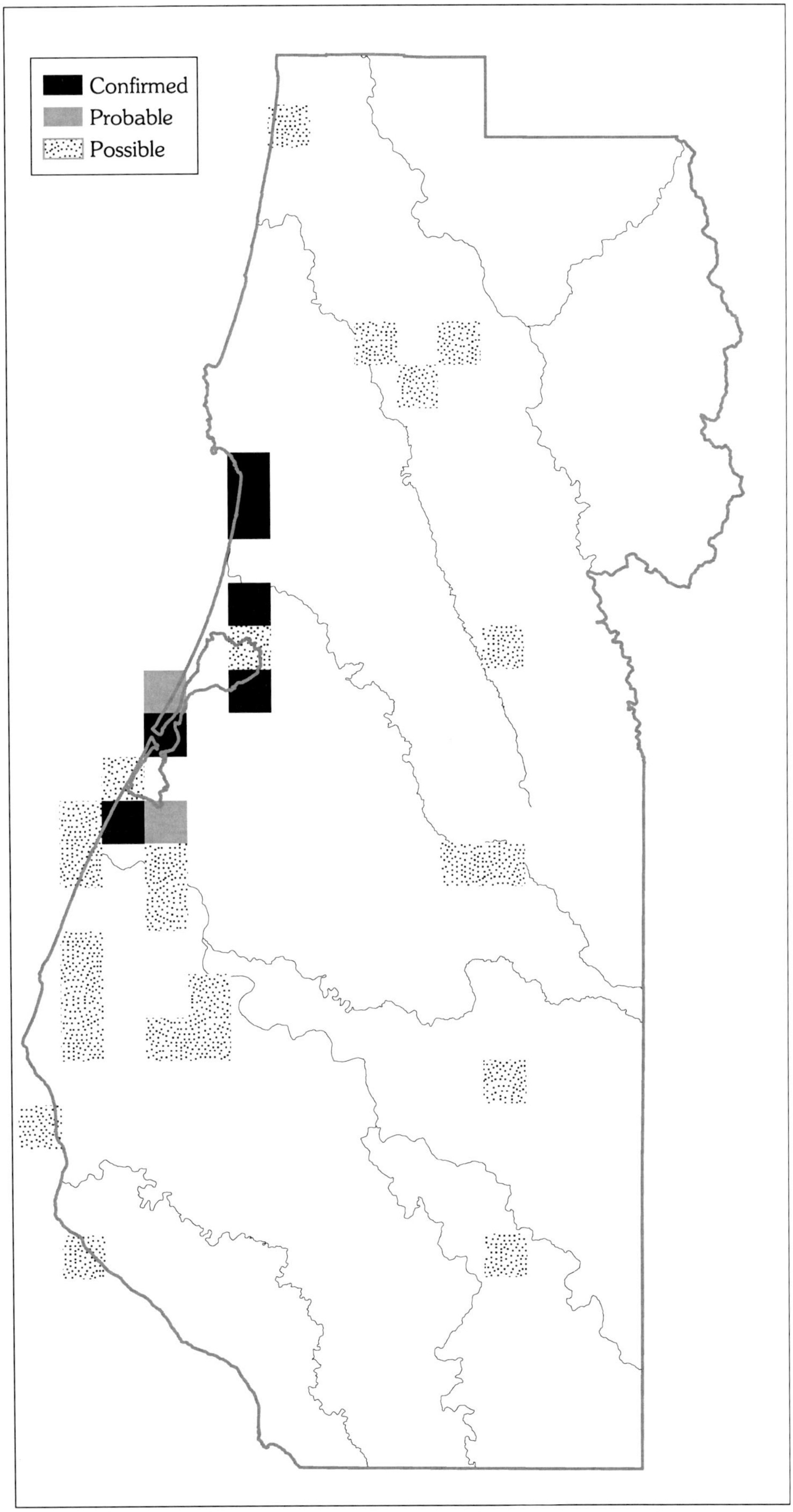

A white rump, slim proportions, and languid, slow-motion style of cross-country flight mark the Northern Harrier. Paradoxically, this raptor is at once among the most distinctive yet potentially confusing small hawks. Seen near at hand coursing low over a meadow, nothing else living in North America looks quite like a Northern Harrier. In full sunlight, the gleaming silvery upperwings and jet black wing-tips of adult males are an arresting and unmistakable sight. However, preconceptions of an odd *Buteo* or *Accipiter* often attend the thoughts of incautious observers witnessing a harrier backlit in a full soar overhead, or moving steadily down a mountain ridge with wings set at half-sail. And well into one's experience with birds, a brown-plumaged harrier stubbornly unmoving atop a fencepost in the heat-hazed distance may recall any number of other hawks, or even an owl.

The Northern Harrier is one of three Western Hemisphere representatives of the genus *Circus*, a group of 10 small- to medium-sized hawks. This species enjoys a tremendous range, spanning the Holarctic and occurring widely in California. It has been formally known as a bird of Humboldt County since the report of Townsend (1887), who stated that the birds were common at Humboldt Bay. Numerous observers working in Humboldt in the past century have made acquaintance with this graceful raptor. Harris (1996) called it a "common migrant and winter visitor; uncommon breeder and summer resident" in northwestern California. He noted that summer birds frequent open areas with marsh or tall grass cover on beach–dune brushfields and overgrown bottomland pastures.

This characterization was echoed by Atlas results. Northern Harriers were not especially widespread in summer, and they were reported during the nesting season in only 7% of all blocks. Nevertheless, confirmed breeding was established in six blocks, all within what has historically been considered the core summer range of this raptor in Humboldt County. Overall, detections were most frequent in the coastal lowlands from Clam Beach through the Humboldt Bay lowlands, the area where each of the breeding confirmations arose. Harriers were also encountered at various points in the Eel River floodplain from near Rio Dell into the lower delta, and in the expansive hilly grasslands shortly inland from the coast south of there. A single bird at Boyes Prairie, Prairie Creek Redwoods State Park (g03) 23 Jul 1998 (D. Anderson) was intriguing. None were noted at this site in the course of frequent coverage in spring during any of the Atlas years, so perhaps this individual was a post-breeding dispersant or wanderer.

Although this species was clearly most numerous in the extensive open country of the central Humboldt coast, it was not strictly confined to that area. During the Atlas period, harriers were recorded in grassy portions of the county at middle elevations in three blocks in the Bald Hills, extending from Childs Hill Prairie (i08) to French Camp Ridge (k08); in upper mid-slope oak-and-grassland mosaic on Titlow Hill (L15) and in similar upland habitat west of Showers Pass (k20, l20); southeast of Bridgeville (L25); and on Mail Ridge west of Fort Seward (L29). All inland observations were of single birds and referred to only possible breeding but, given the date span of 23 Apr to 14 Jul, it is likely that at least some of these birds were summer residents and, indeed, possibly breeding.

During the spring and summer, Northern Harriers in Humboldt County are to be looked for in open landscapes, with trees of any size only a minor element; large interconnected stretches of unmanaged or lightly managed grassland (sometimes with grazing pressure by livestock, sometimes not); scattered hunting, plucking, and lookout perches such as saplings and fence posts; often, a wetland component; and minimal human intrusion. These needs are met spottily across the coastal lowlands and, evidently, to a lesser degree in the interior of the county. It was the experience of atlasers that occurrence and abundance seemed to increase with proximity to the coast. Pairs of Northern Harriers were most expected in the zone of coastal dunes and in the weedier pastureland shortly to the east of that strip. Clam Beach, the Mad River estuary, both of the Humboldt Bay sandspits, and the extensive flat bottomlands of the Eel River delta were consistent hangouts of Northern Harriers.

Comparatively limited birding coverage of the vast mountain prairies of the Bear River Ridge and Cape Mendocino country produced numerous sightings of single birds, suggesting that a significant summering population occurs there. The great amount of mountain prairie habitat in which pairs of harriers might nest leads one to suspect that a concerted survey for these birds might reveal many more than were detected during the course of roadside birding from the limited travel routes available.

Four of the six confirmed breeding reports of Northern Harriers during the Atlas period were of occupied nests. While no details were provided, these were reported 23 Apr 1996 (F11), 26 Apr 1996 (c19), 29 Apr 1996 (f14), and 12 May 1996 (f16; all J. Kidd). Two birds were seen dropping prey into tall grass in the Arcata Bottoms 26 and 29 May 1999 (F11; K. Irwin). An occupied nest was reported 16 Jun 1996 at Clam Beach (F11; M. Marshall), and a single nestling there was photographed (date unknown). An adult carrying food was seen near the south end of the north spit of Humboldt Bay (D17) 28 May 1998 (DFx). Two reports of probable breeding were received, of a pair in courtship 11 Apr 1998 near Manila (d16; JEH) and agitated behavior 12 May 1996 (D19; KMS).

Northern Harriers are least abundant in Humboldt County in late spring through early fall. As is the case with some other birds of prey, abundance increases distinctly in autumn, remains at a higher level through winter, then diminishes in early spring until only scattered breeding pairs persist by April. In winter, not only are breeding areas saturated with wintering birds, but outlying, peripheral, and more marginal foraging sites where the birds are not commonly encountered are used daily. These settings include the smaller and more confined valley bottoms and stream corridors; river gravel bars; heavily grazed dairy pastures; urban-edge habitats such as old mill sites and weed patches; airstrips; and nearly any terrain in which trees are an exception rather than the rule.

Northern Harriers roost communally in winter. Dusk assemblages of from several to 20 birds were noted during each winter of the Atlas period at favored places in the Eel River delta and in the south Arcata Bottoms. Harriers and White-tailed Kites were observed roosting together at Cock Robin Island (c20) and at the Mad River Slough Wildlife Area (F15). This phenomenon has also been noted for more than 20 years at Fern Ridge Reservoir west of Eugene, Oregon (DFx *et al.*, pers. obs.).

Loss of marshlands around Humboldt Bay has probably affected Northern Harriers. Nesting on the ground in open country, harriers are faced with numerous threats. Many poorly timed management activities, including burning, flooding, livestock grazing, mowing, and plowing can destroy active nests. In addition, natural predators, dogs, and human disturbance probably take their toll. Rank old pastures have very high biomass of small mammals (Chrisney 1988), so mowing or burning to enhance introduced Canada Goose habitat is probably done at the expense of harriers and other aerial hunters.

SHARP-SHINNED HAWK *Accipiter striatus*

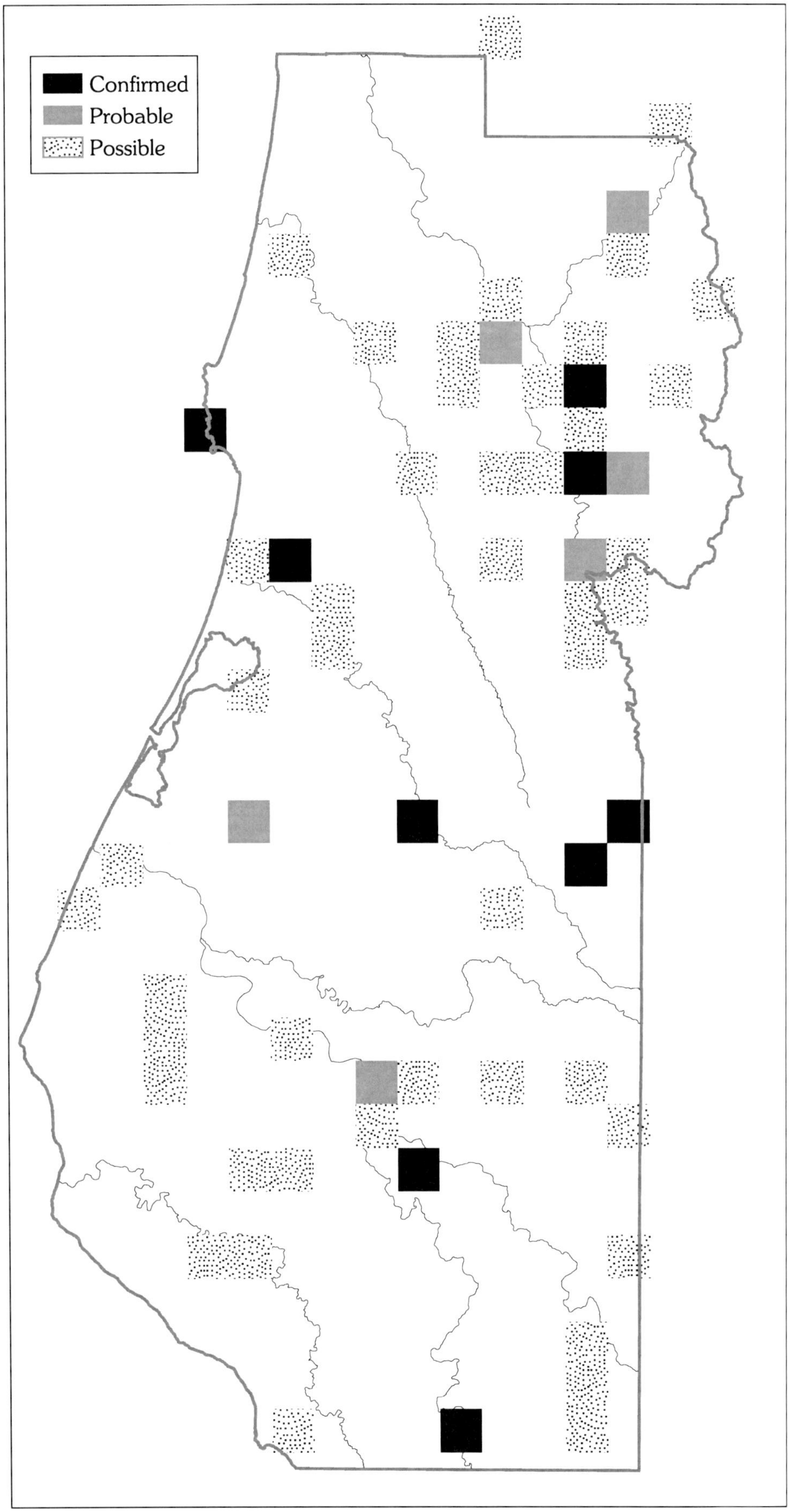

Sharp-shins happen. They occur as interruptions, as rough-cut moments one never quite expects: a heedless blur avoiding yet another windshield; a modest spread-winged form wheeling tightly above a hillside when the air is full; a burst of hectic wingbeats causing juncos to scatter as nothing else can. They are among the most widely seen birds of prey, especially in towns and rural countryside during the colder months, where they haunt any small flock of birds using a backyard feeder or scrap of woods. Early in a naturalist's experience, the first Sharp-shinned Hawk comes as but a minor milepost, lost to distinct memory with one's first woodland orchid, crawdad, and meteor shower.

Sharp-shinned Hawks are comparatively northern breeders. Only a few other North American hawks and falcons have a more generally northern distribution in summer. Southward extensions of the breeding range occur broadly in the West wherever forest cover is sufficient and the birds find site conditions otherwise to their liking. Sharp-shins are known as a nesting species chiefly in the northern portion of California, occurring in semi-open lowlands and arid regions elsewhere in the state from September through April. These hawks have been reported regularly in Humboldt County since the report of Townsend (1887), who stated that it was "not common" at Humboldt Bay in November–December 1885. With general reference to northwestern California, Harris (1996) characterized Sharp-shinned Hawk as a "common migrant and winter visitor; uncommon summer resident and breeder," mentioning only four sites at which nesting evidence had been reported. Perhaps the earliest historical record of breeding in Humboldt was eggs collected 30 May 1935 at McKinleyville (C. Clay; HSU 1085).

While they are certainly not very numerous in summer, it has been apparent that Sharp-shinned Hawks may be found nearly anywhere in Humboldt County during the breeding season. Atlas results indicated that Sharp-shins occurred in a diffuse manner across the county during the 5 years of the survey. The only clustering of detections was in northeastern Humboldt, where the Hoopa Valley Indian Reservation and adjacent lands accounted for nearly one-third of all blocks with reports. Otherwise, nothing approaching a cluster of blocks can be discerned on the map; the balance of detection-blocks was quite thoroughly scattered.

As a rule, Sharp-shins seem to occur most often in the breeding season in areas of contiguous forest cover. Generally, such sites are dominated by immature to mature Douglas-fir mixed with tanoak, California black oak, Jeffrey or ponderosa pine, and other routine forest trees. Sharp-shinned Hawks are probably sensitive to forest fragmentation (Rosenberg and Raphael 1986). As forest cover in the county is abundant and occurs by and large on hillsides at all elevations, it follows that no particular correlation with landform or river courses should be apparent on the distribution map. Although this is surely not the only possible explanation for the pattern of distribution in Humboldt, it would seem reasonable to suspect that this phenomenon contributes to the scattering of blocks with detections.

Although 86% of the 425 total Atlas blocks had no reports of Sharp-shinned Hawks, only a few concentrations of these "no-hawk" blocks fall within regions readily describable with respect to landform or vegetation. It can be noted that industrial timberlands dominate the landscapes of far northern Humboldt, east of Trinidad, and southeast of Eureka. The east-central portion of the county, dominated by oak-and-grassland mosaic, is deficient in dense stands of trees, primarily a result of thin, unstable soils, not human-related activities. As was the case with Cooper's Hawk, reports from interior southern Humboldt County were few. The reason for this is not apparent. Perhaps that area's excessively hot summer weather and spotty representation of preferred nesting habitat are limiting factors.

Nests of Sharp-shinned Hawks are not often located by human observers, as these birds favor thick stands of trees for placement. Only three nests were reported during the 5 years of Atlas surveys: an occupied nest near Pilot Rock about 15 Jul 1996 (n20; S. Bunnell); a nest with young at Iaqua Buttes the same day (J19; S. Bunnell); and another nest with young 15 Jul 1995 (o19; C. Hibbard). The remaining four confirmations were of adults carrying food to young 20 Jun 1998 (k33; JEH) and 26 Jun 1996 (e10; GAS), and fledged young 15 Jul 1995 (g13; J. Haggard), 28 Jul 1998 (N09; *fide* JMH), and 7 Aug 1997 (N11; *fide* JMH). Evidence from six blocks indicated probable breeding status. The most frequent behavior noted among these was an agitated adult, reported 3 Jun 1998 (o11; KMS), 7 Jul 1995 (N13; TWL), and 30 Jul 1996 (l08; JMH). In 75% of the total blocks in which Sharp-shinned Hawks were detected, only possible breeding was reported. The date span of these reports fell mostly outside the expected periods of migration, so it is likely that breeding could have taken place in many if not most of these blocks.

These small hawks are highly migratory. They are among the raptors most commonly seen in passage at coastline points and along mountain ridges, particularly inland from the Eastern Seaboard and in the foothills of western mountains. Sharp-shins are frequently noticed actively migrating through Humboldt County, but no significant concentration points have yet been uncovered by raptor enthusiasts.

COOPER'S HAWK *Accipiter cooperii*

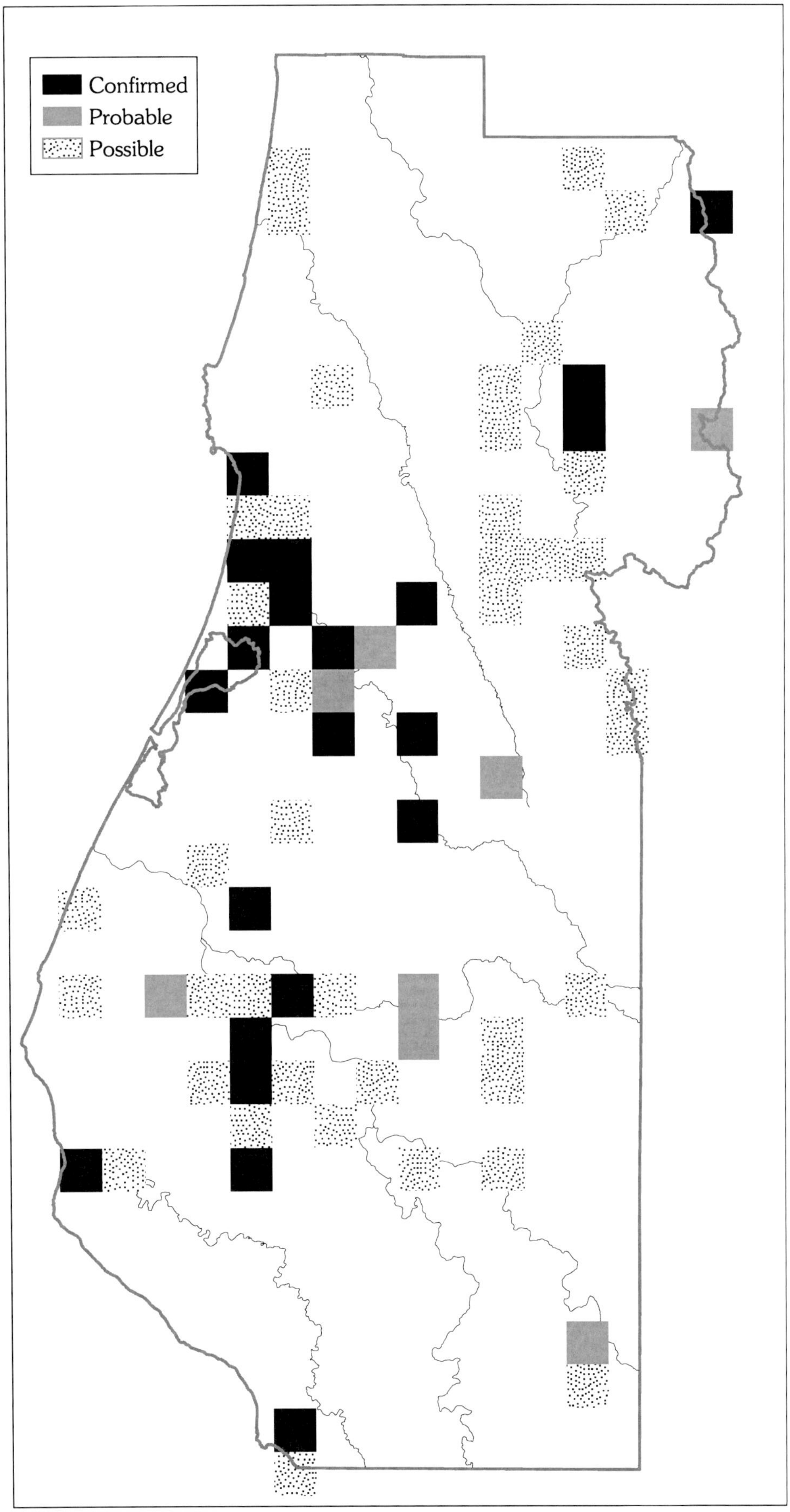

Cooper's Hawks are small to medium-sized raptors of wooded country. They have a less boreal, more temperate distribution than the other two U.S. and Canadian accipiters. They differ also from Northern Goshawk and Sharp-shinned Hawk in habitat—although much overlap with either of the two species occurs—and in foraging behavior. In a general sense, Cooper's Hawks in the breeding season appear less tied to extensive tracts of forest, are less frequently high-mountain birds, and nest in riparian or lowland woodland settings more commonly than the other two. These birds appear to prefer broken forest cover, especially of uneven-aged or mixed-species tree communities. An ability to tolerate unsuspecting human presence has allowed recent increases in some urban areas, such as Sacramento and Victoria, B.C., as well as persistence in woodlots in suburbs of the northeastern United States.

Though classically described as forest birds, Cooper's Hawks are not averse to hunting in semi-open settings. They will hunt small birds in sapling- or shrub-dotted bottomlands, provided sufficient cover exists to allow surprise attacks. Backyard bird feeders are well known as a source of attraction. Cooper's Hawks may be seen soaring over treeless areas at any time of year. An *Accipiter* sitting upon a fencepost in a pasture or on a roadside power pole most often proves to be this species.

With a distribution spanning California at one season or another, Cooper's Hawk has long been known as a bird of Humboldt County. Historically, this species has been encountered in very small numbers throughout the year. Most birders and naturalists see Cooper's most commonly either in migration or in winter, when concentrations of small landbirds fall under their watch. Sightings in late spring and summer are less expected. Nonetheless, occasional individuals or pairs may be seen widely. During the Atlas period this species was detected in

16% of the 425 blocks, compared to 14% of blocks for Sharp-shinned Hawk and 5% for Northern Goshawk. Reports were scattered more or less evenly across the 5 years of the project.

Cooper's Hawks nest high in trees, and attendant adults are stealthy enough to avoid easy detection. Nesting attempts may remain undetected unless an observer inadvertently approaches closely and agitates the adults. The distribution of Cooper's Hawk as a breeding bird in Humboldt was characterized in a cautious vein by Harris (1996), who stated that the species "breeds sparingly throughout"; he listed eight widely scattered stations in the county from which nesting evidence was reported 1930–1995. Results of the Atlas surveys may serve to shed a bit more light on their status. It is clear that this bird was quite well distributed in the breeding season during the 5 years of the study. Records came from essentially each of the four corners of Humboldt and at numerous points in the interior. Indeed, most detections fell within the central portion of the county.

The spread of detection-blocks on the map indicates no strong association with either landform or regional cover type, although there were many more reports of Cooper's Hawks from blocks at lower than at higher elevations. Blocks bisected by the Trinity, Eel, Van Duzen, and Mad rivers and their larger tributaries were well represented. Many reports among those comprising a loose cluster centered around block (F25) represented birds seen in the course of fieldwork by Pacific Lumber Company biologists. Forest composition and extent of tree cover vary considerably among these block clusters.

Cooper's Hawks were only rarely reported from two regions of some size within Humboldt County. The northwestern part of the county appeared to hold little attraction for these birds. Much of this area is commercial timberland under intensive management. As a rule, forest cover here either is lacking or is immature and even-aged; tree species and structural diversity are lower than in many other parts of the county. This region also contains old-growth redwood parks. In extreme southern Humboldt, this species was almost unreported as well, with only four detection-blocks among the southernmost seven tiers of blocks. It is difficult to pose any reason for a seeming scarcity of Cooper's Hawks there, as many of these blocks would appear to offer the broken, variegated forest mosaic inhabited elsewhere in the county. However, other blocks are thoroughly forested with closed-canopy tanoak/Douglas-fir regrowth with limited openings and edge. It is conceivable, though only speculation, that atlasers working from roadsides in heavily wooded surroundings enjoyed fewer opportunities to spot Cooper's Hawks in flight or in trees out of a limited line of sight.

Given the secrecy with which these hawks nest, a surprising proportion (30%) of the total blocks with detections reported confirmed breeding. An adult carrying nest material was reported 7 Apr 1996 (F13; GAS). Twelve occupied nests were reported, spanning 23 Apr 1996 (F11) to 1 Jun 1999 (B27; both J. Kidd). A nest with young found at the Mad River Fish Hatchery (H15) 18 Jul 1995 (PFS, C. Walling) was about 50 feet up in a 16-inch-diameter coast redwood. By 26 Jul three young had fledged, but they were still in the vicinity of the nest and still dependent on the adults (C. Gutiérrez). The feeding of young was also reported 19 Jul 1996 (F15; GCH). As might be expected, confirmed breeding reports involving the carrying of food or the observation of fledged young fell during late July and early August. While it is conceivable that some individual birds reported in blocks with only possible breeding may have been non-breeding birds (such as later spring or earlier fall migrants), birds present from April through July are probably summer residents.

Cooper's Hawks suffered a continental decline during the 1960s and 70s, but their numbers have apparently stabilized. Evidence for this stabilization comes from data gathered during decades of systematic counts at hawk-watch sites at many points in the species' range. Continued protection from shooting and environmental contaminants and the restoration of lowland riparian habitats have also benefited these dashing birds of prey.

NORTHERN GOSHAWK *Accipiter gentilis*

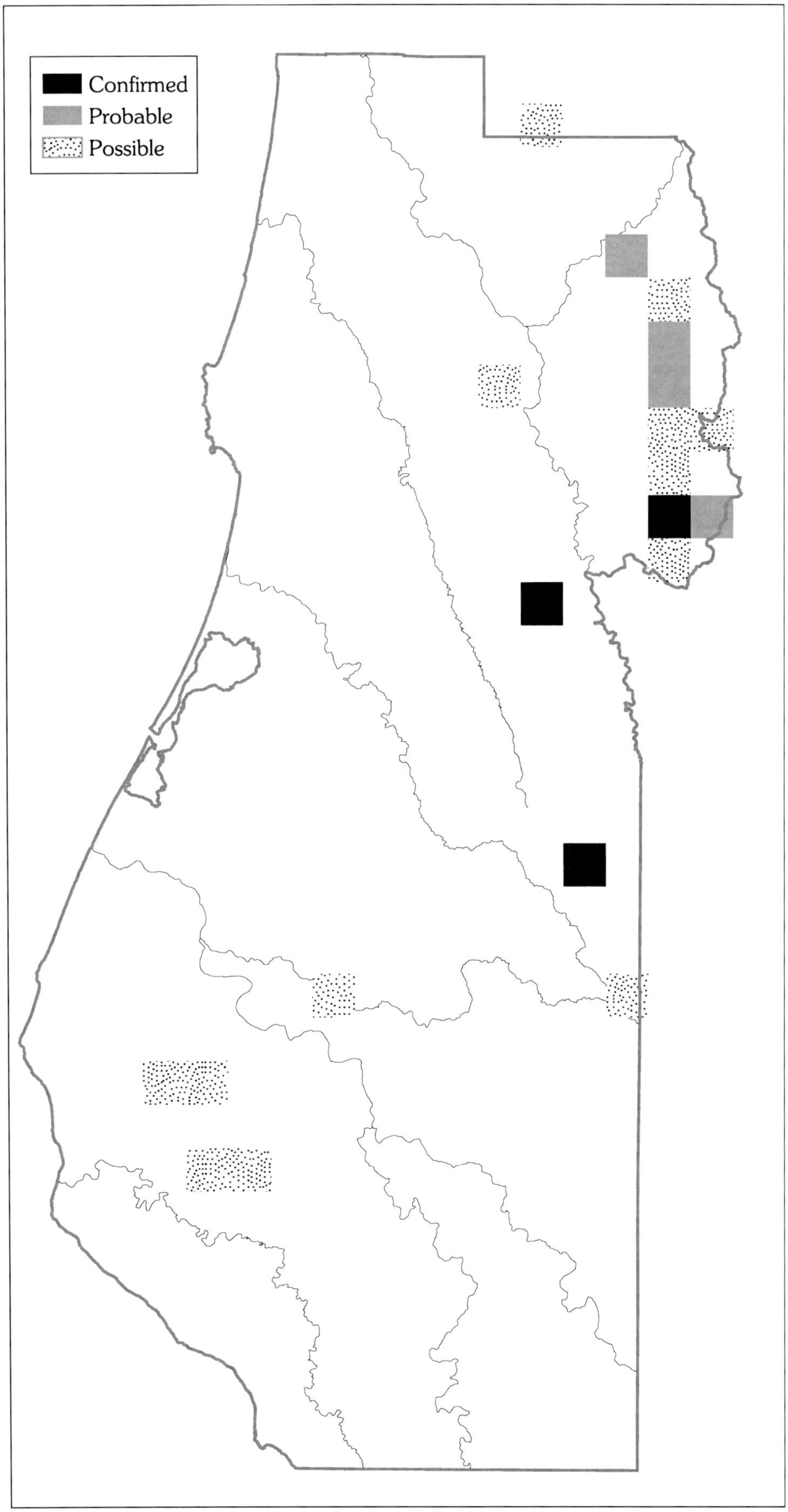

Northern Goshawks are spirited and powerful hawks of extensive forests. Across their wide range in the Northern Hemisphere they inhabit stands that include spruce, larch, pine, aspen, and mixed hardwoods. Regardless of forest type, the birds are most frequently encountered in stands of big trees with minimal understory and a dependable prey base of mammals and birds. Although not limited to wilderness, much of the population occurs in remote country throughout the year. The range of the goshawk in North America extends southward from Alaska and boreal Canada through the mountains of the western United States. They are sparingly distributed at any season in California but are regular nesting birds of mature forests. Goshawks breed in the Cascades, the Sierra Nevada, and the Klamath Mountains, and locally southward in the Coast Range to Mendocino County. This species was considered a rare resident and breeder in northwestern California by Harris (1996).

During the Atlas period, Northern Goshawks were reported in 20 (5%) of the 425 total blocks. Blocks with detections fell in two areas of unequal significance, each of which bears individual discussion. The larger of the two is the far-inland area, which includes middle and higher elevations of the Klamath Mountains and Inner North Coast Ranges. Most of the detections were from the Klamath Mountains north of the Trinity River. South of the Trinity, the forests of the Inner North Coast Ranges (n20 and o23) are seemingly more suitable because of a more open understory. This type of habitat becomes more common, as do goshawks, in adjacent areas of Trinity County. Hall (1984) studied goshawk nesting habitat in this region and found that nest stands were typically composed of dense single-storied stands of younger Douglas-fir with scattered hardwoods, large mature trees, and a park-like understory.

Within the far-inland area, Northern Goshawks were reported from 15 blocks extending from near the Del Norte County line (m03) south to near Dinsmore (o23). Despite the comparatively small number of reports, the proportion of possible, probable, and confirmed breeding reports from that region is similar to the pattern of more numerous or more widespread species encountered during the Atlas period. All three blocks with breeding confirmations fell within this area. A nest with young attended by a pair of adults was observed 5 Jul 1999 near Pilot Rock (n20; S. Bunnell). Two aggressive adults with a fledgling were seen 17 Jul 1996 in true firs along Forest Service road 7N31A (p12; TWL). An adult was observed carrying food at Brush Mountain Lookout (m14) 15 Aug 1996 (M. Smith). Probable breeding was established in four blocks in this region. A bird was watched "fighting" with a pair of Golden Eagles 3 Jul 1998 near the end of Forest Service road 9N03 (p08; J. Ward). Agitated behavior also was reported from southeast of Blacks Lake in the Trinity Alps Wilderness 26 Jun 1998 (P09; KMS), on 20 Jul 1995 (q12; T. Wallace), and a pair acting that way 7 Aug 1998 (o06; D. Kudrna).

The other area of the county in which Northern Goshawks were reported during the study period was Pacific Lumber Company timberlands in the interior south-central portion of Humboldt. During the course of fieldwork—some of which involved the broadcasting of territorial goshawk calls to elicit responses—these birds were detected in five blocks. This loose cluster of blocks with possible breeding evidence is disjunct from the population known from eastern Humboldt, as it lies roughly 50 miles to the southwest. Examples of some goshawk sightings in this area include a bird in Alwardt Creek drainage, a tributary of the North Fork of the Mattole River (D25) 15 Jul 1998 (K. Sundeen), and one heard near Pritchard Creek (e27) 8 Aug 1999 (*fide* S. Chinnici).

The possibility that a small nesting

population of Northern Goshawks occurs in southwestern Humboldt County has been the subject of speculation among biologists and birders. While not validated by probable or confirmed breeding evidence, sightings of individual goshawks during or following the nesting season in 3 years, and in different watersheds, seem to keep the matter an open question. While many of these areas in south-central Humboldt have a significant component of Douglas-fir, coast redwood is also common. The status of the goshawk in redwood stands is even more of a mystery. In redwood-dominated forests, these birds have either greatly declined or were never widespread. An observation of an adult bird in the vicinity of Van Duzen County Park (H23) on 21 Apr 1998 (B. Mauney) was well within the redwood zone, helping to keep that particular discussion alive.

Harris (1996) indicated a long history of Northern Goshawk reports for northwestern California, listing 20 sites in Humboldt County where the species had been reported. Despite this representation, it is plain that these *Accipiters* are very uncommon here in the breeding season and, indeed, throughout the year. Even the most active bird observers in the county rarely see them. Classic migrants obviously on the move through the county have been reported only a few times. Harris (1996) remarked that "post-breeding dispersal by juveniles generates reports in the coastal lowlands in fall and winter some years." During the Atlas period, there were no reliable reports of goshawks in the coastal lowlands despite considerable birding coverage.

While the true status of Humboldt's Northern Goshawk population is not well understood, many biologists familiar with this species have expressed concern for its welfare. Nesting stands may be unfavorably altered by harvesting overstory trees or by fire exclusion, both of which can result in increased density of brush and smaller trees in the understory. Human disturbance around nest sites can cause these high-strung and aggressive birds to abandon them.

RED-SHOULDERED HAWK *Buteo lineatus*

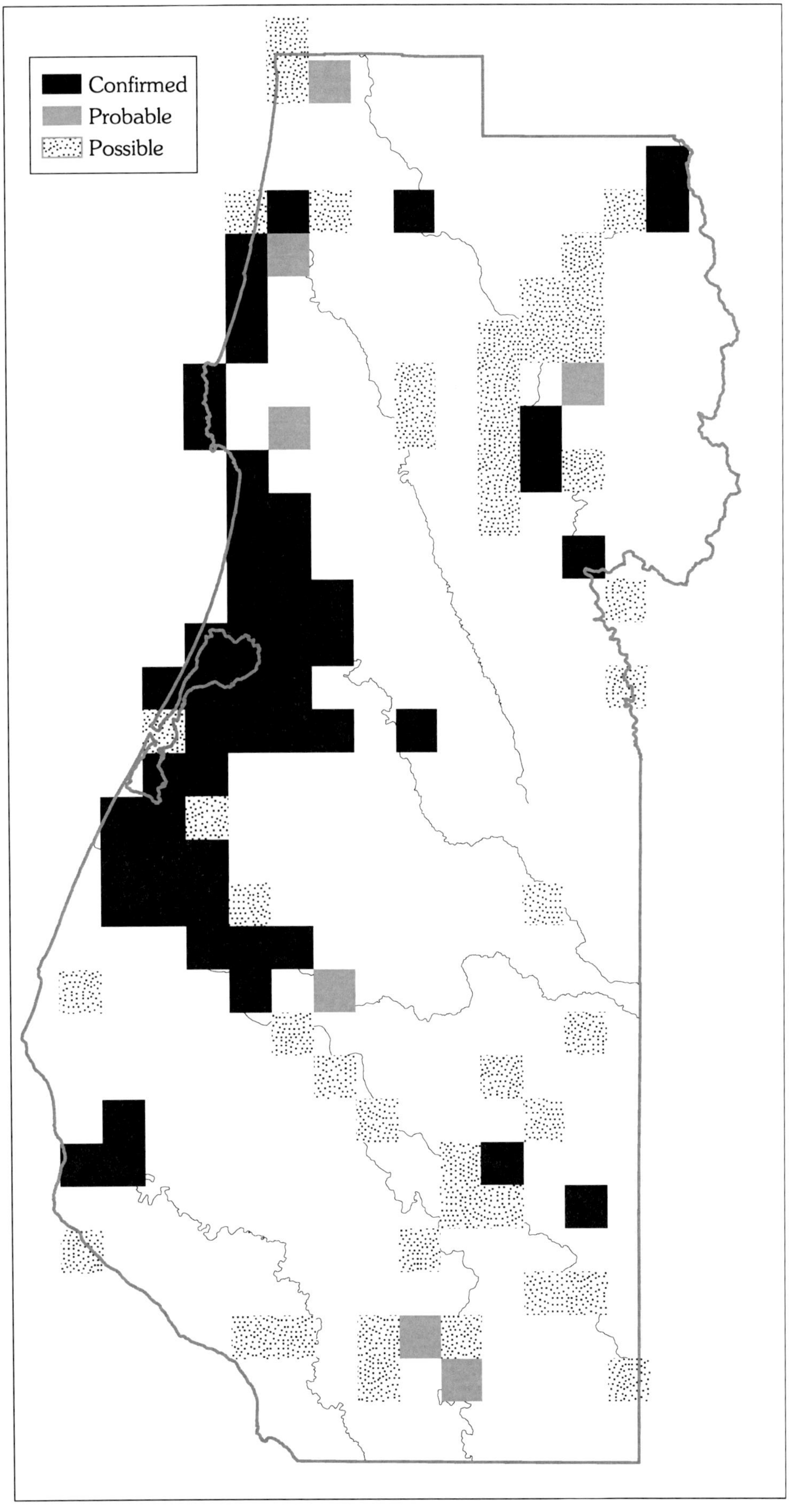

The shrill, measured shouts of adult Red-shouldered Hawks are a familiar sound of valleys and floodplains in Humboldt County. Beginning at sunrise, peaking in mid-morning, and continuing at intervals throughout the day, their exclamations vie with the chatter of swallows as a voice bespeaking the urgency of spring. Red-shoulders vocalize at any time of the year, but most persistently in late winter and spring. These hawks catch the eye through their habit of sitting on roadside wires and fence posts. A curiously hunched and attentive posture while perch-hunting, showy good looks, and a distinct tolerance of human activity conspire to stir comment and admiration, even from persons who would scarcely know one type of hawk from another.

With Red-tailed Hawks, Ospreys, White-tailed Kites, and American Kestrels, Red-shouldered Hawks are among the most frequently seen raptors in Humboldt County. During Atlas fieldwork, Red-shoulders were detected in 24% of all blocks. These hawks were common along and near the coast from the Redwood Creek estuary through the Humboldt Lagoons, all around Humboldt Bay (though not on the south spit), in the lower Mad River drainage, and throughout the Eel River delta. Scattered blocks with detections lay along the South Fork Eel River and the main Eel River, including two confirmed breeding reports in the latter area. Presence in the Trinity River and Klamath River corridors is reflected by the rough arc of nearly contiguous detection-blocks there.

Red-shouldered Hawks were seen only sparingly in southwestern Humboldt County. Habitat in this section of the county is localized, but three adjacent blocks with breeding confirmations along the lower Mattole River suggest a regular population in that area. Only a few blocks along the Van Duzen River hosted Red-shoulders. This is perhaps an effect of the many miles coursed by the river within its incised canyon, where suitable floodplain habitat is discontinuous or altogether lacking. The central Mad River corridor also appeared to hold few Red-shoulders. Great stretches of dry open oak-grassland—and the fact that principal routes through that country occur well away from riparian stretches—may explain the lack of reports in that part of the county.

These hawks are strongly associated with forest edges. Much of the birds' foraging is conducted in expanses of lightly managed bottomland adjacent to riparian areas or timber stands. Neglected pastures or farm fields dotted with saplings and bulrushes attract Red-shoulders, as do city parks, urban gulches, and riparian stringers of cottonwood, alder, and willow bisecting open country. Contrary to the characterization of Red-shoulders expressed in many field guides, those in Humboldt County are not forest dwellers; they hunt to a great extent in somewhat exposed situations. Because these settings exist chiefly on and at the edges of floodplains and in semi-open valley bottoms, it is no surprise that large sections of Humboldt do not support breeding Red-shouldered Hawks. Nearly without exception, there were no reports in blocks dominated by industrial timberlands in the northern and central interior of the county, in rugged interior southwestern Humboldt, and at higher elevations.

It should be noted that, while Red-shouldered Hawks certainly are numerous in the lower Mad River, Humboldt Bay, and Eel River lowlands, the noticeable saturation of those blocks with breeding confirmations was more likely due to an intensive raptor-nest search effort in 1996 (J. Kidd). These data, in combination with Atlas data, illustrate the conspicuous nature of nesting behaviors by Red-shoulders during the nesting season in the part of Humboldt County where they appear to be most common. Red-shouldered Hawks were confirmed breeding in 52 blocks. Nests of various sorts were reported from 1 Mar 1999 (F15; PFS) to 15 Jun 1995 (g13; M. Maxwell); these dates encompassed nearly all breeding confirmations. The earliest fledgling was reported 15 Jun 1996 (e16; GAS); the latest, 4 Aug 1996 (F13; JEH).

Red-shouldered Hawk has long been known as a common California bird, but its status as a locally common resident and breeder in northwestern California is a recent phenomenon (Harris 1996). Hoffmann (1927) indicated that these birds did not occur along the coast north of San Francisco. Harris (1996) presented a detailed record of the expansion this species underwent during the last half of the 20th century, noting that the first report for Humboldt was of one bird seen near Korbel 10 Apr 1943 (W. Anderson). Following several additional near-coastal records, birds at Willow Creek in 1953 (Talmadge 1953b) provided the first evidence of the species away from the coastal lowlands. The first Red-shouldered Hawk detected on the northern Humboldt County coast was seen at Orick in 1957 (AFN 11:374). Harris (1996) stated that sporadic reports of single birds continued through the 1960s, and that it was only in the mid-1970s that the species "suddenly became common" in the coastal lowlands and in the Trinity River valley. The earliest Humboldt breeding confirmation was of a decomposing fledgling found 19 Oct 1974 near Arcata (S. Harris; Yocom and Harris 1975). It was in the 1970s that the expansion began to manifest itself north of California as well. Illustrating a larger regional context, Red-shouldered Hawks have been resident in small numbers in southwestern Oregon since at least the early 1980s (Gilligan *et al.* 1994), and winter visitors in western Washington are no longer a great surprise.

A broad-brush hypothesis explaining this change in distribution is that land clearing during the era of modern settlement increased the amount and variety of bottomland habitat, served to expose previously forested riparian corridors, and produced a mosaic of complex edge habitat at the margins of floodplains. Ultimately this habitat was "discovered" and exploited by the birds. Because Red-shouldered Hawks in California are not strongly migratory, wayward individuals emigrating from the distant core range likely were few. One might speculate that it may have required decades following wholesale transformation of the settled lowlands for a population to finally become established.

RED-TAILED HAWK *Buteo jamaicensis*

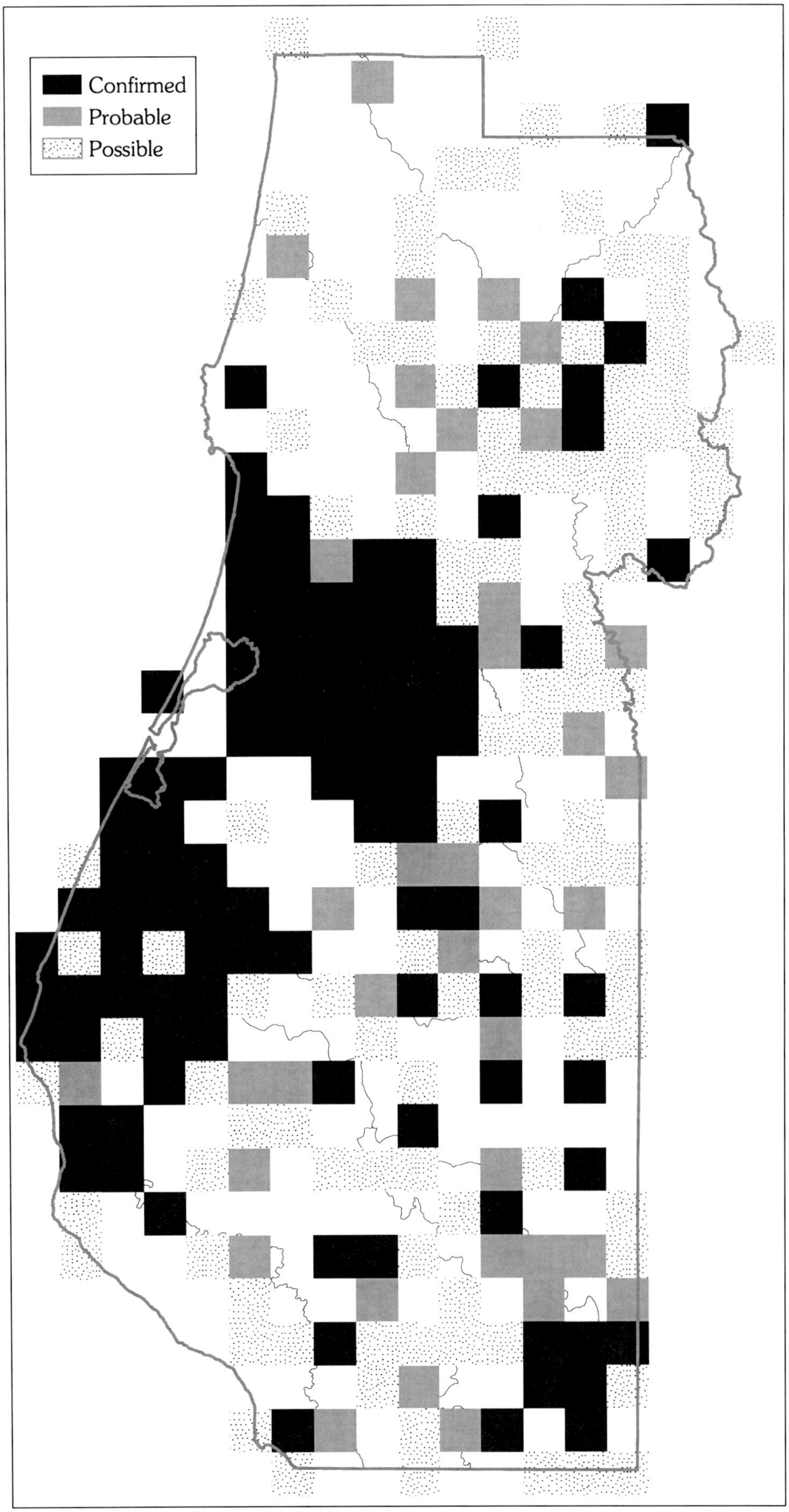

Putting aside quibbling over various regional exceptions, it could be argued that the Red-tailed Hawk is the most frequently seen bird of prey in North America. It is as familiar a raptor as any: common, usually distinctive, inhabiting a great range, and easily spotted perched and in flight. The presence of the species stamps any piece of ground with the wholesomeness of simple countryside, any reach of sky with the beauty of aerial form and function. These birds are at home at the edges of the human footprint as well as in wilderness. Reluctant to give ground at the suburban/rural frontier, pairs of Red-tails make their stand in woodlots, river groves, and scraps of forest in close proximity to human activity. They are a routine sight along highways in open country, where power poles and billboards are favored perches. This is especially true during the colder months, when northern breeders have moved south and augment local populations in valley-bottom and upland habitats of all kinds.

Red-tailed Hawks are habitat generalists. They occur wherever nesting opportunities and freedom from persecution coincide with an abundant mixed prey base of small mammals, reptiles, birds, and large insects. Their ability to find a niche nearly everywhere is reflected in the map generated by the Atlas project. No large section of Humboldt County lacked Red-tailed Hawks. Indeed, the thorough sprinkling of blocks with records of possible, probable, and confirmed breeding indicates that this species was the most widely distributed bird of prey in the county during the Atlas period. Red-tails were found in 58% of all Atlas blocks. With reference to northwestern California, Harris (1996) called the species a "common resident and breeder; becomes more common in winter." Except in Humboldt's near-coastal bottomlands and riparian corridors, this species outnumbers Red-shouldered Hawk in the breeding season.

Noted in the case of Cooper's and Red-shouldered Hawks and attending the present species was an organized nest search effort in 1996 (J. Kidd). Results from that effort, incorporated into Atlas databases, produced a striking concentration of blocks with breeding confirmations (nearly all pertaining to occupied nests) in west-central Humboldt and in the Cape Mendocino–Bear River Ridge area. To be sure, this amazing cluster of confirmed breeding blocks resulted from greater observer effort, not greater number of birds. Throughout the balance of their range in the county, Red-tailed Hawks were confirmed at a far lower density—although not much less than that of some songbirds sharing the same territory! It may be noted that, as is true for many species, slightly more confirmed and probable breeding evidence was obtained in priority than in non-priority blocks, particularly in interior portions of the county.

Red-tails were uncommonly encountered in northern Humboldt County. Ill-defined clusters of no-detection blocks shortly inland from the coast from the Del Norte County line south to east of Trinidad, and also southeast of Eureka, were largely dominated by industrial timberlands. Old-growth coast redwood parks such as Redwood National Park and Prairie Creek Redwoods State Park also are in these areas. The diversity of vegetative structure is somewhat less than that typical of the rest of the county, which is composed of dense regrowth conifer and tanoak stands and cut blocks that revegetate in short order. It is possible that this phenomenon limits Red-tailed Hawks locally. Red-tails were also undetected in a narrow arc of blocks comprising higher elevations and remote sites in extreme northeastern Humboldt County.

Humboldt County offers diverse opportunities for Red-tailed Hawks. Practically every upland habitat with stands of trees, openings, and some grassy or weedy open ground is used at some season of the year. Viewed from a landscape perspective, these hawks seemed to show no distinct preference for one habitat over another during the 5 years of Atlas surveys. This may in part be because Red-tails cruise over sizable territories during the breeding season. These may include a spectrum of habitats, ranging from a patch of valley bottom through mid-slope forest and hillside meadows, or hillsides and adjacent ridge-tops.

Red-tailed Hawks are resident in Humboldt County throughout the year, yet the species is strongly migratory, and the numbers that are present vary by season. They appear to remain in forested and hilly regions all year. In the open lowlands and in the rolling grassy hills of the Cape Mendocino and Bear River ridges, abundance distinctly increases in October with the passage of migrants and the arrival of wintering birds. Far more Red-tailed Hawks occur in open lands around Humboldt Bay and in the Eel River delta during the colder months than are present in the nesting season. As many as a half-dozen may be spotted in a thorough binocular sweep of bottomlands offering lightly managed land and isolated perches, a setting in which the birds are easily seen. By late March, this glut of wintering birds diminishes.

At that time and continuing through spring, pairs of adults are routinely seen performing courtship displays and attending nests or probable nest sites. Courtship was recorded as early as 3 Mar 1996 (B23; L. Bacon-Ogden) and 25 Mar 1997 (D21; S. McAllister). The earliest evidence of actual reproduction was a report of young (presumably nestlings) being fed 23 Apr 1995 (g33; K. Dyer). The 63 occupied nests reported spanned the dates of 12 Apr 1996 (d18, e18, J17, k17; *fide* J. Kidd) to 26 Jun 1996 (I12; JMH). Fledged young were reported from 8 Jun 1996 (J23; E. Gilchrist) to 18 Aug 1997 (i29; JG).

These adaptable birds are probably doing quite well in Humboldt County, aided by land-use practices such as logging and ranching that increase the amount and interspersion of open habitat. Any impact these powerful predators may have on other wildlife populations, especially those not as well adapted to open spaces, can only be surmised (Hunter *et al.* 1993).

GOLDEN EAGLE *Aquila chrysaetos*

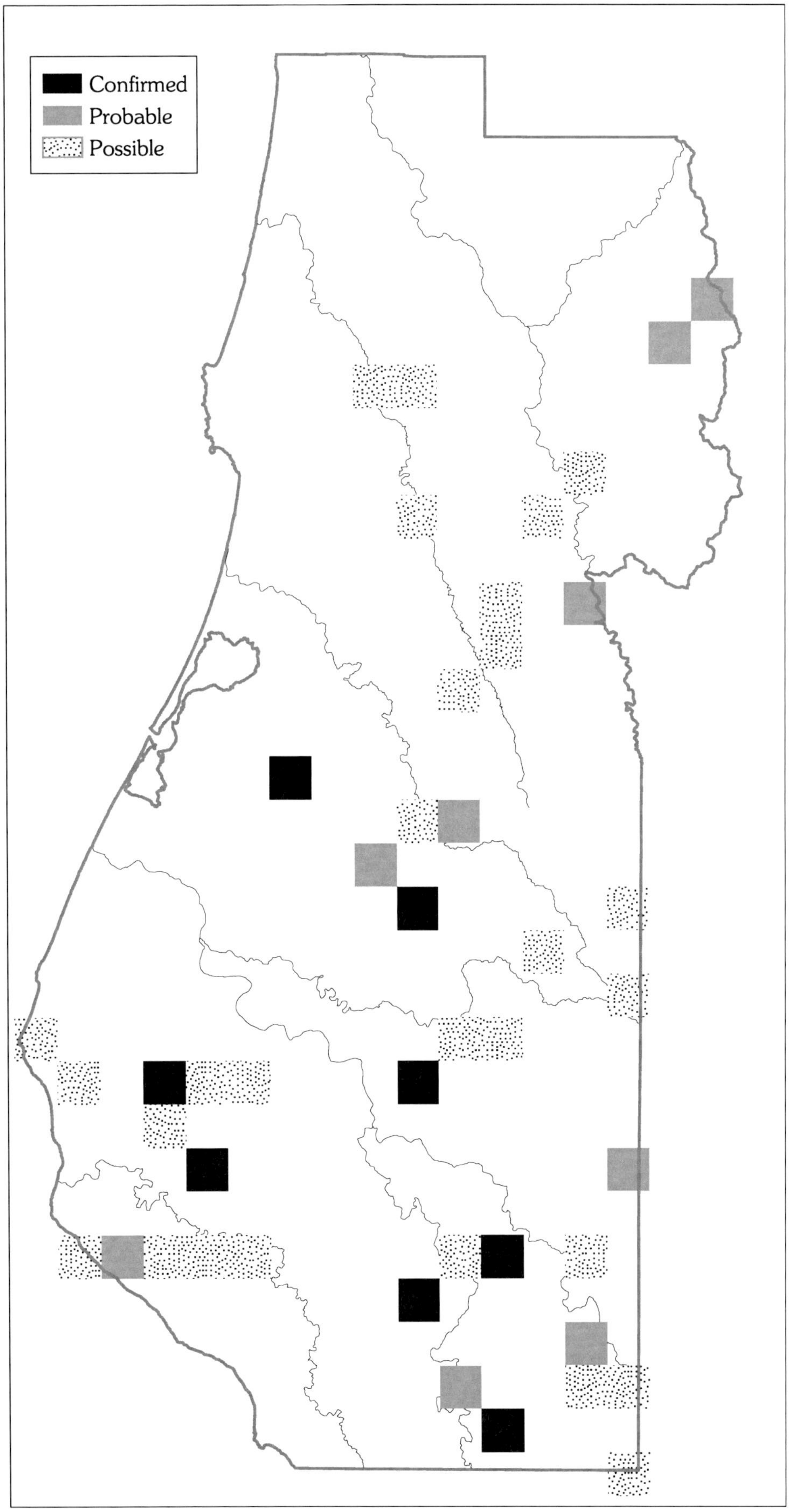

Admired across the ages, Golden Eagles inspire honor and kindle reassurance in the hearts of many. Typically, they live in greatest numbers where the biophysical elements of large landscapes function as they should. Their presence is characteristic of highly rural or wild areas, places in which the human tread remains light. California's mountains, deserts, grasslands, cattle range, and open oak woodlands support a strong population of Golden Eagles. Their range within the state has long been known to include Humboldt County.

In the course of Atlas fieldwork there were numerous reports of Golden Eagles. Most (28) sightings involved single birds, with pairs noted in eight of the 45 total blocks with detections. As a rule, they were seen in parts of Humboldt County where they had been historically known. The general range of Golden Eagle in Humboldt can safely be stated to take in nearly all of the county save the coastal slope and fog belt from north of Cape Mendocino to the Del Norte County line and the lower Klamath River drainage. They are distinctly more numerous and widespread across the southern one-third of the county than elsewhere. Because they establish large territories, they are nowhere common, but the species is seen throughout the year. The extent to which individuals migrate through Humboldt is unknown. Classic fall migrants are rarely reported.

It was understood prior to the beginning of the Atlas period that these birds favor the warmer, drier, more open countryside of southern Humboldt County, extending from near the ocean coast east to the Trinity County line. The more Mediterranean climate and sparse human settlement of the southeastern portion of the county afford habitat similar to that inhabited by Golden Eagles farther inland in California. A noticeable cluster of blocks with detections and several breeding confirmations reflects this. At many places in the hills out of Bridgeville or Alderpoint, good Golden Eagle habitat would appear to extend unbroken to the horizon and eagles are frequently seen. One might intuit or reason that this trend toward more sightings in the southern interior of the county suggests greater numbers to the east and southeast, a situation that seems likely to exist.

As they do at many points farther south in California, Golden Eagles closely approach the coast in the area about Cape Mendocino and in the King Range. Windswept expanses of grass and miles of broken forest canopy seem to offer the birds what they require. They are known to occur rarely out onto the southwestern Eel River delta in winter, but are almost never seen along or on the coast north of there. The distribution of Golden Eagles in Humboldt County clearly is confluent with a larger range that includes much of western and northern North America and, indeed, much of the Northern Hemisphere.

There were eight breeding confirmations during the Atlas period. All but two were in the region of greater numbers in southern Humboldt, the exceptions being an adult carrying a jackrabbit (*Lepus californicus*) above the Mad River between Redwood House Road and Showers Pass (J21) 13 Jul 1999 (DFx, JCP), and an adult feeding young near Kneeland (g18) 27 Jul 1996 (A. Desch). Kneeland stands at the cusp of the fog zone, as do the Bald Hills and Redwood Valley, two other sites at which eagles were observed. Three nests were found during the Atlas period. An occupied nest was reported 11 Mar 1999 about 1 mile west of Bear Buttes (j30; M. Anderson). An unoccupied nest was observed from a helicopter 12 May 1999 in the Larabee Creek area (J25; K. Sundeen, M. Freitas, R. LeValley). Another unoccupied nest was found 1 Jul 1998 along Pritchard Creek (e27; D. Dill, B. Boyd). The remains of an apparent nestling Golden Eagle were later found near that site (S. Chinnici). The remaining three breeding confirmation were of fledglings.

Landscapes used by Golden Eagles vary widely in looks and layout. Wide-open expanses of foraging habitat are used, with rangeland, treeless ridges and slopes, and the more open oak woodlands favored. However, these eagles have been seen hunting above the canopy of mature Douglas-fir and mixed-species forest, vegetation that clothes a great deal of the north-facing ground in the interior of the county. Pair activity centers and the few known nests have been located at or near the crests of ridge knobs or buttes, often with large old Douglas-firs or California black oaks in the vicinity. Golden Eagles in the Bald Hills must regularly cruise a mosaic of coast redwood forest, regrowth timberlands, and fog-shrouded hillsides.

Harris (1996) characterized Golden Eagle in northwestern California generally as a rare to uncommon resident and breeder, "generally seen soaring over open woodlands, forests, and grasslands along the higher inland ridges." The roster of sites he listed included all of the areas in which eagles were reported during the Atlas period. These eagles are at risk from shooting and poisoning by ranchers and others who consider them a threat to livestock.

AMERICAN KESTREL *Falco sparverius*

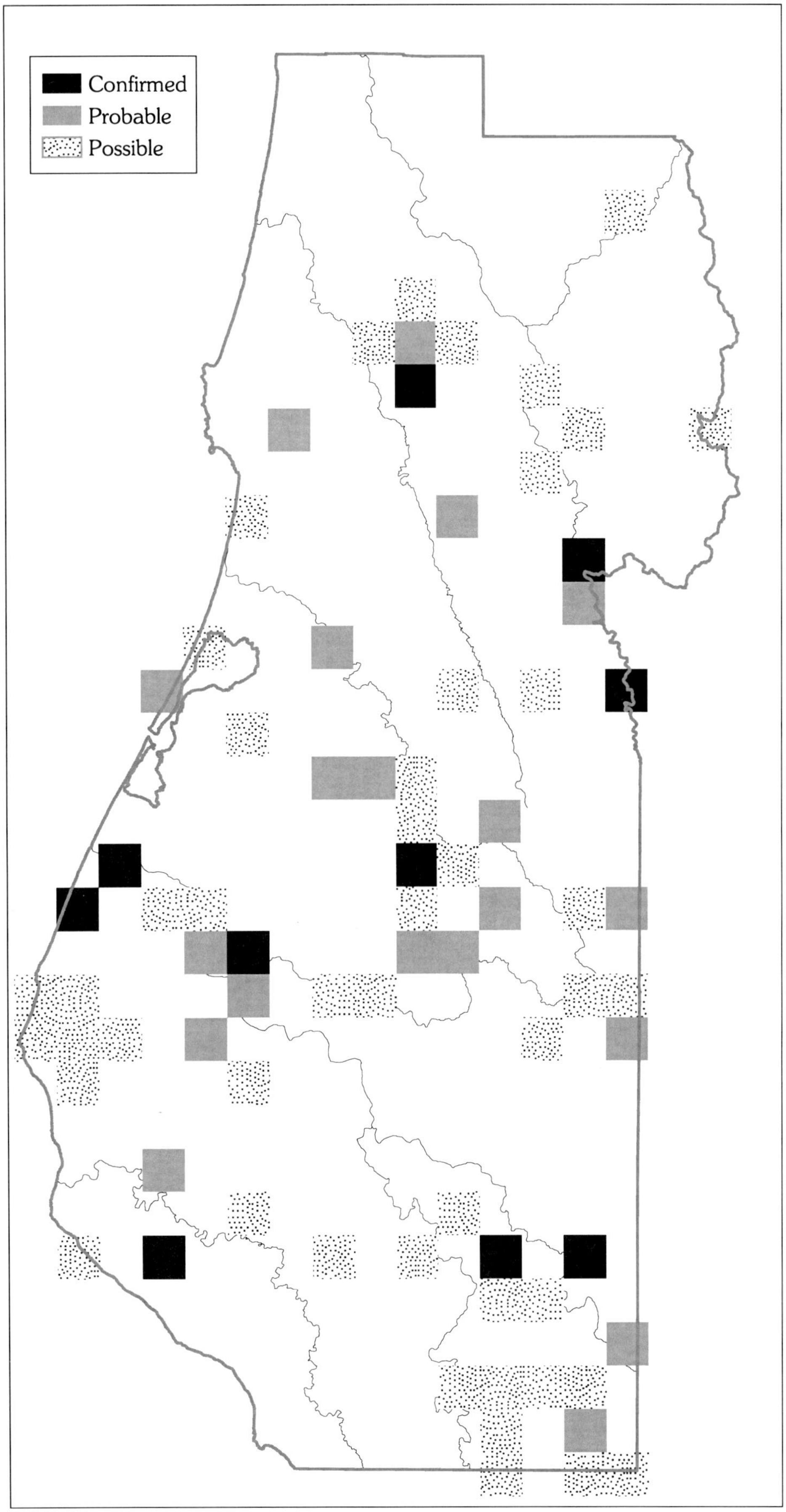

Trim and buoyant, American Kestrels hunt for mice, insects, and other small prey items while perched or hovering. Their routine habitation of settled areas and a proclivity for using roadside wires have made them familiar to many.

The American Kestrel is a common resident and breeding species in northwestern California, more common in winter and migration (Harris 1996). Throughout the year these birds are partial to expanses of unmanaged or lightly managed countryside, which provides habitat for their prey and a measure of freedom from disturbance. Lookout posts such as small trees, power lines, wire fences, and other constructions typically dot preferred haunts, yet the advantage offered by a mere breeze will enable prolonged hovering and dashing sorties over nearly featureless terrain. Atlas results indicated that coastal headlands, agricultural lands, forest edges, and a spectrum of open interior-county sites host small numbers of kestrels during the breeding season.

The distribution of this species over the 5-year Atlas period coincided in a general way with the coastal mountain prairies northeast of Cape Mendocino; the lower Eel River bottoms; hillside grass-and-oak mosaic of the southeastern portion of the county, and in the upper Mad and Van Duzen River drainages; along the Trinity River; and in the Bald Hills. Each of these regions is rural, dominated by expanses of cattle pasture or otherwise open lands.

Kestrels were reported in 18% of all Atlas blocks and were confirmed breeding in 10 well scattered blocks. Half of those were based on reports of food-carrying that spanned the dates of 4 May 1996 (f22; P. Marsh) to 19 Jun 1998 (B21; COg). Feeding of young was observed 5 Jun 1998 near Coyote Peak (J09; TWL). Only two occupied nests were reported: an active nest in one of the orange aircraft-hazard spheres placed on high-voltage power lines crossing the Trinity River at Willow Creek 3 May 1996 (N13; TWL), and another lacking details 24 Aug 1999 near the Todd Ranch (o16; TWL).

Large clusters of blocks without detections occurred in extreme southwest county; along the Eel/South Fork Eel divide; southeast of Eureka; east of Westhaven; and throughout extreme northwestern Humboldt. The last three areas comprise large expanses of dense timberland generally lacking suitable open areas. It was plain to atlasers that, during the breeding season, American Kestrels were to be anticipated in rural settings offering good foraging substrate and a source of nest holes. Over much of Humboldt County, pairs of kestrels seem to be distinctly partial to sites with large old oaks, which afford nesting opportunities created not only by limb breakage with age, but through the excavations of primary cavity-nesting birds.

It is worth noting that a great amount of appropriate breeding habitat sprawls across sparsely roaded portions of Humboldt where American Kestrels are reasonably numerous. Thus, despite the conspicuous behaviors of these birds, they probably occurred in many blocks that received limited coverage by atlasers. A search for the birds aided by unrestricted access in these blocks probably would have resulted in a map indicating a slightly more widespread and contiguous breeding distribution.

Despite the skewing of observer effort toward central-coastal Humboldt, that area had neither an increased density of blocks with detections of American Kestrels nor an increase in breeding confirmations. This suggests that breeding in the coastal pasturelands is relatively uncommon. Yet, in fall and winter these same coastal lowlands appear to host as many, if not more, kestrels than most other areas of the county. This species is quite migratory, and indeed a sizeable jump in numbers takes place in September and October as migrants pass through and wintering birds arrive.

PEREGRINE FALCON *Falco peregrinus*

Since times of antiquity, few birds have held faster hold of the human heart than has the Peregrine Falcon. Renowned for speed and killing prowess, this bird has graced the fist of the privileged falconer for ages. Wild peregrines are enjoyed today by more people than ever before. Their pursuit of winged prey elicits an outcry of admiration from birders on weekend outings—as well as words of a different tenor uttered by shorebird enthusiasts suddenly chagrined behind the spotting scope. New technology allows us to click our way to real-time webcam views of peregrines nesting in scrapes erected for them outside high-rise windows and on urban bridges. The living image is likely to be encountered no farther away than one's nearest National Wildlife Refuge or other expansive wetland.

Those who have gained an interest in birds within the past 20 years may imagine that the frequent sight of a Peregrine Falcon crossing the sky is nothing out of the ordinary. This has not always been the case. The species underwent a critical decline in North America in the 1950s and 60s, to the extent that no known breeding pairs remained in the U.S. east of the Rocky Mountains. Such an alarming decline for such a charismatic species prompted a listing of "endangered" in 1970 under the Endangered Species Conservation Act of 1969, which predated the current Endangered Species Act of 1973. The recovery of this bird required a huge effort that ultimately became one of the most intensive and innovative projects undertaken in wildlife management. It involved the cooperation of fractious groups often at odds over how this essence of wildness was to be managed. Using captive-breeding and site enhancement techniques now widely applied in the conservation of endangered birds, biologists managed to pull the species back from a precarious low of just two known nesting pairs in California in 1970 (Herman 1971). Perhaps the most significant factor in the recovery of this falcon, however, was the ban on DDT and other organochlorine pesticides in Canada and the U.S. in the early 1970s. This resulted in improved reproductive success throughout North America.

The statewide breeding population has rebounded in spectacular fashion. Eighty-two active nest sites were known by 1988 (Jurek 1989), and a remarkable 167 occupied sites were known by 1998 (USFWS 1999b). In contrast to the northeastern U.S., where urban concentrations of nesting peregrines originating from hacked transplants have largely served to swell the ranks, much of the recovery in California ultimately owes perhaps as much to watching nature take its course. The Peregrine Falcon was officially removed from the federal list of threatened and endangered wildlife in 1999 (USFWS 1999b).

Peregrine Falcons have long been known by ornithologists to occur in Humboldt County. Townsend (1887) and Palmer (Fisher 1902b) both talked to local collector C. Fiebig and concluded the peregrine was "one of the commonest hawks about Humboldt Bay" (November–December 1885) and "abundant" (May–June 1889). Although it is reasonable to assume that Peregrine Falcons were numerous more than 100 years ago and have always nested in the county, the historical record is difficult to reconstruct. Perhaps the earliest historical report of confirmed nesting in Humboldt was a collection of eggs made near Trinidad 12 Apr 1926 (G. Atwell; Humboldt State University specimen missing). Yocom and Harris (1975) described the peregrine as having nested in Humboldt County until 1954. There were no further reports specifically indicating nesting for more than two decades—and, most telling, no reported sightings of peregrines in Humboldt County in June and July during the 1960s and 1970s. Harris (1996) indicated that well known historical sites at Castle Rock, Del Norte County and Trinidad Head in Humboldt had not been occupied for decades. It should be noted that many of the eyries discovered in the county have been situated in highly rural areas, well off the beaten track for most birders and biologists. Some sites are on cliff faces miles removed from common avenues of travel, while others in the major river gorges are visible only if one knows where and how to look for them. For these reasons, it is conceivable that some sites may actually have remained occupied during the era of diminished numbers in the 1960s and 70s, either without discovery, or with word of their existence suppressed.

Then, within the past quarter-century, nesting activity increased in Humboldt County, as it had across the historical range in California and the West. Yet, pre-Atlas breeding reports made known outside the falconry community and government agencies were few. Four birds seen together in the Van Duzen River drainage in early June 1978 likely belonged to a family group (S. Harris), and were the first evidence of local breeding known to the general birding community since 1954. Two birds were at a different site up the Van Duzen in early June 1980 (R. LeValley). A nest attended by a pair was discovered along the Trinity River in 1983 (R. Erickson, J. Sterling). This site was active during most, if not all, Atlas survey years. The first report of a nest site along the lower Eel River came in 1986. This eyrie was abandoned when the Petrolia earthquake of 24 Apr 1992 dislodged the nest ledge. By 1990, eight Humboldt sites were monitored for nesting activity (Pagel 1990). This completes the slim roster of reports in late spring and summer prior to the inception of the Atlas project.

Currently, the number of Peregrine Falcons present at any season of the year in Humboldt County varies, as does the birds' distribution, but they are more or less routinely seen by persons on the watch for raptors. During the 5-year Atlas period, more than 260 reported sightings in the county were compiled (S. Harris, unpubl. data). Many of these observations were known or suspected to pertain to the same individuals, especially as near-daily use of the Arcata Marsh and the Mad River Slough Wildlife Area took place during much of the year. Peregrines were also widely reported along the northern and central Humboldt County coast and infrequently inland, mostly from September through early May. This is a period when the small local breeding population is greatly augmented by migrants or winter visitors. Few North American CBCs approach or exceed the totals reported for Peregrine Falcon on both the Arcata and Centerville Beach to King Salmon CBCs each year. This species has been considered an uncommon migrant and winter visitor, and rare and local breeder and summer resident in northwestern California by Harris (1996). Except in spring and summer, far fewer peregrines are seen in the interior of Humboldt County than in the coastal lowlands.

Even in an era when Peregrine Falcons can be secured for falconry through legal and appropriate means, the possibility of criminal acts by the unscrupulous remains a threat. Since the welfare of the birds was top priority in the scheme of the Atlas project, known nesting areas are discussed only indirectly. In addition, the final breeding distribution map for the Atlas period has not been published here and block numbers have been deleted from the final bird observation database. To summarize, Peregrine Falcons were confirmed breeding at 10 sites, each site in a different Atlas block. Six were in the northern one-third of Humboldt, two in the middle, and two were in the southern third. Two sites can be described as coastal, two as near-coastal. Of these four, two were on coastal bluffs and two were near the tops of residual old-growth coast redwoods on private timberland. The remaining six were in the interior of the county. Three of these were along the Klamath River, one along the Trinity River, one along the South Fork Eel River, and one in the main Eel River drainage. Additionally, probable breeding status was recognized in three blocks in extreme northeastern Humboldt, one at Humboldt Bay, and in one block in south-central county. In each of these cases, agitated behavior or birds visiting a probable nest site were observed. Peregrines recorded as possible breeders during the spring or summer occurred in five blocks, four of these in southern Humboldt and one in the northeast. Thus it is possible that 15 or more nest sites may have been active during the Atlas period.

As can be envisioned from this array of coastal, near-coastal, and interior locations, Peregrine Falcons breeding in Humboldt County occupy sites differing in topography, vegetative cover, openness of landscape, and potential prey base. Elevations of occupied nests known during the Atlas period were comparatively low, ranging from shortly above sea level to 2400 feet; however, a cliff above a subalpine lake was the setting for one of the

Continued on page 389

RUFFED GROUSE *Bonasa umbellus*

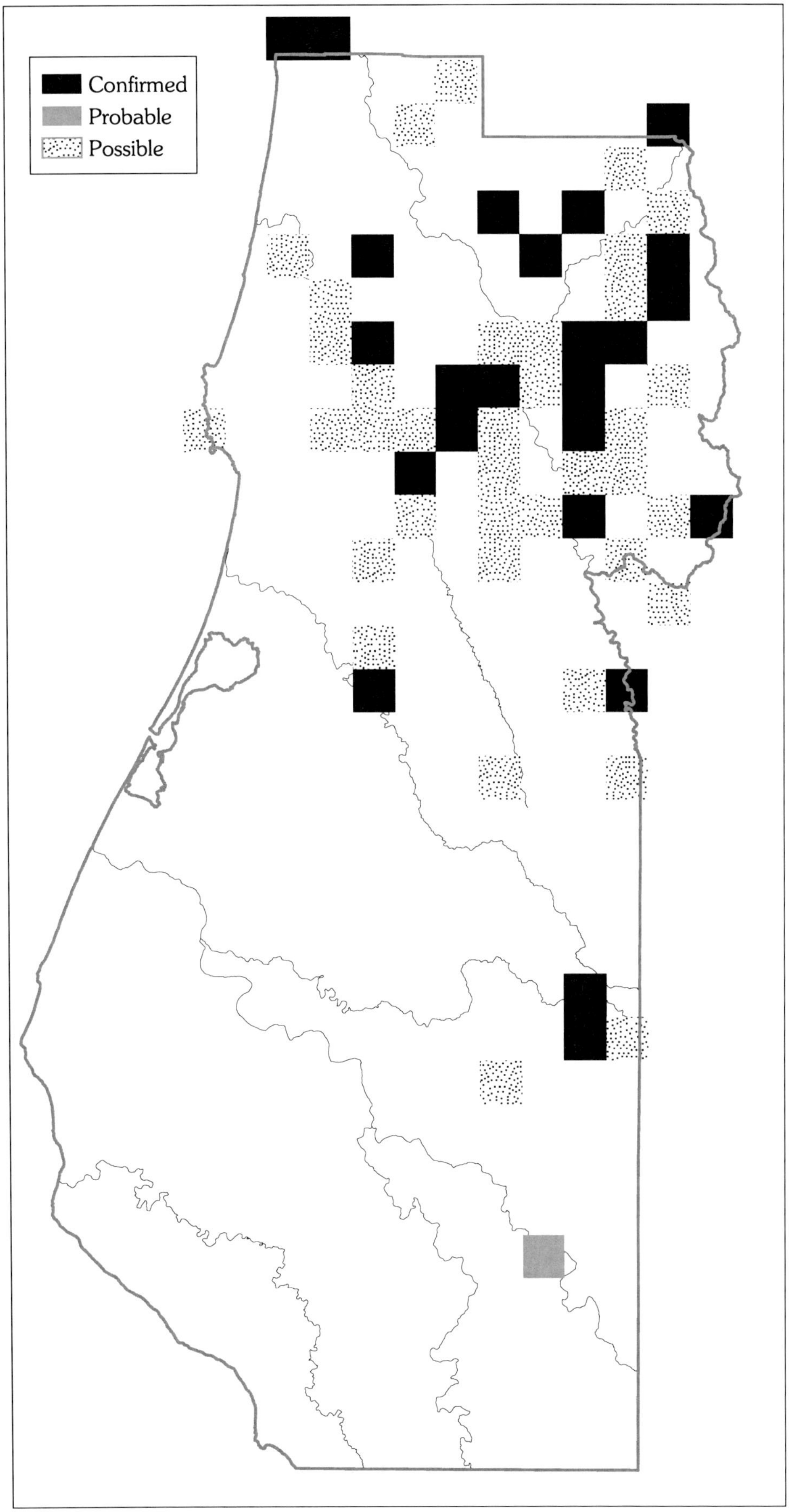

A familiar woodland bird across much of its extensive range, the Ruffed Grouse is encountered only locally in California. Portions of Del Norte, Siskiyou, Trinity, and Humboldt counties support the state's entire population. Ruffed Grouse have been known to occur in Humboldt County since the 1880s (Townsend 1886), and numerous observers have recorded them in the northeastern half of the county over the past century (Harris 1996). Scattered reports from southeastern Humboldt indicate that birds occur to within about 15 miles of the Mendocino County line; the species remains unknown in that county. Extensive coverage of the more remote areas during Atlas survey work resulted in a surprising flush of reports. Although reported more often than was expected by some atlasers, these observations portray Ruffed Grouse as an uncommon species that occurs in detectable numbers across a rather broad portion of Humboldt County. They were found in 62 (15%) of the 425 total blocks, and breeding was confirmed in 24.

Many California birders have traveled to Humboldt County to search for a Ruffed Grouse. Some have come away disappointed. It is a matter of wry understanding that consistent "stakeout" grouse at readily accessible sites are few. Although these birds are not rare, most of their Humboldt haunts are far from human population centers, and they are not often encountered along well-traveled roads. Only those who live or work in their highly rural habitat see or hear them with any frequency. Generally, Ruffed Grouse are found in the wetter zone near the coast and in the Klamath Mountains of the northern part of the county. Far fewer birds are found in the drier mountains of east-central Humboldt, in areas such as Bridgeville and South Fork Mountain.

Ruffed Grouse in Humboldt County are known to favor ravines and seeps with at least a semblance of opening or edge. Down logs, clumps of moist-site brush, berry-bearing shrubs, and a source of water are common denominators. However, many birds have been found on rolling topography, such as benches and hollows. Provided that the habitat meets their needs, birds may be encountered at microsites tucked within the winter snow zone, even to 5000 feet. A broken but mainly forested mosaic of white fir, incense-cedar, corn lily, *Spiraea*, red-osier dogwood, mountain alder, and a low herb layer characterizes sites such as Box Camp in east-central Humboldt near the Trinity County line. These birds are also found in mixed evergreen hardwood forests with a shrub understory and adjacent to riparian vegetation. Although exceptions occur, most sites hosting these grouse have a fairly high forest canopy closure and are on a shady exposure. Melt-off of winter snowpack helps water a good deal of Ruffed Grouse habitat in the county.

A male drumming on territory 6 May 1995 (N23; D. Simmons) and an occupied nest 24 May 1999 (n24; M. Shreeve) represented the early nesting season. Breeding was confirmed by observations of precocial young in 21 of the 62 blocks in which this species was found. Such reports spanned the dates of 19 May 1996 (g01; TWL) to 10 Aug 1997 (L05; P. Herrera). For information on three well described nests with eggs near Willow Creek, see Talmadge (1957).

Grinnell and Miller (1944) stated that in northwestern California, Ruffed Grouse was formerly fairly common locally, "but now rare, having disappeared completely from places where formerly reported (for example, valley of Van Duzen River, *fide* H. E. Wilder)." Yocom (1978) mapped observations of Ruffed Grouse 1954–1977, and his findings did not indicate any further range contractions. However, the Atlas map indicates that no Ruffed Grouse were detected in southernmost Humboldt County in a number of areas in which Yocom (1978) had reported them. Whether this does indeed represent continued range contraction or is an artifact of inadequate Atlas survey coverage is unknown.

There is some concern about the effects of extensive logging on this species. Unlike logging in the Lake States, logging in northwestern California may not result in improved habitat (Yocom 1978). Existing glades favorable to grouse can also be destroyed during logging operations. Some observers, however, have noted that logging of true fir forest stimulates the growth of red-osier dogwood and snowberry, and is therefore beneficial to these grouse. At Ruffed Grouse sites within the coast redwood zone such as along Skunk Cabbage Creek, the removal of old-growth forest and subsequent replacement by red alder also may increase the amount of available forage (T. Leskiw, pers. comm.).

The often highly local distribution of Ruffed Grouse, along with their remarkable lack of fear of humans, could make them vulnerable to hunting pressure. From 1995–1999 there was an estimated annual harvest of 447 Ruffed Grouse in Humboldt County (CDFG 1995, CDFG 1996, CDFG 1997, CDFG 1998, CDFG 1999). The status of this species in northwestern California should be carefully watched. The possibility of a long-term decline, as well as the value of Ruffed Grouse to local ecotourism, should be considered during any sport harvest planning.

BLUE GROUSE *Dendragapus obscurus*

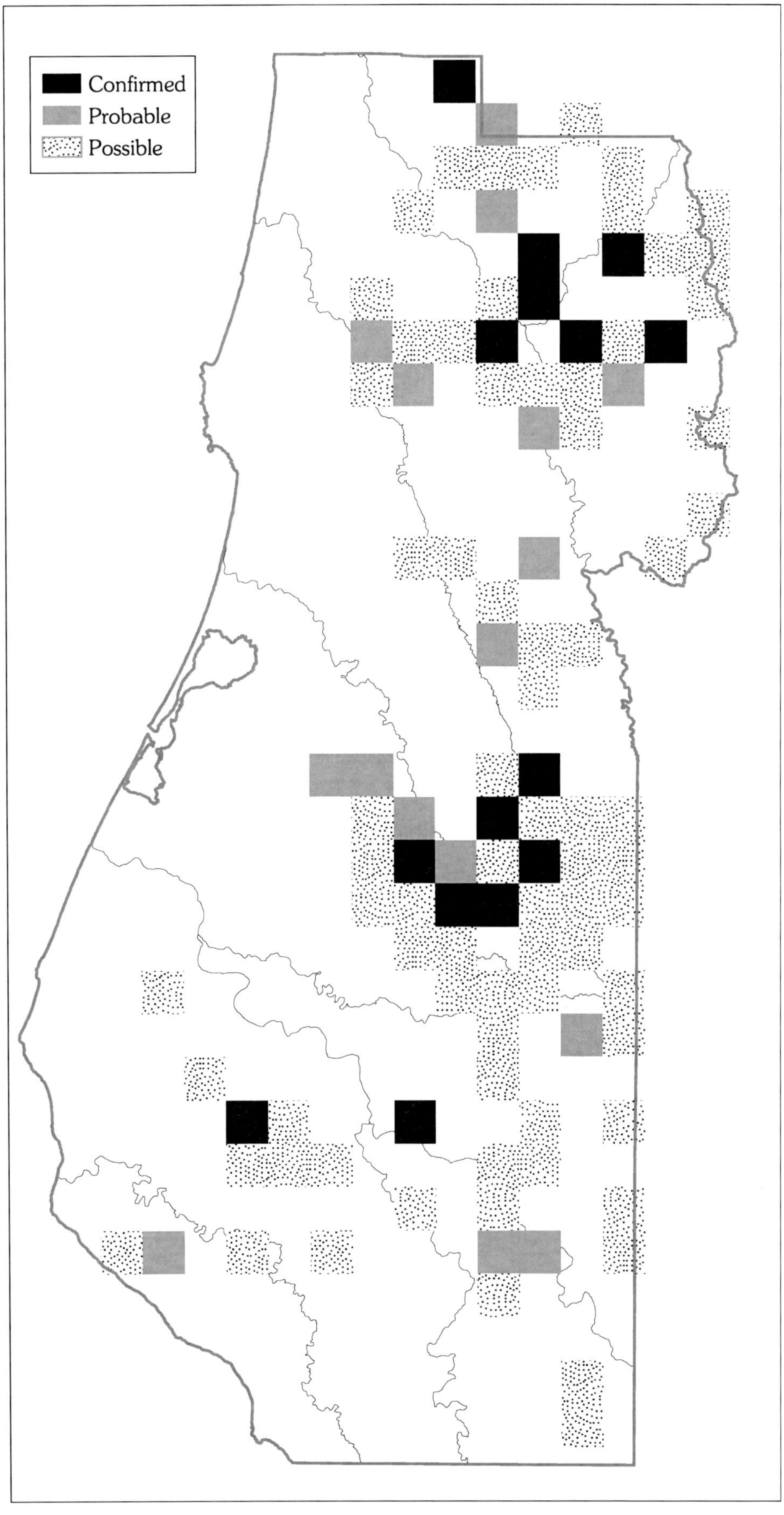

Remnants of winter snowpack persist in the shady hollows of Douglas-fir stands as male Blue Grouse begin the courtship ritual of another spring season in Humboldt County. Deep, penetrating "hoots" carry through the forest, seeming at some moments to come from nearby, yet at other moments to have originated from somewhere downslope. With much effort, perhaps the grouse responsible for the strange noises can be spotted. As often as not, it is nearly overhead on some stout, sheltered bough adorned with epiphytes. Blue Grouse also "hoot" and display on down logs or stumps, occasionally at roadside within easy view of human observers.

Harris (1996) considered this widespread mountain grouse of the West to be common in the timbered expanses across the middle and higher elevations of Humboldt County and rare at lower elevations west of the Eel River. Atlas results suggest that its abundance may be better characterized as uncommon at middle and higher elevations. During the 5-year Atlas period, Blue Grouse were reported in 106 (25%) of the 425 total blocks, and breeding was confirmed in 15 blocks.

Higher ridges and slopes of the northeastern mountains, South Fork Mountain, and Snow Camp Mountain have been among the more reliable places to search for this bird. In the mountains above the Klamath River canyon, around the Bald Hills, on Kneeland, and in the upper Bear River/Mattole country, Blue Grouse penetrate some miles into the summer fog zone. However, they do not appear to use old-growth coast redwood stands. Blue Grouse were conspicuously absent during the study period at low elevations near the coast north of the King Range and in extensively logged areas. Not surprisingly, detections were few in the oak-dominated, lower-elevation foothill country of southwestern Humboldt.

Blue Grouse frequently are encountered on the upper one-third of forested slopes, where they inhabit mature or maturing stands of Douglas-fir, white fir, red fir, and mixed conifer-and-oak near the upper elevation limit of "tree" oaks. As elsewhere within their range, these grouse are strongly associated with Douglas-fir. At least a token few larger conifers with large limbs commonly are noted in better Blue Grouse habitat, although such trees are not always present. The abundance of Blue Grouse shows a strong negative correlation with higher abundance of hardwoods (Ralph *et al.* 1991). A considerable degree of habitat continuity also seems necessary.

Later in the breeding season, family groups can be found foraging in fairly open ridge-top habitats, ungrazed oak stands, clearcuts, old log landings, and roadsides. They may also be encountered in and about scattered stands of open-grown Douglas-fir at the edges of the mountain prairies, as in the upper Bear River area and in the Bald Hills. The open oak woodlands often used in spring and fall are typically adjacent to mature or old-growth Douglas-fir stands (R. Gutiérrez, pers. comm.). In winter, Blue Grouse often move upslope into Douglas-fir stands where they feed on needles. Aside from these seasonal altitudinal movements within watersheds, Blue Grouse are thought to be quite sedentary. Thus, this species probably breeds in most of the blocks in which it was detected.

Breeding was confirmed by observations of precocial young in only 12% of the detection-blocks. This compares to 34% of blocks in which Ruffed Grouse were so confirmed. Precocial young records spanned the dates of 4 Jun 1998 (k02; KMS) to 5 Aug 1998 (m06; D. Kudrna). A single nest was found, which contained eggs when discovered 22 May 1996 (o06; B. Norman). One instance of a distraction display was recorded 8 Jun 1995 (m18; GAS). Males "counterhooting" were recorded 31 Mar 1998 (m10; *fide* JMH).

This species may have declined somewhat from historical numbers due to hunting and fragmentation of older inland coniferous forests, to which they may respond unfavorably (Rosenberg and Raphael 1986). Raphael *et al.* (1988) estimated that Blue Grouse had declined 34% from historical levels in this region as a result of timber harvest. Conversely, it is quite possible that logging in extensively forested areas has improved habitat locally by creating open areas where broods can now feed on a variety of abundant insects and forbs. It is unknown if the annual sport harvest has any impact on local populations.

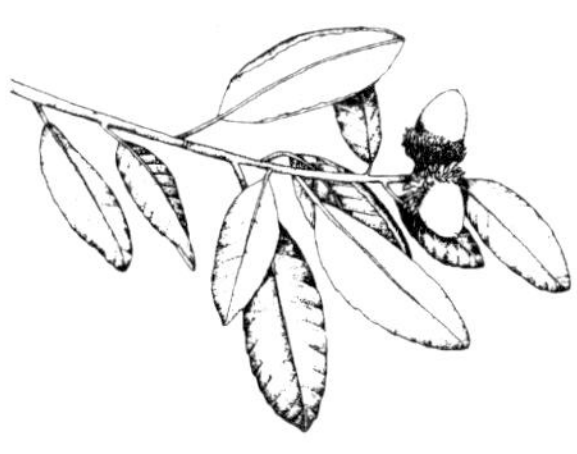

WILD TURKEY *Meleagrus gallipavo*

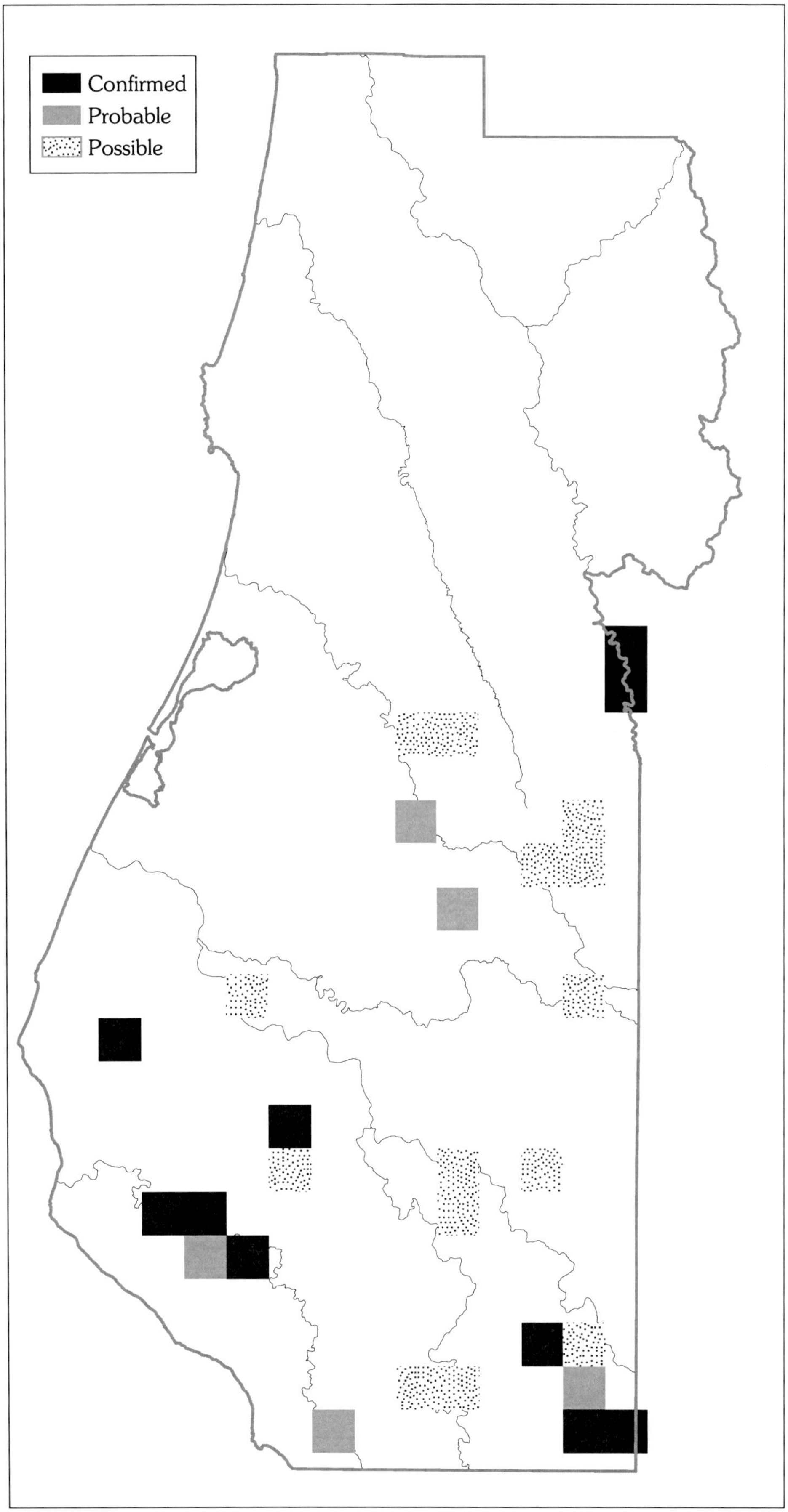

Following a long era of decline and scattered extirpation since the European colonization of North America, populations of Wild Turkeys within their original range have recovered. An expansion into areas previously unoccupied in historical times has been achieved through vigorous introductions of transplanted birds in recent decades. Aside from the fact that the Wild Turkey exists in the fossil record, the species is not "native" to Humboldt County, or to California.

Introductions began as early as 1913 (Grinnell 1915), with intermittent releases (*e. g.*, in 1930; Anonymous 1931) continuing through March 1999, when 42 birds were released just across the South Fork Trinity River in Trinity County near Deep Gulch (o15; J. Lahr). Harris (1996) stated that the population in parts of Humboldt, Mendocino, and Trinity counties derives from stocking efforts that began in the early 1960s, and that the species was by then an uncommon local resident. These introductions have resulted in a scattered but substantial population in southern Humboldt. Hunter and Hazard (1998) provided details on the various subspecies released into Humboldt and adjacent areas of Trinity County. During the Atlas period, Wild Turkeys were reported from 29 (6%) of the 425 total blocks, and were confirmed breeding in 10 blocks.

Hardwood mast (particularly acorns) is a critical component of this species' diet. Because of this, Wild Turkeys are typically, though not always, encountered in or near stands of trees in which oaks are well represented. A woods-and-opening structure characterizes much of their habitat in the county. They are found most commonly at the margins of lightly grazed grassy expanses which are in use as horse or cattle pasture and which are edged by, or interdigitate with, California black or Oregon white oak woodlands. Such sites often grade upslope into drier-site Douglas-fir or pine regrowth with an open forest floor or additional grassy openings, habitats also used by turkeys.

Much, though not all, turkey habitat is distributed along the lower slopes of the larger river drainages. During Atlas work, Wild Turkeys were also encountered near, and even along, ridgelines in better-quality, "clumpy" oak woodland, as in the rolling hills southwest of Showers Pass, around Last Chance Ridge, and in the extreme southeastern corner of the county. The upper extent of either black or white oak would appear to be a limiting factor; habitat would therefore seem to run out above 4000 feet anywhere in the county.

During the Atlas period, Wild Turkeys appeared to be rather common in the central Mattole River valley (vicinity of Honeydew and westward), along Mail Ridge and southeast toward Island Mountain, and in the upper Mad River drainage. None were found north of the Trinity River. Based on gross cover type, far less potential habitat exists in northern Humboldt. Forests and regrowth in that region have much more tanoak than black or white oak. Tanoak reproduces in its own shade, creating dense forest understory with comparatively low botanical diversity. Might it be that edge-poor tanoak stands fail to meet the needs of turkeys? It is of some interest that no Wild Turkeys were reported from a corridor of seemingly suitable habitat which broadly bisects the Hoopa Valley Indian Reservation. Oaks of several species occur commonly both in the Hoopa Valley and on the slopes above. Have no introductions been made here, or is their seeming absence due to intensive hunting pressure?

As appears to be true for many populations of reintroduced Wild Turkey, some flocks habitually frequent pastoral settings surprisingly near the rural fringe, such as livestock corrals and farms. Birds in these places may become tolerant of routine human approach, with hens accompanying poults occasionally seen along dusty roadsides just outside small hamlets. Nine of the 10 breeding confirmations were of precocial young; these records spanned the dates of 1 Jun 1999 (o15; TWL) to 30 Jul 1999 (g26; JG). The remaining confirmation was of a nest with eggs found 20 May 1995 in the vicinity of Chamise Creek (o33; C. Secor). One of the earliest documented breeding confirmations in Humboldt County probably stems from the early era of introductions: precocial young seen on Henry Ridge 12 Jul 1972 (B. German).

The California Department of Fish and Game, in conjunction with private organizations, continues to release turkeys into California upland ecosystems for sport hunting. Impacts these semi-exotic birds may have on native plants or wildlife of the region are unknown.

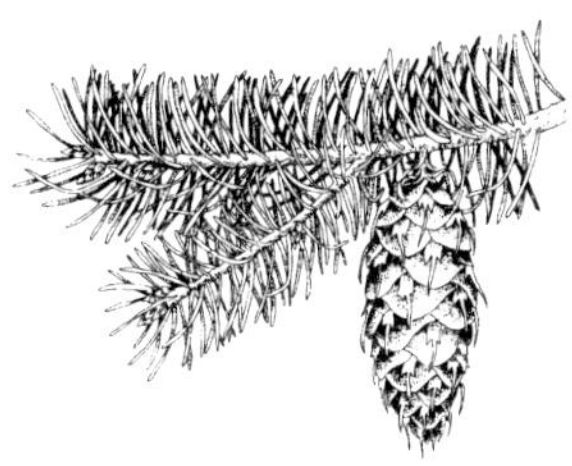

MOUNTAIN QUAIL *Oreortyx pictus*

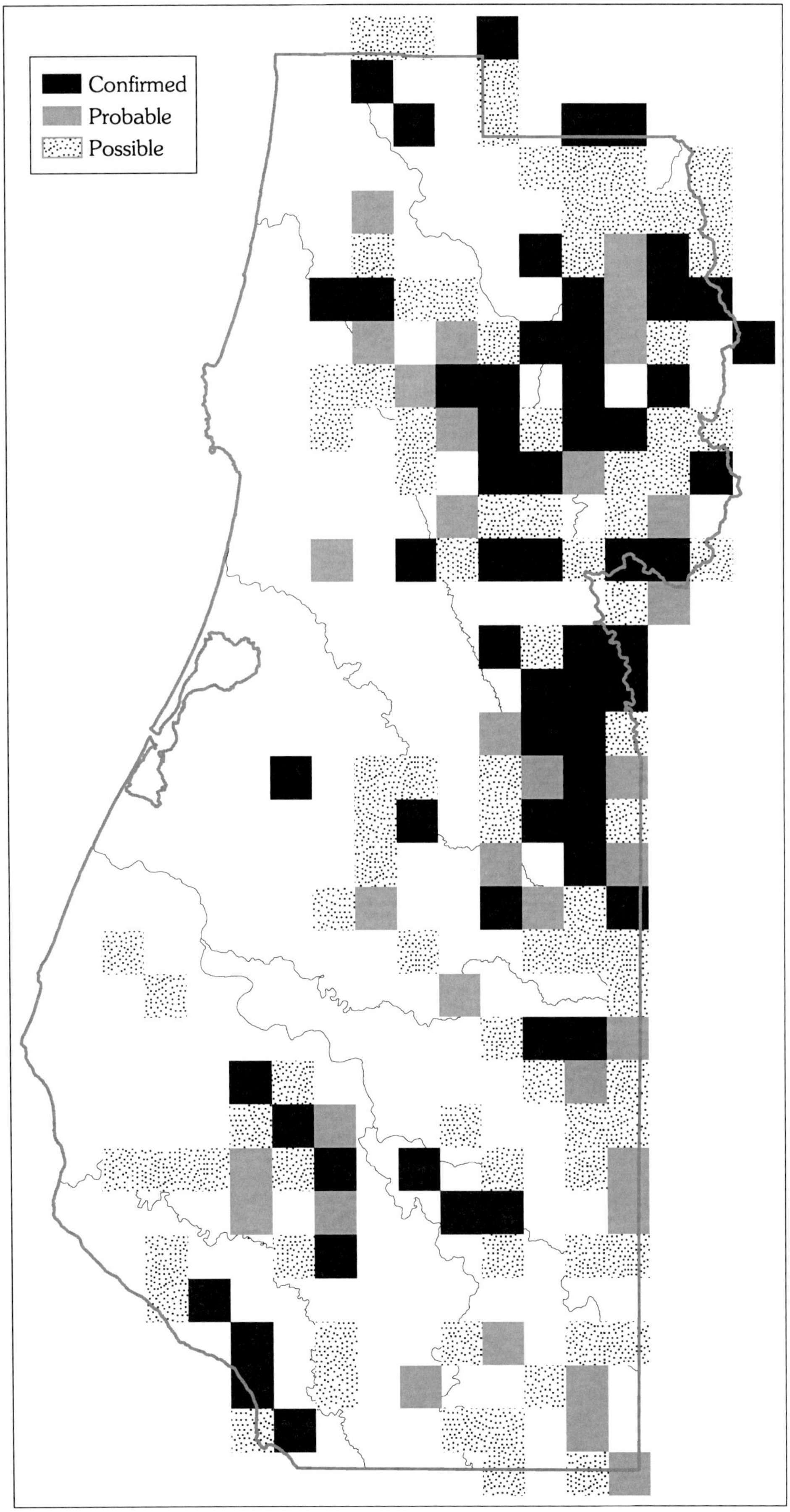

Much of interior Humboldt County is a highly rural expanse of foothill and mountain slopes. Where such places are grown to tall shrubs and conifer saplings, fronted by a ragged forest margin, and broken by grassy clearings and neglected dirt roads, conditions begin to favor the presence—if not the detection—of Mountain Quail. Especially during spring, these birds are more often heard than seen. From about the time of the vernal equinox through June, males deliver a loud whistle-like note with a rising inflection. In an environment of ambient hush, this unique call carries a great distance—especially when broadcast from an exposed boulder or down tree overlooking the head of a rugged drainage. In stretches of good habitat, the calls of several quail may be discerned, some nearby, others at the edge of audibility.

The Mountain Quail has long been recognized as a fairly common resident of Humboldt County. These birds are well known by local birders, naturalists, and sportsmen. Those people living in the country, working in the woods, or routinely driving on back roads at middle and higher elevations often see them. Writing of this species with reference to northwestern California, Harris (1996) characterized it as an "uncommon to locally common resident and breeder in higher elevation brushlands and forest margins." He speculated that the general distribution had remained little changed since settlement days. However, clearcut logging may have actually increased amounts of suitable brush and edge habitat (*fide* R. Gutiérrez). Brennan (1984) and Brennan *et al.* (1987) studied summer habitat ecology of Mountain Quail on Kinsey Ridge, and found a density there of about 20 birds per 250 acres.

Though not abundant, Mountain Quail are quite numerous and are widely distributed within their range in the county. During the Atlas period, they were detected in 42% of the total blocks. In particular, the species was frequently reported in the northeastern section of the county and along the general South Fork Mountain uplift, and the bulk of confirmed breeding reports came from these areas. Presence in the King Range was also well documented. Suggesting a positive correlation between increasing elevation and Mountain Quail occurrence is the unmistakable zone of absence in the coastal lowlands and in the lower Mad and Eel River corridors. Yet these birds reach all the way to the ocean where suitable habitat exists at lower elevations in central coastal California, so their presence in similar situations in southern Humboldt should not be ruled out (R. Gutiérrez, pers. comm.). While clearly most numerous east of the fog zone, Mountain Quail do penetrate toward the coast in hills above the lower Klamath River; in lower Redwood Creek; up the North Fork of the Mad River; at Kneeland (g18); on Wildcat Ridge (c22) and adjacent Bear River Ridge (D23); in hills drained by the Mattole, Bear, and South Fork Eel rivers; and in the King Range.

Based chiefly on the observation of precocial young, confirmed breeding by Mountain Quail was established in 59 blocks. Precocial young constituted 93% of the total confirmed breeding reports, illustrating the comparative frequency with which family groups were seen. These records spanned the dates of 20 Jun 1998 (L15; COg) to 1 Oct 1995 (g33; K. Dyer). The remaining four blocks with confirmed breeding reports included two instances of distraction displays by adults, 4 Jun 1996 (m11; JMH) and 26 Jun 1999 (l28; JG), and—surely among the niftiest discoveries during the 5 years of the Atlas study—nests with eggs 16 May 1996 (m19; D. Carpenter) and 4 Jun 1996 (p06; KMS). The earliest evidence of probable breeding was multiple males singing on territory 1 May 1996 (n32; M. Bailey) and 1 May 1998 (i08; COg).

Mountain Quail once benefited from the brushfields and open forests which originated from unchecked wildfires. In today's era of fire suppression, lightning has found its role as an agent of habitat creation surpassed by the chainsaw. Mountain Quail are at home in the clearcut-and-leave block patchwork created by industrial forestry. Hagar (1960) observed numerous broods using brushy clearcut blocks near Willow Creek, and thought that logging improved breeding habitat. But while they are a predictable element of the bird community in many clearcuts, Mountain Quail are hardly restricted to them. Forested slopes well clothed with jungles of canyon live oak and evergreen mixed-hardwood forests with a shrub understory are inhabited, as are the edges of natural prairies, power line rights-of-way, and naturally occurring brushfields. These quail extend upslope essentially to the highest elevations in the county, as well as spreading out over a great area of mid-slope regrowth forest and brush. They rarely occupy the same territory as California Quail, although habitats may interdigitate locally where the upper extent of pastureland or rangeland meets the lower edges of timber. Individuals or coveys of the two species are sometimes encountered in the same general area, but in Humboldt County they seem never to mix.

Visual encounters with Mountain Quail typically involve glimpses of a few streamlined forms sprinting madly along the road shoulder, then, with scarcely a break in stride, barreling on a diagonal track up the cutbank and into the shelter of the nearest shrubbery. Running with upperparts thrust forward, their narrow head plumes provide exclamation points to their seeming anxiety. Mountain Quail are regarded with exasperation by many birders who have searched for them and come back to town unsuccessful. For those who miss them, or find unfulfilling the call-notes of hidden males, chances of eventual satisfaction increase from July to September, when family groups may be seen along lightly used roads and about old clearcut landings.

CALIFORNIA QUAIL *Callipepla californica*

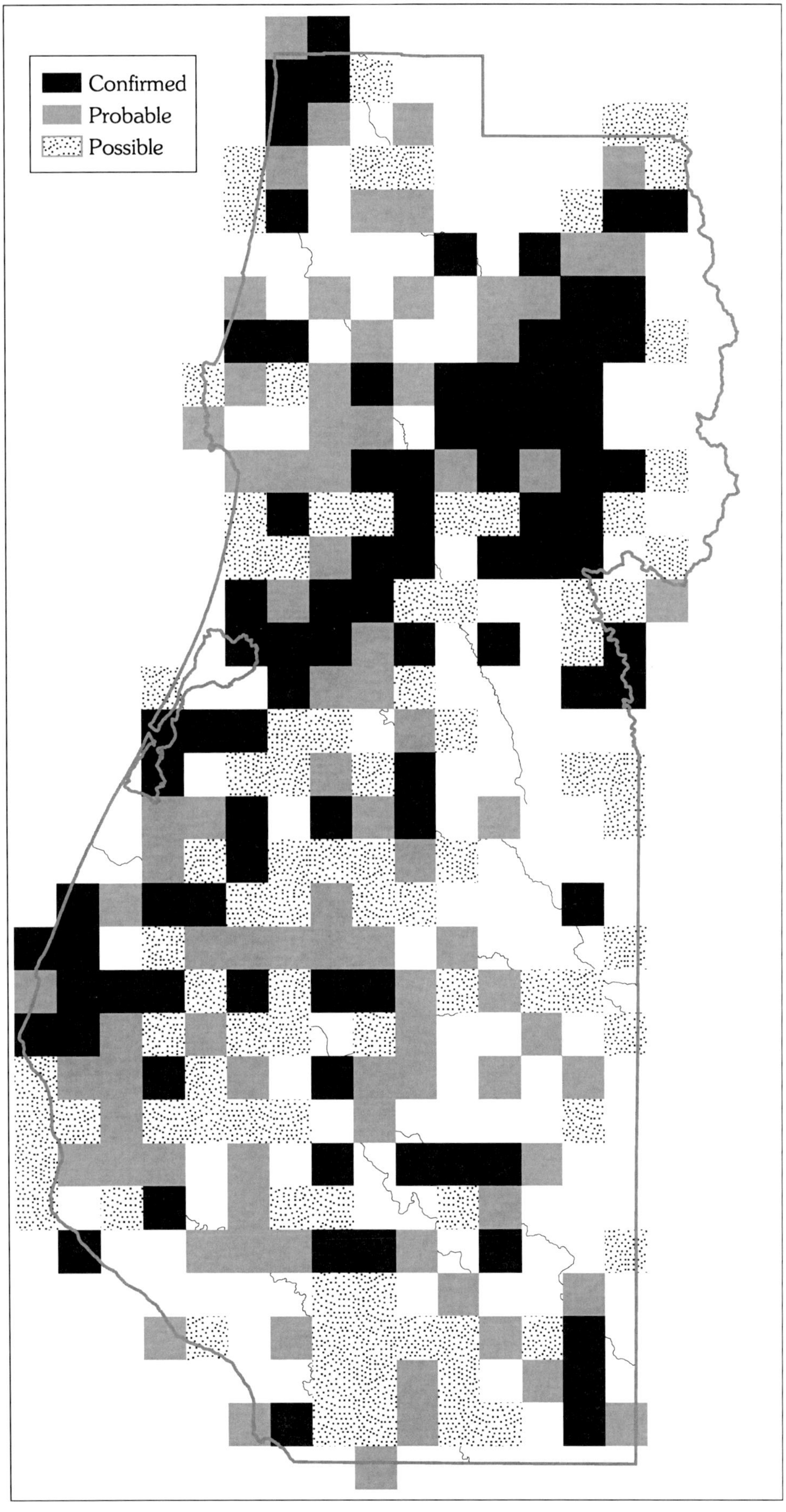

California's state bird is well distributed and variably common outside of the deserts, dense forests, and higher mountains. Within Humboldt County, California Quail are similarly widespread, and in northwestern California have been described as a common resident and breeder (Harris 1996). They are familiar and numerous permanent residents at lower and middle elevations, ranging throughout the county and occupying a fairly broad spectrum of habitat types. Although Atlas data fell short of validating the supposition, they probably bred in nearly all of the 266 blocks in which they were detected during the Atlas period.

California Quail may be encountered in many settings in Humboldt County. They seem to favor a mixture of dense brush and openings, often dominated by grass, annual forbs, and a sprinkling of small- to medium-sized trees. High-quality edge cover within a short flying distance is important to them, although they will also forage a considerable distance from such cover. Places providing these elements are found in abundance over most of the county's settled lowlands. As many of the openings existing today within the original forest and shrub cover have, by and large, been human-made, these quail are chiefly birds of sites that are in some sense or another disturbed. The setback of vegetation to earlier seral stages, a process sustained by timber harvest and the general clearing of land, has created opportunities for these quail where none previously existed.

The urban-rural interface around the cities of McKinleyville, Arcata, Eureka, and Fortuna support coveys of California Quail, as do the long-established brush patches and weedy hollows characteristic of abandoned industrial properties. Moderately grazed pastureland and farmland with sufficient hedgerow cover and a steady source of seeds, insects, and free water are well populated. California Quail also occur regularly in vegetated dunes, and in the ill-defined but extensive dune-pastureland edge along the west side of Humboldt Bay and in the lower Eel River delta. On the outer coastal plain, they use the broken riparian thickets and the great tangled masses of blackberry growing over and about long-abandoned ranch structures.

These quail follow the mosaic of open or semi-open ground well into the middle elevations. They are frequently seen in or about recent clearcuts on private timberland, both within the general zone of coastal overcast and eastward. They are found locally in the larger openings in the coast redwood forests and well toward the heads of small tributary valleys, persisting in such spots as old homesteads, crop rows near streams, neglected blackberry thickets, vacant lots dotted with coyote brush, and around the edges of parking areas and one-store hamlets.

California Quail avoid closed-canopy forest with extensive damp undergrowth. They are uncommon and localized within the summer-parched hillside pasture that sprawls over dozens of square miles of the upper Mad River drainage. The higher mountains of extreme northeastern Humboldt County do not support California Quail.

Atlas fieldwork revealed that, throughout the county, California Quail may be found upslope to the local upper limit of oak woodland, as well as in areas where the landscape is dominated by intensive industrial forestry practices. Logged sites on gently sloping ground, on which logs were tractor-skidded, appear to offer the species the irregular brush-and-clearing structure they favor. Steeper clearcuts or partial-cuts on which logs were yarded by a tower or "high-lead" system tend to cause less compaction, less outright setback of shrubby vegetation, and consequently a more vigorous and thorough regeneration to extensive closed brush. Such a situation is more to the liking of Mountain Quail.

Timber harvest in Humboldt County seems to have aided the spread of California Quail (Harris 1996). Even on steep northern exposures at 3500 feet, these quail can be encountered wherever slides, grass-grown road sidecast spoil, or old log decks provide a sunny spot with some bare ground, small seeds, and ready escape cover. There exists a zone of abutting or overlap of California and Mountain quail where oak woodland or pastureland habitats intergrade with higher forest stands. In such places, it is possible to encounter members of both species within a short distance of one another. California Quail would appear out of place well within the zone of late winter snowpack and montane conifers.

Despite the persistent nesting-season vocalizations and posturings of males, California Quail are secretive during nesting. Chicks scurry about with remarkable celerity soon after hatching, and they begin to fly at a young age. This allows family coveys to forage over larger areas and evade ground predators. The tendency for midsummer coveys to feed in semi-open settings provided atlasers with ready evidence of nesting. Precocial young were the most frequently observed manifestation of breeding.

Of the 89 blocks with breeding confirmations, 87 of those were of precocial young. These records spanned the dates of 26 May 1998 (B23; COg) to 16 Sep 1996 (m13; TWL). The other two confirmations were of nests with eggs 28 Jun 1995 (i09; C. Argello) and 8 Jul 1995 (n12; D. Pagen). Compton (1931) reported an adult with what appeared to be an exceedingly late 10-day-old chick 10 Jan 1931 near Scotia.

Due to extensive logging of redwoods, California Quail may have been most abundant in northwestern California during the late 1800s (Harris 1996). But because logged redwood country is apparently less favorable to these quail than logged Douglas-fir lands, these quail may have actually peaked in numbers during the much later heyday of Douglas-fir logging in the interior of the county (R. Gutiérrez, pers. comm.). Harris (1996) also speculated that recent re-clearing and "cleaning-up" of lowlands had diminished local populations. These quail are harvested in large numbers by sportsmen, but well regulated hunting probably has little effect on populations (Leopold *et al.* 1981). Feral cats, however, can decimate local populations of these ground nesters, and should be eradicated wherever they occur.

VIRGINIA RAIL *Rallus limicola*

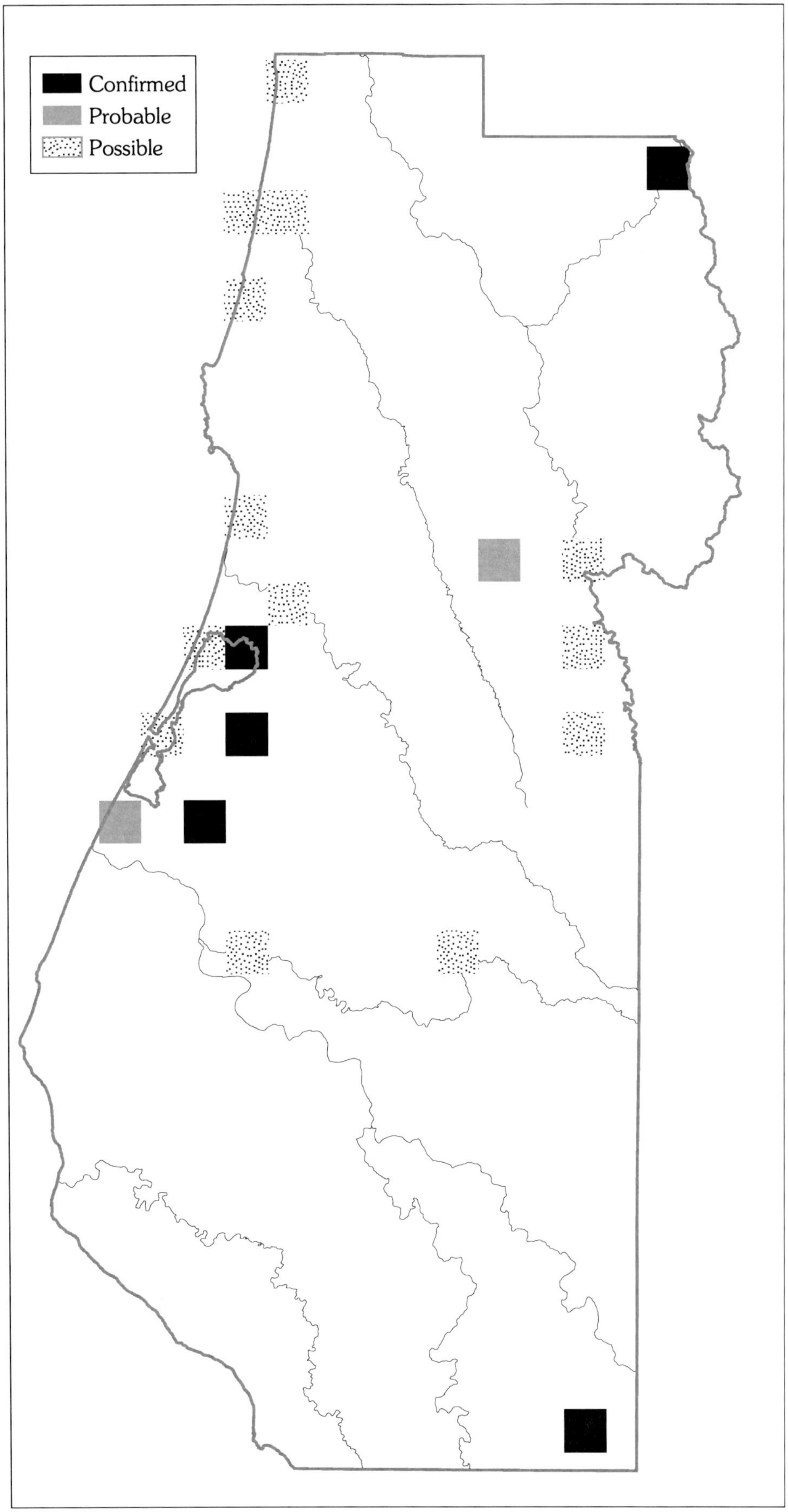

A Virginia Rail spends its life scarcely able to see in front of itself. Yet it is fully at home in dense thickets of low, wet vegetation, hunting tiny animals and eating seeds, seldom venturing into the open. Virginia Rails are long-drawn, gaunt birds, typical of the tribe of "water rails", whose many species worldwide have laterally compressed bodies and slender bills. These birds enjoy a great range across North America, able to exploit northern wetlands in summer while resident over much of their range to the south. Although these birds are not often seen, their voices are distinctive and, once learned, can be recognized at the limit of hearing. Because of this, one comes to understand that Virginia Rails may be expected in many wetland settings with suitable overhead cover.

The Virginia Rail occurs regularly in Humboldt County and has been widely noted by birders and ornithologists. Yocom and Harris (1975) stated that it was a "common resident and breeder, less numerous in summer." Though long known from Humboldt, one of the earliest confirmations of breeding here was of a nest with eggs discovered at Dry Lagoon 30 Apr 1956 (K. Legg). An early inland confirmation was of two adults with five young traveling overland between the Twin Ponds near Oak Knob 28 Jul 1977 (B. Marcot; Marcot 1978). Other locations where breeding has been confirmed over the years include the ponds at Clam Beach and near Arcata and King Salmon (Harris 1996).

While not appearing to be an especially common breeder during the 5-year Atlas period, Virginia Rail was frequently reported, especially considering its habitat requirements. Twenty blocks (5%) among the 425 total blocks had reports, including five with confirmed breeding and two with probable breeding. Like many other species dependent upon healthy freshwater wetlands, Virginia Rail were detected during the breeding seasons of 1995 to 1999 in two parts of Humboldt County. A concentration of reports along the northern and central coast suggests breeding around the Humboldt Lagoons and in the Humboldt Bay lowlands. Three of five confirmed breeding blocks were in the latter area. A sparse smattering of detection-blocks accrued in the interior of the county. Given the scarcity in Humboldt of marshes away from the coastal lowlands, the two breeding confirmations well inland near the eastern boundary of the county stood out.

Coastally, possible breeding by Virginia Rails was reported in shallow bulrush-grown pools at the end of Lanphere Road (e15) 4 May 1996 (DFx); at Aldergrove Pond (g14) 5 May 1999 (GAS); along the Skunk Cabbage Trail in Redwood National Park (g05) 11 May 1997 (TWL); at the Eureka Marsh (D17) 30 May 1995 (J. Haggard); in extensive cattails at Dry Lagoon (F07) 4 Jul 1996 (JEH); near Hydesville (f22) 6 Jul 1996 (JEH, TWL); at the marsh at the end of Hufford Road west of Orick (f05) 8 Jul 1997 (JCP); along lower Prairie Creek (g02) 7 Aug 1996 (M. Kenney); and in the "Rose Patch" at Clam Beach (f12) 15 Aug 1996 (TWL). A pair in the Eel River delta (c19) 4 May 1998 (M. Wachs) provided the only breeding evidence for the near-coastal area classified as probable. Each of three near-coastal breeding confirmations was established with the observation of precocial young. On 18 May 1998 a female with five chicks was seen at a marshy meadow off Old Arcata Road (F17; RJA). An adult and two downy chicks were reported at No Name Pond, Arcata Marsh (F15) 13 Jun 1997 (S. Harris). Precocial young also were reported at Catfish Lake (e19) a few miles southeast of Eureka 1 Aug 1996 (M. Wachs).

Inland, reports of possible breeding were received from Bigfoot Golf Course (N13) 28 May 1998 (TWL); at a pond near Oak Knob (N15) 13 Jun 1999 (TWL); at an isolated marsh along the Kneeland–Bridgeville Road (k22) 3 Jul 1997 (DFx); and at Big Lake (N17) 24 Jul 1998 (GAS). The only probable breeding evidence inland was of an adult heard "singing" 5 Jun–4 Jul 1997 in the Three Creeks drainage (L13) 4 Jul 1997 (TWL, S. Petersen). The two confirmed breeding reports in the extreme interior of Humboldt County were of precocial young at the pond at the junction of Ishi Pishi Road and Forest Service road 11N34, 2.5 miles north of Orleans (p04) 8 Jun 1998 (S. Robinson), and out Island Mountain Road (N33) 21 Jun 1996 (C. Secor).

Habitats used by Virginia Rails in Humboldt County include the array of sites mentioned above, as well as brackish marshes and saltmarshes in fall and winter. Remnant and degraded saltmarsh all around Humboldt Bay and in the Eel River estuary is inhabited. This species is detected more frequently than Sora in such places as roadside ditches, wet pastures, and small ephemeral marshes.

Virginia Rails are well recognized for strong migratory movements that occur twice each year over most of the species' range. Hapless members of this species, along with Soras, have long been the source of odd tales of encounters at backyard ponds, large parking lots, in phone booths, shopping malls, garden planters, or in the jaws of house cats. Migration is well evidenced in Humboldt County, not by sightings of individuals in unusual places—though this is sometimes reported—but by an obvious increase in fall, and greater abundance and more thorough distribution within the region in winter. This is followed by an ill-defined decrease that takes place at some point in late winter or early spring. Virginia Rails may be anticipated at nearly any marsh worthy of the description in the colder months, when permanent wetlands are at full pool and ephemeral or marginal microsites become widely available across the lowlands and in stream corridors. Harris (1996) remarked upon this phenomenon, relating that the species "seems to be absent or scarce in saltmarsh in midsummer, but with the arrival of fall migrants in late August through September, these areas fill with birds."

SORA *Porzana carolina*

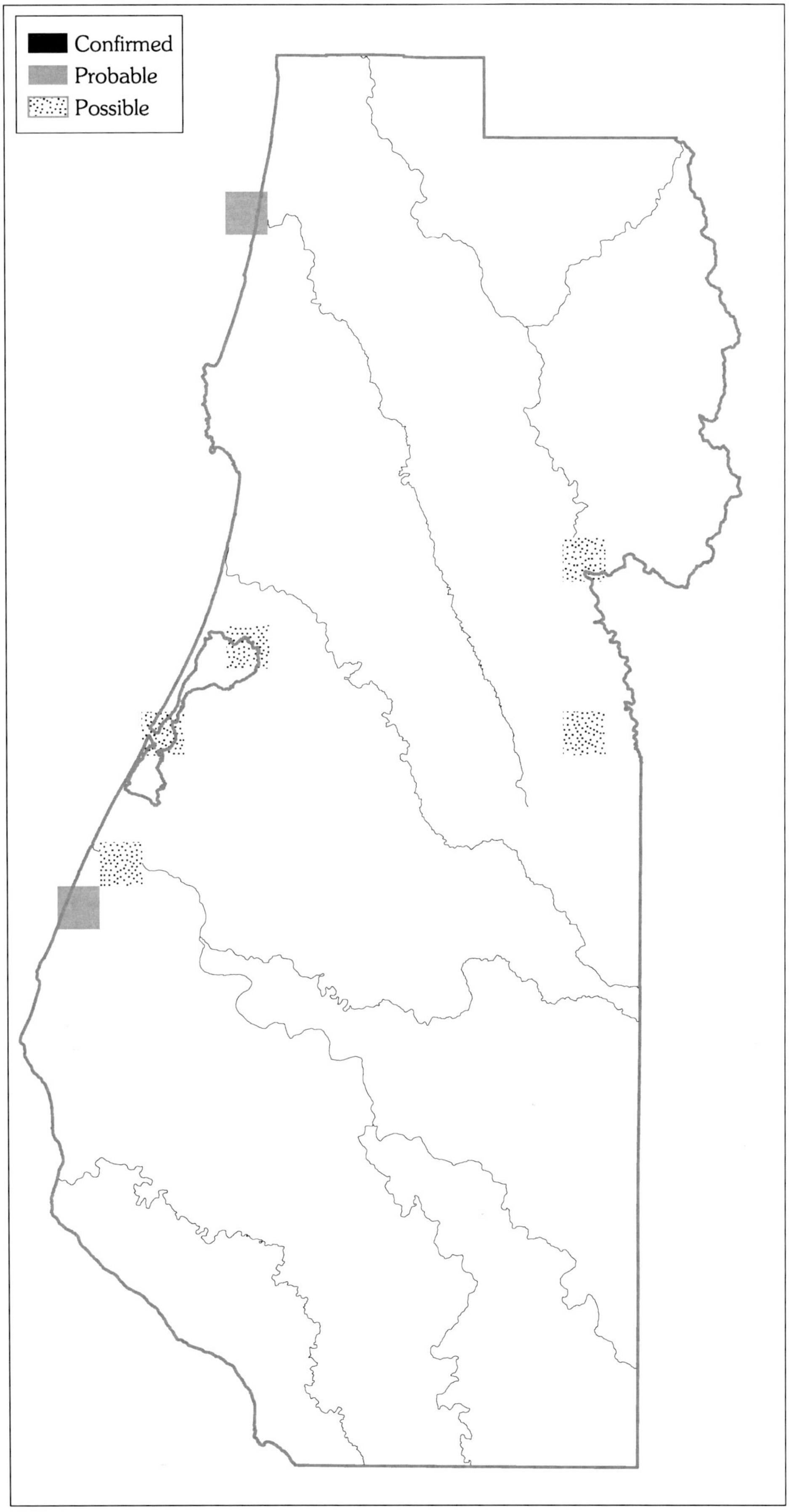

Soras are small, short-billed rails most often detected in freshwater marshes. Their breeding range extends across much of Canada and the northern U.S. Northern regions are vacated during the colder months. Cattail beds, rushy lake shores, and persistently wet meadows support nesting Soras in spring and summer. Small size, a laterally compressed body, and a streaky plumage pattern resembling the stems and shadows of its chosen environment allow the Sora to hunt insects, snails, and other invertebrates, and to glean plant material while remaining out of sight of both predators and humans. A seeming aversion to direct sunlight is another trait of this species to which most birders would readily attest—the whinnying "song" and querulous call-notes of this species are heard far more often than the birds themselves are seen. Shy and secretive though it is, the Sora is sufficiently numerous that regular visits to one or another of the better freshwater marshes in Humboldt County will ultimately show it to be present. In northwestern California, the Sora is generally considered an uncommon migrant and winter resident, and is casual to rare in summer (Harris 1996).

This species has long been known to occur in Humboldt County (Townsend 1887). The great majority of Sora records are of migrants and wintering birds (from mid-August to early May) in the marshes and ponds of the coastal lowlands, with few early summer breeding season reports. Since fall migration begins quite early, observations of apparently independent young Soras can pose a puzzle. Harris (1996) indicated that he had "no positive evidence of nesting." However, after seeing Soras at Espa Lagoon, Gold Bluffs Beach through the summer, G. Strachan reported a juvenile there 3 Aug 1981. This report, along with others—such as an adult with "youngsters" at Moxon Road in May 1982 (J. Sterling, K. Rosenberg)—provided sufficient evidence to confirm breeding in Humboldt.

During the 5 Atlas years, most reports of Soras fell within the established pattern. They were observed sparingly but regularly from August into April at such sites as the Arcata Marsh, the Arcata sewage ponds, Humboldt Bay National Wildlife Refuge, Eel River Wildlife Area, and similar venues in the coastal lowlands. Some examples of early fall records that could mistakenly suggest local breeding include four birds, including one flying juvenile only two-thirds as large as an adult, in No Name Pond at the Arcata Marsh 16–24 Aug 1997 (S. Harris), and two juveniles judged to be capable of sustained flight in that same pond 9 Sep 1995 (GAS). However, reports believed possibly to indicate local breeding came from seven Atlas blocks. In two blocks, probable breeding was noted, but there were no breeding confirmations.

During late April through 5 Jun 1995, a Sora was heard "singing" on dates a week or more apart in a freshwater marsh along Hufford Road in the lower Redwood Creek estuary (f05; DFx). The same breeding evidence was recorded in a small marsh west of Ferndale (B21), with a latest date of 19 Jun 1998 (COg). The presence of singing birds in what appeared to be suitable nesting habitat on dates well past the expected period of spring migration indicates that they probably represented breeders.

One or more Soras were detected in suitable breeding habitat in five additional blocks. Coastally, birds were reported 20 May 1995 (D17; GCH), 25 Jul 1998 at the Arcata sewage ponds (F15; S. Harris), and 11 Aug 1996 along the lower Eel River (c20; JG). Only two breeding season reports were from the interior of the county: one from Bigfoot Golf Course 4 Jun 1997 (N13; TWL), and one from Big Lake just northeast of Sugarloaf Mountain 19 May 1996 (N17; L. Bryan). While some of these possible breeding records may have involved migrants, they were nevertheless retained in order to document sites that should be watched in the future.

Soras are most frequently encountered in larger or more homogeneous parcels of marsh than are Virginia Rails, although many exceptions can be noted. This species is perhaps less likely than the Virginia to occupy smaller, linear, or ephemeral wetland habitats such as roadside ditches, wet pasture edges, very small marshes, and those that are highly overgrown with brush or overarched by trees, and is encountered less often and in lower numbers than Virginia Rail.

AMERICAN COOT *Fulica americana*

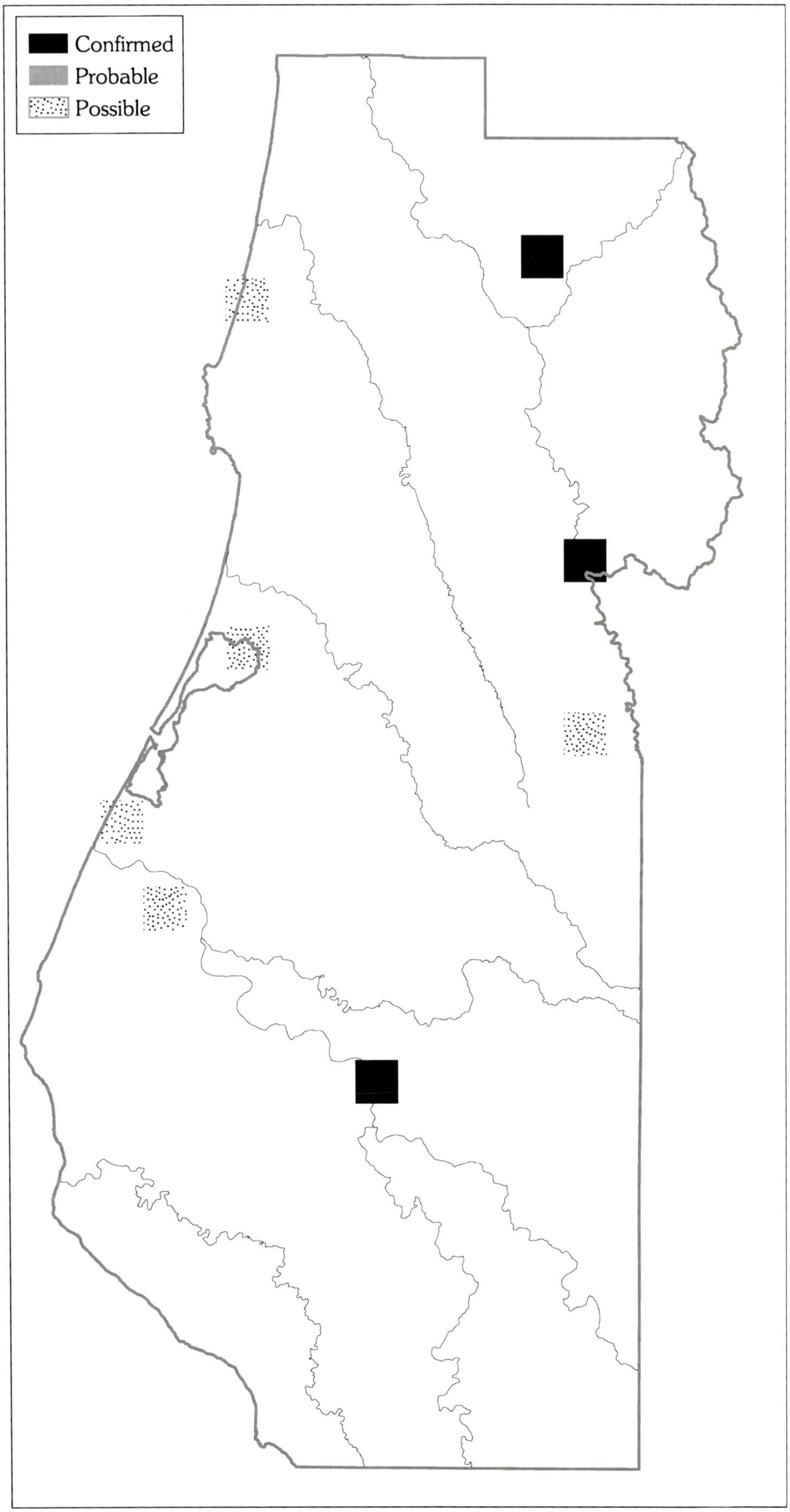

The identity of stout gray waterbirds with white bills, lobed toes, and a distinctly amphibious lifestyle has caused puzzlement for many. The question, "Are those ducks?" accompanied by gestures toward a flock of American Coots, is sometimes asked of birders. A habit of grazing on golf courses, in city parks, and along urban greenways has made this species visible to many. This holds most true during the colder months, when coots have moved out of their breeding marshes and concentrate in large numbers on open water, or on lawns and pastures adjacent to water. In spring and summer they are less conspicuous, retiring to reedy lake shores and extensive freshwater or alkaline wetlands to breed.

American Coots were very uncommon in summer in Humboldt County during the Atlas period. Reports in the nesting season came from only eight (2%) of the 425 total blocks, and in only 4 of the 5 project years (none in 1997). Yet these were widely scattered both along the coast and inland. There were breeding confirmations in three blocks. Adults feeding young were reported at Fish Lake (m07; E. Elias; no date). At the marshy water feature at Bigfoot Golf Course near Willow Creek, a bird was seen carrying nest material 13 May 1995 (N13; JEH), and precocial young were observed there 4 May 1996 (DFx, JCP, E. Elias, S. McAllister). Two occupied floating nests were in plain sight on a cattail-fringed pond at Larabee Ranch near Shively 26 Jun 1999 (i25; DFx, JCP, JEH). Reports given possible breeding status were of birds in the Ferndale Bottoms 9 Apr 1995 (D21; L. Karsteadt); at Big Lagoon 13 Apr 1996 (F07; JEH); along the "V Street Loop" (Old Samoa Road) in the Arcata Bottoms 17 Jun 1995 (F15; JEH); and at Eel River Wildlife Area 21 Jul 1995 (c19; S. McAllister).

That all three blocks with confirmed breeding were in the interior of the county—with no confirmed or even probable breeding reports from the coast—may seem surprising. However, American Coots seemed consistently scarce in spring and early summer at each of the major wetlands in the county, such as the Arcata Marsh, Humboldt Bay National Wildlife Refuge, and Eel River Wildlife Area. Yocom and Harris (1975) characterized American Coot as a "common to abundant migrant and winter visitant, uncommon in summer". Those authors stated that they knew of 12 records of broods for northwestern California: four from the Arcata area, one from Willow Creek (Bigfoot Golf Course), and the rest at Lake Earl, Del Norte County. Harris (1996) noted that most of the population departs in April–May, and that "a few breeding pairs or non-breeders remain on freshwater ponds through the summer." He mentioned breeding at Bigfoot Golf Course and at the Arcata Marsh prior to the Atlas period, as well as the earliest historical Humboldt County breeding record of "half grown young" somewhere around North Humboldt Bay 23 May 1943 (W. Anderson).

Habitats used by breeding American Coots in the county have been small- to medium-sized wetlands, either ponds with marshy edges or marshes with open water in the middle. Stands of cattails occurred during the Atlas period at the Bigfoot Golf Course pond, site of one of the three confirmed breeding reports. Emergent vegetation, occurring both as edges and as tiny islets or clumps, grew at the pond on Larabee Ranch, where the two occupied nests were found.

American Coots are far more abundant in Humboldt County from September through March than during the balance of the year. The period of influx in autumn appears protracted and is visibly incremental, evidenced by frequent increases in the number present at such sites as the Arcata Marsh and adjacent sewage ponds over the course of the season. In each of the 5 Atlas years, typically one to several summering visitors were augmented by a slight increase in late July, to dozens in August, hundreds by October. Flocks commonly numbering in the hundreds and locally into the thousands of birds assemble and linger for months at favored wintering sites. Freshwater Lagoon, various ponds and impoundments in Arcata and Eureka, wet dairy pastures, tidelands of South Humboldt Bay, and the Eel River estuary host flocks of coots each winter. During year-round studies at the Humboldt Lagoons, American Coot was the most common species, with peak numbers occurring during the winter and early spring (Devlin-Craig 1992). Between 1995–1999, high counts for coots were 2407, 3273, and 125 on the Arcata, Centerville Beach to King Salmon, and Willow Creek CBCs, respectively. These numbers indicate that the abundance of American Coots in winter is much greater along the Humboldt County coast than inland.

SNOWY PLOVER *Charadrius alexandrinus*

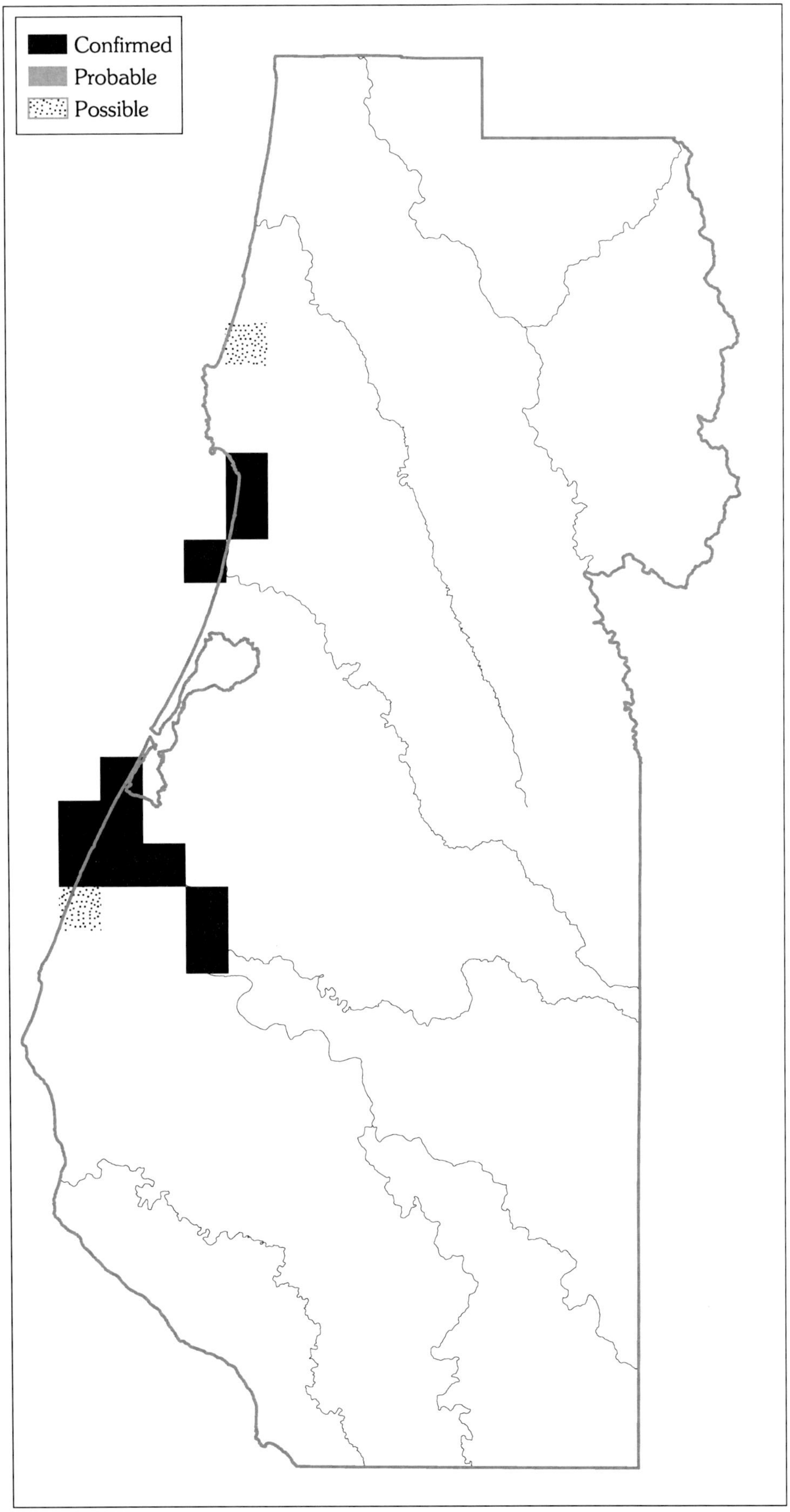

Following on the hind toes of the Spotted Owl and Marbled Murrelet, the Snowy Plover and its struggle to survive made headlines in Humboldt County during the 1990s. Human use and abuse of coastal beaches and the invasion of European beachgrass has diminished breeding success of Snowy Plovers here. Because of problems in Humboldt—and at nearly every other place where this species breeds on the West Coast—the Western Snowy Plover (*C. a. nivosus*) was federally listed under the Endangered Species Act in 1993 (USFWS 1993). A greater understanding of the biology and conservation needs of this species in Humboldt County has been achieved through the work of California Department of Fish and Game, Humboldt State University, Mad River Biologists, U.S. Fish and Wildlife Service, and other governmental and non-governmental organizations.

The concerted effort to study and safeguard the species' future in the county began in earnest in 1998. Much of what has been learned of the biology and population dynamics of the Snowy Plover in Humboldt is a result of that work. These efforts have included an annual winter survey; regular breeding-season surveys, including intensive nest monitoring; a public awareness campaign; banding of as many birds as possible; habitat studies; sharing of data and techniques among plover biologists; erection of exclosure fences around nest sites; and a somewhat heightened law enforcement profile which, however, remains insufficient to deter much of the inadvertent or purposeful disturbance and destruction of nests.

The Snowy Plover has long been recognized as a bird of Humboldt County beaches. Harris (1996) mentioned two sets of eggs collected from the ocean beach near Samoa 27 Apr 1902 (M. & J. Davis). Before introduced plants stabilized the dunes, it seems likely that these birds have nested on or near the sandy ocean beaches for ages. Yocom and Harris (1975), summarizing the occurrence of this species in northwestern California, stated that Snowy Plover was an uncommon migrant and winter visitant and a rare breeder. They listed it as expected regularly on Gold Bluffs Beach, Dry Lagoon Beach, Big Lagoon Beach, Clam Beach, the north and south spits of Humboldt Bay, Centerville Beach, and the beaches from Cape Mendocino south to the Mattole River area.

By the summer of 1977 this situation no longer persisted. During that year 18 Humboldt beaches were checked, and a total of 54 adults and 18 nests were found only on Big Lagoon Beach, Clam Beach, Mad River Beach, Lanphere Dunes Beach, the north and south spits of Humboldt Bay, and Elk River Spit (Page and Stenzel 1981). By the summer of 1999, more extensive and site-intensive beach surveys counted about 30 plovers and 17 nests or broods only on Clam Beach, the south spit of Humboldt Bay, and the beaches at Eel River Wildlife Area (R. LeValley, unpubl. data). Thus, by 1999 many of the historical nesting beaches in Humboldt County were no longer being used. For example, no evidence of nesting was found at Big Lagoon (not since 1989), Lanphere Dunes Beach (not since 1977), the north spit of Humboldt Bay (not since 1977), Elk River Spit (not since 1977), and Centerville Beach (not since 1994). However, while there had been a clear decline in plovers on Humboldt beaches over the past three decades, in 1999 it appeared that the population had been fairly stable for the previous few years (R. LeValley, pers. comm.).

European beachgrass was first introduced to the West Coast in 1898 in an attempt to stabilize dunes. The plan succeeded, and this exotic plant subsequently spread along the coast from British Columbia to southern California. Today the upper waveslopes of many Humboldt County beaches grade into formerly open dunes well-clothed with beach grass, as well as with bush lupine and, locally, ice plant. Stabilization of the dunes by beach grass has impeded the dynamic by which dunes might otherwise be perpetuated, imperiling unique flora and fauna and greatly reducing the amount of potential breeding substrate available to Snowy Plovers. As the breadth of the unvegetated beach/dune zone has narrowed, so has the zone in which plovers can nest. Pressed between the lower waveslope and the grassy foredune, the birds inhabit the upper beach. Visitor use at ocean beaches used by plovers for nesting, such as Clam Beach, occurs daily and is intense in summer, especially on Independence Day. With no patch of sand safe from the risk of foot traffic, unleashed dogs, horses, and off-road vehicles, Snowy Plovers are increasingly at risk in their efforts to reproduce.

The discovery that Snowy Plovers nest on gravel bars of the lower Eel River caught the ornithological community by surprise in 1996, when four nests or broods were discovered breeding in the vicinity of Fernbridge (G. Lester; Tuttle *et al.* 1997). For the record, it should be noted that a "family group" had been reported 5 Jul 1983 just below Fernbridge (J. Sterling, B. Garrison). Snowy Plovers along the Pacific Coast had not been known to nest along rivers more than one-half mile upstream from their mouths. Further investigations revealed not only that there were plovers nesting on the gravel bars, but that there were many of them—scattered from above Cock Robin Island to the confluence of the Van Duzen River. Surveys of Eel River gravel bars in 1999 found 19 plovers and six nests or broods on Fulmor, Fernbridge, Worswick, Mercer/Fraser, and Sandy Prairie bars, and on Canevari Island (R. LeValley, unpubl. data).

Plovers nesting along the Eel use broad, uniform expanses of river cobble, each of which consists of particles from sand- to cantaloupe-sized, but simulating the look of a gravel beach. Areas favored for nesting are typically well up from the waters edge

Continued on page 389

KILLDEER *Charadrius vociferus*

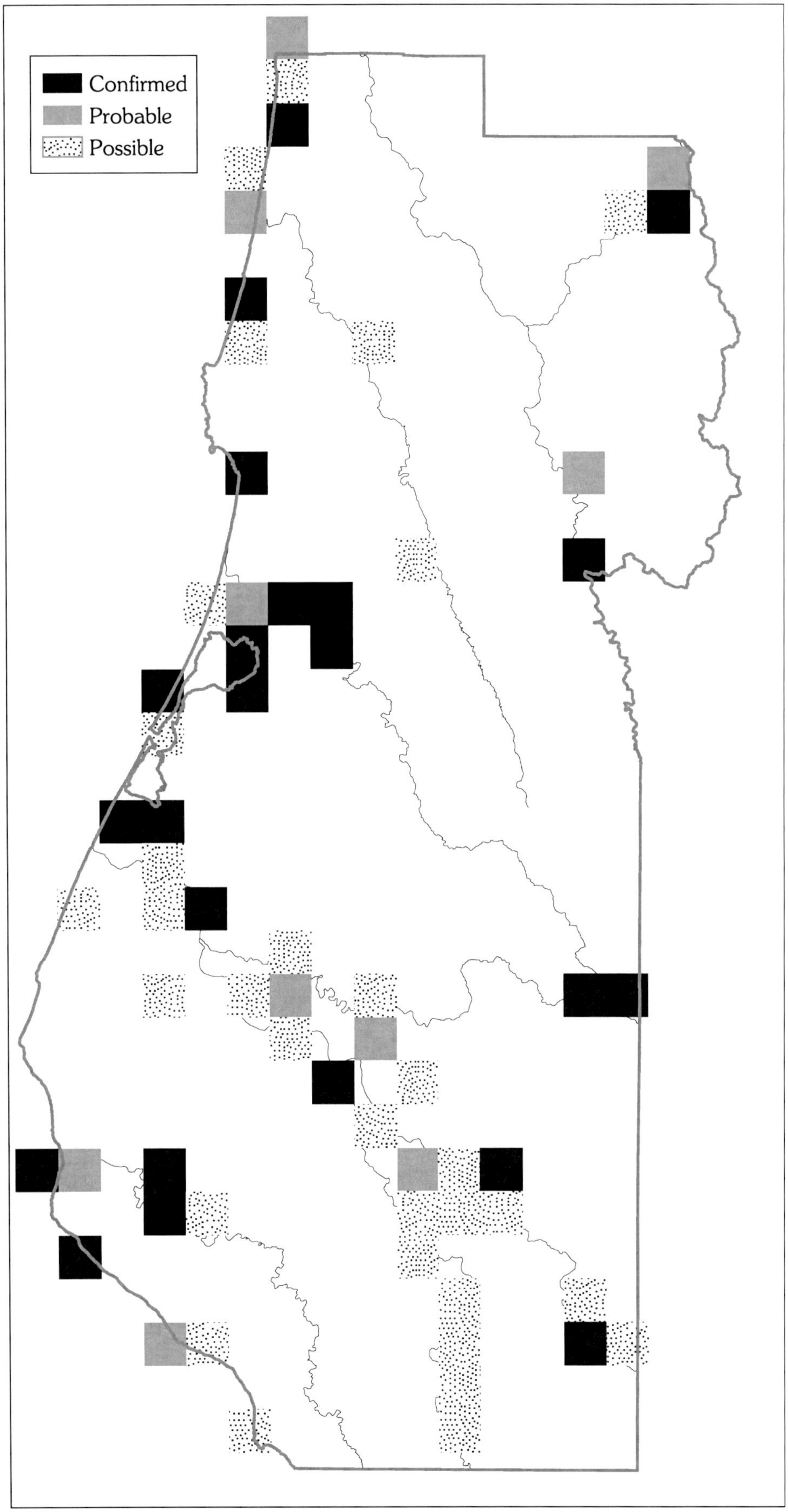

Killdeer are upland shorebirds, partial to open country. They become familiar to many people because of their commonness on athletic fields, farm pastures, and city lawns. Killdeer are wary birds, quick to give voice to unease. Adults nesting in areas of human activity seem haunted by chronic anxiety. This vigilance is fanned to a frantic pitch in dramatic "hurt wing" displays, performed by adults distracting predators from the vicinity of the eggs or precocial young. This species is widely distributed across temperate North America. In northwestern California, the Killdeer is a "common to abundant migrant and winter visitor; common summer resident and breeder. Widely distributed on freshwater habitats and some upland areas throughout the region" (Harris 1996).

Killdeer were found nearly to the four corners of Humboldt County during Atlas fieldwork. They were detected in 65 (15%) of 425 blocks, both near the coast and in numerous places in the interior of the county. Along or near the coast, Killdeer were reported from Gold Bluffs Beach (g01) south through the Humboldt Lagoons, Mad River bottoms, and the Humboldt Bay area to the Eel River delta. Killdeer were reported from three blocks at the seaward base of the King Range, as well as at Shelter Cove (f33). Oddly, none were reported from the nearly unbroken grassy expanse immediately inland from the ocean beach between False Cape (a22) and the mouth of the Mattole River (a27).

A noticeable pattern of blocks with detections indicates the general corridor of the Eel/South Fork Eel River (south to k33). Elsewhere away from the coast, Killdeer were found breeding in settled and agricultural land about Orleans (P05); along the Trinity River (N11, probable) and near Willow Creek (N13); in adjacent blocks along the Van Duzen River (N23, o23); in the middle Mattole River country (d28), and along the upper main Eel River near Fort Seward (N31). Sightings ranked as only possible breeding were reported at Childs Hill Prairie in the Bald Hills (i08) and at Lord-Ellis Summit on Highway 299 (J13).

During the breeding season in Humboldt County, Killdeer are seen and heard most often around the edges of towns and cities, in pasturelands, and along rivers. They are found in short grass habitats such as large lawns and heavily grazed dairy pasture; on compacted, weed-dotted old mill sites; at industrial parks; along graveled service roads and on dikes, and around the edges of shallow ponds.

Among the 65 blocks in which Killdeer were detected were 23 with breeding confirmations. These came from nearly every general location at which the birds were encountered. One-half of the total confirmations were established by the observation of distraction displays. The earliest report was 7 May 1996 (N13; TWL) and the latest was 25 Jul 1998 (d28; TWL). Precocial young were the best evidence in eight blocks, spanning 5 May 1999 (d16; GCH) to 10 Jul 1996 (D19; K. Foerster). Somewhat surprisingly, only three active nests were reported during the 5 years of the Atlas. Two were occupied nests, 10 Apr 1995 (F11; TWL) and 23 Jun 1997 (g14; T. Bowden), and the other was a nest with young 13 Jul 1996 (P05; T. Kirk).

The seasonal movements of Killdeer are somewhat difficult to perceive. Beginning at some point in autumn, numbers assemble in riverine habitats such as gravel bars, islands, and beaches, as well as in plowed fields. The vast open expanses of the Humboldt Bay and Eel River bottomlands attract not only breeding pairs but flocks numbering dozens, sometimes many hundreds, in the non-breeding season. Most wintering Killdeer in Humboldt County are found in the coastal lowlands, with far fewer lingering along or near the larger rivers. Harris (1996) noted that in some years, thousands winter in wet lowland pastures of the Mad and Eel River bottoms. It is assumed that migrant Killdeer must pass through Humboldt County with some regularity, yet flocks appearing to be actively migrating by day are not seen—and the voice of no more than the occasional nighttime flyover is heard by birders who listen for flocks of migrating shorebirds.

BLACK OYSTERCATCHER *Haematopus bachmani*

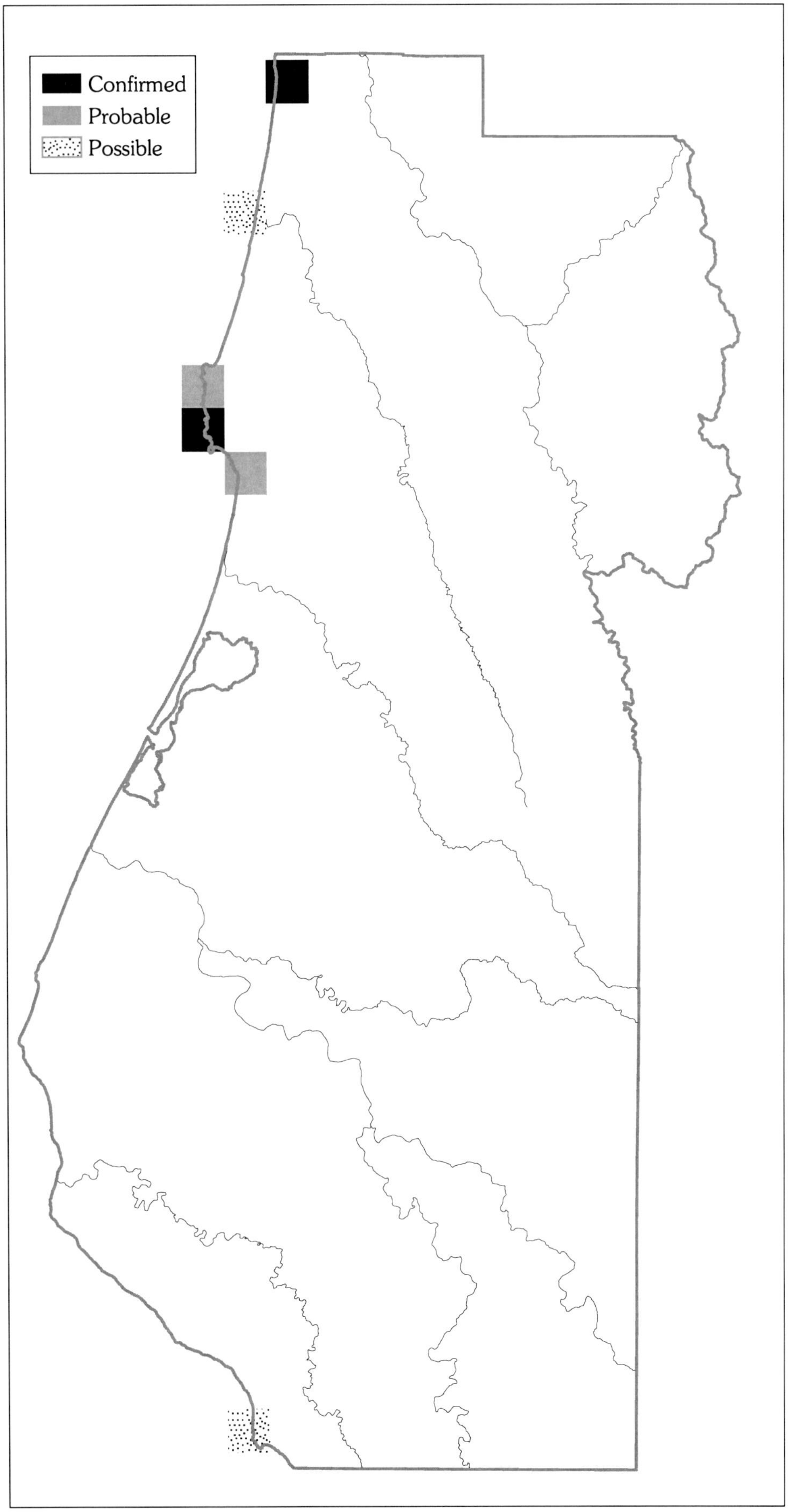

Black Oystercatchers are specialized birds of rocky North Pacific shores. Using long, thin, laterally flattened bills, they extricate and consume the soft flesh of unwilling mollusks living in the intertidal zone. Their deep sooty-brown plumage and methodical inspection of wave-swept mussel beds causes them to be difficult to spot when foraging. When alarmed or on the wing, however, Black Oystercatchers typically are heard before they are seen. Even in the lathered chaos of winter storms, their shrill whistles penetrate through the white noise of breaking waves and the gales that accompany them ashore.

This species is much more numerous farther north in its range than it is along the coast of California. Nevertheless, they populate stretches of suitable rocky coastline in modest numbers wherever it occurs in the state south to the Channel Islands, with fewer birds on the mainland shores of southern California. Black Oystercatchers are to some extent resident, although considerable apparent migration among more northerly populations has been demonstrated (Andres and Falxa 1995). Little is known of seasonal movements of oystercatchers in Humboldt County.

This species has been known in Humboldt County since C. Clay reported seeing several at Flatiron (Off-Trinidad) Rock 16 Jul 1911 (Harris 1996). Although frequently reported from reliable sites along the coast, such places are comparatively few. Hence, the species was considered an uncommon resident and local breeder by Harris (1996). Atlas fieldwork resulted in detections of Black Oystercatchers in only six blocks, all of them along the outer coast. Confirmed and probable breeding were each noted in two blocks, with simple presence in appropriate habitat reported in the remaining two.

The stretch of coastline supporting Humboldt County's summering Black

Oystercatchers is well defined. The birds have been most commonly seen along the outer coast from Patricks Point southward through the Trinidad area to Moonstone Beach near Westhaven. The roughly 10 miles of shore favored by the birds features extensive rocky or cobble beaches, untrampled tidepools, California mussel (*Mytilis californianus*) colonies, and protected shelves of rock above the range of normal high tides. These expanses of rocky shoreline enclose pocket beaches, the sands of which are dissected by little freshwater outflows preferred by oystercatchers for bathing, loafing, and occasional foraging. Use of such sites is partly dependent upon the minimal presence of people and unleashed dogs. Helbing (1977) provided interesting details (and photos) on nest sites, prey, activity budgets, breeding behavior, and other aspects of the life history of five pairs nesting near Trinidad. Because of the fragmented distribution and somewhat limited extent of these habitats in the region, the total population in Humboldt and Del Norte counties is not especially great, and is thought to be only about 100 nesting birds (Sowls *et al.* 1980).

Evidence of confirmed breeding by Black Oystercatchers in Humboldt County during the study period was scant. Precocial young were seen at Elk Head (e10) 22 Jun 1995 (JEH). Of interest was the only other breeding confirmation, a sighting of precocial young 7 Aug 1996 in limited but evidently suitable habitat near the north end of Gold Bluffs Beach (g02; M. Kenney). This observation may have been sufficiently late in the breeding season to remain ambiguous with respect to point of natal origin as, in a summary of nesting chronology, Andres and Falxa (1995) indicate that early August is a period during which fledged young may accompany adults from the breeding territory to remote feeding locations. It is conceivable that the young birds seen at Gold Bluffs Beach may have hatched in more suitable, more extensive rocky shoreline habitat either to the north or south. Courtship display witnessed near Patricks Point (e09) 23 Apr 1996 (GAS) strongly suggested nesting in the vicinity, as did the presence of a pair of adults near Luffenholtz Beach (F11) 21 May 1995 (JEH).

A single adult seen at the Redwood Creek mouth (f05) 8 Jul 1997 (DFx) would appear to have been associated with the rocky coast shortly to the north of that estuary. Another lone bird at Shelter Cove (f33) 21 Jun 1995 (D. Roberson) causes one to wonder if enough habitat exists around the low, broken headland of Point Delgada to attract a pair or two; no further reports from that isolated patch of potential nesting habitat were received. Although it is possible that a few birds may inhabit the intertidal zone about Cape Mendocino, no reports were received from there. This area is difficult to reach and was little visited during the Atlas period, so a definitive statement on whether or not oystercatchers occurred there cannot be made.

AMERICAN AVOCET *Recurvirostra americana*

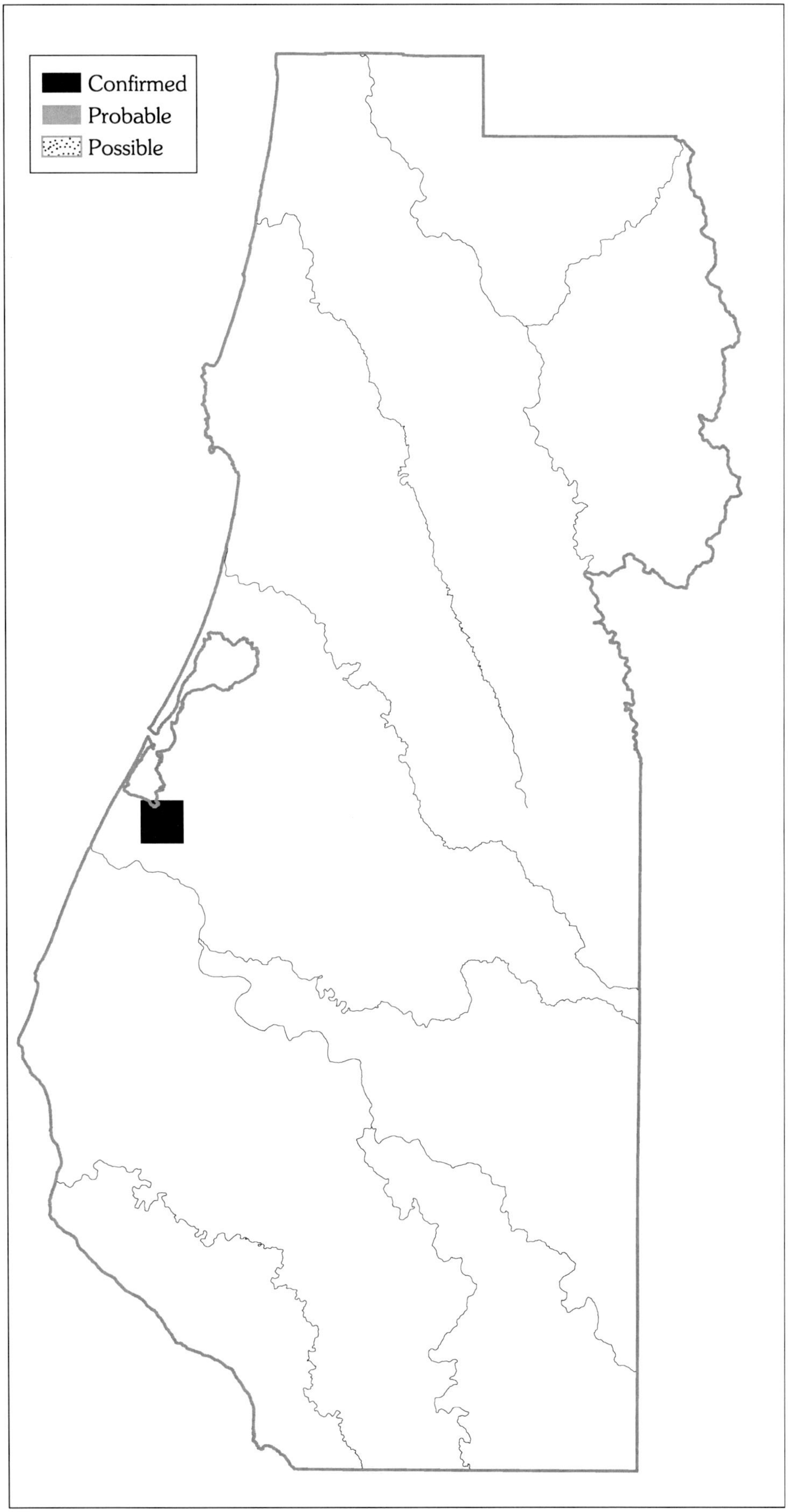

At high tide, an observer of Humboldt Bay can see no mud. During the long hours of low water, one sees almost nothing *but* mud. Forced ashore by a returning tide, migrant and non-breeding shorebirds abandon the bay to preen and rest nearby while it mirrors the day's skies. Following every freshly exposed inch of the ebb, they then feed anew, their ranks of thousands ultimately dispersing into the misted corrugations channeling water throughout the breadth of the empty tidal basin.

Among the many waders whose daily routines are profoundly molded by this tidal scenario are American Avocets. These birds are attuned to the availability of an invertebrate food resource most vulnerable to exploitation during approaching and receding high tide. At the dynamic frontier between shallow water and fluid mud, flocks of these appealing shorebirds "scythe" with a neat sideways flick of the bill, reaping macroplankton generated by nutrient-rich outflow charging the bay. Favored stretches of estuarine shoreline around the margins of Humboldt Bay and the lower Eel River provide habitat sufficient to support hundreds of avocets during the non-breeding season. With the exception of 1995–1996, discussed below, during the Atlas period the numbers of birds present typically dwindled to nearly none by May, and then increased after the arrival of post-breeding adults in midsummer. The main influx of fall migrants and wintering birds in the county took place in September and October. In northwestern California, Harris (1996) classified the American Avocet as a "common, local migrant and winter visitor; casual breeder and summer resident."

Not all who are accustomed to seeing flocks of American Avocets on Humboldt Bay may be aware that their expansion into the region is a fairly recent phenomenon. It probably is related both to a larger-scale expansion witnessed in the San Francisco

Bay area and to the incidental creation of suitable habitat in Humboldt Bay. Unknown to Humboldt's earliest ornithologists, this species was first reported from the county 17–18 Aug 1935 when two were collected in Eureka (Davis 1939; HSU 4800, HSU 4801). Talmadge (1946), Evans (1988), and Harris (1996) provided information regarding some of the 16 subsequent Humboldt Bay records from 1943 through March 1961.

The construction of the Arcata sewage treatment plant in 1957 was possibly the event that provided prospective wintering birds with preferred conditions. A regular wintering population was established by the early 1960s. As many as 800 wintering avocets were counted by 1982–1985 (Evans 1988, Harris 1996). These birds concentrated their activity at, and on mudflats opposite, the sewage ponds. It is believed that avocets prospered because of increased freshwater flows and elevated biological activity in the bay near the sewage plant (Evans and Harris 1994). The San Francisco Bay experienced a similar, but larger, expansion of avocets during roughly the same time period as did Humboldt Bay (Shuford 1993).

During the establishment of the wintering population, American Avocets remained strongly restricted to the northeastern corner of North Humboldt Bay. The number of birds increased to a high of about 1000 by the early 1990s. By the late 1990s the wintering population had decreased to perhaps 50% of its former size, but at the same time avocets began to appear at sites at which they previously had been unknown or were rarely seen (Colwell *et al.* 2001). Persistent attendance at Manila, Eureka Slough, Indian Island, and southward along the east side of South Bay was noted by Colwell and others. Tidal flats and brackish channels around lower Hookton Slough were especially well used. The large open marsh on the south end of the Salmon Creek Unit of Humboldt Bay National Wildlife Refuge became established as a roosting and, in short order, nesting site for avocets. Use of Eel River Wildlife Area and estuarine reaches of the Eel River was also documented. For nearly the first time, avocets began to forage in pasture rainpools after storms. An anomalous first historical Humboldt County breeding record for American Avocet was made with the sighting of a pair of adults accompanying a single three-quarters-grown juvenile inside the Eel River entrance at Crab Park in June 1980 (B. Slocum; Harris 1996).

American Avocet was confirmed breeding in Humboldt County during the Atlas period only in 1995 and 1996. Those were the only Atlas years in which any breeding activity was reported. Fourteen adults and two chicks were seen at the refuge during the summer of 1995 (K. Foerster), and a pair attending a nest with eggs at the southernmost marsh on the Salmon Creek Unit was also recorded 22–31 May 1996 (D19; GAS, K. Foerster). But these reports came as no great surprise, as nesting by a very few avocets had been observed at this site in 1993–1994. The outcome of the 1996 nesting attempt was undetermined. Elsewhere in 1996, avocets were recorded at the Arcata Marsh (F15) 5 May 1996 (KMS) and 13 Jul 1996 (JEH). None appeared at the refuge during the 1997 breeding season. About 50 avocets lingered at their favored south Salmon Creek Unit wetland 19–22 Apr 1998, but there were no further reports that year before the arrival of expected migrants or winterers at the site. Stragglers lingered there until 19 May 1999 (S. Harris, M. Colwell).

A study carried out at numerous sites on Humboldt Bay (Danufsky and Colwell 2003) indicated that avocet occurrence was positively correlated with substrates characterized by fine particles (as comprise mud), rather than by coarser materials such as sand. During the relatively wet 1995–1999 period, increased sediment loading of the Humboldt Bay tidal system may have occurred as a result of logging and consequent runoff of soil and silt.

SPOTTED SANDPIPER *Actitis macularia*

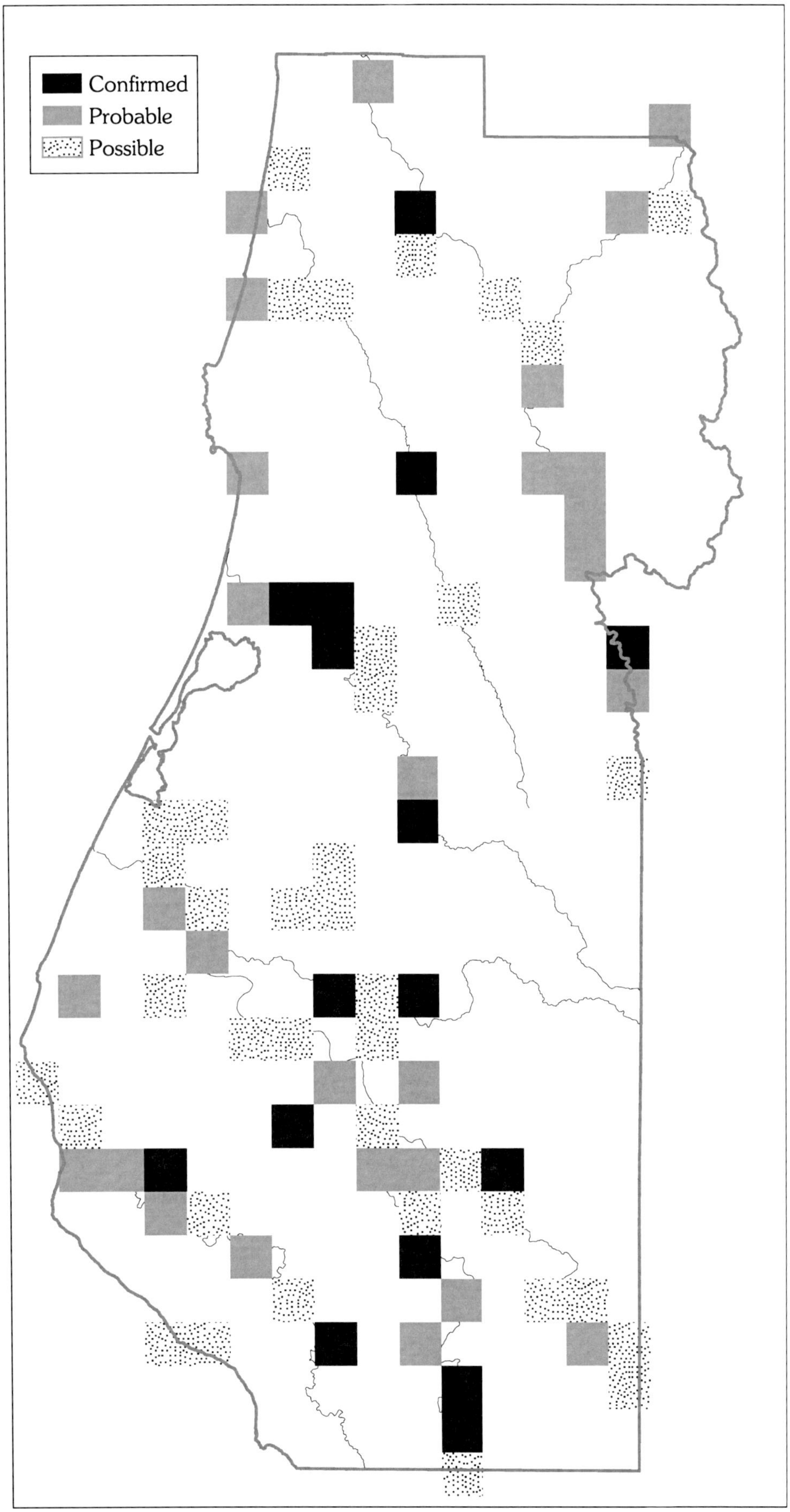

Appealing and easily seen, the Spotted Sandpiper typically is among the first shorebirds to come to the notice of those with a growing awareness of birds. Their distinctively stiff-winged flight, loud calls, "teetering", and propensity for inhabiting waterways also used by people cause them to be one of the most conspicuous species of streamsides and inland shores.

Seeming to demand no more than a minimally disturbed pebbly beach washed by clean waters and supporting a vigorous invertebrate fauna, Spotted Sandpipers enjoy a wide range, breeding nearly throughout the temperate and boreal United States and Canada. Atlasers found these small waders along every one of Humboldt County's principal rivers and streams and, where suitable conditions existed and access permitted, on a number of smaller tributaries. The birds were encountered in 82 (19%) of the 425 total blocks and were confirmed breeding in 16 of them.

Not surprisingly, distribution of Spotted Sandpipers in the county closely coincides with the larger watercourses. They were found on the Bear, Eel, South Fork Eel, Klamath, Little, Mad, Mattole, Trinity, South Fork Trinity, and Van Duzen rivers. Nests have also been found in the dunes at the Little River mouth (*fide* M. Colwell). Creeks along which the birds were found during Atlas surveys included Big Flat (King Range, drains directly into the Pacific Ocean); Blue (into the Klamath River); Bull (into the South Fork Eel River); Davis (south of Cape Mendocino, drains into the Pacific); Larabee (into the Eel River); Lawrence (into Yager Creek); Prairie (into Redwood Creek); Redwood (drains into the Pacific), and Yager (into the Van Duzen River). The most concentrated clustering of blocks with detections was along the Eel River and South Fork Eel River, with consistent presence noted along the nearby Mattole as well. It was clear that the lower Mad and Trinity rivers also supported these birds along much of their lengths.

It can be noted, too, that no Spotted Sandpipers were encountered in much of interior northeastern Humboldt, nor in the east-central part of the county. These regions are characterized by middle-elevation ridges and plunging draws drained by short, steep streams. Watercourses in these sections may not afford appropriate habitat. Though many run cool and unsullied, their flows typically diminish by late spring. Further, the banks of many are highly incised, with the semi-open cobble bars and gravel beaches so favored by Spotted Sandpipers generally small and uncommon. While insufficient habitat may well be the cause of no detections, it is important to note that atlasers often could not enjoy access to many reaches of sizable streams along which some habitat may exist. This held true for significant reaches of the larger rivers as well. Hence, these negative data are best viewed in a conservative light.

During the nesting season, May to July, Spotted Sandpipers are frequently seen in pairs; nearly three-quarters of all the observations of probable breeding were of pairs. The earliest that courtship was recorded was 4 May 1997 (k30; JG). Successful nestings detected during the Atlas period were most commonly evidenced by the sight of semi-dependent fledglings toddling about with a parent; 80% of all breeding confirmations were of precocial young. Such records spanned the dates of 30 Jun 1998 (h14; KMS) to 5 Aug 1995 (H15; PFS). Nests with eggs were found only twice: 21 Jun 1995 (H31; D. Roberson) and 29 Jun 1996 (J29; RJA).

Suitable breeding habitat for Spotted Sandpipers is found along perhaps several hundred river-miles of Humboldt County waterways. If several pairs summer along each mile of river that offers suitable habitat, a rough estimate of the county-wide breeding population would be hundreds of pairs. With respect strictly to Humboldt County, the Atlas results would appear to affirm the statement of Harris (1996), who summarized Spotted Sandpiper in part as an "uncommon resident and breeder" in northwestern California.

Populations of Spotted Sandpipers using the Eel River in summer may be affected by factors arising within historical times that have led to this river being one of the most degraded in northern California. Much of the river has become silted as the unstable Franciscan parent material that underlies soils within the watershed has slumped and washed into the main river course. Logging and road construction have greatly exacerbated this natural phenomenon. Annual diversion of up to 85% of the upper Eel River flow into the Russian River system via the Potter Valley Project in neighboring Mendocino County has conspired with other human impacts to alter the very nature of the river. These effects are manifested in exceedingly low flows during summer and fall, during which time large gravel bars and cobble beaches are persistently exposed. While these changes have certainly diminished habitat for many species, it is possible that Spotted Sandpipers may have benefited from them. Conversely, it is probable that increased human access and recreational use of the immediate river channel and gravel bars by off-road vehicles and unleashed dogs may reduce nesting productivity. In addition, the occasional large volume of water released into major rivers as "maintenance flows" during the nesting period undoubtedly washes away some active nests.

WESTERN GULL *Larus occidentalis*

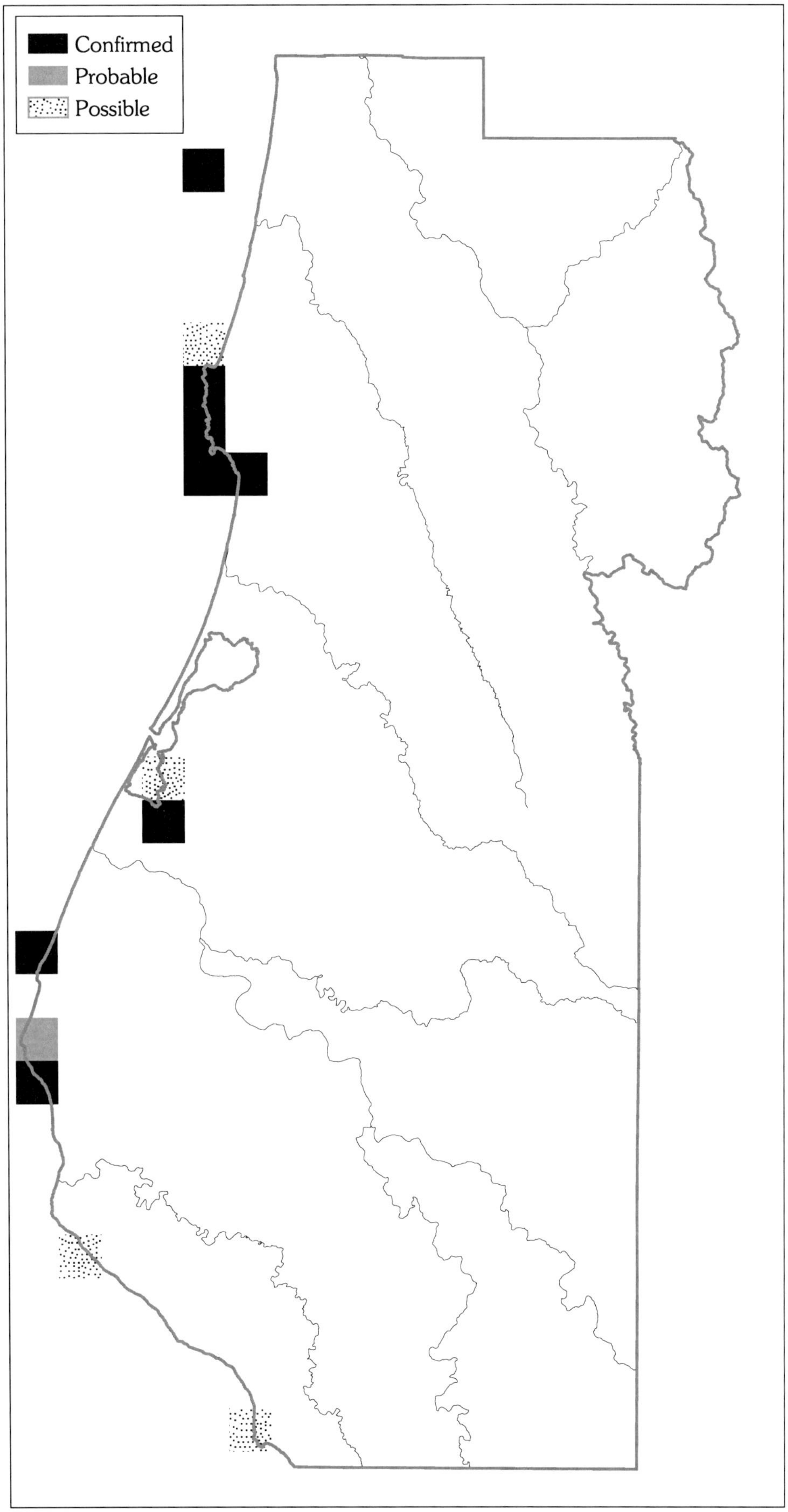

Western Gulls live only along the coastline of the eastern North Pacific Ocean from British Columbia to Baja California. Rare stragglers visit southeastern Alaska and appear inland in the western U.S. Most of the population breeds in Oregon and California. This is the only large, pink-footed, dark-backed species of gull resident or otherwise occurring commonly along the California coast.

Few birds to be encountered along the coast of Humboldt County are as conspicuous as Western Gulls. The dark gray upper wings and back, no-nonsense bill, and fearlessness of adults make them stand out both when perched and in flight. They are able to fly long distances with ease and can sustain themselves on nearly any organic material—traits that have served them well in the era of modern settlement. Fish-packing plants, industrialized waterfronts, and garbage dumps have afforded Western Gulls new foraging opportunities. In some locations, pairs have nested on channel markers in harbors or on urban rooftops.

The entire extent of the Humboldt County coast is inhabited by Western Gulls throughout the year. Use of different stretches of coastline varies with season, with food supply, and with respect to suitability for nesting. Western Gulls have long been known as a breeding species in the county. Referring to northwestern California, Harris (1996) characterized Western Gull as a "common resident and breeder." While adults continue to occupy much of the outer coast while foraging and commuting, far greater numbers occur in spring and summer in the sections of the county where most nesting takes place.

During the Atlas period, Western Gulls were confirmed as breeding in eight blocks scattered in several areas along the northern and central Humboldt coast. In those eight blocks with confirmed breeding, only one report was of a nest with eggs, at Teal Island (D19) 22 May 1996 (K. Foerster). There were three reports of occupied nests, at Wedding Rock (e09) 17 May 1998 (JEH), at Steamboat Rock (a25) 10 Jun 1998 (DFx), and at False Cape (a22) 23 Jun 1995 (DFx). Nests with young were reported from an offshore rock (e11) 19 May 1995 (JEH) and from "Little Head", near the base of the Trinidad pier (e10) 22 Jul 1995 (JEH). Finally, dependent fledged young furnished the best evidence in two blocks, at Elk Head (F11) 22 Jun 1995 (JEH), and at Redding Rock (e04) 31 Jul 1998 (J. Jacobsen, DFx, JCP, JEH). Visitation by adults at a probable nest site at the Cape Mendocino Lighthouse (a24) was recorded 6 Jul 1996 (JEH, TWL). In four other blocks, adults were present in places offering appropriate nest substrates, but no evidence of nesting was observed.

With one exception, Western Gulls nested during the study period at locations historically recognized as breeding sites. In 1996, at least seven pairs nested on Teal Island in southeast South Humboldt Bay. This island also supported a colony of Double-crested Cormorants. The denuding of the once dense coyote brush at this site by cormorant excrement and the consequent creation of semi-exposed terrain may have created a situation attractive to the gulls.

Unlike other parts of California, Humboldt County is not known for supporting large colonies of these gulls. Much of the population in the county occurs as pairs nesting in loose association, in accordance with the availability of ledges and nooks on the coastal cliffs and offshore rocks. Nests are generally located on the upper slopes of such rocks, or on cliffs well above the splash zone. The nest is built either on bare ground, in grass and forb cover, or on a rocky ledge. Where there is one pair of nesting Western Gulls, more are usually nearby, anywhere from several feet to some distance away. Harris (1996) stated that "nests are known from most larger and many smaller seastacks." There were an estimated 229 breeding pairs of Western Gulls in Humboldt County in 1969 (Ainley and Whitt 1973).

The list of locations where breeding was detected in 1995–1999 may fall short of reiterating the complete historical roster of known sites, but it illustrates the patchy distribution of nesting habitat. Places for pairs of gulls to build nests are lacking or scarce along the coast from the Del Norte County line south to Patricks Point, from Little River through the Humboldt Bay and Eel River estuaries, and along the seaward base of the King Range, where there are no protected coastal cliffs or offshore rocks.

Although portions of the Humboldt County coast are without nesting birds in summer, subadult (and perhaps some non-breeding adult) Western Gulls remain present. These birds are easily overlooked, but contribute significantly to the midsummer gull population. They feed in the intertidal zone in estuaries and to a considerable degree offshore, where Western is among the most commonly seen gull species. They may be seen loafing in small flocks at creek outfalls and other gull gathering sites. In late spring and early summer, Western Gull is the most common gull species in Humboldt, but during migration, California (*L. californicus*), Glaucous-winged (*L. glaucescens*), and Herring (*L. argentatus*) gulls may outnumber it. Clearly, the many thousands of Mew Gulls (*L. canus*) present in winter would appear to surpass the number of Westerns in the area at that season. This is particularly true during extended storms, when great numbers of Mew Gulls forage in flooded pastureland as well as along the coast.

Although Talmadge (1953b) reported this species along the Trinity River during salmon runs, more recent observers have not witnessed them there. Otherwise, Western Gulls are quite restricted to saltwater habitats. Their zone of abundance essentially includes all of the outer coast and waters of the continental shelf and slope. Daily use is made of Humboldt Bay and the lower Eel River throughout the year. The birds use the Humboldt Lagoons proper only sparingly. Although a familiar sight along the Eureka waterfront, Western Gulls are much less numerous around Arcata. Visits to athletic fields, flooded pastures, and large parking lots are more frequent in the colder months, as are overflights of other habitats removed from the outer coast.

COMMON MURRE *Uria aalge*

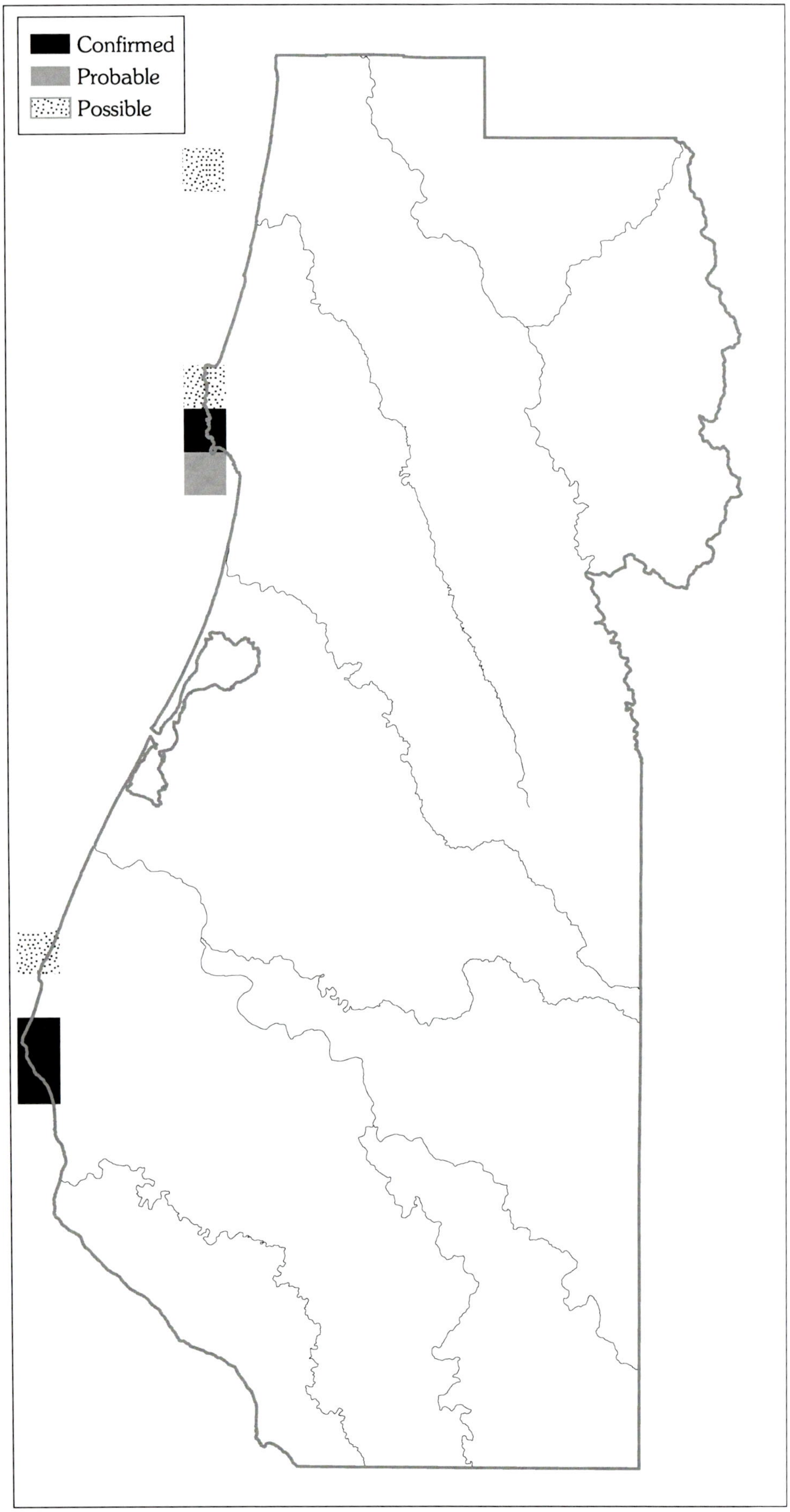

Common Murres are highly social alcids, breeding in colonies ranging from a few dozen birds to tens of thousands of pairs. They are widespread along North Pacific coasts, extending southward along the West Coast to Monterey County as nesting birds and farther south in winter. This seabird has long been known to nest on many offshore rocks along the Humboldt County coastline; a set of eggs was collected from an unknown site in the county 2 Jul 1897 (A. Shields; MVZ 5791). Harris (1996) characterized Common Murre in northwestern California generally as a common resident and locally abundant breeder.

Historically, numbers of nesting Common Murres have declined in California, with loss of attendance at some sites. This phenomenon has not been as clearly evidenced in Humboldt County as it has elsewhere. Sowls *et al.* (1980) reported the following numbers at Humboldt County nesting colonies in 1979–1980: Redding Rock, 2100 birds; White Rock, 2600 birds; Green Rock, 55,000 birds; Flatiron (Off-Trinidad) Rock, 24,000 birds; Blank Rock, 1000 birds; Pilot Rock, 2500 birds; False Cape Rocks, 8000 birds; and Steamboat Island, 4500 birds. In 1995, additional counts were made for Flatiron (25,494 birds) and False Cape Rocks (12,426 birds; Manuwal *et al.* 2001). Additional murre count data for Humboldt sites can be found in Osborne (1972) for the period 1969–1971, and Manuwal *et al.* (2001) for the period 1979–1995.

Atlas-period confirmations of breeding include occupied nests on Sugarloaf Island off Cape Mendocino (a24) 4 Jun 1995 (D. Ogden) and near Trinidad 29 Jun 1995 (e10; DFx). Nests with young were reported on Steamboat Rock (a25) 28 Jul 1995 (S. Levy). About 1000 murres were seen visiting probable nest sites on Pilot Rock (e11) 26 Mar 1995 (JEH).

Counts at Redding Rock (e04) from 1989 to 1995 indicated a decline, and there is fear that this colony could be extirpated in the future by chronic disturbance by the U.S. Coast Guard (Manuwal *et al.* 2001). No murres were found nesting on Redding Rock when atlasers (J. Jacobsen, DFx, JCP, JEH) visited it on 31 Jul 1998. This finding during their single visit came as no surprise, since that survey was conducted during an ongoing El Niño event. However, great numbers of sea lions (mostly California sea lions, *Zalophus californicus*) used Redding Rock as a haul-out during the Atlas period. About 200 were counted that day, from the waterline to the crest of the rock. This behavior clearly disallows use by nesting Common Murres. Also on 31 Jul 1998, hundreds of murres were on Flatiron Rock (e10), but none were seen on Green (e10) or Pilot (e11) rocks, and only one juvenile was seen on the water.

It should be noted that adult Common Murres commute many miles each day in the course of feeding their single nestling. Such feeding flights made to and from colonies caused this species to be noted in 35 additional outer-coastal blocks. Since these blocks lacked offshore rocks of size and structure suitable for breeding sites, these observations were not considered evidence of possible breeding and were not mapped.

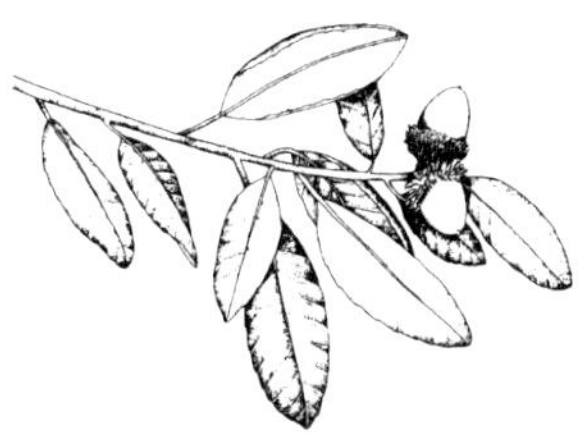

PIGEON GUILLEMOT *Cepphus columba*

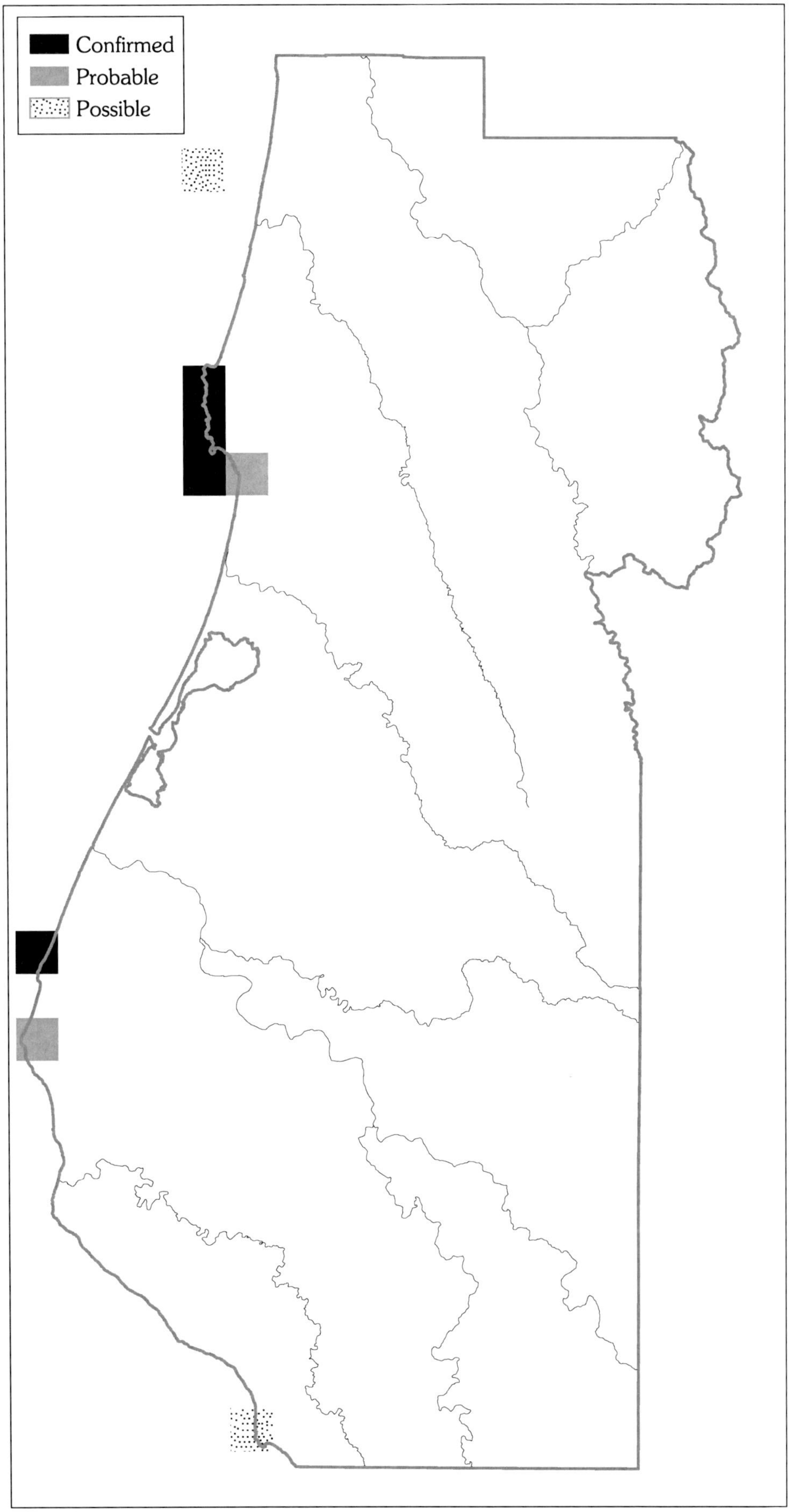

Pigeon Guillemots are medium-sized alcids partial to nearshore ocean waters throughout the year. They are common and widely distributed along the California coast from early spring through mid-autumn, with few detected during the rest of the year. Guillemots nest in potholes, crevices, and other recesses of offshore rocks and on rocky coastal headlands. This species has long been known to occur in appropriate habitat along the coast of Humboldt County. An early Humboldt breeding record is of eggs collected 14 Jun 1926 near Trinidad (G. Atwell; HSU 1733). Harris (1996) summarized the status of Pigeon Guillemot on the far northern California coast as a "common summer visitor and breeder; casual in winter," adding that fewer than 10 winter reports for Del Norte, Humboldt, and northern Mendocino counties were known to him. Ainley and Whitt (1973) estimated the entire Humboldt breeding population to be 129 pairs.

During the 5-year Atlas period, Pigeon Guillemots were found commonly from Patricks Point State Park southward through the general Trinidad area to Little River Rock. This stretch of shore is dotted with rocky islands and sea-stacks. Coastal cliffs of Wedding Rock and Trinidad Head provide nesting sites for many pairs of guillemots. Atlasers noted birds on or near numerous rocks within the northern cluster of mapped blocks; however, no attempt was made to census this species. Sowls *et al.* (1980) indicated that birds were present at Wedding, Sea Gull, Sea Lion, White, Green, Puffin, Flatiron (Off-Trinidad), Blank (Sugarloaf), Pilot, Prisoner, and Little River rocks, on numerous islets of the "Trinidad Bay rocks", and on Trinidad Head. None of these locations supported more than an estimated 60 nesting birds. During the Atlas period, occupied nests were reported on or near Little River (Camel) Rock 19 May 1995 (e11; JEH) and at Wedding Rock (e09) 31 Jul 1998 (DFx). An adult carrying a fish to hidden young was observed near Trinidad (e10) 22 Jul 1995 (JEH), and another visiting a probable nest site was recorded from Luffenholtz Point (F11) 21 May 1995 (JEH).

North of this area, Pigeon Guillemots at Redding Rock (e04) 31 Jul 1998 (J. Jacobsen, DFx, JCP, JEH) were suggestive of local nesting, but they were seen only on the water, with no pairing, "singing", or flybys noted. The heavy use of Redding Rock as a sea lion haul-out may limit occupation of the site by guillemots, although Sowls *et al.* (1980) estimated that there were eight nesting birds there in June 1980. Other than this offshore location, the far northern Humboldt coast offers no rocks of sufficient height, surface complexity, and protection from high surf events to support nesting pairs. The tall, sandy faces and steep slopes that front the ocean at the base of Gold Bluffs Beach have few nooks and crannies and are at some distance from the ocean, and would thus seem to offer limited nesting opportunities.

In addition to the breeding sites in the Trinidad area, two sites in the vicinity of Cape Mendocino, False Cape Rocks and Sugarloaf Island, hosted Pigeon Guillemots during the Atlas period. Both sites appear to support large numbers of guillemots, and were also recorded by Sowls *et al.* (1980), who counted a total of 390 nesting birds. Atlas records for these two islands were of at least one occupied nest 23 Jun 1995 on False Cape Rocks (a22; DFx), and adults visiting probable nest sites 6 Jul 1996 on Sugarloaf Island (a24; JEH). Sowls *et al.* (1980) also indicated that only two birds were breeding at Steamboat Rock, not far south of Sugarloaf Island. Neither shore- nor boat-based observers detected Pigeon Guillemots on or in the vicinity of this particular rock during the Atlas period. A guillemot on the ocean opposite Point Delgada (f33) 18 Jun 1995 (K. Dyer) could have been summering locally, perhaps nesting. This headland's low elevation and nearly complete exposure to inundation by surf would seem to explain the near absence of Pigeon Guillemots from south of Cape Mendocino to Shelter Cove.

Pigeon Guillemots are known to nest under docks and wharves in lower-estuarine settings at many points within their range, including in the Crescent City harbor, Del Norte County, until 1987, when the docks and their nesting birds were set on fire and destroyed (Harris 1996). Use of these types of sites has not been documented inside Humboldt Bay or elsewhere in Humboldt County, but the extensive wharves of a bygone era, nearly all fallen to time, conceivably could once have supported an estuarine nesting population of guillemots.

PIGEON GUILLEMOT *Cepphus columba*

MARBLED MURRELET *Brachyramphus marmoratus*

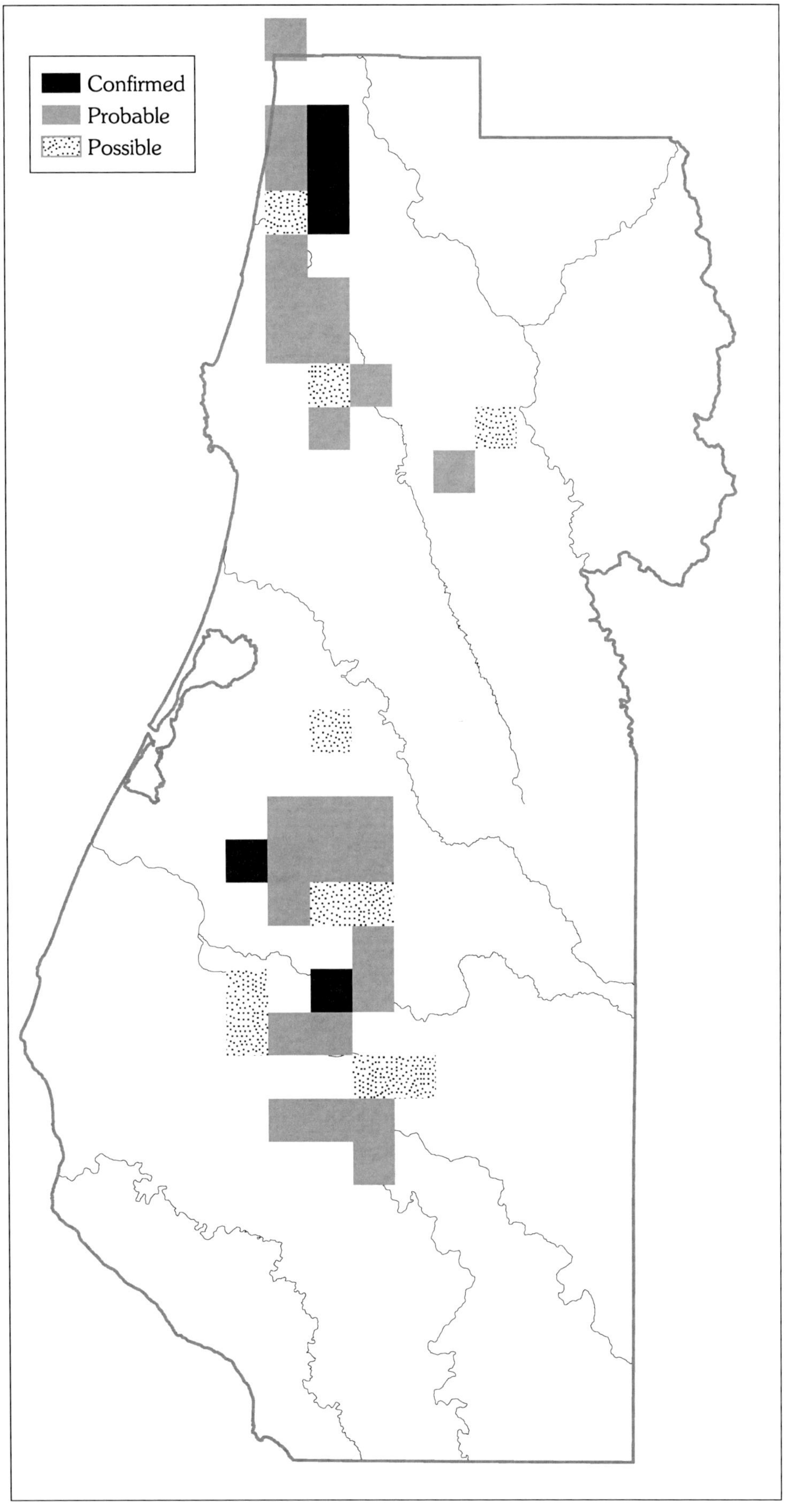

No native of Humboldt County's old-growth forest received more attention in the 1990s than the Marbled Murrelet. In a given week, anyone within reach of a wildlife journal, the Eureka *Times-Standard*, or a television remote might learn of the latest events in an unfolding story of controversy and discovery. Agency biologists, environmentalists, industrialists, consultants, politicians, reporters, investors, lawyers, loggers, scientists, and others have all joined the fray at one point or another. Why all the hubbub? Simple: these seabirds nest almost exclusively in large old conifers. As a result of the extensive cutting of coastal old-growth forests, the Marbled Murrelet was listed as "threatened" in California, Oregon, and Washington in 1992 (USFWS 1992).

As is true for the Spotted Owl, much of the attention has resulted in reasonably good local information having been learned about murrelet distribution, biology, and habitat associations. While knowledge about the murrelet is better than for many other species, it was the last regular breeder in the U.S. and Canada to have its nest discovered. In 1974, a tree trimmer near Santa Cruz discovered a downy murrelet chick on a large limb in a Douglas-fir (Binford *et al.* 1975), thereby ending more than 100 years of speculation regarding where these birds might nest. Humboldt's first historical breeding confirmation was of a downed fledgling found in Eureka 4 Jul 1924 (C. Clay; HSU 2557; Carter and Sealy 1987).

These North Pacific alcids like cold marine waters, and are found breeding from about Santa Cruz north to the Aleutian Islands. While at sea, these birds are seldom found more than a few miles from the shore, a characteristic that makes them particularly vulnerable to oil spills and, in some areas, gill netting. In Humboldt, Marbled Murrelets feed opportunistically on a variety of small crustaceans and fish such as the northern anchovy

(*Engraulis mordax*; Burkett 1995), and are most abundant directly offshore from the last remaining areas of coastal old-growth forest (Miller *et al.* 2002). They are known to forage far from breeding stands; during at-sea murrelet surveys in 1995–1999, birds were observed foraging in 30 coastal blocks in Humboldt County where evidence did not suggest breeding (DFx *et al.*). The habit of nearshore feeding provides researchers in boats with a reasonably good chance to count murrelets and their young, and it is from this work that population estimates have been made. The resulting approximations should be viewed with some caution, however, since sampling and analytical methods are variable and require further refinement. Yet the numbers provide a sense of how many murrelets may be present. On the basis of data from 1989–1993, about 4120 birds were estimated to be present between the Klamath River and Cape Vizcaino (Ralph and Miller 1995); 1995 data from Strong *et al.* (1997) suggest that perhaps 3400 birds were distributed from the Oregon border south to Cape Mendocino; in 1999 about 3600 birds were possibly present from the Oregon border to Trinidad (Ralph *et al.* 2000); and in 2000 there were an estimated 4876 murrelets from Coos Bay, Oregon south to Shelter Cove (Bentivoglio *et al.* 2002).

Conversely, detecting, let alone counting, these birds when they visit inland nesting areas is far more difficult because of their crepuscular habits, cryptic plumage, rapid flight, secretive behavior near the nest, and the high observer skill required (Hunter and LeValley 1996). Accordingly, a standard inland survey protocol has been developed (Pacific Seabird Group 1994), under which countless predawn surveys have been conducted locally for timber management, population monitoring, and research purposes. It is from these survey results that much Atlas data was gleaned. Under the survey protocol, observations of murrelets flying through or below the tree canopy or circling above suitable habitat were generally considered evidence of "occupancy"; these observations were considered equivalent to visiting a probable nest site under Atlas methodology. Records of birds present at inland sites, but not exhibiting these highly suggestive behaviors, were afforded possible breeding status.

During the Atlas period, Marbled Murrelets were detected in fewer than 10% of all blocks. Possible breeding was recorded in 10 blocks, which included observations at Dean (F23), Freshwater (H17), Owl (i21), Yager (H21), and Bridge (H09) creeks. All 26 probable breeding observations were of "occupied behaviors" recorded during inland surveys and ranged from 6 Apr 1995 (g04; JEH) to 3 Aug 1995 (k11; J. Price). Of the five confirmations, two were observations of birds carrying fish to presumed nestlings at Lost Man Creek (H05; no date; *fide* S. Miller) and 17 Jul 1996 at Boyes Prairie (h04; JEH). Murrelet eggshell fragments were found 2 Aug 1995 at Cheatham Grove (H23; P. Herrera) and also on 2 Aug 1995 along Little South Fork Elk River in Headwaters Forest (f20; M. Perry; WFVZ 164715). The only nest discovered during the Atlas period was one found 24 Oct 1998 by a researcher climbing around in the canopy of a redwood appropriately named the Atlas Giant at Prairie Creek Redwoods State Park (H03; S. Sillett). The few Atlas records above fit with what is known of the breeding chronology in California, where incubation starts about 24 Mar and ends 13 Aug, with the nestling period running from about 23 Apr to 9 Sep (Hamer and Nelson 1995a).

As indicated by the distribution map, murrelets in Humboldt occurred almost exclusively in forests dominated by coast redwoods. While there have been a few exceptions to this rule, most of the Douglas-fir stands where birds have been found around the southern fringes of their distribution in Humboldt contain at least some redwoods. Examples of these areas include stands on Larabee Creek near

Continued on page 390

CASSIN'S AUKLET *Ptychoramphus aleuticus*

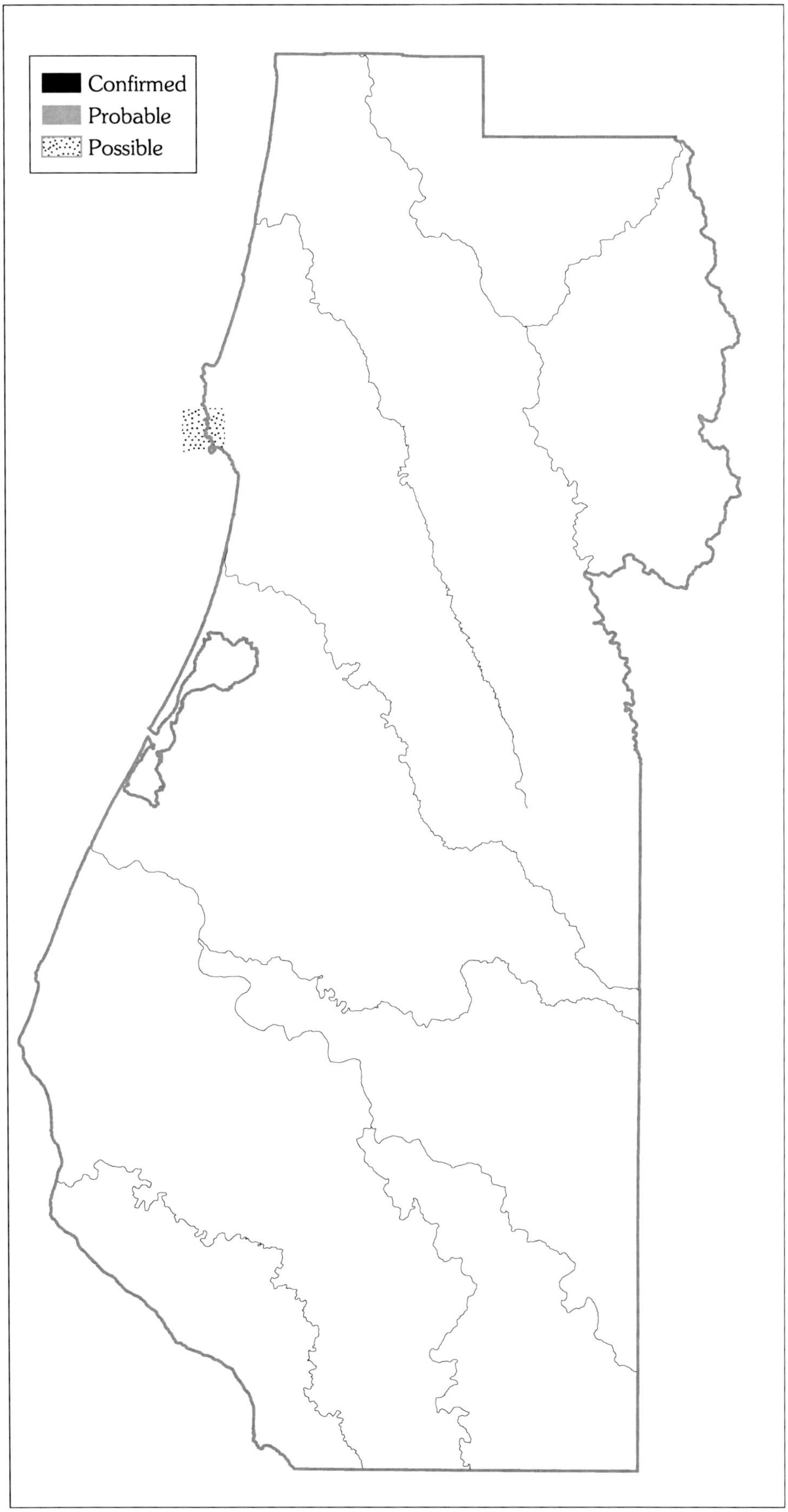

Cassin's Auklets are found in cold marine waters on each side of the North Pacific Ocean. These small alcids breed colonially in burrows on sea islands, and are almost entirely nocturnal near land during the breeding season. Only a comparatively small proportion of the large North American population breeds as far south as California. Of California's estimated 131,170 birds, 80% nest at the Farallon Islands and only 3% in northwestern California (Sowls *et al.* 1980). There are an estimated 25 and 1000 breeding pairs in Humboldt and Del Norte counties, respectively (Ainley and Whitt 1973). While Harris (1996) characterized Cassin's Auklet in northwestern California as a common resident and locally common breeder, in Humboldt this species is a very local and somewhat rare breeder.

These birds are found throughout the year in ocean waters off Humboldt County but are not commonly seen near shore. Historically, observers on boats out over the continental shelf have regularly reported Cassin's Auklets in numbers ranging from a few birds to many thousands. Because of restrictions of opportunity and weather, most such reports are made in summer and fall. Birders doing seawatches from shore points have seen apparent north- and southbound migrants in spring and fall. More spectacular evidence of migration has been provided by counts of from 5000 to more than 10,000 birds during those seasons by observers on boats (Harris 1996). Cruises 10 to 50 miles offshore between September and February usually produced counts of 50–200 Cassin's Auklets (Yocom and Harris 1975). It appears that this species is probably present in winter every year.

Humboldt's earliest historical breeding confirmation was made 22 May 1910 when about 20 pairs were found nesting in a small sandy flat at the north end of Flatiron (Off-Trinidad) Rock (C. Clay). Clay (1913) subsequently collected and artificially hatched an auklet egg from there. Osborne (1972) estimated that no more than 25 pairs nested at the county's only other known Cassin's Auklet breeding site, Green Rock. During the Atlas period, Cassin's Auklets were detected in only a single block containing a plausible breeding site (e10; Green Rock). One such report was of birds on the water in the vicinity of that rock 6 Jul 1999 (E. Elias). Although no direct evidence of nesting was obtained, this report was retained as evidence of possible breeding at a long-established nesting site.

Difficult to access, seldom visited, and appearing immune to human-related alteration, seabird breeding rocks would seem unlikely to have undergone significant recent changes. However, while auklets nested on Flatiron Rock as late as 1934 (C. Clay), soil erosion has resulted in the apparent extirpation of the breeding colony (Osborne 1972). Increasing use of offshore rocks by Double-crested Cormorants and other birds has become a concern of seabird biologists. From an historic low in the 1960s and early 1970s, Brown Pelicans have rebounded greatly. Their heavy annual use of offshore islands as loafing sites during post-breeding incursions into northern coastal California carries with it the potential for degradation or loss of the delicate sod caps vital to burrowing nesters such as Cassin's and Rhinoceros auklets, Tufted Puffins, and Fork-tailed and Leach's storm-petrels. Damage, if any, to the limited sod cover at Green Rock by pelicans or cormorants has not been demonstrated. However, the establishment of marine nesting colonies of Double-crested Cormorants in Humboldt County since the 1970s has definitely caused loss of vegetative cover at Little River Rock (S. Harris, pers. comm.), a process leading directly to loss of soil. Breeding by Double-crested Cormorants at Green Rock was not noted during the Atlas period, but, given present trends, might reasonably be expected to occur in the future.

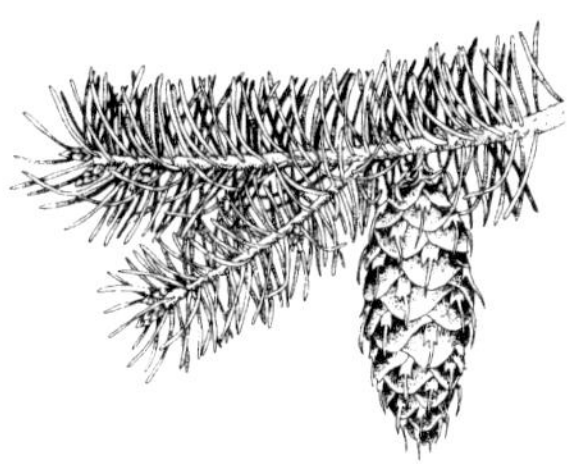

RHINOCEROS AUKLET *Cerorhinca monocerata*

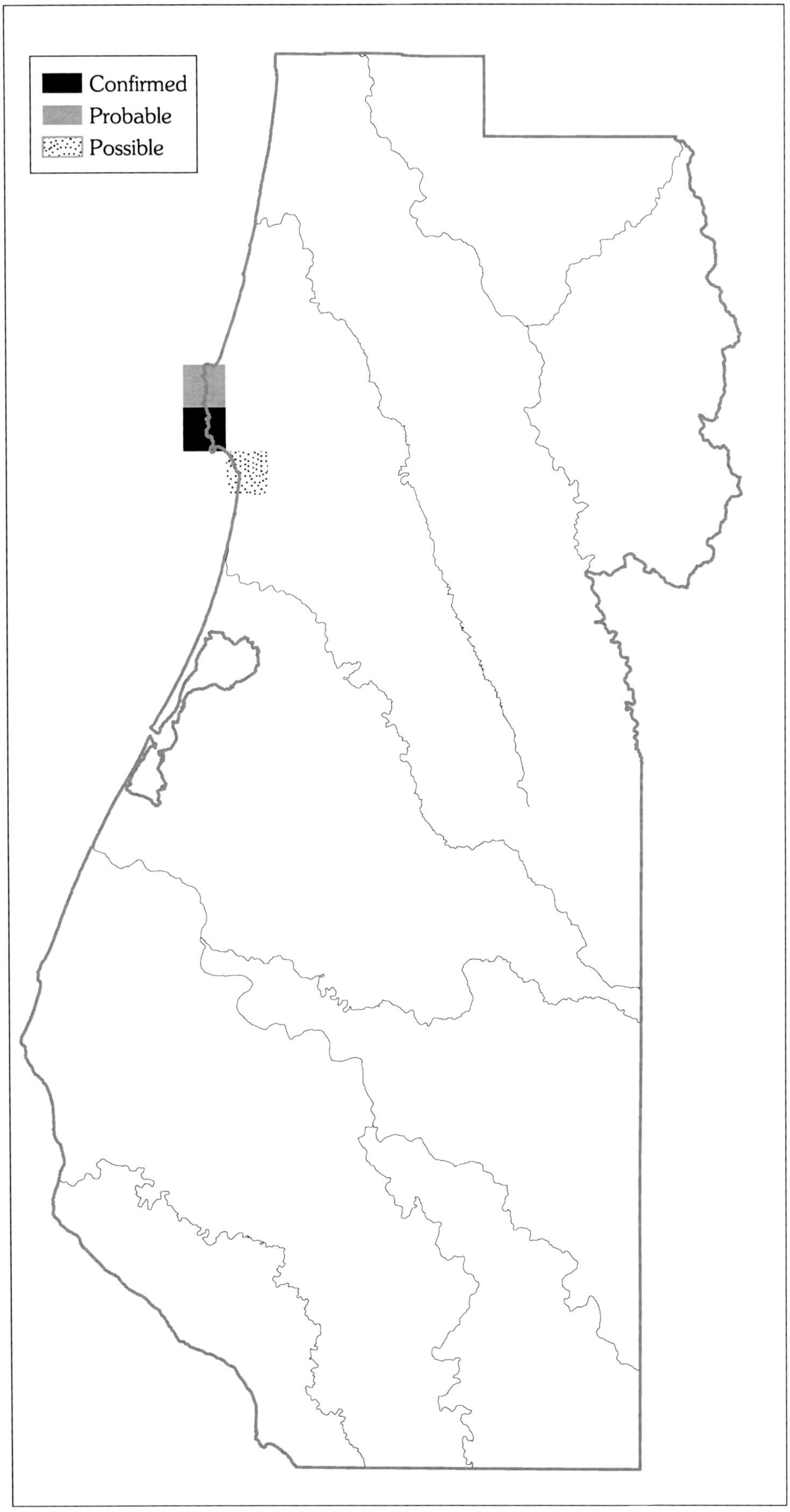

Rhinoceros Auklets occur in some numbers in Humboldt County's offshore waters during much of the year. Their general avoidance of protected inshore settings and their nondescript appearance cause them to be known to few persons other than birders, marine biologists, a few fishermen, and sea-kayaking naturalists. While nearshore incursions of these auklets occasionally bring dozens into the Crescent City harbor in Del Norte County, this phenomenon does not take place within Humboldt Bay. When observed here, "Rhinos" are most often spied through a spotting scope trained upon the ocean beyond the surf. Individuals, pairs, and the occasional small flock may be located with patience and luck. Apparent migrants are sometimes detected flying north or south. They are known to be present in winter (Harris 1996), but are rarely reported at that season, especially from shore points.

Despite observations like those reported by Sowls *et al.* (1980) of multiple birds on and near Green Rock, including some entering burrows, Harris (1996) considered breeding by Rhinoceros Auklets in Humboldt to be unproven. Breeding has also been suspected at Little River Rock (Harris 1974). Considering the historical information and Atlas reports from the vicinity of Green Rock (e10) of apparent pairs on the water 24 Jun 1998 (M. Morris) and 10 Jul 1995 (JEH), of birds entering burrows 4 May 1998 (L. Lester) and 22 Jul 1995 (DFx), and of a juvenile Rhino on the water 22 Jul 1995 (DFx), this species was upgraded to a confirmed Humboldt breeder. While there was limited opportunity for focused seabird surveys during the Atlas period, it was thought that at least a small population was breeding on Green Rock and, possibly, on other rocks and sea-stacks in the Trinidad area during the late 1990s.

Rhinoceros Auklets were once considered extirpated from California (Grinnell 1926), but apparently recolonized their historical range within the past 30 years. They were long believed to breed at only a few locations south of Washington. This situation—or perception—changed during the 20th century after surveys and studies of numerous seabird colonies in Oregon and California. During that time it was discovered that birds had both reoccupied long-abandoned sites and shown up as nesting birds at rocks where they were formerly unknown (Sowls *et al.* 1980).

The dynamic of this species in Humboldt County remains unclear. Sowls *et al.* (1980) noted that nesting by Rhinoceros Auklets was confirmed along the far northern California coast at Castle Rock, Del Norte County, in 1917. Osborne (1972) estimated 100–150 auklets at that site. It is conceivable that at least a segment of the population probably nesting in Humboldt during the Atlas period had become established as a reflection of the general trend of increasing numbers along the northern and central California coast. Harris (1996) mentioned Sugarloaf Island, just off the south end of Cape Mendocino, among a short list of Humboldt sites at which the species was believed to breed. No attempt was made to search for this species at that rock, as a close approach could not be made.

As do Tufted Puffins, Rhinoceros Auklets nest in burrows excavated in sod or soil on offshore rocks. In recent years, concern has been expressed over the ability of large social waterbirds, specifically Brown Pelicans and Double-crested Cormorants, to degrade and denude this substrate while loafing or, in the case of cormorants, nesting. It is hoped that any such loss of soil on islands important to colonies of auklets, puffins, and storm-petrels will ultimately prove not to seriously reduce their regional populations. Like those species, Rhinos also are at risk from oil spills and human disturbance at nest sites.

TUFTED PUFFIN *Fratercula cirrhata*

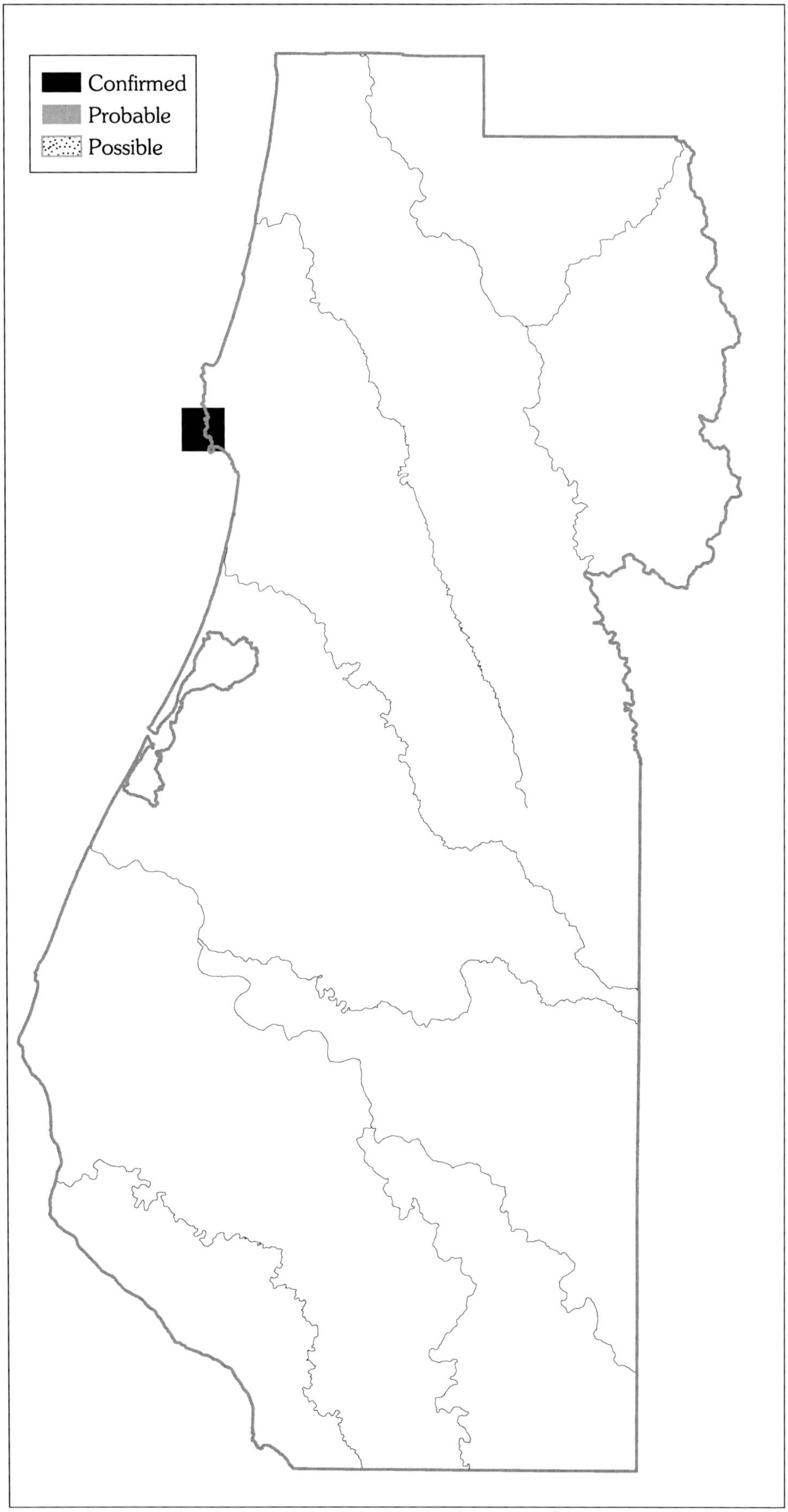

Unforgettably striking, the Tufted Puffin causes all who see it well to marvel at its wispy head plumes and heavy ornamented bill. This bird is a "signature species" of colder North Pacific waters. It is a highly adapted and distinctive member of the Alcidae, a seabird family whose species exhibit a great diversity of size, shape, color, diet, and breeding behavior. This puffin is a characteristic nesting bird of subarctic and northern temperate shores and islands. The southern end of its eastern Pacific range encompasses the rocky coasts and offshore islands of northern California. Here it is a scarce and localized summer resident, occurring in small numbers at traditional sites year after year.

Harris (1996) summarized the puffin in northwestern California as a "locally rare to uncommon summer resident and breeder; perhaps declining." During the Atlas period, confirmation of breeding was established at a single Humboldt site, Green Rock, which stands shortly beyond the surf zone northwest of Elk Head. Green Rock had been known to support a tiny population prior to the beginning of the study. There were no Atlas reports from other sites, although puffins have been reported historically on White, Puffin, Flatiron (Off-Trinidad), and Blank (Sugarloaf) rocks (Clay 1916) and Sugarloaf Island (Sowls *et al.* 1980). Little River (Camel) Rock has also been suspected as a nest site (Harris 1974).

Within the past 150 years, the population of Tufted Puffins in California has diminished. Many places at which they once were known no longer support them. The same may be said for Humboldt County. Certainly, it is generally understood among the Humboldt ornithological community that puffins were observed with greater frequency, and at more sites, in the past. The Tufted Puffin population in the Trinidad area has undergone a decline to near-extirpation. Humboldt's earliest historical breeding confirmation was of a bird removed from a burrow on Flatiron Rock 22 May 1910, where these puffins were described as "breeding in great numbers on Offshore Rock. They breed regularly on the cliff under the Trinidad Light House" (C. Clay; Harris 1996). By 1980, there was an estimated single pair still nesting on Flatiron Rock and none on Trinidad Head (Sowls *et al.* 1980). None have been reported on Flatiron since the early 1990s (G. Lester, pers. comm.). Puffins have also apparently disappeared from Blank Rock (Sowls *et al.* 1980) and Puffin Rock (Harris 1996). Additionally, Harris (1996) noted that the 18 Tufted Puffin eggs in the Davis Collection, procured in 1924–1925, clearly point to a larger population in the Trinidad area in the past.

Osborne (1972) found five pairs of Tufted Puffins still nesting on Green Rock 18 Jun 1970. Marbled Murrelet surveyors working at sea, as well as shore-based observers armed with scopes—and with some familiarity with the species' early-rising habits—produced only a few sightings at or near Green Rock (e10) during the Atlas period. However, they were reported each year from this site. A bird was seen poking its head out of a nest burrow 23–24 Apr 1995 (JEH). One to three puffins were seen there 8–10 May 1996 (S. Harris, R. Ridenhour). Birds were reported 12 and 21 May 1997 (TWL). An occupied nest was reported 4 May 1998 (G. Lester). And finally, in 1999 a bird was seen on the water and flying back and forth by the rock 17 Apr (M. Wachs), and one was observed 6 May popping out of a burrow, flying around, and reentering the burrow (E. Elias, J. Sooter).

At one other historical colony site, Sugarloaf Island, just off the southern terminus of Cape Mendocino, a team of atlasers intent on scanning for puffins in July 1999 met with poor visibility. They could not inspect the rock closely and, in the absence of other reports, it was not determined whether puffins still remained at this location. Judging from views gained from the nearby mouth of Singley Creek, a moderate amount of suitable sod cover appeared to persist on the rock.

The reason for the decline of Tufted Puffin as a breeding bird in Humboldt County is perhaps never to be fully understood. Sowls *et al.* (1980) and Harris (pers. comm.) suggested that loss of soil atop the coastal islands and sea-stacks may be a contributing factor or indeed the main factor. The trampling of fragile turf-and-rootlet caps by the burgeoning numbers of Brown Pelicans that loaf on Humboldt's offshore islands from midsummer to late fall has been implicated. Concern has arisen as well over recent establishment of marine-nesting colonies of Double-crested Cormorants, another species that has rebounded greatly since the domestic ban on DDT was enacted in 1972. While Green Rock was not known to support a Double-crested colony during the Atlas study period, it may be only a matter of time before this occurs. This species' excrement kills terrestrial plants, and heavy rain washes the exposed soil away. Other threats to this species include human disturbance at nest sites, oil spills, and a reduced prey base due to overfishing. Decline of the sardine (*Sardinops sagax*) fishery may be another factor in puffin declines in California.

ROCK DOVE *Columba livia*

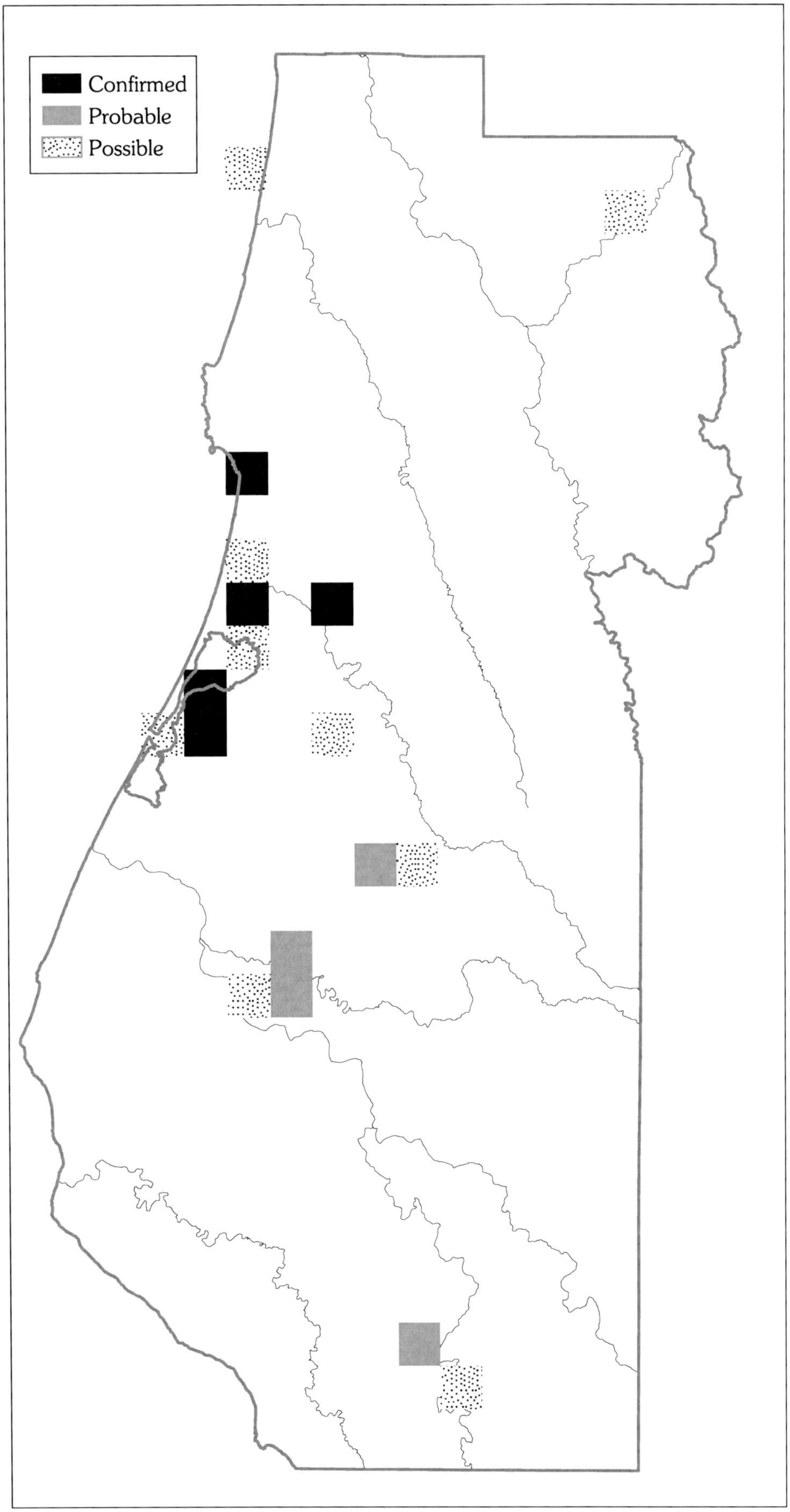

Feral Rock Doves of European ancestry are widespread throughout temperate North America, relying upon the hand of man to persist and flourish. When Rock Doves arrived in Humboldt County is unknown (Harris 1996), but the species was first recorded nesting on Little River (Camel) Rock in June 1970 (S. Harris, pers. comm.; Harris 1974). Pigeons from one or another source can be assumed to have been present since modern settlement and commerce began in the mid- and late nineteenth century.

Rock Doves were noted at several widely scattered locations in Humboldt County during the Atlas period. The map may suggest a more or less evenly distributed population around Humboldt Bay, yet this is not the case. Birds were found to be concentrated in large numbers only in and about Eureka. Flocks of dozens may be seen there, assembling at favored feeding sites such as the Nilsen Company feed store and around the sidewalks and waterfront of Old Town. In the latter area, the small plaza at 2nd and F streets daily attracts an outsize share. Flocks also resort to backyard feeders in Eureka where allowed to do so. The many corniced and decorated Victorian buildings offer suitable nest sites for pigeons. These birds also nest and shelter in nooks beneath the Samoa Bridge. Although the Eureka population of Rock Doves is not large—perhaps some hundreds of birds—its members are constantly in sight, imparting the feel of a larger city center. Rock Doves in these urban situations seem to occupy habitats that are usable by only a few native species, and they help feed the local Peregrine Falcons.

Outside Eureka, Rock Doves are only occasionally noted in urban settings. A few do breed in Arcata; for example, a nest with young was at Humboldt State University (f14) 17 Mar 1995 (L. Doerflinger; earliest confirmation). However, they seem unable to establish a major presence there, as they are not seen on sidewalks or even on the central plaza in that city, nor in McKinleyville. Somewhat surprisingly, no reports came from Fortuna, a town seemingly large and old enough to sustain pigeons. Lone Rock Doves are not a surprising sight anywhere in the outlying coastal lowlands, especially farther from the coast and near farms, dairies, and ranches.

Atlas fieldwork also revealed that Rock Doves were present in several more rural locations. They were detected in or near the smaller communities of Blue Lake (h14), where a nest with young was reported 1 May 1997 (K. Nelson); near Carlotta (g23), where birds visiting a probable nest site in a bridge over the Van Duzen River were observed 1 Jun 1997 (JEH); and at Redway, Garberville, and Orleans. Pigeons seen at Kneeland and to the southeast of there were associated indirectly with livestock operations. A group of several dozen was commonly seen at the Moxon Dairy (f14) in the Arcata Bottoms. In such places, they may have benefited from the protection and nest sites afforded by people, even from actual pigeon cotes or lofts built for them. Rock Doves were noted as continuing to breed at outer coastal sites. Nesting was detected in the rocks above Moonstone Beach (F11) 22 Apr 1995 (JEH), and atlasers observed a pair on Little River Rock (F11) 31 Jul 1998 (JEH, JCP, DFx, J. Jacobsen).

Some of the dates above suggest that Rock Doves initiate breeding early in the season. The latest breeding confirmation during the study was a nest with young 4 Jul 1995 (g14; KMS). Courtship may be witnessed at almost any time of year. Examples of breeding in Eureka (e16) included nest building 12 Apr 1996 and 2 May 1995 at 123 F Street (L. Bacon-Ogden), and an occupied nest 27 May 1999 in the attic of the old Daly Building (G. Lester).

BAND-TAILED PIGEON *Patagioenas fasciata*

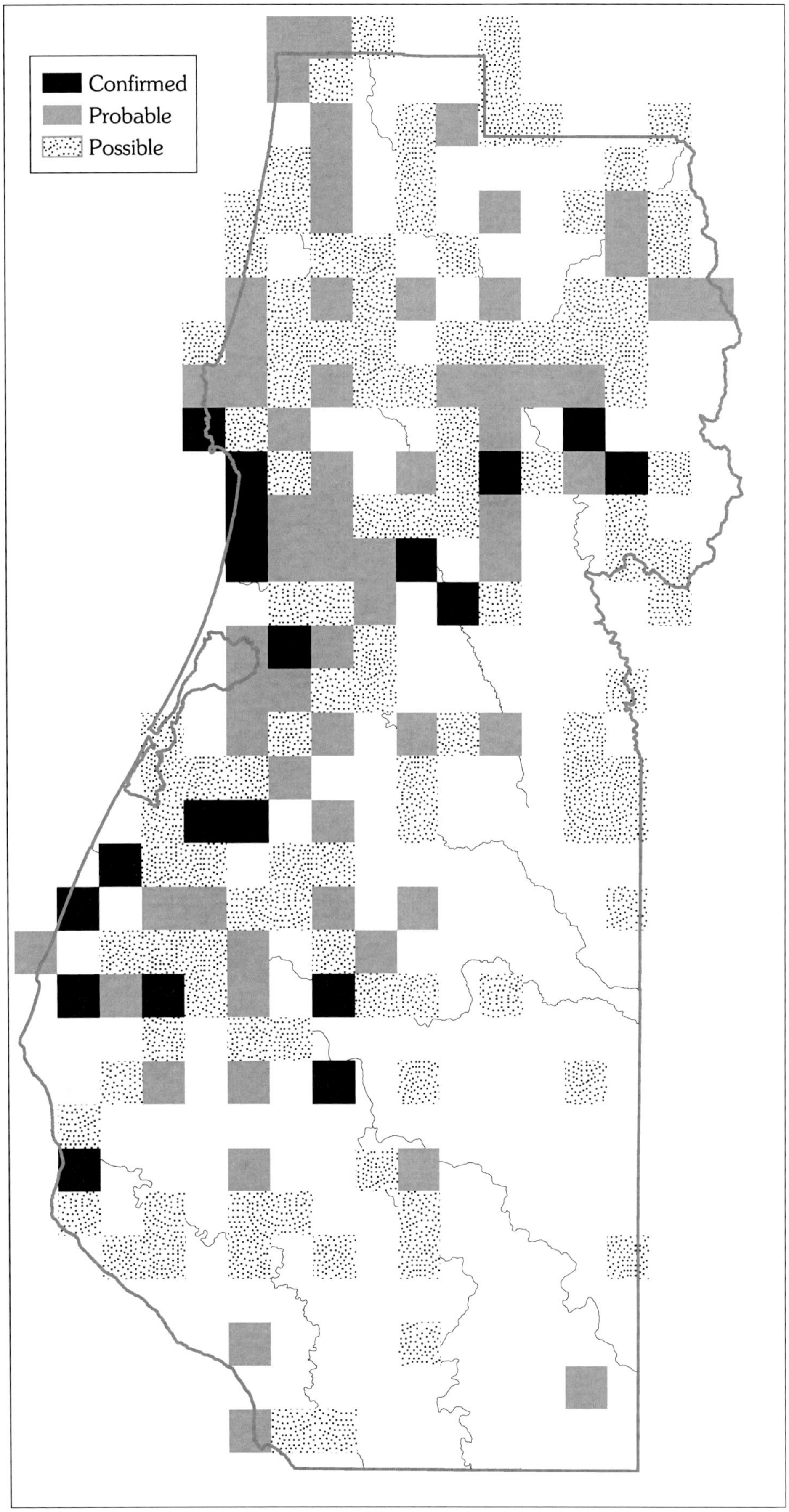

Band-tailed Pigeons are relatively common in many portions of the western U.S., and occur more sparingly south into South America. These chunky, purplish-gray birds are regularly seen speeding across forested canyons of the Pacific Slope in search of berries, hardwood mast, and other seeds. Shy, nervous, and vigilant, Band-tails are typically "backwoods" birds. In California, they are less commonly seen in lowland cities and occur in residential neighborhoods only locally, as where a frontier of unmanaged countryside meets the outer suburbs or where there is an especially attractive bird feeder. They are most numerous in rural foothills and lower mountain canyons forested with oaks, madrones, and scattered tall conifers. Considering their bulk, these pigeons are remarkably inconspicuous, flocks often feeding undetected overhead until the muffled slap of wingbeats is heard. Were it not for the resonant cooing of males perched high in treetops, this species would be even less readily detected.

Northwestern California's extensive and diverse forests, including regrowth timberland and closed-canopy oak woodlands, support their share of Band-tailed Pigeons. In Humboldt County, this offering comprises both conifer- and hardwood-dominated tree communities. Atlas surveys revealed that, during the 5 years of the project, Band-tails could be generally expected in most of the northern third and the western sections of the county. Forests and managed timberland in the fog zone were well inhabited, and this zone of fairly thorough occurrence spilled eastward into the middle elevations of the interior. The Klamath, Trinity, South Fork Trinity, and main Eel/South Fork Eel river drainages were well represented. Altogether, Band-tailed Pigeons were encountered in 45% of all blocks. They were confirmed as breeding in 10% of the blocks with detections and established as a probable breeder in 33%. Of interest is that no confirmations came from the northern one-quarter of the county, nor from interior east-central or southeastern Humboldt, both regions where comparatively few birds were detected overall.

Among the 63 blocks with probable breeding status, the observation of pairs was the highest-ranking evidence obtained in 39. Courtship displays or copulation were reported as the highest evidence in 10 Atlas blocks and spanned the dates of 11 May 1997 (F09; M. Morrissette) to 29 Jul 1995 (e10; JEH). The displays, in which males launch themselves from an exposed high perch and circle on quivering wings, uttering a scratchy call, are an arresting sight. Yet many observers may misinterpret these short flights as movements made while foraging. Only two nests were reported, probably because these pigeons usually nest in dense foliage high in trees. One nest had eggs 22 Jul 1997 (n10; *fide* JMH), and the other was occupied 8 Sep 1997 (o11; *fide* JMH). The latter record and observations of courtship into late July indicate that this species has a prolonged nesting period in Humboldt. When seen clearly, young birds are readily distinguished by their dark-tipped bills and the lack of a white band across the nape. This played an important role in establishing breeding by this species, as fledged young accounted for 15 of the confirmations. Fledged young were seen as early as 19 May 1997 (B23) and as late as 24 Aug 1997 (F19; both COg).

Band-tailed Pigeons are resident in Humboldt County throughout the year. However, they are clearly migratory, since they are much more numerous from March into October. After the extended breeding season, these pigeons gather in flocks that assemble daily, sometimes for weeks, in the vicinity of stands of madrone, oak, red elderberry, Pacific dogwood, and cascara. At those times, individuals, pairs, and flocks may be seen flying across canyons and along ridgelines as they commute to and from foraging sites.

Historical and recent winter records in Humboldt County suggest patchiness. Withdrawal of birds from areas not rich in concentrated food resources is indicated by seeming absence from many places where they are found in summer. Single birds, pairs, and small flocks occur in lower-elevation oak and madrone stands, particularly in the Willow Creek area during years of high madrone berry production. Harris (1996) stated that, although Band-tailed Pigeons are generally present in scattered small flocks in winter, large concentrations sometimes are found. He cited as examples reports of 2000 near Thorn Junction 10 Jan 1975 (R. Sutherland) and 2000 on the 1982 Willow Creek CBC. Elsewhere, few are reported, other than from favored backyard bird feeders in the near-coastal lowlands.

With general reference to northwestern California, Harris (1996) characterized Band-tailed Pigeon as a "common migrant, summer resident, and breeder; rare in winter. Appears to have declined in recent years." However, assessing population trends is difficult given the species' highly nomadic nature. Atlas results suggest that this species may be considered an uncommon to fairly common breeding species over most of Humboldt County, except for the drier, more open oak-grassland landscapes of the southeastern portion of the county. Since Band-tails show a positive association with hardwoods (Ralph *et al.* 1991), monoculture forestry practices that remove hardwoods in favor of conifers probably have had an unfavorable effect on this species in northwestern California. Glover (1953a), Houston (1963), and Drewien *et al.* (1966) studied the biology of this species in Humboldt and provided extensive details on breeding chronology, physiology, behavior, food habits, and reproductive output. Davis (1938) documented some early nesting records for Humboldt, including the earliest historical breeding confirmation, a nest 23 May 1923 south of Eureka (J. Davis).

MOURNING DOVE *Zenaida macroura*

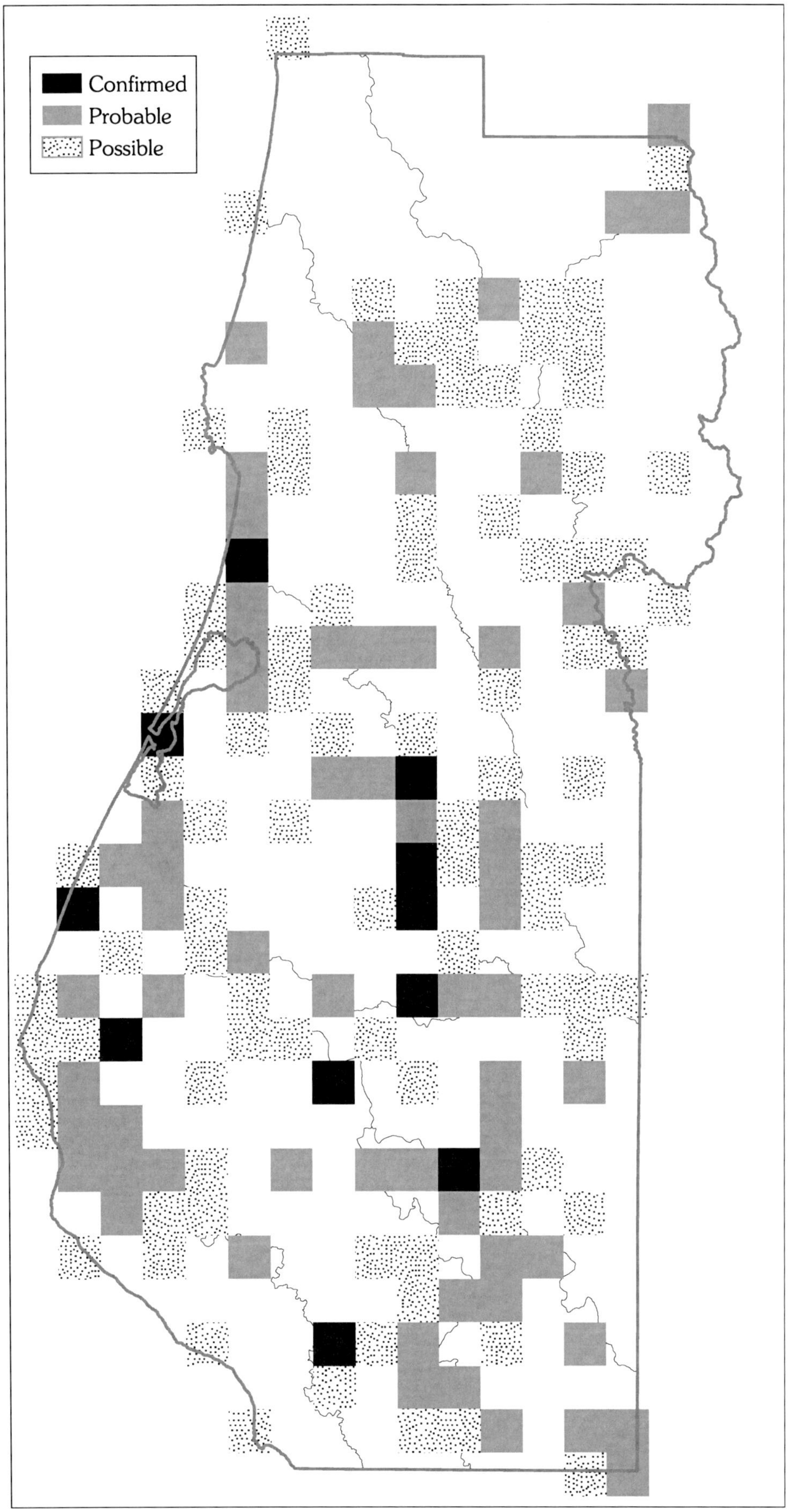

Open spaces providing the seeds of grasses, forbs, or cultivated grains would seem to be nearly all that is necessary for Mourning Doves to prosper. The clearing, simplification, or outright replacement of natural habitats by humans across the past several centuries has caused a host of native North American birds to diminish, but this species is not among them. They are common or abundant (at least locally) in each of the lower 48 states and in much of southern Canada, particularly in the Prairie Provinces. Able to nest on the ground or in trees, they accept treeless terrain and, though drawn to water for drinking, may be encountered in abundance where it is scarce. Mourning Doves offset a small clutch size (two eggs) by nesting as many as five or even six times per year, this occurring more commonly in the southern U.S. Such high fecundity supports a bountiful harvest of many millions by hunters each fall.

Offering great expanses of open or semi-open countryside, nearly everywhere rich with both native and alien grasses and forbs, California hosts huge numbers of Mourning Doves. Yet sizable sections of the state, such as the higher mountains, are comparatively "dove poor." Humboldt County might be described as falling somewhere in between these two situations. Although certainly well represented in Atlas results, Mourning Doves exhibited a somewhat patchy pattern of occurrence. Most of Humboldt seemed to support at least some Mourning Doves, but there were areas within the county from which there were only scattered reports, and in extreme northern Humboldt this species was virtually undetected. Atlas surveys revealed the presence of Mourning Doves in 39% of the 425 total blocks, with slightly more reports from the southern portion of the county. Near-absence of doves was most evident in the industrial forestlands near the northern end of the county and in the heavily forested mountains of northeastern Humboldt.

Mourning Doves were encountered in a wide variety of habitats. Indeed, to describe each of them might become an exercise in listing all significant non-forested habitat types in the county. Grassland of any description was well inhabited, ranging from wind-whipped slopes within earshot of the ocean to sere, sun-baked uplands dotted with occasional oaks or rock outcrops at middle elevations in the interior of the county. Birds were noted in river corridors, on gravel bars, in coastal dunes, in farmland, ranches, towns, cities, oak woodland, and forest edges. These birds were more numerous during the Atlas in the less forested sections of the county, becoming less common and less anticipated where conifer forest is the dominant cover. Mourning Doves seem distinctly out of place both in the big redwood parks and in commercial timberland. They occur only sparingly in clearcuts, perhaps as a result of the typically rapid succession of logged sites to dense brush and sapling trees, which quickly precludes the growth of the seed-producing herbs and forbs upon which these birds depend.

Considering that this species is conspicuous, easily identified, multiple-brooded, and reasonably common over much of Humboldt County, it is intriguing that breeding confirmations were reported in only about 7% of blocks with detections. All 11 blocks with confirmations were in the southern or western part of the county. Ten involved observations of fledglings, ranging from 27 May 1997 (k27; JG) to 23 Aug 1998 (j18; C. Keil). The remaining confirmation was of a nest with eggs on the ground beneath lupines near the cypress grove at the base of the north jetty of Humboldt Bay (D17) 27 May 1996 (GAS). In contrast to the paucity of confirmed breeding reports, probable breeding was reported in 65 blocks, three-quarters of these based on observations of pairs. Other than a few pair reports, the earliest probable breeding report was of courtship 14 May 1995 (H15; PFS). Reports of probable breeding were quite thoroughly distributed within the general mosaic of detection-blocks.

The Mourning Dove has long been recognized as a common resident of Humboldt County, most numerous in the warmer months, less common and localized near food sources in winter after many birds have migrated south. Referring broadly to northwestern California, Yocom and Harris (1975) termed the species an "uncommon migrant, summer visitant and breeder; rare winter visitant," found in "agricultural lands, cutover timberland, grassy ridges inland, and riparian bottomlands." In an expanded summary treatment, Harris (1996) characterized it as a "common summer visitor and breeder; locally rare to uncommon in winter. Scattered pairs breed in open and woodland habitats, most commonly in the drier, southern inland valleys supporting oaks."

BARN OWL *Tyto alba*

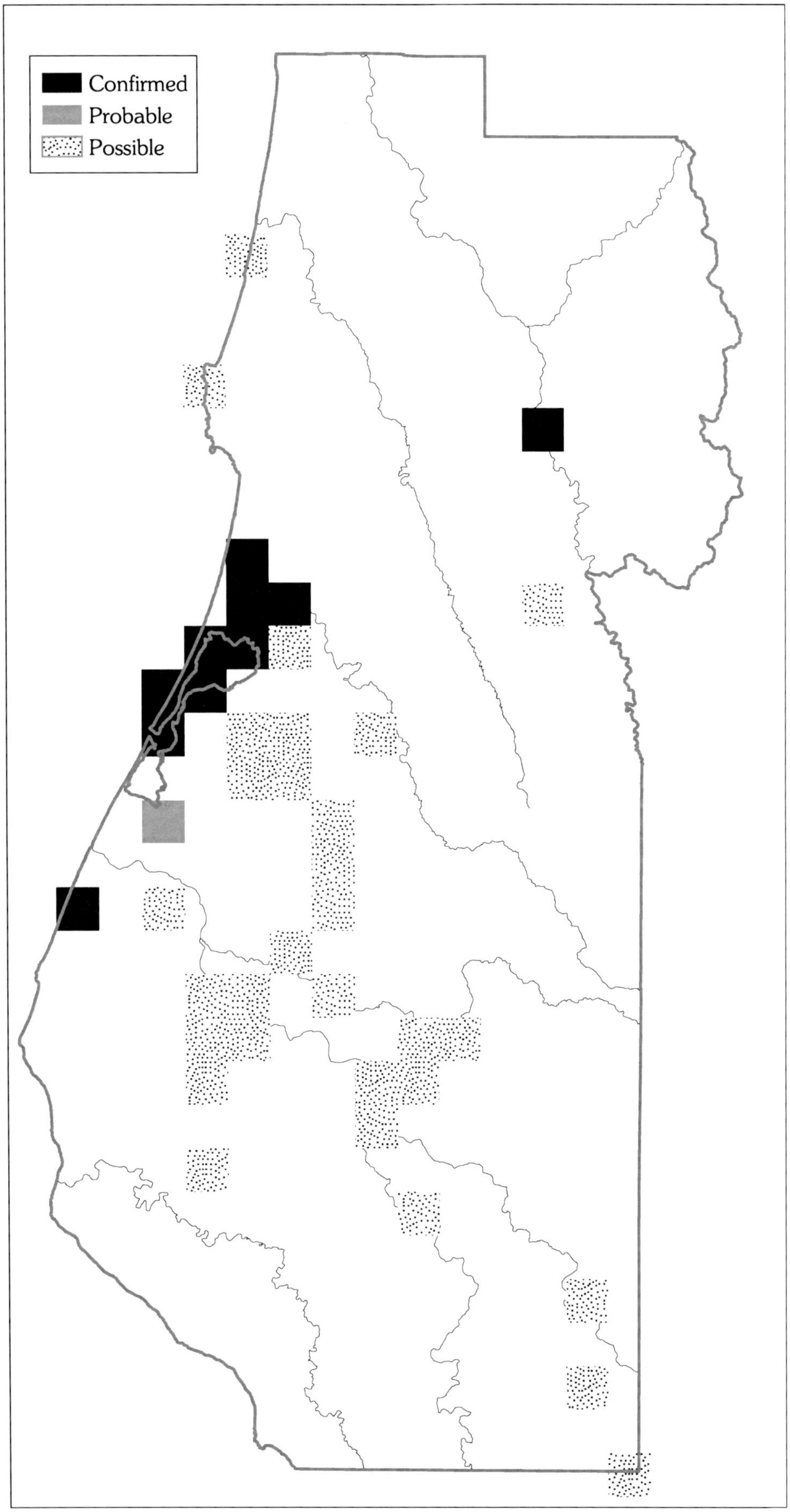

Barn Owls of one species or another are found throughout most of the world's temperate regions. They occur widely across the United States and locally in southern Canada. Found commonly over much of lowland California, they are best represented in open country. Densely forested areas and the higher elevations appear to be avoided. The ghostly form of a Barn Owl about to flush in apprehension from a telephone line or floating wraith-like across an old pasture is no great surprise to those who drive rural roads in the bottomlands of Humboldt County. Their unique call-note is frequently heard at any hour of the night as individuals commute to and from foraging sites, passing over cities, neighborhoods, and pastoral countryside. Foraging occurs over pastures, farmland, oak woodland, and on bald ridgelines, as well as within industrial sections of cities, around the edges of residential neighborhoods, and in outer-coastal dunelands.

Barn Owls were detected during the Atlas period chiefly around Humboldt Bay, along the tributaries of the major streams draining into the lower Eel River, in the broken forest/grassland mosaic of the upper Bear River-Mattole River country, and in the mid-elevation oak woodland of extreme southeastern Humboldt County. Breeding was confirmed in eight near-coastal blocks extending from McKinleyville to Centerville Beach, with a distinct cluster centered on North Humboldt Bay. For example, one nest with young was found 17 Jul 1997 in a hole in the bank of the Mad River at the end of School Road (F13; M. Wilbourn) and a used nest was found under the Samoa Bridge (e16) 1 Sep 1998 (T. Ash). Among only two detections in the northeastern portion of the county was a confirmation near Hoopa, where fledglings (the earliest reported) were seen near the nest under the Highway 96 bridge over the Trinity River (m10) 26 Jun 1997 (JMH). Like other owls, these start breeding relatively early in the season:

the earliest record of confirmed breeding was of an occupied nest 23 Apr 1996 (F13; J. Kidd). Although some detections, particularly in the dairylands, were made by searching old barns, most records of possible breeding were nighttime detections.

Barn Owls appear to be either absent or in very low numbers in the commercial timberland of interior northwestern Humboldt as well as in the mountains. Their seeming absence in these places, as well as in the rugged and largely forested hills of southwest county, may be explained by the comparative scarcity of old barns and similar structures. However, a great many Atlas blocks outside areas intensively surveyed for Spotted Owls did not receive sufficient searching for Barn Owls, as in most instances this was impractical. One intuits that scattered pairs of Barn Owls probably remained undetected in many blocks where habitat exists.

Historically, this species has been well known in Humboldt County. Wilder (1916a) considered it "rather common in the dairy region of Humboldt County." The first recorded breeding was a nest with eggs in Eureka 19 April 1924 (J. Davis). Harris (1996) characterized Barn Owl in northwestern California as a "common resident and breeder in open habitats, particularly associated with agriculture and pasture lands." In the course of modern settlement of the region, extensive clearing of lowland forest cover and subsequent establishment of the dairy industry surely increased potential habitat for Barn Owls. Barns and outbuildings associated with dairies and local agriculture provide nesting and roosting sites for these birds. Abandoned or unmanaged structures are favored, as they offer access holes, a selection of roosting eaves, and the dim recesses of rough-cut haylofts built in a bygone era.

Because of their large size, gradual "owl-friendly" decay, and location among or near productive foraging grounds, Humboldt's many old barns continue to serve the owls. With the passage of years, an increasing number of the county's big barns have become dilapidated, have met a fiery end, or have been storm-thrown. Barns built with modern materials and techniques would appear to afford fewer opportunities for Barn Owls.

FLAMMULATED OWL *Otus flammeolus*

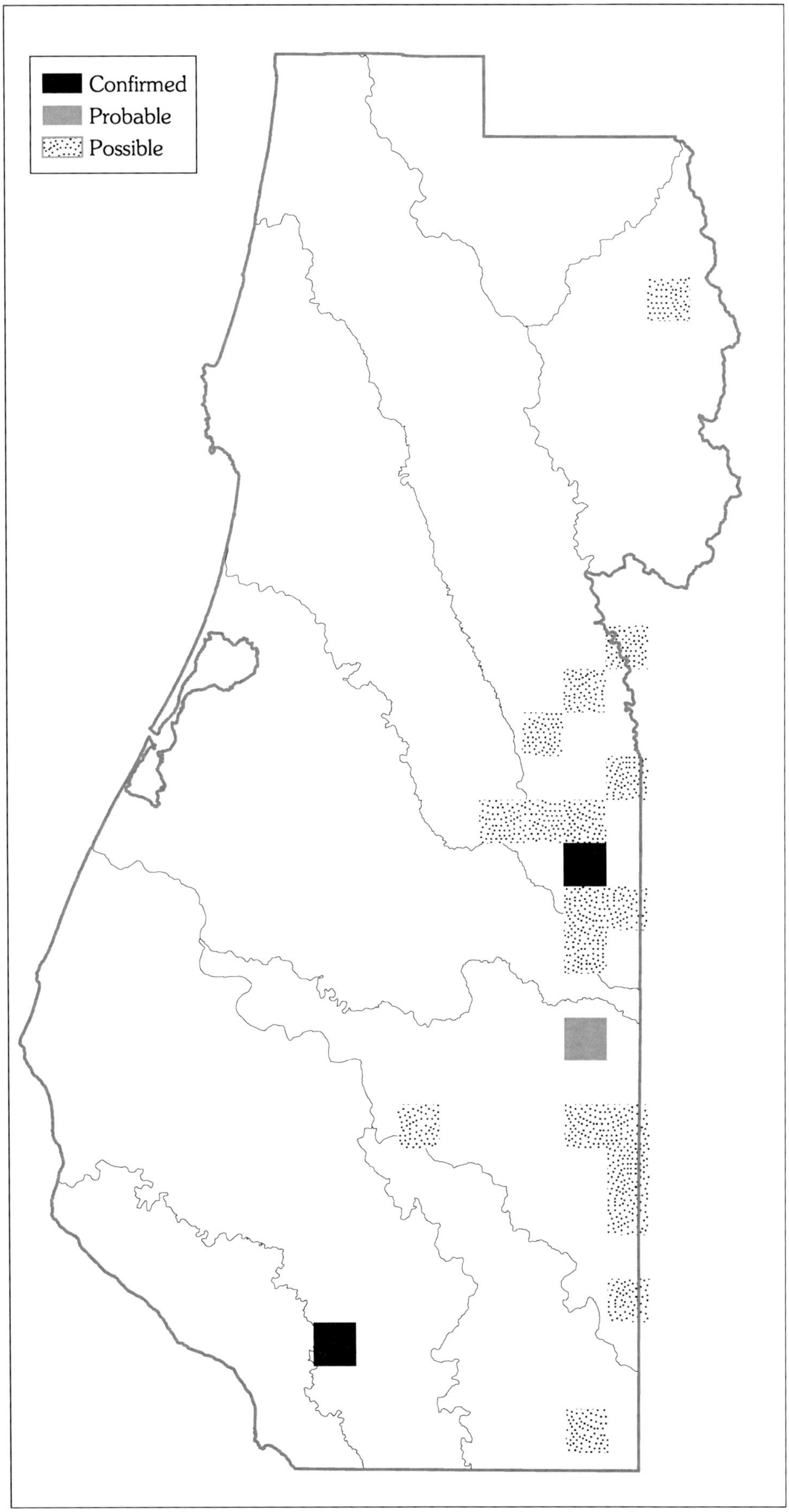

The voice of the Flammulated Owl is an odd night sound of western mountains. Beginning well after sunset, one becomes aware of a soft but persistent hooting, consistently at the same low tenor note: *Oop . . . er, Oop . . . er, Oop. . . .* The grace-note and announcement amount to but a muffled punctuation of the breath of air through tall pines, yet it is a sound bringing satisfaction to those who are aware of these reclusive owls.

"Flams" are small, dark-eyed gnomes cloaked in a complex mixture of warm and somber feathers. They were once considered quite rare. But thanks to the efforts of researchers and birders over the past several decades, we now perceive the Flammulated Owl to be a variably common summer resident in most of the major mountain ranges of the western U.S. and locally in southern British Columbia. Much of what is known about the distribution and comparative abundance of Flammulated Owls in northwestern California is due to the original work of Marcot and Hill (1980), who compiled records and conducted surveys for these birds in Humboldt and Trinity counties. They tallied 38 territories in Humboldt County. They further reported that forest structure—an open understory and down logs, and the presence of either Jeffrey/ponderosa pine, California black oak, or either pine along with oak—were consistent habitat elements in their study area. Abundant nocturnal insects such as adult boring beetles, crickets, and large moths, as well as the presence of disused woodpecker nest or roost holes, are also important components of good habitat. Flammulated Owl was considered a rare to uncommon, local summer resident and breeder in northwestern California by Harris (1996).

During the Atlas period, small numbers of Flammulated Owls were detected in appropriate forest habitat at middle and higher elevations near the eastern edge of Humboldt County.

Forests inhabited by the owls exist chiefly at higher elevations on Snow Camp Mountain, on the northern one-half to one-third of South Fork Mountain, in and around the Lassics, and locally on Island Mountain. Outlier detections were made up the North Fork of Red Cap Creek in the Klamath River canyon south of Orleans, on Fruitland Ridge, and at Briceland. Additional surveys aimed specifically at Flammulated Owls would probably show that a greater portion of interior southern Humboldt County is inhabited by Flams than indicated by the distribution map. This owl is known to occur in summer as low as 1800 feet elevation in somewhat xeric oak-pine stands near Willits, in northern Mendocino County (DFx, pers. obs.). Whether similar sites within the great array of mixed oak-and-conifer forests scattered across the warmer regions of Humboldt County support Flammulated Owls remains unknown.

On South Fork Mountain and its many significant side ridges most of the trees are white fir, Jeffrey pine, incense-cedar, and Douglas-fir. Blake Mountain, which is actually part of South Fork Mountain, contains some of the largest stands of Shasta red fir in Humboldt County. California black oak is an abundant tree in the drier forests and woodlands at middle elevations across much of southern Humboldt County, but little is known of the possible use of oak-dominated stands by Flams there. From this westernmost portion of its California range, suitable habitat would appear to extend locally, but more or less continuously, eastward into adjacent Siskiyou and Trinity counties.

If it is true that Flammulated Owls cannot endure prolonged separation from their arthropod prey, then the reason for their winter absence from the mountain forests where they breed is apparent. They are known to migrate south from the U.S. and their limited range in Canada in fall, returning in April and May. Little is known of the seasonal life history in Humboldt County. During the 5 years of the Atlas project, most detections were of single birds at night, many of which were recorded during the course of Spotted Owl surveys. The span of dates for detections of possible breeders (18 blocks) during the Atlas period was 9 May 1998 (N33; DFx) to 2 Aug 1995 (N19; K. Baker). As these owls are occasionally encountered at desert oases in the Great Basin through much of May, it appears reasonable to suspect that some continue to arrive or pass through Humboldt during that time.

Probable breeding was recorded in only one Atlas block: copulation was observed at dusk in the low branches of a white fir on top of Buck Mountain (n24) 9 Jun 1995 (KMS). Breeding was confirmed in only two blocks. A nest with at least three young in a mixed-species montane conifer stand was observed 7–8 Aug 1996 near Severin Opening on Pilot Ridge (n20; COg, J. Bettaso). The nest cavity was at 3900 feet elevation, about chest high in a black oak. This nesting occurred within the expected zone of summer residency of Flammulated Owls. A juvenile owl foraging in an atlaser's garden near Ettersburg (H31) 14 Aug 1995 (R. Sutherland) was a surprise. Since this site is only about 10 miles inland from the ocean coast, some distance west of more regular Flam haunts, this bird may have actually been some distance from its natal area. However, since a nest was found not many miles to the northeast near Phillipsville in 1988 (Harris 1996), this record was retained.

Much of the mountain and higher foothill forest inhabited by Flammulated Owls in summer in Humboldt County lies on the Six Rivers National Forest. "Sanitization" of forests through removal of dead or dying trees, salvage logging, and exposure of white fir snags to the elements through the creation of forest edge poses possible long-term threats to these and other birds.

WESTERN SCREECH-OWL *Otus kennicottii*

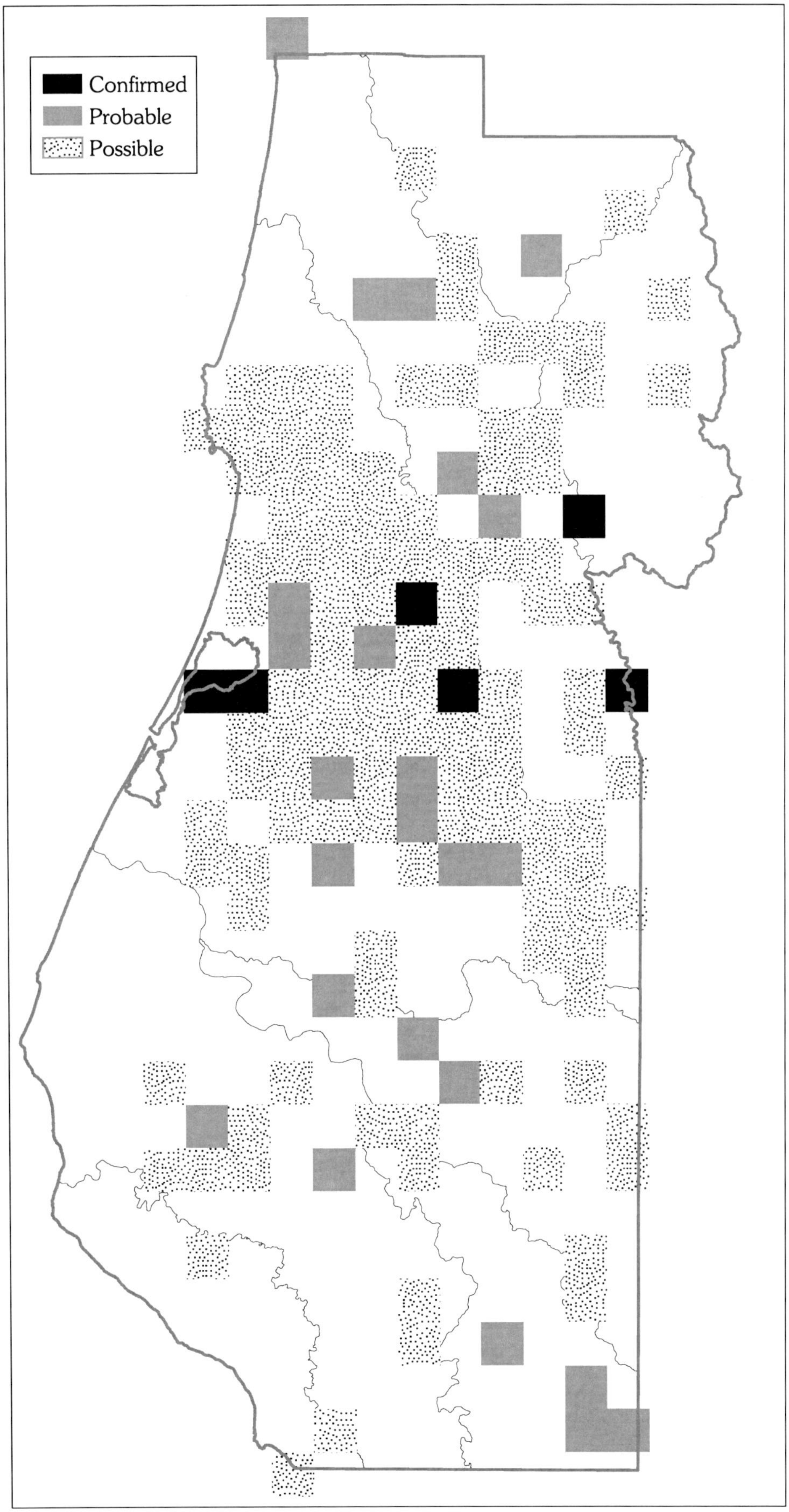

Small owls of many genera have found environmental niches throughout most of the world. Some are rare and local and remain poorly known. Others are common, widespread, and well studied. Clearly to be included within the latter group is the Western Screech-Owl, a nocturnal predator resident from southeastern Alaska to central Mexico. This species has been known to ornithologists as a bird of northwestern California since the late nineteenth century, and is considered a common resident and breeder (Harris 1996). There can be no doubt that these birds were well known to Native Americans long before that time, as their tremulous hooting is a characteristic nighttime sound of broken forests, woodlands, and oak groves throughout much of Humboldt County. During the Atlas period, Western Screech-Owls were detected in 141 (33%) of the 425 total blocks. Breeding was confirmed in six blocks.

Like most owls, Western Screech-Owls are difficult to locate during daylight hours. Although some were encountered during the Atlas period by investigating the uproar created by mobs of Steller's Jays first on the scene, most detections involved single birds and pairs heard at night. Because visits to most Atlas blocks by Eureka- or Arcata-based observers could be made in a day's time and required no camping, little owling was done. However, sufficient effort was undertaken to illustrate that screech-owls are found widely in Humboldt County.

The prominent cluster of blocks with detections of Western Screech-Owls centered in north-central Humboldt can largely be attributed to numerous incidental detections of screech-owls during Spotted Owl surveys by Simpson Timber Company and Pacific Lumber Company biologists. The screech-owls reported by these biologists were heard or seen without deliberate attempts to detect them. Nearly all of these reports were of single owls. Rather than representing the results of extensive search for screech-owls, the contiguous cluster of blocks with detections suggests how commonly the birds may be found—even when not specifically looked for. The region covered at night by Spotted Owl surveyors might appear to offer some special appeal to Western Screech-Owls, but assuredly this is not so. In reality, it is altogether a fairly nondescript collection of habitats, characteristic of much of Humboldt at lower and middle elevations. Sites where screech-owls were reported on managed forestlands comprise a spectrum of vegetation types, seral stages, slope position, and exposure. Both conifer- and hardwood-dominated stands were represented.

It is the opinion of the more active local owlers that the map does not represent the actual distribution of Western Screech-Owls in the county. It is likely that there are screech-owls in nearly all of the blocks at middle and lower elevations which support stands of trees edged or interspersed with grassy or weedy open ground. These owls are habitat generalists, getting along well in both managed and unmanaged landscapes, provided there are trees with cavities of size and shape sufficient to offer nesting opportunities. Both conifer- and hardwood-dominated forest edge and woodlands are believed to be widely occupied in Humboldt. This perception was supported when a deliberate nighttime attempt to locate screech-owls on 23 Jun 1998 resulted in the detection of 23 birds in both coast redwood- and oak-grown habitats from Korbel to Bald Mountain (DFx).

General owling along back roads throughout the county prior to and during the period of the Atlas study has shown that Western Screech-Owls appear to saturate available habitat. Prime oak-and-grassland mosaic may support several calling owls at various points of the compass simultaneously audible from one spot. With this borne in mind, it becomes an inescapable assumption that the population within the county probably runs well into the thousands of birds. There is some evidence that these screech-owls have increased in California over the past 100 years (DeSante and George 1994).

The earliest male recorded calling on territory for at least a week was on 3 Apr 1996 (n32; M. Bailey). Two nests with eggs were found in Wood Duck boxes, but no details were provided (e16, f16; P. McNeil). An occupied nest was reported 15 May 1998 (k16; K. Nelson, G. Bloomfield) and a nest with two young was observed 14 Jul 1997 in a California black oak at Horse Linto Campground (n12; TWL). Fledglings were reported 23 Jun 1998 (j14; DFx) and 14 Aug 1995 (o16; P. Carlson).

GREAT HORNED OWL *Bubo virginianus*

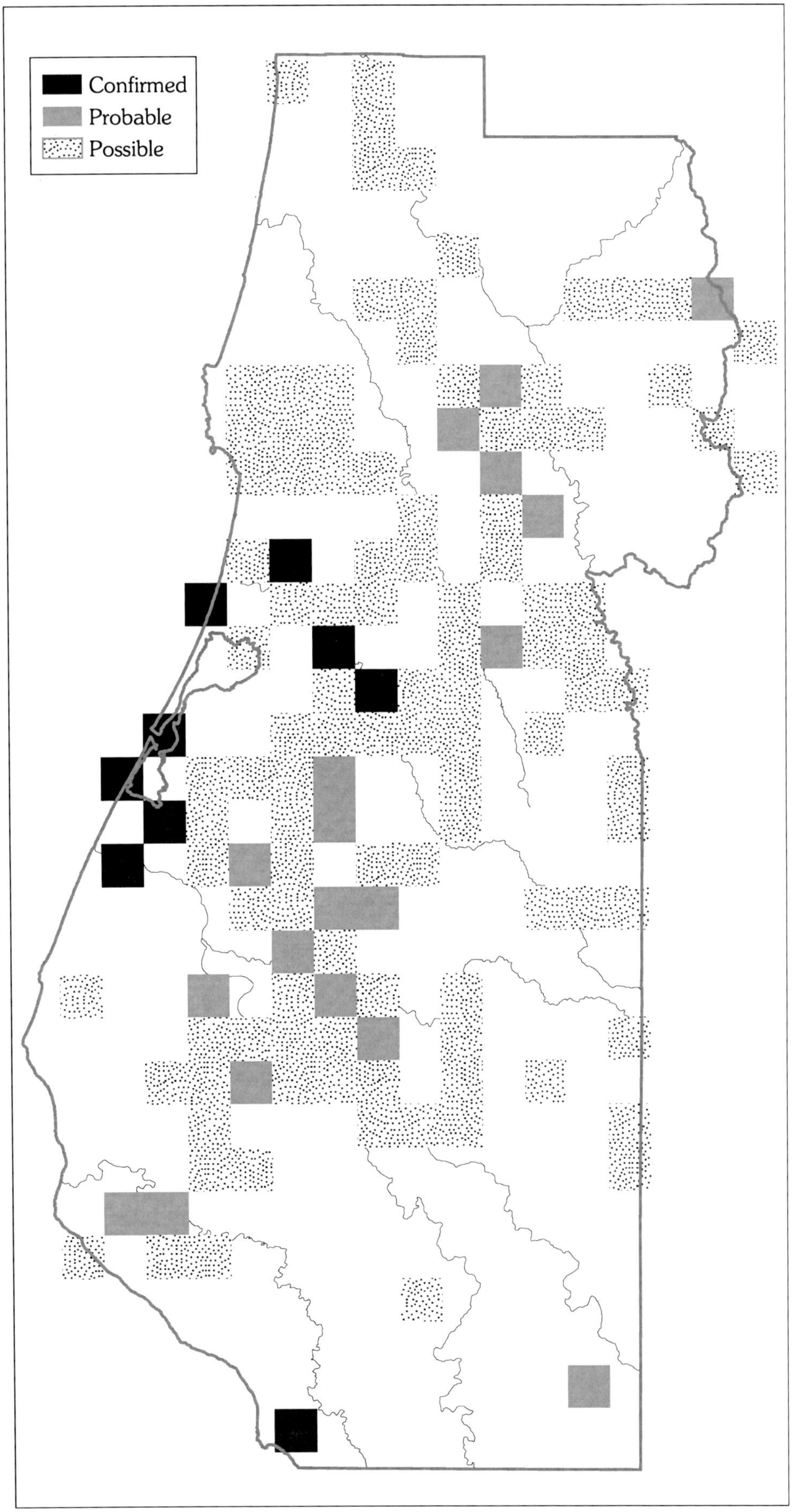

A powerful build, nasty talons, near-fearlessness, and the ability to take a great variety of prey species in any terrestrial habitat has spelled success for the Great Horned Owl. This species has one of the greatest ranges of any owl, extending from the fringes of the Arctic south through the Americas to Tierra del Fuego. Forests, grasslands, deserts, and city parks all host Great Horned Owls, suggesting the spectrum of cover types exploited by these birds. This species has been recognized as a regular resident of Humboldt County since the beginning of ornithological investigation in the area and—while not seeming especially abundant—is nowhere unexpected. Harris (1996), referring to northwestern California in general, stated that Great Horned Owl was a "common resident and breeder." In the course of Atlas fieldwork, Great Horned Owls were encountered in 32% of all blocks, with breeding confirmed in nine blocks and evidence of probable breeding obtained in 19 others.

As with nearly all other owl species, most detections of Great Horned Owls during the survey period were made by Spotted Owl surveyors. The distribution map indicates a pattern of occurrence broadly similar to that of some of the other owls, with a disproportionate share of reports from private timberland in extreme northern and west-central Humboldt County, the Six Rivers National Forest along the central-eastern fringe of the county, and on or near the Hoopa Valley Indian Reservation. Scattered reports, both from within the general region of timberland owl surveys and well outside it, were contributed by "owling birders" or those fortunate enough to encounter the birds by day. Comparatively few overnight Atlas field trips to areas of the county remote from Arcata and Eureka were made, so encounters with nocturnal owls by atlasers were more by happenstance than by intention. Absence or scarcity of reports from sections of Humboldt such as the near-coastal "mountain prairies", the northeastern corner, and across the far southern part probably is largely due to this. Because this owl is widely recognized as occurring in nearly all terrestrial habitats, there is no reason to suspect that the "white areas" on the distribution map reflect actual absence of these birds.

Given the disproportionate amount of nighttime coverage in some parts of the county, the breeding distribution map may appear as misleading as it is enlightening. Surely, there is nothing remarkable or amazingly attractive about the habitats in which these owls were recorded on private, tribal, or federal timberland. Indeed, much of this habitat is, if anything, less diverse than that in outlying sections; oak-dominated woodlands and riparian bottomlands, in particular, are places not commonly surveyed for Spotted Owls. One might therefore reason that, were similar survey effort applied throughout the county, Great Horned Owls would be found about as widely as they were in the intensively surveyed sections of Humboldt—or, perhaps, even more frequently.

Eight of the nine blocks with confirmed breeding were clustered in the central-coastal lowlands and near the lower Mad River. Confirmation in five blocks was established through the discovery of occupied nests, which spanned the dates of 12 Apr 1996 (e18) to 29 Apr 1996 (D17; both J. Kidd). There was a single report of an adult feeding young 31 May 1998 (H15; PFS). There were three reports of fledged young: 22 Jun 1996 (g13; GAS), 2 Jul 1995 (c20; JEH), and the unusual date of 1 Oct 1995 (g33; K. Dyer).

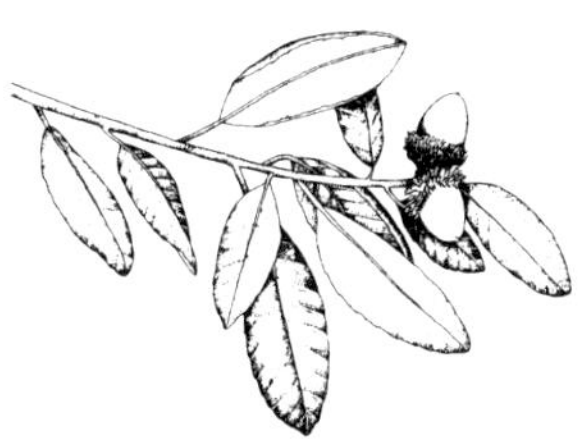

NORTHERN PYGMY-OWL *Glaucidium gnoma*

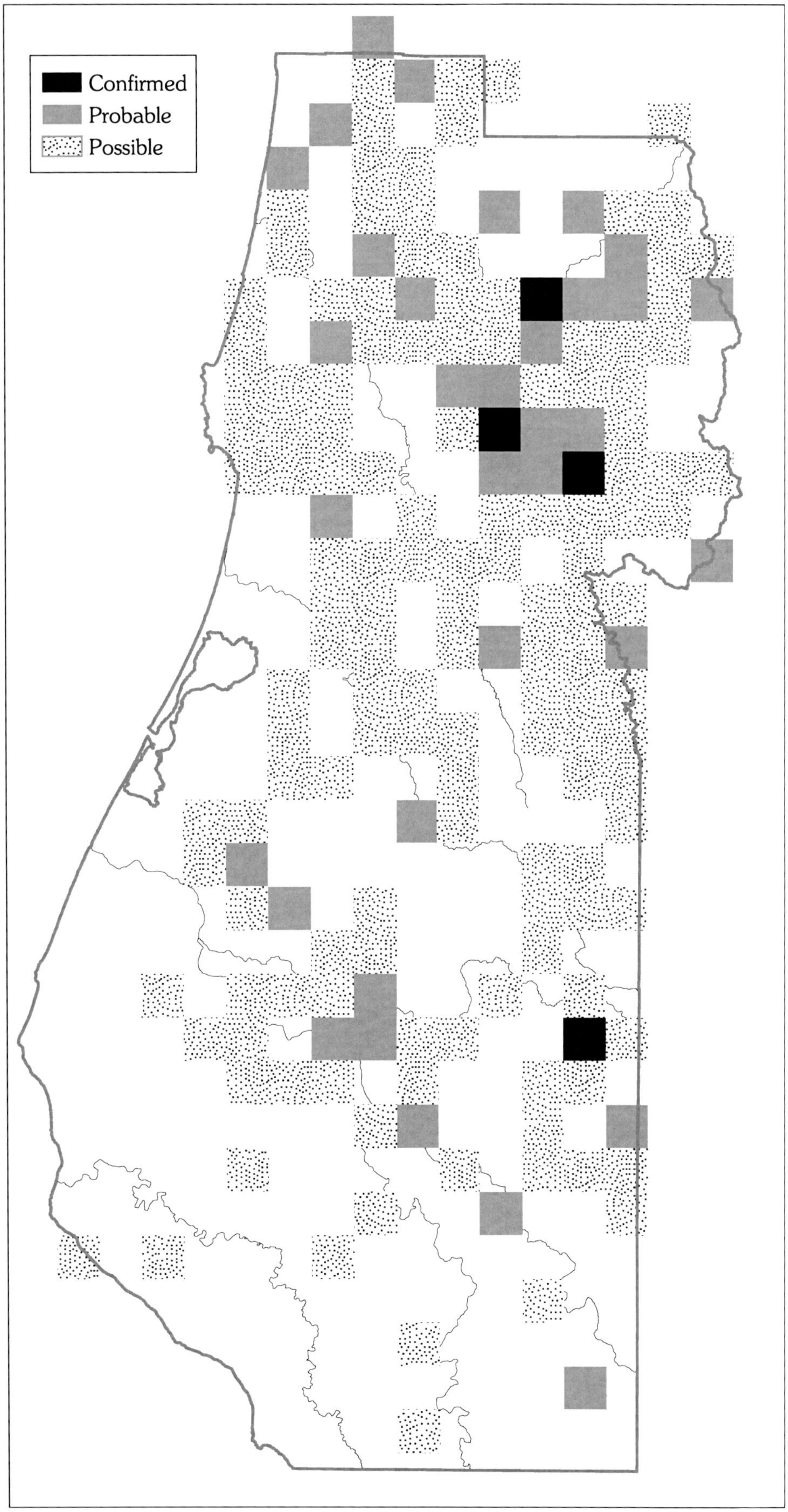

Forests composed of conifers and oaks with plenty of old woodpecker holes and an abundant prey base of small songbirds appeal to Northern Pygmy-Owls. These tiny owls are resident across a broad range in western North America and are quite widespread in the mountains of California. Humboldt County's extensive and variegated forests provide habitat for a well distributed population of these birds. They do not seem especially numerous; pairs or apparently single birds probably occur in small numbers in suitable habitat. They are routine in the sense that much habitat exists, they may be encountered at any time of year, and a pygmy-owl is never wholly unexpected in any well-wooded setting. And they are distinctly birds of a well-forested landscape: where there are Pygmy-Owls, trees typically either stretch to the horizon or obscure it. Although exceptions to the rule take place, notably in semi-open lowlands of the intermountain West in winter, few of these owls are found far from the nearest patch of forest.

Northern Pygmy-Owls have been well recognized as occurring routinely in Humboldt County. There are countless historical and recent reports from the county, and this species continues to be seen or heard frequently by Humboldt's biologists, naturalists, hikers, and others attuned to the outdoors. Harris (1996) stated that the pygmy-owl in northwestern California was an "uncommonly seen resident and breeder . . . thinly scattered in all types of forests or woodlands containing conifers throughout the region." His perception was one probably shared by most local observers.

Efforts to gather information on Northern Pygmy-Owls during the Atlas period benefited from ongoing Spotted Owl surveys on private, tribal, state, and federal lands. Detections of pygmy-owls figured large among the ancillary data on other species of owls encountered by surveyors. Formal surveys for Spotted Owls took in reaches of Humboldt County in which the presence of pygmy-owls has been well established. Apparent densities in commercial timberlands during the Atlas period were higher than had been imagined by some at the outset of the project. Outside regions of Humboldt where organized Spotted Owl surveys were conducted, reports of these birds were generated more commonly by atlasers. This was particularly true in the redwood parks, on the Six Rivers National Forest, and in portions of southern Humboldt. However, the timing of most one-day trips made from Humboldt Bay-area communities into far-flung parts of the county probably put atlasers in suitable habitat outside the most advantageous hours of dawn and dusk.

During the Atlas period, Northern Pygmy-Owls were found in 41% of all blocks. The map suggests greater abundance, detectability, or Spotted Owl survey effort in the northern one-half of the county. Reports were few in the pastoral countryside of the upper Mad River corridor and across southern Humboldt. A zone of seeming absence, coincident with minimal and fragmented forest cover, hillside prairies, and treeless bottomland, was apparent from Cape Mendocino north to Westhaven and some miles inland.

Northern Pygmy-Owls in Humboldt County live in nearly all closed-canopy forest types, from near sea level to above 5000 feet. Old-growth coast redwood stands are well known to support them. Older stands of Douglas-fir or pine, with or without oaks, and second-growth stands dominated by tanoak and conifers constitute much of the local habitat at middle elevations. This owl seems not especially partial to northern exposures, with south-facing or rolling ground and ridgelines often inhabited. At higher elevations, pygmy-owls are found most frequently in older forests featuring large trees and a strong white fir component. Protected gulches and stream corridors provide pockets of habitat in heavily cutover landscapes.

There were only four confirmed breeding reports during the Atlas period. The drama surrounding a nest in Burr Valley (n24) was well documented in 1997 (M. & J. Shreeve). Birds there were first detected calling 11 Mar. Members of a pair were observed allopreening 13 Apr, and entering a probable nest hole in an oak 18 Jun. On 3 Jul a fledgling was seen leaving the nest hole, harassed by a Bullock's Oriole. The owls were not seen in the area again until 18 Sep. The three remaining confirmations involved the observation of fledglings: 11 Jun 1996 (N11; H. Brown), 30 Jun 1999 (L11; JMH), and 10 Jul 1998 (m07; JEH). The earliest report of probable breeding was of territorial behavior 21 Mar 1997 (i01; C. Hibbard).

The extent to which Northern Pygmy-Owls undertake seasonal movements within Humboldt County is unknown. Appearances by these birds well away from forest cover, including the settled and agricultural areas of the central-coastal bottomlands, are rare. This masks any seasonal movements. Pygmy-owls remain unrecorded at nearly every one of the well covered lowland birding sites surrounding Humboldt Bay, shun even high quality floodplain riparia, and are only lightly reported from the forest/residential fringe in Eureka. Certainly the species is resident throughout the year, as indicated by CBC data and other reports in December and early January.

Simplification of forests through logging and the establishment of even-aged plantations harvested in rotation makes one wonder how much habitat for Northern Pygmy-Owls may have been lost. As secondary cavity users, pygmy-owls are thought to be highly dependent upon woodpeckers for their share of disused living quarters. To varying degree, standard silvicultural treatments of nearly any description generally diminish woodpecker populations, indirectly affecting the fortunes of species such as the pygmy-owl. It is hoped that the retention of snags and "legacy" or "structure" trees in managed timberlands—as well as a growing appreciation of the bird itself—will help ensure the future of the Northern Pygmy-Owl in Humboldt County.

SPOTTED OWL *Strix occidentalis*

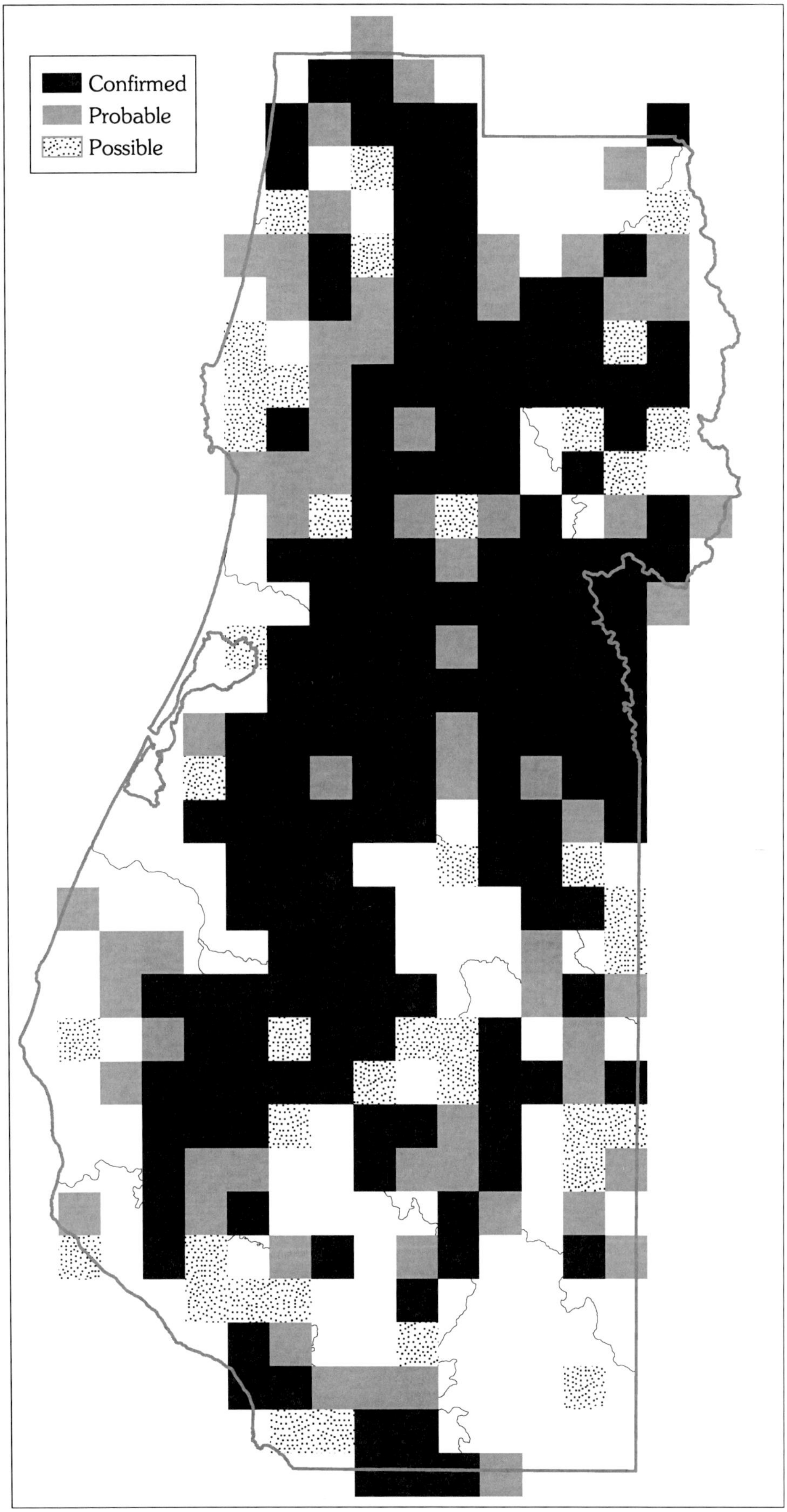

The Northern Spotted Owl (*S. o. caurina*) has come to represent many things to many people. It is, by and large, a bird of the backwoods, inhabiting places where few people live or visit. In imagining the form and presence of a Spotted Owl, it becomes plain that few North American species so completely conjure a vision of the creature's environment. In one's mind, there can be seen the shafts of mild sunlight. The odors of leaf mold and vital dampness seem nearly real. Shreds of valley fog burn off in tendrils combed by the tops of the tallest trees. For those hundreds of biological technicians who have "moused" Spotted Owls and captured them for banding in the 1980s and 90s, to think of owl work is to recall all of this—as well as to remember the report forms and mapping of detections, the mind games played out along the old spur roads after midnight, and the husbandry of unwitting white mice. Yet for thousands of others, the very name of the bird draws expressions of contempt and a railing at the efforts of "environmental extremists". As the 20th century drew to a close, the "timber wars" had simmered down, yet the owl remains an enduring icon of land-use controversies that continue to affect rural communities, and which have polarized citizens concerned with these issues.

Living in forests supporting conifers upon which a regional economy has been based, Spotted Owls have been threatened by the logging of old growth and the management of regrowth timber harvested in short rotation. Thanks to restrictions on the harvest of nest stands and other management mandated after the listing of the Northern Spotted Owl as "threatened" under the Endangered Species Act in 1990 (USFWS 1990), these birds persist across most of their original range, with numbers still comparatively high in many areas. Offering extensive forests of varying species and structure, maturing regrowth, and plenty of habitat connectivity, Humboldt County supports many Spotted Owls. While Spotted Owls appear to be well distributed in Humboldt, much uncertainty exists about long-term population trends and viability, especially if these birds were somehow to lose the protections they are now afforded. Harris (1996) termed the bird an "uncommon resident and breeder" in northwestern California.

The distribution map reflects the huge amount of information generated by the extensive Spotted Owl surveys required for the approval of timber harvest plans (and for research) that were undertaken during the five Atlas years. Thousands of hours of concerted surveys in known and suspected habitat on private, state, federal, and tribal lands has produced a good portrait of where Spotted Owls live within Humboldt County—as well as where they appear not to live. During the Atlas period, they were detected in 64% of all blocks. They were also confirmed breeding in 167 blocks, more than any other Humboldt breeder. It should be noted that many blocks in which these owls were detected probably support multiple territories, since a high percentage of the detection-blocks are well-forested and contain sufficient breeding habitat for more than one pair. One study in the Douglas-fir zone found that Spotted Owl territory centers averaged about 1 mile apart (Hunter *et al.* 1995). Indeed, by 1995 there were 755 documented Spotted Owl sites in Humboldt County (G. Gould, unpubl. data). It should be noted, however, that many of these 755 sites were probably not actually occupied in 1995, while other Humboldt sites may have gone unreported.

The range occupied by Spotted Owls in Humboldt County corresponds well with the general zone of forested land, and they occur from nearly sea level to almost 5000 feet. Overlaying the land cover map on the owl map reveals the extent to which this is true. Both second-growth and old-growth coast redwood forests and mature or older Douglas-fir/tanoak stands are well populated with these birds. Originally believed restricted to old-growth forests, Spotted Owls are now widely known to also use stands subjected to prior timber harvest—provided that requirements for prey, thermal cover, protection from predators, and suitable nest sites are available.

In Humboldt County, the primary prey of the Spotted Owl is the Dusky-footed Woodrat (*Neotoma fuscipes*; Hamm 1995, Ward *et al.* 1998). Woodrats reach high densities in early seral stage clearcuts (Sakai and Noon 1993), which is why some degree of clearcutting probably enhances the owl's reproductive output (Thome *et al.* 1999, Franklin *et al.* 2000). However, as a northern forest species, Spotted Owls seek cool, shaded areas for daytime roosting and for nesting. In the moist and cool redwood zone, Spotted Owls appear to tolerate younger and more exposed stands, while in the summer-hot Douglas-fir zone they favor denser and older stands typically located in drainage bottoms (Ting 1998). Other reasons that Spotted Owls occur more often in young redwood stands than in young Douglas-fir stands could be that second-growth redwood stands have relatively rapid tree regrowth, higher woodrat productivity, and often contain remnant large trees. Dark and dank nest groves may

Continued on page 391

BARRED OWL *Strix varia*

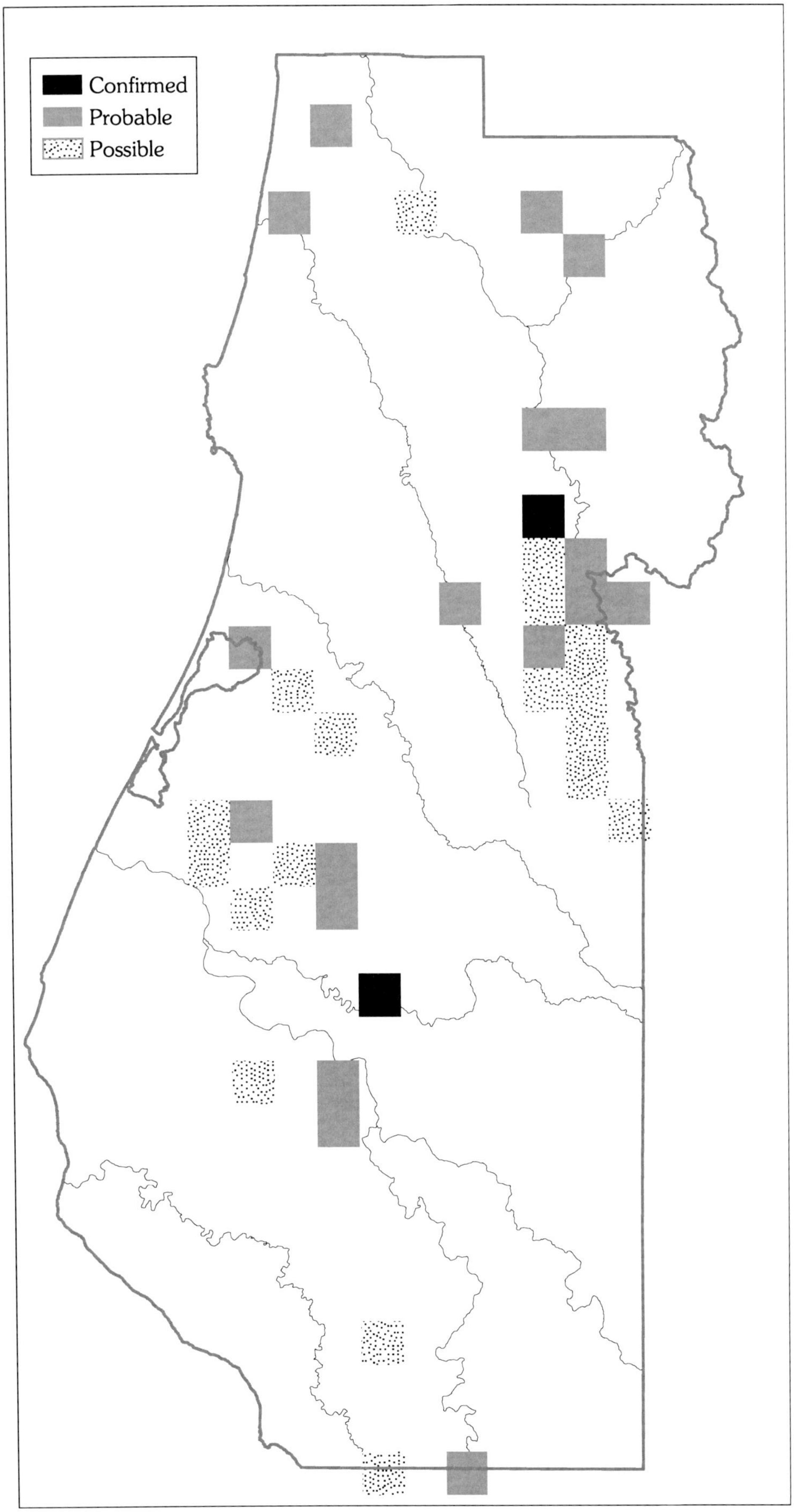

In a past era, any medium-large forest owl with dark eyes seen in California would certainly be a Spotted Owl. This is no longer the case, as the last few decades of the 20th century saw a dramatic appearance by the Barred Owl in the Pacific Northwest, including parts of California (Dark *et al.* 1998). The establishment of Barred Owls in northwestern California mirrors the pattern of their appearance and increase in Oregon, Washington, western Canada, and southeastern Alaska. This species was first detected in California at Salyer, Trinity County, in 1981, and then in Del Norte County the following year (AB 37:221). Suggesting that these occurrences were part of a larger, emerging scenario was Humboldt County's first known Barred Owl, a calling bird along Patterson Road near Willow Creek 30 Apr 1983 (R. LeValley), followed by another near Orleans 13 Jun 1983 (D. Solis). Since the time of the first report in the state, Barred Owls have rapidly expanded into northwestern California. By 1995, many were known in the three counties mentioned above.

The Atlas "snapshot" has captured, in part, an image of an event unique in the annals of North American ornithology: the wholesale invasion of the range of one owl by the ranks of a related and similar second. That a large forest owl should undergo a major range increase has spurred considerable local and regional interest among birders and naturalists. Not only have Barred Owls become widespread and quite numerous in the region, but biologists have come to hold concerns that the species poses threats to the welfare of the indigenous Spotted Owl. Time will reveal the unfolding outcome of perceived competition between these owls. The effect of possible usurpation of Spotted Owl habitat by the more aggressive Barred Owl is a topic of much speculation by those concerned with the welfare of Spotted Owls. A concrete example of the outcome of conflicts between these owls came on 11 May 1997. On that day, observations of a Barred Owl along Skunk Cabbage Creek in Redwood National Park (g05) strongly suggested that it had killed and partially consumed a Spotted Owl (TWL, S. Petersen; Leskiw and Gutiérrez 1998). Hybridization between these two owl species is also a concern. The female widowed at Skunk Cabbage Creek in 1997 apparently paired with the male Barred Owl and produced young the following year (*fide* K. Schmidt). An adult hybrid was also seen near Willow Creek in 1994 (A. Franklin).

At the beginning of the Atlas period, Barred Owls had become more routine than novel in Humboldt County. By the end of the project, the species had been detected in 9% of all blocks. The great majority of reports involved individual adults or pairs heard in the course of Spotted Owl surveys. The distribution of Barred Owl in Humboldt as it appears on the map is probably incomplete, as it reflects the results of deliberate search for owls in several discrete regions of the county. Survey effort elsewhere was typically less well distributed. It should be borne in mind that habitats used by the owls are by no means limited to the areas in which most detections originated. Therefore, these owls may have been in a number of blocks in which there were no reports, perhaps even in blocks well removed from the clusters of reports mapped.

Barred Owls were reported during the Atlas period in three blocks within the coast redwood zone in northwestern Humboldt; in three blocks above the lower Klamath River; in three blocks extending from Arcata to Kneeland; in 12 blocks on and around private timberland southeast of Humboldt Bay; in 16 blocks within the combined Trinity and South Fork Trinity River watersheds; and in three blocks in extreme southern Humboldt. These owls appeared to have adopted a range of older coniferous forest types within the county during the Atlas period. An informal survey of government, private industry, and tribal biologists revealed that at least 37 Barred Owls were in Humboldt County by the fall of 1998 (DFx; NAB 53:101).

During the Atlas period, confirmed breeding by Barred Owl was documented in two blocks, each in sections of the county with many detections. The most noteworthy of these was Humboldt's first breeding confirmation, which occurred 11 Aug 1995 when a pair with fledged young were detected at Brannan Mountain (m12; C. Miller). Subsequent to that observation, two fledglings were observed and one was banded 10 Jul 1998 in timberlands just north of Grizzly Creek Redwoods State Park (i23; P. Carlson). That only two instances of confirmed breeding came to light during the project surely owed in part to lack of concerted daytime follow-up effort invested in Barred Owls, since Spotted Owls were the primary focus of surveys. In addition, Barred Owl nests and fledglings appear to be more difficult to locate than those of Spotted Owls.

The reasons behind the expansion of the Barred Owl into the Pacific Northwest may never be fully understood. It has been suggested that an increasing population of reproductively vigorous owls in the Pacific Northwest set the stage by establishing pioneering outpressure, with individual dispersants and pairs steadily popping up in new areas. Second, it is generally thought that the establishment of a patchwork of older and younger stands as a consequence of logging has created a habitat mosaic favorable to this species. This broad hypothesis is open to question, but nevertheless has been proposed. Though the "truth" is unavailable, these phenomena may have worked in tandem to open the doors of the Golden State to these opportunistic birds.

LONG-EARED OWL *Asio otus*

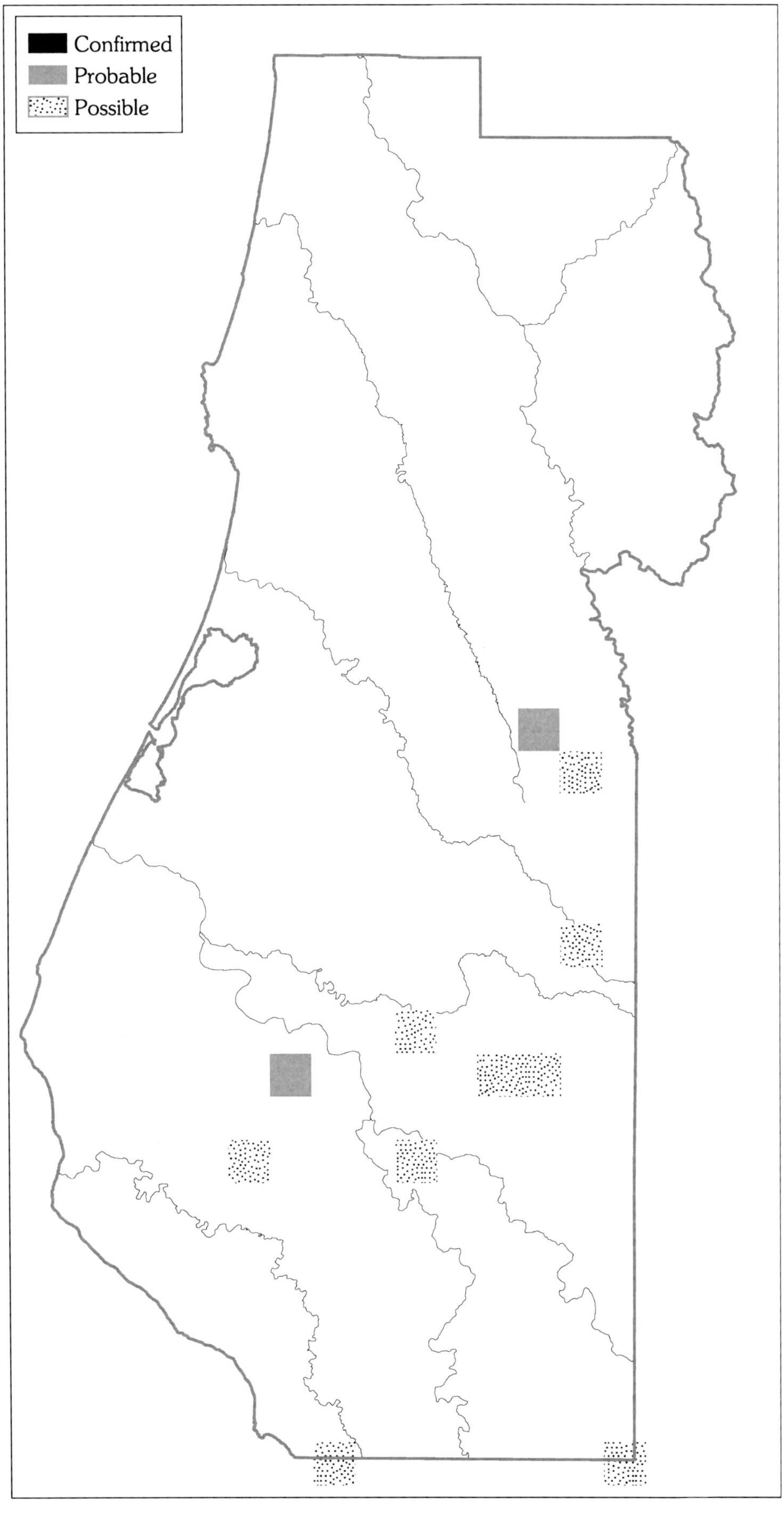

Few species believed to be resident throughout the year and known to breed in Humboldt County remain as enigmatic and poorly understood as the Long-eared Owl. Put plainly, not a whole lot is known about this bird in Humboldt. Weekend birdwatchers and veteran owl surveyors alike are apt to scratch their heads when asked what they think they know about this species' status in the county. It can be labeled an "owl of mystery" or, for those with little familiarity with their calls, an "owl of uncertainty" in Humboldt. Encounters with it invariably are few and far between. This is hardly a matter of shame, as it is true for most birders living within its range. Slim, dark, cryptically patterned, active chiefly during hours of darkness and not especially vocal, Long-eared Owls are well suited to avoid detection except by the few who have come to know them well.

During the Atlas period, Long-eared Owls were reported in spring or summer in 11 blocks, and in each of the 5 Atlas years. Reports were well scattered, though all blocks with detections were confined to the southern half of the county. Breeding was not confirmed in the course of the project, but was considered probable in two blocks. In the remaining nine blocks with detections, breeding status was classified as possible. Some representative possible breeding observations include: birds calling 5 Mar 1999 and 17 Apr 1999 (it is unknown if the same territory was involved both nights) just north of Whitethorn (h34; S. Pagliughi); one 28 May 1995 near Butte Creek (m25; P. Carlson); a bird calling 28 May 1999 from near Island Mountain in Trinity County (o34; JEH); and one near Sims Mountain 17 Jul 1995 (n18; P. Carlson). The two probable reports were of a pair calling 13 May 1997 (g25; D. Dill) and a male vocalizing on territory from 16 Jun–22 Jul 1997 near Grouse Mountain (m17; P. Carlson).

Atlas results provided no specific details on habitat use by Long-eared Owls. However, since most reports were from southern and interior Humboldt, it is likely that drier, mixed stands of conifers and hardwoods are used. Edges or openings such as meadows, prairies, and roadsides are fairly common where Long-eared Owls have been historically encountered in Humboldt. Harris (1996) related that most of the spring/summer reports from plausible breeding sites were of owls heard calling or seen foraging along edges of mountain prairies such as those at Blake Mountain. Another habitat used by these birds in Humboldt is open Jeffrey pine forest, such as near Horse Mountain, where a bird was recorded 18 Aug 1991 (K. Irwin, TWL, JEH). Other pre-Atlas locations where possible breeders have been noted include Kneeland, the Bald Hills, Pilot Creek, Ettersburg, and Snow Camp Mountain.

This species was not mentioned in the early literature, which reveals how little was known of the Long-eared Owl in northwestern California prior to the recent era of increased birding coverage and owl surveys. Yocom and Harris (1975) termed it "casual in fall and winter," listing only four records from the region. By the time of his first edition of *Northwestern California Birds*, Harris (1991) described Long-eared Owl as a "casual fall migrant; casual breeder." He cited an unconfirmed report of three chicks brought to the Humboldt Wildlife Care Center 12 Jun 1987 from a nest south of Honeydew (M. Dickman), and a nest with two young near King Peak in June 1981 (B. Woodbridge). Sufficient additional records accrued during the first half of the 1990s that, in his second edition, Harris (1996) modified the status characterization to, "Rare. Appears to breed at higher elevations in open, edge habitats; moves downslope to lowlands in fall/winter."

Harris (1996) also summarized records known to him of 32 individuals from the coastal lowlands, and indicated that they were clearly of non-breeding or wintering birds. Many of the winter birds were found roosting in patches of conifers or willows; it is unknown if any of these winter birds are the same individuals that nest in the interior, or (more likely) if they are migrants from out of the county. The irregular occurrence of Long-eared Owls in winter in the Humboldt County lowlands may be evidence of the somewhat nomadic behavior these birds will exhibit in response to changes in prey abundance.

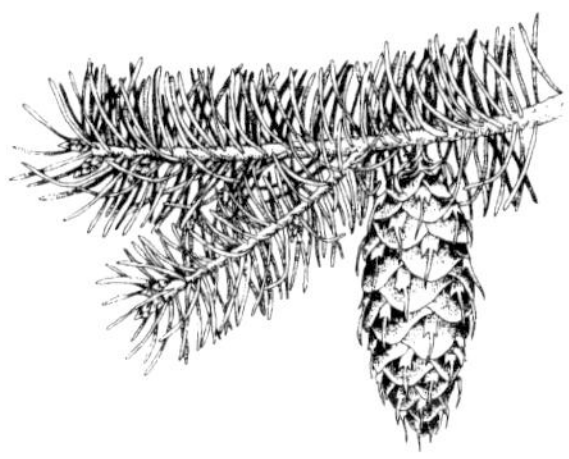

SHORT-EARED OWL *Otus flammeus*

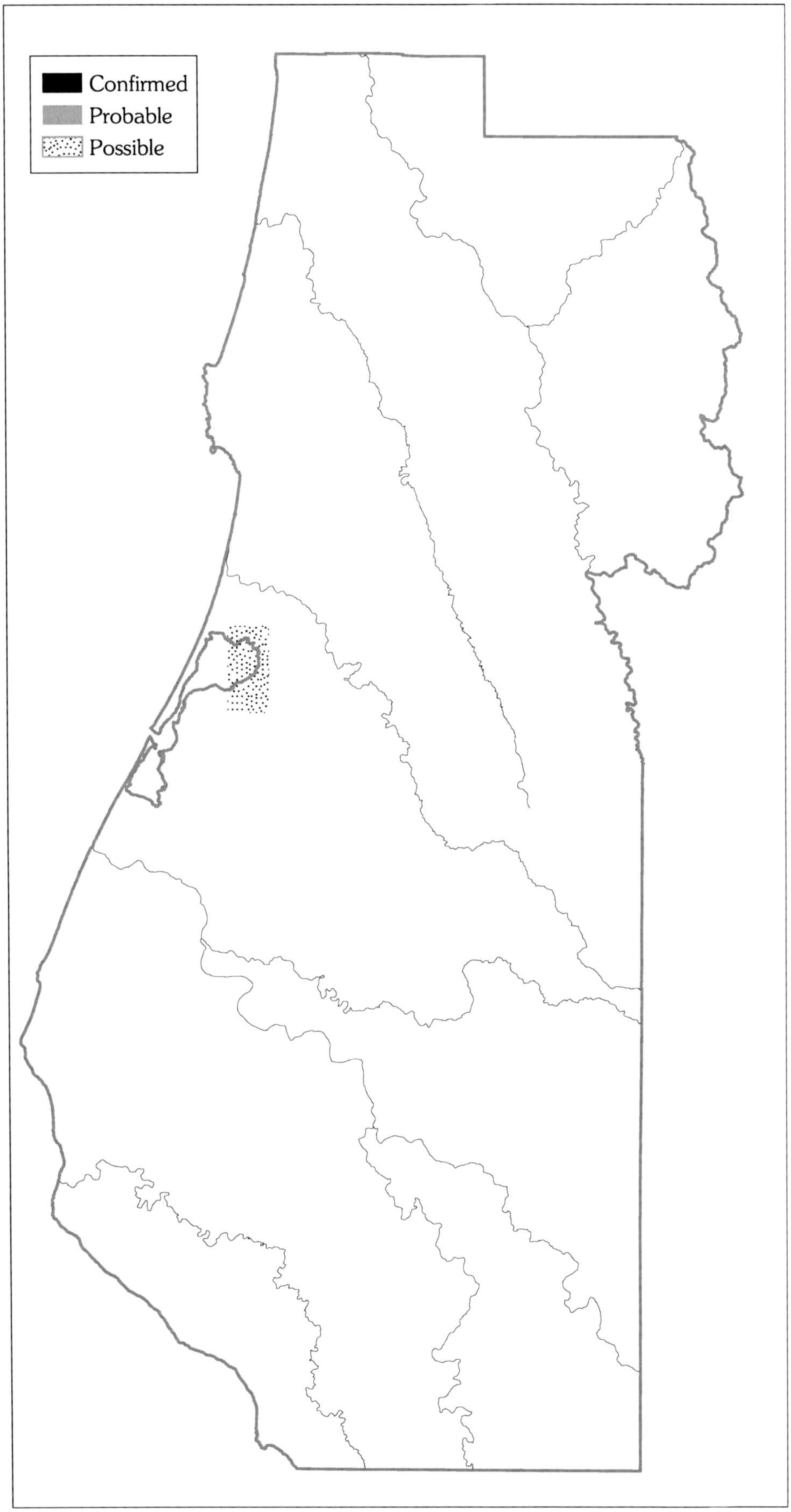

Flight made to look easy distinguishes the Short-eared Owl at any distance when on the wing. Deep, slow, uneven wingbeats carry the bird on a wending path across foraging grounds in open countryside. The stir-and-wriggle of a vole among last year's stiffened grass stems elicits a pouncing dive; the sky is suddenly emptied of the tawny ghost-bird it held, what had animated it now unseen within the wind-dabbed sweep of meadow. In the presence of Short-eared Owls, more often than not the roar of urban surf is absent, and one can take pleasure in the hush and color of a careless moment. A person may forget his or her first Short-eared Owl, but the feeling lingers. This experience is available to a great many people, as the species occupies portions of every temperate continent as well as some remote Pacific islands.

Short-eared Owls are partial to treeless or tree-dotted country heavily grown to grass or marsh vegetation. As can be imagined given the great range of this bird, habitat varies in looks and in layout. Natural prairie or grassland parcels, disused or sparsely grazed pastures, sapling-dotted wet meadows, marshes, and undeveloped industrial sites attract them. Patches of appropriate habitat of size or quality sufficient to host these birds are scarce in the rugged interior of Humboldt County, with correspondingly few reports of Short-ears. Nearly all of the regularly inhabited sites supporting one or more of these owls lie in little-disturbed open country of the central-coastal lowlands.

Within the record of ornithology in Humboldt County, Short-eared Owls have been recognized to occur annually in suitable habitat, typically in very small numbers, in fall, winter, and early spring. Concentrations of up to 30 owls have been encountered in some winters, probably in response to booms in prey populations. Humboldt's only historical breeding record followed one such winter, when up to 24 owls were seen 7 Apr 1990 (F. Broerman) near what is now the Mad River Slough Wildlife Area. Three adults were still present there 3 Jun 1990, one of which was apparently feeding two fledglings (F. Broerman; Harris 1996). Such episodic nesting appears to be quite rare in Humboldt County and was not recorded during the Atlas project.

Each winter of the Atlas period, Short-eared Owls were encountered at any of several spots, at each of which they were more or less expected. These included the Mad River Slough, Fay Slough, and Eel River Wildlife Areas, tidelands of the Eel River estuary, and coastal sand-spits. Making a distinction between wintering birds, migrants, and prospective summering birds in early spring was not simple. At the Mad River Slough Wildlife Area in the Arcata Bottoms (F15), three owls were seen 17–19 Apr 1998 (KMS, DFx *et al.*), and in 1999 two were seen 13 Apr (K. Irwin) and one on 14 Jul (B. Cariss). At Fay Slough Wildlife Area (f16), three birds were seen 23 Mar 1999 (TWL) and one was reported 19 Apr 1998 (KMS).

Around Humboldt Bay, Short-eared Owls commonly prey on voles, which comprise about 70% of their diet (Dunk and Cooper 1994). Since these small rodents are virtually absent when grazing reduces the vegetation to less than 4 inches in height (Dunk and Cooper 1994), it is easy to understand why these owls show a preference for fairly tall, rank grasses, sedges, forbs, and similar vegetation that is not impacted by grazing. Ungrazed, old fallow ground and undisturbed dunelands are localized, with practically all of the few truly sizable parcels existing during the Atlas period named within the discussion above. Such places are threatened by planned return to active pasturage, off-road vehicle use, and natural succession. Of particular concern locally are taxpayer-funded burning and grazing schemes that favor waterfowl while compromising habitat for Short-eared Owls and other raptors of ungrazed grasslands. Declines have been documented in many portions of this species' range.

NORTHERN SAW-WHET OWL *Aegolius acadicus*

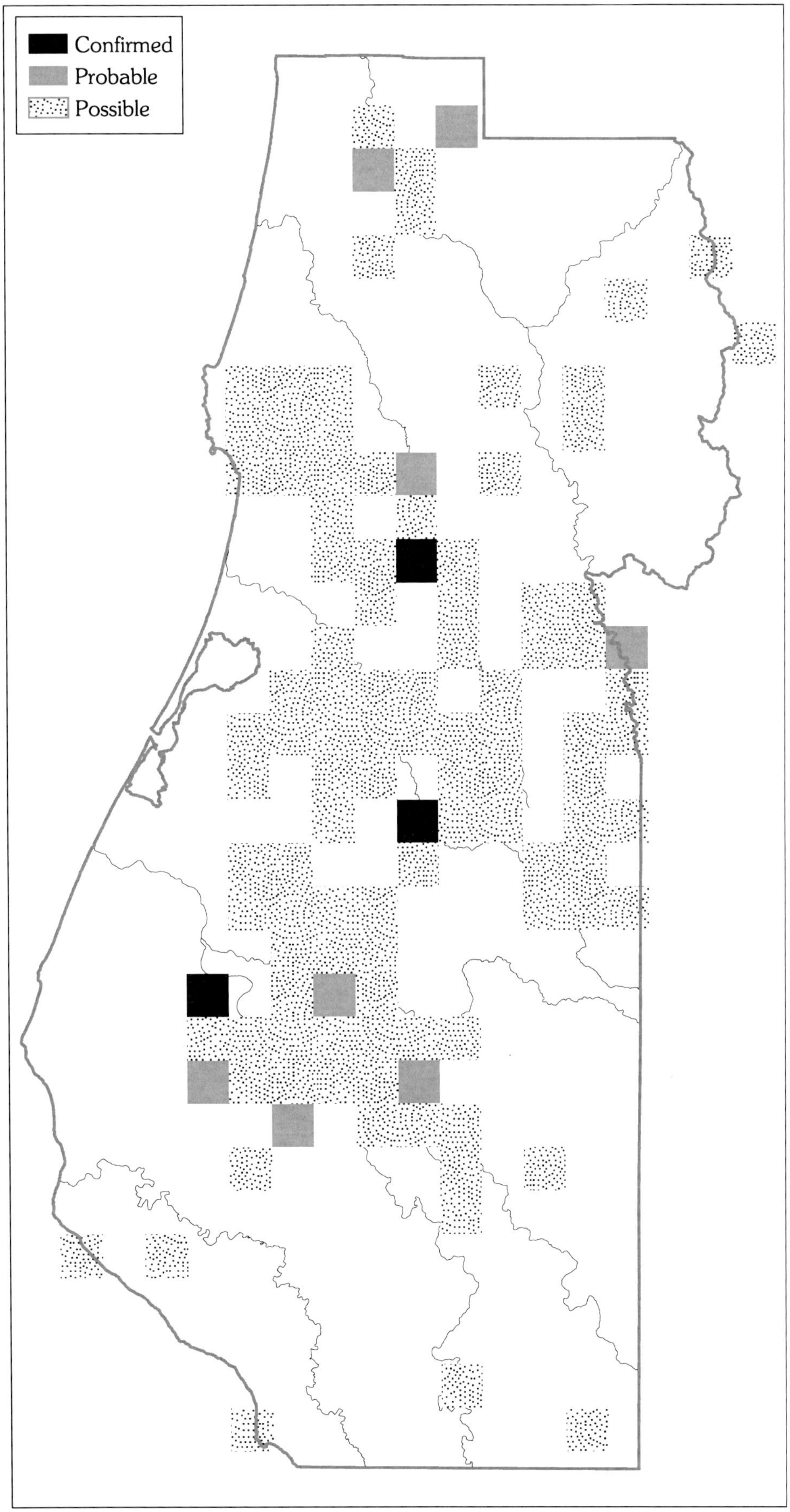

Northern Saw-whet Owls are minuscule and reclusive night hunters with a fondness for dense stands of trees. Saw-whets breed across much of the northern tier of the United States, through boreal Canada, and in Alaska. They are resident across most of forested northern California but are partially migratory throughout the state. Because they are strictly inactive during the day and do not call until nightfall, they are unknown to many naturalists and birders. Northern Saw-whet Owls are occasionally spotted at night by motorists alert to their top-heavy form and their propensity for hunting along rural roadways. Often what is seen is but the flash of a chubby brownish apparition betrayed by headlight beams, lit brightly from beneath and gone. It is unfortunate that an understanding of their seasonal presence has been based to some extent on the discovery of birds struck by vehicles.

Reports of Northern Saw-whet Owls in Humboldt historically have come from widely scattered places throughout the county, and from all months of the year. Harris (1996) characterized Northern Saw-whet Owl in northwestern California as an uncommonly-recorded resident and breeder. He noted that the Humboldt State University collection had 32 mostly roadkill specimens of Saw-whets, compared to 15 skins of Western Screech-Owl and 14 of Northern Pygmy-Owl. The high ratio of Saw-whet to other small owl skins at Humboldt State University is probably evidence of both a greater tendency to hunt along roadsides and greater seasonal movements undertaken by this particular owl. However, the true degree to which this species is migratory in Humboldt County is unknown.

Northern Saw-whet Owls were detected in 111 (26%) of all Atlas blocks. The pattern of reports strongly suggests that Saw-whets are to be expected in forested areas wherever concerted surveys of owls are conducted. The host of incidental nighttime detections by Spotted Owl surveyors made it obvious that Saw-whets are distinctly well represented, and may be quite numerous on private timberland in the central portion of the county. These holdings comprise a fairly large region in central Humboldt, stretching from the upper Mattole River country in the southwest through the lower Mad River/North Fork Mad River corridor and some distance northwestward. This belt spans a significant moisture gradient, and tree species composition varies considerably from area to area. In a general sense, though, much of this country is similar in supporting low amounts of old-growth forest and a patchwork of recent clearcuts, regenerating cuts, stringers, and stream corridors typified by uneven-aged or fragmented stands, and minimal to moderate hardwood understory and edge. Dense stands of unthinned Douglas-fir/tanoak are known to be used by Saw-whet Owls in Humboldt County during spring and summer. As this habitat is abundant across a large part of the county, it stands to reason that intensive survey of areas throughout Humboldt's second-growth conifer/oak forest—impractical during Atlas work—might bring to light many more of these owls. Were this to prove the case, then the skewed pattern of detections in the heavily surveyed commercial timberland represents not a concentration of Saw-whets, but rather an artifact of systematic Spotted Owl survey effort.

Because calling commences in winter and continues for the most part only until late spring, when the birds fall silent, it is likely that most atlasing fell during those months when Northern Saw-whet Owls are either decreasing their night calling or have altogether ceased giving unsolicited calls. This is believed to have diminished opportunities to find Saw-whet Owls during the Atlas period. Nevertheless, three confirmations of breeding were made during the Atlas. In June 1995 a logger working on Monument Ridge west of Rio Dell (e23) discovered a cavity containing a brood of nearly-fledged nestlings in a recently felled tree (*fide* S. Chinnici). An adult was observed feeding a fledgling in Redwood Valley (J13) during the early morning hours of 27 Jun 1997 (JCP). A dead fledgling was found 16 Sep 1995 near Iaqua Buttes, southeast of Kneeland (J19; S. Bunnell). All eight records of probable breeding were of either vocalizing pairs or males on territory for at least a week. The earliest recorded pair was heard 16 Mar 1999 (e25; *fide* S. Chinnici), and the latest male calling on territory was recorded 23 Jun 1998 (o15; P. Herrera).

COMMON NIGHTHAWK *Chordeiles minor*

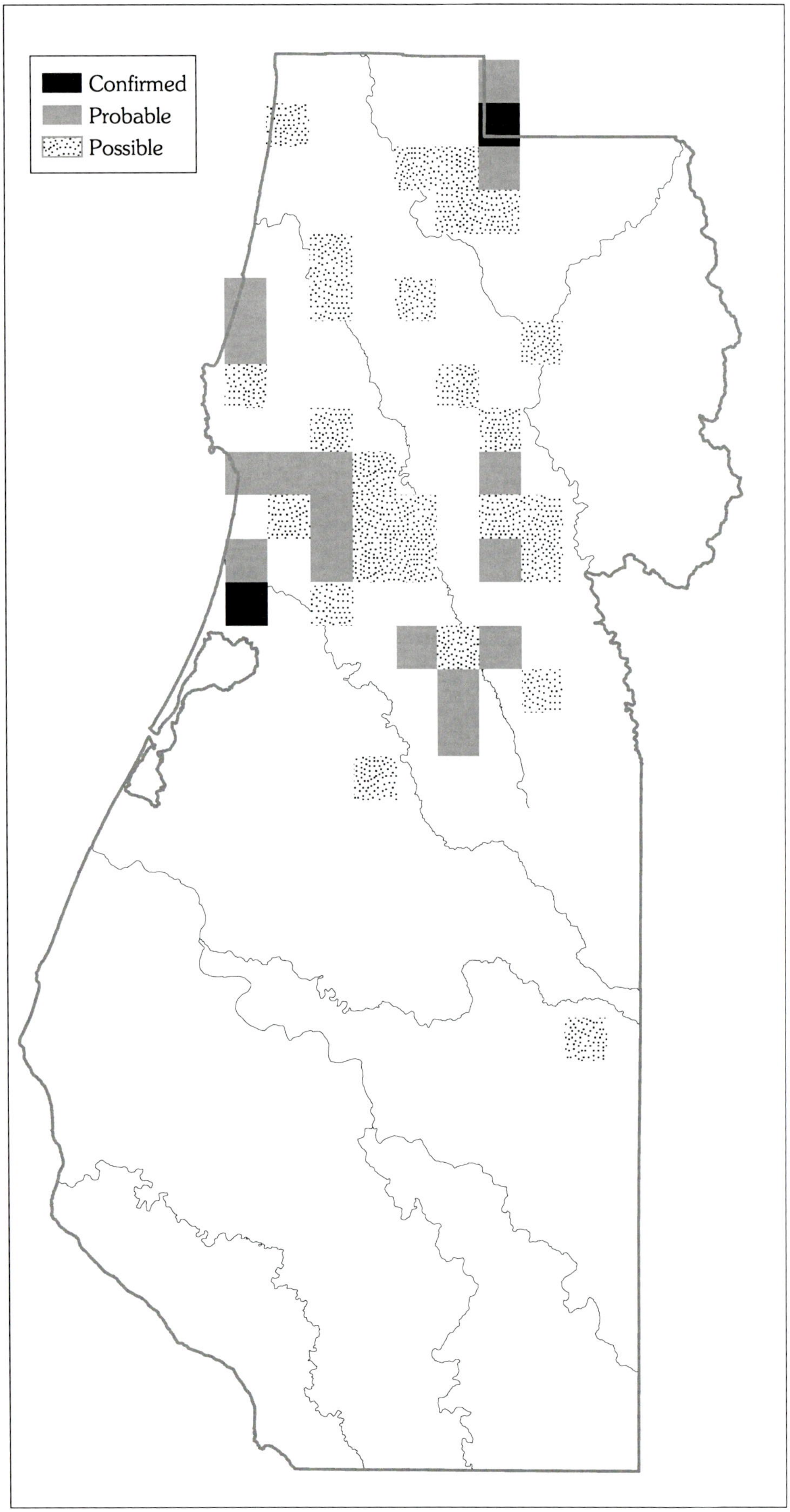

Unique call-notes, heard every few moments overhead, herald evening twilight for summer visitors to mountain campgrounds and forested retreats in northern California. So, too, are Common Nighthawks listened for by those who know them as city birds. This unusual voice can be associated with rural landscapes as well as the sound of late-night traffic. Common Nighthawks are found from late spring through early fall in northern California in the mountains, in the Great Basin, and sparingly in the lowlands with the exception of the Central Valley. This species is the only member of the nightjar family that occurs regularly in Humboldt County. Humboldt stands at the cusp of the breeding range in the state. South of the area, this species is known chiefly as an uncommon to rare migrant in near-coastal districts.

Common Nighthawks were reported from 11% of blocks during the Atlas period, and were encountered in each of the 5 years. The Atlas surveys make clear that the distribution of Common Nighthawk in Humboldt County during the project was restricted to northwestern and north-central portions of the county, with the exception of one report near Buck Mountain (n24) 25 Jul 1995 (KMS). The birds were found from near the Del Norte County line (l02) south to the top of Friday Ridge on South Fork Mountain (m16); at Lake Prairie near Snow Camp Mountain (k17); on Kneeland (i18); at Blue Lake (h14), and in Arcata (f14). The western edge of the area from which reports came also included Westhaven (F11), Patricks Point (f08), and Prairie Creek Redwoods State Park (g03). Enclosed within this perimeter, blocks with detections were well scattered, but seeming concentrations occurred near Onion Mountain (l03); near Fieldbrook (g13) and north to Big Lagoon (F07); from the southwest corner of the Hoopa Valley Indian Reservation (l12) south to Willow Creek (m13), and on the Snow Camp Mountain ridge (k16). It is likely that nighthawks could have occurred undetected in many intervening blocks. A very few reports came from Arcata (f14), and none from Eureka.

Common Nighthawks can be observed over a wide variety of terrain and vegetation. Elsewhere within their range, they are summer residents of open pine forests, river gravel bars, sagebrush communities, and grasslands. This flexibility has been noted to some degree in Humboldt County, where they have been seen foraging and interacting over mountain forest (as at South Fork Mountain; L15), redwoods and oak with grassland openings (Bald Hills; h06 and H07), over sandspits and dune systems (F07, f08), and over cities. Probably the most important factor influencing the breeding distribution in Humboldt is the availability of a bare substrate, upon which they lay their eggs. Many of the observations in Humboldt County during the Atlas were made in landscapes dominated by intensively managed commercial timberlands. It has been suspected that, as is the case in similar habitat elsewhere in the Pacific Northwest, the isolated, seldom-disturbed old landings and disused logging roads everywhere in evidence may provide nesting sites suitable for nighthawks. Such an environment would appear to be free from all but passing human disturbance. Ten nighthawks displayed in late morning above just such habitats near Lake Prairie (k17) 2 Jul 1998 (DFx, JCP).

In northwestern California, Common Nighthawks are most frequently encountered in areas with significant amounts of unvegetated ground (such as serpentine soils), like those in the Onion Mountain area. This relationship is more apparent in portions of Del Norte County, where serpentine soils are even more common, as are these particular birds. Along the coast, nighthawks often have been associated with dune systems or gravel-covered roofs of buildings, both of which afford the exposed substrate required.

Because they build no nest and rely upon their cryptic plumage pattern and stillness for concealment, it is difficult to confirm breeding for Common Nighthawks. Among the blocks with detections there were only two with confirmations. Each of these reports illustrates the disparate environments used by the rural and urban populations. At Blue Creek Mountain (l03), a remote area in southernmost Del Norte County, both members of a pair were observed engaged in distraction displays 24 Jul 1997 (S. Thiesen). In a much different setting was a nest with one young nestling on the roof of Gist Hall, Humboldt State University, Arcata (f14) 10 Aug 1996 (S. Dark, J. Dunk). Unfortunately, this nestling was found during a re-roofing project, and was removed to a wildlife care facility.

Sixteen blocks produced reports of probable breeding. These were widely scattered, but most fell in the western and southern portion of the species' limited range in Humboldt County. In nine of these blocks, courtship, exhibited by "booming" flight displays, was evidence of probable breeding. Probable breeding was established along the outer coast when a pair was seen at Big Lagoon sand spit 7 Jun 1997 (f08; COg), and in F07 when an adult was at a probable nest site 11 Jun 1997 (R. Hamlin, L. Roberts). A nightjar, probably a Common Nighthawk, was reported in similar beach habitat on 15 Jun 1996. That bird was roosting on the ground among driftwood at Fern Canyon Beach (g03; D. Diehl).

Historically, Common Nighthawk in northwestern California has been recognized as a "common summer resident and breeder about towns and cities . . . normally arrives in late May. . . usually leaves in August" (Yocom and Harris 1975). Sixteen years later, Harris (1991) characterized it as a "locally uncommon summer resident and breeder. Seems to be restricted to the vicinity of towns and cities." Yet later in the second edition of *Northwestern California Birds*, this statement was amended to, "Locally rare summer resident and breeder. Most often recorded in the vicinity of towns and cities" (Harris 1996). In that work, Harris noted that "This species has certainly declined in numbers, at least near Humboldt Bay since the early days. Most modern breeding records are from rooftops in urban areas. Thinly scattered "booming" birds (breeders?) have been recorded widely in the interior mountains." He listed such reports in Humboldt County as having come from northeast of Orleans, at Willow Creek, in mountains south of Willow Creek, on Pecwan Creek, and at Weitchpec.

Fisher (1902b) called the species "very abundant" in May–June 1899 in the Redwood Region. Between 1913 and 1929, early collectors took eggs on adjacent Woodley and Daby islands in Humboldt Bay, on the Elk River sandspit, at Bucksport, and along the Mad River gravel bars at Blue Lake (Harris 1996). Harris (1996) related that breeding was no longer known at any of these sites. As late as the early 1990s, however, these birds still occurred in summer in

Continued on page 392

COMMON POORWILL *Phalaenoptilus nuttallii*

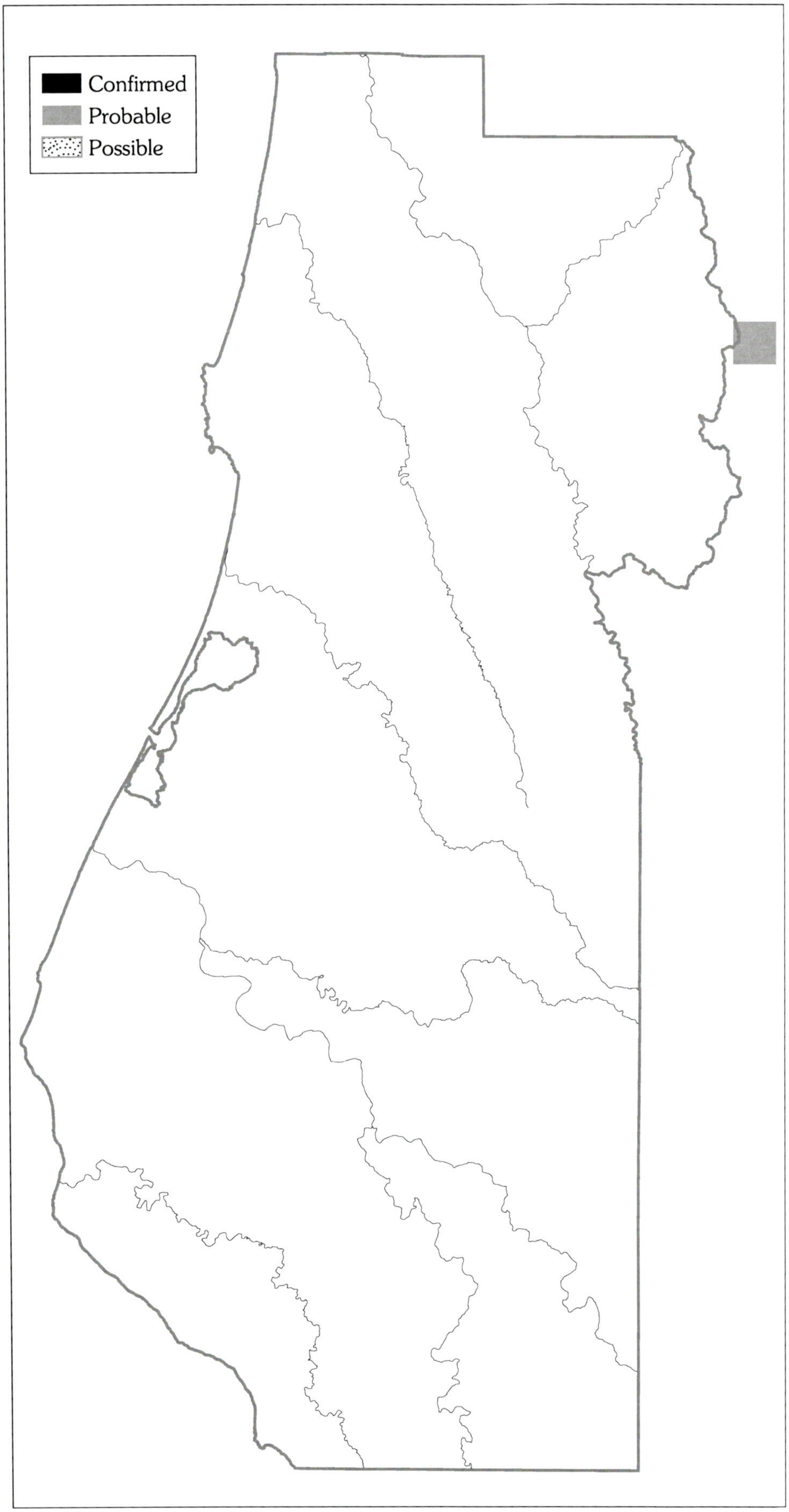

Arid slopes with sparse grass and with few or no trees host Common Poorwills across much of the western U.S., extreme southwestern Canada, and northern Mexico. These nocturnal birds are occasionally flushed from back roads at night but are seldom encountered during the day. They are superbly camouflaged by cryptic, mottled plumage and are given to remaining motionless on the ground while the sun is overhead. They would be even less often perceived were it not for their odd two- or three-syllable cry, delivered at intervals of a few seconds from somewhere in the night. In good habitat one may hear a half-dozen poorwills, some nearby, some at the edge of perception.

The Common Poorwill is at best a peripheral or marginal breeding species in Humboldt County. It is known to summer in western Siskiyou (AB 34:927), Trinity (Hunter and Hazard 1998) and eastern Mendocino counties (Carter 1998), but is unrecorded in summer in Del Norte County. Near-absence in Humboldt is due to lack of the semi-arid, shrub-dotted expanses favored by these birds. Here, there are no classic, open "hard" chaparral patches like those found to the southeast in California, nor does sagebrush occur, nor xeric grassland. The historical record of this bird in Humboldt County is short. It comprises a handful of reports of spring and fall migrants in the coastal lowlands, where Poorwills are entirely unexpected, as well as occasional reports in spring and summer in the interior mountains, where they are less unexpected but always quite rare. Records falling within the latter group have typically come from higher-elevation forest edge, serpentine openings, and clearcuts.

Breeding confirmations for Common Poorwill in Humboldt County are lacking, but this species can be considered a probable breeder here based on scattered pre-Atlas reports of males singing on territory over an extended period, such as one near Petrolia in the summer of 1975 (B. Clow), two or three near Whiteys Peak 3–10 Jun 1987 (G. Lester, L. Lester, R. Erickson; AB 41:1484), and two birds near Horse Mountain–Friday Ridge Road 13–16 Jul 1992 (DFx, JEH, G. Lester, L. Lester, B. Deuel). Other pre-Atlas summer locations have included Kneeland and Hoopa. The Lassics area along the border of Humboldt and Trinity counties was once a semi-reliable place where local birders could go to hear poorwills. Suggesting a local decline, however, is the fact that the most recent report from there was of four birds near Black Lassic 8 Jun 1986 (PFS; Hunter and Hazard 1998). During the Atlas period, the only report of this species was of a bird singing east of Salmon Summit, Siskiyou County (r08) from 21–30 Jul 1998 (GCH, JEH). This high-elevation site (about 5000 feet) had much exposed rock and soil and was somewhat reminiscent of the Whiteys Peak and Lassics sites.

In northwestern California, Harris (1996) termed this species an "uncommon summer resident and breeder." Yet the Humboldt breeding population must be tiny in the best of years, and is probably nonexistent in many. It is noteworthy that no poorwills were reported during the Atlas period in the course of a considerable amount of Spotted Owl surveys in the interior mountains, nor by the few active birders working the open hills of interior Humboldt at night. It remains for future field workers to confirm breeding by this species in Humboldt County.

VAUX'S SWIFT *Chaetura vauxi*

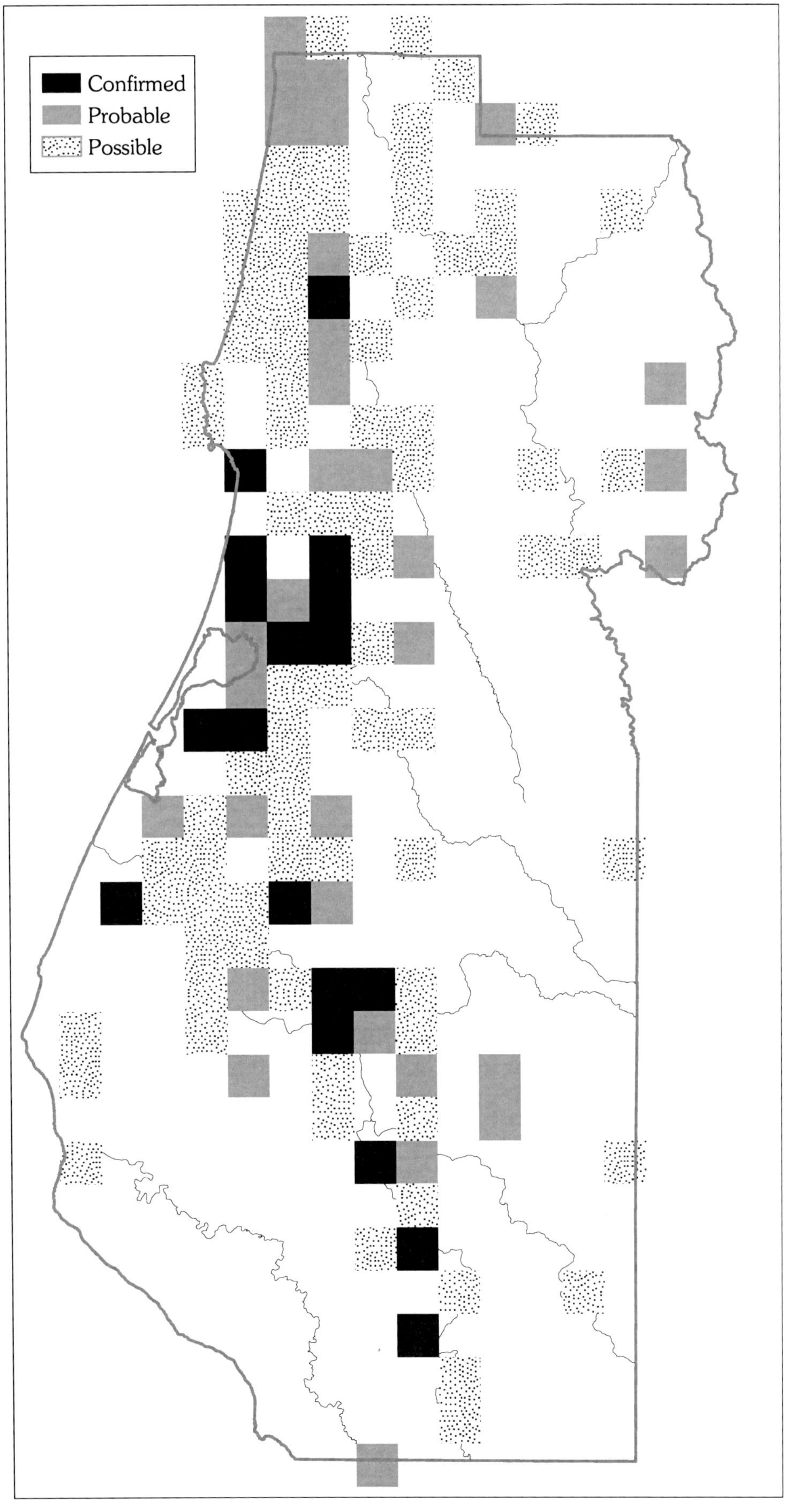

Spirited chippering notes delivered in short, rapid bursts advertise the presence of Vaux's Swifts overhead. This is a sound of the Pacific Northwest in the warmer months, once to be heard only in or near coniferous forests but nowadays as likely to be detected over towns and older residential neighborhoods. Brick chimneys of homes, churches, and schools attract nesting Vaux's Swifts in many cities. While it has not led to any significant expansion of the overall range of the bird, this broadened spectrum of opportunities may have partly offset widespread losses attributed to logging. Destruction of most of California's old-growth redwood forests and their replacement with younger managed stands deficient in hollow or cavity-laden trees has undoubtedly resulted in declines in this species.

The species has long been recognized as a regular spring and fall migrant and summer resident in Humboldt County, and it is reported frequently each year. Harris (1996) termed Vaux's Swift a "common summer resident and breeder; casual in winter" in northwestern California. These birds were detected in 129 (30%) of all Atlas blocks. Vaux's Swifts arrive from their neotropical wintering range in April and the first half of May. Typically only a few birds are seen during the first half of April, but by late in the month the skies over much of Humboldt County are well populated. During early May, concentrations numbering from hundreds to as many as several thousand may assemble over the near-coastal lowlands. This occurs to greatest extent within the redwood zone, and particularly over the Redwood Creek estuary and the Humboldt Lagoons. A drop in abundance of swifts follows as spring migrants clear out, leaving the more modest summering population.

Considering the many conifer species and great variety of tree communities within the range of Vaux's Swift, it is surprising that only a portion of this array of forests is commonly inhabited. In California, most of the breeding population of Vaux's Swift occurs within the redwood zone, probably because of the presence of the large nesting cavities they require (Sterling and Paton 1996). These swifts are encountered in significantly greater numbers in old-growth redwoods than in younger redwood stands (Hazard 2002). In Humboldt, the species is most commonly detected within the redwood zone, and all 18 of the blocks with breeding confirmations reported during the Atlas period were from this portion of the county. Few birds found in Humboldt County exhibit to such a degree this apparent preference for one vegetative region. Yet these birds were reported in other forest types and probably breed there too, but in much lower numbers. Outside of Humboldt, these birds have been found breeding in a variety of forests including those dominated by Douglas-fir, grand fir, western hemlock, or (from Oregon northward) western larch or western red cedar. The importance of forests other than redwood to Vaux's Swifts in Humboldt is unclear, but it is probably not great; lands managed by the U.S. Forest Service for late-seral Douglas-fir may be of limited value to this species.

These swifts are to be expected throughout older forests of the fog zone from the Del Norte County line south to Ferndale, extending 40 miles up the Klamath River, in the Redwood Creek and lower Mad River watersheds, and in much of the hilly wooded country shortly inland from Humboldt Bay. The range appears to be continuous from there southeastward to the Mendocino County line in a corridor well removed from the coast. Vaux's Swifts were also reported from a scattering of blocks near the eastern edge of Humboldt County, most of those from lower and middle elevations of the Trinity River. This loose cluster of inland detections included three priority blocks in which pairs were observed 22 Jun 1998 (P11; KMS), 22 Jun 1999 (P13; GCH), and 22 Jul 1999 (P09; GCH). Other blocks outside the main redwood zone with evidence of possible breeding were at two high-elevation sites, on Blake Mountain (o20) and in the Lassics (o27), and in the lowlands along the main Eel River at Alderpoint (n30).

Humboldt County's earliest historical breeding confirmation was described by Taylor (1905), with subsequent descriptions of Humboldt nests provided by Dawson (1923) and Bent (1940). Almost all of these early nests were found in burned-out hollow redwood snags or stumps. More recently, these swifts have been recorded nesting in fire-created basal hollows or "goose pens" in large living redwoods, as well as in a few other types of cavities in redwood trees (Hunter and Mazurek 2004).

It is interesting that, of the 18 Atlas confirmations, only two were nests in trees: nests with young were reported 25 Jul 1996 in the Gould Grove of Humboldt Redwoods State Park (i27; M. Bailey) and 15 Aug 1995 in Cheatham Grove of Grizzly Creek Redwoods State Park (H23; C. Powell). Both of these trees were more than 12 feet in diameter and 230 feet tall. An occupied nest was also reported in a stump 23 Jul 1999 (g21; B. Mauney). The nine remaining confirmations pertaining to nests were in chimneys or other human-made structures. Chimney nests with young were reported 16 Jul 1995 on Fickle Hill (g15; PFS), 26 Jul 1998 in Ferndale (c21; GCH), and 27 Jul 1995 in Blue Lake (h14; observer unknown). There were five reports of occupied nests in chimneys, spanning the dates of 15 May 1995 in Arcata (f14; D. Anderson) to 24 Jul 1998 in Eureka (e17; L. Bacon-Ogden). More than one chimney with a swift nest was reported from within many of the above blocks.

An unusual nest site was reported at the Blue Lake Fish Hatchery (H15). In 1995, two pairs nested there in a water-filter gallery which consisted of an underground concrete corridor. The nesting chronology for one of the pairs was as follows: they were observed copulating 26 May and carrying nest material 27 May; one egg

Continued on page 392

WHITE-THROATED SWIFT *Aeronautes saxatalis*

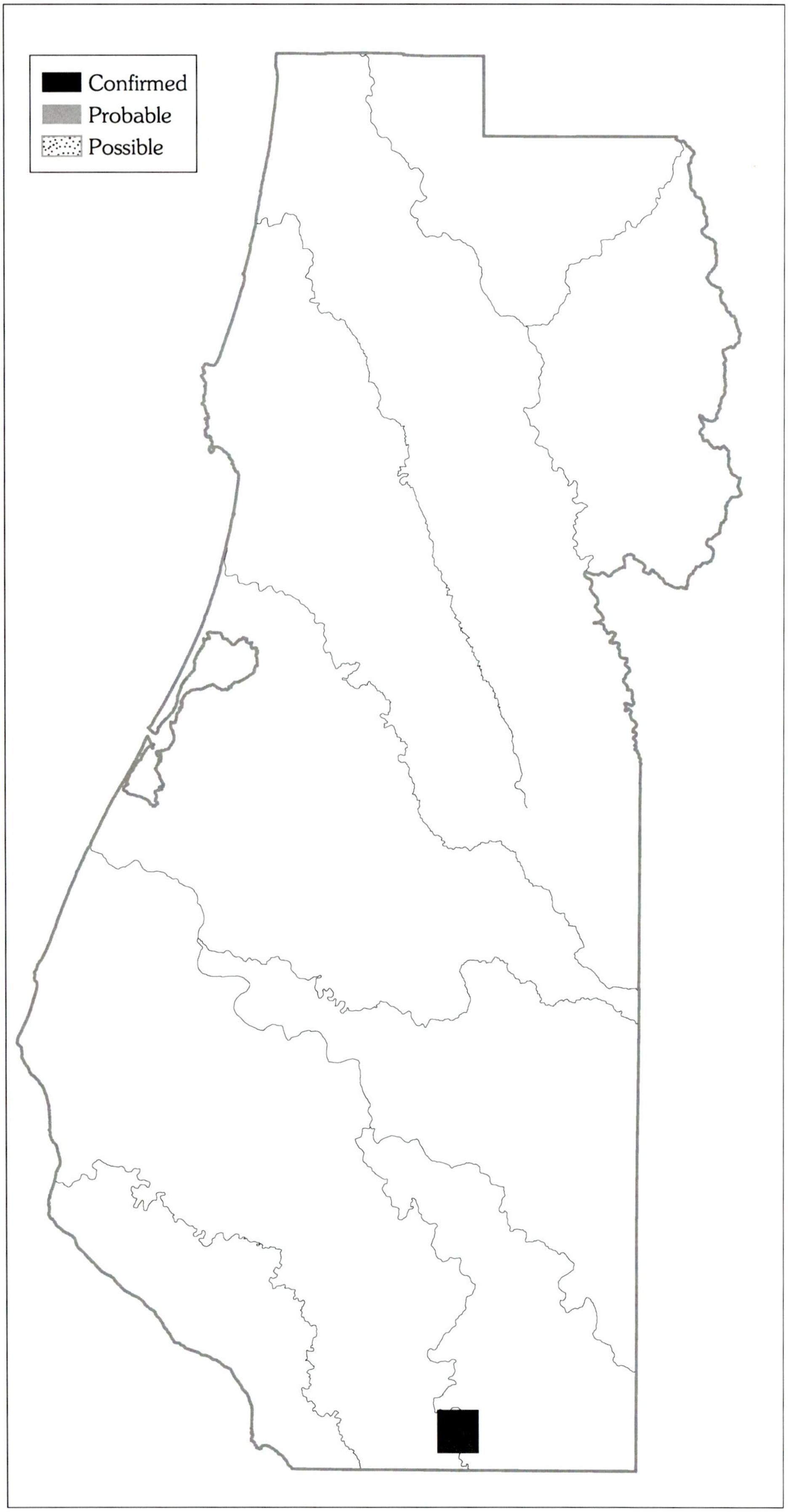

White-throated Swifts are among the fastest-flying birds in North America. The sight of their scimitar forms coursing past battlements of rock—or mocking city traffic far above concrete canyons—is in either case a vision of an animated sky. These swifts are characteristic breeding birds at cliffs, rimrocks, and rock pinnacles across much of the interior western U.S. and locally in southwestern Canada. They have acquired a taste for city life in some places, (*e. g.*, Sacramento), where the birds find nesting sites in cavities of buildings and in expansion joints and drain holes in freeway overpasses.

At the beginning of the Atlas period, White-throated Swift had been recorded only twice in Humboldt County: two near the Weitchpec Bluffs along the lower Trinity River canyon between Hoopa and Weitchpec 4 Jun 1977 (B. Claypole), and one with Violet-green Swallows near Orleans 6 Jul 1977 (R. Escaño; Harris 1996). It therefore came as a great surprise when a small colony of White-throated Swifts was discovered 19 Jun 1998 beneath the Edward G. Wagner Memorial Bridge on Highway 101 (k33) 3 miles south of Benbow (DFx, JCP).

The discovery of breeding White-throated Swifts in Humboldt County owes more to serendipity than to application of talent or effort. Atlasers sipping shade-grown coffee were en route to a meeting with a private landowner living on Reed Mountain. Nature called before reaching the gate, and they pulled off beneath the Wagner Bridge carrying Highway 101 across the South Fork Eel River (k33). As soon as the car doors were opened, the loud chatter of a group of White-throated Swifts flying about the bridge came as an arresting announcement of only the third sighting of the species in the county. That observation certainly suggested the birds were nesting there; subsequently, adults were observed arriving at and departing from a few select drain or ventilation holes beneath the bottom of the span. At least five pairs and 15 birds (highest count reported) were present (DFx, JCP). The formal report establishing confirmed breeding was made 27 Jun 1998, when the voices of multiple swifts were heard emanating from a hole upon arrival of an adult (GCH). Birds were reported at the Wagner Bridge in 1998 as late as 26 Jul (J. Sterling, W. Widdowson *et al.*). Swifts were noted at this site by many observers during the final Atlas year (1999) as well.

The Wagner Memorial Bridge is a typical California Department of Transportation structure of reinforced concrete perhaps 1500 feet long and crossing the South Fork Eel River with plenty of fresh air beneath. Observation of swifts was typically made from a frontage road accessed from the south side. Birds also were occasionally seen at various times of day by distracted birder-drivers traversing the bridge. The South Fork Eel canyon in this area is fairly well incised, grown to pockets of coast redwood near the stream but otherwise clothed in drier-site Douglas-fir, tanoak, and canyon live oak. Vigorous forests and associated riparian habitat probably provide a highly varied potential insect prey base for swifts foraging at and about this site.

The colonization of Humboldt County by White-throated Swifts fits nicely within the range expansion noted in other areas of northwestern California. Mendocino's first historical breeding confirmation was reported 11 Jun 1996 near Hopland (C. Vaughn, R. Kieffer; NASFN 50:993). Two swifts were also seen 20 Jun 1998 not far south of the Wagner Bridge, in adjacent Mendocino County 1 mile south of Piercy where another large bridge crosses the Eel (JEH, TWL). Trinity County's first historical breeding confirmation was recorded 5 May 1997 at the Zenia Bluffs (C. Secor; FN 51:924). Additional nesting sites farther north should be looked for in the future.

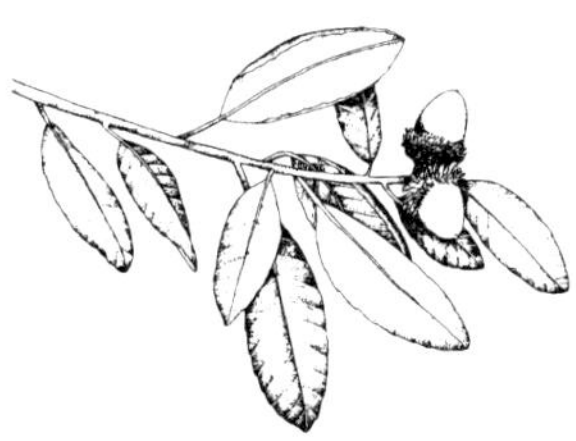

ANNA'S HUMMINGBIRD *Calypte anna*

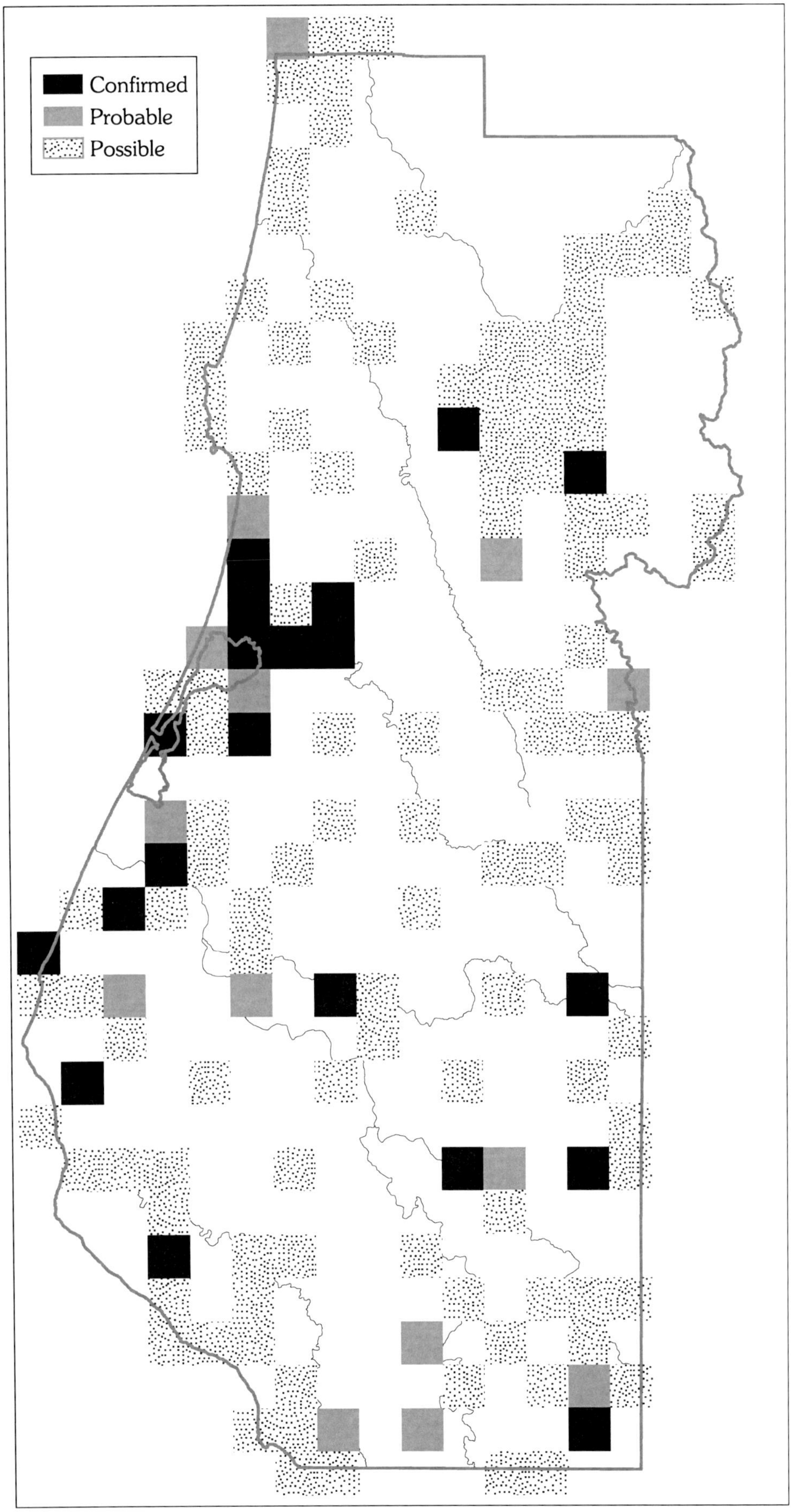

Attractive and common, Anna's Hummingbirds are one of the "gateway birds" for those developing an expanded interest in birds. Hummingbird feeders serve to make their presence obvious, as individuals vie for dominance and chase-flights abound. Adult male Anna's Hummingbirds delight not only with a glittering rose-red head and throat, but with their odd song: an extended, deliberate series of scratchy notes, suggesting a cricket in the shop for a tune-up.

This species has an intriguing history in northwestern California as well as farther north in lowlands of the Pacific Slope. One early Humboldt County observer found it "rather a surprise" that Anna's was "a regular winter resident in limited numbers" (Wilder 1916a). Despite the abundance of the species today, some birders will likewise be surprised to learn that the earliest breeding record for the county was of a nest located along the Mad River at Blue Lake in May 1940 (Talmadge 1953a). Harris (1996) provided additional information supporting the assertion that Anna's Hummingbird has become more abundant in northwestern California in the past 20 years. This was also a period of apparent population increases in Baja California, western Oregon, western Washington, and extreme southwestern British Columbia. Although where they now occur in summer they generally are found throughout the year, Anna's are known to "migrate" in some sense, insofar as scattered birds appear out of the breeding range each winter in such places as the Southwest and the Gulf states. Owing to the species' widespread occurrence and presence throughout the year, the extent to which Anna's Hummingbird in Humboldt County is migratory is almost impossible to determine, although numbers are lower in winter. Harris (1996) considered the species to be a common resident and breeder in northwestern California.

In the experience of those who confine most of their birdwatching to urban and otherwise well-settled areas, Anna's Hummingbird may be perceived as a "town bird", expected chiefly in residential areas and specifically at feeders. This perception might change after a long look at the distribution map. Most of Humboldt County is very rural, yet this species was detected in a rather remarkable array of blocks distributed across most of the county. Anna's Hummingbirds were reported in 35% of the 425 total blocks. Nearly every describable subregion within the county was represented. Detections appeared most frequent in the central-coastal lowlands, perhaps betraying a hint of the extra results provided by "backyard birding". There were also numerous reports on the Hoopa Valley Indian Reservation, in the Mattole River valley, in southwestern Humboldt, and in the far southeastern corner of the county. Detections also came from scattered points along and near the north coast, in the main stem as well as tributary watersheds of all of the major rivers, and at a few locations well into the hills. There were nearly no detections in blocks taking in the higher mountains of northeastern Humboldt.

It was clear that this hummingbird could be expected almost anywhere in the lowlands during the breeding season. Habitats occupied by Anna's varied considerably. In general, towns and residential neighborhoods seemed almost a sure bet. Since prolonged hard freezes in the bottomlands and foothills are infrequent in Humboldt, the county's towns and cities are able to maintain many exotic shrubs, trees, and flowers, a food source seemingly very attractive to these birds. Increases in ornamental plantings (and feeders) in settled areas also may have something to do with increased numbers of Anna's Hummingbirds during winter. Many Anna's also were encountered in streamside woods; this species is a regular inhabitant of riparian zones containing a diverse assortment of trees and shrubs. Abandoned mill sites grown to coyote brush, willows, and alien vegetation hosted them, as did hillside tangles of poison-oak, deciduous oaks, and wildflower patches. These birds have probably increased in abundance inland as well, where they are most abundant in the late shrub/sapling stage of clearcuts (Marcot 1984). Raphael *et al.* (1988) estimated that as a result of timber harvest, Anna's Hummingbirds have increased more than 200%.

Confirmed breeding was noted during the Atlas period in 20 blocks. Anna's Hummingbirds are well known for initiating nesting early in the year. The earliest of four reports of adults carrying nest material was 29 Jan 1999 (D17; GAS). The three other reports were 1 Apr 1996 (F15; GCH), 18 May 1997 (D29; TWL), and 1 Jun 1996 (N33; C. Secor). Nests of this species are not particularly difficult to locate. Somewhat surprising, therefore, is that only five were specifically reported during the course of the project. Three nests with eggs were noted, 23 Mar 1999 in north Arcata (f14; K. Irwin), on 19 May 1996 (H23; C. Powell), and 19 May 1997 (N11; *fide* JMH). The female incubating eggs in north Arcata was seen to fly from another nest—where she had been feeding two young that were about to fledge! Another nest with young was reported in the low boughs of a Douglas-fir near Blue Lake Fish Hatchery (H15) 16 May 1996 (C. Gutiérrez). An occupied nest was found 4 Jul 1995 (g15; RJA). The most commonly reported confirmation evidence was of fledged young, reported from 13 May 1997 at Blue Lake (h14; K. Nelson) to 10 Aug 1997 (k27; JG). One of these reports was in May, none were in June, four were in July, and two fell in August.

CALLIOPE HUMMINGBIRD *Stellula calliope*

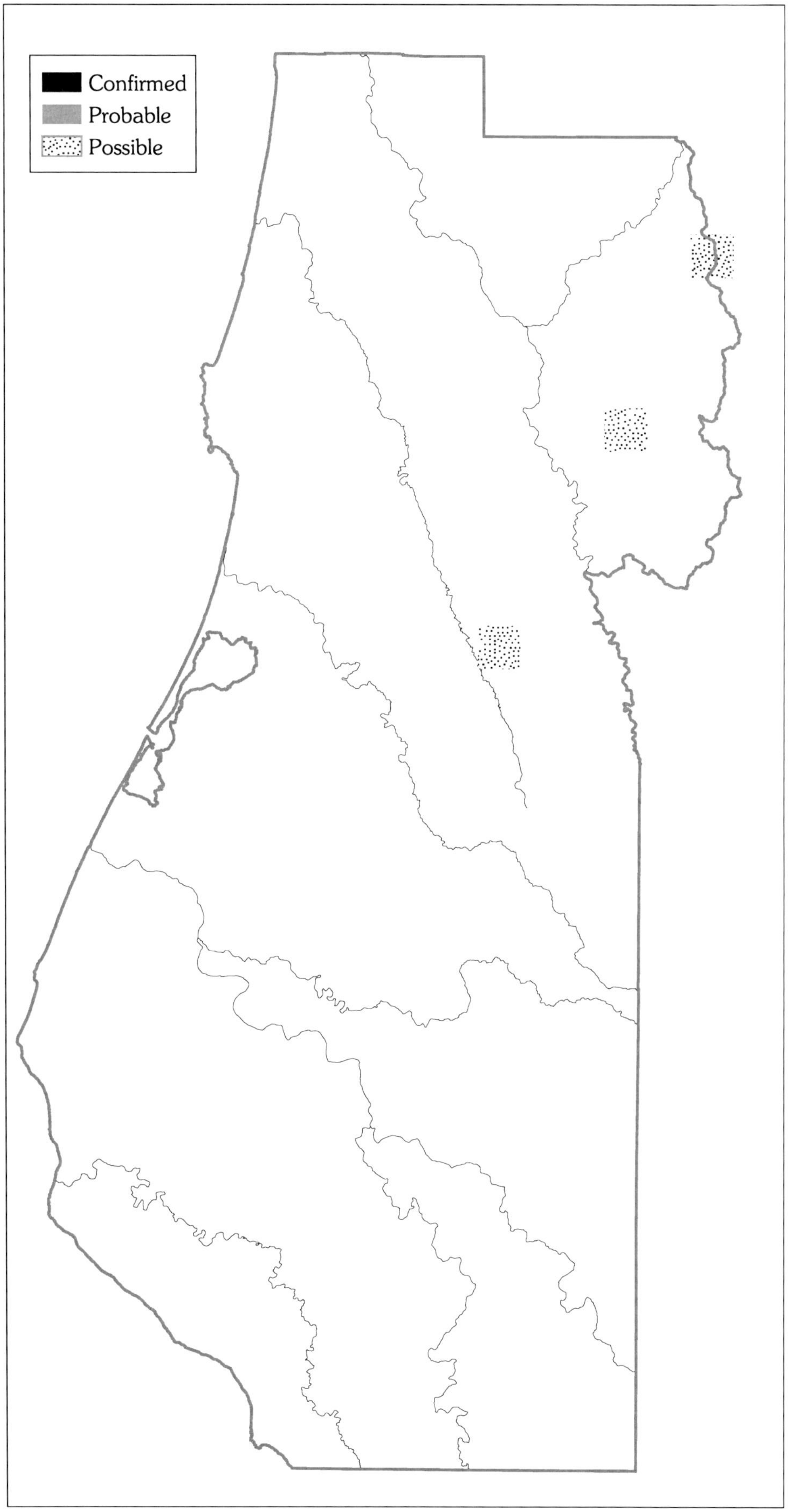

Calliope Hummingbirds are tiny, delicate birds, 10 to the ounce and none but precious. They know the full charms of a subalpine meadow in midsummer, bucking afternoon breezes in flights from one clump of penstemon to another. Set against a backdrop of raw-boned peaks, their small size and endearing looks surely strike a contrast in mass and texture. While many Calliope Hummingbirds inhabit parched brushfields in forest clearcuts, others live in moist mountain glades. Some can be found in places of great beauty, as witness their presence in many of the classic national and provincial parks. In passage through the lowlands, the occasional spring migrant enjoyed at a backyard feeder brings the satisfaction of hosting a high-elevation specialist for a day or two before it moves on.

The extensive summer range of the Calliope Hummingbird in the West includes the higher mountains of interior northwestern California. In this sense, the species is one among a large suite that is characteristic of the Cascade Range, the Sierra Nevada, and northern forests in general and that occurs locally in the mountains of eastern Humboldt County. Calliope Hummingbirds, however, have been reported only infrequently in the county, with reports scarcely annual. This species has come to be recognized as a "casual to rare spring migrant in coastal lowlands (Apr–May) and along higher inland ridges (Apr–Jul)" in northwestern California (Harris 1996). Although no site is known to host them reliably, this may be as much a result of the remoteness of most potential breeding habitat as to the scarcity of the birds.

Hiking trips through possible Calliope Hummingbird habitat purely in search of this species remained the stuff of wishful atlaser daydreams. The need to complete blocks during available birding time dictated a less thorough approach. In the future, the pursuit of these hummingbirds after the close of the spring migration in likely-looking places in the mountains of northeastern Humboldt County may reveal greater presence than is now perceived.

During the 5 years of the project, sightings were made in three blocks in the mountains, and a few lowland spring migrants were noted as well. In each case, no more than possible breeding status was established by the mere presence of birds in appropriate habitat in the nesting season. These reports were of single male Calliope Hummingbirds displaying in brushfields along Big Hill Road (o10) 28 May 1996 (JMH) and at Mud Springs (also o10) 2 Jun 1998 (KMS), and single birds 4–5 May 1996 along Red Cap Road (q06; G. Friedrichsen) and 20 Jun 1998 along Friday Ridge Road (L15; COg).

While Grinnell and Miller (1944) indicated breeding by Calliope Hummingbird on South Fork Mountain in nearby Trinity County, the details do not support conclusive breeding there (Hunter and Hazard 1998). Within Humboldt County, pre-Atlas records such as a "courting male" 9 Apr 1985 on Waterman Ridge (J. Sterling), a female 8 Jun 1984 at Mt. Lassic (J. Sterling, K. Rosenberg, L. Doerflinger), and displaying single males along Big Hill Road 30 Apr 1994 (DFx, RHw) and 3–24 May 1992 (GCH, RHw, G. Lester) suggested probable breeding. Thus, it would appear that the mountains of eastern Humboldt County lie at the extreme western edge of the regular summer range of this species.

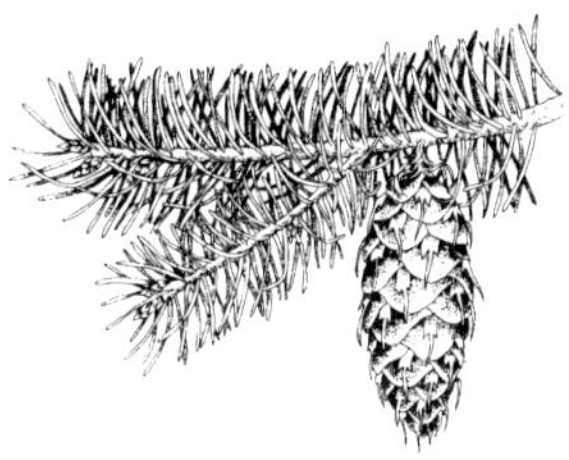

ALLEN'S HUMMINGBIRD *Selasphorus sasin*

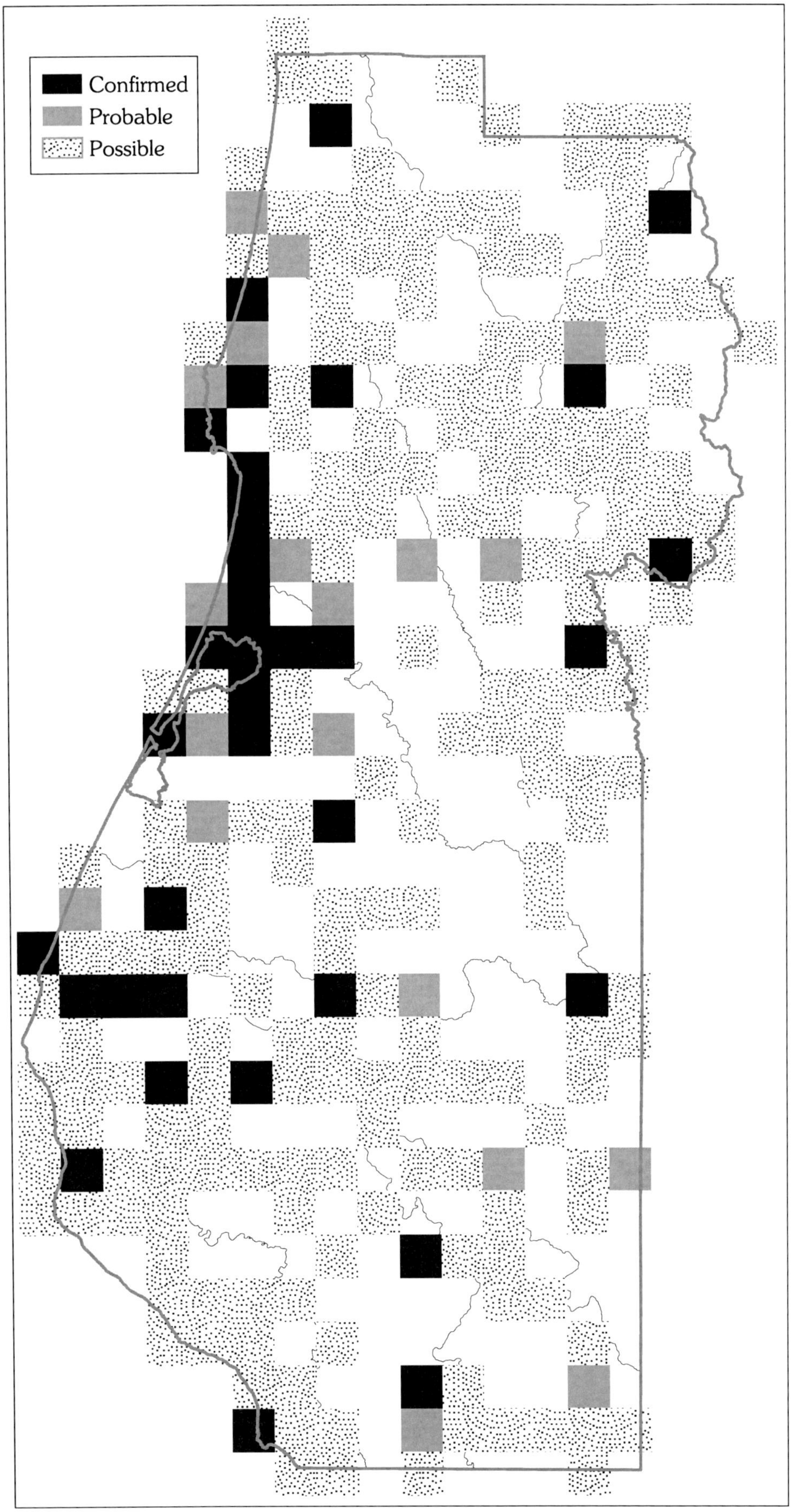

Allen's Hummingbird is a common migrant and abundant breeder in Humboldt County (Harris 1996). Early ornithologists recorded this species often. A nest with eggs was collected as long ago as 8 Apr 1926 in Eureka (G. Atwell; specimen apparently later lost). Because it has been well recorded historically, is numerous, is widely seen at hummingbird feeders, and often nests in suburban backyards, one might think that a great deal should be known about the status, movement patterns, and population dynamics of the species here. Surprisingly, however, much remains unknown about Allen's Hummingbird in Humboldt County.

Atlas fieldwork resulted in a great many detections of this species—or birds believed to be this species. Yet it did little to make clearer the precise breeding distribution within the county, nor to shed light on seasonal movements made by the breeding population and progeny. This was due not to any shortcomings by atlasers, but rather to an inability to draw inference from what was observed. Questions linger because, locally, adult male Allen's Hummingbirds are uncommonly seen after late spring; female and immature Allen's/Rufous hummingbirds, which cannot be readily separated in the field, constitute most of the population from late spring into early autumn. This matter is complicated by the broad south- and eastward movement of male Rufous Hummingbirds after nesting takes place, as early as late May. Most of this movement occurs at higher elevations, where the birds appear to pursue the peak period of flowering as the general zone of blooming retreats upslope throughout the West. A scenario then develops annually in which the countryside at middle and lower elevations is full of hummingbirds that, strictly speaking, are unidentifiable.

Recognizing from the outset of the project that broad-brush Atlas methodology could not achieve the fine strokes necessary to "solve" this problem, most records of non-adult male *Selasphorus* hummingbirds were treated as Allen's. This decision was made to conform to the existing understanding that Allen's is the common breeding *Selasphorus* hummingbird wherever birds of that genus occur during the period March through May. The portrait of the breeding distribution of *Selasphorus* in Humboldt County should accordingly be interpreted with caution. During the Atlas period, birds treated as described above were detected in 240 (56%) of the 425 total blocks. Breeding was confirmed in 28 blocks.

Classically, Allen's Hummingbird has been considered a breeding bird in California west of the Sierra Nevada and away from the Central Valley floor. Some published sources stress that the nesting population is concentrated along and near the coast. If this is so, and assuming that all reports of confirmed breeding refer to Allen's, then a more coastal bias is supported by the Humboldt map. However, as was the case for many of the county's more common breeders, everyday "backyard birding" effort around Humboldt Bay probably skewed the results. In any case, the most unambiguous reports of confirmed breeding came from coastal areas. Examples include birds carrying nest material 7 Apr 1996 (f12; TWL), 13 Apr 1996 (F07; JEH), 12 Jun 1999 (F09; JEH), and 23 Jun 1996 (B27; GCH); nest building 22 May 1996 (c23; COg); a female with a brood patch killed by a house cat 17 May 1996 (F13; JEH); a nest with two eggs (full clutch size) 9 Apr 1995 (F11; JEH) and nests with eggs 25 Apr 1996 (g15; JCP) and 14 May 1995 (e10; GAS); a nest with young 10 May 1996 (H15; PFS); and the feeding of young, seen 11 Jul 1998 (D25; JCP, DFx). The span of dates for these activities suggests a somewhat extended nesting period along the coast, possibly involving multiple broods.

Assuming greater significance with this coastal bias taken into account are the widespread and scattered confirmed and probable breeding reports well inland in both northern and southern Humboldt. Examples of Atlas records from inland locations include copulation witnessed 13 Apr 1996 in the vicinity of Three Creeks (L13; GAS); the carrying of nest material in the vicinity of Mill Creek of the lower Trinity River 8 Apr 1997 (N09; *fide* JMH) and 18 Apr 1995 near Oak Knob (N15; P. Carlson); nest building 22 May 1998 at Groves Prairie (P13; KMS); and fledglings 19 Jun 1996 near Dinsmore (N23; GAS) and 25 Jun 1996 near Orleans (P05; KMS). Yet the question remains: are there appreciable numbers of Allen's Hummingbirds nesting in the interior of the county? If not, what might be limiting factors? These questions may be answered by future investigators.

Irrespective of the species question, it is plain that *Selasphorus* hummers are to be expected nearly everywhere in the county during the warmer months. Suggesting local scarcity was a cluster of no-detection blocks in the midst of intensively managed commercial timberland in extreme northern Humboldt and two clusters situated in the oak-grassland mosaic of east-central county and northeast of Garberville.

Described as casual in winter (Harris 1996), Allen's Hummingbirds are essentially absent until the first northward-moving adult males appear in early March. Occasional pioneers are reported earlier. Since none are known to breed north of the Coquille River on the southern Oregon coast, it seems reasonable to assume that some of the birds arriving in spring are not passing through, but are local breeders. While Allen's Hummingbirds nesting to the north of Humboldt must also pass southward through the county, their true status in fall is greatly confused by the presence or assumed presence of Rufous Hummingbirds.

BELTED KINGFISHER *Ceryle alcyon*

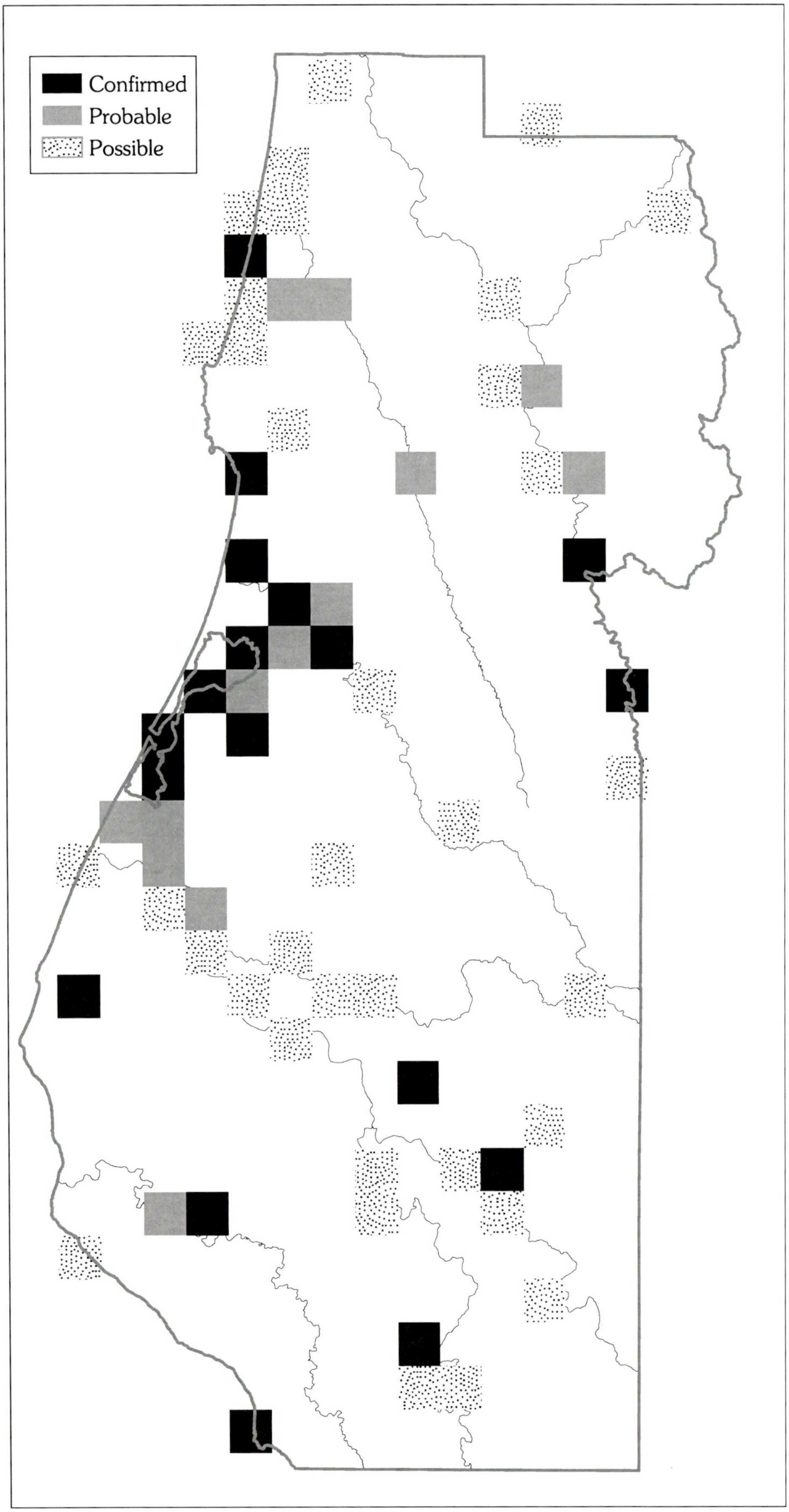

Clear streams, rivers, lakes, and estuaries throughout much of North America provide habitat for Belted Kingfishers. The sight of a kingfisher perched at attention on a piling or flashing past with irregular wingbeats adds immediate appeal and a stamp of wholesomeness to the surroundings in which we see them. Nervous and vigilant, these birds appear ever watchful, scanning the water for small fishes foolish enough to closely approach the surface. During spring and summer they concentrate their activity in the vicinity of steep earthen banks and bluffs, in which pairs excavate a burrow terminating in a nesting cavity. This is a creature so widespread and so recognizable that many birders who can recall any number of first encounters with "life birds" might remember their first kingfisher only as a dim memory from pre-birding days of childhood.

Belted Kingfishers are familiar residents of Humboldt County. Harris (1996) perfectly characterized Belted Kingfisher in northwestern California as a "common resident and breeder throughout the region along rivers, lagoons, lakes, bays, and coastlines . . . nests in widely scattered clay or sand banks." While not abundant, they are numerous and occur widely along the coast and inland. Atlas surveyors found them in 66 (16%) of the 425 total blocks. The map furnished no surprises, indicating that these birds can be expected along and near the northern two-thirds of the Humboldt coast, at isolated ponds, and in association with the county's larger rivers and many of their tributary streams. As might be expected of a species so closely tied to shores and river corridors, the map reveals a somewhat linear pattern of Atlas blocks, with reports corresponding to estuarine shores, the Humboldt Lagoons, and the Klamath, Trinity, Mad, Van Duzen, Eel, and Mattole rivers and some of their tributaries. Straightforward access to lands within the general Eel River drainage, combined with the size of that river system, may have contributed to a fairly thorough stippling of Atlas blocks with detections in south-central Humboldt County. The South Fork Eel River, in particular, cuts its way through steep and unstable banks of soft sandstone, affording much potential nesting substrate in proximity to fish-bearing waters.

Because a great amount of hilly countryside at middle and higher elevations is not dissected by large streams, broad expanses of Humboldt County are essentially uninhabited by kingfishers. The rugged timberlands of extreme north-central Humboldt, the northeastern mountains, the incised (and largely private) foothills of interior southwest county, and the open grassy hillsides of the upper Mad River drainage are kingfisher poor. The coast from Cape Mendocino southward is watered with small, short streams typically lacking steep, exposed banks; the Lost Coast from Punta Gorda southward is especially inhospitable. Pockets of suitable breeding habitat may be safely assumed to occur at lower elevations along clear streams within the county where no detections were reported; limited or forbidden access during Atlas fieldwork precluded full investigation.

Although Belted Kingfishers seldom, if ever, forage over the open ocean off Humboldt County, the birds nevertheless are a frequent sight in the breeding season along portions of the outer coast. Exposed earthen banks beneath the sod caps of small promontories are known to be used for nesting, as at the south end of Freshwater Lagoon beach and just east of Clam Beach. In such settings, kingfishers forage at the nearest body of fresh water. Roadcuts may also provide nesting possibilities. Banks suitable for nesting need not be especially commanding, but they must provide some protection from potential predators. Certainly, lack of appropriate nesting substrate becomes a limiting factor in generally flat terrain, such as around the shores of North Humboldt Bay.

Considering how conspicuous Belted Kingfishers are when carrying food, it is not surprising that 12 of the 18 breeding confirmations were based on this behavior. Observations of adults carrying food spanned 20 Apr 1998 (d18; M. Wachs) to 20 Jul 1996 (L27; D. Diehl). All four reports of fledglings were in August, the earliest 3 Aug 1995 (H15; PFS).

ACORN WOODPECKER *Melanerpes formicivorus*

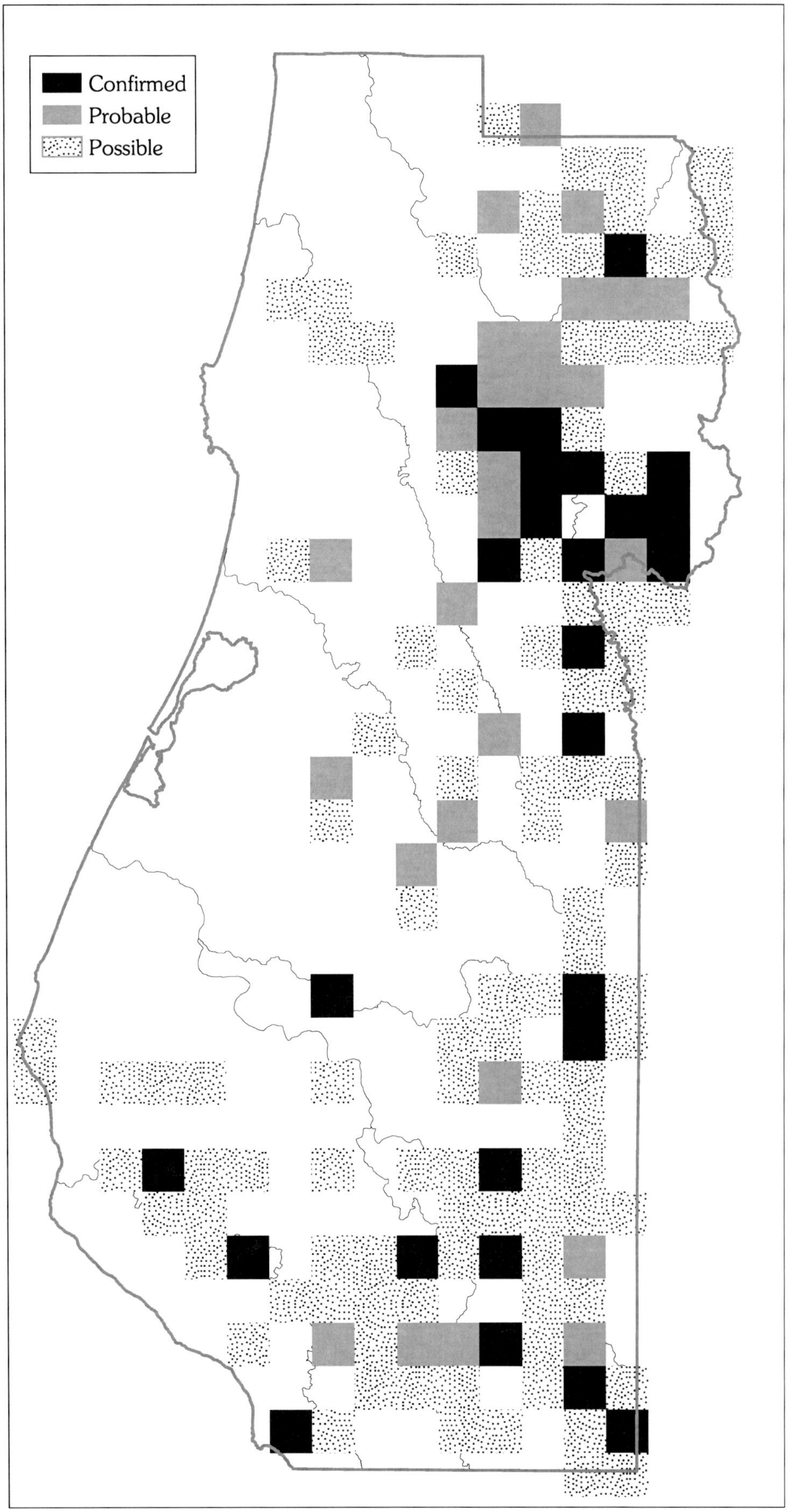

Think of Acorn Woodpeckers, and what picture comes to mind? In the same instant as one conjures an image of the bird there comes, too, the upper canopy of nearly any oak. Stick with the dreamscape for a moment, and there might now be three or four of the birds, posturing, gazing warily about, fluttering a few feet to a new perch. Their habitat, and the manner in which these woodpeckers pass their lives, is distinctive. Even a first encounter with their kind leaves the observer with an impression that is vivid and persists through the course of further experience.

This species is well named, for across its extensive western North American range it is closely tied to oaks. As California supports many oaks, Acorn Woodpeckers are abundant and widespread in the state. Throughout the lower and middle elevations west of the Cascade–Sierra Nevada axis occur a great many forest and woodland communities rich with many species of oak. The presence of even one oak species in stands otherwise characterized by conifers appears to attract them. In northwestern California the Acorn Woodpecker is a common resident and breeder (Harris 1996). During the Atlas period, birds were encountered in 156 (37%) of the 425 total blocks. Breeding was confirmed in 27 blocks and was considered probable in 28.

Valleys and foothills host the greatest numbers of Acorn Woodpeckers found in Humboldt County. With oaks absent or of only spotty occurrence in Humboldt's foggy northwestern section, it is in the interior of the county that this species comes into its own. Beginning just beyond the zone of persistent daytime overcast and extending eastward lies an expanse of countryside with abundant oaks, a region overarched by sunny summer skies. It is within this broad belt of warmer, drier forests and groves that most of the Acorn Woodpeckers found in the county live throughout the year.

Acorn Woodpeckers are variably common in the foothills of the middle elevations, where they are routine forest birds wherever California black oak is found. Provided oak at any reasonable density is present, these woodpeckers may occupy any slope aspect. They may occur in somewhat harsh places such as just beneath ridge-top openings at the western edge of the range of tanoak in the county, where only a sparse and ragged edge of Oregon white oak fronts wind-whipped meadows. In the incised and hot Trinity and Klamath River canyons, canyon live oak seems important to them. North of Cape Mendocino, Acorn Woodpeckers were conspicuously scarce or absent along the coast and in coastal bottomlands, where oaks, as a rule, do not occur.

These birds are less abundant in thick stands of Douglas-fir/evergreen hardwood forest than in the more open oak woodlands and savannahs. In denser stands of trees, shade-intolerant "true" oaks do not prosper, and it is clearly the presence of tanoak that is of particular importance to these woodpeckers. Given its ability to reproduce in its own shade, tanoak offers a food source otherwise lacking in these situations. Especially favored by the birds are broken or vertically diverse conifer stands well grown to tanoak and other oaks and in which remnant tall, dead-topped or snag Douglas-firs or pines poke up above the canopy. The latter trees are persistently used as "granaries" wherein Acorn Woodpeckers typically pound hundreds, even thousands, of acorns into neat holes for later retrieval.

In areas where tanoak more closely approaches the eastern edge of the fog belt, this woodpecker maintains at least a low density presence as far west as the trees are numerous. The importance of tanoak is easily noted in lower Redwood Creek, a fog-swaddled drainage supporting some of the world's tallest trees. Despite the dominance of redwoods, habitat exists for Acorn Woodpeckers in tanoaks interspersed amongst the giant conifers.

With their tie to oaks well understood, the absence of Acorn Woodpeckers from the highest mountains of Humboldt County is not surprising when one considers that no species of large oak extends to such places. Shrub oaks of several species, usually growing to a height of no more than 4 or 5 feet, are characteristic of snow-zone forests and forest edge above 4000 feet. While locally abundant, they appear to offer no attraction for the birds. Perhaps their sparse, small acorns—growing within the lunging range of terrestrial predators—are a food source these woodpeckers disregard. Although the location of many Atlas blocks with detections in the northeastern and east-central parts of the county might suggest presence of acorn woodpeckers at the higher elevations, their presence there is largely a result of incursions of California black oak along southwesterly ridges up into some of these higher blocks. Following these stands, the birds do climb to about 4000 feet in northeast county and to 4500 feet (sparingly) on South Fork Mountain, but occur there in only modest numbers.

Intensively managed expanses of industrial forestland in extreme northern and interior northwestern Humboldt offer little habitat. It is conceivable that removal of snags and large hardwoods during logging has affected Acorn Woodpeckers in these areas. Because of the low density and diversity of oaks, numbers of these woodpeckers may have been historically low there.

Since development pressure in interior Humboldt is comparatively light, it would appear that this species' future is secure. However, the recent spread of a fungal pathogen, *Phytophthora ramorum*, which causes Sudden Oak Death Syndrome, threatens a startling spectrum of indigenous hardwoods, oaks being the most significant among these because of their sheer volume. It is yet unclear to what extent any manifestation of the disease might affect populations of oak-dependent wildlife, but dramatic declines in oaks would clearly exert profound effects on Acorn Woodpeckers.

Acorn Woodpeckers may begin breeding quite early if conditions are favorable. A good description was submitted of begging young in a nest cavity 1 Mar 1996 at Hoopa (m10; J. Biggin). However, during the Atlas period most breeding activity started in earnest in April. The earliest fledglings were reported 22 Jun in 1997 (P11; K. & E. Bentler) and 1998 (L31; S. Rovell). The latest active nest was one with young 2 Aug 1999 (N11; JMH).

RED-BREASTED SAPSUCKER *Sphyrapicus ruber*

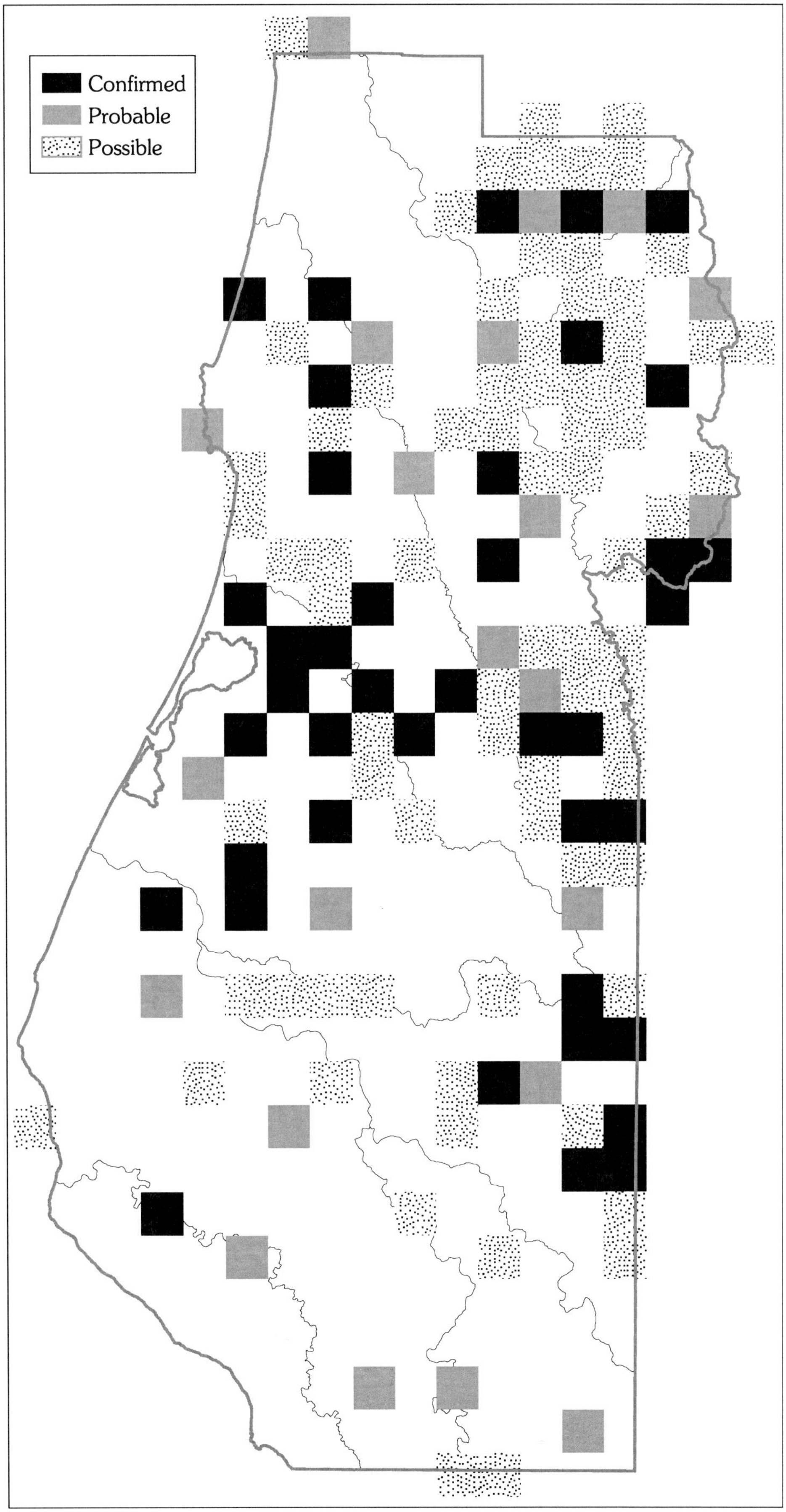

Leisurely behavior and bold plumage patterns cause Red-breasted Sapsuckers to be admired by nearly all who notice them. Given to a lifetime of drilling and revisiting sap wells chiseled into tree cambium, they raise the concern of uninitiated observers worried for the welfare of favorite fruit or ornamental trees. Will the dozens—or hundreds—of holes opened in the bark invite infection and lead to the tree's demise? Should the incessant tapping and drilling be discouraged? Such concern may be allayed through observing any number of older, healthy trees thoroughly tattooed with healed sapsucker wells—some surely the work of birds long since gone from the scene. Scores of North American tree species have coevolved with the four species of sapsuckers, and the birds have not been known to be the chief agent in the demise of any. The constant outpressure of sap in a healthy tree would appear to repel most pathogens. On the credit side, sapsuckers in general have been recognized as "keystone species", whose activities create opportunity for wasps, warblers, hummingbirds, and other creatures able to exploit the flow of nutritious fluids and the insects ensnared within (Ehrlich *et al.* 1988).

Long regarded as a distinctive race of Yellow-bellied Sapsucker (*S. varius*), the Red-breasted Sapsucker is currently granted full species rank. However, birds of this form are known to hybridize with Red-naped Sapsucker (*S. nuchalis*) in the Cascades, and, adding to the potential for confusion, there are two subspecies. In Humboldt County, the southerly form *S. r. daggetti* is an uncommon to locally fairly common resident. Red-breasted is the only species of sapsucker regularly occurring in the county, with the two above-mentioned species and Williamson's Sapsucker (*S. thyroideus*) only rare visitors.

This woodpecker has long been recognized as a regular element in the Humboldt County avifauna

(Townsend 1887). Summarizing the species' occurrence in northwestern California, Harris (1996) termed Red-breasted Sapsucker an "uncommon resident and breeder in riparian and mixed woodlands throughout the region. Nests in cutover second-growth forest, forest edges, riparian, and residential areas." The experience of atlasers supports this characterization. During the 5 Atlas years, Red-breasted Sapsuckers were detected in 137 (32%) of the 425 total blocks and were confirmed breeding in 40.

In the breeding season, Red-breasted Sapsuckers are birds of forests and unmanaged woodlots, and, to a lesser degree, riparian woodlands. It has been the impression of Humboldt birders that, in spring and summer, sapsuckers are most frequently encountered where forest cover is extensive. The Atlas map suggests a broad association with conifer forests. Of interest is the vaguely two-parted distribution exhibited by this species. Within about 20 miles of the eastern border of Humboldt lie several significant, more-or-less contiguous clusters of detection-blocks. This region, extending from north of the Klamath River south into the Lassics, corresponds to the range of white fir dominated montane conifer forests. It would appear either that abundance of Red-breasted Sapsuckers within this zone is greater, or that they are more readily detected. Surely, it is in the mountains of eastern Humboldt County where this species most closely approaches "routine" status in the nesting months. The prevalence of dense stands of white fir in this section of the county provides a boon for woodpeckers (as well as secondary cavity users) in the form of stressed, damaged, dying, or dead trees. This generally mountainous area of the county also supports such higher-elevation forest birds as Hammond's and Dusky flycatcher, Mountain Chickadee, Townsend's Solitaire, and Fox Sparrow. Indeed, if one disregards the more scattered detection-blocks occurring in the western portion of the species' range in the county, the remaining area closely resembles the maps for each of those species.

On the west side of the county, occurrence of Red-breasted Sapsuckers during the study period lay largely within the fog belt. In this section of Humboldt, the birds were encountered in, or at the edges of, forests of coast redwood, Douglas-fir, grand fir, bigleaf maple, and red alder. The old-growth redwood parks evidently held no particular attraction for them, for comparatively few detections were made in such places. Rather, they were found more commonly in older unmanaged timber, no-cut stream buffers, and at forest edges. Sapsuckers routinely nest and forage in or near clearcuts, provided a good number of snags, dead-topped trees, or live "cull" trees are left standing (Marcot 1983). The stretch of countryside from near Big Lagoon south through the low hills inland from Humboldt Bay, an area peppered with breeding confirmations, is the section of the county in which grand fir is most abundant. This unheralded species is in many ways an ecological analog to white fir, and probably provides similar opportunities for cavity excavation. Protected, wind-firm snags of various sizes are often numerous in unmanaged stands containing this tree.

The earliest reported probable breeding observations were of courtship 25 Apr 1999 (m25; JEH, TWL) and nest building 1 May 1998 (e10; TWL). Among the 40 blocks with confirmed breeding by Red-breasted Sapsucker, nest with young was the most frequently reported code establishing this status. Twelve such reports all fell during a 6-week span from 7 Jun 1996 (F17; H. Brown) to 21 Jul 1998 (P13; S. Taylor). All five occupied nests were reported in June. Fledged young were noted as the best evidence of breeding in 11 blocks, from 22 Jun 1996 (F7; JEH) to 10 Aug 1997 (L05; P. Herrera). Not surprisingly, the eight observations of adults feeding young were also made during a narrow date range, from 8 Jun 1998 (p14) to 22 Jul 1996 (P05; both KMS).

Red-breasted Sapsuckers generally are considered resident in northwestern California, yet the appearance of individuals away from breeding habitat in fall, winter, and early spring indicates that at least some movement takes place. This species is known to increase in abundance in lowlands of the Pacific Northwest during cold and snowy weather. In Humboldt County, it is typical for individual Red-breasted Sapsuckers to appear in residential yards of the near-coastal lowlands. A child's first sapsucker, discovered industriously working an old fruit tree on a sunny morning, is a sight capable of inciting astonishment and wonder. Those with a great concern for the welfare of individual trees, and birders who sometimes throw up their hands at the identification puzzles posed by hybrids and backcrosses, might tire of these birds. On all other counts, they are widely enjoyed.

DOWNY WOODPECKER *Picoides pubescens*

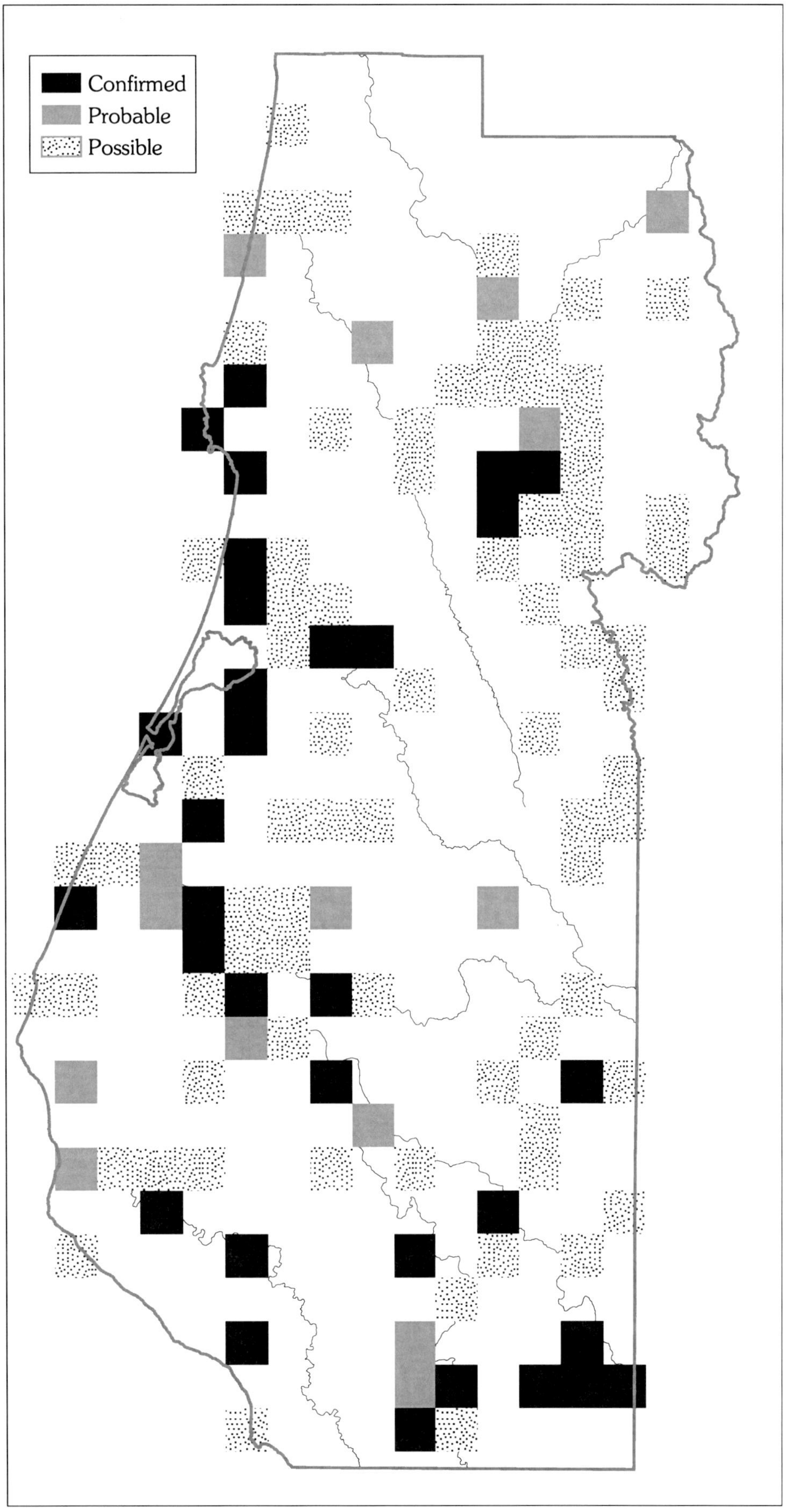

Petite and active, Downy Woodpeckers charm many to whom most small birds remain anonymous or unknown. An ability to maintain a strong presence in the tread of the human footprint has allowed it to become among the continent's most familiar backyard birds. Despite its small size and the leafy shadows in which it forages, it is a conspicuous species because of its frequent appearance in yards, parks, orchards, riparian groves, and any manner of nondescript regrowth woodland along rural roadsides.

Harris (1996) characterized the Downy Woodpecker in northwestern California as a common resident and breeder, most frequent in lowlands and inland valleys. Atlas results were consistent with this description. Downy Woodpeckers were detected in 29% and confirmed breeding in 8% of all Atlas blocks. Surveys revealed that they are to be expected over most of the lower elevations of Humboldt County, especially along the courses of the Trinity, lower Mad, and Eel/South Fork Eel rivers. The mapped breeding distribution is interrupted somewhat by a north to south swath of no-detection blocks in central Humboldt County, possibly because atlaser access was limited there to dense hillside tanoak-and-Douglas-fir forest interspersed with meadows. Another possible explanation is that much of this swath is composed of intensively-managed coniferous timberland with few bottomland riparian corridors.

Downy Woodpeckers in Humboldt County inhabit groves, woodlots, streamside trees, and forest edge at lower elevations. Hardwood trees and shrubs predominate over conifers, or are present exclusively, wherever these woodpeckers are numerous. The vigorous red alder regrowth widespread in logged areas within the fog zone, particularly around and north of Humboldt Bay, is commonly occupied. These woodpeckers were found during 11% of all point counts within riparian habitat along the Eel River during 1999 (R. Hewitt, unpubl. data). Inland, much of the population occurs along streams, with black cottonwood and the ubiquitous alder affording much of the foraging substrate and opportunities for cavity excavation. During the Atlas period, Downy Woodpeckers were occasionally noted in oak woodlands well upslope from the nearest creek, but they seemed absent from closed-canopy forest at middle and higher elevations. It was clear that much of the great expanse of general hillside forest across interior Humboldt County was rich in Hairy Woodpeckers and poor in Downies.

The earliest probable breeding evidence gathered was of a bird excavating a nest cavity 20 Mar 1996 (D17; GAS). The only courtship behavior was observed on 7 May 1998 (D21; L. Karsteadt), and the first breeding confirmation was of adults feeding fledged young 14 May 1996 (e19; M. Wachs). Active nests provided the highest-ranking evidence in seven blocks and spanned the dates of 28 May 1995 (F31; KMS) to one with young 11 Jul 1998 (H15; KMS). Adults were still feeding young as late as 11 Jul 1999 up the Little Van Duzen River (N25; DFx).

HAIRY WOODPECKER *Picoides villosus*

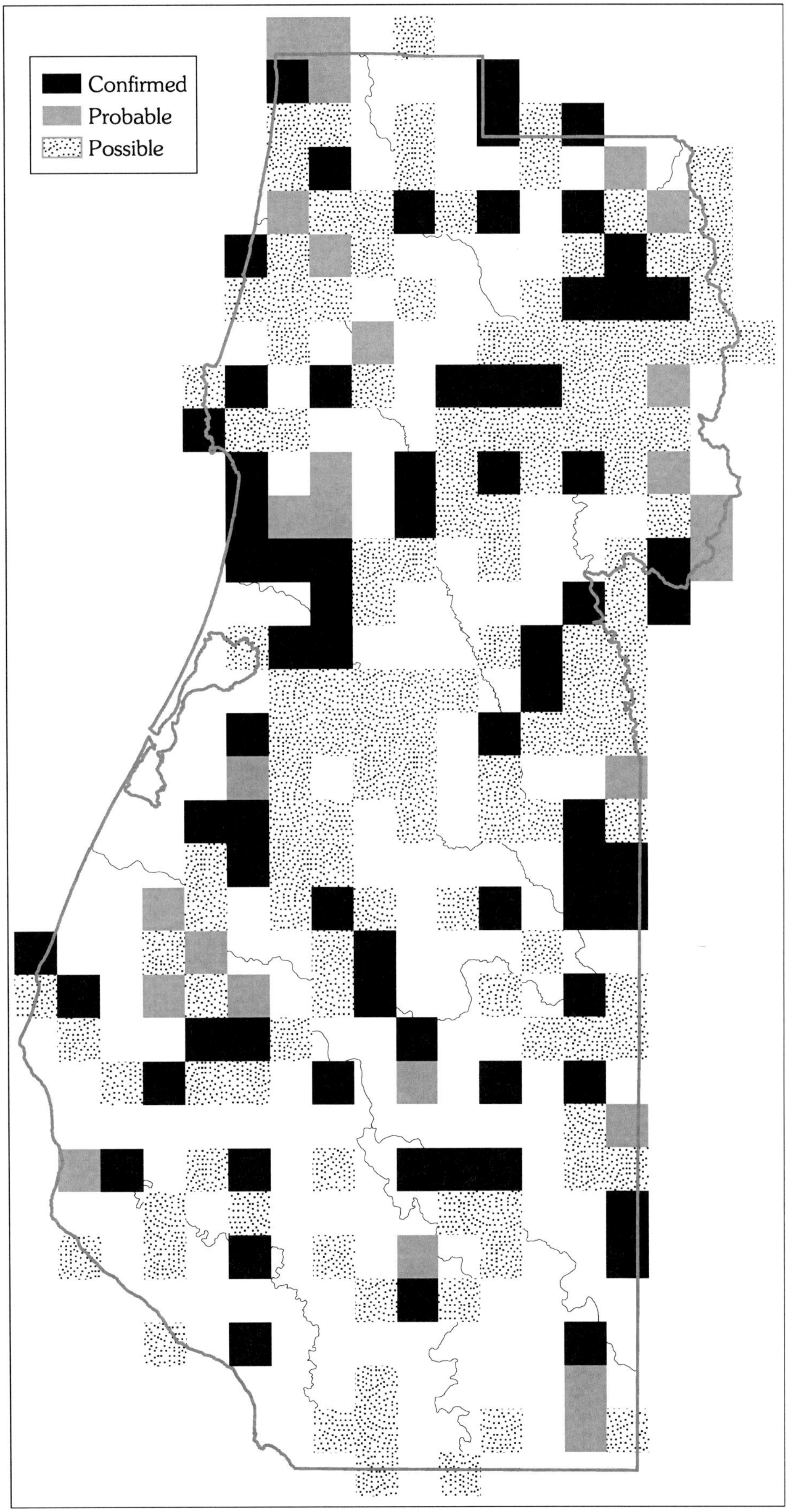

Sharp, slowly repeated *peek!* calls coming from a stand of large trees announce the presence of a Hairy Woodpecker. These are small woodpeckers, yet powerful enough to secure a home in nearly any kind of forest. Big limbs, dying understory trees, snags, and down logs are common in the places they favor, offering opportunities for any bird able to exploit the bark-beetle larvae they support. With their heavy pointed bills and ability to forage in differing forest structures, Hairy Woodpeckers have found success. The species inhabits a vast range, from the northern edge of the boreal forest south across North America to Panama. The great sprawl of forested country in Humboldt County hosts these birds in large numbers.

Hairy Woodpeckers were commonly and widely seen in Humboldt throughout the Atlas period. The call-notes, rattle series, and drumming of advertising males were often detected, as were many birds seen flying across roadways, perched on the sides of exposed snags, or glimpsed in backyard trees. Truly, this woodpecker was well represented in the county, with reports from 54% of all blocks. No section of the county was clearly without Hairy Woodpeckers, although there were small clusters of contiguous blocks in north-central, western, and southern Humboldt that lacked detections. Perhaps the most conspicuous of these zones corresponds to the open coastal lowlands from the Arcata Bottoms southwestward through the lower Eel River delta. Harris (1996) thought that "this may well be the region's most widely-distributed woodpecker, given the abundance of suitable habitat" and categorized Hairy Woodpecker as an "uncommon to common resident and breeder" in northwestern California.

These birds were encountered in a rather remarkable array of forest types during the Atlas period, although nearly all areas have at least some conifer component. Near the coast, they were found in old-growth coast redwoods; in stands of Douglas-fir, grand fir, Sitka spruce, and western hemlock; and in cutover land grown back to unmanaged red alders with some conifers. Along rivers and smaller streams, the riparian zone of black cottonwood, willow spp., white alder, bigleaf maple, and similar trees was used. At middle elevations in the interior of the county, the broad and variable zone of Douglas-fir, tanoak, and California black oak was widely inhabited. Forests and uncut blocks in the higher mountains within the white fir zone were also well populated; presence of Hairy Woodpeckers was obvious along most of the eastern fringe of the county.

While Hairy Woodpeckers are birds of the forest, they appear routinely in burns and clearcuts. They are attracted by forestry practices in which snags or live "cull" conifers are retained on a logged site (Marcot 1983). They may situate their nest cavity well out in the open in the side of a barkless snag, or in a green tree with a long unhealed scar or rotten heart. An adult was watched carrying food to a nest only several feet above the ground on the lower side of a big Douglas-fir stump on a steep slope at Snow Camp Mountain (L17) 12 Jun 1999 (JCP, DFx). Hairy Woodpeckers breeding in these settings forage on snags and down woody material within the clearcut as well as commuting across the opening to adjacent timber.

Adult Hairy Woodpeckers are not reticent at their nests, and their dependent nestlings call loudly from the cavity as they approach fledging. Family groups remain conspicuous for at least a short time after fledging. These behaviors and others assisted atlasers in establishing many breeding confirmations. Confirmed breeding was recorded in 31% of the 229 blocks with detections. Of these, five were occupied nests, spanning the period 29 Apr 1995 (n14; RHw) to 2 Jul 1996 (f06; DFx); one of the remaining three was in mid-May and two were in mid-June. Nests with young furnished the highest breeding evidence in 21 blocks. The earliest was 10 May 1996 (j12; D. Kriger). Three others were reported in May, eight during the first half of June, four in late June, and five in July, with the latest report 14 Jul 1997 (f20; D. Garrison). In 10 blocks, Hairy Woodpeckers were confirmed breeding via the observation of adults feeding young, with dates spanning about a 1-month period from 13 Jun 1995 (o28, o29; GCH) to 15 Jul 1997 (L21; PFS). In 25 blocks confirmations were made by the observation of fledged young. These reports fell from 15 Jun 1998 (h14; KMS) to 7 Aug 1996 (g02; M. Kenney). Eight of the 23 were in late June, nine were in early July, five were in late July, and only one (noted above) was in August.

WHITE-HEADED WOODPECKER *Picoides albolarvatus*

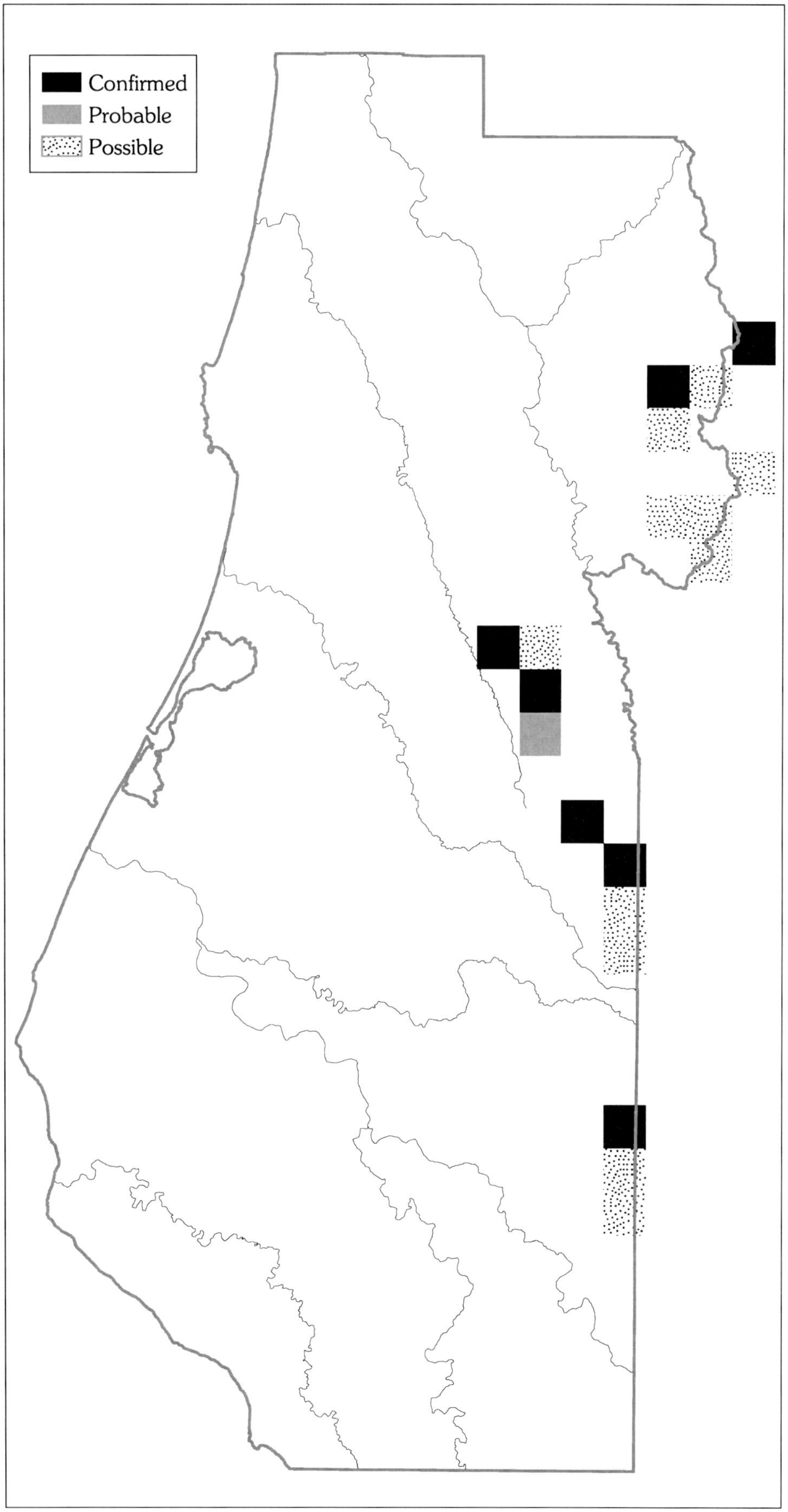

Mountains of the interior Pacific Northwest support a distinctive suite of birds found most commonly in pine and fir forest. Among these species is the White-headed Woodpecker. Its striking appearance and restless behavior while foraging are further enhanced by the scenic appeal of the places in which these birds live. Add to this the species' regionality, its typically Western bird associates, and memories of past summers in the mountains: the sum is a woodpecker that has about it an aura of charisma, keenly held in the minds of many birders acquainted with it.

White-headed Woodpeckers have long been known to occur in modest numbers in the higher forested mountains of northwestern California, where they have been considered rare to uncommon (Harris 1996). Within the cooler true fir- and pine-dominated stands above about 4000 feet, singles, pairs, and occasional family groups may be met with. Nearly all White-headed Woodpeckers seen or heard in eastern Humboldt County are encountered during late spring and summer, when observer interest runs highest and roads on north exposures well into the mountains are snow-free. The birds are considered to be present year-round, based on the lack of migration within their overall range and on occasional sightings at Horse Mountain (K. Lester, G. Lester) during the Willow Creek CBC. During the Atlas period, White-headed Woodpeckers were detected in 19 (4%) of the 425 total blocks. They were confirmed breeding in seven blocks.

Results of fieldwork indicated that, in a general sense, White-headed Woodpeckers were present in the same places where they had been historically known. Harris (1996) listed Trinity, Blake, Salmon, Horse, Board Camp, and Snow Camp mountains, Box Camp, Groves Prairie, and the Lassics as some known Humboldt locations. An exception was Snow Camp Mountain, the westernmost station for the species in the county (Harris 1996), where none was encountered by atlasers despite considerable searching. No truly new or unexpected breeding locations came to light. Also unchanging was that most of these birds were seen or heard along lightly traveled main routes and spur roads on Six Rivers National Forest.

Stands where White-headed Woodpeckers are most often detected are within the winter snowpack zone and have either Jeffrey pine, white fir, or Shasta red fir and a multi-storied canopy in which numerous dead or dying firs or other conifers persist, along with minimal silvicultural entry. Although not the case farther south in California, hardwood trees of any size are nearly absent—a smattering of poor California black oaks or tanoaks at their upper elevational limit may intrude. The shrub layer is sparse, the forest floor largely open. Canopy closure is sometimes quite high, but White-headed Woodpeckers do not shun edge, nesting and foraging to a considerable extent at forest's edge or in partial-cut openings in which snags or residual live trees persist after logging.

White-headed Woodpecker distribution in Humboldt County is confined to the interior, high-elevation ridgelines. The total span of blocks with detections actually describes a vague arc delimiting the westernmost extent of the high forest type preferred by White-headed Woodpeckers. This encompasses a region extending irregularly beyond Humboldt through Siskiyou and Trinity counties and extreme southeastern Del Norte County to the east and northeast, respectively, and southeast through the higher elevations of eastern Mendocino and adjacent counties.

Within Humboldt County, White-headed Woodpeckers are seemingly restricted to three areas containing pine-and-fir forests: northeast of Willow Creek and east of the Hoopa Valley Indian Reservation; along the South Fork Mountain ridge from Horse Mountain to Blake Mountain; and in the Lassics to the southeast. The three patches are somewhat broadly dissected by two corridors of unsuitable lowland gaps created by the Trinity and South Fork Trinity rivers in the north and by the Mad and Van Duzen rivers in the south. While surely crossable by individual woodpeckers, the extent of foothill and valley-bottom to be traversed in moving from one of these ridge-tops to another would appear sufficient to isolate nearly all of the birds within their islands of habitat. However, suitable habitat confluent with these "patches" in Humboldt extends on an increasingly broad front east of the county, so these patches are better described as "fingers".

The nesting season of White-headed Woodpeckers in Humboldt County appears to be relatively short. The earliest (and southernmost) confirmation was an adult carrying food 15 Jun 1996 near Black Lassic (o26; GAS), and the latest (and northernmost) confirmation involved the feeding of young 30 Jul 1998 near Salmon Mountain (r08; GCH). Only one nest with young was reported, 26 Jun 1996 (N19; GAS), and the earliest fledgling was seen 10 Jul 1998 (m16; COg).

Threats to this species have not often been directly discussed. Many stands of white fir, red fir, incense-cedar, Jeffrey pine, and other species were rather heavily logged in parts of Humboldt County through the 1980s. While loss of these charmed places through clearcutting has not been calamitous, degradation remains a threat in the form of salvage and sanitation cuts and other harvest treatments. Prior to "new forestry", reforestation of sites naturally supporting white fir and other trees of low economic value may not have resulted in reestablishment of the tree community previously occupying those sites. In the past such "low value" species were not commonly planted and were routinely removed during thinning. While this practice is now much less common than in previous decades, it did alter the species composition of some high elevation forests. The welfare of Humboldt County's limited high mountain forests and their birds cannot be championed too loudly.

NORTHERN FLICKER *Colaptes auratus*

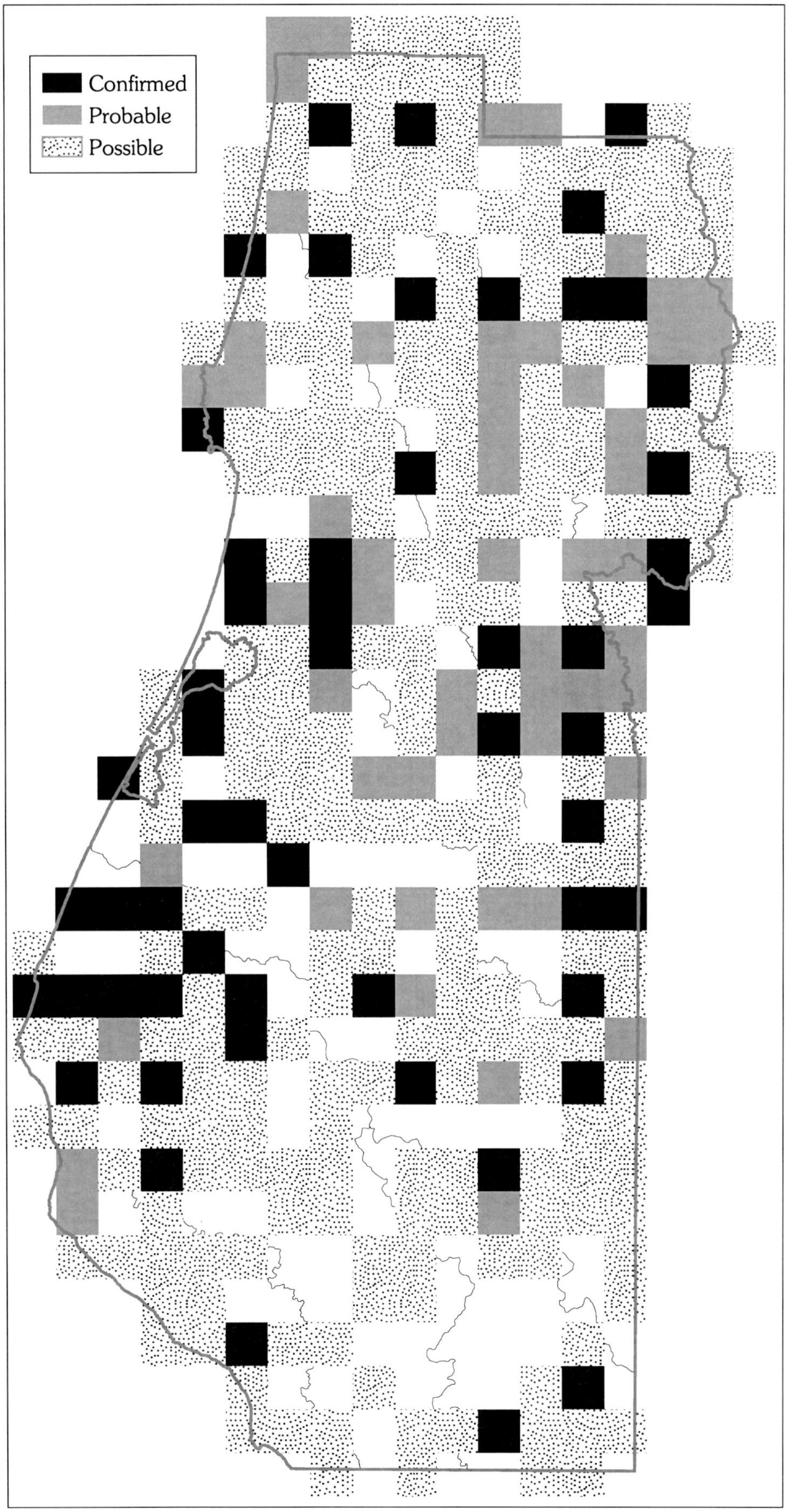

The bird known as Northern Flicker comprises two forms, the "Yellow-shafted" and "Red-shafted" Flickers. In the western U.S. north of the Colorado Desert (except in Alaska) the Red-shafted Flicker is the only representative of the genus normally found in the breeding season. Flickers are large, they use exposed perches, have loud voices, and show little aversion to human-altered habitats—provided there is sufficient open ground on which to forage and dead or dying trees in which to nest. These habits cause them to be widely seen and heard.

The Northern Flicker is a common breeder that is present throughout the year in Humboldt County, as it is throughout northwestern California (Harris 1996). The species is migratory, and a noticeable influx of migrants and wintering birds occurs in fall. Spring migration is little perceived. Flickers have long been known to be numerous and thoroughly distributed in Humboldt County. As the distribution map attests, they are among the most widespread breeding birds in the county. During the Atlas period, flickers were found in 76% of all blocks. Confirmed breeding was noted in 55 blocks, probable breeding in 54.

The foraging behavior of Northern Flickers differs from that of most other woodpeckers in that ants, a preferred food, are obtained not only from standing and down trees but from the ground. Although flickers may breed in extensive tracts of forest, generally some edge or semi-open country is nearby. Sites with low or spotty grass and disturbed or unstable earth, such as streamsides, roadsides, livestock pasture and rangeland, woodlots in residential areas, and lightly developed urban acreages are inhabited in spring and summer. Unlike some woodpeckers that are restricted to forests, Northern Flickers will nest in shade trees in parks, in the eaves of buildings, in burrows excavated in roadcuts, and in even the poorest stringer of riparian habitat provided there are potential nest trees of appropriate size and condition.

For these reasons, flickers can be expected from near sea level to the highest elevations in Humboldt County, in all types of tree communities, and in forests and woodlands of widely varying structures. Atlasers commonly encountered them in and near settled areas, in and about large expanses of forest or regrowth, and in clearcuts, where they nest in snags left on the site. Across the county, distribution is quite thorough. Confirmed and probable breeding blocks were well scattered, although less frequent in southern Humboldt County. In northern Humboldt there were no clusters of blocks without detections. In southern Humboldt, no-detection blocks were about three times as numerous, yet only in the extreme southern-interior part of the county was there a suggestion of somewhat restricted occurrence.

The earliest probable breeding record was of courtship behavior 16 Mar 1997 (F23; L. Bacon-Ogden, D. Ogden). The only nest with eggs reported was one in a Wood Duck nest box 18 Apr 1995 (f14; B. Oppenheim). Occupied nests accounted for 27% of breeding confirmations, ranging from 29 Apr 1996 (p14; TWL) to 17 Jul 1998 (L33; JCP). Of these, the eight in June were twice the total of July, with two in May and one in April. Fledged young accounted for the highest proportion of breeding confirmations (31%) and spanned the dates of 22 May 1996 (c23; COg) to 20 Aug 1996 (o07; H. Brown). It is clear from these reports that the local breeding season extends from at least early April through July.

In autumn, numbers of Northern Flickers appear to increase throughout the county. Beginning in October, birds with various plumage characteristics of the "Yellow-shafted" Flicker move into the area. Such birds are noted throughout the winter, although it is entirely likely that many seen in fall are replaced by others. These intergrades or backcrosses are quite common, to the extent that one might readily detect one or several such individuals in the course of an outing. Conversely, extremely few birds seem to be unalloyed Yellow-shafteds. When examined closely, nearly all flickers with yellowish underwings show telltale features of the other form.

PILEATED WOODPECKER *Dryocopus pileatus*

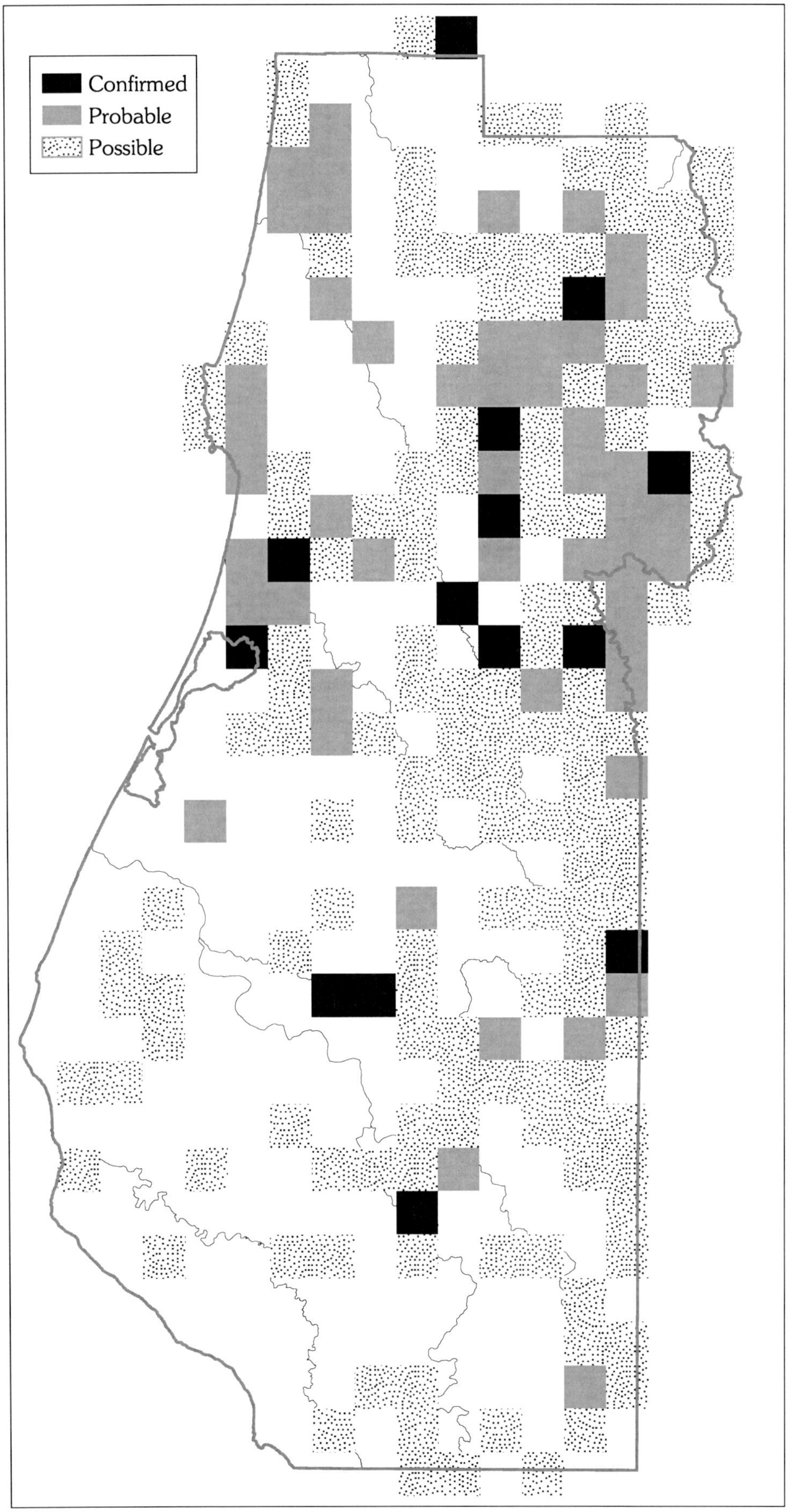

The Pileated Woodpecker is on nearly everyone's list of favorite birds. Dramatically large and dark and with a bright red crest, these woodpeckers draw comment from non-birders and bird enthusiasts alike. Pileated Woodpeckers are forest birds. Large standing trees, fallen trees, and snags are common elements of high quality habitat. They accept a great variety of tree communities across their range. Boreal forests of the North, foothill and mountain conifer forests in the West, mixed hardwoods in the East, and cypress swamps in the South are all inhabited by these woodpeckers. Offering a sprawling sweep of forested countryside, Humboldt County hosts many Pileated Woodpeckers throughout the year and in many forest types. The bird is numerous enough that an occasional wanderer makes its way onto "yard lists" in towns and suburbs of the coastal lowlands.

During the Atlas period, Pileated Woodpeckers were detected in 206 (48%) of all blocks. Confirmed breeding was reported from 14 blocks, probable breeding in 51 others. The majority of reports were from the northern half and the eastern side of the county. Distinctly few detections were made in southwestern Humboldt, nor in excessively open sections of interior county. Owing in part to much fieldwork in the mountains of northeastern Humboldt, many probable breeding observations were made in that area. The range of Pileated Woodpecker in Humboldt County takes in mature and old-growth forests of the northern coast; a mosaic of managed timberland at lower and middle elevations through much of the county; higher-elevation forests of fir and pine; and remnant stands of old Douglas-firs and tanoaks scattered through the grassy prairie country inland from Cape Mendocino. The extra effort expended in priority blocks did not result in a greater detection rate than in non-priority blocks, suggesting that Pileated Woodpeckers are less numerous than most other species of forest birds in Humboldt County.

A fairly distinct zone of blocks without detections occurred slightly inland from the coast in northern Humboldt County. Much of this area is intensively managed industrial forestland supporting mostly immature stands with few large trees and snags, limited down woody material, and low structural diversity. Clearcuts are numerous in this section of Humboldt, and altogether the area probably supports fewer Pileated Woodpeckers than parts of the county that have more intact forests. These woodpeckers were also unreported from the outer coastal lowlands from Little River south to near the mouth of the Mattole River; in varied private timberland or grassland-dominated blocks in interior southwest Humboldt; in some blocks within the oaky countryside of the central Mad River drainage; and in numerous blocks of mostly oak-dominated hillsides and openings in southern Humboldt. Reports from the King Range were notably scarce.

During the Atlas period, Pileated Woodpeckers were encountered in nearly all of the forest types found in Humboldt County, except for the drier, more open oak woodlands. As a rule, this woodpecker is not closely associated with oaks in this area, although California black oak, canyon live oak, and tanoak are common components of many forests in which the birds may be found. Clearly, most of the better and more contiguous habitat in Humboldt exists in the northern and interior portions of the county. Forests containing large conifers and lots of standing dead and down trees in a generally forested expanse are most attractive to these birds. On the Six Rivers National Forest, these birds forage in stands of all ages, but more so in older stands and on larger diameter trees (Boleyn 1997). On the coastal slope of northern Humboldt County, they may be found in older, unmanaged stands of regrowth red alder with patches or pockets of Sitka spruce, Douglas-fir, grand fir, and western hemlock. They follow near-coastal forests nearly as far west as would be possible; they were noted in Prairie Creek Redwoods State Park each year of the Atlas, and were resident in the Arcata Community Forest.

Nest building was observed 5 Apr 1996 in the Arcata Community Forest (g14; K. Sundeen). Reports of active nests spanned the dates of 19 May 1996 (F15; GCH) to 29 Jul 1998 (P11; S. Taylor). The very latest breeding confirmations were all on 30 Jul: fledglings in 1996 (j28; COg) and 1999 (i23; GCH), and adults feeding young in 1998 (N07; E. Elias).

Harris (1996), referring to northwestern California, characterized Pileated Woodpecker as a "rare to uncommon resident and breeder . . . Thinly scattered over the entire region. A few reports from the coastal lowlands in fall of apparent post-breeding wanderers or birds displaced by logging." It can be argued that a label of "uncommon" may best describe the population in Humboldt County, which probably numbers hundreds of birds. Raphael *et al.* (1988) estimated that this species may have declined 38% because of logging. While these woodpeckers likely respond negatively to logging (Rosenberg and Raphael 1986), they may be holding their own in Humboldt because they use a fairly wide range of forest stands.

OLIVE-SIDED FLYCATCHER *Contopus cooperii*

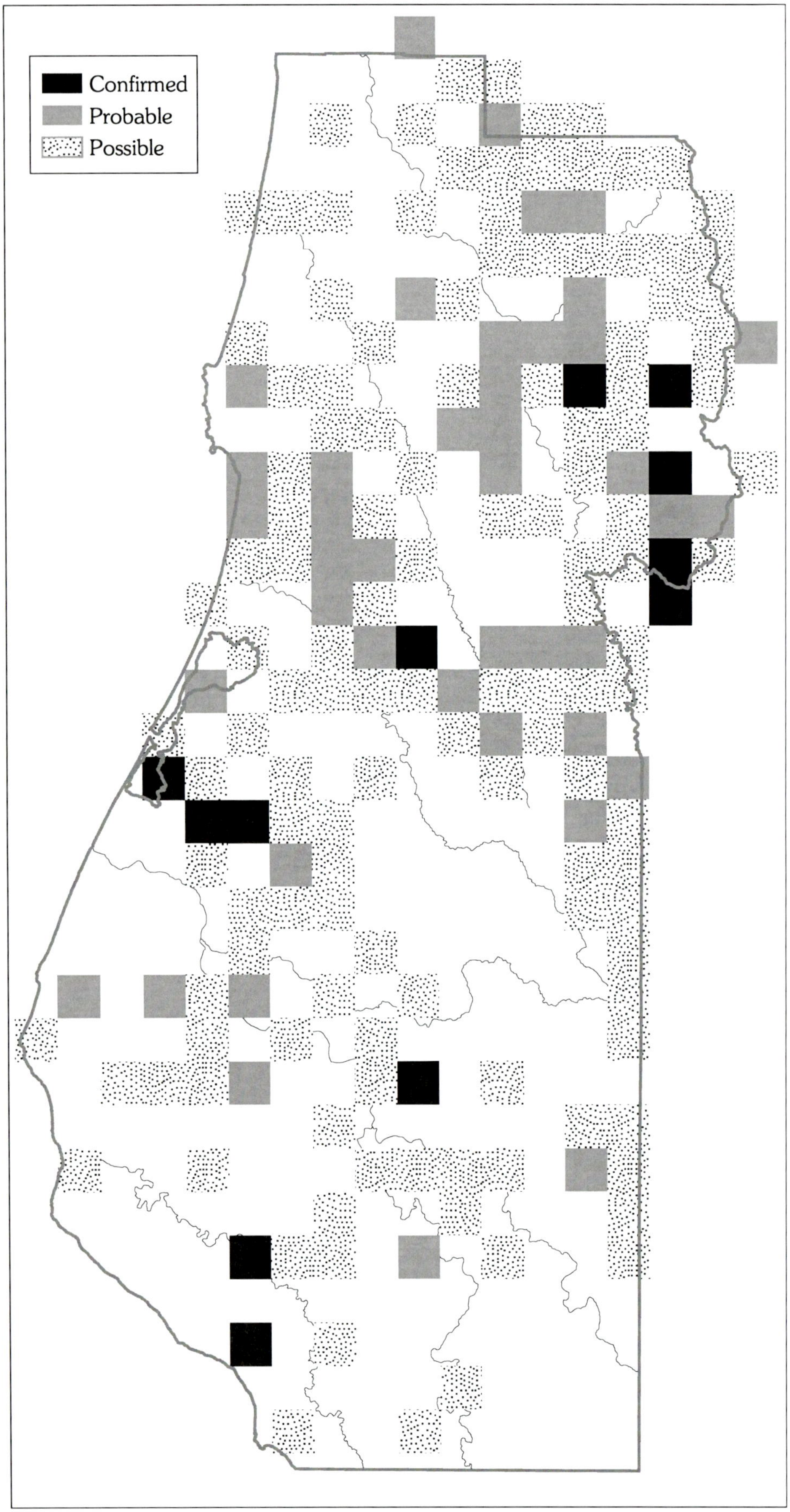

Olive-sided Flycatchers are short-tailed, bull-headed flycatchers of coniferous and mixed forests. They are most frequently seen perched at the tip of a tall tree or snag in wooded settings. These flycatchers are actually wood-pewees; the name Boreal Pewee has been proposed as an alternative to the present vernacular. Like other birds in this genus, Olive-sided Flycatchers are long-distance migrants. Returning from wintering grounds in the mountains of northern South America, they breed across a vast sweep of forests in Alaska, Canada, the northeastern U.S., and in foothills and mountains of the West. Humboldt County, with much conifer forest, clearly offers considerable habitat for Olive-sided Flycatchers.

During Atlas surveys, Olive-sided Flycatchers were found throughout much of Humboldt County. They were reported from 184 (34%) and confirmed breeding in 12 of the 425 Atlas blocks. The map shows a distinctly greater abundance of detection-blocks in the northern one-third of the county than elsewhere. They were also reported from many blocks in interior southwestern Humboldt County. In general, portions of the county that are well forested with mature Douglas-fir, true fir, pine, and mixed conifer-hardwood communities supported Olive-sided Flycatchers.

Virtually no reports came from blocks within the more open, oaky, less conifer-dominated landscapes of the central Mad River drainage and in extreme southeastern Humboldt. For Olive-sided Flycatchers, a mix of tree ages and a certain degree of edge seem to be an attractive element in northwestern California (Brandy 2001). The dearth of reports from the King Range was somewhat surprising. These flycatchers also were unreported from the mostly treeless Eel River delta, as well as in some blocks composed chiefly of grassland inland from the coast in the vicinity of Cape Mendocino.

Sites used by Olive-sided Flycatchers in Humboldt County vary. Typically the birds are encountered in large stands of trees either dominated by conifers or exclusively coniferous, featuring a mix of sizes, snags or dead tops, and often with an abrupt edge or opening. Brandy (2001) found territorial birds in narrow stream buffers in commercial timberland in Humboldt County. Coast redwood, Sitka spruce, Douglas-fir/grand fir, Douglas-fir/tanoak, and the true fir and pine forests of the higher elevations are all inhabited by at least modest numbers of these birds. During the Atlas period, it appeared that the foothills and mountains supported more of them than did the coastal lowlands and river canyons. However, they were found in some blocks with forested slopes shortly east of the coast through the Humboldt Bay lowlands and elsewhere. This species was quite consistently reported from the mountains along the eastern and northeastern edges of the county.

In a 1998–1999 study conducted in managed redwood timberlands in Humboldt and Del Norte counties, Brandy (2001) found that Olive-sided Flycatchers used habitat in proportions different from those generally available in the environment. He noted that males were commonly found at or near the top of a tree on a perch without live foliage. Snag density was not important in heterogeneous landscapes dominated by mature forest. Greater forest edge, a juxtaposition of stand size classes, or both, appeared to be factors contributing to greater numbers of Olive-sided Flycatchers. Brandy (2001) also found that occupied locations contained more young forest and less regenerating forest than random locations, and suggested that greater abundance of insect prey in recently harvested sites may have induced birds to establish territories in such sites. He hypothesized that a decrease in the structural complexity of stands from decades of fire suppression may have lowered habitat quality on a landscape level within the region. Brandy (2001) suggested that logging operations may increase attractiveness of sites through the establishment of a mix of tree sizes and newly-created "hard edge", but that a loss of size class diversity might take place as an area is repeatedly logged. Other studies in redwood forests found more Olive-sided Flycatchers in younger age classes (Hazard and George 1999).

Olive-sided Flycatchers arrive late in the spring landbird migration. Although occasional migrants are heard as early as April, these are exceptional; it is not until well into May that the rural foothills ring with the birds' spirited songs. Indeed, much of the population may not appear in the area until late in May or even June. Apparent migrants appearing outside expected nesting habitat—such as in lowland city parks, patches of spruce along coastal bluffs, and in clumps of residual trees well out in the open in recent clearcuts—may be anticipated through Memorial Day and into the first two weeks of June. For this reason, it is conceivable that some reports of birds in early June may refer to individuals singing or otherwise detected in passage. In 29 of 130 blocks reporting possible breeding, the reports establishing this status fell in May. Adding to the potential for overestimating the breeding distribution is that migrants tend to select well-treed habitats, many approximating that found at their breeding locations. Given the abundance of reports from mid-June and July, though, it is obvious that numbers of the birds call Humboldt County home in summer.

It was evident that nesting by Olive-sided Flycatchers began at least by early May, while others must have begun considerably later. Nest building was observed 27 May 1995 (F31; KMS). Adults carrying nest material were seen 7 Jun 1996 (P13; TWL) and 14 Jun 1997 near the Honeydew Creek recreation site (F29; JEH). An adult was flycatching with an already full bill at Mill Creek Lake (P09) 20 Jul 1995 (JEH). The evidence most frequently used to establish confirmed breeding was fledged young, reported in eight of the 12 blocks with confirmations. All of these reports spanned a 5-week period from 22 Jun 1997 (P11; K. & E. Bentler) to 28 Jul 1995 at Humboldt Hill (d18; P. Spruance). Harris (1996) noted that fledglings had been seen as early as mid-June in Arcata.

Referring to northwestern California, Harris (1996) termed Olive-sided Flycatcher a "common summer resident and breeder in woodlands and forests with a conifer component." Atlasers had the impression that, while numerous, Olive-sided Flycatchers do not approach the abundance of most other flycatchers resident in Humboldt County in summer. Other than the ephemeral association shown by family groups shortly after fledging, they do not occur in flocks, instead dispersing at low density in suitable habitat. These birds seem to cease singing while nesting is still underway, making their detection more difficult. The summer population appears to melt away in August. Apparent fall migrants well out of breeding habitat are rarely encountered in Humboldt in late August and early September. The latest report for the county, and the only one for October, was of a single bird 25 Oct 1955 on Fickle Hill above Arcata (C. Yocom; Harris 1996).

This species is possibly declining in California. As noted above, forestry and fire management activities have great potential to affect this species by reducing variability in both horizontal and vertical habitat structure. The birds favor edge, and snags and other tall trees are needed as song and foraging perches. Habitat loss and other factors on the wintering grounds may also be to blame for the decline.

WESTERN WOOD-PEWEE *Contopus sordidulus*

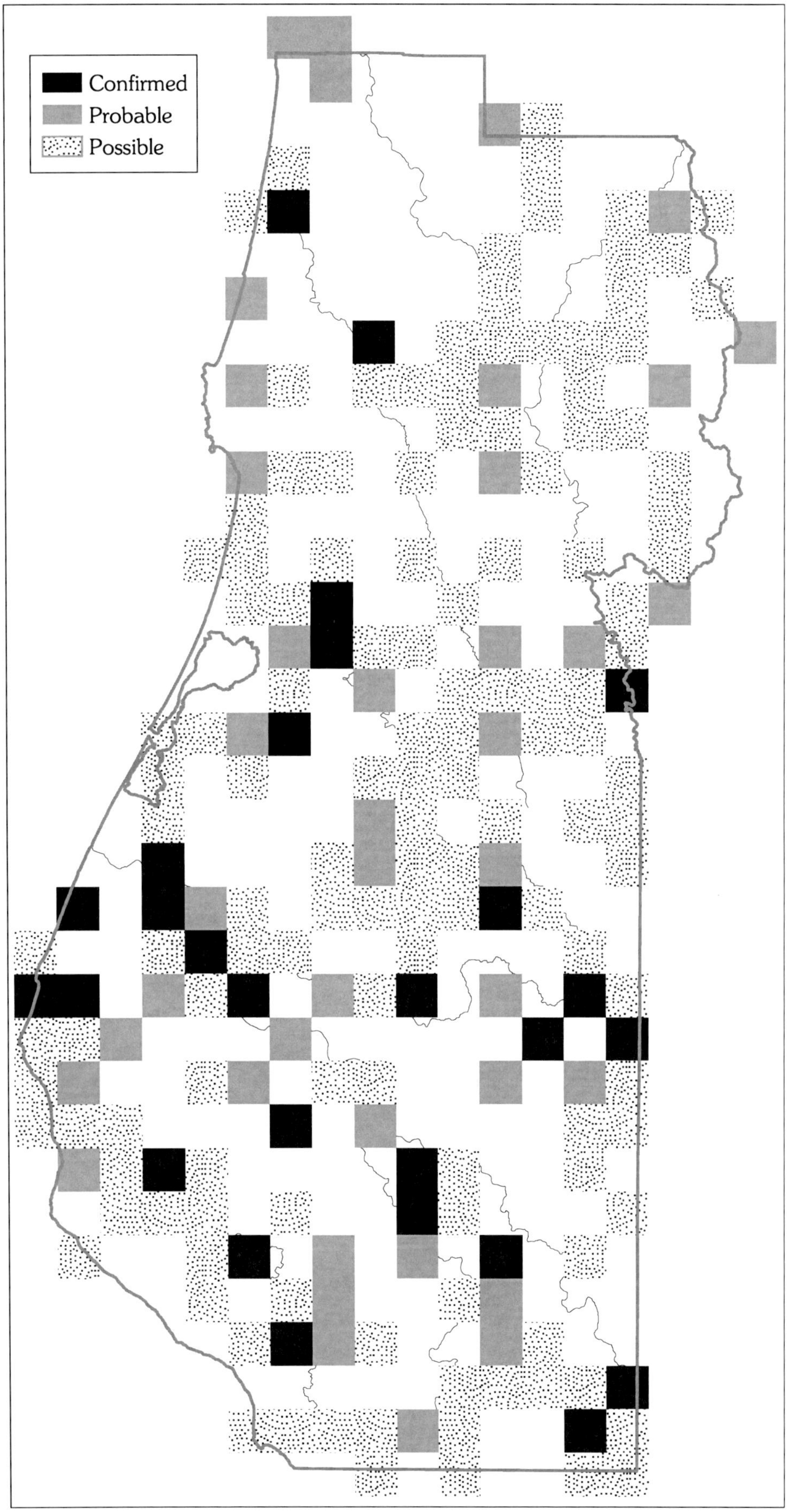

Various open forest and woodland settings across Humboldt County appeal to Western Wood-Pewees from late April into September. These flycatchers are fairly common throughout most of the area, though nowhere are they especially abundant. The tireless early summer chant of territorial wood-pewees is a sound many come to associate with a leisurely and bucolic ambience—bringing to mind generic images of nameless back roads crackling with grasshoppers, smelling of cut hay, and loud with sun. Harris (1996) considered the Western Wood-Pewee a "common summer resident and breeder in dry, mixed woodlands and riparian areas throughout the region . . . more common inland than along the coast." During the Atlas period, Western Wood-Pewees were reported from 48% of the blocks and were confirmed breeding in 6% of all blocks.

Western Wood-Pewees are not truly birds of the forest interior, instead preferring broken, open, or uneven-aged stands containing a considerable degree of exposed "edge". A large proportion of their foraging efforts consists of open-air sallies of varying duration initiated by flying from semi-sheltered flycatching perches at, or near, the outer mid-canopy of trees, whether coniferous or broadleaf, evergreen or deciduous. Somewhat exposed limbs with dead tips offer preferred perches. Open, maturing and older conifer-dominated stands, forest edge, riparian corridors, oak woodland, and higher fir and pine forests appear to fulfill their requirements in Humboldt County. Riparian corridors and meanders grown to alder, black cottonwood, Oregon ash, and large willows are also used.

Unlike some other species routinely found in oak woodlands, Western Wood-Pewees also occur well within the summer fog belt. However, there was no obvious clustering of detections in well-peopled central-coastal Humboldt. This suggests that numbers in the coastal lowlands and on the immediate coastal slope are not particularly great. These birds were encountered during the Atlas period from shortly above sea level to above 5000 feet. In the mountains, they occur about the margins and meadow-edges of conifer forest above the zone of California black oak, often inhabiting uneven-aged stands with an understory of white fir, in which they nest. They do not seem to linger commonly in urban or suburban areas in the summer, but will do so wherever a sizable patch or stringer of forest edge ties the fringe of human settlement to an expanse of habitat beyond. Wood-pewees appear to tolerate at least moderate undergrazing by cattle in the lowland and foothill sites they occupy in Humboldt. Small clearcuts or "leave-tree" cuts often support pairs of these birds, but expansive areas of intensively logged redwoods appear to attract few of them.

Atlas blocks in which confirmed or probable breeding status was established were slightly more numerous in the southern one-third of the county. Ready birder access to tree-dotted valley bottoms and edges in the Eel, Mattole, and upper Mad River corridors may have contributed to this apparent skewing. It is possible that, on a landscape level, the sprawling oak woodlands, broken or "edgy" mixed-species second-growth stands, and unmanaged riparian habitat common within that region may support wood-pewees in higher numbers than the intensively managed coniferous timberland and deep-shadowed tanoak stands of northern Humboldt.

Harris (1996) stated, "Logging may have increased habitat by creating a mosaic of open areas and snags with early second growth." Raphael *et al.* (1988) estimated that these wood-pewees may have increased 78% as a result of logging, and Marcot (1984) also indicated that population size or distribution increases have probably occurred as a result of clearcutting. During surveys in Douglas-fir sawtimber, mature, and old-growth stands, Western Wood-Pewees were detected in 29–38% of all plots (Raphael 1984). During 1999 point counts in riparian habitats along the lower Eel River, they were detected during 20% of surveys (R. Hewitt, unpubl. data).

The earliest breeding confirmation was of the carrying of nest material 29 May 1995 (D21; L. Karsteadt), and the earliest nest with eggs was discovered 14 Jun 1997 (N33; GAS). The earliest fledgling was reported 20 Jun 1996 (N23; GAS). The latest nest with young was observed 30 Jul 1999 (g26; JG), and the latest breeding confirmation was of adults feeding young 9 Sep 1998 (d20; *fide* RHw). The earliest spring arrival reported for Western Wood-Pewee during the Atlas period was 18 Apr 1996 (L27; D. Diehl).

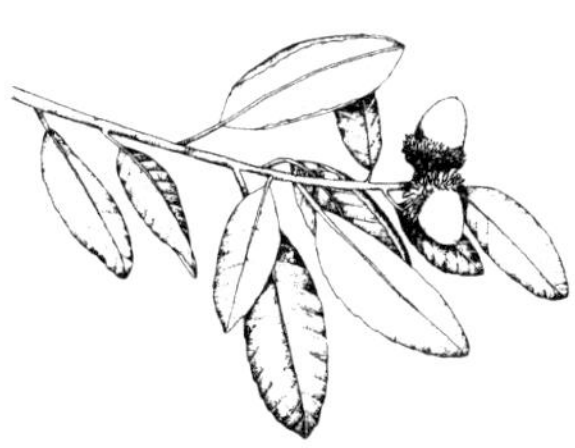

WILLOW FLYCATCHER *Empidonax traillii*

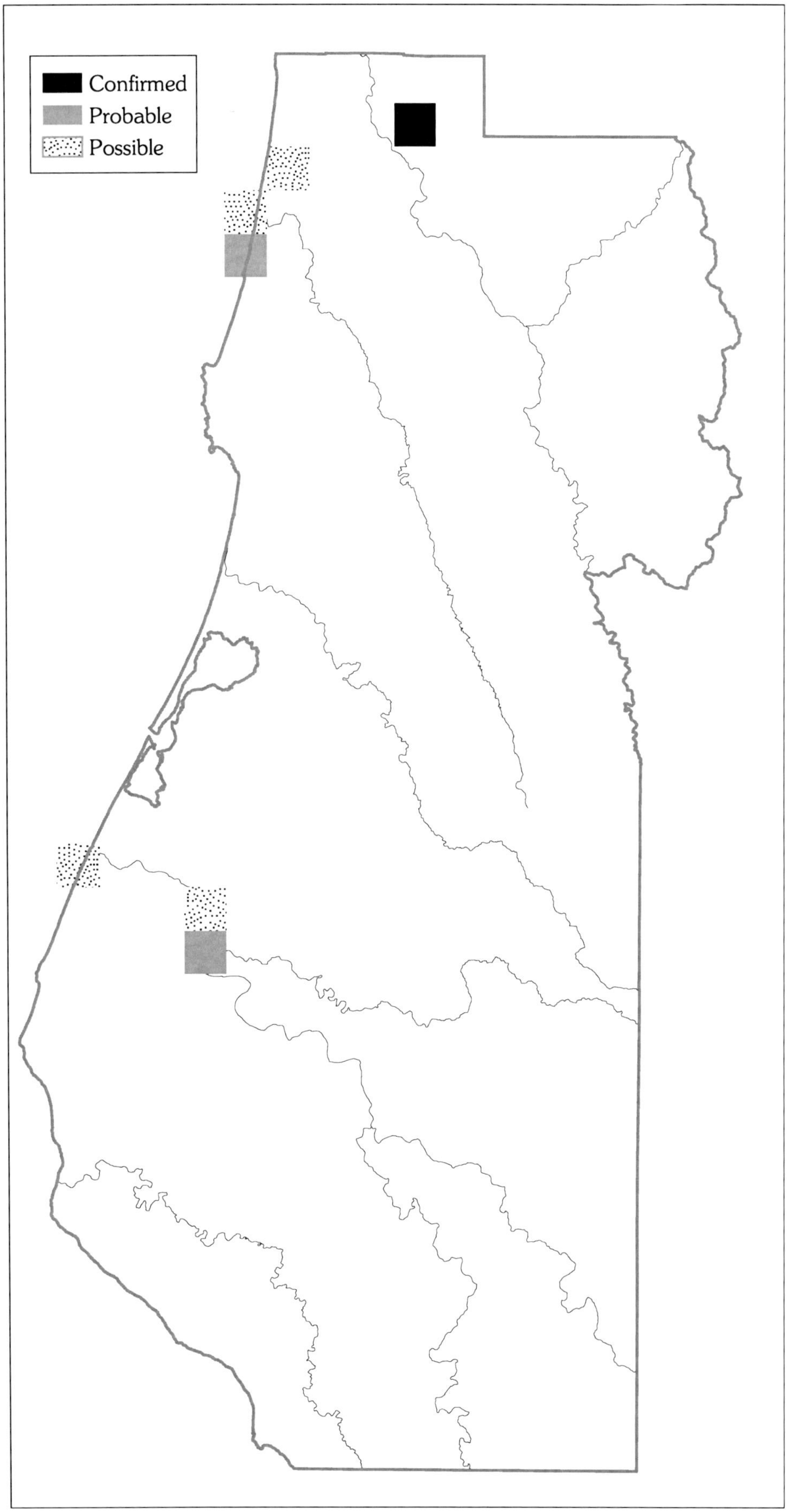

Originally called Traill's Flycatcher, the creature now considered the Willow Flycatcher comprises five subspecies. Three of these occur within California, with *E. t. brewsteri* (the Little Willow Flycatcher) occurring annually both as a spring and fall migrant and casual summer resident and breeder in northwestern California. All subspecies breeding in California are listed as "endangered" by the State of California, yet only the Southwestern Willow Flycatcher (*E. t. extimus*), occurring in southern California, is federally listed as "endangered" (USFWS 1995). The Little Willow Flycatcher is an uncommon migrant through Humboldt County during the period mid-May to mid-June and can be fairly common at times from mid-August through September. The status of this flycatcher in the breeding season in Humboldt is not thoroughly understood. Presently, summering by Willow Flycatchers in the county appears to be a rare and localized phenomenon.

That Willow Flycatchers are among the latest-migrating, latest-arriving passerines in spring causes difficulty in determining the status of given individuals not tracked over days or weeks. This is a fly in the ointment in attempts to tentatively assign birds a seasonal label. To illustrate, while a Pacific-slope or Dusky Flycatcher singing or otherwise detected in appropriate breeding habitat in Humboldt County any time after early May can be reasonably assumed to be summering, a Willow Flycatcher in May or early June in a setting that looks suitable for nesting could be a bird still on the move. For this reason, special attention was paid during Atlas data-checking to ensure that a conservative approach was taken in addressing the occurrence of this species in Humboldt. Considered of distinct interest were sightings in seven (<2% of 425 total) blocks. In two of these blocks probable breeding was established, and in one, a well-documented breeding confirmation came as one of the highlights of the Atlas project.

Three other Atlas records of Willow Flycatchers that likely involved summering birds were clustered along and just inland from the northern coast. Observations of single, singing birds were made in forest/meadow edge at the south end of Boyes Prairie in the main "elk meadow" of Prairie Creek Redwoods State Park (g04) 20 Jun 1998 (G. Lester), and in a shrubby marsh beyond the end of Hufford Road in the lower Redwood Creek estuary (f05) 8 Jul 1997 (JCP). While these reports were considered evidence of possible breeding, a bird singing 25 Jun–8 Jul 1997 in moist-site shrubs, forbs, and sapling red alders in a lake shore opening along the east side of Freshwater Lagoon (f06; K. Irwin *et al.*) suggested probable breeding.

In the lower Eel River corridor, a bird giving position calls but no songs along the lower Salt River at the end of Port Kenyon Road (b20) 10 Jul 1998 (DFx), and another bird farther up the Eel in riparian cottonwoods and willows at Sandy Prairie opposite Alton (e21) 28 Jun 1998 (P. Herrera) were both assigned possible breeding status. Up to two males sang on territory 28 Jun–14 Jul 1998 in well developed riparian woods near the Hauck/Hansen Bar at the confluence of the Eel and Van Duzen rivers (e22; P. Herrera), strongly indicated probable breeding in that area. Before the Atlas period, Humboldt's only historical breeding confirmations were from farther up the South Fork Eel River near Burlington and Miranda, where three egg sets (labeled "Traill's Flycatcher") were collected in June of 1930 and 1931 from willows on gravel bars (Harris 1996). Despite the current abundance of Brown-headed Cowbirds in the riparian habitat along the Eel River, additional breeding activity should be looked for in these areas. Some atlasers believed that an increase in breeding-season reports from lowland areas indicated that these riparian habitats were being recolonized in Humboldt County (R. Hewitt, pers. comm.)

On 26 Jul 1998 a *whit* call-note from an unseen *Empidonax* in a recent clearcut in upper Bear Creek at 2300 feet in hills south of the lower Klamath River (J03) proved to be made by an adult Willow Flycatcher—carrying a large insect in its bill! It flew out of sight behind sapling tanoaks and Douglas-firs, where a moment later a quivering, begging juvenile Willow Flycatcher was observed sitting atop a shrub (DFx, JCP). In all, two adults and one fledgling were seen. The clearcut was about 10 years old and was dominated by crowded healthy young Douglas-firs (3–12 feet tall) and low, dense, and mostly uniform brush with a few willows. This was the first breeding confirmation in Humboldt County since the egg sets were collected in 1930–1931, and an unexpected find by atlasers.

The nesting of Willow Flycatchers in northern California in a clearcut with almost no willow and dominated by regrowth forest trees and seral shrubs was a shock to many Humboldt birders and biologists. However, in the Coast Ranges of Oregon and Washington, Willow Flycatcher is a fairly common breeding species in regenerating clearcuts (B. Altman, G. Gillson, M. Hunter, D. Irons, T. Mickel, T. Rodenkirk *et al.*, pers. comm.). Those clearcut sites have been described as generally ranging from about 5–15 years old, 1000–3000 feet elevation, somewhat moist, with young conifers, and typically containing a strong deciduous hardwood component (mostly willow or alder). No doubt contributing to the suitability of such montane sites as habitat is that fewer cowbirds are found in landscapes dominated by coniferous trees, and cattle are seldom allowed to degrade commercially valuable conifer regeneration as they are riparian habitat. Treating clearcuts with herbicides to kill all non-conifer vegetation undoubtedly renders such areas unsuitable for this species.

The region of regular breeding of Willow Flycatcher within clearcuts extends southward at least to the point where the southern Oregon Coast Range dissolves into the northwestern extent of the Klamath Mountains (T. Rodenkirk, pers. comm.). Early-seral vegetation in the Coast Range north of California consists of more species favoring moist sites and fewer favoring dry sites, a condition that, in Humboldt County, persists locally within the area of industrial forestry falling under overcast summer skies of the fog belt. Other observations in northwestern California have hinted at the possibility that these birds use clearcuts. In 1993 two singing birds lingered into early July in similar habitat not many miles away in upper Cappell Creek of the lower Klamath River (DFx), and in 1997 a bird sang on territory 18 Jun–2 Jul in a clearcut on the Smith River National Recreation Area in Del Norte County (KMS, JEH). Predating those records was one of a bird "singing wildly" in a regrowing clearcut on Waterman Ridge from at least 26 May–10 Jun 1983 (R. Erickson).

Willow Flycatchers arrive in Humboldt County beginning in mid-May, with a few exceptionally early reports from early May (Harris 1996). In most years very few are detected in spring, but in others a perceptible pulse of birds is reported, this ordinarily taking place in the last week of May and the first two weeks of June. Fall migrants fly south through a broader window, and are sometimes rather numerous in both inland riparian areas and coastal willow patches from mid-August to mid-September. More than 50 were banded along the Klamath River near Orleans in late August and early September 1992 (S. Sniado). There are very few records of Willow Flycatchers in northwestern California after the end of September (Harris 1996).

HAMMOND'S FLYCATCHER *Empidonax hammondii*

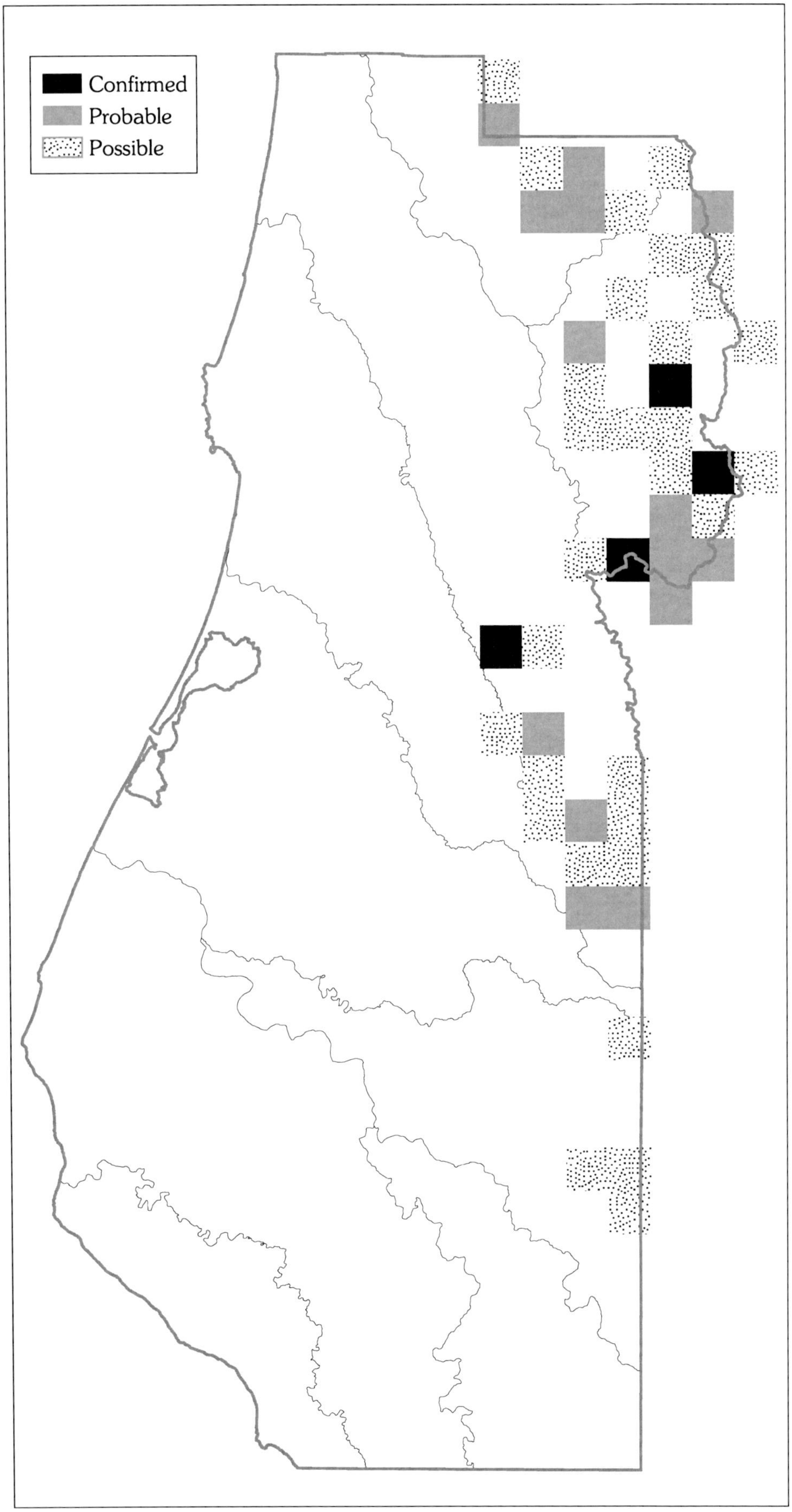

Hammond's Flycatchers are tiny, retiring, soft-voiced birds of inland mountain forests, miles removed from urban centers. As a result, this species is poorly known by many observers. The thin, flat, distinctly "spiritless" song could lift no heart but a birder's—yet it is a sound so characteristic of the buggy months in the high forests that those who learn it savor it. Heard well only when close by are the unincisive call notes which advertise its presence. Yet despite the success of the search, Hammond's Flycatchers are difficult to observe in detail. They may "closet-up" for many minutes within the deep noonday shadows, or else flit and chase rapidly through eye-hurtingly bright forest sunbeams—all the while remaining dozens of feet overhead.

This species was reported in 48 (11%) of the 425 total blocks in the course of the Atlas period. Hammond's Flycatcher is one of many species favoring the true fir zone of the higher elevations in summer. These birds are uncommon to common in unharvested or lightly harvested stands of white fir and Shasta red fir above 4000 feet in eastern Humboldt County. They also are present, but less abundant, in older Douglas-fir forests. In these Douglas-fir stands, the abundance of Hammond's Flycatchers is inversely correlated with the presence of hardwoods (Ralph *et al.* 1991). Pacific-slope Flycatcher tends to replace this species as one moves downslope into more temperate stands with better developed hardwood midstories. Sakai (1988) described nesting chronology, nests, nesting success, and other aspects of breeding biology at some Humboldt sites. Sakai (1987) and Sakai and Noon (1991) studied nest site characteristics in Douglas-fir/tanoak stands and found nests in stands with high canopy closure and age greater than 90 years. Most nests were in tall, large-diameter green trees.

Since they breed at higher elevations, Hammond's Flycatchers arrive on breeding grounds later than do Pacific-slope Flycatchers (Sakai 1988). Some do show up by late April at 4000 feet and above, at which time atlasing at higher elevations was generally precluded by snow-covered roads. Some early possible breeders in suitable habitat were recorded 3 May 1996 (o05; K. Baker) and 4 May 1996 (n10; DFx). The earliest probable breeding records were of multiple males singing 8 Jun 1998 (p14; KMS) and territoriality 15 Jun 1996 (l03; P. Herrera). No nests were found during the Atlas period, and only four breeding confirmations were reported: adults carrying food at Water Dog Lakes (P09) 8 Jul 1995 (JEH, TWL), at Grizzly Camp (q11) 12 Jul 1997 (JEH), and on Horse Mountain (L15) 29 Jun 1998 (COg), and feeding young on Waterman Ridge (o13) 2 Jul 1998 (KMS). Limited confirmations may be due in part to identification difficulties encountered by atlasers.

This species has undoubtedly been affected by several decades of intensive logging of old growth at middle and upper elevations in Humboldt County, but trend and population numbers are unavailable. Responses to single-tree selection, salvage logging, and other uneven-age harvest prescriptions are unknown. Raphael *et al.* (1988) estimated that numbers of Hammond's Flycatchers may have declined by about 40% due to intensive timber harvest. Clearcutting generally converts Hammond's to Dusky Flycatcher habitat. Given the species' apparent preference for the forest interior, it is likely that road construction and other activities that fragment stands affect this species.

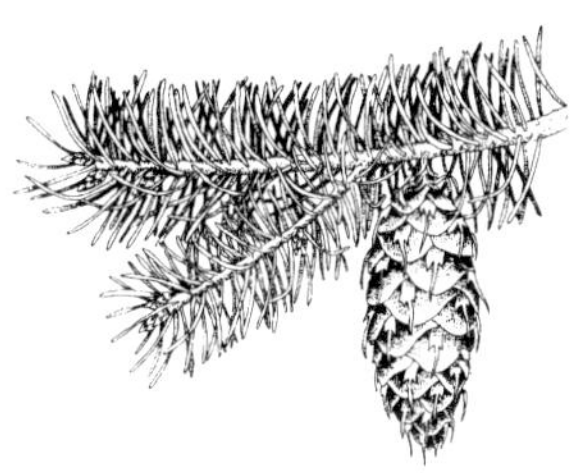

DUSKY FLYCATCHER *Empidonax oberholseri*

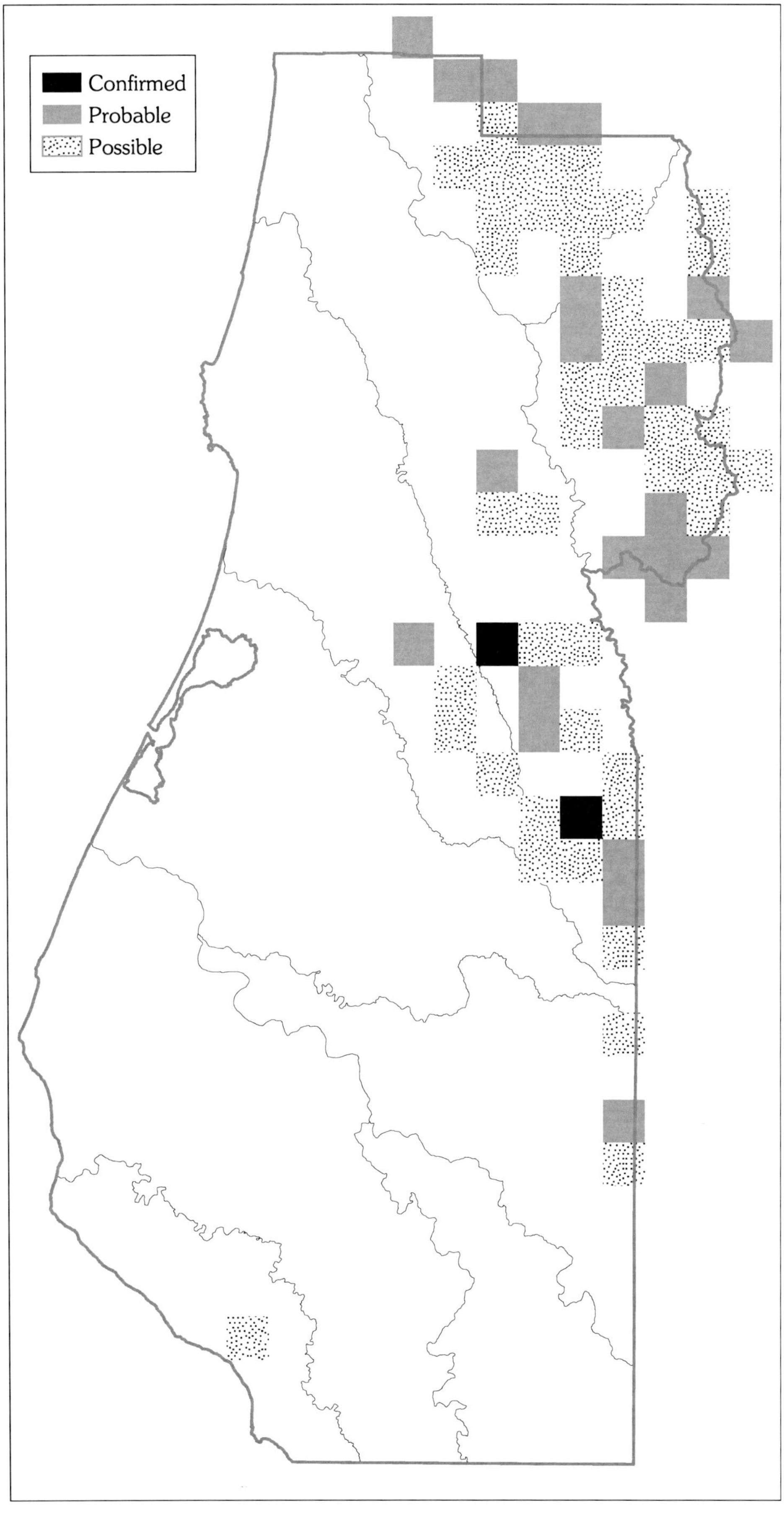

Soft, plaintive songs with a rhythmic lilt come to an attentive birder's ears throughout the hot hours of early summer in mountain brushfields of the West. Among the most challenging small songbirds to identify strictly by its looks, the Dusky Flycatcher is more readily distinguished by its chant-like *dew . . . dew-hick* midday song. Unlike many *Empidonax* flycatchers, Duskies exhibit no pronounced aversion to direct sunlight; they typically perch on or around the tops of shrubs and saplings in preparation for an aerial sally after flying insects. Eschewing the closed-canopy forests so preferred by its look-alike congener, Hammond's Flycatcher, the Dusky favors sunny openings well-grown to snow-zone brush and scattered conifers. It routinely replaces Hammond's Flycatcher throughout the altitudinal range of that species when forest is cut and the resulting opening grows back to brush. One may hear each species from the same spot along forest edges within a landscape mosaic of timber-and-clearcut.

In addition to its notorious similarity to Hammond's and Gray (*E. wrightii*) flycatchers and a few other small, plain birds, several circumstances render Dusky Flycatcher a potentially under-detected species. Its breeding range in Humboldt County is restricted to the hinterlands, far from cities and towns. Spring and fall migrants are rarely encountered, nor are birds often identified away from known or plausible nesting sites. Judging from the mere handful of reports, Dusky Flycatchers are apparently rare in the coastal lowlands, and they are virtually unknown as "vagrants" at the well-birded outer-coastal willow thickets. Daunting as this low profile may be, Atlas surveyors experienced in finding interior Humboldt's summer birds encountered them frequently. Over the 5 Atlas years, these birds were reported in 16% of all blocks.

Harris (1996) stated that Dusky Flycatcher was a "common summer resident and breeder in mountain chaparral, shrubby cut blocks, and open pine stands throughout the interior mountains" of northwestern California, that migrants arrive by mid-April, and that summer residents are on territories by mid-May. In Humboldt, the breeding range of Dusky Flycatcher lies chiefly at higher elevations in the northeastern and north-central portions of the county. This distribution is contiguous with a far more broad-ranging occurrence throughout the mountains of northwestern California and the Cascades, Sierra Nevada, Rockies, and other western ranges. In the Klamath Mountains, this flycatcher is expected at elevations from about 2600 to 5400 feet (Sniado 1999).

During the Atlas period, Dusky Flycatchers were encountered widely across the middle-to-higher elevations. The area in which they were found extends from the Del Norte–Humboldt County line generally south-southeast along higher slopes and ridges to the Lassics, at the Trinity–Humboldt border. A brief but incised break in the contiguity of the species' distribution occurs in the lowlands of the Trinity River canyon, where lower-elevation brushfields do not support them. Aside from the exception discussed below, the most westerly occurrences reported were near South Red Mountain (j01)—only a dozen miles from the coast and at the edge of the marine fog belt—and at Fernwood (J15), just south of Bald Mountain. Fernwood was also the lowest elevation at which the birds were found during the study, at only 2800 feet (territorial males 2 Jul 1998; DFx). Intriguing was a report of a bird in seemingly suitable habitat in an old burn in the King Range (F31) 2 Jul 1999 (KMS). The limited amount of survey in that remote section during the nesting season leads us to suspect that the birds may breed somewhere among the peaks and higher ridges within this area. Whether Dusky Flycatchers occur regularly in the King Range—within several miles of the ocean—is a mystery left to energetic bird surveyors of the future.

Dusky Flycatchers were not detected in the Bald Hills. It appears that at least some appropriate habitat should exist in the ridge-top clearcuts on private timberland east of Schoolhouse Peak. Vigorous "stump-sprout" tanoak originating from the second timber rotation characterizes brushfields there, and is a habitat in which Dusky Flycatchers have been noted on Snow Camp Mountain.

Habitat chosen by summering Dusky Flycatchers in Humboldt County is similar to that selected elsewhere in the Klamath Mountains bioregion and in northwestern California. These birds are strongly partial to sun-exposed brushfields that are moderately- to thickly grown to such mountain species as green-leaf manzanita, snowbrush, bitter cherry, oceanspray, snow-suppressed tanoak, and *Ribes* spp. Shrub height typically runs from 4–15 feet tall. Sites at which the birds are numerous often are clearcuts dotted with scattered emergent conifer saplings and post-logging residual trees. These elements may be clumped, irregularly distributed, or in stringers. Patches of open ground and piles of logging slash are common features. Slope ranges from flat ground to steep slopes; rolling or moderately sloped sites are common. Birds were noted at all slope aspects. Dusky Flycatchers will colonize some clearcuts or partial-cuts as soon as 3 years post-harvest, though typically 5 or more years passes before the generation of a diverse shrub stratum sufficient to attract breeding birds. While some brushfields in the mountains of east county would appear to persist over many years, seral succession, typically characterized by increasing dominance of conifers and diminishment of brush, causes habitat at many sites to become ultimately unsuitable. Thus, site for site, habitat for Dusky Flycatchers here is more ephemeral than that of most other locally breeding songbirds.

At higher elevations, Dusky Flycatchers were also commonly encountered in unmanaged or lightly managed pine and fir stands undergrown with huckleberry oak, as along the higher portions of the Gasquet–Orleans (G–O) Road and on Horse Mountain. This species was found to be associated with *Pinus* spp. and lower total tree cover, as well as occurring most frequently within the white fir zone (Sniado 1999).

What would appear to be naturally occurring habitat is not often encountered, as wildfire—certainly once the foremost agent of creation—is typically suppressed before a large area has been burned. Instead, Dusky Flycatchers occur here largely because of logging within the mountain forest zone. Although few baseline data exist, the amount of habitat for these birds has been greatly increased since the 1950s by intensive logging at middle and higher elevations of the Six Rivers National Forest and private inholdings (Marcot 1984). Raphael *et al.* (1988) estimated that numbers of Duskies may have increased by more than 200% as a result of extensive logging and suggested that they may have expanded downslope. However, in the most recent years, increased attention to non-timber resources on federal lands will probably reduce the size of new clearcuts. Since those clearcuts already in existence are bound to grow into conifer regeneration, it is conceivable that the number of Dusky Flycatchers nesting in the county will decline in the future.

While these flycatchers were detected in 68 Atlas blocks, breeding was confirmed in only two. Both confirmations were based on observations of adults carrying food, one 20 Jun 1998 (L15; COg) and the other 14 Jul 1995 (N19; GAS). Probable breeding observations, recorded in 23 blocks, spanned the dates 16 May 1996 (p12; TWL) to 30 Jul 1998 (r08; GCH).

PACIFIC-SLOPE FLYCATCHER *Empidonax difficilis*

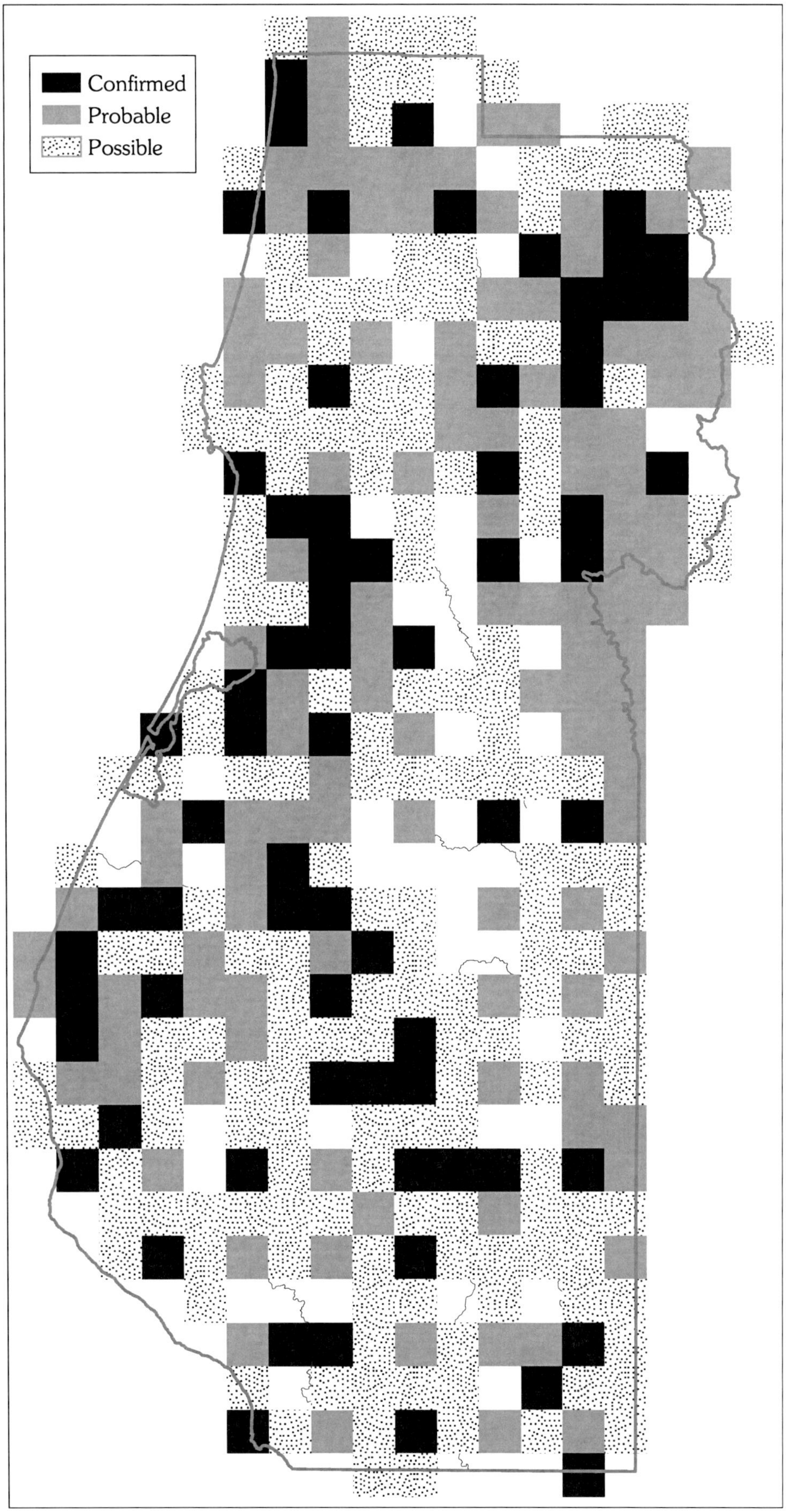

As birds of dense forest undercanopy and persistent shade, Pacific-slope Flycatchers find the great wooded expanses of Humboldt County much to their liking. From mid-April to August this species ranks among the most abundant birds of forests, managed timberland, riparian corridors, and the neglected second-growth that has followed the general clearing and settlement of land. Observers performing morning chorus counts on BBS routes have consistently recorded large numbers of "Pac-slopes", and in a broad spectrum of timber types. Raphael (1984) found them in 94–100% of his plots in 50- to 350-year-old Douglas-fir stands. So many individuals may be heard singing before sunrise in stretches of preferred habitat that song elements continually overlap, frustrating attempts to count the number of birds actually giving voice. Harris (1996) reported Pacific-slope Flycatcher to be a "common to abundant migrant, summer resident, and breeder" in northwestern California. Other than possibly a few Willow Flycatchers, this is the only *Empidonax* that regularly nests in the lowlands of Humboldt County.

Pacific-slope Flycatchers may be found in practically any well-wooded situation whether evergreen or deciduous, provided that extensive low limbage affords deep shade. Both old-growth and regrowth conifer forest appeals to them, though they tend to be more abundant in older stands (Ralph *et al.* 1991, Hazard and George 1999). Stands of unmanaged tanoak and Douglas-fir, so abundant across middle elevations east of the fog belt, are thickly populated. Pac-slopes breed from just above sea level to about 4500 feet, the approximate lower elevational limit for the congeneric Hammond's Flycatcher. At about this elevation oak and most other hardwood tree species become sparse, and fir-and-pine forest becomes prevalent.

Thinned commercial timberland appears to meet their needs, as long as canopy closure is sufficient and a consistent understory of tall shrub-trees persists. In Douglas-fir forests, the abundance of these flycatchers is strongly correlated with the canopy volume of tanoak (Raphael 1987). Even warm southern exposures grown to canyon live oaks support these birds wherever canopy cover is dense and a comparatively shaded, humid microsite may be found. In some inland areas a strong association with riparian habitat has also been noted (Bardolf 2001). Sakai (1987, 1988) and Sakai and Noon (1991) provided considerable detail on the breeding biology of this species in local Douglas-fir forests.

The Atlas effort resulted in detections in 83% of blocks, clearly indicating that the Pacific-slope Flycatcher is to be expected virtually anywhere and that it is the most widespread breeding neotropical migrant species in Humboldt County. This species probably bred in every one of the blocks in which it was encountered. The map suggests that the birds may be a bit less detectable in east-central county. If so, such a situation might owe to distinctly restricted patches of habitat in this oaky open country, and to incomplete sampling because of the comparatively few public roads from which atlasing could be conducted.

While on its breeding grounds, the Pacific-slope Flycatcher is the most difficult songbird within its range to spot—and to keep in view. No other diurnal bird is heard so continually, yet seen as infrequently. That these birds are so surpassingly common makes this all the more noteworthy. With this in mind, it is a testament to the perseverance of Humboldt atlasers that so many probable and confirmed breeding observations were noted. Nests were detected among outbuildings, along roadcuts, and tucked beneath the eaves of cabins. Young in a nest under an awning near Blue Lake (H15) were still being fed 21 Jul 1995 (C. Walling, C. Gutiérrez). These birds waste no time after their appearance each April. The carrying of nest material was observed 18 Apr 1998 (H23; M. Wachs), and nest building was observed 30 Apr 1998 (g15; DFx). A nest with young was found 2 May 1995 (H31; R. Sutherland). All five nest-with-eggs records were from June. The earliest fledgling was reported 14 Jun 1997 (k23; JG), and the latest feeding of young was seen 3 Aug 1998 (H17; T. Roelofs).

These flycatchers typically appear in detectable numbers in spring during the first days of April. By mid-month they are widely reported, yet it is not until late April that forests everywhere within their extensive Humboldt County range are occupied. By early August, breeding is completed for the year, and by mid-month numbers would appear to plummet. Although Pac-slopes are inconspicuous and largely silent when in migration, small numbers of passage birds are detected each autumn in deciduous hardwood cover of the coastal lowlands. Reports after mid-September are unusual.

BLACK PHOEBE *Sayornis nigricans*

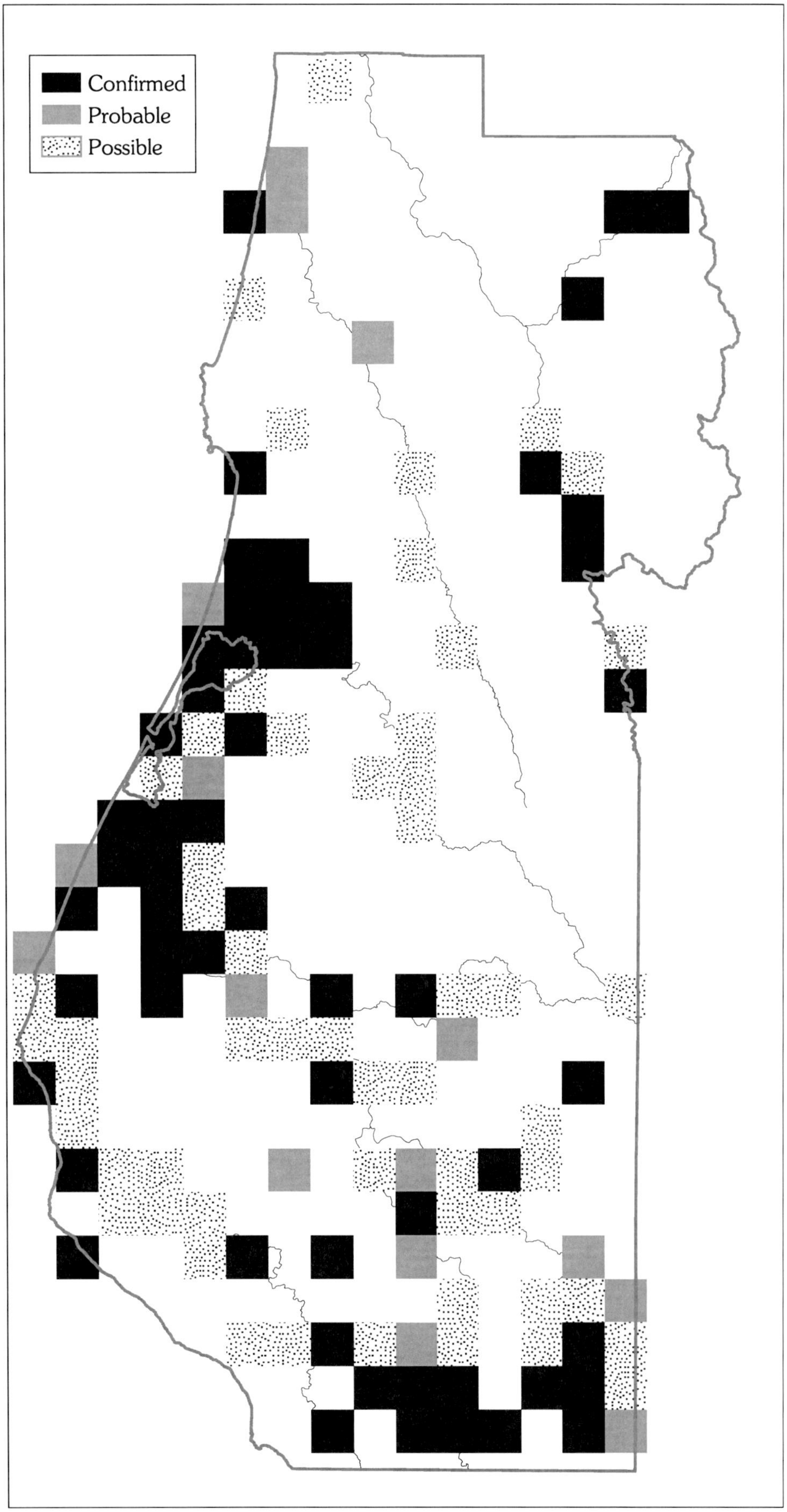

Spirited and confiding, Black Phoebes animate streamsides, pasturelands, and residential settings with a sweeping flash of charcoal-and-white. Their sweet call-notes penetrate above the murmur of waters and carry above the bustle of city traffic. Bold appearance, an impressive force of numbers, and a tendency to select perches in the open create a high "detectability" profile for this flycatcher among birders and other persons attuned to the outdoors.

Throughout their extensive range, Black Phoebes show a distinct fondness for water. Creeks of nearly any size attract them, and even a cattle trough can be enough to lure in a bird. A mix of streamside vegetation and semi-open country, steep earthen banks, and sparsely grown cobble bars are common habitat elements wherever these birds are routine during the breeding season. Humboldt County's major rivers support breeding pairs wherever protected nest sites are afforded by the shelter of a protruding root mass, a rocky canyon wall, or an abandoned outbuilding. Black Phoebes commonly nest around dairies and farms, and at the rural/urban edge created by the mingling of residential yards with tangled ravines or adjacent pastureland. These birds are well known for building nests in the shadows of eaves, under bridges, on the shady sides of old sheds, and on barns—altogether, it would seem the hand of man is well tolerated. Irwin (1985) provided more information on the biology of this species in Humboldt County.

During the Atlas period, Black Phoebes were found in 30% of all blocks. While it is a common resident species in northwestern California (Harris 1996), some seasonal movement occurs, although the exact nature of these movements is unknown. Detections during the Atlas period were vaguely clustered along the county's larger rivers and in blocks where road or foot access to tributary streams was possible. Black Phoebes are quite common throughout the lowlands surrounding Humboldt Bay, east to Blue Lake, south across the Eel River delta, and along the entire reach of the combined Eel/South Fork Eel River system within the county. Some numbers also occur around the Humboldt Lagoons and along lower Redwood Creek.

The distribution map indicates that, over much of interior Humboldt County, these birds were either absent or occurred only in very small numbers during the study period. However, it can be reasonably assumed that appropriate sites for nesting Black Phoebes probably exist in exposed creekbottoms, at stream confluences, about isolated residences, and in other situations where ready access by atlasers was not possible. Certainly, the great expanses of up-and-down, mountainous terrain well inland in Humboldt offer little habitat for these phoebes.

Along Island Mountain Road (N33), a nest with the first brood of young was observed 30 May 1998, and a second brood from the same nest was reported 15 Jul 1998 (both C. Secor). Records that represented the early part of the breeding season included adults carrying nest material 15 Mar 1997 (g15; DFx) and an occupied nest 2 Apr 1998 (F15; DFx). The earliest fledgling reported was seen 18 May 1996 (F13; GAS). The tail end of nesting activity was represented by a nest with young 24 Jul 1995 (H15; KMS), adults carrying food 26 Jul 1996 (k32; JEH), and adults feeding young 7 Aug 1998 (g15; S. Levy). Fraser (1931) found a pair feeding young at a nest near Benbow on the late date of 28 Sep 1930.

Glancing at either the map presented here or at a field guide, a student of western bird distribution might wonder whether Black Phoebes occur only sparingly to the north of Humboldt County. Historically, this was indeed the case. Within the past several decades, these flycatchers have become more common both in adjacent Del Norte County and in southwestern Oregon, where a large population is now resident in the Coquille Valley southeast of Coos Bay, roughly 175 miles northeast of Humboldt Bay. Clearing of land, rural settlement, and the establishment of the dairy industry may have created additional nesting and foraging opportunities.

ASH-THROATED FLYCATCHER *Myiarchus cinerascens*

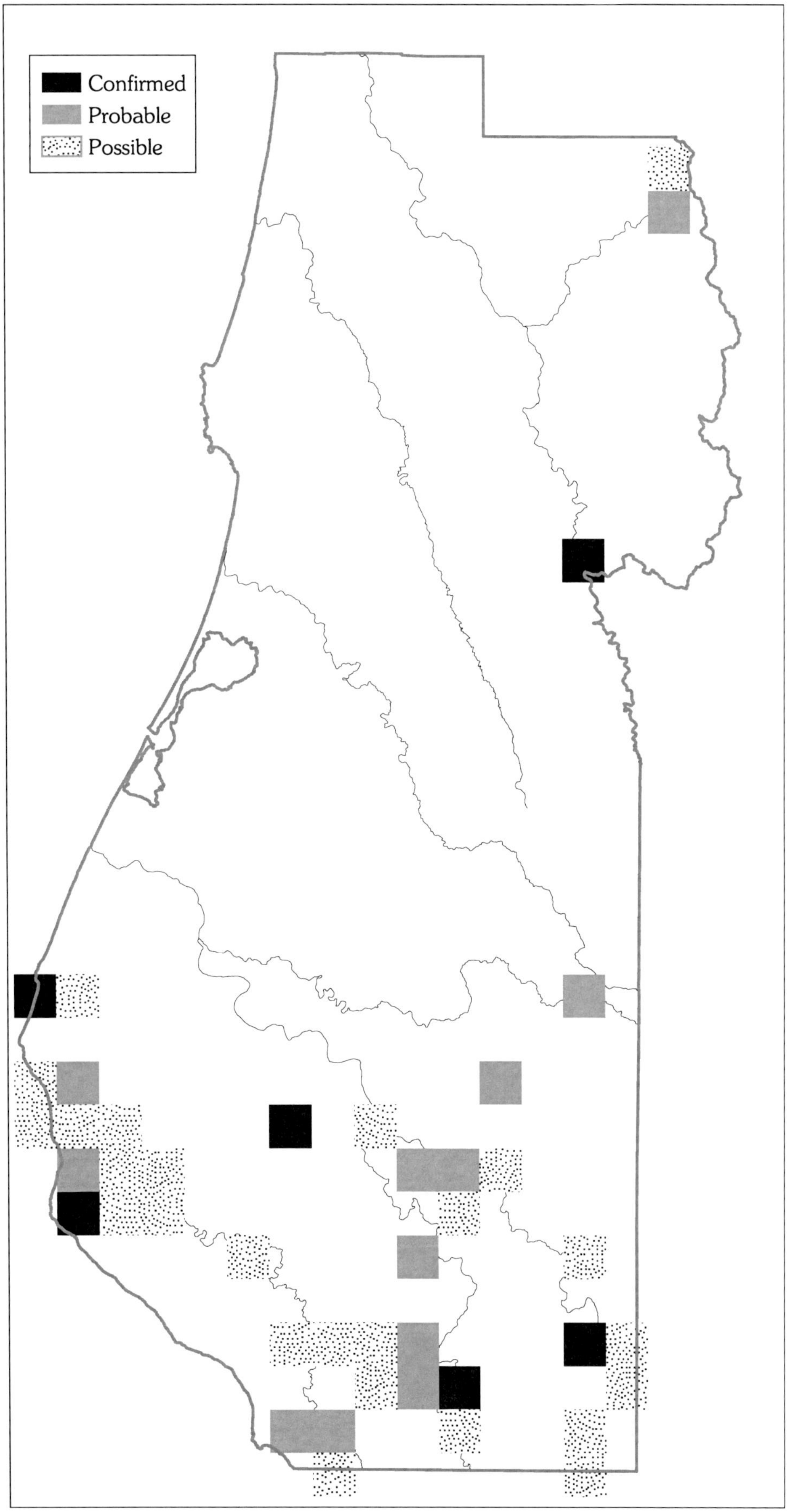

Although it breeds sparingly as far north as south-central Washington, the retiring Ash-throated Flycatcher reaches the northwestern limit of its extensive California distribution in Humboldt County. Here it occurs chiefly within the mosaic of oak woodland and drier Douglas-fir/oak regrowth across the southern third of the county. Considered a locally common migrant, summer resident, and breeder in northwestern California (Harris 1996), this species is reasonably numerous and routine in pockets and corridors of better habitat. In the Garberville area, the main canyon and lower tributary drainages of the Eel River, and along the Mattole River, the burry call-notes of these large flycatchers are a minor but expected element of the morning chorus from late April into July. These birds were found in 43 Atlas blocks. Breeding was confirmed in six of these.

During the Atlas period, Ash-throated Flycatchers were detected most frequently in oak woodlands, at the edges of forests with a significant oak component, and in drier-site riparian corridors at lower and middle elevations within the Eel, South Fork Eel, and Mattole river drainages. The map clearly indicates that the Ash-throat is another among many species whose preference for drier habitats limits its occurrence in Humboldt County to southerly and interior locations. However, a male singing on territory through 18 Jun 1996 (KMS) well to the northeast at Orleans (P05), and nest building observed 27 May 1996 near Willow Creek (N13; JMH) suggest an irregular northward sprinkling of birds. These flycatchers closely approach the outer coast at Shelter Cove and from Punta Gorda to near Cape Mendocino, but are absent from the coastal slope north of there. They are distinctly averse to mountain forests, and were not detected at higher elevations.

Throughout their extensive range, Ash-throated Flycatchers are birds of canyons and foothills. Oaks of some size are nearly always present wherever numbers of these birds are found in the breeding season within the interior of Humboldt County. Because they often contain numerous natural cavities, oaks afford secondary cavity-nesters opportunities for nesting that are not as easily come by in tree stands dominated by conifers. Some degree of departure from this rule is exhibited by birds summering in the seaward portion of southern Humboldt. On the sprawling slopes above Shelter Cove, these birds reside in a rich mosaic of tall chaparral dominated by blue blossom which, while deficient in oaks, is dotted with large Douglas-fir snags. Within this area, it is conceivable that the numerous conifer snags, well worked by woodpeckers, offer sufficient nest cavities.

Regions of the county in which tree stands are rich in tanoak but lacking large true oaks appear not to support Ash-throated Flycatchers. Much of the northern portion of Humboldt County at middle elevations is clothed in tanoak. Ash-throated Flycatchers were undetected across nearly this entire belt during the Atlas survey effort. The shade tolerance of tanoak results in dense and humid stands with a distinctly different "feel" from true oak woodland—in which trees characteristically are open-grown, and are somewhat separated from one another. In addition, the struggle by tanoaks to reach sunlight may produce slender trees less likely to develop cavities.

Just back of the coast from north of the mouth of the Mattole to Cape Mendocino, these birds show up along tortuous riparian corridors densely grown to wind-firm old white alders. Left unmanaged, the crowded alders and other riparian trees include the ranks of the dead and dying—a resource not overlooked by woodpeckers, whose industry helps the efforts of the flycatchers to gain a foothold in such confined locations—often surrounded by miles of windswept, nearly treeless coastal prairie.

The seeming absence of Ash-throats in the Trinity River canyon is intriguing. For some miles in the vicinity of Hoopa there is a considerable amount of habitat in which true oak is mixed lightly with dry-site Douglas-fir, tall manzanita, and even some gray pine. Summer skies are clear, and summer daytime temperatures are very warm. Are the birds present in the Hoopa Valley, but perhaps distributed so spottily or at such low density that they may have been missed by atlasers?

The earliest probable breeding evidence obtained during the Atlas period was a pair on 23 May 1999 (k27; JG). All carrying of nest material or nest building was seen in late May, all food-carrying or occupied nests were in mid- to late-June, and the latest confirmation provided by an adult carrying food was 24 Jun 1995 (b28; GCH). No fledglings were reported.

WESTERN KINGBIRD *Tyrannus verticalis*

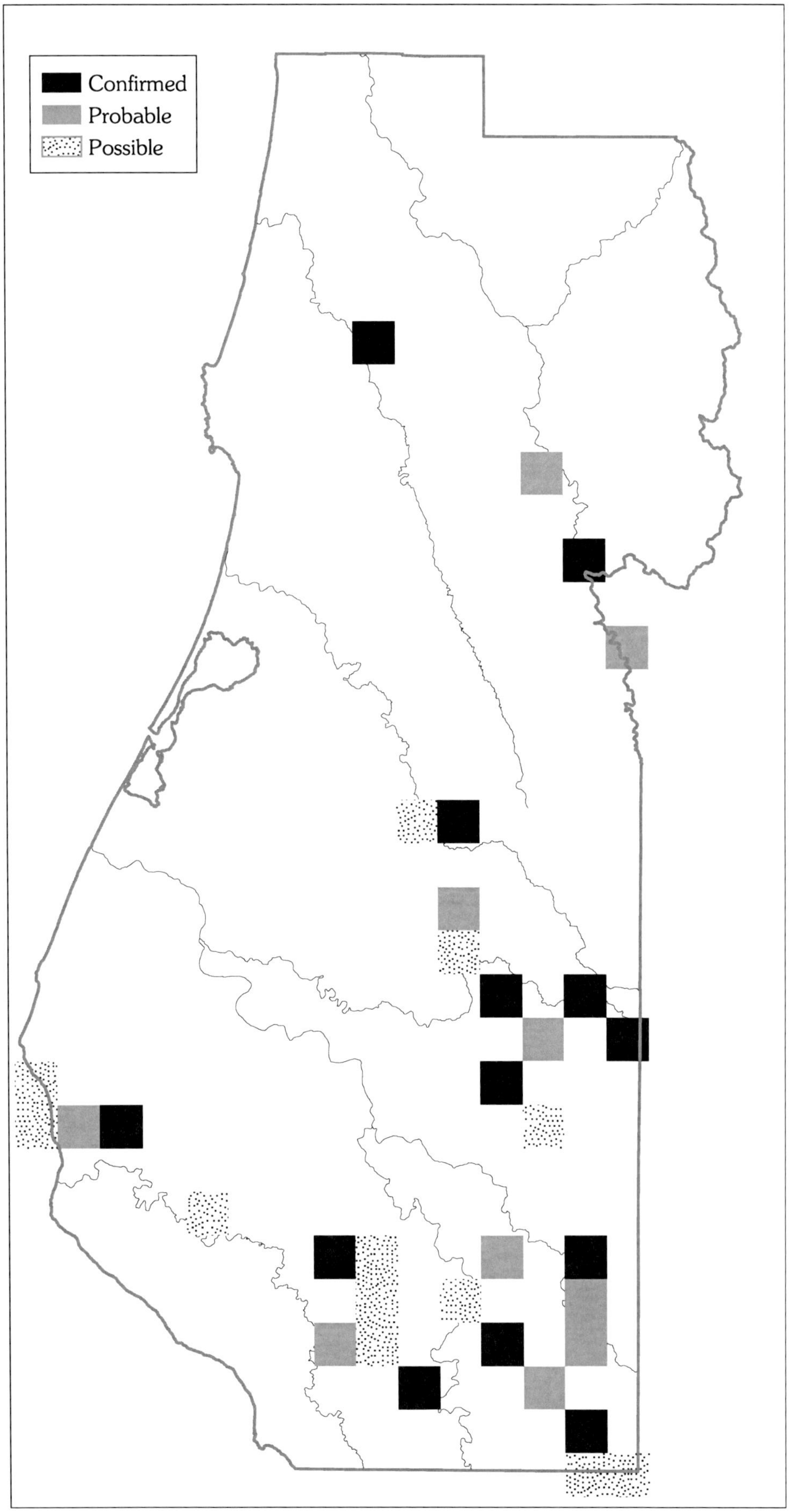

The Western Kingbird is a common and familiar breeding bird across most of California at lower and middle elevations. The species' range within the state extends northwestward into portions of Humboldt County, where the birds occur patchily and in modest numbers where expanses of preferred habitat exist. Kingbirds plainly appear to meet some sort of environmental barrier here. Much seemingly appropriate habitat in the county is unfrequented. The range of Western Kingbird in Humboldt is similar to that exhibited by other species of passerines that typically are more broadly distributed and more abundant to the southeast of the region. Harris (1996) summarized the status of Western Kingbird in northwestern California as an "uncommon migrant in coastal lowlands; uncommon to common summer resident and breeder in drier valleys and open fields of the interior and southern portions of the region." During the 5 Atlas years, these kingbirds were encountered in 35 (8%) of the 425 total blocks. They were confirmed breeding in 13 blocks.

Western Kingbirds are best represented during summer in the warmer, drier country of southeastern Humboldt County. Many detections during the study came from a broad swath extending northwestward and eastward from Garberville. This concentration of blocks takes in the Salmon Creek watershed, the Alderpoint area, and intervening country extending southeastward along the general Island Mountain ridge.

The smattering of detections extended north into portions of the Van Duzen River and tributary drainages and the central Mad River corridor northward to east of the Iaqua Buttes. The farthest west in the Van Duzen River drainage atlasers found this species was near Bridgeville (L23). A few kingbirds were found in the near-coastal pastureland south of "The Wall" (south of Cape Mendocino) and up the Mattole River valley.

A small population of Western Kingbirds along the Trinity River corridor provided a confirmed breeding report as well as two probable reports. Although Fisher (1904) recorded the bird as "common" in the Hoopa Valley from 29 May to 7 Jun 1899, the only report received from there during the Atlas period was of a pair on 21 May 1996 (m11; *fide* JMH).

Finally, a kingbird carrying food in a region of disjunct habitat—the sere and grassy Bald Hills (i08)—provided the northernmost nesting evidence obtained during the Atlas period 11 May 1996 (JEH, GAS). This record was also the earliest kingbird breeding confirmation during Atlas fieldwork, and no other evidence was reported from the Bald Hills. All other breeding confirmations fell from early June through 17 Jul 1998 (H29; JCP), with the earliest fledgling reported 22 Jun 1998 (L31; S. Rovell).

Western Kingbirds are characteristic of open, rolling, rural or lightly-settled situations across most of their range in California. Humboldt County is no exception. Here they are most often encountered in sunny, wide-open valley bottoms or lower-foothill pasturelands well grown to grasses and dotted with full-crowned oaks. Much of the habitat occupied by Western Kingbirds in Humboldt is private ranchland, upon which grazing or haying usually occurs. Flat or nearly flat ground on upper-river floodplains and adjoining lower foothill slopes are most frequently inhabited. With respect to slope aspect, comparatively little pastureland in Humboldt occurs on the more well-watered northern exposures capable of producing commercial forestland; therefore, most Western Kingbird territories lie on the sunnier slopes inhospitable to conifer reproduction and growth. These flycatchers sometimes appear in younger clearcuts during spring migration.

During the Atlas period, five nests were discovered. Typical of these was an occupied nest east of Bridgeville (L23) 2 Jul 1996 that was built about 10 feet below the top of a large oak standing at the edge of a grazed pasture, with a house nearby (M. Morris). Given the sparse human settlement in the entire range of Western Kingbird in Humboldt, it stands to reason that more nests in the county are built in the crowns of oaks or other larger living trees than are tucked into recesses on power poles, where they are also commonly seen.

Small and variable numbers of Western Kingbirds were reported each spring during the Atlas period from the coastal lowlands, particularly about Humboldt Bay and in the Redwood Creek estuary. Historically, such appearances have been nearly annual. The earliest report of a Western Kingbird in Humboldt during the Atlas period was of a probable migrant along Monument Road 14 Apr 1996 (e24; COg). Scattered individuals or even loose flocks were also encountered in other open country settings that somewhat resembled nesting habitat. Although some of these reports were of pairs of kingbirds and occurred as late as mid-May, these birds were also considered to be migrants moving through the area. Western Kingbirds are recognized as having a lengthy spring migration. Birds are regularly noted at places where nesting is unknown into early and even mid-June from the southern Oregon coast into southern British Columbia.

CASSIN'S VIREO *Vireo cassinii*

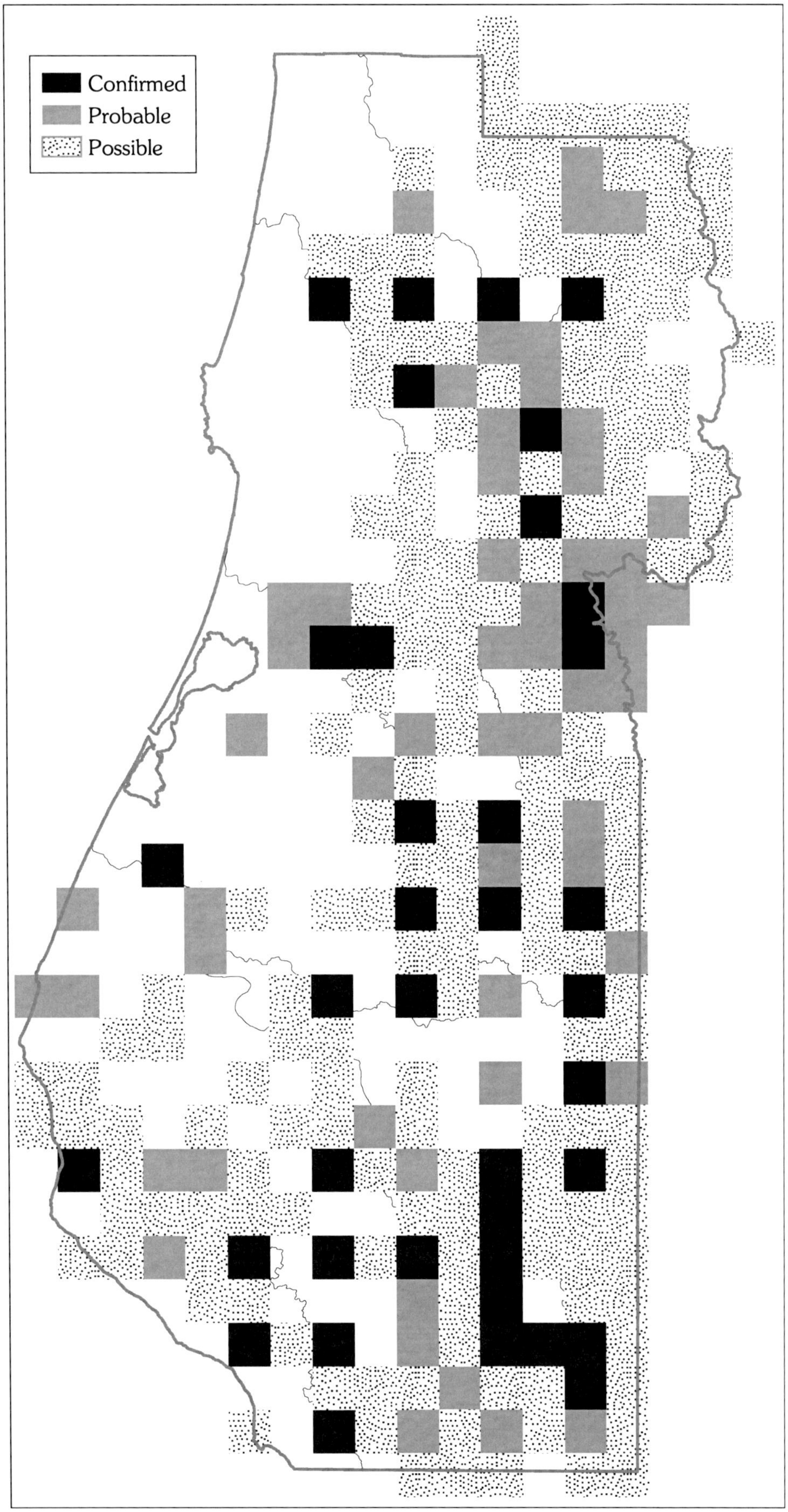

Cassin's Vireo is a common summer resident and breeder that is casual in winter in northwestern California (Harris 1996). Small size, subdued colors, and slow foraging movements cause these birds to be easily overlooked. As is true for vireos as a group, Cassin's is much more frequently heard than seen. Throughout the breeding season, males sing a deliberate song of brief, incongruous phrases separated by moments of silence: *see me . . . De-troit . . . sur-real*. Well into July, one may detect widely scattered Cassin's Vireos carrying forth during hot afternoon hours when singing by other species has diminished or ceased. Females betray their presence by uttering growling *sneer* calls when an observer unwittingly approaches a nest or dependent fledglings.

Cassin's Vireos are among the most widespread and common spring and summer birds of nearly all of Humboldt County outside the summer fog belt. During the Atlas period, they were detected in 256 blocks (60% of all blocks). Confirmed breeding was reported from 38 blocks. The extreme northwestern corner of the county was nearly without detections, and in only two blocks on the windward slope of the King Range were these vireos encountered.

While generally avoiding pure stands of redwood, Cassin's Vireos otherwise inhabit quite a range of forest and woodland types within the county. Most though by no means all preferred habitats contain at least a few oaks. Stands containing any of the five species of tree-size oaks occurring locally (California black, Oregon white, canyon and interior live oaks, and tanoak) are routinely selected. Drier-site conifer-and-oak forests and forest fragments, brushy hillsides with clumps or stringers of trees, and cutover timberland coming back in nondescript hardwood regrowth are all widely used. Well-developed gallery forests, undisturbed stringers of black cottonwood, and mature alder on and around the floodplains of rivers and larger creeks also typically support at least a few Cassin's Vireos during the breeding season. Harris (1996) noted that breeders have been seen in riparian areas as close to the coast as the Mad River bottoms near Blue Lake, a situation that persisted during the Atlas period.

In vigorous younger industrial-forest regrowth with a great deal of dense tanoak at middle elevations, these birds are not particularly abundant but are well scattered and occupy all slope aspects. Like a number of songbirds that may do best in situations offering greater stand diversity, stands of young intensely managed forest in extreme northern Humboldt were found not to support Cassin's Vireos. Within the fog belt, much of this country has regrown to dense immature stands of wet-site conifers in which red alder is the common hardwood. Such sites seem not to meet with the birds' favor. At higher elevations Cassin's Vireos are less common but regularly detected in unmanaged or lightly managed stands thoroughly undergrown with uneven-aged white fir and interspersed with sunny brushfields. They appear to summer essentially to local "timberline" shortly below the highest ridges in the county.

Cassin's Vireos are among the earliest arboreal neotropical migrants to make it north to Humboldt County in spring. They begin to arrive in numbers in early April, concurrent with the appearance of the first Warbling Vireos, Nashville and Black-throated Gray warblers, and newly territorial Hermit Thrushes, with which they share many summer hangouts. Multiple males (i29; JG) and countersinging males (L13; TWL) were reported as early as 24 Apr 1997.

Atlas results suggested that the peak period for nesting within the county is from mid-May into mid-July. The earliest confirmation was an occupied nest 1 May 1996 (m32; M. Bailey), and the first fledglings were reported 19 May 1996 (H23; C. Powell). The latest confirmation was of adults feeding young 21 Jul 1999 (N07; GCH). Atlasers discovered that some degree of perseverance is usually required to locate nests. However, Cassin's Vireos often place their nests at surprisingly low heights, often in sparse understory limbage near eye level. The preponderance of confirmed breeding in priority blocks suggests that the entirety of Humboldt County, except for the northern coastal strip, is occupied.

After breeding, the appearance of family groups of Cassin's Vireos gives way to a seeming near absence by mid-August. The presence of lingering summerers or early migrants in the hushed treetops is betrayed by infrequent *sneer* notes or half-hearted snatches of song. These vireos are fairly common migrants across the county, but so much of the landscape is clothed in suitable breeding habitat that this is perhaps perceived by many birders more as a strong intuition than as a tangible phenomenon. During the 5-year Atlas period, obvious migrants well outside breeding habitat were encountered in small numbers in the coastal lowlands, more commonly in fall than in spring. A very few Cassin's Vireos are encountered as late as mid-October in most years, with an occasional bird lingering as late as the time of the coastal CBCs.

HUTTON'S VIREO *Vireo huttoni*

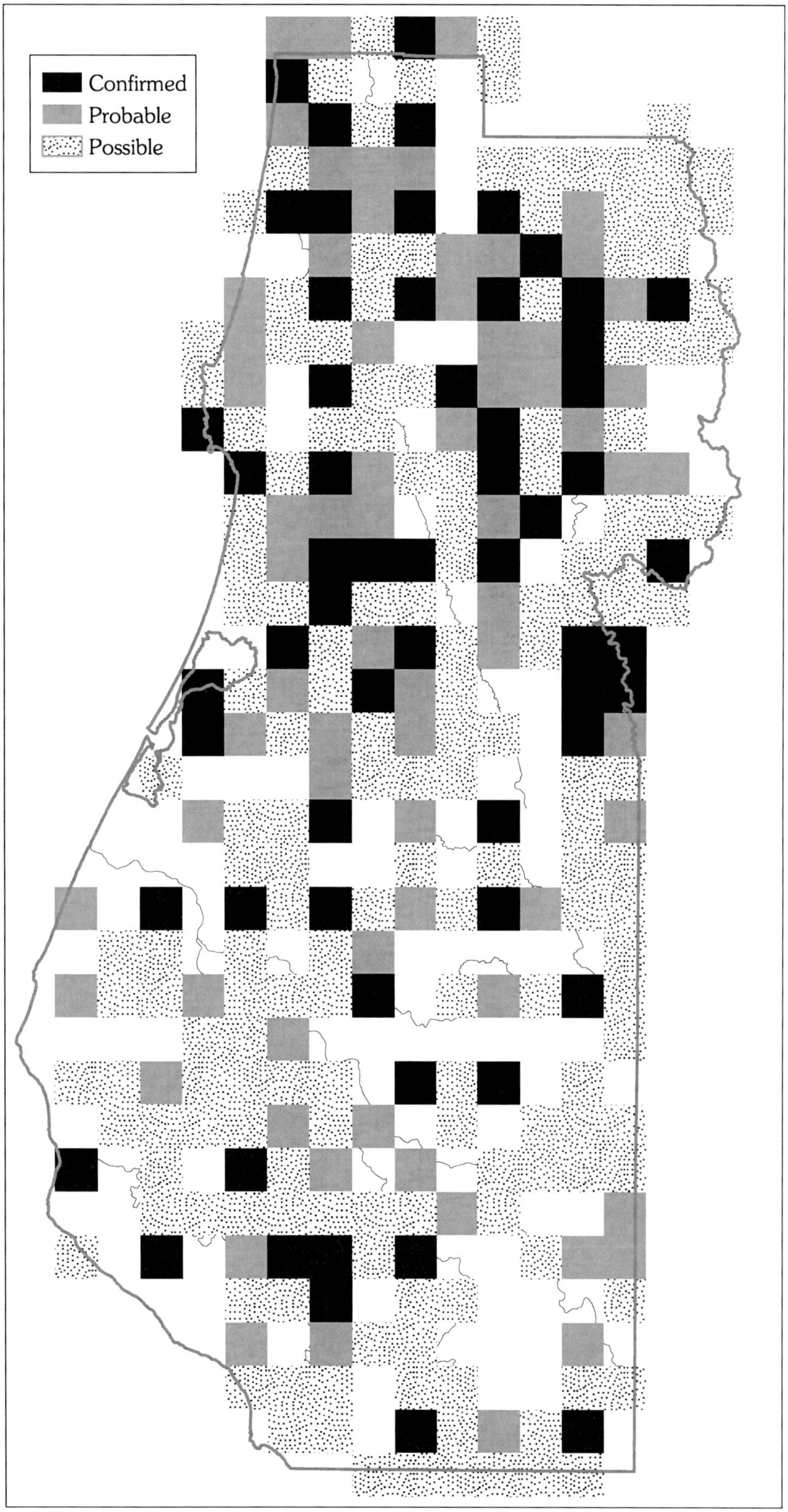

Among the least conspicuous birds wherever it is found, the tiny Hutton's Vireo makes up for dull plumage and indistinct field marks with an insistent voice. The simple, shrill song drifts from the understory of dense stands of trees at all hours of the day from late February into late summer. Outside of the breeding season, infrequent snatches of song, as well as the Hutton's peevish calls, can often be detected among the lisps, chatterings, and chip-notes of the chickadee-and-kinglet flocks with which these vireos routinely forage. Patience on the part of the inexperienced observer is required to firmly associate the sneezy noises made by Hutton's Vireos with the diminutive olive-drab creatures from which they may ultimately be found to emanate.

Despite this low profile, there were nevertheless a great many detections of Hutton's Vireos during the Atlas period. These birds were encountered in 295 (69%) of the 425 total blocks. That the distribution map for this small forest bird is as well spangled with confirmed and probable breeding reports as are the maps for more conspicuous songbirds suggests that this species is truly common in Humboldt County. Indeed, across northwestern California, this is among the more frequently detected forest passerine species, and it is considered a common resident and breeder (Harris 1996).

A student of bird distribution with no specific knowledge of the habitats occupied by Hutton's Vireo in Humboldt County might intuit from the map that these birds can be encountered across a spectrum of forest types. Such a perception would be accurate. During the Atlas period, they were detected in nearly every closed-canopy forest type outside of the highest mountains. Irrespective of species composition, a high degree of understory complexity and tree-to-tree connectivity is a structural element common to sites at which the birds are most numerous. The array of tree communities in which Hutton's Vireos are regularly found in Humboldt includes Sitka spruce/red alder; various Douglas-fir dominated forests; tanoak dominated stands; and a great variety of situations in which true oaks (*Quercus* spp.) are present. Canyon and foothill slopes heavily forested with evergreen hardwoods such as canyon live oak and madrone appear particularly favored. Even old-growth redwood forests support them if a midstory layer is present.

These birds are quite common not only in unmanaged oak/conifer stands but also in regrowth commercial forest, a great deal of which occurs in the north-central portion of Humboldt County. Provided that no more than light thinning has been performed, these vireos will occupy regrowth stands as young as 20 years. Mid-seral and maturing mixed conifer-hardwood stands in which high natural pruning of the canopy has yet to take place are commonly used. Such older regrowth stands comprised a considerable portion of the forest cover in the county during the Atlas period.

Over much of their range, Hutton's Vireos are associated with drier-site forests in which one or more species of oaks is well represented. Conversely, in Oregon, Washington, and British Columbia, they commonly inhabit well-watered conifer stands in which no oaks occur. In a general sense, these disparate regional habitat predilections dovetail in Humboldt County. Thus, the birds occupy forests with few oaks wholly within the coastal fog zone as well as oak-rich forests in warmer, drier places. In the Klamath Mountains, their abundance is positively associated with hardwoods and negatively associated with conifers (Ralph *et al.* 1991).

A distinct pattern of more confirmations in priority blocks than in non-priority blocks is evident on the distribution map, suggesting that Hutton's Vireos breed in most of the blocks in which they occur.

No strong indication of absence over any significant area within the county can be discerned from the map. However, several small regions in which Hutton's Vireo may have been less numerous or less thoroughly distributed can be noted. They were seemingly absent from the mountains of northeastern Humboldt, the Humboldt Bay and lower Eel River bottomlands, and grassy, mostly treeless coastal blocks from Cape Mendocino to Punta Gorda. A vague pattern of no-detection blocks scattered in east-central and southeastern Humboldt may be related to the largely open, grassy, oak-dotted landscape traversed by the roads from which most atlasing was conducted. Almost certainly, scattered patches of habitat support Hutton's Vireos on inaccessible private land in these areas.

Evidence that these resident vireos initiate breeding early in the season was provided by records of multiple males singing on territory 25 Feb 1995 (i05; JEH JCP, RHw, D. Roberson, R. Carratello) and an adult carrying nest material 19 Mar 1997 (e10; M. Morrissette). Only three active nests were reported, the earliest 29 May 1995 (D21; L. Karsteadt) and the latest 17 Jul 1998 (e17; L. Bacon-Ogden). The earliest fledgling was reported 23 May 1999 (g29; JCP). Adults were seen feeding young as late as 7 Aug 1996 (g02; M. Kenney).

Although Hutton's Vireos are resident in Humboldt County, some degree of migration or dispersal evidently takes place during the non-breeding season, when it is typical to encounter one or two birds well away from breeding habitat in flocks of chickadees and kinglets. Appearance of a very few Hutton's in the isolated dune-hollow willows at Fairhaven during periods of landbird migration suggests that some movement of these birds may be underway. Riparian thickets in the Humboldt Bay lowlands and Eel River delta supported at least a few of these vireos during the winters of the study period.

WARBLING VIREO *Vireo gilvus*

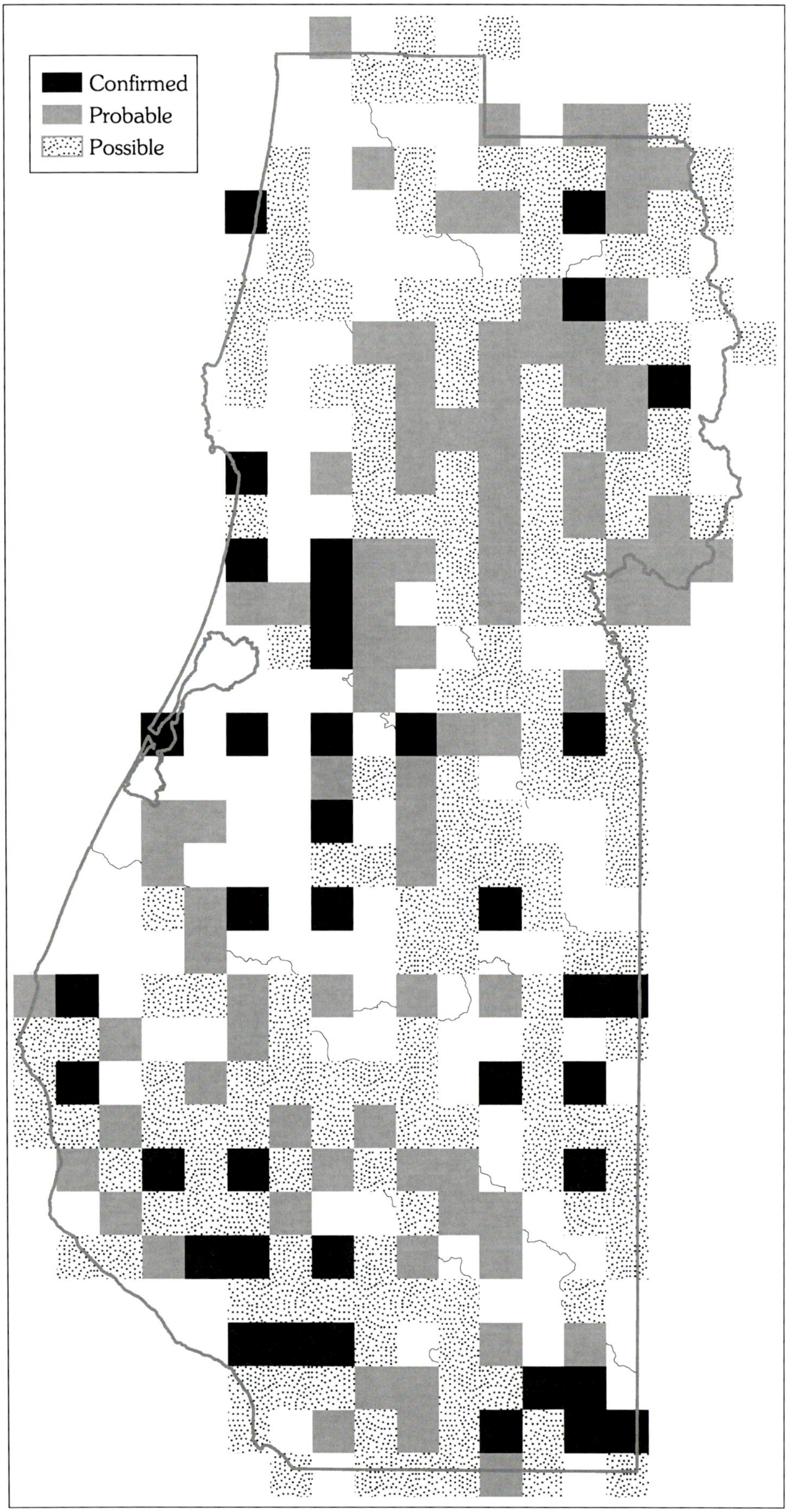

Much of the Humboldt County landscape hosts Warbling Vireos in abundance from mid-April into late summer. Locally, they are among the more common songbirds of many wooded habitats, occurring from sea level to above 5000 feet. Harris (1996) stated that the Warbling Vireo in northwestern California was a "common migrant, summer resident, and breeder in mixed forests, riparian areas, and oak woodlands." In many sections of Humboldt that are well wooded with nondescript regrowth timber, the Warbling Vireo ranks among those songbirds most frequently heard during the general period of spring and summer singing. During the Atlas period, this species was detected in 287 (68%) of the 425 total blocks. Breeding was confirmed in 38 blocks.

In the minds of many birders, Warbling Vireos are associated with leafy bottomland corridors, aspen parks, and similar settings. While riparian areas are certainly important to them, Warbling Vireos in Humboldt County are not partial to streamside vegetation or stands of deciduous hardwoods. Most of the population occurs not along streams but on wooded slopes, often at some distance from the well-watered haunts of Yellow Warblers and Bullock's Orioles. They were detected at only 30% of systematic point counts in lower Eel River riparia in 1999 (R. Hewitt, unpubl. data), but in more than 65% of old-growth Douglas-fir plots (Raphael 1984).

Providing habitat for Warbling Vireos across much of the county are unmanaged stands of second-growth Douglas-fir thoroughly undergrown with evergreen hardwoods, particularly tanoak. In Douglas-fir forests the abundance of this vireo is strongly positively correlated with the canopy volume of tanoak (Raphael 1987). The spectrum of habitat also encompasses red alder regrowth near the coast; white alder riparia (inland, and near the coast south of Cape Mendocino); stands of black cottonwood and willows along rivers; higher-closure oak woodland; regenerating moist-site clearcuts with thickets of vigorous shrubs and small trees; and, at higher elevations, mixed coniferous and deciduous growth at forest's edge and around the edges of meadows. In the Eel River delta, Warbling Vireos favor alder/cottonwood/willow forest with high densities of alder (Kelly 1987).

During the Atlas period, no region of the county was wholly without detections, although an irregular smattering of blocks with no reports can be discerned on the map. These "white" areas coincide roughly with intensively managed commercial redwood timberland (northwest and central county); the higher mountains near the boundaries Humboldt shares with Siskiyou and Trinity counties; open grassland; coastal and near-coastal pastureland and dunes; and conifer-clothed slopes of the seaward King Range. Restricted survey access to private lands, particularly in interior Humboldt, certainly left areas of suitable habitat unchecked. Because of the distribution that was revealed, along with the fact that Warbling Vireos here are not restricted to one or a few distinct vegetation types, many atlasers concluded that these vireos were actually present during the summer in most inland blocks.

Territoriality was observed as early as 21 Apr 1998 (N11; *fide* JMH). Nest building was observed 9 Jun 1999 (L21; DFx), and the carrying of nest material was recorded 22 Jun 1995 (g31; D. Roberson) and 22 Jun 1998 (H29; JCP). Only seven nests were found. The earliest occupied nest was found 31 May 1999 (o23; C. Cool). The earliest fledgling was seen being fed 24 Jun 1996 (F13; GAS). Near Blue Lake (H15), an adult sang while sitting on the nest 18 Jul 1995 (PFS), and was feeding young 4 Aug 1995 (C. Gutiérrez); this was the latest report of confirmed breeding during the Atlas period. While this species should be watched for possible declines, Humboldt's breeding population of Warbling Vireos appears robust.

GRAY JAY *Perisoreus canadensis*

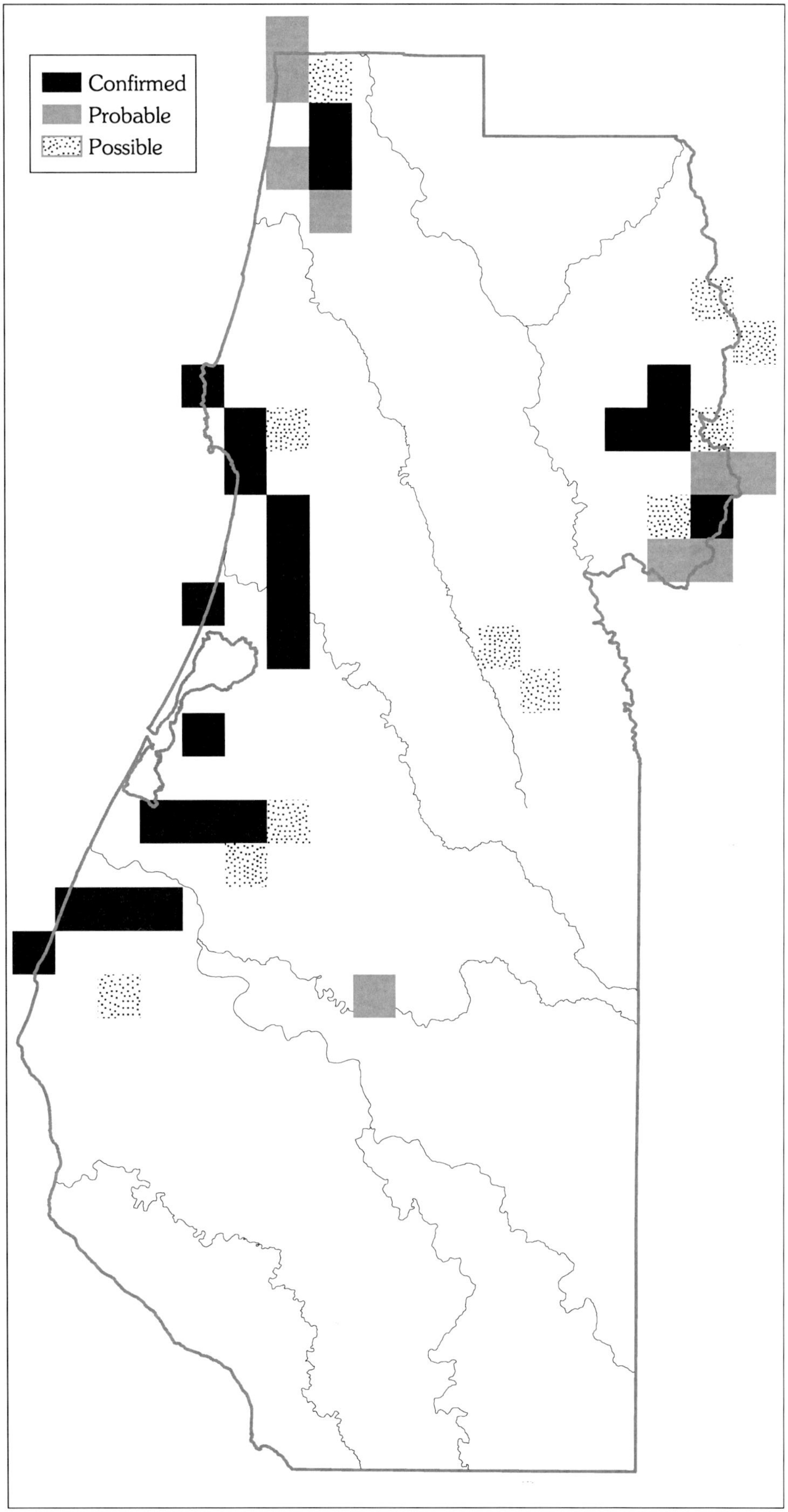

Wraiths of the cool conifer forests, Gray Jays appear seemingly from nowhere to investigate hikers along trails and at campsites. Apart from a few whistles and chattering notes, little notice is given of their arrival. One by one, a flock materializes from the shadows of the mid-canopy, individuals slipping from higher tree limbs to lower and closer perches, their curiosity impelling them onward. With an alternation of seemingly ineffectual flapping and wavering glides, birds within the group soon assemble nearly overhead and just beyond arm's reach at eye level. Should a handful of nuts or raisins be offered on an outstretched palm, it is probably a matter of moments before the impromptu corvid subsidy program is accepted. First one, then another jay boldly plucks the goodies—each bird remaining on one's hand no longer than it takes to tickle the fingers with a delicate grip, cock a bright eye upward, snatch the handout, spring away with a flutter of wings, and retire to the nearest bough to devour the item. This scene of goodwill and brazen trust is played out regularly in parks and forests of the North, as well as in select locations in Humboldt County.

California lies at the southwestern edge of the range of the Gray Jay. This species extends into the state in the Warner and Klamath mountains, in the Cascades, and in near-coastal forests of Del Norte, Humboldt, and (very sparingly) Mendocino counties, south rarely to the Gualala River (Burridge 1995). The distribution of the jay in Humboldt is limited, occupying two sections of the county: the coastal and near-coastal conifer forests from the Del Norte County line south to Centerville, and the northeastern mountain forests. This distinctly dual range was perceived by many birders prior to the initiation of Atlas surveys, and was well documented by the end of the project.

During the Atlas period, Gray Jays were encountered in 10% of all blocks. Confirmed breeding was noted in more than half of these, and probable breeding in nine others. In the western segment of the range within Humboldt, these jays were detected in Prairie Creek Redwoods State Park and adjacent Redwood National Park (cluster from g01 to H05); at Patricks Point State Park (e09); shortly east of Trinidad (f10, g10) and in Westhaven (F11); near Fieldbrook (g12, g13); on Fickle Hill both within and outside the Arcata city limits (g14, g15); at Lanphere Dunes (e14); on Humboldt Hill (e17); from near Loleta east and southeast into industrial forestland (D19 to g19, f20); along the south side of the Eel River delta from just south of Centerville Beach (B21) to Wildcat Ridge out of Ferndale (D21), and near Malfunction Junction (c23).

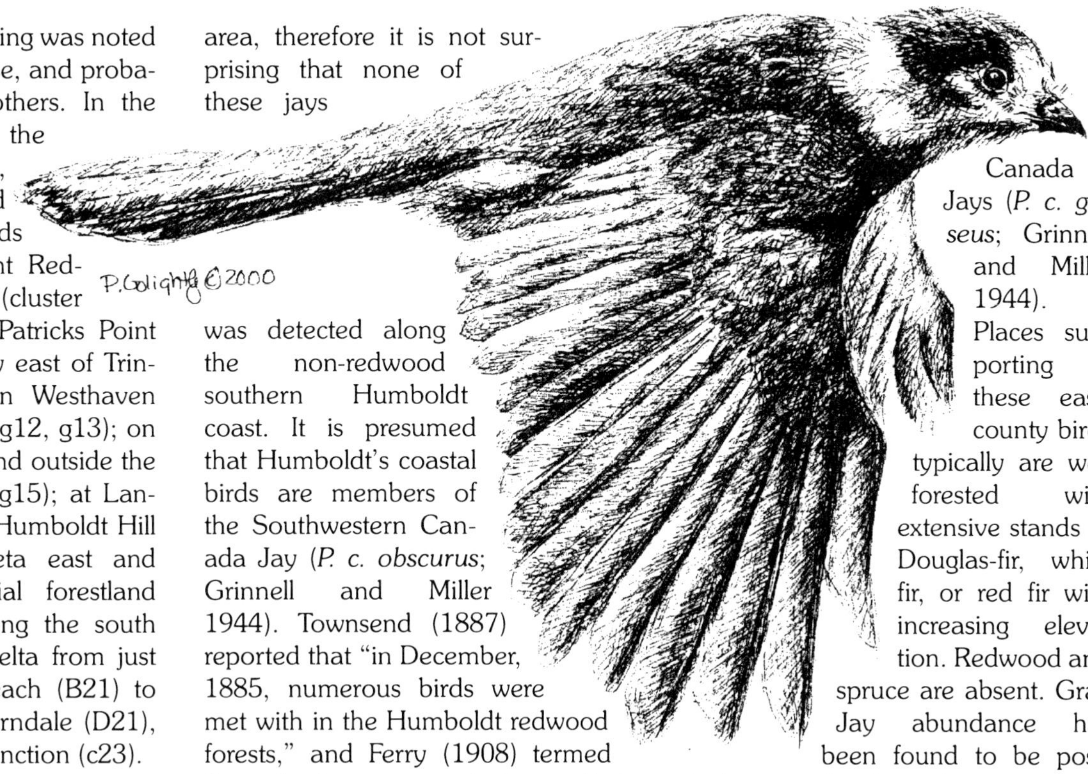

Gray Jays inhabiting the coastal and near-coastal portion of their Humboldt County range tended to be encountered in three kinds of settings. Each is conifer forest characterized by high canopy closure, stand uniformity, and minimal, or no, hardwood component. First, they were found in coast redwood-dominated old-growth and regenerating logged forest of Redwood National and State parks and nearby industrial timberland. These stands also contain much Douglas-fir, Sitka spruce, grand fir, and western hemlock. Second, birds were noted in maturing redwood-dominated second-growth forest of the Arcata Community Forest and adjacent stands at the eastern outskirts of Arcata. Third, they were found in mature, mid-sized Sitka spruce woods at Patricks Point as well as near Ferndale. While considerable forest exists in the former location, the stands inhabited in north-facing draws above the Eel River delta are typically on the tatty side, with some little more than long stringers through which one can see daylight on the opposite side. Homogeneous or nearly homogeneous stands of spruce appear to attract these birds. Such habitat quickly diminishes south of the Centerville area, therefore it is not surprising that none of these jays was detected along the non-redwood southern Humboldt coast. It is presumed that Humboldt's coastal birds are members of the Southwestern Canada Jay (*P. c. obscurus*; Grinnell and Miller 1944). Townsend (1887) reported that "in December, 1885, numerous birds were met with in the Humboldt redwood forests," and Ferry (1908) termed them "quite common in heavy Redwood timber at Orick," suggesting they were formerly more abundant here.

A notable outlier report was of a bird seen in Grizzly Creek Redwoods State Park 30 Jul 1999 (i23; GCH). It is interesting that only one sighting was recorded at Grizzly Creek, and none at all in the much larger Humboldt Redwoods State Park. Might there be a population of Gray Jays in Humboldt Redwoods simply overlooked by atlasers?

In the higher mountains of northeastern Humboldt County, Gray Jays were encountered in 14 blocks. These reports fell within a cluster of a dozen, with two blocks lying not many miles to the southwest of that concentration. Birds were found in the general area of the Humboldt/Siskiyou county line and, farther south, the Humboldt–Trinity county line, from near Whiteys Peak (q07) southward to the vicinity of Onion Camp (q13). Gray Jays living in the mountains of northeastern Humboldt are restricted to forests between 4000 and 6000 feet in elevation and usually, though not always, occur on northerly exposures. These inland jays are thought to be Gray Canada Jays (*P. c. griseus*; Grinnell and Miller 1944). Places supporting these east-county birds typically are well forested with extensive stands of Douglas-fir, white fir, or red fir with increasing elevation. Redwood and spruce are absent. Gray Jay abundance has been found to be positively correlated with stand age (Raphael *et al.* 1991).

The extensive montane conifer forests clothing much of the South Fork Mountain ridge appeared not to support breeding Gray Jays, with the exception of possible breeders noted at Horse Mountain (L15) and between Friday Camp and Spike Buck Mountain (m16). That these jays would be rare on at least the Humboldt portion of South Fork Mountain was anticipated. Before the Atlas period there had been only a handful of reports of apparent transient Gray Jays on Blake Mountain and points farther south in Trinity County (Hunter and Hazard 1998).

The earliest probable breeding record was of courtship observed near Lost Man Creek (h04) 15 Apr 1998 (W. Widdowson). Gray Jays were observed building a nest near Catfish Lake (D19) 20 Apr 1998 (M. Wachs). The nest was constructed of spruce twigs, and was located in an alder surrounded by spruce. This was the only Gray Jay record submitted that pertained to an actual nest site. In the 22 blocks in which confirmed

Continued on page 393

STELLER'S JAY *Cyanocitta stelleri*

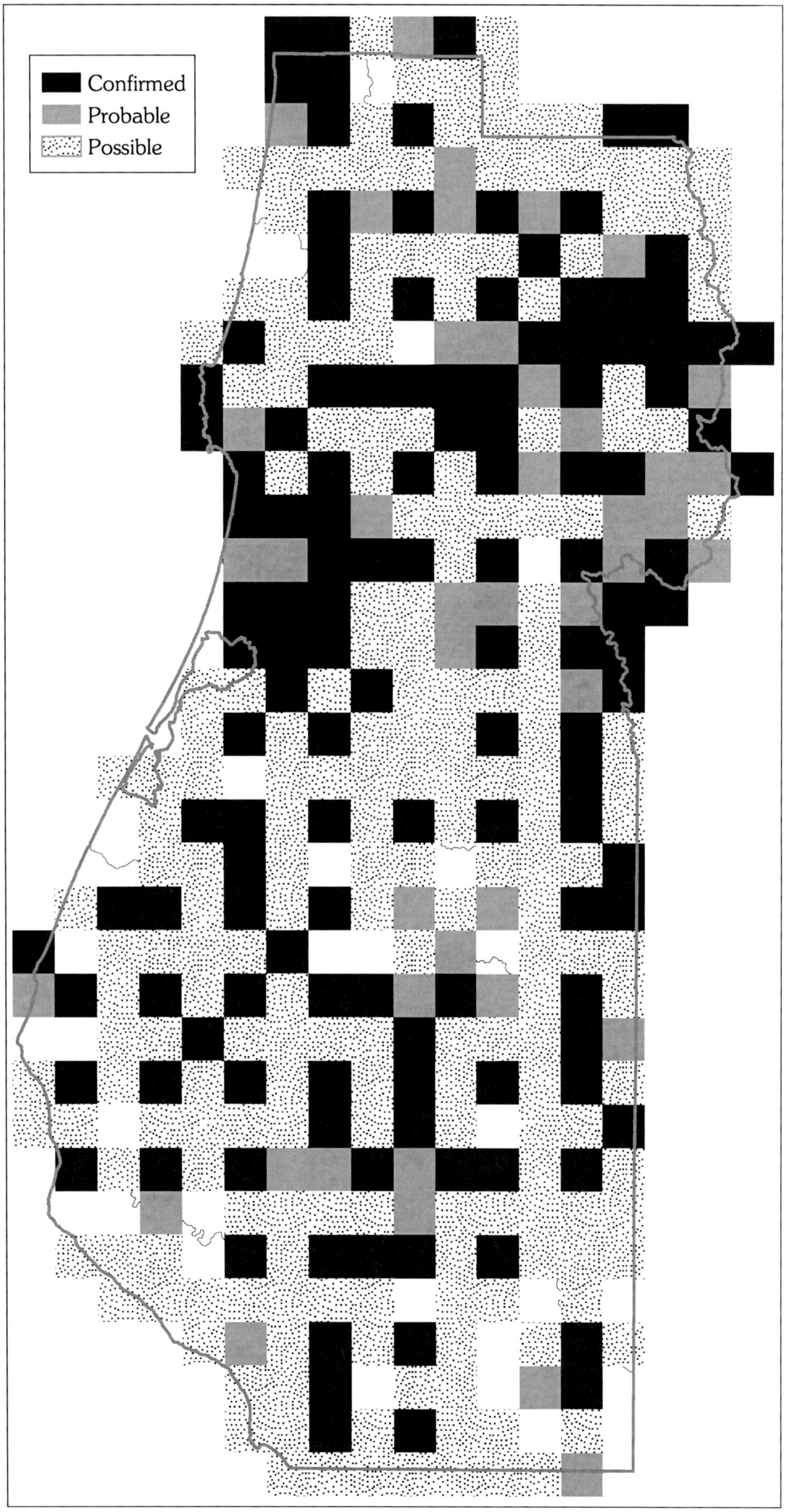

Striking appearance, bold behavior, and clannish ways make Steller's Jays among the most visible of Western forest birds. Although much of the time they are stealthy and voiceless creatures of shadow's edge, any cause worth making a fuss about may quickly lead to a noisy assemblage of jays. Inciting one another to a frenzy of harsh calls, the hush of the woods may be broken for long moments as corvid complaints are delivered. Across its extensive range this jay is typically a common resident wherever it occurs. A broad spectrum of forest types, characterized by conifers, is inhabited.

Steller's Jay has long been known as a bird of Humboldt County. Since Townsend (1887) wrote that this species was "moderately common among the redwoods," naturalists have found these birds to be almost continually in earshot and frequently in sight within the forest environment. In northwestern California, Harris (1996) considered this species "a common resident and breeder in forests and woodlands with a conifer component." In fact, Atlas data suggest that this is the most widely distributed bird species in Humboldt, detected in 380 (89%) of the 425 total blocks. This jay was somewhat more widespread than the second most-commonly reported species, the Common Raven, which was found in 83% of all blocks.

In no region within the county is this species unusual. A comparative sparseness of confirmed and probable breeding reports in the southern half of the county hints at either fewer birds or—perhaps a more likely explanation—less observer effort in the rural hinterlands. Possibly an effect of a less-forested landscape, confirmed and probable reports were also less frequent in the pastoral open country near the geographical center of the county. Coastal and near-coastal blocks lacking detections typically offered little or no habitat. Habitat for Steller's Jays in extreme southeastern

Humboldt is largely confined to northerly exposures. In blocks where limitations of access precluded sampling of such sites, the species was unrecorded. Particularly in the southern half of the county, extra observer effort invested in priority blocks appears to have paid off with increased reports of breeding.

Steller's Jays occupy many forest types in Humboldt County, provided that conifers are present. The birds inhabit coast redwood- and Sitka spruce-dominated old-growth; regrowth spruce mixed with red alder; Douglas-fir old-growth and regrowth timber; forests dominated by evergreen hardwoods but with a modest or patchy conifer component; the Jeffrey pine/white fir/incense-cedar forests of the snow zone; Shasta red fir stands at the highest elevations within the county, and pine-oak associations with few short-needle conifers. Brand (1998) categorized Steller's Jay as an edge species in redwood forests, since it exhibited higher densities near edges than in the forest interior. These jays are ordinarily absent during the breeding season from treeless expanses, dunelands, open oak woodland, and riparian woods with no conifers. They are also largely absent from shore pine forests (Sferra 1988). Numerous birds appear in winter and early spring at bird feeders in towns, or in the smaller woodlots where they are not encountered during the breeding season.

As would be expected of a widespread and somewhat abundant resident species, breeding confirmations occurred over a comparatively wide span of dates. The carrying of nest material was reported as early as 12 Apr 1996 (F15; GCH), and food-carrying was reported 21 Aug 1998 (f20; M. Wachs). A single nest with eggs was reported 29 May 1996 (N09; JMH), and the earliest fledgling was seen 19 May 1996 (g01; M. Kenney).

Populations of Steller's Jays are thought to be increasing in California, although the increases apparently are not as pronounced as those witnessed among other corvids (Liebezeit and George 2002). Because this species is a known nest predator, such increases are a cause for concern, especially as they relate to threatened or endangered species. For example, predation of Marbled Murrelet nestlings by Steller's Jay has been documented (Singer *et al.* 1991).

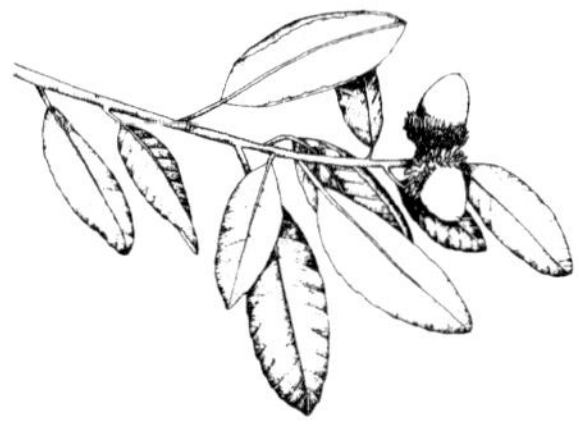

WESTERN SCRUB-JAY *Aphelocoma californica*

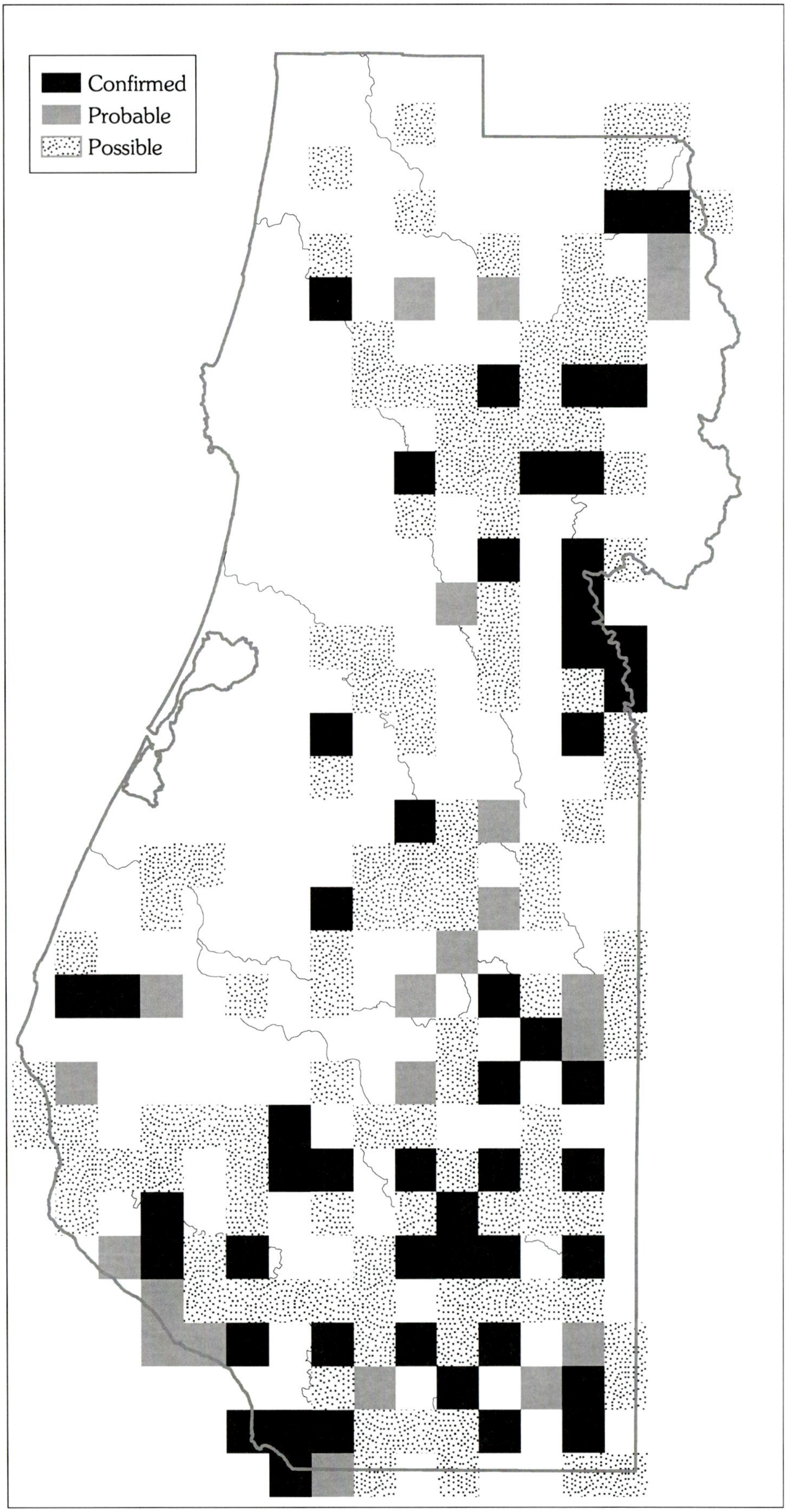

Crestless jays of the genus *Aphelocoma* are indigenous to middle and temperate North America. Spurred by evolving knowledge of each species' respective genetic relationship to the others, as well as greater understanding of their differing life histories, the taxonomy of this group has recently been revised. Five species are currently recognized. Four of these occur in the U.S., two of which are resident in California. The Island Scrub-Jay (*A. insularis*) of Santa Cruz Island is found nowhere else, while the Western Scrub-Jay is widespread at lower and middle elevations across much of the state. This is a bird of both town and country, visiting backyard feeders and nesting in hedges, as well as inhabiting rural or wild lands where no handout is forthcoming.

The Western Scrub-Jay has long been known as a common resident in Humboldt County. The early annals of ornithology in the region contain many references to this species. Fisher (1904) reported it to be abundant in the Hoopa Valley in 1899, with subsequent Humboldt reports provided by Mailliard (1916) and Clay (1917). Yocom and Harris (1975) described it as a common resident and breeder in northwestern California. Harris (1996) repeated this, adding that the birds were most common inland and in the southern portions of the region and that they occurred within 10 miles of the coast on Fickle Hill (southeast of Arcata) and at Fortuna.

During the Atlas period, the Scrub-Jay remained a common bird in Humboldt, and survey results indicated no noticeable change in its distribution. Western Scrub-Jays were found in 43% of all blocks. They were found in much of Humboldt County with the exception of the coastal fog belt and the higher mountains. Most of the county east of the fog belt from the Klamath River to the Mendocino County line is inhabited to some degree, but Scrub-Jays are much more conspicuous in the county's southern one-third, from the coast to the base of the Lassics. Vague clusters of blocks with jays in the central Mad River and along the combined Klamath-Trinity River corridor can also be discerned.

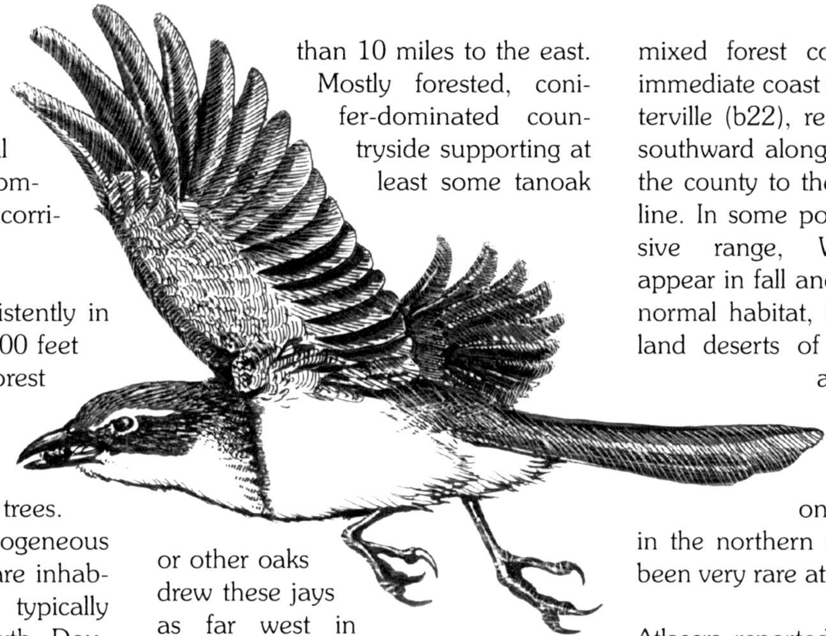

Scrub-jays occur most consistently in Humboldt County below 3500 feet in woodland or broken forest containing California black, Oregon white, canyon or interior live oak, or any combination of these trees. Although some nearly homogeneous stretches of oak woodland are inhabited, scattered conifers are typically present. Patches of regrowth Douglas-fir, Jeffrey/ponderosa pine, or incense-cedar, mixed irregularly with oaks, commonly occur where Western Scrub-Jays live. These preferred habitats are found in settled valleys and stream corridors, lower foothills, and rolling semi-open country. Additionally, extension upslope nearly to the beginning of the mountain conifer forests has been noted where broad bands of oaks march into the middle elevations. Power line rights-of-way and clearcuts supporting oak also provide habitat. Disturbed ground regrown to clumps of trees and brush attract them, particularly if such places are within ready flight distance of oaky forest edge or woodland.

Western Scrub-Jays extended westward during the Atlas period to the edge of the fog belt and intruded locally into the zone of coast redwoods—although redwood stands *per se* were not used. This penetration into the region of persistent summer morning overcast was almost certainly enabled by the presence of tanoak and the westernmost vestiges of California black and Oregon white oak on the ridges. The line separating acceptable and unacceptable habitat seems to be a fairly precise one. These birds historically have been nearly unknown in coastal towns and cities from Loleta north, but they occur regularly fewer than 10 miles to the east. Mostly forested, conifer-dominated countryside supporting at least some tanoak or other oaks drew these jays as far west in northern Humboldt as just east of the redwood parks (h04, h06, H07, i08, and i09); in Redwood Valley (J11, j12); on the "back side" of Fickle Hill (H15), and at Kneeland (H17, h18). A good example of these coastward encroachments was the report of adults feeding young 10 Jul 1996 near the Bald Hills (H07; GCH).

No scrub-jays were encountered along the coast from the Del Norte County line south through the Eel River delta. This is a region completely lacking the acorns so important in their diet. However, again mirroring the historical distribution, Scrub-Jays reappeared in coastal-slope chaparral and mixed forest cover quite near the immediate coast in the vicinity of Centerville (b22), remaining conspicuous southward along the western edge of the county to the Mendocino County line. In some portions of their extensive range, Western Scrub-Jays appear in fall and winter outside their normal habitat, for example, in lowland deserts of the Southwest and along stretches of the Pacific Coast. Yet, in Humboldt, the occurrence of even one Western Scrub-Jay in the northern coastal lowlands has been very rare at any time of the year.

Atlasers reported breeding confirmations in 51 blocks and probable evidence in 22 others. Despite the fact that this is a resident species, it appeared that the onset of nesting did not take place especially early in the season. Various reports of breeding activity, however, occurred into August. While all but five confirmations were reported after May, some early ones included the carrying of nest material 24 Apr 1995 (g33; K. Dyer), a nest with young 11 May 1995 (H33; K. Dyer), and food-carrying 22 May 1996 (c23; COg). The earliest fledglings were seen 6 Jun 1995 (n32; M. Bailey). Of the three nests found, a late one with young 7 Aug 1995 (J11; D. Kriger) was among the few August confirmations.

AMERICAN CROW *Corvus brachyrhynchos*

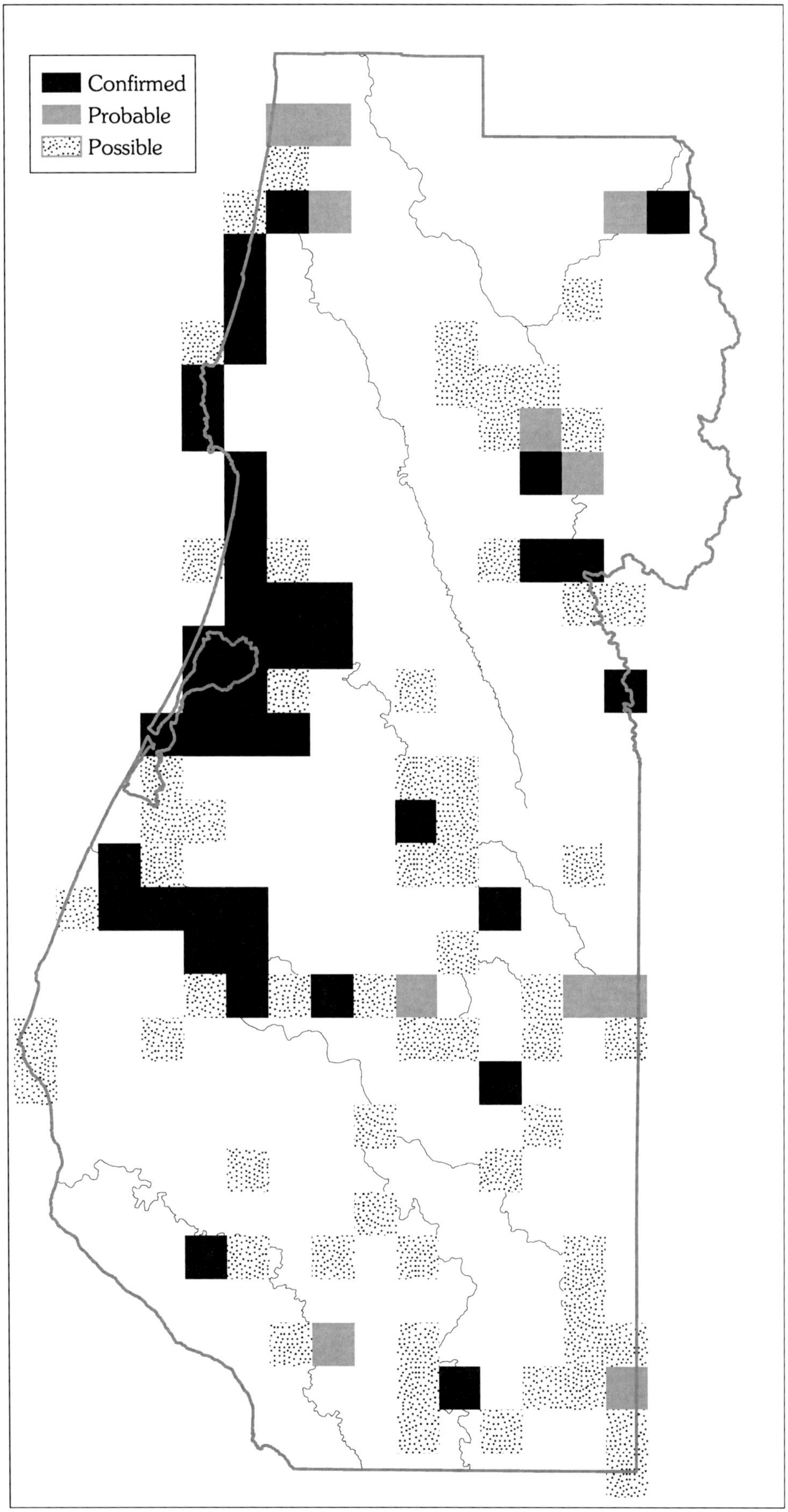

Few birds are as conspicuous, familiar, and steeped in lore as American Crows. Their cleverness has endeared them to many, but application of this trait has brought exasperation and some measure of financial loss to many a farmer. Their clannishness extends nearly throughout the year; gaining protection and other vital benefits from the flock, a crow by itself and strictly on its own is a crow down on its luck. With the clearing of the Eastern hardwood forest and the breaking of land by the plow, American Crows have done well in the presence of humans.

One attuned to birds quickly becomes nearly inured to the presence of American Crows. They are frequently seen in and around cities and suburbs and in outlying farmland. Atlas results provided no surprises; rather, the pattern of detections supported a widespread preexisting understanding that where people live, crows also live. The map indicates both that crows are found in and around settlements and that they generally do not occur in sections of Humboldt County in which human residents are scattered and few. This rule of thumb holds true throughout the county but, as it is a rule, a few exceptions may be expected.

American Crows were reported in 27% of all blocks. The intense clustering of blocks with detections of crows—chiefly with breeding confirmations—in the coastal lowlands between lower Redwood Creek and the upper Eel River floodplain suggests that these birds are numerous there. Indeed this is the case, as crows were found to be routine breeding birds around the Humboldt Lagoons, in the urban/suburban expanse from McKinleyville to Fortuna, and in the dairy pasturelands south and southeast of Humboldt Bay. During the nesting months, the combined forces of nearly continual observer effort, large numbers of birds, and the easily noted breeding behaviors of American

Crows abetted the intensity of confirmations in this portion of the county.

A clustering of mixed possible, probable, and confirmed breeding observations from the vicinity of Willow Creek (N13) northward through the Hoopa Valley (m11) and thence shortly northeast around Orleans (P05) indicates the presence of crows about settlements and in broken riverside country in the Trinity–Klamath river corridor. Similar, though less well defined, strings of blocks fell within the Mad, Van Duzen, and Eel river valleys. No such distribution was apparent along Redwood Creek, which is a largely forested and deeply incised drainage, nor along the Mattole River, although crows were confirmed breeding near Honeydew (e29). Crows were absent, or seemingly so, throughout the great expanses of commercial timberland in the central and north-central portions of the county, in the mountains of the northeast, along the South Fork Mountain ridge, about the Lassics uplift, in the King Range, and in the grassy "mountain prairie" country inland from Cape Mendocino.

Atlas data indicate that American Crows began nesting, or at least courting, in late March. Reports of adults carrying nest material spanned the dates 28 Mar 1995 (e16; JEH) to 16 Jun 1996 (F13; JEH). Nest building was reported 7 Apr 1997 (f06; DFx) and 15 May 1995 (f14; RJA). The sole report of a nest with eggs fell on 3 May 1996 (F13; GAS). Feeding of young spanned the dates 13 Jun 1998 (H23; JEH) to 28 Jul 1998 (f06; D. Anderson). A nest in a coast redwood near Blue Lake (H15) was observed under construction 17 May 1996 and fledged three young on or by 13 Jul 1996 (C. Gutiérrez).

Historically, the American Crow has long been known as a common resident bird in the county. Harris (1996) stated that the crow is a "locally common resident and breeder . . . mostly associated with riparian areas and agricultural lands, but recently has become an urban species in Arcata, even visiting backyard feeders." Harris (1996) noted that winter population estimates made for select northwestern California sites by Emlen (1940) were smaller than population sizes inferred from recent CBCs conducted at these locations. Indeed, 1959–1999 CBC data and 1966–1999 BBS data indicate statistically significant increases in populations throughout California (Liebezeit and George 2002).

The extent to which American Crows migrate through Humboldt County, and the degree to which population size may fluctuate seasonally, are little understood. Serving to mask any transient element in the local population is the fact that, unlike many other smaller birds, flocks of crows are seldom encountered well away from expected haunts such as productive foraging sites or roosts. Flocks which appear to be actively migrating north or south are almost never reported. After the fledging of young, family groups begin to commingle into larger late summer and autumn flocks. These concentrations establish colder-season communal roosts in stands of trees along rivers or on ridgelines above valley edges. Such roosts are used nightly and may support hundreds of birds. Regular processions involving discontinuous, straggling lines of crows may be observed commuting toward a roost site shortly before sunset or leaving it at daybreak.

Given the propensity of crows for exploiting human-related food resources, it seems likely that increasing conversion of homogeneous parcels of native habitats to suburbs and commercial development will continue to benefit this species. Any actual effects of the increase in crows on other bird populations are undocumented in Humboldt County. However, predation by crows on Snowy Plover eggs and young has been well documented elsewhere in California and Oregon (Liebezeit and George 2002). Thus, beach-going family groups of crows may be cause for concern. Such apparent groups were frequently seen throughout the Atlas period feasting on garbage at Dry Lagoon State Beach (F07). Breeding by crows was confirmed there when fledglings attended by adults were seen 30 Jun 1996 (JEH).

COMMON RAVEN *Corvus corax*

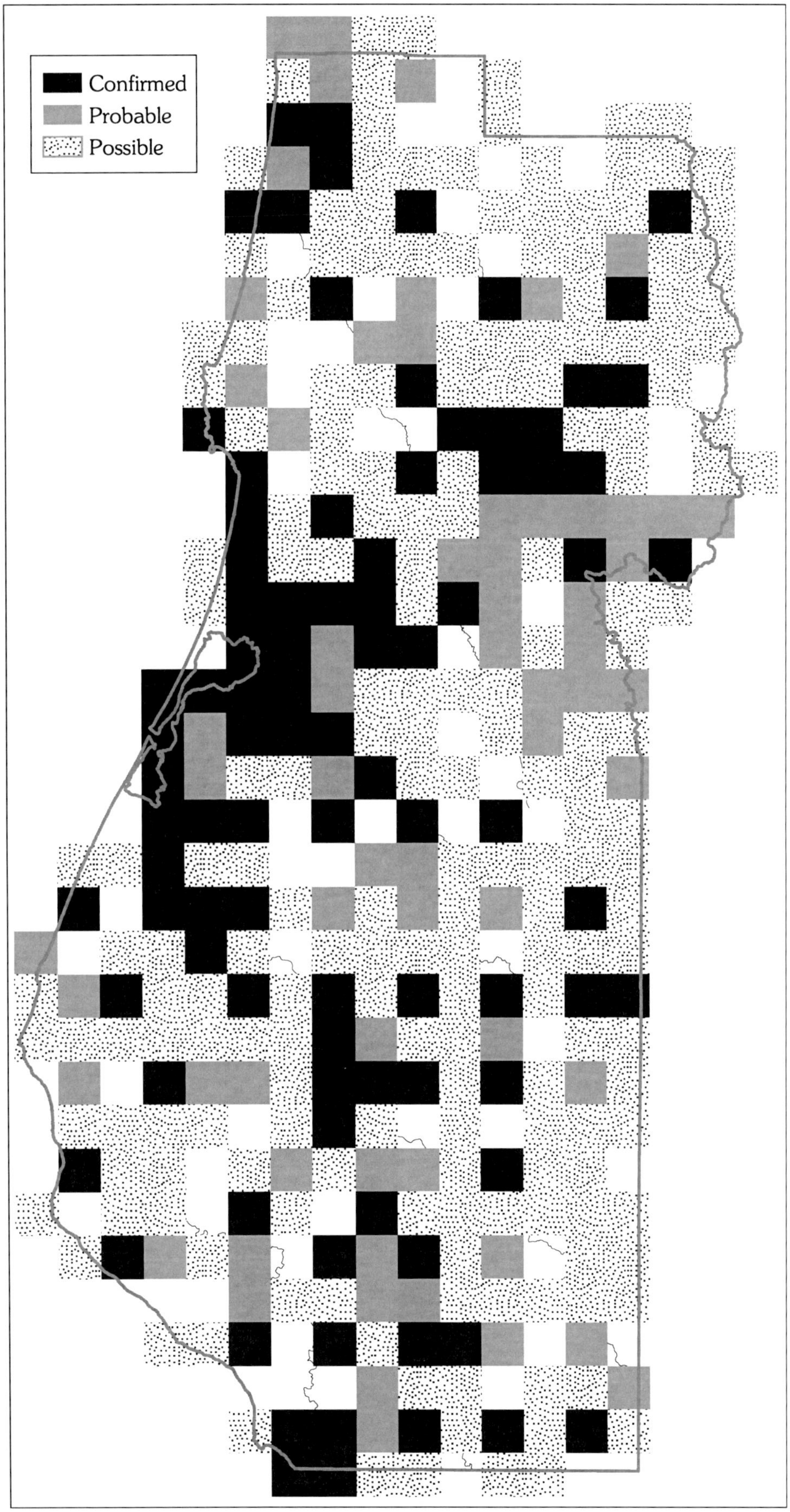

The raven is a bird of many attributes, so well known and so widely immersed in myth that it is of the same menagerie as lion, serpent, eagle, salmon. Long acquaintance with the Common Raven, and an assimilation of its wiles into lasting cultural reference, has been shared by peoples scattered throughout its range, for Raven has attended the hunt and forespoken of grave events across most of the Northern Hemisphere since the dawn of man. Today it still occupies nearly all of the North American range it was known to originally inhabit, with local contractions in some regions offset by gains made elsewhere. The chief hallmark of Common Ravens is perhaps their ability to thrive in a great array of settings and to exploit all available opportunities when foraging, selecting nest sites, and avoiding persecution. Tundra, boreal forest, mountains, steppe, desert, and shores support ravens. Greater presence in recent years in urban and industrial areas tells of a capacity to flourish amid the challenges of an increasingly humanized world.

Common Ravens are resident across almost all of California with the exception of the Big Sur coast of Monterey County (Roberson and Tenney 1993) and the Central Valley. In northwestern California, they are resident and are found in all habitats (Harris 1996) and can be anticipated nearly anywhere. During the Atlas period, no corner of Humboldt County was unoccupied, with ravens present essentially across the entire study area from sea level to above 6000 feet. Common Ravens were detected in 352 (83%) of the 425 total blocks. This ranked it as the second-most widely distributed breeding species in the course of the Atlas study, surpassed only by Steller's Jay. The 73 blocks without ravens seemed to occur without respect to predominant elements of cover type or terrain, although sparse human settlement in those blocks may have played a role.

Blocks without detections generally were non-contiguous, with no more than three in a row anywhere.

Common Ravens patrol foraging grounds on a landscape level, eyeing ridgelines, slopes, valleys, stream corridors, and shores for food. Crossing canyons and flying along highways, they are widely seen. In particular, their habit of scanning roadways for vehicle-struck animals—ranging in size from bumblebees to large mammals—places ravens in the public eye. Highly intelligent, these birds actively investigate human activity, often waiting patiently at a safe distance for their chance to have a closer look. As in the case of Turkey Vultures, the fact that these birds are so continually visible may cause one to imagine that their numbers are greater than they actually are. Hence, an appropriate characterization of the species in Humboldt County would be "fairly common."

Seeming abundance of Common Ravens within the county varies with location and, locally, with seasonal abundance of food sources. One noteworthy example is a record of 100+ in one flock, and another 100 nearby in scattered groups, feeding on grasshoppers in open grassland on Bear River Ridge 24 Sep 1975 (S. Harris; Harris 1996). Garbage dumps, dead livestock, and other rich food sources can attract large concentrations of ravens as well. Although atlasers did not make estimates of abundance in blocks, it was nevertheless apparent that ravens were encountered in greatest numbers in the county's central coastal bottomlands and in outlying foothills. Single and paired ravens were almost continually in sight at some point of the compass in the Mad River, Humboldt Bay, and Eel River lowlands; in summer, family groups of adults and juveniles were widely noted in these and other areas. Common Ravens were once thought to be declining in California because of human persecution (Grinnell and Miller 1944). However, since then both CBC and BBS data indicate increases in most of the state, including northwestern California (Liebezeit and George 2002).

Because cliffs are scarce in most of Humboldt County, Common Ravens are suspected to nest chiefly in trees. These birds exhibit a great contrast in behavior while nesting: stealth and secrecy are employed at or near the nest, even though overt aerial behaviors and vocalizations may be commonplace at some distance from the site. Given the numbers of ravens in and around the cities of the central coastal lowlands in spring and summer, it is an inescapable conclusion that they must routinely breed in woodlots and forest edges close to humans. A raven nest situated 180 feet up in a Douglas-fir in a residential woodlot at Willow Creek was observed for 60 hours from 30 Apr to 4 Jul 1977 (Wyckoff 1983). Observations revealed that eggs were laid before the study began and that young fledged 26 Jun, a fledge date consistent with those noted during the Atlas period.

Breeding was confirmed in 92 blocks. The carrying of nest material was the highest evidence recorded in nine blocks and ranged from 26 Mar 1995 (F11; JEH) to 15 Jun 1998 (k14; TWL). There were no reports of nests with eggs, but seven nests with young were observed, spanning the dates of 25 May 1995 (f16; C. Hibbard) to 3 Aug 1995 (h14; C. Keil). Food-carrying reports spanned the dates of 23 Apr 1996 (N13; TWL) to 28 Jul 1996 (F21; JEH). The earliest fledglings reported were on 2 Jun 1997 (e10; DFx) and the latest were on 5 Aug 1996 (J15; JCP). Reports of fledged young accounted for 40% of all blocks with breeding confirmations. This was the most frequently reported evidence of confirmed breeding. Juvenile ravens were readily distinguished by their lack of wing and tail molt and, when unseen, by their falsetto begging calls.

HORNED LARK *Eremophila alpestris*

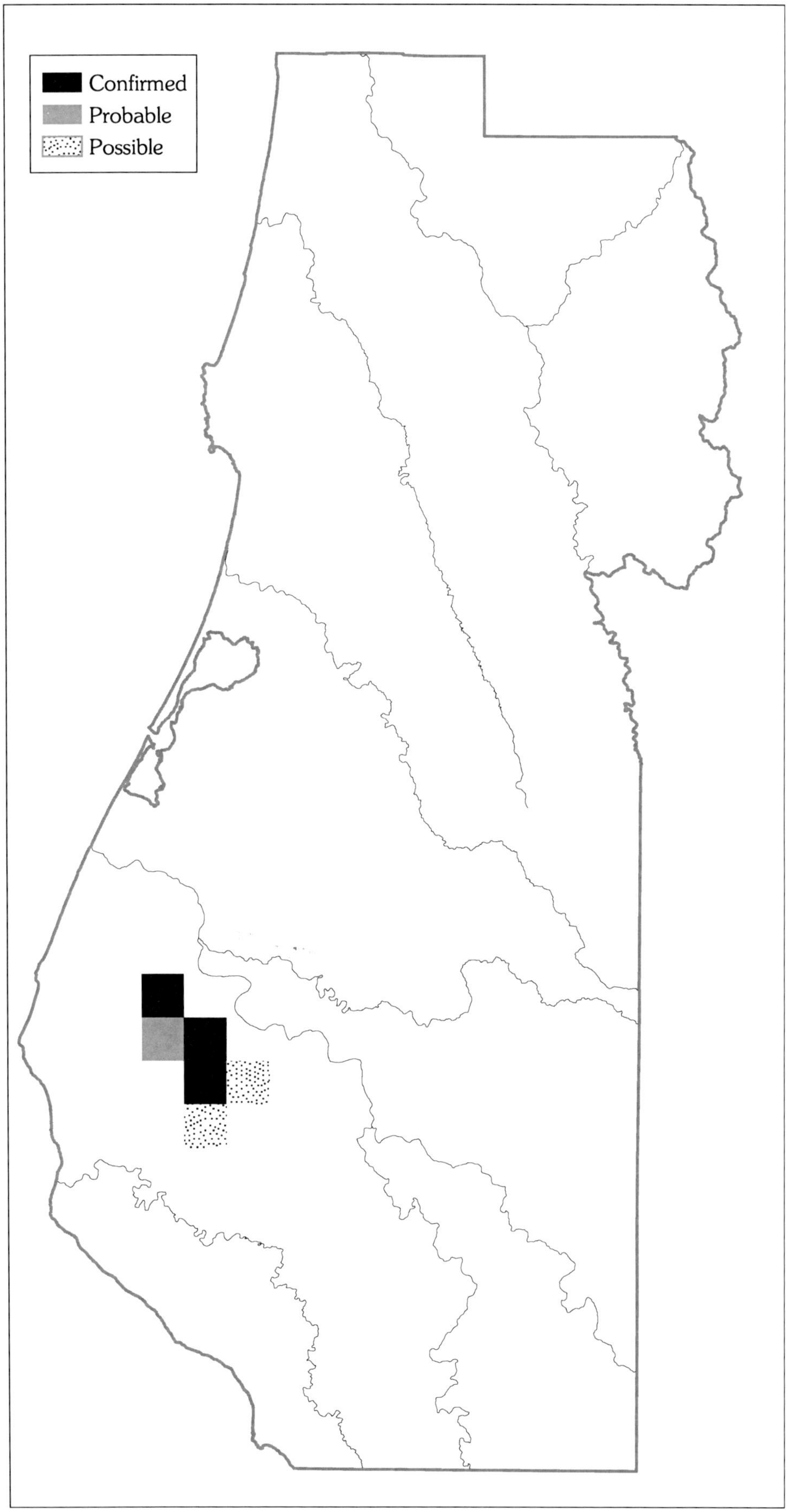

Horned Larks are small ground birds of striking pattern, their faces emblazoned with a decurved black mask and with two tiny recurved "horns" of delicate feathers atop their heads. Often heard before they are seen, Horned Larks flush with lilting *see dirt* notes, a mnemonic aid to identification that recalls an element of their habitat. Their song, commonly delivered in flight, is as spare and bittersweet as an autumn frost—sharp tinkling notes that begin in seeming hesitation, then ascend to a jumbled, high-pitched finish. Treeless, short grass habitats around the Northern Hemisphere host these birds. Barren, plowed, hard-grazed, or windswept expanses in which few other species persist in numbers may attract great flocks of them. In regions that are forested, shrubby, or heavily settled, Horned Larks may remain nearly unknown.

Only one area within Humboldt County has been known to support Horned Larks throughout the year: the mountain prairies of the Bear River and inner Cape Mendocino ridgelines, about 10 or 12 miles inland from the coast and up to about 3000 feet in elevation. Larks were first reported from this region in 1927 (G. Atwell; Harris 1996; HSU 4829) and were first documented breeding when a male and a nest with four fresh eggs were collected 9 May 1929 (Grinnell 1931; MVZ 53976). G. Atwell collected additional study skins there 2 Jun 1929 (Grinnell 1931; MVZ 53981, MVZ 53982, and MVZ 53983). They have been noted by many observers in the decades since, and the population is well known to Humboldt County birders and naturalists. Unknown as a breeder in northwestern California outside of Humboldt County, here they have been considered a very local resident and breeder (Harris 1996). Winter status in local breeding areas is unclear.

Atlasers succeeded in confirming breeding in three blocks, observing probable breeding in one, and possible breeding in another two. Indicating the localized and circumscribed nature of Humboldt's Horned Lark breeding habitat is the fact that these six blocks form a cluster. The range of nesting Horned Larks in the county trends roughly northwest to southeast, in accordance with the major ridges in the area. The distribution takes in several large ridgeline and upper mid-slope prairies. Atlasers found access to a significant portion of potential habitat impossible because of private property restrictions. It is conceivable that Horned Larks may have nested in several other blocks in the vicinity, but nevertheless the breeding distribution of the species in Humboldt County has probably been well determined at the landscape level.

Horned Larks were confirmed breeding in two blocks along the main Bear River Ridge, between the junction of Wildcat Ridge Road (D23; 10 Jun 1997; COg) and Monument Road (e24; 7 Jul 1996; L. Bacon-Ogden). Both confirmations were of adults feeding young. A pair was also seen in the intervening Bear River Ridge block (d24; 1 Jun 1996; COg). Another confirmation was made 11 Jul 1998 at Windy Nip (e25), an aptly-named saddle on Rainbow Ridge, separating the Bear and North Fork Mattole river drainages. On that day at Windy Nip, at least five Horned Larks were seen near the Doppler radar site, including fledglings and territorial males (JEH, TWL *et al.*). Possible breeders were reported on Monument Ridge east of Mt. Pierce (F25) 17 Jun–11 Jul 1998 (M. Morris) and on Rainbow Ridge about 1 mile southeast of Windy Nip 21 Apr 1995 (e26; JEH). There were no detections of larks during the breeding season in outlying blocks.

Sites inhabited by Humboldt County's breeding Horned Larks in spring and summer are all in treeless corridors atop tall ridges. These places are sky islands of sorts, delimited by shrubby or wooded regions occurring generally throughout lower elevations below the ridgeline country. Although today these areas receive perennial grazing, they may have existed as grasslands before the earliest human settlement. It could be argued that the highest portions of the Bear River and Cape Mendocino ridges experience the region's harshest weather. The conspiring elements of thin soils, winter storms, and desiccating summer winds thwart the growth of anything but grass, which is kept sparse and low by the grazing of beef cattle. This condition plainly appeals to Horned Larks, which are ordinarily not difficult to locate once one has traveled well above the limit of trees.

Elsewhere in the county, Horned Larks are known only as occasional migrants or winter visitors. Harris (1996) indicated that lowland occurrences were generally of one or two birds, with most reports in November and December. This species is rarely detected on local CBCs. Reports from January through March are few, suggesting that truly wintering birds anywhere away from the mountain prairies are rare.

Grinnell (1931) initially considered the Humboldt birds to be "Streaked" Horned Larks (*E. a. strigata*), but subsequently Grinnell and Miller (1944) included them with "California" Horned Larks (*E. a. actia*). This change was probably based on Behle's (1942) analysis of the MVZ specimens, which he considered to be California Horned Larks. A recent cursory examination of the MVZ specimens supported placing this population with *actia* rather than with *strigata* (C. Cicero, MVZ, pers. comm.).

PURPLE MARTIN *Progne subis*

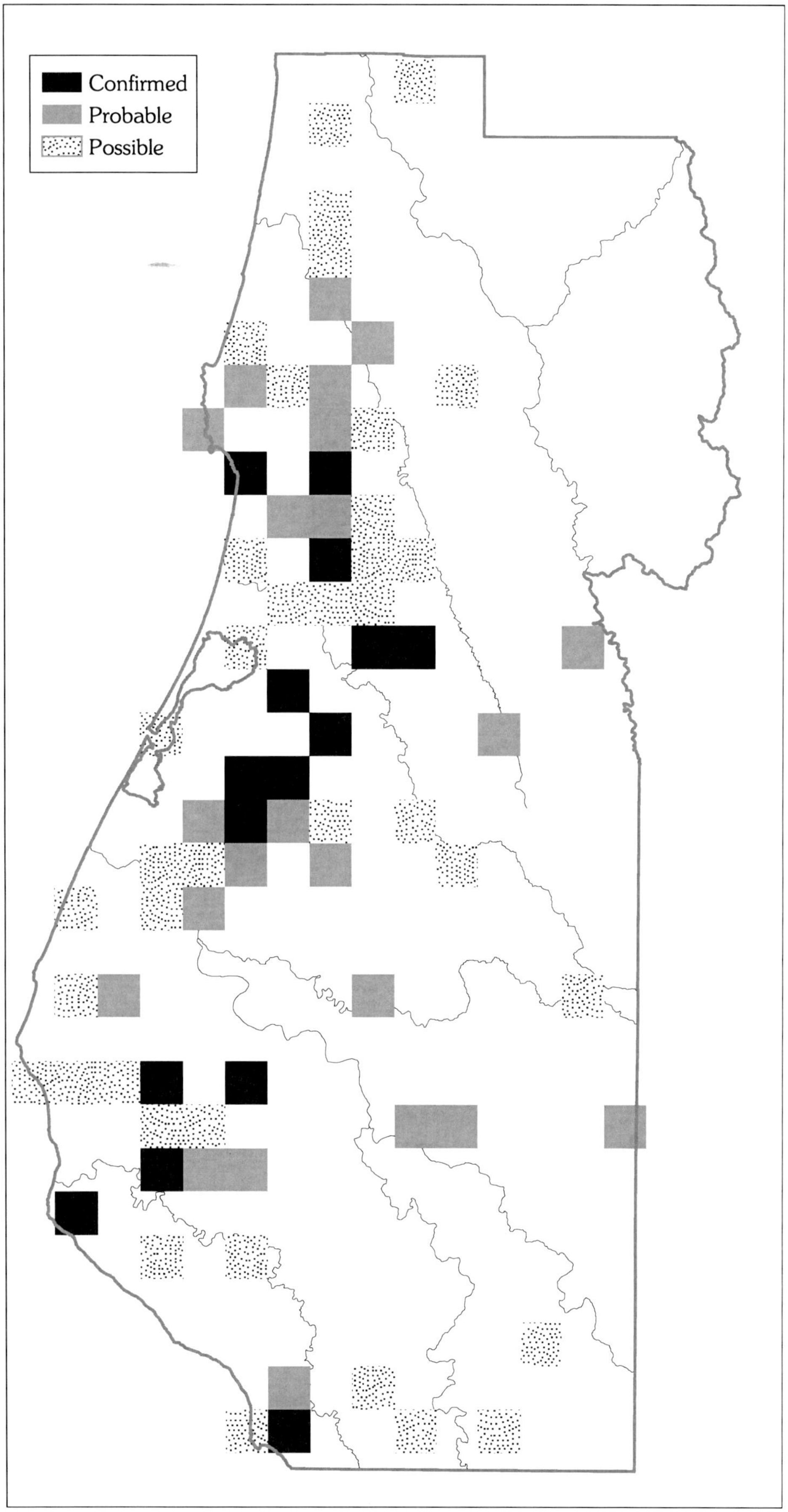

Purple Martins are large, gregarious swallows, appealing in appearance and behavior. Adult males are dressed in deep purplish feathers that shimmer with iridescence in strong sunlight. Females and young males lack the flash of older males, yet all cut a distinctive image with their broad wings, slow-motion circling, and penetrating voices. They originally nested in natural cavities, but in the eastern U.S. and southern Canada they have been aided by widespread placement of apartment-style nest boxes designed to accommodate their colonies. Thousands of martins benefit from this largesse, which can significantly increase populations where such a program is maintained. However, in the western U.S. almost all attempts to establish colonies in apartment-style nest boxes have been unsuccessful.

In the West, Purple Martins are localized and generally uncommon to rare. In many sections they are almost unknown even as migrants. Many pairs or small colonies occur in lowlands and foothills of the Pacific States, almost entirely west of the Cascade–Sierra Nevada divide. Humboldt County's share of martins is considerable, although recognition of this might elude a given birder working only a small section of the county. In line with the regional pattern of distribution exhibited both to the north and south of the county, single birds, pairs, and small colonies were found at widely scattered locations during the 5 Atlas years. In contrast to the situation on the Oregon coast, few appeared to breed near estuaries. Purple Martins were recorded in a somewhat surprising 18% of all blocks.

Roughly three-quarters of the blocks with detections of martins occupied low and middle-elevation foothills within the coast redwood zone in the west-central part of the county. Most of these blocks were somewhat removed from the immediate coast. Blocks constituting this loose cluster extended almost continuously from lower Redwood Creek to the south side of the Eel River delta. Many of the birds were encountered on managed private timberland at lower and middle elevations. In particular, concentrations of detection-blocks east of Westhaven and southeast of Humboldt Bay are believed to reflect a greater abundance of Purple Martins in cutover lands in which remnant redwood snags persist.

Outside the vague arc of blocks in west-central Humboldt, Purple Martins were found most frequently in the hills extending inland and southeast from Cape Mendocino. In that area they occurred along or near ridgelines in a sweep of country characterized by grassland and patches of mature Douglas-fir forest. A typical nesting situation there involves multiple pairs using old woodpecker holes in a snag located on an exposed or grassy ridge. Such places, often in profoundly rural settings, may also offer partial or complete freedom from competition by European Starlings for nest sites.

Observations of Purple Martins elsewhere in the county during the Atlas period were comparatively few and were quite scattered. Most birds were seen or heard in forested places at middle elevations, often, though not always, in the vicinity of dead-topped trees or snags. Purple Martins were historically rather numerous and regularly seen or heard in the hills not far above Shelter Cove (f33, g32, g33), and this situation continued during the 5 Atlas years. Many fire-created snags persist in the brushfields of the Shelter Cove Highlands (g33), and an occupied nest was reported in one of them 29 May 1998 (K. Dyer). No Purple Martins were detected in the northeastern portion of the county.

Purple Martins were confirmed breeding in 15 of the 76 blocks with detections. The bulk of confirmations came from those sections of the western part of the county where, historically, the birds have been most frequently seen. There were no confirmations in extreme northern Humboldt nor in the eastern part of the county. The confirmed breeding report that was farthest inland was of an occupied nest near Fernwood (J15) 26 Jul 1996 (JCP). Many of the blocks farther inland with detections also probably hosted nesting martins. Historically, apparent spring migrants have been almost completely unreported from outside Humboldt's coastal lowlands, thus birds inland would appear to be prospective summerers.

The earliest Atlas report of a migrant martin was 11 Apr 1998 (h14; C. Keil). All but seven of the 23 probable breeding reports were from June. The most frequent evidence of confirmation recorded was occupied nests (12 blocks), which ranged from 28 May 1999 (g18; JEH) to 26 Jul 1996 (J15; JCP). Only two nests with young were reported: 21 Jun 1998 (F19; M. Wachs) and 30 Jun 1997 (F25; D. Garrison). The only fledgling reported was seen 24 Jun 1995 (b28; GCH).

Purple Martins are present in Humboldt County from April into September each year. A variable appearance by spring migrants takes place at or near the Arcata Marsh. In some years, scores of martins may be present over a period of days in early May. That other wetlands may attract concentrations during spring migration is supported by the observation of 104 birds at Eel River Wildlife Area 5 May 1994 (M. Fisher; Harris 1996). No early-fall staging spectacles like those witnessed on the East and Gulf coasts have ever been noted in Humboldt County, and indeed the local martins are little reported after the fledging of young and dispersal of family groups.

Harris (1996) categorized the Purple Martin as a long-known uncommon summer resident and breeder in northwestern California. Williams (1998) estimated that there were 80–180 nesting pairs in Humboldt County. Northwestern California appears to be a comparative stronghold for Purple Martins in the state, perhaps because of fewer European Starlings in many of the more rural situations and the greater availability of large trees. The status of this species here is in sharp contrast to some other parts of the state where numbers have declined markedly. The favorable situation in Humboldt might be maintained if foresters safeguard the existing snags and dead-top trees martins favor, and allow the "recruitment" of future snags. One would be hard-pressed to design a forestry practice worse for this species than the archaic tradition of cutting snags on ridgetops, purportedly for "fire safety". The clustered single-pair nest box approach has helped sustain populations of martins at historical sites in western Oregon and could potentially play a role in conserving local populations of this swallow. Although some cities, such as Sacramento, host nesting Purple Martins, by and large this is a bird of rural or out-of-the-way places in which establishing a nest box program may be impractical.

TREE SWALLOW *Tachycineta bicolor*

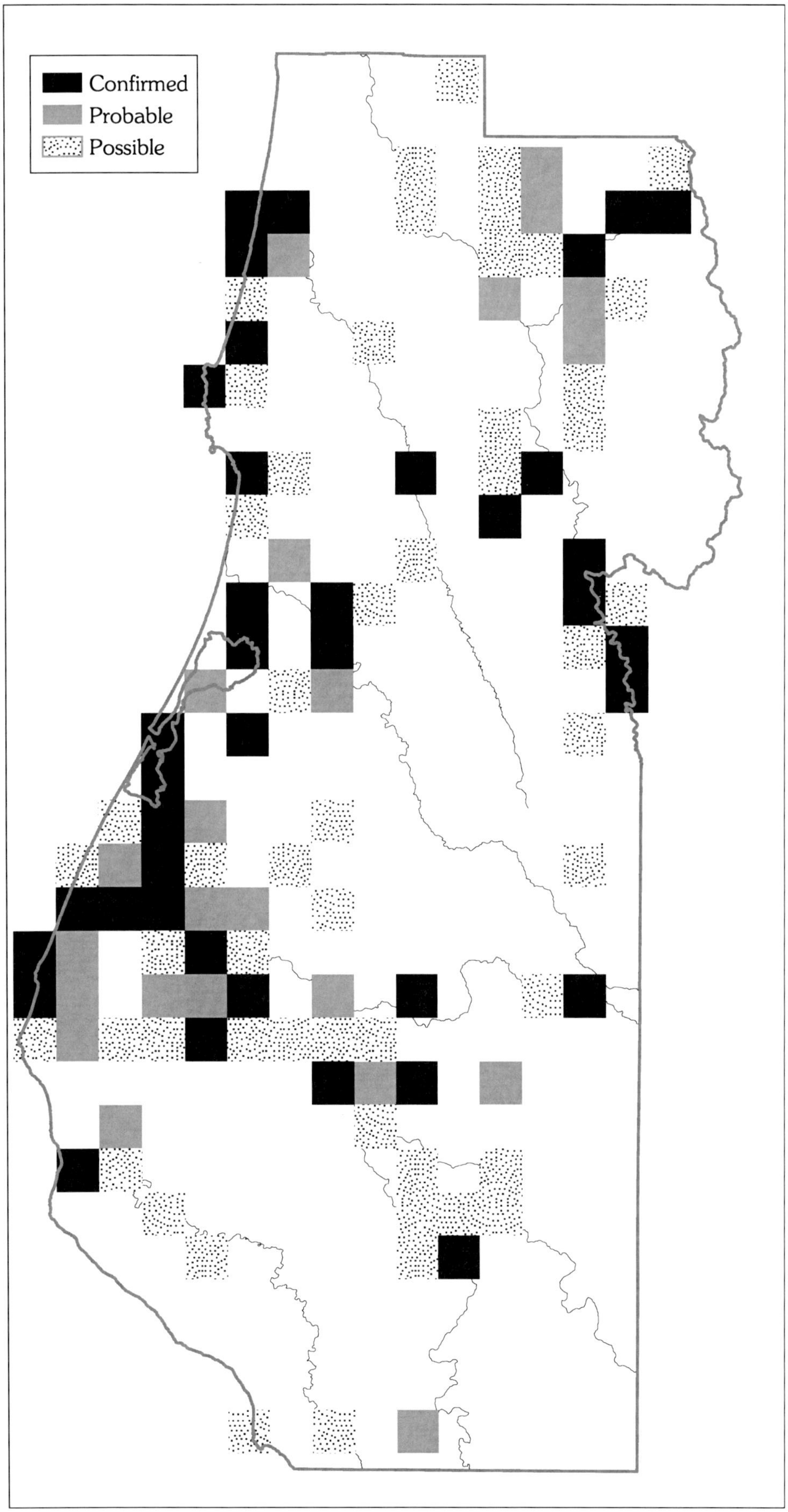

Tree Swallows are highly aerial birds, enlivening wooded streamsides and lake shores with glitter and grace. Set sharply against snowy underparts, the cobalt-blue upperparts of adult males furnish dead alder limbs and old woodpecker holes with accent notes of spring's arrival. Tree Swallows are typically among the earliest migrant landbirds in the U.S. and Canada, enduring spring snowstorms and late freezes as they push north from a more temperate wintering range. As well, they complete their nesting efforts by mid-summer and, as the species is single-brooded, an unheralded exodus takes place beginning as early as late July. The southward migration is extended, however, and many swallow flocks lingering in late fall in northern areas are chiefly of this species.

The Tree Swallow is well represented throughout much of California, variably as a breeding species, a migrant, a winter visitor, or all three, depending upon location. In Humboldt County, Tree Swallows have long been recognized as common migrants as well as routine spring and summer residents in suitable habitat. In most years, migrants appear widely in lowlands and around fresh water or grassy habitats of all descriptions in late February and increase in March, with subsequent restriction of birds to appropriate breeding sites. These are usually, though not always, at, over, or near water. Referring to northwestern California generally, Harris (1996) termed Tree Swallow a "common migrant, summer resident, and breeder; casual to rare in winter."

Atlas results indicated that Tree Swallows are widespread in Humboldt County during the nesting season, yet their distribution is somewhat localized. They were reported in 27% of blocks during the Atlas period. Tree Swallows were present along much of the coast or in the near-coastal lowlands from the Redwood Creek estuary (f05) through the Humboldt

Lagoons, and from the Little River estuary (F11) southward in a scattering of blocks taking in areas around Humboldt Bay. The Eel River valley upstream to about Salmon Creek (J29), the Eel River delta, and the blocks east of Cape Mendocino formed the greatest cluster of detections in Humboldt.

Tree Swallows were also found in blocks of the Van Duzen River corridor; around Blue Lake; and along the Klamath, Trinity, South Fork Trinity, and Mattole rivers and some of their tributaries. Outlier blocks with detections were on the southern Humboldt coast at Shelter Cove (f33); near Whitethorn (H33); on Sprowel Creek southwest of Garberville (J33); in Redwood Valley (J11); at Childs Hill Prairie in the Bald Hills (i08), and south of Last Chance Ridge on South Fork Mountain (n20).

The map clearly suggests that large sections of Humboldt County are without suitable breeding habitat for Tree Swallows. Uplands well away from water, such as forested slopes and ridges and the oak-grassland country, offer no attraction to these hydrophilic swallows. East-central and southern Humboldt were particularly poor in reports of these birds, comprising "no-detection zones" of considerable size.

Habitats used by Tree Swallows nesting in Humboldt County are characterized by the presence or proximity of fresh water—often a river or creek, but sometimes a lake or pond—and by holes in trees, in which the birds nest. As secondary cavity nesters, they are largely reliant upon the work of woodpeckers, whose disused nesting or roosting cavities are regularly adopted by the swallows. Stands or stringers of riparian trees provide most of the breeding habitat in Humboldt. Black cottonwood, alder (red alder in the fog belt, white alder inland), or tall willows are usually found where Tree Swallows occur in summer. Dead, dying, or live trees with cavities are often numerous in quality Tree Swallow habitat.

The comparative instability of riparian stands may enhance opportunities for these birds. Flood scouring, drought, and the inevitable replacement of one tree community by another all work to afford a supply of stressed trees for Downy and Hairy woodpeckers and Northern Flickers, each of which probably contribute their share to the pool of potential Tree Swallow nest sites. These birds also use nest boxes and other human-provided structures; *e. g.*, a nest near Blue Lake (H15) in a 3-inch pipe had young birds in it 28 Jun 1995 (B. Hawes).

Tree Swallows were confirmed breeding in 39 blocks. These were thoroughly scattered through almost all sections of the county that had detections. Occupied nests were reported in 11 blocks, ranging from 20 Apr 1995 (F15; G. Friedrichsen) to 26 Jun 1998 (o15; *fide* RHw). There were 12 blocks in which nests with young was the highest-ranking breeding behavior, spanning a comparatively brief period from 4 Jun 1996 (D19; K. Foerster) to 23 Jul 1996 (F17; PFS). Five reports of adults feeding fledged young ranged from 25 Jun 1997 (H15; T. Bowden) to 11 Jul 1995 (e22; *fide* RHw).

Protection of riparian vegetation, providing for the "recruitment" of new dead and dying trees, and minimizing human disturbance around active nests will help ensure the future of the Tree Swallow in Humboldt County.

VIOLET-GREEN SWALLOW *Tachycineta thalassina*

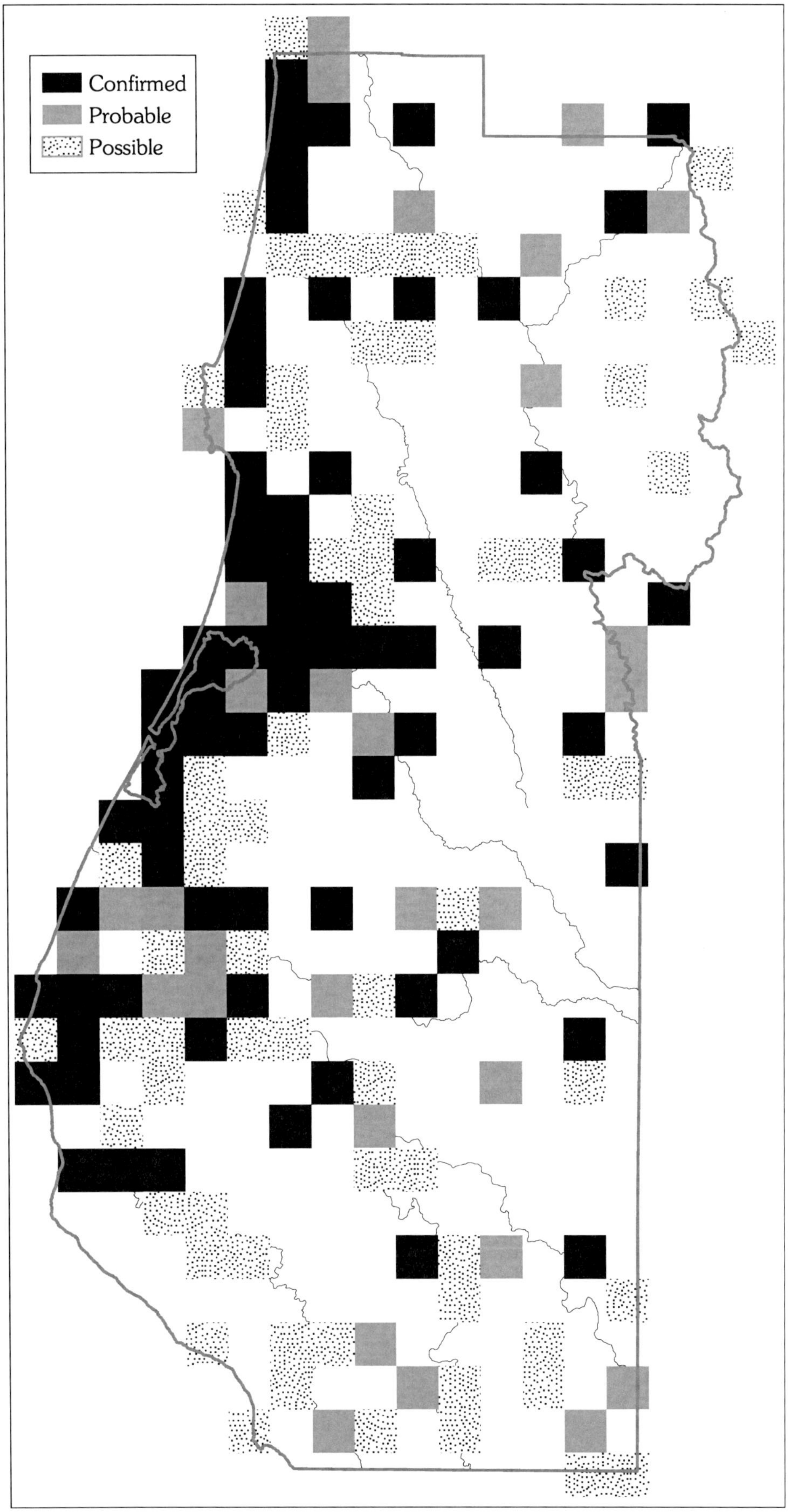

The exuberant chatter of Violet-green Swallows foraging high overhead is a sound familiar to birders living in the western U.S. and Canada. These small white-bellied swallows are common in many habitats, especially in forests with snags and dead-top trees, or at rock faces with crevices and holes. As breeding birds, they are numerous across most of California outside of desert regions. In Humboldt County, Violet-green Swallows are to be expected during summer in forested areas, in riparian woodlots, in cities and towns, and around cliffs. They also occur in the mostly treeless bottomlands wherever old buildings with suitable nest crevices are found. Migrants overfly essentially all parts of the county at some time during spring or fall and are nowhere unexpected. Swallows of this species may fly exceptionally high, with foraging birds sometimes so high overhead that they are not detectable with the unaided eye, nor can their calls be heard.

Violet-green Swallows were widely encountered during the Atlas period. Although this species was detected in 39% of blocks, Atlas efforts indicated that the distribution of Violet-greens within Humboldt County was irregular. Most coastal and many near-coastal blocks were inhabited, but scores of blocks in the interior of the county apparently were without them during the nesting season. This patchiness is explainable to some degree by scarcity of trees (and cliffs) in many blocks, particularly in the grasslands of the central Mad River, yet seemingly appropriate habitat exists in many other blocks where no swallows were detected.

The region of Humboldt County with the highest detection rate for and the greatest abundance of Violet-green Swallows during the Atlas period was the coastal lowlands and foothills from the Del Norte County line to the Mattole River valley. A fairly high proportion of blocks with confirmed breeding occurred along the coast south to near the Mattole River mouth and in the lower Mad River watershed, attesting either to comparatively large numbers of swallows or to their readily witnessed breeding efforts. Occurrence of Violet-greens was somewhat coincident with stream corridors, but this trend was far less apparent than that exhibited by the related Tree Swallow. Violet-green Swallows clearly are not restricted to riparia, although they are found in it routinely. Of restricted extent in a highly corrugated landscape, riparian tree communities constitute but a small proportion of the habitats available for these swallows. Indeed, Violet-green Swallows were encountered in many blocks characterized by hilly terrain, and through which none of the larger rivers flow.

During the Atlas period, Violet-green Swallows were noted in old-growth conifer forests of the fog zone; in varied structure and settings in the managed timberlands, typically where snags persisted; in fragmented forest dominated by oaks; in the montane true fir zone; along streams with mature alders or cottonwoods; and about human dwellings and outbuildings. As secondary cavity nesters, these swallows are largely beholden to woodpeckers for the disused nest and roost cavities they provide. Forests and woodlots supporting Violet-green Swallows commonly contain snags or exposed dead-topped trees. Along stream corridors, tall cottonwoods and the abundant, comparatively short-lived white alders provide nesting opportunities. Limited use of birdhouses was also noted during the 5 years of Atlas surveys.

In unforested sections of their range, Violet-green Swallows rely to a great extent on tall rock faces for breeding sites. In Humboldt County, cliffs with suitable crevices and niches are uncommon and localized, yet the birds were regularly found using such situations where this habitat exists. Cliffs along the lower Trinity, Eel, and South Fork Eel rivers hosted Violet-greens, as did the few commanding ridge-top and upper mid-slope rocks standing in these drainages. Steep banks created by blasting during road construction also provided sites for these swallows. Among the more unusual nests was one found 24 Jul 1997 that was only 18 inches above the ground in a cavity in a large piece of driftwood on the nearly featureless Big Lagoon spit (F07; TWL).

Violet-green Swallows typically arrive in Humboldt County in late February and March, with birds widespread by the middle of March. The earliest Atlas record of a bird in suitable breeding habitat was 2 Mar 1997 (o15; TWL). While the arrival of Tree Swallows usually precedes that of Violet-greens, these two species dominate the swarms of migrant swallows over wetland habitats in the coastal lowlands

during a period of several weeks in March and early April. Prospecting pairs appear to inspect many possible nest sites before selecting one; the earliest such observation was 21 Mar 1996 (o15; TWL). The earliest breeding confirmation was of an adult carrying nest material 29 Mar 1995 (g14; JEH). Violet-green Swallows generally nest in exposed situations. Consequently, their breeding activities are readily observed. Among the 69 blocks with breeding confirmations, occupied nests supported this status in 28 (41%), with the earliest report 23 Apr 1995 (d16; JEH) and the latest 23 Jul 1998 (g03; D. Anderson). Only one occupied nest was discovered in April, six in May, 15 in June, and six in July.

This species is single-brooded. Juveniles become conspicuous in early July. Fledged young were the highest-ranking code in 20 blocks, ranging from 23 Jun 1996 (B27; GCH) to 13 Aug 1995 (F15; C. Meyers). Two of these were in June, 16 in July, and two in August. Adults feeding young were reported as late as 8 Aug 1996 (J13; JCP). Beginning as early as late July, flocks of from dozens to hundreds of Violet-green Swallows assemble in foothill and mountain clearcuts and over near-coastal grasslands. Unlike Barn Swallows, which are commonly seen migrating southward by day, Violet-green Swallows seem not to undertake such apparent movements. Large flocks may fill the sky one September afternoon—and none may be present the next. Late influxes following periods of seeming absence or scarcity have been noted in early October in some years. Harris (1996) stated that a few stragglers remain in coastal lowlands through some winters. None of these were seen during the Atlas period.

Summarizing the status of Violet-green Swallow in northwestern California, Harris (1996) stated that the species was a "common migrant, summer resident, and breeder; casual to rare in winter." He also noted that this species is far more likely than Tree Swallow to nest in residential areas, a statement that held true during the Atlas period.

NORTHERN ROUGH-WINGED SWALLOW *Stelgidopteryx serripennis*

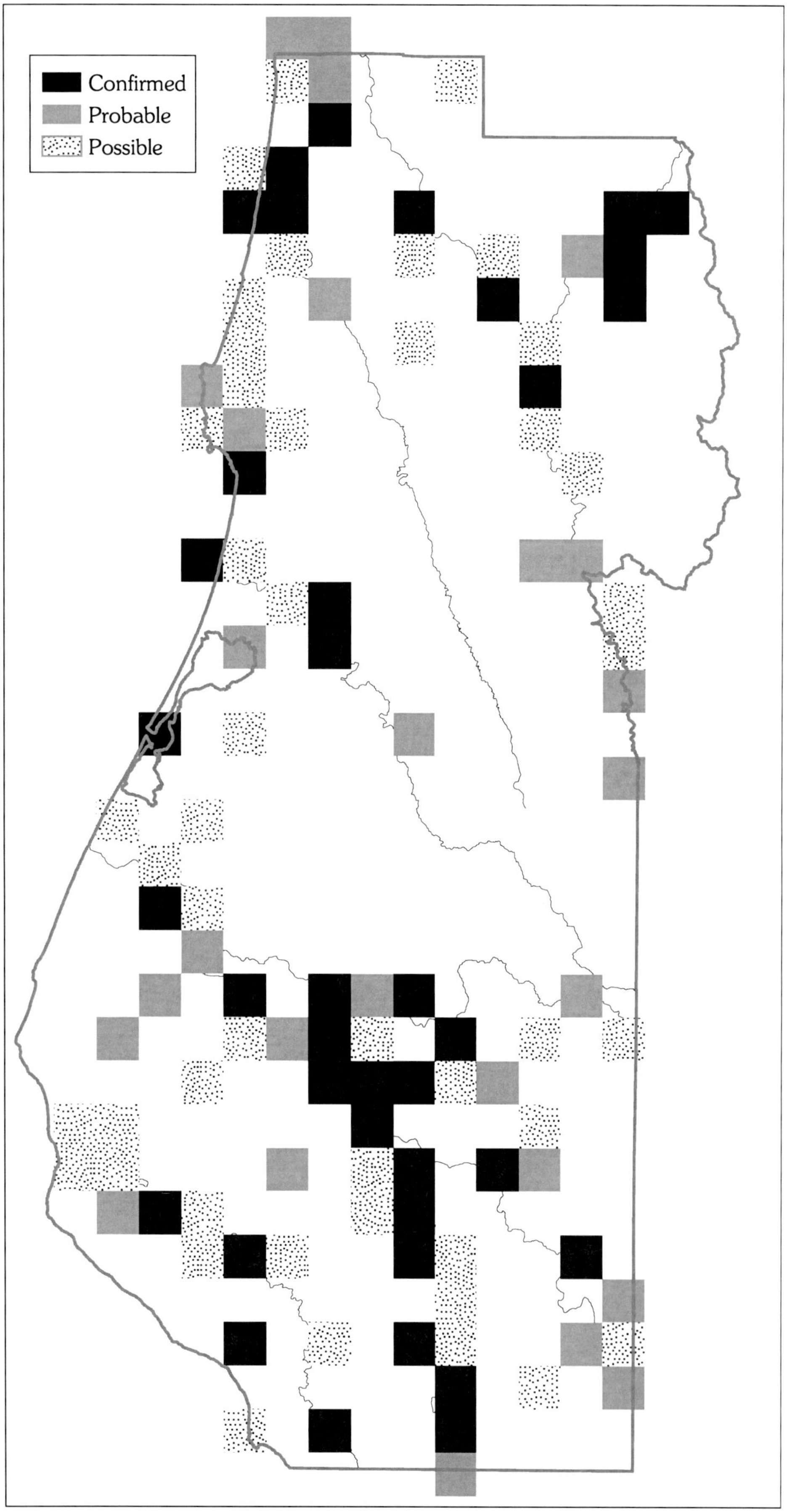

Common and widespread, Northern Rough-winged Swallows are yet little known to most people. Lacking the iridescent colors and affiliation with humans exhibited by many other swallows, they draw no special notice. Few persons other than birders and knowledgeable fishermen, boaters, and trampers-of-riverbanks come to recognize them. Their earth-toned plumage, dry voice, and habit of occurring in only modest numbers at any given site cause Rough-wings to have a distinctly lower profile among the public than such species as Barn, Cliff, Tree, and Violet-green swallows and Purple Martin, each of which may nest in close proximity to dwellings and other constructions. Nevertheless, the arrival of these birds each spring is well marked by those who have gradually gained a fondness for them.

The Northern Rough-winged Swallow has long been known as a bird of Humboldt County. The earliest historical report was of birds along the Mad River May–June 1889 (T. Palmer; Fisher 1902b). Harris termed the species a "common migrant, summer resident, and breeder" in northwestern California. This status persisted through the Atlas period. In the course of the project's 5 years, Northern Rough-winged Swallows were detected in 26% of the total blocks. Confirmed breeding was established in 38 blocks, probable breeding in 27 others. The species was reported in all 5 Atlas years and was encountered widely.

Atlasers reported detections of Northern Rough-winged Swallows most commonly along streams, at bridges with drainage holes, at exposed roadcuts, and about low coastal cliffs. They were found along the coast consistently from the Del Norte County line south to Trinidad and sparingly southward to the lower Eel River. The only outer-coastal blocks with Rough-wings south of there were around the mouth of the Mattole River (b26, B27) and at

Shelter Cove (f33). Detections were frequent along the Klamath, Trinity, lower Mad, Eel, South Fork Eel, Van Duzen, and Mattole rivers and their tributary streams. It appeared that much of the population breeding in Humboldt County occurs in association with larger creeks and rivers. The coastal component would appear smaller; however, these birds are quite common at Gold Bluffs Beach (g01, g02); along Prairie Creek and lower Redwood Creek (f04, g04, g05); throughout the Humboldt Lagoons (f05 to f08); and from Patricks Point (e09) south to Little River (F11). Birds nesting in this stretch of coast do so chiefly in the numerous low cliffs of soft, uplifted marine sediment facing the ocean.

A remarkably "white" swath of no-detection blocks marks nearly all of central Humboldt County away from the coast. This zone stretches from east of Big Lagoon in the northwest and upper Redwood and Willow creeks in the northeast to the confluence of the Van Duzen and Eel rivers in the southwest, and then everywhere that is not in the immediate Van Duzen River corridor in the southeast. The disjunct block with probable breeding reported takes in the vicinity of Maple Creek (J17), along the Mad River some miles above Blue Lake. The apparent absence of Northern Rough-winged Swallows across so much of the county has some basis in the lay of the terrain and its cover type. In these middle elevations, a great deal of the landscape accessible by road traverses slopes that are forested or have been clearcut and are regrown, or that are grassy expanses little broken by cuts or interruptions in the uniformity of the ground. As well, much of the surface rock in Humboldt County is graywacke sandstone, a friable material prone to slumping; consequently, most upper road cutbanks slough into talus, which then becomes vegetated. Thus, comparatively few cutbanks on moderate slopes provide appropriate steep, exposed earthen bluffs suitable for breeding Rough-wings. The small bridges along the minor streams in this section of Humboldt appeared not to attract these swallows.

More to the birds' liking are a persistent, steep bank of soil or loose rock. Settings preferred by these birds are hardly limited to roadways. Streams scarcely large enough to wet one's feet in crossing may support Northern Rough-winged Swallows if there is even a modest stretch of "porous verticality" rising from the stream, in which they can excavate burrows. In an extreme case, breeding was confirmed when an adult was seen carrying a fecal sac from the entrance of a nest burrow 18 inches above the ground in a gravelly rim only 2 feet high along the South Fork Eel River at McCann (J27) 24 Jun 1999 (JCP). A pair of Rough-wings was watched scouting for a prospective nest site along the inner lip of the dike separating Klopp Lake (F15) from North Humboldt Bay 28 Mar 1995 (DFx); this was also the earliest report of breeding activity of any kind during the Atlas period.

The routine hydrodynamics of creeks and free-flowing rivers provide both destruction and renewal of Northern Rough-winged Swallow nesting sites from year to year. Along the outer coast, soil and soft rock faces typically slough off sections during high rainfall and heavy surf, perpetuating nesting opportunities. Excavations made for buildings, the periodic retrenchment of ditches bisecting dairy pasture, and the removal and piling of aggregate by the local gravel-extraction industry all are agents of change acting to varying degree on the fortunes of these swallows.

Occupied nests were the most commonly recorded evidence of confirmed breeding by Rough-wings, and this code was employed nearly five times as often as the next most frequently observed one, which was fledged young. The earliest such record was of a bird apparently nesting in an old Belted Kingfisher burrow near Little River Beach (F11) 22 Apr 1995 (JEH). Six occupied nests were noted in May, four in the first half of June, 10 in the second half of June, and two in July; the latest was found 15 Jul 1998 (L07; JCP). In only a single block (H03) did a nest with young rate highest (19 Jun 1999; JEH). Fledged young, recognized by their cinnamon-buff wingbars and occurrence in family flocks, were the best evidence of breeding in five blocks, with dates spanning 15 Jun 1997 (N29; RJA) to 27 Jul 1998 (J05; DFx). Confirmations in three blocks were based on adults carrying food, spanning 16 Jun 1997 (J31; RJA) to 27 Jul 1995 (F29; JEH).

Northern Rough-winged Swallows are among the early-arriving spring migrants. Although preceded each year by Tree and Violet-green swallows, they show up in many places just as the general northward movement of warblers, vireos, and flycatchers begins. In some years, first reports for Humboldt County fall in late March; in others, early April. By mid-April these swallows become numerous. At the initiation of nesting, pairs and groups of Rough-wings engage in rapid chases, vocalizing constantly and generally filling the air with the full feel of spring. At sites where they are numerous, these antics can be among the most overt and persistent courtship behaviors displayed by small birds in the area. Although Rough-wings occur in modest assemblages in spring and make appearances during that time at sites where they may not breed, generally they are well outnumbered by other species in places used by swallows as a group, such as wetlands and pastures. The breeding population fades from the scene in August, and there is no discernible fall migration. Only occasional apparent migrants are encountered in September. The latest accepted sighting from Humboldt County was 13 Oct 1968 (T. Eley; Harris 1996). There have been no winter reports.

CLIFF SWALLOW *Hirundo pyrrhonota*

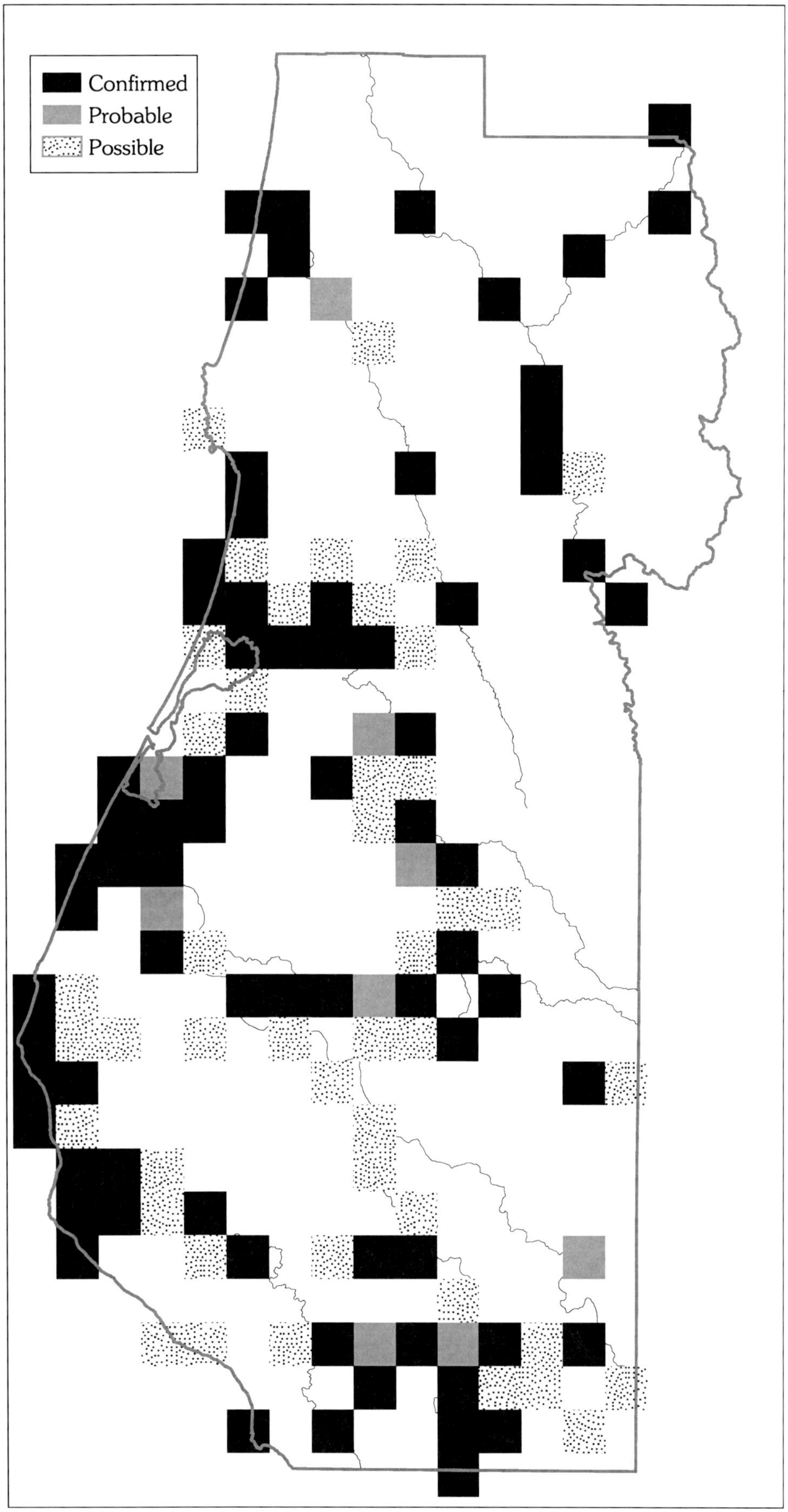

Cliff Swallows are best known for their habit of nesting in colonies on barns, houses, freeway overpasses, bridges, and other human-made verticalities. Mixing wet earth with saliva, they build straw-lined mud nests gobbet by gobbet. Placed side-by-side and over-and-under, nests may number in the hundreds at favored sites. Prior to human settlement, these birds must have relied upon rock faces and overhangs, a substrate still in wide use wherever it occurs. Many colonies of Cliff Swallows nest on rock cliffs scattered irregularly across Humboldt County, the bulk of those in interior and southern Humboldt. They probably have done so for ages. With respect to total numbers, however, it is clear that most of the birds present during the breeding season within the county use human constructions.

Atlas surveys indicate that, in a general sense, Cliff Swallows are common nesters wherever people live. The combination of tall buildings, easy availability of mud suitable for nest construction, and insect populations typical of rural or urban-edge settings has spelled success for this swallow in Humboldt's expansive ranching and agricultural lands. They are locally abundant throughout most of the coastal lowlands. That they are most numerous where sources of mud are common is reflected in the pattern of confirmed blocks approximating the courses of the Klamath, Trinity, Mad, Van Duzen, Mattole, Eel, and South Fork Eel rivers and Redwood Creek. At lower and middle elevations near water they are to be expected everywhere. Harris (1996) deemed them a "common migrant, summer resident, and breeder" in northwestern California. One hundred twenty-eight (30%) of the 425 total blocks held Cliff Swallows during the 5 Atlas years. The range of this species in Humboldt County is similar to that of the Barn Swallow.

Barns, bridges, and other buildings attractive to breeders are sufficiently numerous on the landscape that concentrations of Cliff Swallows, as well as single or dispersed birds, are routinely seen from early April through August. An old barn or a cluster of outbuildings in the last pasture up a settled creek hollow may support a thriving colony, as could an isolated maintenance yard, firehouse, or radio facility. As a rule, only small numbers colonize well-built-up neighborhoods and urban core areas. Small colonies of a few pairs frequently persist around the urban/rural perimeter where they are not discouraged from doing so. Those portions of Humboldt lacking Cliff Swallows are generally dominated by extensive forests.

During the Atlas period, Cliff Swallows were reported nesting on rock cliffs along or above the Klamath, Trinity, Mad, Van Duzen, and Eel rivers. In most cases, such habitat is not extensive. However, many isolated outcrops or rock heads above the upper Mad River, on portions of the general Island Mountain ridge, and just above the coastal bench south of Cape Mendocino were noted to host colonies. Exposed mid-slope rocks amounting to little more than great boulders may support Cliff Swallows if there is sufficient verticality—and a well-trampled stock pond nearby.

Where numbers of Cliff Swallows are encountered in spring and summer, breeding evidence is readily obtained, as their colonies are usually conspicuous. Not surprisingly, nest building, nests with young, occupied nests, and used nests accounted for 60 of the 74 breeding confirmations. Nest building was reported as early as 9 Apr 1995 (D19; S. McAllister) and a nest with young was reported as late as 15 Aug 1997 (b29; H. Taylor). All but one 29 Jun 1996 (i29; JG) report of fledglings were from July. This extended breeding season allows ample time for at least two broods.

After winter spent in South America, Cliff Swallows begin to appear in Humboldt County in late March, becoming common by mid-April. Adults forage over pastures, ranchland, marshes, and fields, sometimes flying low in the manner of Barn Swallows but often quite high, routinely mixing with Violet-green Swallows near sites where both species breed. No great numbers are perceived to occur as pure migrants, although reason suggests that many Cliff Swallows seen among the general swarms of swallows in early fall must have originated from outside the area.

BARN SWALLOW *Hirundo rustica*

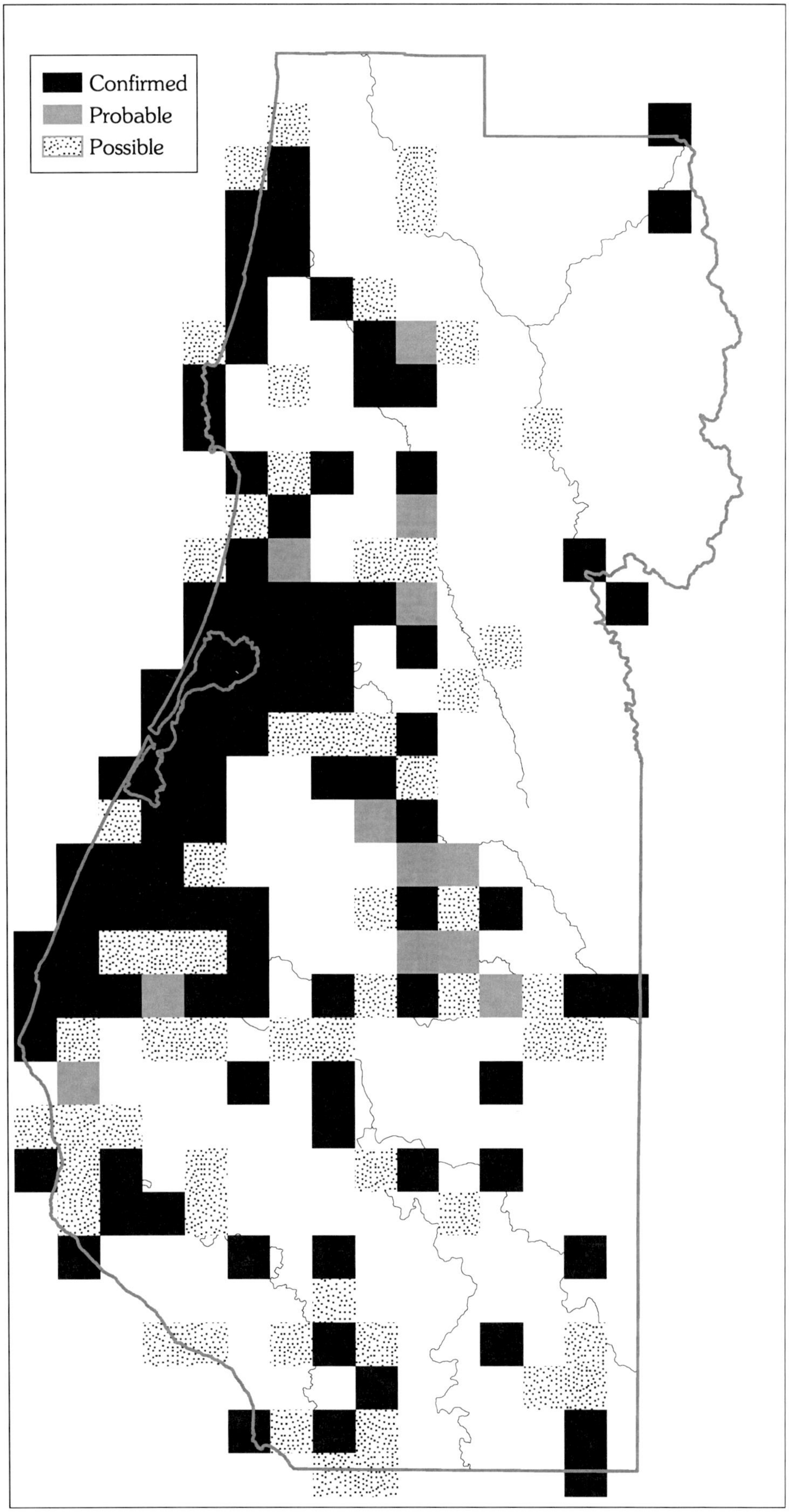

Vernal meadows touched by the shadows of cumulus puffs running before a breeze set the stage for spirited foraging flights of Barn Swallows. Familiar across temperate North America, Barn Swallows prosper wherever expanses of grass or semi-open countryside lie in proximity to human constructions. Their reliance upon porch eaves, open outbuildings, barns, bridge beams, and similar situations for nest placement—together with confiding behavior and an exuberant song—has earned them a place in mythology, poetry, and pastoral portraiture.

During the Atlas period, Barn Swallows were encountered in 163 (38%) of the 425 total blocks and were confirmed breeding in 93. In northwestern California this species is generally a common to abundant migrant, summer resident, and breeder, and is casual in winter (Harris 1996). This holds true for Humboldt County, where Barn Swallows are variably abundant and thoroughly routine summer birds across most of the western portion of the county. A distinct clustering of blocks with confirmations extends from near Trinidad south to Cape Mendocino. As was probably the case with many widespread species frequently detected by atlasers, a measure of the seeming abundance of Barn Swallows in the central-coastal portion of the county must have been due to the "backyard birding" phenomenon. However, the vast sweep of dairy pastures with outbuildings, areas of dispersed human residence, and the neglected constructions of a past era of settlement and industry all suit this species well, and it is truly abundant on the coastal slope.

Away from the zone of coastal abundance, Barn Swallows were found frequently throughout the lower and middle elevations of all larger river drainages, with the exception of the Trinity–Klamath corridor. Human settlement in that area is restricted largely to Willow Creek, Hoopa, and Orleans. The mapped detections reflect this.

There were confirmations of breeding at Willow Creek (N13) and Orleans (P05), presence only at Hoopa (m10), and no detections in the intervening blocks. It seems reasonable to assume that many pairs of Barn Swallows occurred at isolated homes in many blocks in which there were no detections.

Not surprisingly, expanses of commercial timberland in west-central and north-central Humboldt, which have few human constructions and little appropriate foraging habitat, were found to be unoccupied by breeding Barn Swallows. Similarly, the more extensively forested mountains in the northeast section of the county were without detections. The long stretch of unoccupied blocks along the eastern edge of the county is broken in only three places: in the northeast by virtue of the town and the truck farms of Orleans (along the Klamath River), farther south at and near the town of Willow Creek (Trinity River), and in open lowland country at Dinsmore (Van Duzen River).

Tolerant of persistent close human presence, Barn Swallows exhibit the full spectrum of reproductive behaviors before all eyes caring to witness. They are surely among the easiest birds to confirm as breeding. As a result, a significant proportion of all detections during the Atlas period involved proof of nesting. Of the 93 confirmations, 75% were either occupied nests, nests with eggs, or nests with young. Fledged young accounted for 15% of the total of confirmations. The remaining five confirmed categories that were reported (*i. e.*, feeding of young, nest building, used nest, carrying of food, and carrying of nest material) together comprised only 15% of the total. Further evidence of the ease of confirming breeding by this species is the fact that only 7% of the highest evidence obtained in all 163 Barn Swallow blocks was of probable breeding behaviors.

Pairs of Barn Swallows commonly raise two broods per year in Humboldt County. Do they raise three? No direct accounts of such protracted breeding came to light during the Atlas. However, considering the earliest report of nest building on 25 Apr 1995 (g04; JEH) and the latest report of a nest still occupied 27 Sep 1996 (e16; D. Ogden), it would seem likely that sufficient time might be available for particularly hardworking parents to bring off three broods in a good season. The earliest report of a fledgling was 19 May 1995 (D17; JEH).

Because of a mass arrival of Barn Swallows in early April and the initiation of nesting soon after, the fledging of first broods across the near-coastal lowlands is nearly synchronous each year. This results in the appearance of large numbers of juvenile swallows in June. These youngsters remain dependent upon adults for but a brief period, thereafter concentrating about uncut hay meadows, pastures, freshwater ponds and marshes, and along estuary margins, all places where abundant flying insects may be pursued. No such pulse of young birds resulting from second nesting attempts is noted in Atlas data. As the nesting season winds down, numbers of Barn Swallows appear to dwindle. However, a strong and conspicuous movement of southbound daytime migrants, foraging as they fly through, takes place each year in early September. Apparent migrants occasionally are noted well inland, but most are encountered along the coast. Typically, by October only a few stragglers remain in the county.

BLACK-CAPPED CHICKADEE *Poecile atricapillus*

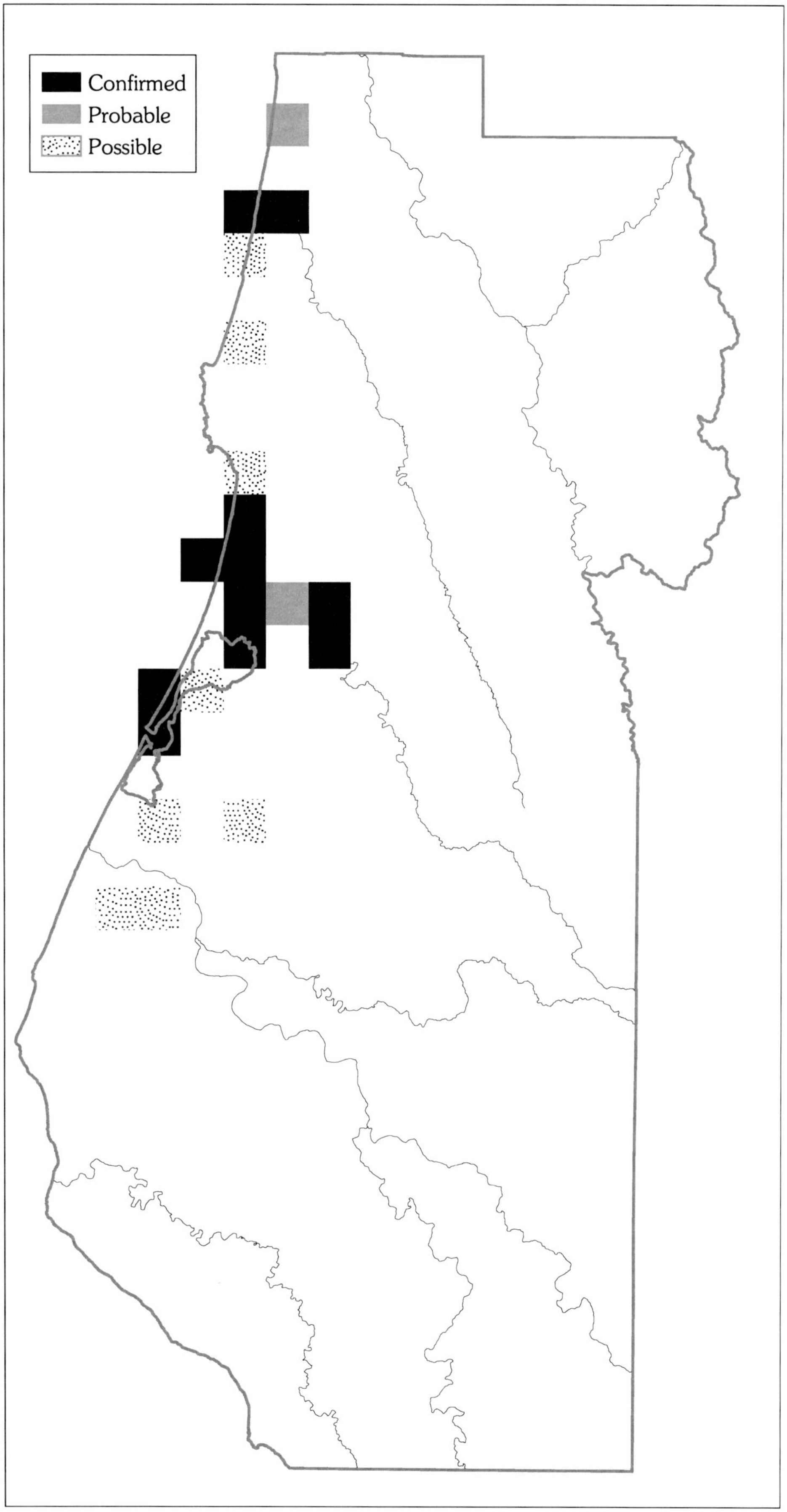

The Black-capped Chickadee is one of the most familiar birds within its extensive range in North America. These hyperactive creatures are among the most warmly regarded "watchable wildlife". They are everyday visitors to an enormous number of backyard bird feeders during the colder months, providing feathered cheer in regions of the continent where the winter avifauna may be reduced to but a few dozen species. Simple verse and homecrafted artwork pay them tribute. Their image is emblazoned upon bags of birdseed, the covers of mail-order catalogs, and a blizzard of holiday greeting cards.

In contrast to its ubiquity elsewhere, the Black-capped Chickadee has a restricted profile in the California avifauna. It is known from only a few counties in the state and is relatively localized where it does occur. It is only in the northernmost 100 miles of coastal lowland of Del Norte and Humboldt counties and in parts of Siskiyou County that these chickadees are regularly encountered. During the Atlas period Black-capped Chickadees were detected in 21 (5%) of the 425 total blocks. They were confirmed as breeding in 11 blocks and considered a probable breeder in two others.

Virtually the entire population of Black-capped Chickadees in Humboldt County is found in damp deciduous thickets in the coastal and near-coastal bottomlands. The range of the species in Humboldt is notably narrow west-to-east—scarcely 1 mile to perhaps 10 miles wide—and is of limited altitudinal extent, from just above sea level to the bases of the first hills east of the coast. This distribution is irregularly configured along about one-half of the north-south extent of the county's western fringe. They were found somewhat patchily, from the northern portion of Prairie Creek Redwoods State Park south to the Eel River floodplain.

No truly inland population was discovered in the county, despite observer awareness of the possibility of this species' occurrence along major rivers. Although unconfirmed sightings of Black-capped Chickadees have been reported from alder-dominated riparian habitat along the Klamath River within Humboldt County, no birds were found there during Atlas surveys despite some focused searches.

Black-capped Chickadees in Humboldt County are partial to red alder woodlots, well-developed hardwood riparian corridors, and extensive willow thickets. A great deal of red alder habitat exists along and near the coast of northern Humboldt County. Much of it is a consequence of extensive land clearing, while some of it occupies perennially damp or unstable ground unsuitable for conifer growth. Streams dominated by black cottonwood and alluvial flats afford habitat for these chickadees near Orick and along the lower Mad and Eel rivers. Willow patches containing trees large and complex enough for possible chickadee nest sites are scattered through the bottomlands throughout the species' range here, and such patches are well used. During the past 20 years, Black-capped Chickadees have established a population in the various discontinuous willow thickets in dune hollows of Humboldt Bay's north spit.

Irrespective of stand structure or degree of undergrowth, sites consistently supporting Black-capped Chickadees in Humboldt County are characterized by a deciduous hardwood canopy and, typically, the presence of willow. Older, unmanaged alder stands in which competitive mortality has resulted in much standing dead material are favored for nesting. The activities of woodpeckers attracted to these stands may augment naturally occurring cavities developing in older, declining trees. Alder thickets which are thoroughly grazed, roaded, or have been heavily cut for firewood tend to retain fewer protected snags and appear less hospitable to these chickadees.

As it is a target species for California birders and is sought locally, reports of Black-capped Chickadees in well-birded Humboldt County have been of some interest within the birding community. This interest has helped spur thorough documentation of a recent southward coastal range expansion and greater abundance in the county within the past 30 years. The earliest historical Humboldt records include specimens collected in Eureka 31 Mar 1920 and 23 Oct 1924 (Davis 1924, Davis 1940). The first direct evidence of breeding in the county was an occupied nest in June 1974 in Fern Canyon (G. Strachan; Yocom and Harris 1975). Harris (1996) noted that regular reports of birds from Humboldt Bay locations and Ferndale began only in the late 1980s. While breeding was not confirmed along the Salt River in the Ferndale Bottoms during the Atlas period, birds were reported there 5 May 1996 (D21; L. Karsteadt) and 17 Jun 1995 (c21; JEH). This stretch of riparian habitat is probably the southernmost location in the state where these birds can be regularly found. Black-capped Chickadees appear to conclude breeding relatively early in Humboldt County. Nest building was seen 17 Apr 1998 (h14; M. Morrissette) and the latest young being fed was recorded 3 Jul 1995 (g05; GCH). The earliest fledglings were seen 19 Jun 1997 (f05; JCP).

MOUNTAIN CHICKADEE *Poecile gambeli*

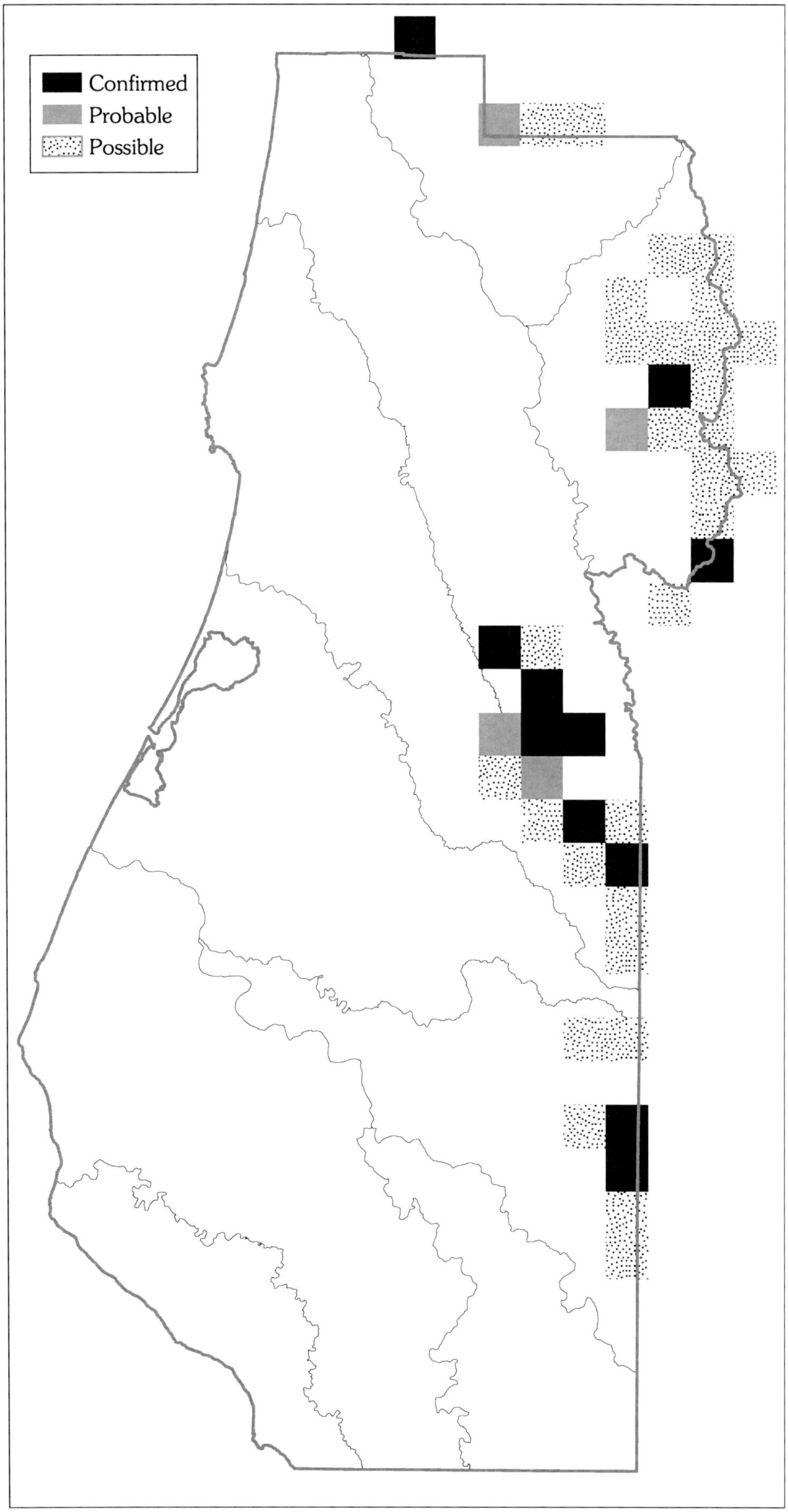

Mountain Chickadees are well named. They are resident in mountain forests of the West from northern British Columbia and Alberta southward in the Rockies, in the Great Basin ranges, along the Cascade–Sierra Nevada axis, at higher elevations in southern California, and in northern Baja California. Throughout this extensive range, they are associated with conifer forests which are typically, though not always, dominated by pines. No other small bird of mountain forests makes itself so quickly known. As the car doors are thrown open at a piney campground, there comes to the ears the distinctive, rough-cut *chick-a-dee* calls or whistled songs which announce the presence of from one to many Mountain Chickadees. Bold and assertive where the possibility of a human handout avails, Mountain Chickadees learn to take scraps from the picnic table, and may land on the outstretched palm to pluck pieces of trail mix. Adapted to the rigors of a snowy winter, they remain throughout the colder months at elevations as high as treeline.

This species has long been recognized as occurring in interior northwestern California, where it has been considered a common resident and breeder in high elevation forests, occurring irregularly in fall and winter in the lowlands (Davis 1933, Harris 1996). These chickadees are better represented to the east of the area, but they nevertheless maintain a consistent presence in the mountains of eastern Humboldt County. At one serpentine site, however, they extend to within 10 miles of the coast, as represented by a notable westerly Atlas breeding confirmation from the South Red Mountain area (j01) shortly across the Humboldt line into Del Norte County (adults carrying food, 1 Jun 1997; KMS). Mountain Chickadees were detected in 10% of the 425 total Atlas blocks.

Mountain Chickadees in Humboldt during the project were correlated with higher-elevation conifer forests. Without exception, blocks in which chickadees were found supported the pine and true fir forests that this species seems to require nearly everywhere within its range. Where white fir begins to appear in abundance within the Douglas-fir forest, and tanoak and other hardwoods diminish, Mountain Chickadees find their lower altitudinal limit. This threshold occurs at roughly 4000 feet, but varies greatly depending on exposure and other site characteristics. They are uncommon to locally common in a variety of higher forest stands, including white fir, white fir/Douglas-fir, Jeffrey pine, and mixed stands with incense-cedar, sugar pine, or western white pine.

Outside the northeastern portion of the county where such forests are extensive, suitable habitat tends to be isolated in more or less continuous strips running along the higher ridgelines. The map shows Mountain Chickadees present during the study from Horse Mountain (L15) southeastward past Grouse Mountain (m17), along Pilot Ridge (n20), and continuing along South Fork Mountain to just south of the Lassics (o29). Poorly indicated (owing to block size) is their absence in intervening lowlands, such as in the canyon of Redwood Creek, which separates birds on Snow Camp Mountain (L17) from those on South Fork Mountain by a few miles of lowland corridor. The distribution of Mountain Chickadee created by the ridge-and-river landscape is similar to that of many other interior-mountain birds found in the county during the Atlas period. Species such as Hammond's and Dusky flycatchers, Townsend's Solitaire, and Green-tailed Towhee occupy preferred ridgetop sites in much the same fashion.

Confirmed breeding was reported from 11 blocks, probable breeding from another four. The earliest evidence of probable breeding reported was of a pair 28 May 1996 (o10; JMH). All but one of the confirmations was in June, with the only post-June confirmation being a late outlier of food-carrying reported 28 Jul 1996 (m17; JCP). These data suggest either a later breeding season for these high elevation birds, or delayed access by atlasers to these areas. There were four reports of nests with young, ranging in date from 10 Jun 1999 (m16; GCH) to 7 Jul 1995 near North Trinity Mountain (P09; JEH). The latter nest was typical: it was in a hole about 6 feet up in a 20-inch diameter "buckskin" fir snag. The earliest fledglings reported were some being fed 18 Jun 1995 (q13; *fide* TWL). Considering how conspicuous fledglings were, it was surprising that only one other report of fledged young was received, 28 Jun 1998 at Horse Mountain (L15; M. Wachs).

Mountain Chickadees in Humboldt County are generally thought to be resident. It is most unusual to encounter even a single bird in lowlands well downslope from ridgeline breeding habitat during spring and summer. However, true lowland incursions, usually involving very small numbers of chickadees, have taken place in fall and winter. At such times, from one to a few birds may be encountered nearly anywhere from the interior valleys to the coast. Yocom and Harris (1975) and Harris (1991) listed some such records from the coastal lowlands. Mountain Chickadees have also been recorded in winter in lowland oaks in the Hoopa Valley. During the Atlas period, one was in a grove of exotic eucalyptus and Monterey cypress on the north spit of Humboldt Bay at Fairhaven 4 Oct 1996 (D17; DFx *et al.*).

Along with other secondary cavity-nesting birds, Mountain Chickadees rely upon the labors of woodpeckers for opportunities to take over disused roosting and nesting holes in trees. Forestry practices which reduce the number of dead or dying trees also diminish numbers of woodpeckers and chickadees. Mountain Chickadees also excavate their own cavities in soft wood, such as is found in dead or dying white firs and other trees (Ehrlich *et al.* 1988). It is hoped that the application of "new forestry" techniques, which aim for retention of snags and other structure, will ensure the protection of habitat for secondary cavity nesters in Humboldt when these higher elevation forests are logged.

CHESTNUT-BACKED CHICKADEE *Poecile rufescens*

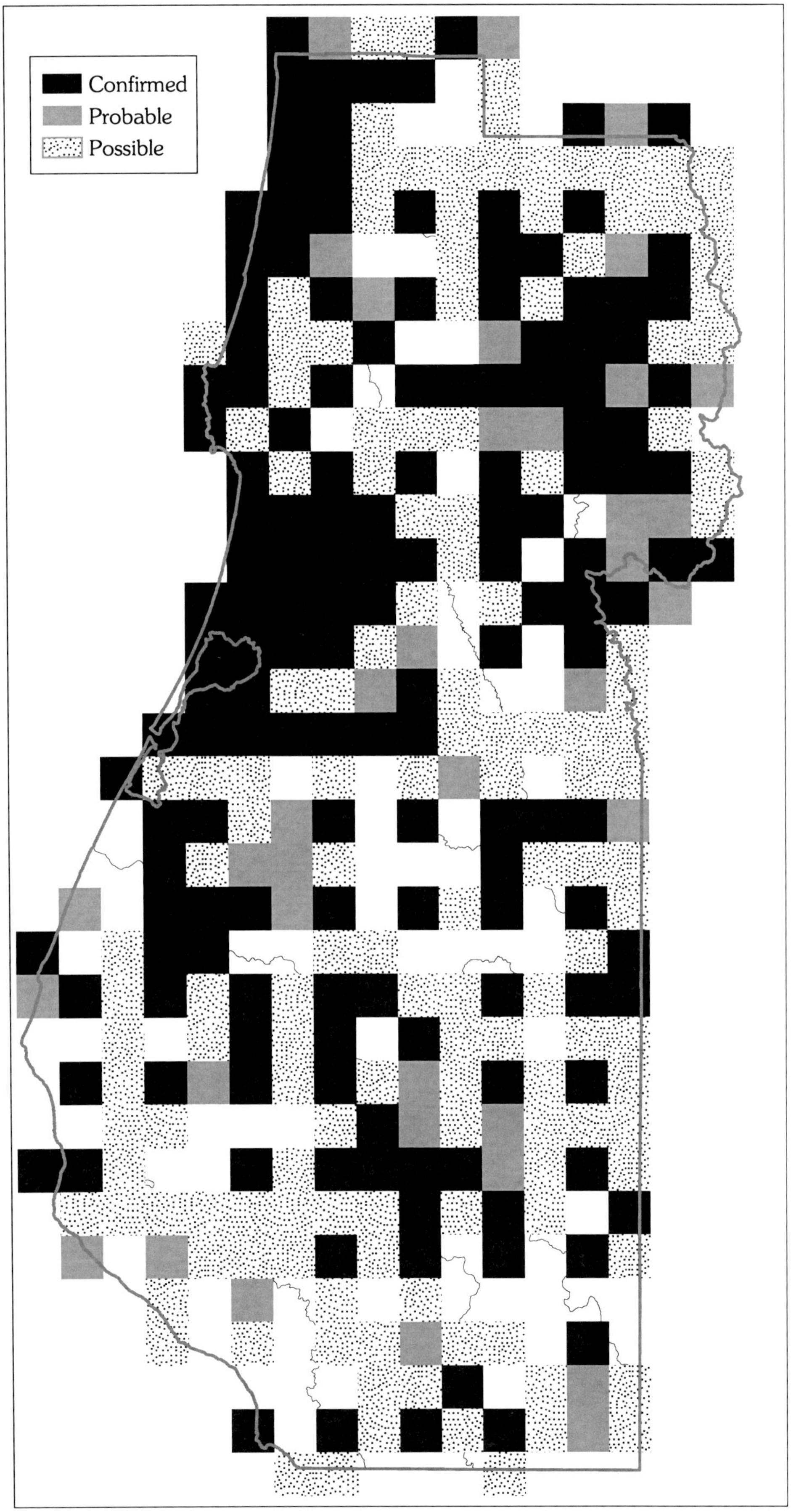

It seems that just about anywhere there are trees in Humboldt County, there are Chestnut-backed Chickadees. As simplistic as this may sound, the rule is quite a good one. With the exception of the highest-elevation pine stands and the most open oak woodlands, no forest type in the county appears to lack these birds. Although they are secondary cavity nesters somewhat dependent on woodpeckers, they can make do with the smallest naturally occurring crevice or cranny. This trait serves them well in today's industrial forest, where woodpecker holes can be uncommon, and where these chickadees maintain a surprisingly strong presence despite intensive timber harvest. They may be encountered at times in the most isolated groves. Coast redwood, Sitka spruce, Douglas-fir, Douglas-fir with hardwoods, true fir, and oak-dominated forests and woodlands all support them, whether they are old growth, recent regrowth, or some mix of tree ages. These birds are found from the forest interior to the forest's edge wherever even the rattiest stands of trees occur, and are commonly seen in towns and residential neighborhoods. Humboldt County's forested expanse altogether hosts huge numbers of these tiny birds. Truly, this is a contender for the most abundant resident bird in the county. Chestnut-backed Chickadees were detected in 79% of all Atlas blocks, ranking as the fourth most widespread breeder in Humboldt.

Referring to northwestern California, Harris (1996) described the Chestnut-backed Chickadee as a "common resident and breeder in moist forest and woodland types throughout the region, but generally absent from dry, inland mountain pine stands or pine–chaparral brush fields. Most common in humid coastal strip." During Atlas fieldwork, no clear association between occurrence and vegetation cover type within the county was noted. Of the few blocks that lacked detections, many were along the outer coast or in non-priority blocks that received minimal survey effort. Mirroring a vague but oft-repeated pattern exhibited by many forest birds, blocks with detections were a bit more evident across the northern one-third of the county and a little less so elsewhere, probably in accordance with general extent of forest cover.

Confirmed breeding was reported from 154 blocks, making this species the second ranking confirmed breeder, trailing only the intensively monitored Spotted Owl (confirmed in 167 blocks). Breeding confirmations were reported at variable density throughout the county. Prominent on the map are clusters of confirmations from the greater Humboldt Bay/lower Mad River area, the northern redwood parks, and the Hoopa Valley Indian Reservation—all sections of the county receiving much survey coverage.

Confirmations were supported by no fewer than seven different behavior codes, the earliest of which was a report of an occupied nest 11 Mar 1997 (D17; GAS). The most frequently reported evidence of confirmed breeding was the observation of fledged young (being fed or not), reported in 69% of the total blocks with confirmed breeding. It is worth noting that, at the outset of the Atlas project, one experienced atlaser (D. Roberson) offered the wry advice that Chestnut-backed Chickadees could be "confirmed" while driving through a block (with vehicle windows rolled down, naturally). While surely not advocating a nonchalant approach to citizen science, that advice proved quite correct. The distinctive begging calls of dependent young Chestnut-backs quickly became well perceived by active atlasers. Fledged young (in these instances, being fed) were reported on dates ranging from 11 May 1996 (e10; TWL) to 22 Sep 1999 (n10; *fide* JMH). Such a date span suggests an extended breeding season, as one might expect of a multiple-brooded songbird such as this one.

This species is among the most frequent inhabitants of bird boxes placed on houses and in yards, particularly where a scrap of woodsy cover adjoins. Many of the nests with young reported were probably from just such situations; nests with young spanned the dates of 11 May 1997 (f16; H. Brown) to 15 Aug 1995 (e16; D. Ogden).

When not nesting, Chestnut-backed Chickadees are social creatures, found in flocks of from several individuals to 20 or more birds. From early autumn through the winter, they associate a great deal with Golden-crowned Kinglets, Hutton's Vireos, Red-breasted Nuthatches, and Brown Creepers. At lower elevations beyond the constant press of freezing weather, their flocks are commonly joined by Ruby-crowned Kinglets (*Regulus calendula*), Townsend's (*Dendroica townsendi*), and Yellow-rumped warblers and, in the northern coastal lowlands, by Black-capped Chickadees—and perhaps, as well, by the odd Nashville or eastern vagrant warbler toughing out a Humboldt County winter. This species is thought to have increased in abundance and distribution in many areas of California from 1944–1985 (Brennan and Morrison 1991).

OAK TITMOUSE *Baeolophus inornatus*

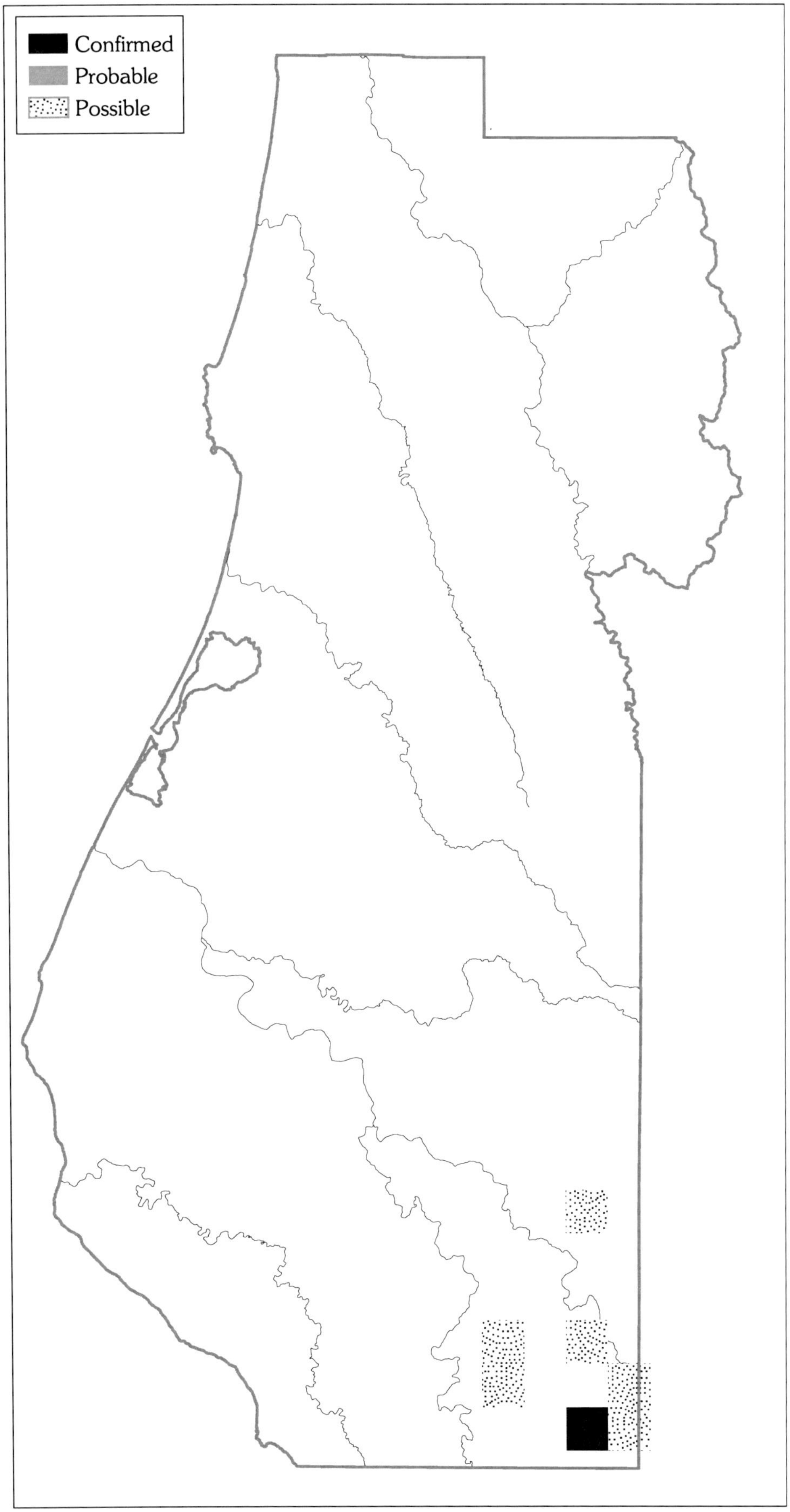

Few birds within its range are as plain as the Oak Titmouse. Indeed, the two taxa now "split out" as Oak and Juniper (*B. ridgwayi*) titmice were once classified as a single species aptly-named Plain Titmouse. Drab, small, and given to remaining in the shadows of dense oak foliage, these birds can be easily overlooked. They would be even less often noticed were it not for the frequent and varied chickadee-like songs and calls they deliver throughout the day. During the hush of a summer afternoon, while many other birds have fallen silent and little but the buzzing of annual cicadas is heard, still the oaks are given voice by the conversational notes of Oak Titmice. This species is distinctly tied to the oak-and-chaparral woodlands of California, Baja California, and interior southwestern Oregon from Grants Pass southward. Where hillsides at lower elevations are grown to a welter of heat-tolerant hardwoods, lizards skitter for cover, and the leaf litter crackles beneath one's feet, these birds can be anticipated. Oak- and sycamore-grown riparian corridors and residential neighborhoods also attract them.

As a characteristic bird of Mediterranean oak-and-brush communities, the Oak Titmouse is another among a small suite of species for which the southeastern corner of Humboldt County represents a frontier. Historically, little was known of the titmouse in the county other than that it seemed resident in small numbers in a restricted area. Harris (1996) related that his files contained 40 reports from east of Highway 101 and south of Highway 36, including nesting records, although the proportion of these originating from Humboldt County was not indicated. The first breeding record for Humboldt was of an active nest near the town of Harris in the late 1970s (S. Harris). Oak Titmouse is a common resident 50 miles northeast of Humboldt at Scott Valley, Siskiyou County (Harris 1996), in Mendocino County to the south, and in portions of Trinity County to the east it is a fairly common resident. Thus, although Harris (1996) termed Oak Titmouse a "rare resident in the southeast corner of the region," this description is best applied only to Humboldt County proper. For most Humboldt birders, a desire to see this species has been challenged by the necessity of a long drive from the coastal cities, uncertainty of success, and some degree of mystery.

While the Oak Titmouse was almost certainly resident in Humboldt before ornithological record keeping, it is intriguing that the first actual record for the county was of a bird at Bayside 19 Dec 1965 (R. Gennelly). Several other dramatically extralimital Oak Titmouse have been recorded within the county, including single records from Orleans and Trinidad. While it is intuitive to consider that these birds may have originated from well to the southeast, it is conceivable that some or all may have come from more northern-interior locations. Prior to the Atlas period, Oak Titmice had been reported from the vicinity of Benbow Inn, along the South Fork Eel River south of Garberville. None were reported there during the Atlas period, although visits to the site were infrequent, and it is possible that titmice, if present, could simply have been missed.

Surveys in extreme southeastern Humboldt County in the course of Atlas fieldwork succeeded in finding the Oak Titmouse in a relatively small cluster of seven blocks. Reports in all but one block suggested only possible breeding. The single Atlas-era confirmed breeding report was of two begging fledglings chasing an adult around in a thistle patch and an oak along Island Mountain Road (N33) 24 Jun 1995 (D. Roberson, R. Carratello). Habitat in this area was typical of that found routinely on lower hillsides along stream corridors in this portion of Humboldt: a patchwork of immature and somewhat older stands of Douglas-fir and oak mixed with other hardwoods, shrubs, and grassy openings, along with scattered residences, outbuildings, gardens (both legal and otherwise), rural roads, and other human doings.

Reports of Oak Titmouse in the remaining six blocks extended north to near Blocksburg (n28) 6 May 1997 (JEH, TWL, S. Harris). They were reported westward in two adjacent blocks along the Alderpoint road between Garberville and Dyerville Loop Road (L31, l32) 23 Jun 1998 (S. Rovell), and in another two adjacent blocks 21 Jun 1998 (o33; JEH) and 22 Jun 1996 (o32; C. Secor). Another bird was heard along the Eel River 1 mile south of Alderpoint (N31) 26 May 1997 (DFx).

Woodlands and forest-edge habitats in which Oak Titmouse has been found in Humboldt County vary from clumps or stringers of oaks and other trees, such as along the Alderpoint road, to oak-dominated riparian habitat and near-riparian hillside stands with occasional openings along the Eel River. Of interest is that woodlands in which interior live oak is common are a repeated theme at lower elevations along the Eel River. The northwestern limit of this oak species is in the vicinity of Fort Seward, not many miles south of Blocksburg (n28), site of the northernmost report of titmice during the Atlas period. It is doubtful that interior live oak, in and of itself, means anything special to the birds; rather, the tree's presence signifies the northern limit of a habitat subtype more widely represented in adjacent southwestern Trinity and interior northern Mendocino counties, and in which Oak Titmice are frequently encountered.

A few out-of-town atlasers with extensive general birding experience well within the range of this species (*i. e.,* D. Roberson, R. Carratello, S. Rovell) contributed greatly to the Atlas titmouse dataset. Is it possible that the various minor call-notes of titmice might have been overlooked at times by Humboldt-based observers? In any event, Atlas results suggest that Oak Titmice are at least as widespread in Humboldt County as has historically been supposed. Additional work in other blocks with limited access in the southeastern portion of the county would probably have resulted in a few more birds and additional breeding confirmations.

BUSHTIT *Psaltriparus minimus*

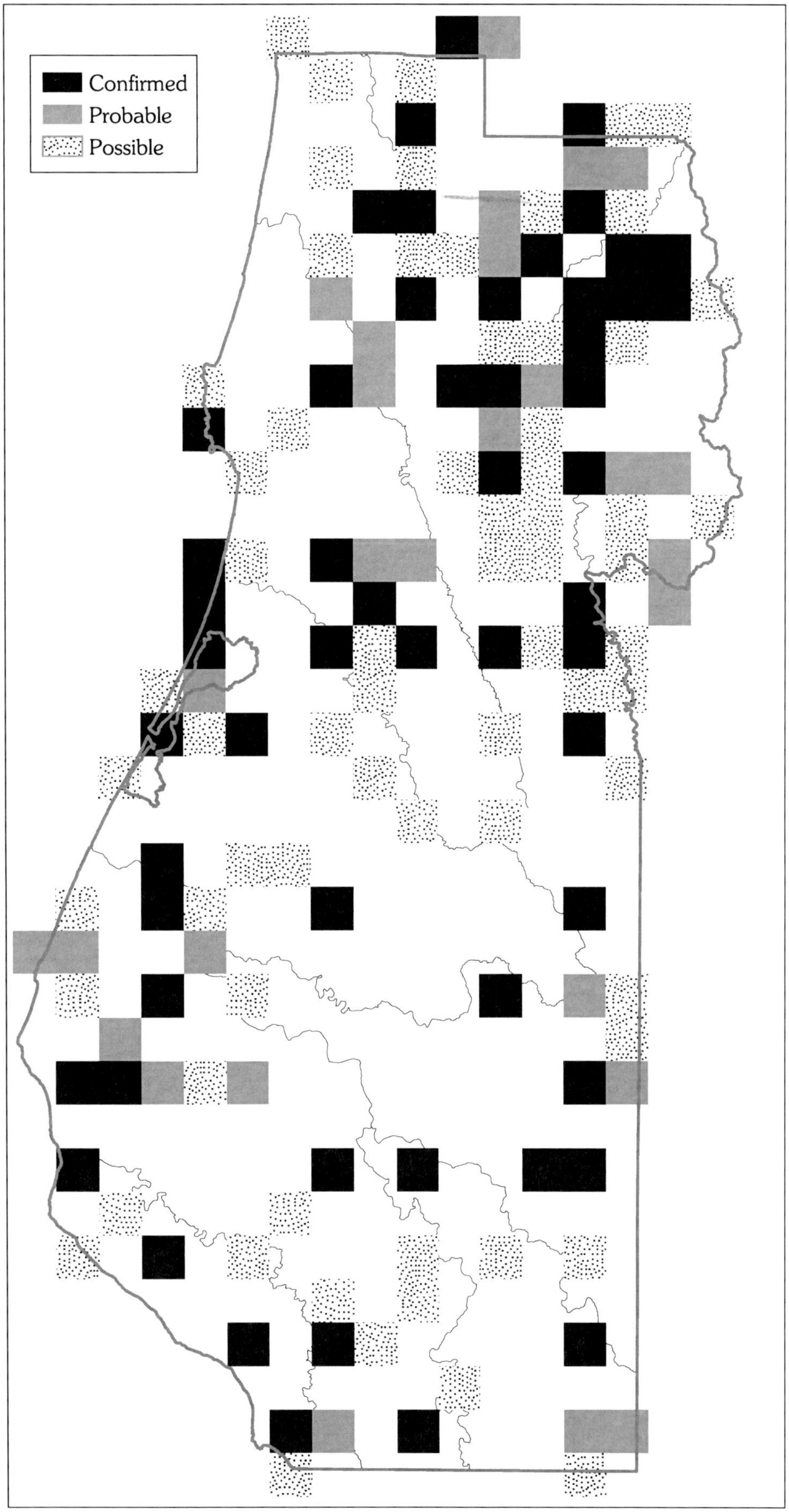

Dating from one's first acquaintance with Bushtits, they are to be remembered for the energetic, fluid manner in which their flocks forage. As the first vanguard of arrivals in a thicket begins gleaning leaves and twigs, more pile on, inciting the band of "birdlets" to forage in higher gear. Many then pulse beyond into the next bush, tossing insistent call-notes over their shoulders, compelling the laggards to hurry in their stead. Ten Bushtits may venture across a rural road, then half as many more. After a long moment, several others do so, and it may seem that all have made the bold crossing and the show is over. But then another half-dozen cluster in the top of a willow on the near side, talk it over, take their aim. In seeming turn, each Bushtit joggles weakly through the air as if being tugged like an elfin kite in a failing breeze. And yet two more appear at the jumping-off spot—now all bets are off.

In only a few sections of Humboldt County would Bushtits appear to be truly out of place. Bushtits are an uncommon resident and breeding species in northwestern California (Harris 1996), and were detected in 35% of Atlas blocks. The distribution map suggests a somewhat random occurrence nearly throughout the county, perhaps a reflection of the success of a habitat generalist. Northeastern Humboldt was detection rich, probably owing in part to great amounts of suitable cover and in part to comparatively high observer effort by Hoopa Tribal Forestry personnel and others. No semblance of a large cluster of detections is otherwise discernible. It is reasonable to suspect that patches of appropriate breeding habitat exist in portions of many Atlas blocks in private ownership. Limited access granted to these blocks often dictated a straightforward traverse along roads bounded by rangeland or other surroundings unsuitable for these birds.

A small region without detections occurred in south-central county. The northern part of this area takes in the rolling open country of the central Mad River watershed, where much of the landscape is tree poor. Some of the blocks on the western side are in Humboldt Redwoods State Park. However, the southern part offers a mixture of lower-elevation timberland and the untidy oaks and second-growth edges of the main Eel River corridor—a section of the county hardly to be described as lacking Bushtit habitat. This "white hole" may be simply an artifact of the vagaries of survey coverage.

Parts of Humboldt County where the absence of evidence may be evidence of absence include sizable tracts of intensively managed timberland centered in extreme northern Humboldt, inland from Trinidad, and southeast of Eureka. Silvicultural practices in these areas usually preclude a semi-open shrub structure, favoring ranks of vigorous young conifers. Hazard and George (1999) found them most abundant in shrub-stage stands of redwood. Bushtits were not reported from blocks supporting extensive montane forest in the far northeastern part of the county, on much of South Fork Mountain, and at the Lassics.

Outside of managed conifer plantations and stretches of dense forest, habitat for Bushtits would seem to lie close at hand nearly everywhere in the wooded sections of Humboldt County. Bushtits can be encountered in many different settings, from just above sea level to nearly 5000 feet. They seem most common in sprawling reaches of nondescript brush. Large clearings grown back to a mid-seral successional stage, in which shrub size and height are commonly diverse and irregular, seem to support Bushtits routinely. Sunny-site species such as oceanspray, coyote brush, *Ceanothus*, toyon, poison-oak, live oak, and young madrone are routine at many of the places where these birds are found. Taller trees with pendant lower limbs, often used for nest placement, are typically present. In the Klamath Mountains, these birds may have increased by over 200% as a result of extensive timber harvest (Raphael *et al.* 1988).

During the Atlas study, Bushtits were encountered along rural roadsides and in riparian woodlots, lower- and middle-elevation clearcuts, power line rights-of-way, and elsewhere at the ragged edge of forest regrowth. Oak woodlands with sufficient canopy closure were also inhabited. Where contiguous wooded or shrubby cover extends nearly to the outer coast, Bushtits follow. They may be found along the immediate coastal slope at such places as Gold Bluffs Beach and Elk Head and are regularly seen in the wax myrtles and shore pines just inland from Lanphere Dunes. They occur year-round in small numbers in the wooded parks and gulches of Eureka and Arcata. They occasionally are seen about the edges of openings in the redwood parks but are not numerous in such places. Each of these situations supported Bushtits during the Atlas period.

Confirmed breeding was noted from throughout the species' range in the county. By a wide margin, July was the month during which most fledged young were seen. Only one such report fell before June, 23 May 1997 at Mad River County Park (e13; GCH), and the latest two were in early August. Observations of adults carrying nest material 17 Mar 1996 (e10; M. Hinton) and 18 Mar 1998 (B27; C. Grant) indicate that these residents start nesting early in the season. The earliest of three occupied nests reported was found 8 May 1997 south of Blue Lake (PFS). Records representing the end of the breeding season were of a nest with young 16 Jul 1996 (N27; GAS) and adults feeding dependent young 23 Jul 1997 (n08; JMH).

Pointing to diminishing totals on regional CBCs, Harris (1996) suggested that a decline in the population of Bushtits in northwestern California might have taken place since the 1970s, but cautioned that further documentation was warranted.

RED-BREASTED NUTHATCH *Sitta canadensis*

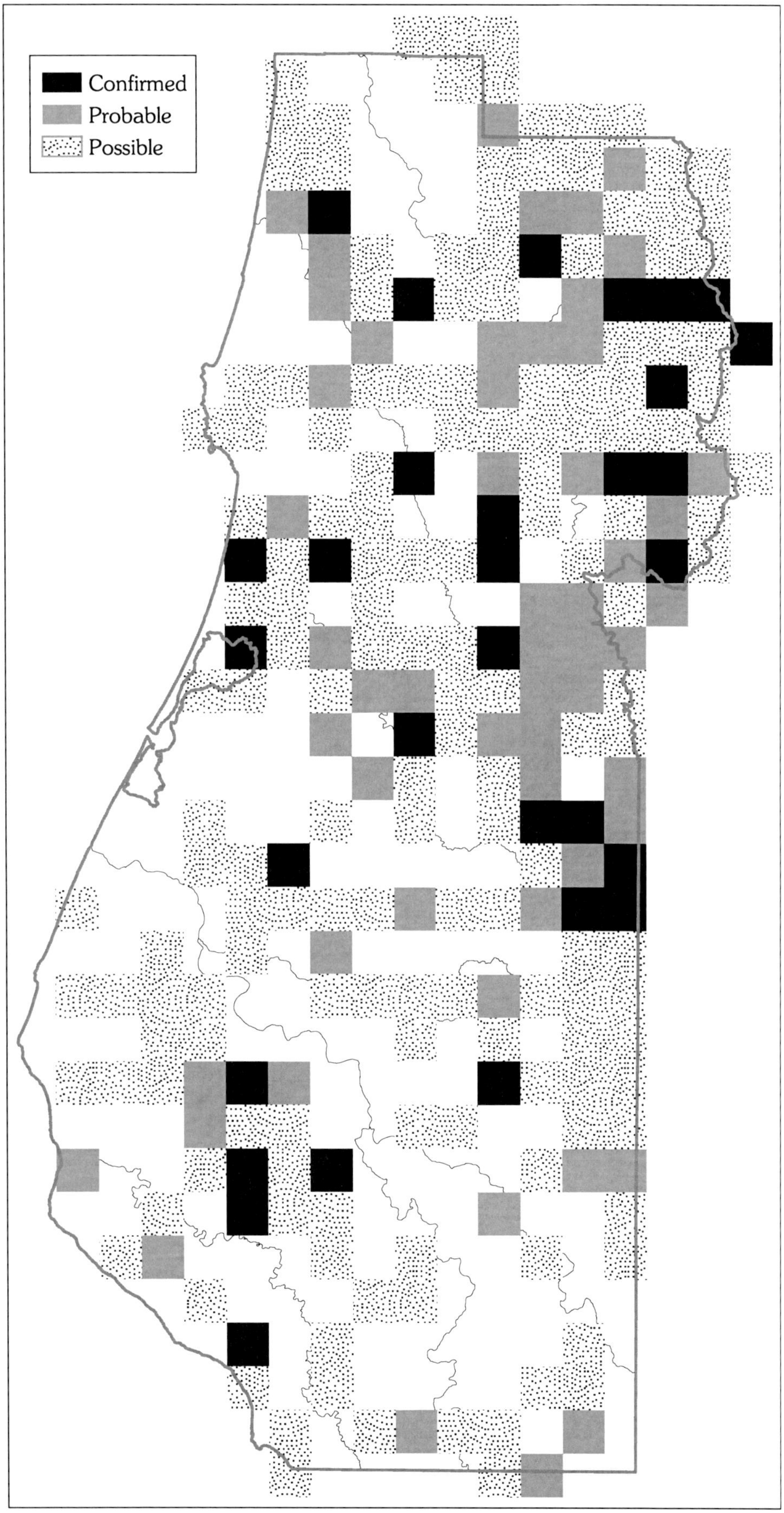

Birders who checked their growing awareness of birds against characterizations in the pages of the late Roger Tory Peterson's field guides found that his description of the voice of Red-breasted Nuthatch as sounding like a "tiny tin horn" rang true. This sound can be heard across a huge stretch of the North American continent. Red-breasted Nuthatches are present in summer, winter, in migration, or are resident from the northern limit of closed boreal forest in Alaska and Canada southward through the U.S., and irregularly into northern Mexico.

This is a routine and common resident species in appropriate habitat throughout forested northwestern California. Unlike some birds of higher-elevation forests, Red-breasted Nuthatches are not restricted to true fir or pine stands, occurring widely in Douglas-fir and mixed conifer/hardwood forests extending downslope locally to just above sea level. Thus they enjoy a broad range in Humboldt County. During the Atlas period, these nuthatches were detected in 58% of all blocks.

In spring and summer, Red-breasted Nuthatches are most frequently encountered in conifer forests. These birds increase as conifer cover increases and decrease as hardwood cover increases (Ralph *et al.* 1991). Atlasers found them in an array of conifer-dominated forests, including stands with coast redwood, Sitka spruce, Douglas-fir, grand fir, and montane forests of Jeffrey pine and white fir. It was the impression of some atlasers—and of Humboldt birders whose experience preceded theirs—that these birds may be most numerous and most dependably found in white fir and Douglas-fir forests of the higher mountains. It therefore came as no surprise that the Atlas map showed the bulk of the breeding population to occur in the northern and eastern parts of the county. Although Red-breasted Nuthatches are not a "mountain specialty," a vague positive correlation between occurrence and altitude was suggested. Similarly, Sniado (1999) found that these nuthatches have an affinity for higher elevations, white fir, and sites with larger trees. Down on the coast, Hill (1977) found them strongly associated with shore pines, but he did not find them nesting in those stands. Sections of Humboldt County in which detections were few or lacking took in the heavily logged industrial forestland of extreme north county, most blocks along the immediate coast and open coastal lowlands, and in blocks supporting a great deal of oak/grassland mosaic in interior and southern Humboldt.

While generally considered birds of coniferous forests, Red-breasted Nuthatches select many tree communities that counter this notion. Stands dominated by red alder or tanoak, or which have considerable bigleaf maple, are inhabited. Such trees, which generally occur in the mid-canopy, may increase in value as a habitat component as a stand matures and overshading induces decline and mortality among the less competitive hardwoods. As cavity nesters lacking powerful bills, Red-breasted Nuthatches ordinarily excavate nest cavities in soft, easily worked rotting wood. Dead branch stubs of alders, tanoaks, or maples frequently are selected as nest sites, even though conifer snags may be available within the birds' nesting territories.

Harris (1996) described Red-breasted Nuthatch in northwestern California as a "common resident and breeder in higher elevation inland conifer forests supporting pines and firs. Generally absent from the redwood belt except as a fall-winter vagrant." Results of Atlas fieldwork were at odds with this characterization. It was found that these nuthatches were present and breeding within the redwood zone, in some cases inhabiting stands containing old-growth redwoods. A representative record for coastal breeding was of an adult feeding two fledglings 4 Jun 1998 in a Sunny Brae yard (earliest fledglings; F15; S. Harris). Elevation *per se* would not seem to be a limiting factor; rather, some aspect of conifer forest structure and/or tree species composition likely plays an important role in determining the breeding distribution of the Red-breasted Nuthatch.

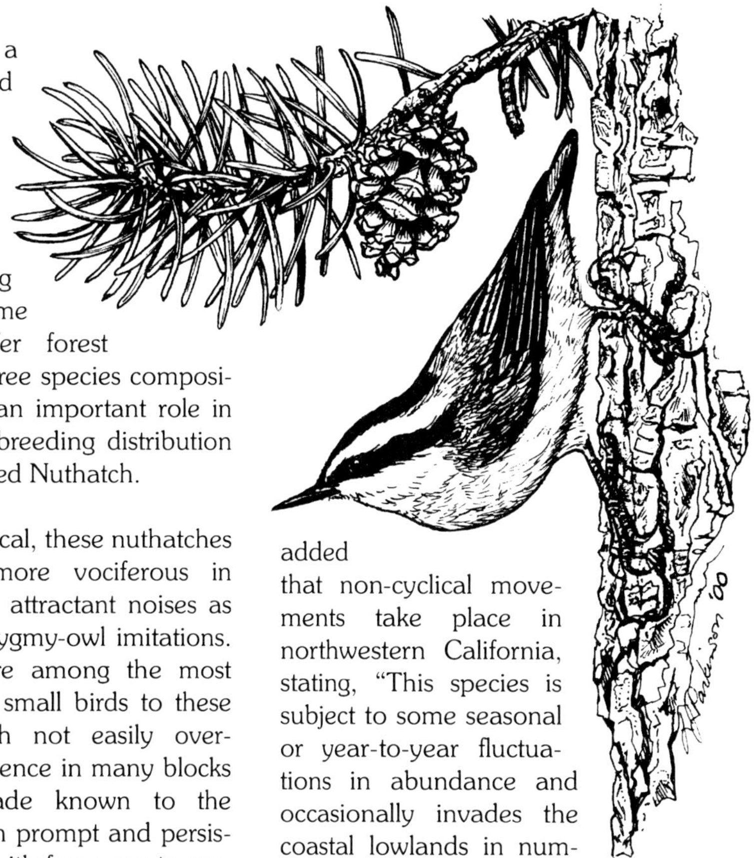

Typically quite vocal, these nuthatches become even more vociferous in response to such attractant noises as "spishing" and pygmy-owl imitations. Indeed, these are among the most responsive of all small birds to these sounds. Although not easily overlooked, their presence in many blocks was initially made known to the observers through prompt and persistent complaints, with from one to several "tiny tin horns" heard close at hand in the forest canopy. Confirmed breeding was reported from 31 blocks, probable breeding from 55. The earliest confirmation was of an occupied nest 8 Apr 1998 (J11; D. Kriger). The latest active nest was one with young 23 Jun 1998 (q07; GCH), and the latest confirmation was of food-carrying 30 Jul 1997 (m19; COg).

While present every day of the year in Humboldt County, Red-breasted Nuthatches are thought to be migratory as well. Surely, the variable fall influxes noted in regions of the U.S. where the species does not normally breed would seem difficult to detect in heavily forested Humboldt County. Nevertheless, the appearance of individuals and very small groups of these birds out of expected haunts in Humboldt has been remarked upon. Yocom and Harris (1975) stated that this species was an "uncommon summer visitant and breeder" but a "common migrant and winter visitant" in northwestern California. Those authors also noted that the species ranges from scarce or absent to abundant in fall and winter. Harris (1996) added that non-cyclical movements take place in northwestern California, stating, "This species is subject to some seasonal or year-to-year fluctuations in abundance and occasionally invades the coastal lowlands in numbers, Sep–Oct." During the Atlas period, a few nuthatches were seen in places such as the isolated cypress and willow thickets on the north spit of Humboldt Bay, in shore pine stands of the coastal dunelands, in exotic Monterey pines at the Arcata Marsh, and elsewhere out of expected breeding habitat in the bottomlands. To what extent the population is actually migratory through the county is not well understood.

As is the case for other birds dependent chiefly upon dead or infirm trees for nesting sites, the way in which forests are managed bears directly on the welfare of Red-breasted Nuthatches. While still relatively abundant, this species may have decreased in number due to extensive timber harvest (Marcot 1984). Salvage, "sanitation", and thinning cuts that remove trees in the canopy and understory of maturing timber stands diminish nesting opportunities. Accordingly, retaining non-crop trees during timber harvest probably will help maintain habitat suitable for this species.

WHITE-BREASTED NUTHATCH *Sitta carolinensis*

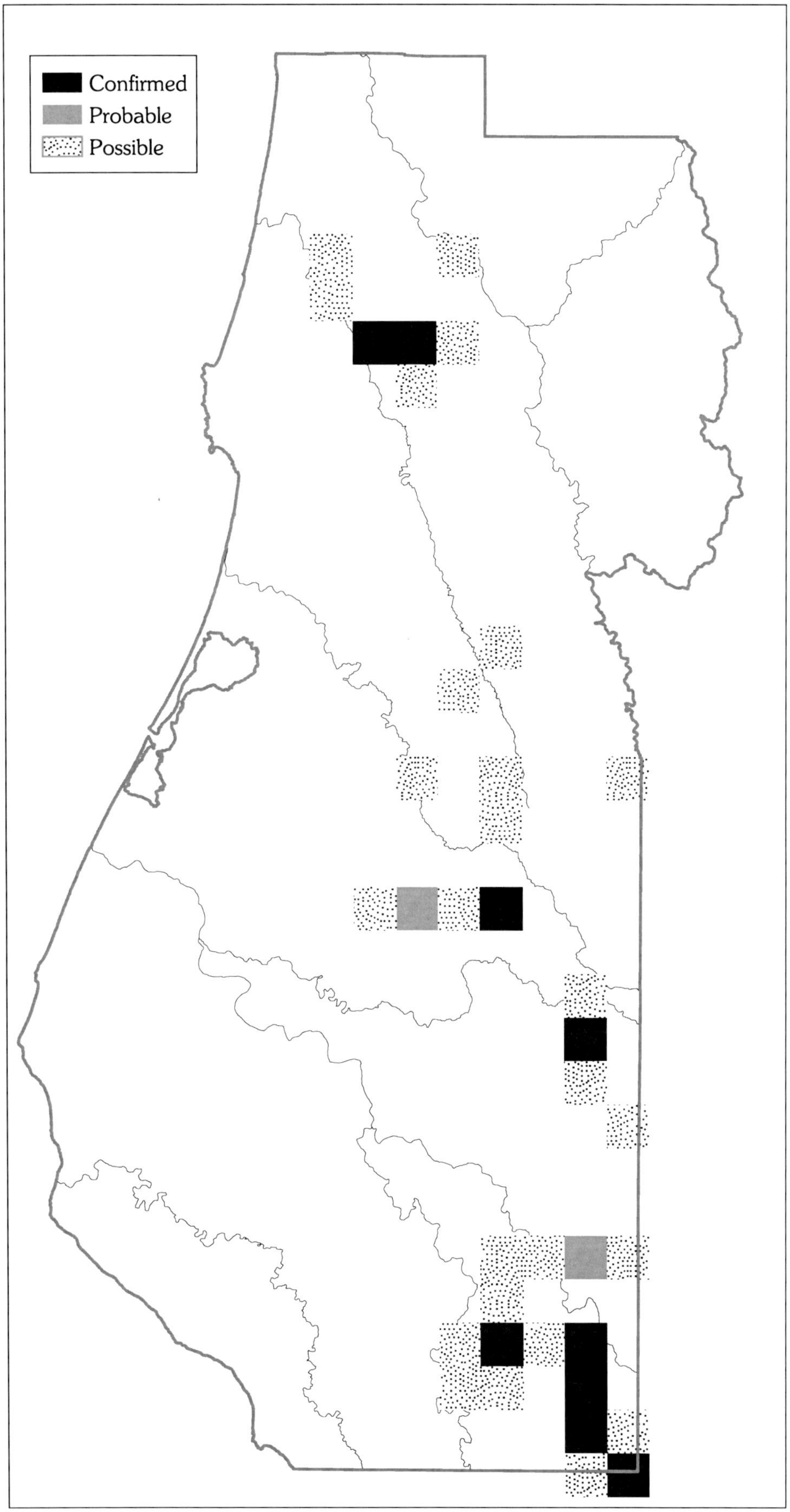

Only a person with the hardest of hearts could watch a pair of White-breasted Nuthatches work over an old oak and fail to feel a sense of fondness for them. The sight of their chubby, wry-necked forms, jinking incrementally down a rough-barked trunk or along gnarled limbs, causes one to marvel at the very different manifestations beauty in living beings can take. As form follows function, so do the compact shape, snubbed bill, short legs, and delicately curved claws best equip these tree-dwellers for a life spent exploring bark crevices for insects and spiders. Throughout their range, White-breasted Nuthatches are most frequently encountered in open forests, parklands, woodlands, and groves characterized by large trees with full crowns and well developed limbs. In California, they are resident in stands of various pines and oaks, as well as in Fremont cottonwoods along streamsides in the Central Valley.

Within the extensive and variegated forest that clothes so much of Humboldt County, suitable habitat for White-breasted Nuthatches occurs only spottily. The closed-canopy coast redwood and Douglas-fir/tanoak forests clearly do not attract them. Rather, it is in the zone of California black and Oregon white oak woodlands in the lower and middle elevations of interior Humboldt that the birds may be encountered most reliably. Large, mature oaks are present at each of the sites supporting these nuthatches.

During the Atlas period, White-breasted Nuthatches were detected in 37 (9%) of the 425 total blocks. They were confirmed breeding in nine blocks and were established as probable breeders in but two others. While this species is probably migratory within some portions of its northwestern California range (*e. g.,* in Redwood National and State parks; R. Erickson, pers. comm.), the possible breeding records suggest that resident birds may nest in those blocks. Areas within Humboldt supporting White-breasted Nuthatches comprise a discontinuous but reasonably extensive collection of sites, stretching from well northwestward in the county through portions of the interior and into the southeastern corner. Spread across a considerable area, these places differ somewhat in appearance.

In the Bald Hills, a disjunct region of wind-blown oak woodlands in the southeastern part of Redwood National Park, nuthatches were found mostly on southern exposures on the north side of Redwood Creek that are grown almost exclusively to Oregon white oak. While impressive and well developed, these oak groves stand within the coastal fog belt and face encroachment by Douglas-firs, which readily seed-in from surrounding conifer forests. In recognition of this, the National Park Service has, in recent years, undertaken a program of oak woodland enhancement by attempting to eliminate Douglas-fir saplings. White-breasted Nuthatches were detected in seven blocks in the general Bald Hills region, and were confirmed breeding in two blocks. An occupied nest was noted shortly south of Schoolhouse Peak 3 May 1998 (j08; TWL), and adults carrying food were seen a few miles to the west at Childs Hill Prairie 11 Jun 1998 (i08; COg). A bird in the westernmost portion of the Bald Hills (h06) 27 Jun 1995 (M. Maxwell) furnished the most westerly report during the Atlas period, only about 6 miles inland from the coast.

Oak woodlands occupying rolling upper mid-slopes and ridges in the central Mad River and lower Van Duzen River drainages were found to support these nuthatches, in an area totaling 13 blocks extending from Titlow Hill (L15) southwestward to ridges above Middle Fork Yager Creek (i21) and southeastward up the South Fork Van Duzen River (o26). Two breeding confirmations were also received from this loose cluster, with adults feeding young 17 Jun 1997 (L21; PFS) and in block n24 (no date; M. Shreeve). A pair of adults noted along Kneeland–Bridgeville Road shortly north of its junction with Redwood House Road 13 Jul 1999 (J21; DFx) furnished one of only two probable breeding reports. Somewhat near these blocks but in the South Fork Trinity River watershed was a report of possible breeding evidence 6 May 1998 from southeast of Sims Mountain (o18; GCH).

It is in the summer-hot, oak-drenched southeastern corner of Humboldt County that these nuthatches were most consistently encountered. In this section of Humboldt suitable habitat is widespread and well developed, comprising any number of variations on a theme. Here, large old oaks stand either in open woodlands, in groves or clumps, or at the edges of mixed-species forests. Within this area was a cluster of 16 blocks with detections, including five with confirmed breeding: an occupied nest 17 May 1997 along the Alderpoint road (N31; M. & M. Bailey) and another 21 May 1995 east of New Harris (N33; C. Secor); adults carrying food on the upper slopes of Island Mountain 28 May 1999 (o34; JEH); adults feeding young 1 Jun 1996 east of Harris (n32; M. Bailey); and a nest with young 22 Jun 1998 along the Alderpoint road (L31; S. Rovell). Finally, a pair seen at Fort Seward 15 Jun 1997 (N29; RJA) provided the only other Atlas-period record indicating probable breeding by this species.

Stands of ponderosa and Jeffrey pine in Humboldt County do not appear to attract breeding White-breasted Nuthatches. An occasional winter bird is reported from the Jeffrey pine stands atop Horse Mountain during the Willow Creek CBCs; *e. g.*, a bird was seen at about 4600 feet near Horse Mountain on Titlow Hill Road 26 Dec 1993 (G. & L. Lester). However, no summering birds have been found there, nor in other pine woodlands in the county. Similarly, while cottonwood and valley oak riparian habitat farther into California's interior is well known as nuthatch habitat, that association does not occur here. Although plenty of black cottonwood occurs along alluvial flats and in stream corridors locally within Humboldt, this tree also would seem to hold no attraction for these birds.

Harris (1996) described White-breasted Nuthatch in northwestern California as an "uncommon to common resident and breeder in the southern and eastern portions of the region. Generally associated with woodlands having a substantial oak component." Atlas results indicate that this characterization continues to apply to interior Humboldt County. It is of interest that the first mention in the ornithological record for this species was of a specimen collected in November or December 1885 in Eureka (Townsend 1887), where this species would seem decidedly out of place.

No threats from human activity would seem to attend the future of White-breasted Nuthatches in Humboldt. However, the potential spread and mortal effects of Sudden Oak Death Syndrome are of concern to many regional biologists. Whether or not this disease will ultimately infect oaks and other trees over a broad area of Humboldt County remains to be seen.

BROWN CREEPER *Certhia americana*

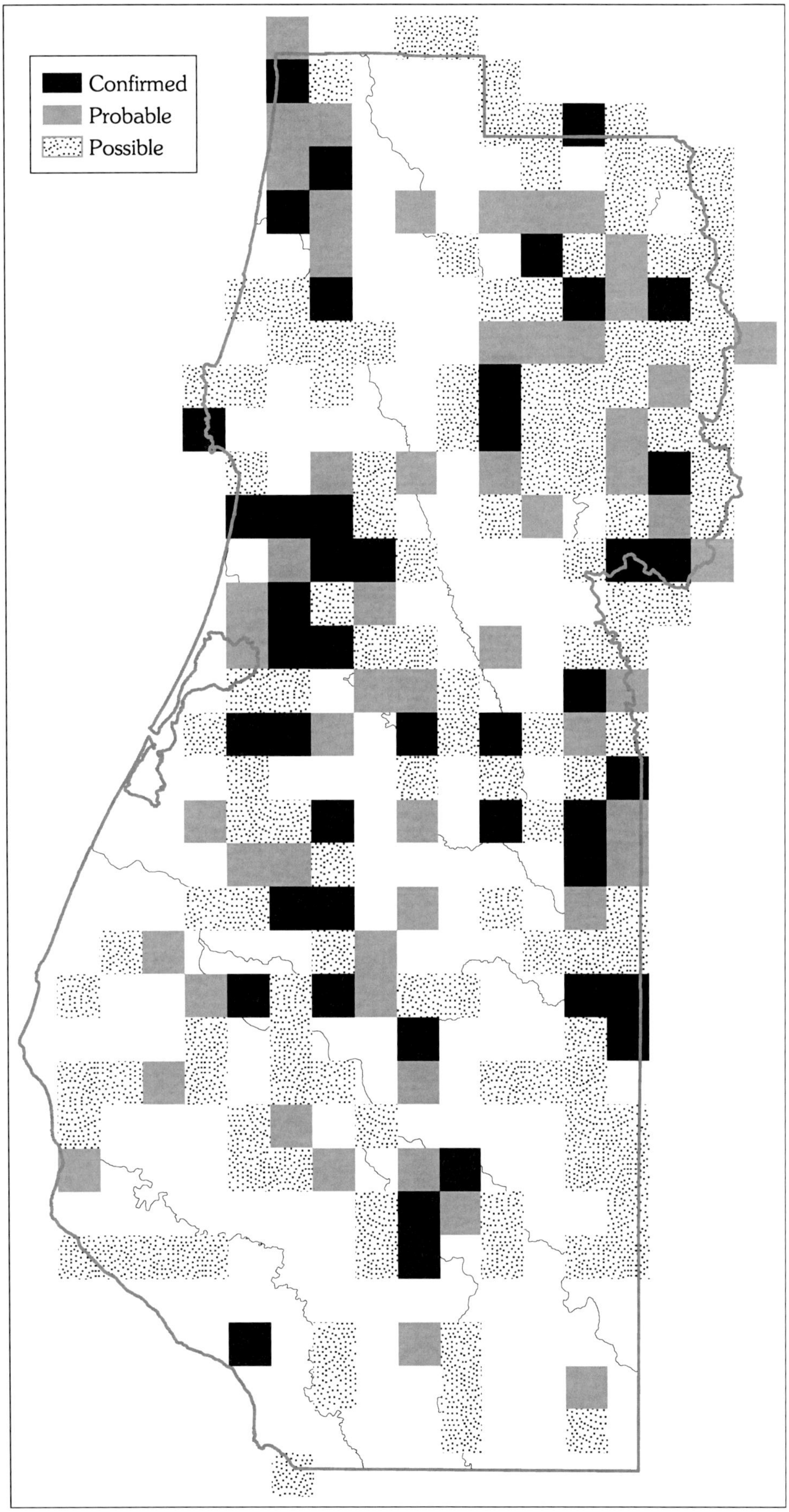

The Brown Creeper is an uncommon resident and breeding species in northwestern California (Harris 1996). It is widely distributed across most of Humboldt County. Because so much of the county is forested, it follows that these tree-hugging birds might be encountered nearly anywhere that stands of trees exist. Indeed, Atlas fieldwork indicated that creepers are present in many of the tree communities recognized within the county—as typifies the species' habitat choices at other points within its extensive North American range. Atlas blocks in large expanses of commercial timberland, both privately and publicly held, contributed a distinctly large proportion of detections, as did the redwood parks. During the Atlas period, Brown Creepers were reported in 225 (53%) of the 425 total blocks. Breeding was confirmed in 44 blocks and considered probable in 56 others.

Habitat occupied by Brown Creepers varies, but stands of mature or maturing trees of reasonably large diameter are preferred. Standing volume of timber is probably as good a predictor as any for the presence of Brown Creepers. Ordinarily, conifers dominate forests in which creepers are found; however, hardwood stands dotted here and there with conifers are also used. Old-growth, mature, and unmanaged or lightly managed stands of coast redwood and Douglas-fir commonly are inhabited at lower and middle elevations. Coniferous forest with white or red fir, incense-cedar, and Jeffrey pine is used at higher elevations. These birds are also resident in Eel River delta riparia, where densities are higher in areas with larger trees (Kelly 1987).

Brown Creepers are present throughout the year in Humboldt County. However, some migration or seasonal shifts occur in September and October (Harris 1996). From about late September into March, small numbers of these birds may be encountered in riparian woods, the more open or structurally simple oak woodlands, orchards, small woodlots, neighborhood parks, and backyards—all places in which they are not typically found nesting. In such situations, one to several birds keep loose company with flocks of chickadees, kinglets, warblers, Red-breasted Nuthatches, Hutton's Vireos, and Downy Woodpeckers.

Creepers forage on the trunks of trees, hitching their way steadily upward, inspecting the bark surface and crevices for adult insects, spiders, and other invertebrates as well as their eggs and pupae. They appear distinctly partial to trees whose bark is corrugated; trees of any size with comparatively featureless bark seem to hold but brief appeal for them. Though creepers are not averse to feeding on dead trees, they are not often seen searching for food on barkless or "buckskin" snags.

Atlas results indicate a consistent presence of Brown Creepers across the more thoroughly wooded sections of Humboldt County. Fewer blocks with detections occurred in the extreme north-central, interior east-central, southeastern, and interior southwestern parts of the county. The seemingly patchy distribution in far southern Humboldt suggests that less habitat exists in that warmer, less heavily forested, more oak-dominated countryside than in the well-watered northern and mountain areas. In redwood stands, Brown Creepers occur at significantly lower densities near urban edges than in forest interiors (Brand 1998). The open pasturelands of the Eel River floodplain, the Humboldt Bay lowlands, and the coastal dunelands and sandspits offer little habitat, nor do the expanses of oak-dotted grassland of east-central county. A cluster of blocks without detections in the extreme north-central portion of Humboldt takes in a region of recently clearcut and immature regrowth timber too young to support these birds. Creepers have probably decreased in number and distribution in Humboldt as a result of clearcutting (Marcott 1984).

Considering the inconspicuous nature of this species, Atlas fieldwork brought to light a surprising number of confirmed and probable breeding reports—especially as a proportion of the total detections. Brown Creepers were confirmed or probable breeders in 44% of the blocks in which they were reported. In contrast to some other forest birds for which patience and time are required to observe breeding behaviors, there is no distinct pattern of increased breeding evidence attributable to more intensive survey effort in priority blocks. The earliest confirmations were nest building 22 Apr 1998 (I10; *fide* JMH) and an occupied nest 25 Apr 1996 (P13; TWL). The first fledgling, being fed by an adult, was reported 18 Jun 1996 (g17; RJA). The latest breeding confirmations were made by observations of adults feeding young 26 Jul 1997 h04; GCH) and carrying food 10 Aug 1996 (J17; C. Keil).

ROCK WREN *Salpinctes obsoletus*

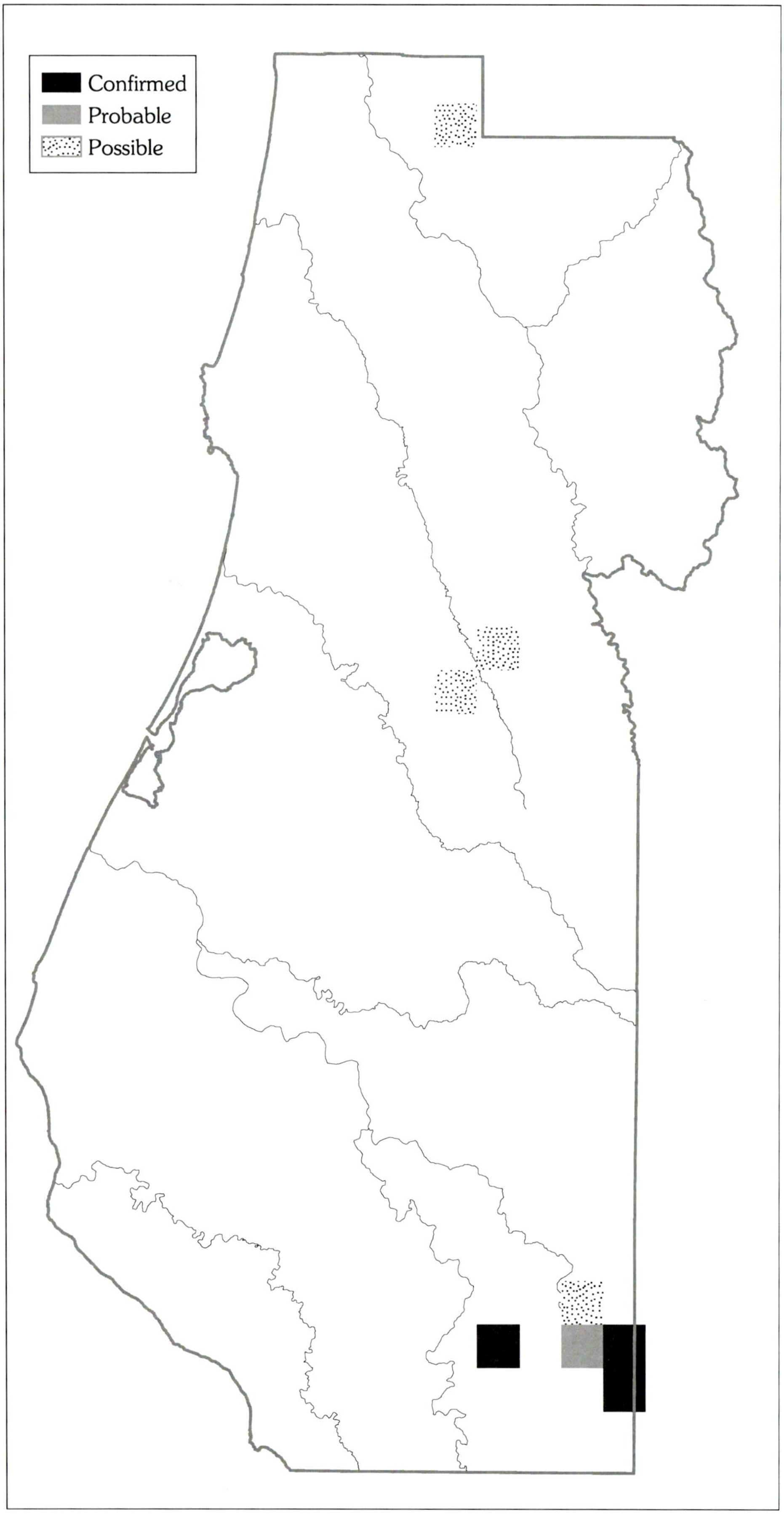

The sixty-some wrens of the family Troglodytidae are a remarkable bunch. Dressed mostly in shades of brown or russet, they make up for plain-Jane looks with peppy vitality and, in the case of some species, voices of extraordinary beauty. They are chiefly birds of the New World tropics. These birds have attained specializations of form and behavior enabling them to live in a great variety of terrestrial and wetland habitats. Among their ranks, the Rock Wren is notable for inhabiting landscapes where fractured surface rock offers what it requires to get along in the world—with whatever vegetation that may be present appearing to afford little more than green scenery. Badlands, mesas, rimrocks, talus slopes, and other hardscrabble terrain characterized by stony or otherwise exposed ground offers a special appeal to them. This is a fairly common species in parts of California offering suitable habitat. In sections lacking it, Rock Wrens are, as a rule, rare or unknown.

Nearly all of Humboldt County lacks appropriate habitat for Rock Wrens. However, there are several places where these birds can be looked for with a reasonable chance of success. In the course of Atlas fieldwork, they were detected in eight (2%) of the 425 blocks. Given the local scarcity of this species, it was a stamp of achievement for atlasers that breeding was confirmed in as many as three blocks.

It is only in southeastern Humboldt that one has a reasonably good chance of encountering Rock Wrens. Even there, the birds are very uncommon and appear restricted to specific places providing quality rocky ground. Well known to birders in the county are the informally- and appropriately-named "Rock Wren Rocks" (L31), a few miles east of Garberville on the road toward Alderpoint. Breeding had previously been confirmed there when a nest and young were seen during May and June 1974 (R. Erickson *et al.*; Yocom and Harris

1975). By the 1990s this location had become recognized as the best and most accessible spot at which to search for them. Breeding at this site was confirmed by an adult carrying nest material, then food, into a nest at the main rock, shortly above the Alderpoint road, 7 May 1995 (JEH, TWL, GCH, JG, K. Dyer *et al.*). Habitat at the Rock Wren Rocks comprises several mid-slope outcrops of various sizes, each shortly separated from the next and set in a sweeping expanse of grassy hillsides. Not surprisingly, slope aspect at the site is southerly. The open nature of the place also attracts numbers of Lark and Grasshopper sparrows, both of which commonly use the main rocks and satellite boulders as song perches.

Habitat in a different setting occurs some miles to the east along the Eel River. Breeding confirmations were made in two adjacent blocks through the observation of an adult carrying a fecal sac 24 May 1998 (o31) and another carrying food 30 May 1998 (o32; both JEH *et al.*). Along this same stretch of river another Rock Wren was noted singing on territory 24–30 May 1998 (N31; JEH *et al.*). These wrens had previously been confirmed breeding along this section of river (Humboldt's first historical breeding record), when four singing birds and one adult feeding a juvenile were recorded 3 Jun 1973 (R. Willmarth, T. Schulenberg; Yocom and Harris 1975). Wrens noted in these blocks were in the shrub-dotted riverbank boulders, shelf outcrops, and small scree slopes that enfold the main stem Eel River for many miles in this area. It was apparent to the atlasers that fair numbers of Rock Wrens probably were present throughout this reach of streambank, extending upriver to and beyond the nearby Trinity County line. Such habitat also continues downstream from the Alderpoint area discontinuously to at least the vicinity of Eel Rock (*fide* G. Bloomfield). This suggests that intensive searches may show that the species is rather well represented along the Eel River within perhaps a 20-mile corridor within Humboldt County. Access to nearly all suitable habitat is restricted; either a river float trip or a grueling trek along disused Northwestern Pacific Railroad tracks offers the only possible means to explore this fascinating corner of the county. The other Atlas record from southern Humboldt was of a single bird detected 15 Jun 1997 near Alderpoint (n30; RJA).

Elsewhere in the county, single Rock Wrens were found at three sites, each in isolated scraps of habitat dozens of miles north of the "SoHum" cluster. One was in marginal, very shrubby habitat at the end of the longest spur off Forest Service road 6N36 east of the summit of Horse Mountain (L15) 1 Jul 1997 (DFx). This site supported sparse montane pine forest with an understory of huckleberry oak, silk tassel, *Ceanothus*, and several species of manzanita, interspersed with small patches of exposed red rock. While giving call-notes, the bird did not sing nor evince other suggestive behaviors.

Another Rock Wren was at a very small ridge-top outcrop adjacent to a cattleguard framed by two fence-jacks filled with rock on Snow Camp Mountain (k16) 11 May 1997 (RJA). As indicated by the chronology of the confirmed breeding reports collected during the Atlas period, the date is well within the nesting season. However, it falls within a period during which an extremely few migrant Rock Wrens have appeared out of expected haunts in the county, as well as across the forested Pacific Northwest. Harris (1996) remarked upon this phenomenon, mentioning eight records of spring vagrants 17 Apr–21 Jun in atypical habitats such as logjams, barn roofs, parking lots, *etc.*, so perhaps the individual on Snow Camp represented a bird still on the move. A bird that was almost certainly a migrant was encountered 23 Apr 1995 along Hufford Road in the Redwood Creek estuary (f05; DFx, JCP *et al.*).

The remaining block with possible breeding in north county involved a Rock Wren singing from down trees and upended rootwads 24 Jul 1999 in a recent clearcut in hills well above the lower Klamath River (k03; DFx, JCP). The clearcut had been prescription-burned, but it burned too hot, with loss of preexisting organic material smaller than log- or stump-sized; much reddish soil was left exposed. It is worth noting that a few pairs of Rock Wrens have nested in recent clearcuts in the Cascades and Coast Range of Oregon (Gilligan *et al.* 1994) and, occasionally, Washington during the modern logging era. These areas are burned to facilitate planting of conifer seedlings. Such places are typically not inherently dry or stony; their attraction springs from the removal of vegetation and the consequent exposure of large down woody material, stumps, open ground, and the generally minimal surface rock. Slope aspect is often, but not always, sunny. The site in block k03 faced northwest and lies within the zone of summer overcast. Indeed, in the case described above, the brief observation period ended when the intruding marine layer lowered, enveloping the two atlasers in fog!

Harris (1996) called Rock Wren a "locally rare summer resident and breeder in the southeast portion of the region, east of Highway 101 and south of Highway 36. Casual vagrant elsewhere in spring, fall, and winter." He stated that the birds were known to breed east of Garberville and along the Eel River from Alderpoint southward (as discussed above), as well as at rock outcrops along Mail Ridge near the town of Harris. The above-mentioned sightings in central and northern Humboldt are of interest, suggesting summering or breeding well north of Highway 36, but otherwise no change to the status described by Harris (1996) was indicated by Atlas results.

CANYON WREN *Catherpes mexicanus*

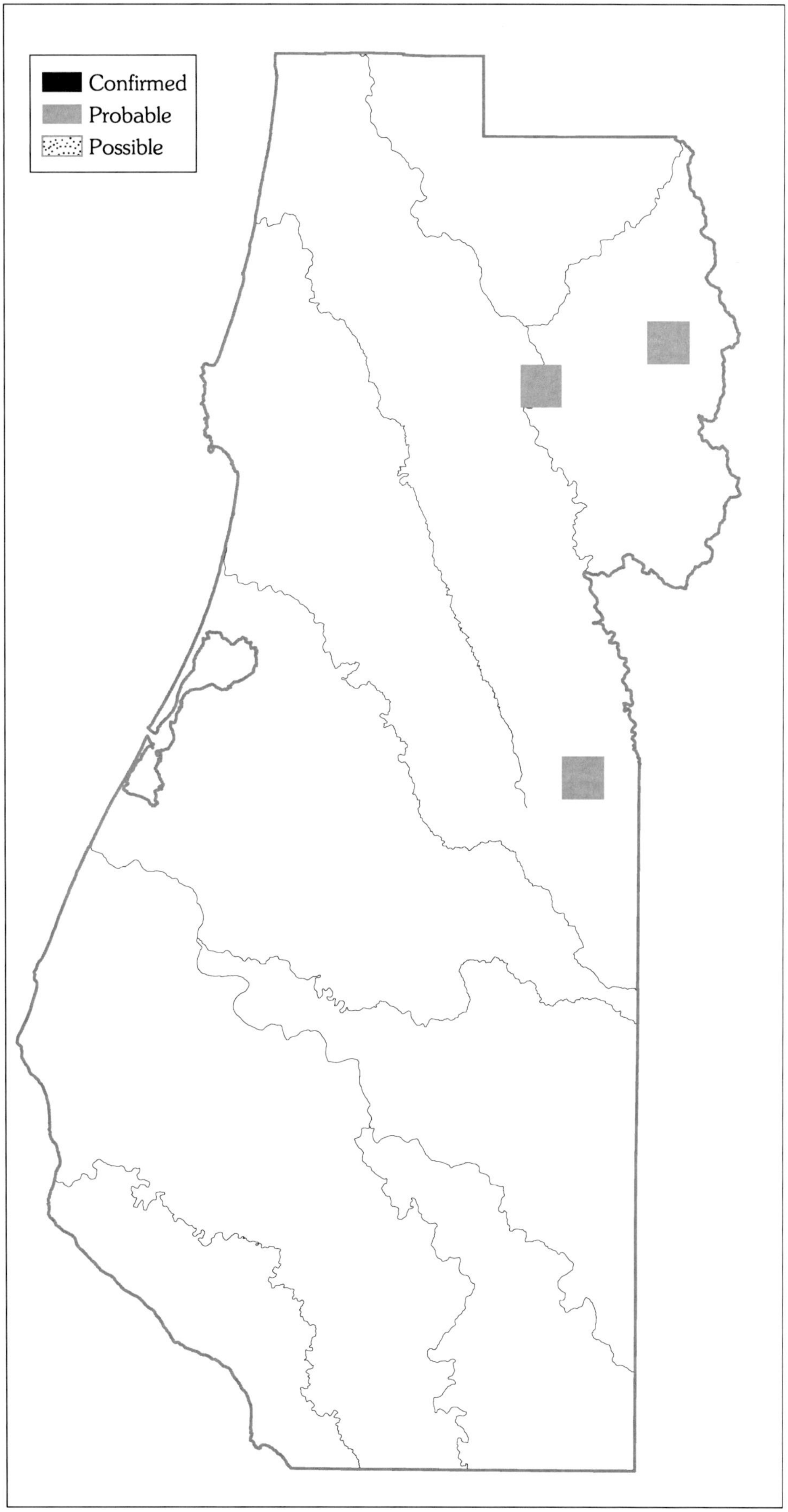

One's first awareness of a nearby Canyon Wren is seldom achieved with vision. Where there had been no wren seen, suddenly from the canyon wall one announces its presence with "clear curved notes tripping down the scale" (R. T. Peterson). After a long moment there comes another song, and then a Dipper-like *zink* note is perceived nearer at hand among jumbled and crowded boulders. Flat but oddly penetrating, the noise would seem as likely to be the call of some squirrel as that of a bird. A tossed fragment of orange peel flipped into the deep shadows between two rocks startles the unseen forager. As it pops into view and flits briefly to the crest of a talus block, the source of both sounds is revealed to be a flat-crowned little bird with a slender bill, colored like old madrone bark—save for its shining white breast. A last call-note is heard as the bird's curiosity impels it into another shaded grotto and a resumption of its search for invertebrates.

Canyon Wrens are widespread in Mexico, the western U.S., and southwestern Canada, but because suitable habitat is scarce they are seldom encountered in Humboldt County. The rock underlying much of the area generally does not project from slopes in massive fashion, instead poking above the soil as outcrops that are generally small and isolated. Truly large cliffs or vertical outcroppings are scarce, and nearly all of those are difficult or impossible to get close to. Among the various common bird species whose normal ranges barely intrude into the county, Canyon Wren is among the rarest and most difficult to locate. Only a very few have actually been seen in Humboldt. Nearly all detections have been of songs given by birds well out of sight on cliffs or down in some inner gorge.

Humboldt County's first and only confirmed breeding record of Canyon Wren was of a bird seen carrying food to a nest 10 Jun 1987 in the Weitchpec Bluffs (R. Erickson, G. Lester).

Other pre-Atlas breeding season locations within Humboldt include Whiteys Peak, Bluff Creek, and along the South Fork Trinity River across from Todd Ranch. In addition, there are records of a Canyon Wren at the mouth of Horse Linto Creek along the Trinity River 29 Aug 1973, four birds at that location 15 Sep 1974, and a bird in the Trinity River canyon four miles north of Willow Creek 24 Aug 1974 (Yocom and Harris 1975). Harris (1996) reported a seeming spring overshoot or emigrant along the coast near the mouth of Redwood Creek 9 Apr 1988 (M. Morris), and summarized the Canyon Wren as a "locally rare summer resident and breeder; less common in winter" in northwestern California. Other than the Weitchpec Bluffs site, no reports were received during the Atlas period from the other historical locations.

Capitalizing on a readily accessed possibility, atlasers made infrequent visits to the top of a cliff extending downward out of sight to the Trinity River from Highway 96, several miles north of Hoopa (m09). This site—locally known as Weitchpec Bluffs—had become recognized during the 1990s as the best place in the county to try to find these wrens. However, as it was impossible to see more than a very small portion of the available habitat from the road, another approach was taken. On 30 Jun 1999 a group of atlasers (TWL, D. Pagen, JEH) took a raft trip down the tortuous gorge of the Trinity to immediately below the cliff in question. Alas, their hopes of seeing some evidence of nesting activity while passing the site were left unrealized. However, probable breeding status had been established there on the basis of a male singing 29 Apr–4 May 1995 (B. Deuel, JEH), 28 Apr–11 Jun 1996 (DFx, JMH *et al.*), and 3 May 1998 (W. & M. Widdowson, KMS, RHw).

There were only two other Atlas period reports of Canyon Wrens in Humboldt County, and both of those also indicated probable breeding on the basis of birds singing over a period of more than a week. One was heard 3 May–3 Jul 1998 at a rocky outcrop on Packsaddle Ridge (p08; J. Ward, J. Osborn) and another was heard 20 May–26 Jul 1998 along Grouse Creek (n18; P. Herrera). Both sites represented new locations for this species in Humboldt.

BEWICK'S WREN *Thyromanes bewickii*

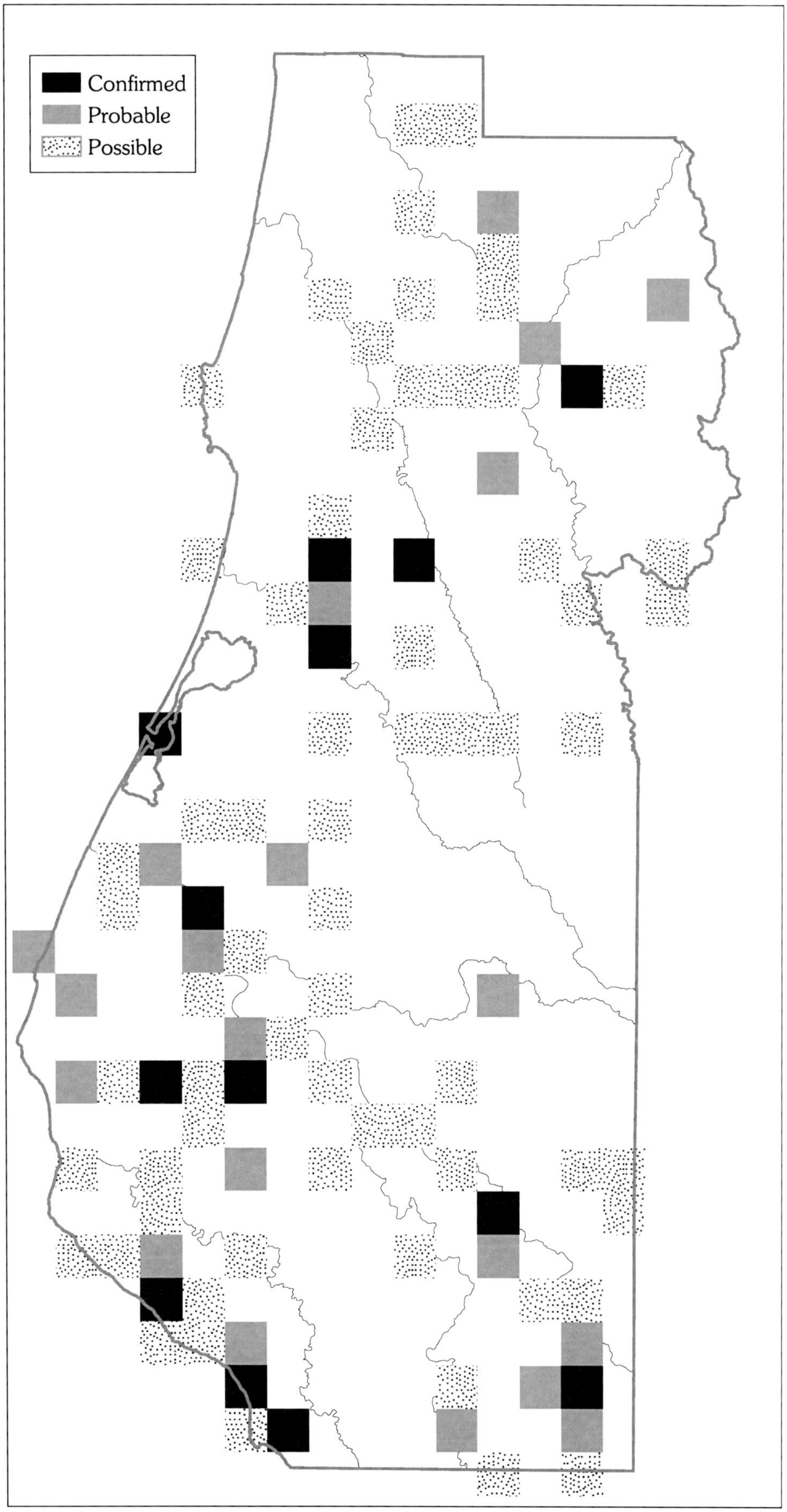

While agitated scolding or boisterous singing often marks the presence of a Bewick's Wren, most birders can attest that skulking in dense tangles is also part of its behavioral repertoire. Although it is an uncommon to common resident and breeder in northwestern California (Harris 1996), Bewick's Wren distribution does not appear to follow a consistent geographic pattern. This may be due to the somewhat local distribution of structurally suitable habitat. During the Atlas period, these wrens were reported from 99 (23%) of the 425 total blocks. Breeding was confirmed in 13 blocks and was thought to be probable in 21 others.

Bewick's Wrens typically are found in bushy places that have been either untouched or neglected by the human hand. Such sites range from coastal lupine–willow dune hollows through an array of semi-open lowland situations, eastward into the oak woodlands and Douglas-fir forests, and up into the industrial timberlands at middle elevations. Extensive stretches of semi-open shrubby growth dotted or edged with taller trees and undergrown with continuous low cover are often used. It has been the general impression of active birders in the county that habitat patches of irregular configuration and with a high edge-to-interior ratio appear to support higher densities of Bewick's Wren.

In contrast to many other species, the heavily sampled coastal lowlands did not show high detection levels during Atlas surveys, suggesting low abundance in agricultural and urban habitats found there. However, during the Atlas period, Bewick's Wrens were common in bushy coastal habitats composed of coyote brush, *Ceanothus*, willows, and lupine. Cottonwood- or alder-lined streams and floodplain riparian habitat in which rambling low brush or clumps of young trees are extensively undergrown with blackberry were widely populated, especially when protected from grazing cattle. These wrens were also often found at bushy disused millponds.

At lower and middle elevations, Bewick's Wrens colonize clearcuts with suitable brush regrowth until saplings choke out the early-seral brush species. A study in northwestern California found that these wrens were more common in clearcuts where snags were retained (Marcot 1983). In the Klamath Mountains, this species may have increased as much as 242% as a result of logging (Raphael *et al.* 1988). Bewick's Wrens were few in the northern coast redwoods, coniferous forests in the mountains of northeast county and along the South Fork Mountain ridge, and oak savannahs dominated by grasses in the east-central portion of the county. However, outside of the redwood zone, atlasers accessed many of these areas chiefly by ridge-top or sidehill roads, which often left bottomland habitats in these blocks unsurveyed. Bewick's Wrens were typically not found in mountain brushfields where winter snow persists.

The rather modest representation of blocks where breeding was confirmed (13% of all blocks with detections) suggests that these birds were not especially easy to confirm, possibly owing to the effort required to observe an unagitated Bewick's Wren more than momentarily in the dense tangles they prefer. Indicating an extended breeding season in Humboldt was nest building observed 3 Apr 1996 (n32; M. Bailey) and an occupied nest 30 Jul 1998 (g33; TWL). The earliest fledgling reported was seen 29 May 1996 (e21; P. Herrera).

HOUSE WREN *Troglodytes aedon*

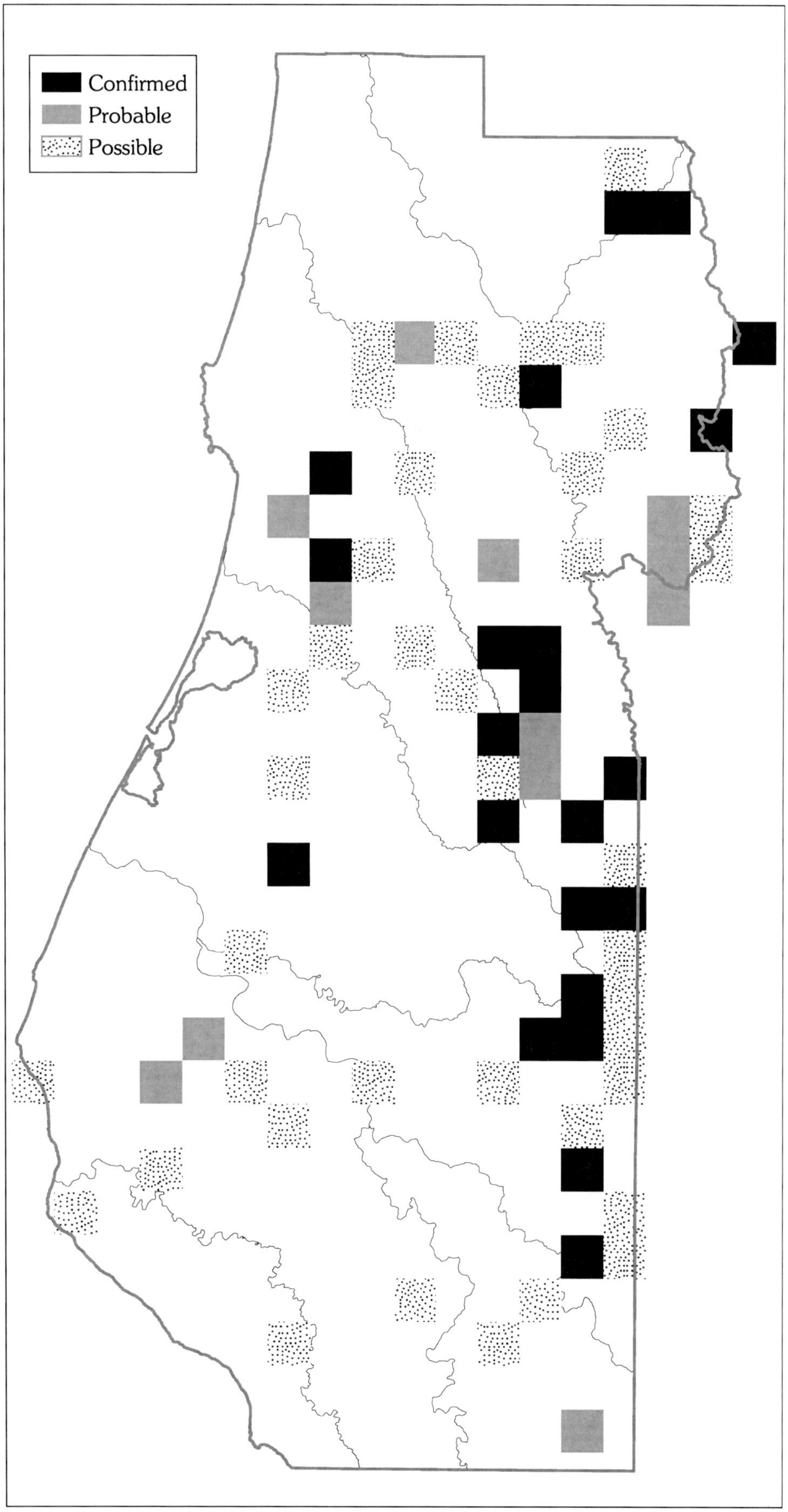

House Wrens are fairly common breeders in Humboldt County, yet their songs are not often heard because of the exceptionally rural nature of their preferred habitat. In contrast to many regions of the West, House Wrens typically are not backyard birds here and are generally unknown, even as migrants, in the more developed lowlands. During the Atlas period, they were found in 74 (17%) of the 425 total blocks.

This wren is distributed in Humboldt chiefly in the higher foothills and mountains well east of the coastal fog belt. It is often found in clearcuts with ample large woody debris and snags in the Douglas-fir and true fir forest zones. Such situations can be found in a broad area extending from northeast county south along the Horse Mountain–South Fork Mountain ridge and adjacent ridges and upper slopes. During extensive point counts in the Klamath Mountains, the white fir vegetation series had the highest proportion of sites with presence of this species (Sniado 1999).

In Douglas-fir clearcuts, Marcot (1984) found House Wrens at their highest densities in the grass/forbs stage. The widespread use of clearcuts by these wrens may be due to two factors. First, clearcuts often afford large expanses of low, patchy, semi-open, and minimally vegetated ground, a situation that the wrens favor locally. Second, the snags and other woody material that often remain after logging provided nest sites for this secondary cavity nester (Marcot 1983). Suitable habitat resulting from timber harvest may be diminished, however, through minimal retention of standing green trees, rapid revegetation, or both. Numbers of this wren may have increased more than 200% because of timber harvest in interior northwestern California (Raphael *et al.* 1988).

Scattered pairs of House Wrens also summer in brushy openings at middle elevations not far east of the zone of overcast. They were noted during the Atlas period in the upper Bear River/ Mattole area, in parcels of intensively managed private timberland southeast and northeast of Humboldt Bay, and sparingly in rural open country elsewhere. Physical structure at these sites was similar to that found in the higher mountains. These wrens were unreported in the northern coast redwoods and in outlying commercial timberland in the northernmost portion of the county. They were generally absent from redwood clearcuts; Hazard and George (1999) did not record any in those they surveyed. Some of the county's House Wrens spend the breeding season in open oak woodland and around the margins of overgrown hillside pastures in the interior. In contrast to their preferences in other regions of the West, they appear to show no perceptible preference for riparia in Humboldt County.

A House Wren exploring a probable nest site 25 Apr 1996 (P13; TWL) was one of the earliest Atlas records; another doing the same 24 May 1996 (N33; C. Secor) was temporally more expected. Most of the 22 breeding confirmations were in June and July. Temporal outliers were an occupied nest 28 May 1997 (m09) and a bird feeding young 17 Aug 1996 (q10; both *fide* JMH). The earliest fledgling was reported 23 Jun 1995 (N19; GAS).

The arrival of migrant or summering House Wrens is an unheralded event, taking place sometime from mid-April to early May. Most of the summer residents have departed by September. Apparent fall migrants well away from breeding sites are almost never reported. There have been about a dozen winter reports for northwestern California (Harris 1996).

WINTER WREN *Troglodytes troglodytes*

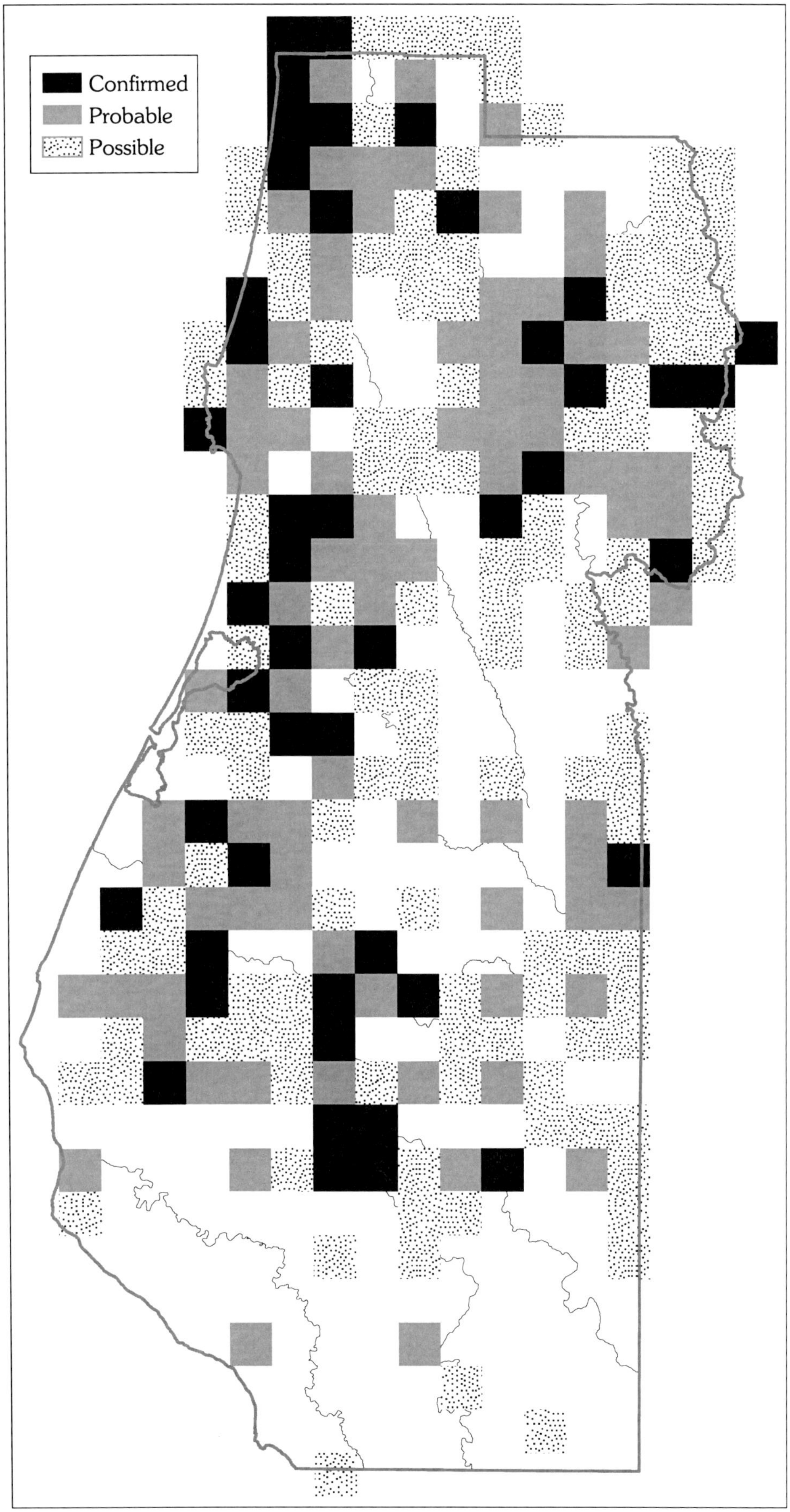

The exuberant songs of Winter Wrens are widely heard across much of forested Humboldt County in spring. For nesting, this wren demands dense overhead cover and a complex forest floor mosaic of down trees and interconnected tangles of shady-site brush. Luxuriant mosses and ferns, columns of sunlight, and the sound of a nearby stream typify locations at which Winter Wrens are commonly encountered. This species is generally common wherever it occurs in the breeding season, although in many places habitat is somewhat localized with respect to landform and slope aspect. Over the 5 Atlas years, Winter Wrens were found in 248 (58%) of the 425 total blocks and were confirmed breeding in 47.

Winter Wrens have long been known to be abundant in the coast redwood zone, especially in less-managed forests and regrowth of the fog belt, as well as inland in appropriate habitat. This wren becomes familiar to each new generation of naturalists tramping the wet woods of the north coast. Townsend (1887) wrote that it was "found everywhere about the redwood logging districts of Humboldt County," and Fisher (1902b) reported that it was "the commonest bird in the great redwood forests, where it is thoroughly at home and sings frequently." Harris (1996) stated that this species was a "common breeder and resident in deep, moist woods; especially numerous in lowland coastal forests."

Atlas fieldwork served to determine firmly that, while the greater portion of central and northern Humboldt County supports them, the southern one-quarter of the county and interior east-central Humboldt do not. Generally, the range of Winter Wren in spring and summer includes both protected and managed timberland from the coast or coastal foothill slopes extending broadly into northeastern Humboldt (notably on the Hoopa Valley Indian Reservation and on the Six Rivers National Forest) and from south of Humboldt Bay southeast through the parks and commercial timberlands of the Eel River drainage. Birds also were regularly encountered at higher elevations in east-central county, such as along slopes of the Snow Camp and South Fork Mountain ridges and around the Lassics.

Near the coast and locally inland, forests of redwood, Sitka spruce, Douglas-fir, and grand fir are thoroughly inhabited by Winter Wrens. In the redwood region they are most apparent in older stands (Hazard and George 1999), but they also occur widely in regrowth stands with remnant woody material on the forest floor and a sufficient salal or huckleberry shrub layer. Allen (2002) found that nest success, territory size, and density were similar within the interior of redwood forests and at redwood forest/urban edges. Within the extensive zone of drier Douglas-fir and tanoak of interior Humboldt, these birds favor humid northern exposures, and show a preference for older stands with more conifer tree cover and less hardwood tree cover (Ralph *et al.* 1991). Inland, they can be found along stream headwaters, in perennially damp ravines and in stands well grown to sword fern, and they exhibit a strong preference for riparian areas (Bardolf 2001). They are also a common year-round resident in bottomland cottonwood forests (Kelly 1987) and other riparian habitats. Winter Wren habitat at higher elevations is afforded by dense thickets of young white fir, especially in shady hollows or gulches with low shrubs and constant moisture.

Plainly, the summer-hot interior of Humboldt County's southern tier and the grasslands of the central Mad River drainage offer little or no habitat for breeding Winter Wrens. A few birds were probably present in summer in some of the blocks without detections, although in many others there truly exists no suitable habitat.

Breeding begins early for these residents. Males singing on territory for at least a week were recorded as early as 3 Mar 1995, and multiple males singing within a block were noted as early as 5 Apr 1998 (both g14; both JEH). There were no reports of the carrying of nest material, but nest building was reported 9 Jun 1995 (N23; P. Morey). Reports of food-carrying spanned the dates of 12 May 1996 (c21; TWL) to 20 Jul 1996 (L27; D. Diehl), feeding-young reports spanned 19 May 1996 (H23; C. Powell) to 22 Jul 1997 (f08; GCH), and occupied nest reports spanned 19 May 1996 (g01; M. Kenney) to 3 Jul 1997 (P09; TWL). One report each was received of an adult carrying a fecal sac on 20 Jun 1995 (H05; B. Devlin-Craig) and one of a nest with young 8 Jul 1997 (f20; D. Garrison). The earliest fledgling was seen 18 May 1998 (f14; observer unknown). Davis (1918) briefly discussed the chronology and other aspects of a Winter Wren nest; although the location was not specified, it was probably within Humboldt County.

Many Winter Wrens are found in fall, winter, and early spring outside of breeding habitat. That there must be an annual movement of birds downslope, southward, or both is apparent in the considerable numbers detected in riparia, blackberry patches, dune hollows, and residential neighborhoods from mid-September into March.

MARSH WREN *Cistothorus palustris*

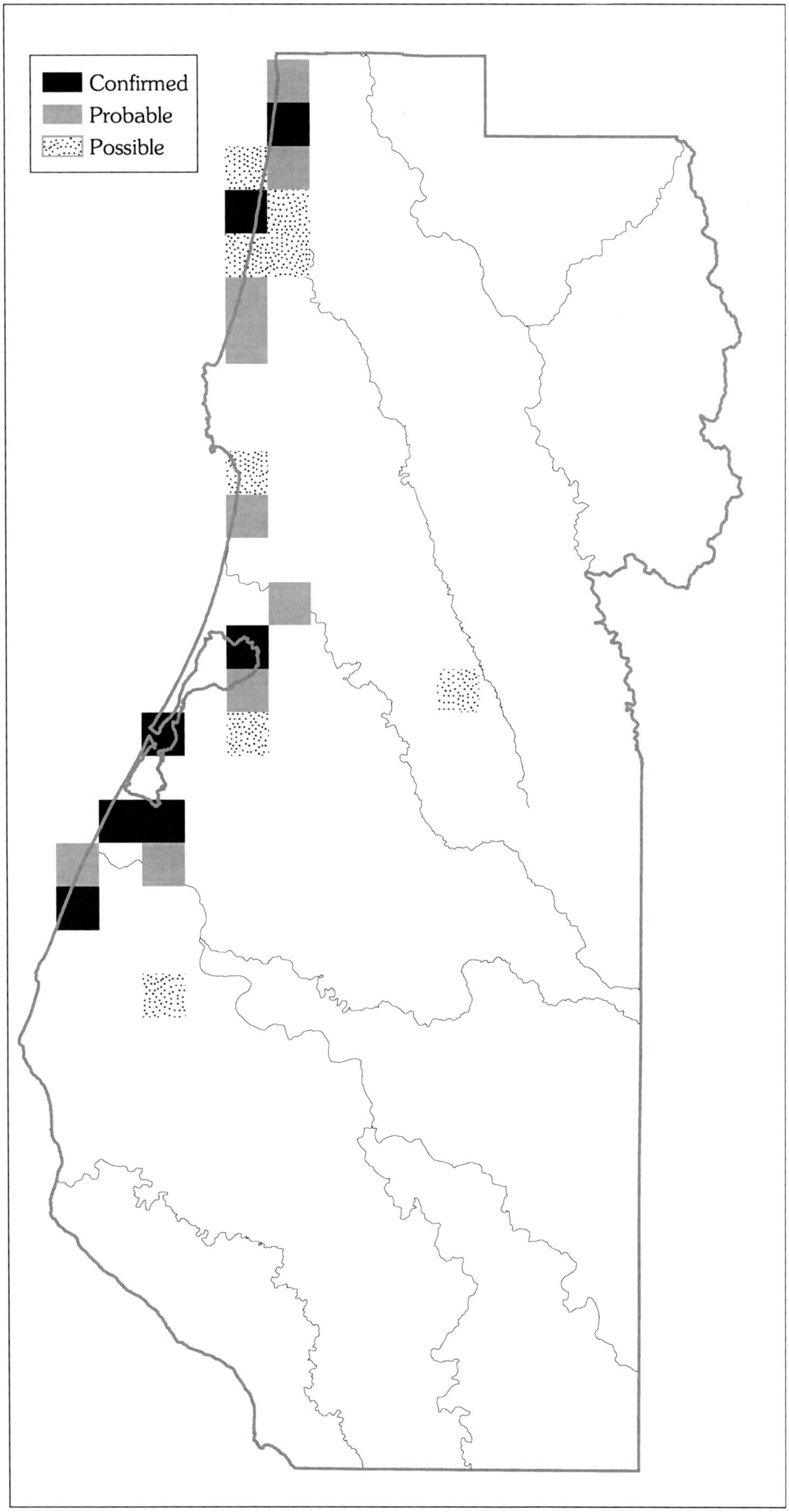

In the marsh, there is stillness. Then a reed twitches. Two reeds now twitch...yet neither begins to bend, nor to betray the nearly weightless fluff of a brown bird in motion among concealing shadows. Always on the go and typically out of sight of predators and people alike, Marsh Wrens signify a piece of ground demanding rubber boots for human passage. Few persons who pass by a marsh on a spring morning are left uncharmed by the exuberant singing of males on adjoining territories.

Harris (1996) considered this wren to be a common resident and breeder in suitable habitat throughout northwestern California. Marsh Wrens may be found in numbers in Humboldt County only in the central-coastal bottomlands. The summering population occurs along sloughs in the Eel River delta; along the shore of Humboldt Bay; around Eureka and Arcata; at a few spots on the north spit of Humboldt Bay; in the Arcata Bottoms and slightly to the northeast; around the edges of each of the larger north-county coastal lagoons; in the pasture edges of the Redwood Creek estuary, and in swales extending from lower Prairie Creek to the Davison Road area.

The near absence of Marsh Wrens away from the coastal plain is due to the notable scarcity of freshwater bodies—rather than fast-flowing streams—in the county's corrugated and incised interior. Breeding was not confirmed inland. A single bird in a roadside marsh on Bear River Ridge 14 Apr 1996 (D23; COg) and one at Lake Prairie north of Snow Camp Mountain 10 May 1997 (k16; E. Elias) were the only truly inland reports.

Dairylands of the Humboldt County coastal lowlands, unmanaged ditches, rushy swales, sapling-dotted vernal wetlands, and persistently wet or swampy corners of bayside pastures are sites at which Marsh Wrens may be found during the breeding season.

Much of this habitat is linear, patchy, and somewhat marginal. In contrast to this are the large stands of broad-leaved cattail and other emergent vegetation at the Arcata Marsh, the breeding metropolis for Marsh Wrens here.

It has been the impression of many observers that Marsh Wrens are far more numerous in the coastal lowlands during fall and winter than they are in the breeding season. At that colder time of year, the birds inhabit not only marshes of the classic definition but also occur widely throughout ungrazed pastures, in wet blackberry thickets, in riparian edge, and even (in low density) in expanses of European beachgrass in protected dune hollows. Most apparent during the colder months is the extent to which Marsh Wrens use saltmarshes. Total numbers of these birds about the margins of the combined Humboldt Bay and lower Eel River shores must be large, as it seems one is never more than a few steps from a wren when tramping a stretch of estuarine marshland.

Marsh Wrens were found in 24 blocks and confirmed breeding in seven. They were recorded nest building as early as 12 Apr 1998 (D19; DFx), and a surprisingly early fledgling was noted 5 May 1998 (F15; DFx). Only a few active nests were reported. One with young was found 20 May 1996 (D17; JEH), and nests with eggs were recorded 1 Jun 1996 (D19; GAS) and 6 Jun 1996 (D19; K. Foerster). The latest breeding confirmation was of an adult feeding young 27 Jul 1997 (g03; G. Lester).

AMERICAN DIPPER *Cinclus americanus*

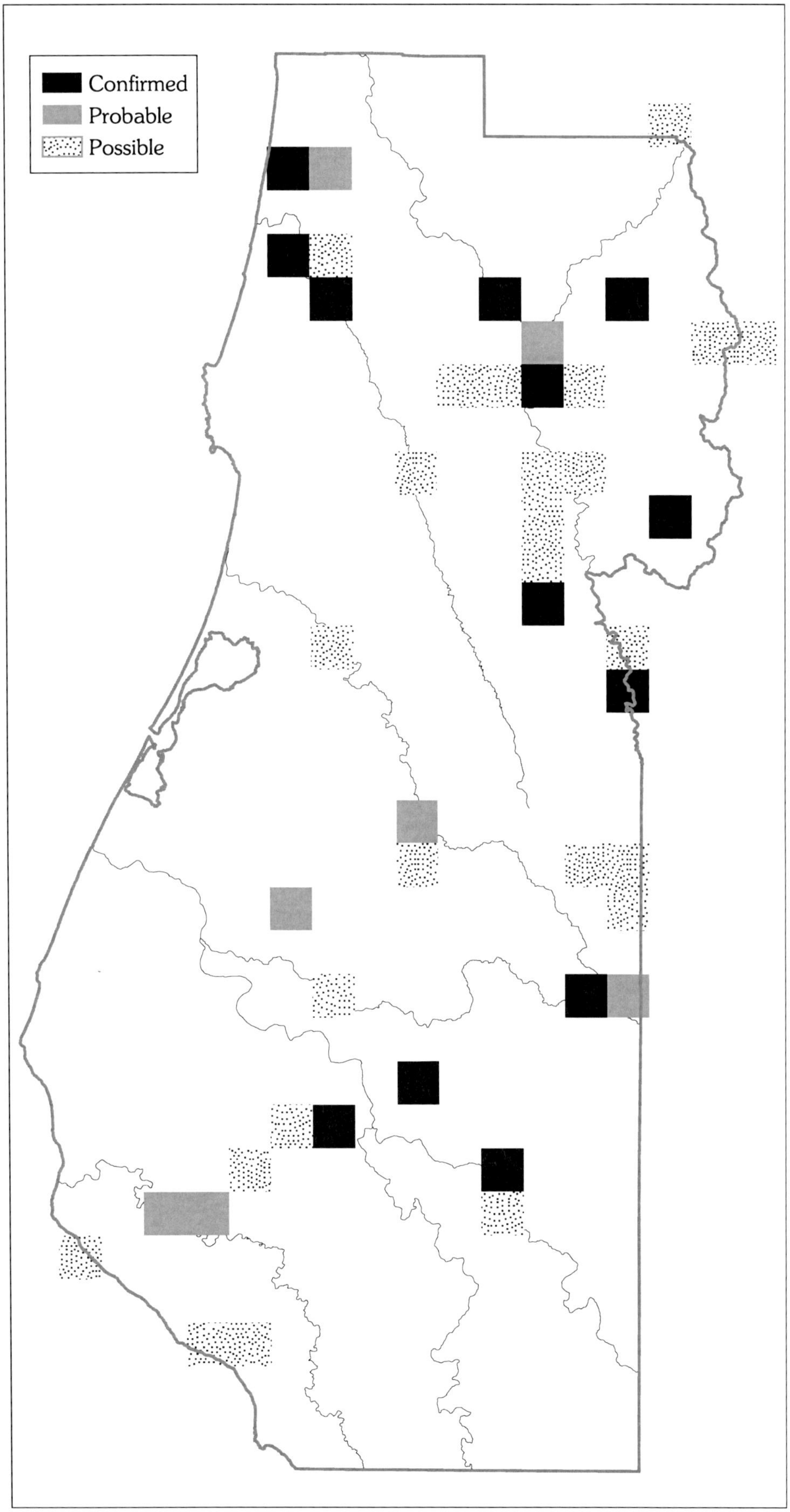

Throughout much of western North America, the voices of American Dippers may be heard above the white noise of fast-flowing streams. They are seldom found far from running water, foraging along the shores and beneath the surface of swift creeks and rivers. Truly aquatic, they have evolved an enlarged preen-oil gland from which they dress their feathers to assure protection from water. They tuck their domed, mossy nests into niches in spray-moistened streamside boulders, in the shadows beneath bridges, and behind waterfalls.

The American Dipper is an uncommon resident and breeding species in northwestern California (Harris 1996). The distribution expands slightly in winter to include lower-gradient watercourses. Patient searches for this bird in appropriate habitat will generally, though not always, reveal their presence. The words "search" and "habitat" may conjure up an image of a stalwart atlaser clad in raingear on a sunny June morning, negotiating frothy torrents atop fern-draped logs in order to find a dipper in the next block over. Such an effort was not mounted. Nearly all detections were made opportunistically from bridges or riverside trails. It is difficult to imagine any other diurnal bird of such intrinsic detectability whose population remains so tough to survey. A great amount of potential habitat comprising clean, high-gradient streams with rocky banks (on both public and private lands) could not be explored owing to time and travel constraints. American Dippers must be more widespread in the county than Atlas results indicate.

The distribution of Atlas reports suggests that American Dippers may be expected in appropriate habitat across much of Humboldt County. They were reported in 45 (11%) of the 425 total blocks. Breeding was confirmed in 13 blocks and was thought probable in seven others. Dippers were encountered during the breeding season along suitable reaches of the main-stem big rivers in the county, mostly well up in their canyons where streamflow is confined and protected nesting substrate is common. However, the tributary streams of these rivers probably support the bulk of Humboldt County's dipper population, as their aggregate length far exceeds that of the larger rivers. Along the Trinity River south of Hoopa, breeding dippers appear to be restricted to tributaries and their confluences with the main stem. Scarcity of protected nest sites on the margins of the Trinity and other "big water" streams may limit opportunities for nesting.

The species plainly is not a "mountain specialty" here. American Dippers may be encountered as far upslope as minimum acceptable stream size allows (locally to 4000 feet in northeast Humboldt), yet they are also noted regularly along lower reaches of small streams near the outer coast. Fern Canyon, Lost Man Creek, and the short turbulent streams finding outflow at the seaward base of the King Range support their share of dippers. Northeastern Humboldt's comparatively higher elevation, greater relief, and (usually) considerable winter snowpack fosters innumerable streams that provide suitable American Dipper habitat.

While American Dippers are known to initiate nesting in February and March to the north of us (Marshall *et al.* 2003), the earliest confirmed breeding reported during the Atlas period was of an adult carrying nest material 26 Apr 1998 along Prairie Creek (g04; DFx). The earliest actual nest observed was one with young 15 May 1996 along the Van Duzen River (J25; JCP). The well known dipper nest near East Fork (of Willow Creek) Campground (m14) appeared quiet 13 May 1995 (JEH, TWL) but had young inside 26 Jun 1998 (RHw). The latter part of the breeding season was represented by occupied nests found 25 Jul 1997 (g06; DFx and h26; L. Bacon-Ogden). Used nests were reported 26 Jul 1997 (H07; C. Grable) and 16 Nov 1995. The latter nest was along Red Cap Creek (o07), and was described as a little cave or hole in the side of a bedrock face that had green moss packed all around the entrance (C. Kent).

The greatest portion of the dipper population within Humboldt County occurs within forested lands managed primarily for lumber. Siltation, warming, loss of pool/riffle/run dynamics, and even wholesale "blowout" of fast-flowing streams in the county have long been associated with clearcutting and road-building. That such activity degrades water quality and aquatic life cannot be disputed. Despite the controversy surrounding these unfortunate phenomena as they relate to anadromous fisheries, never has the case for protecting American Dippers during implementation of forestry practices been considered. The state of dipper populations might serve as an indicator of the health of aquatic invertebrates. If so, adding to the knowledge of the birds' whereabouts may contribute to greater protection for this as well as other aquatic species.

GOLDEN-CROWNED KINGLET *Regulus satrapa*

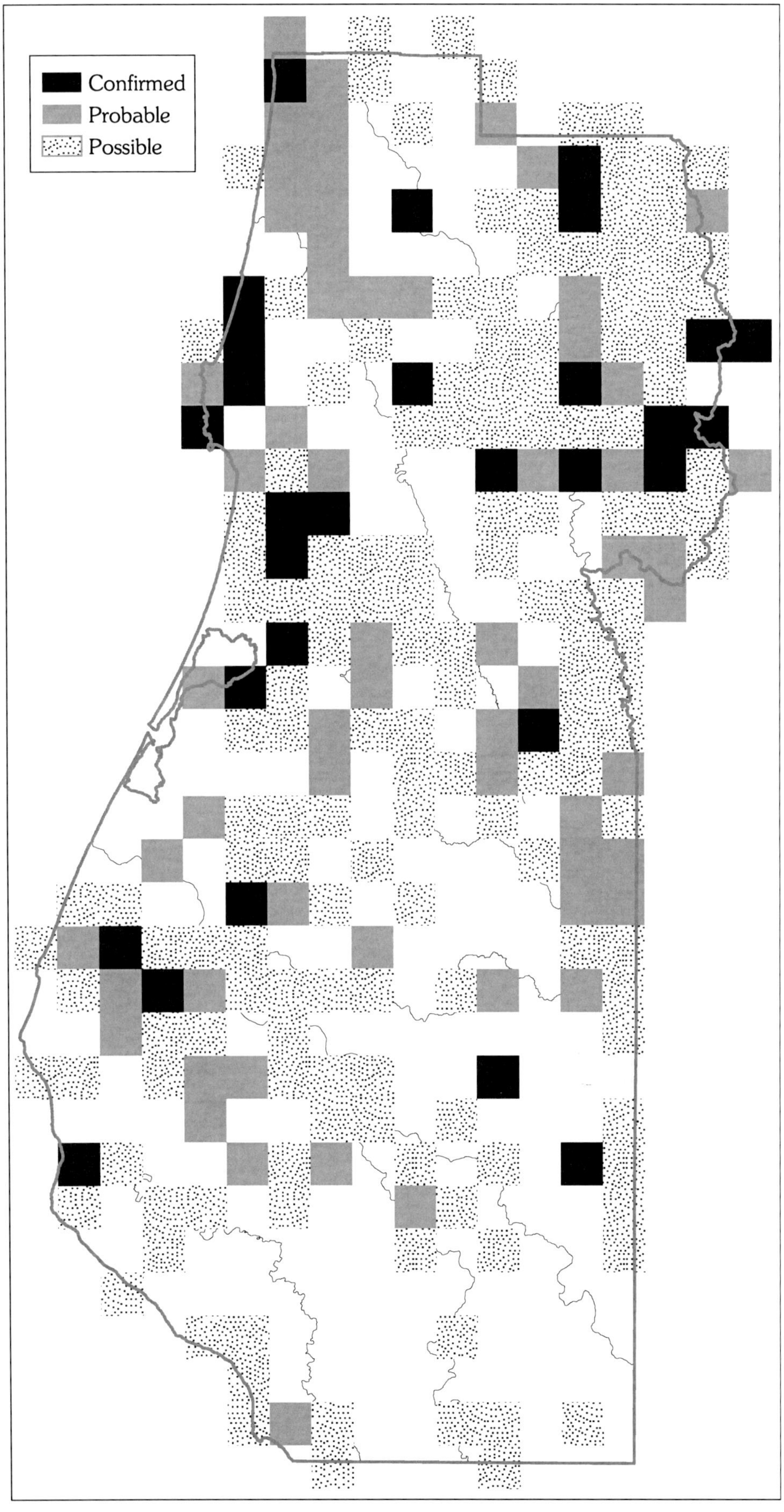

Like a whisper of breeze through the fir-tops when the air is yet still, the triple lisps of Golden-crowned Kinglets give breath to forests throughout the year in Humboldt County. During the spring and early summer, their soft voices are all but lost against the songs of summering neotropical migrants. After the late summer exodus of the "yellow birds" has left the conifer stands without a complex chorus, it is the persistent voices of Golden-crowned Kinglets that break the sunburnt hush of late August and bring a bit of liveliness to sodden weeks in February. In keeping with the hyperactivity that rules the legions of the feathered and tiny, these kinglets rove ceaselessly through conifer sprays in search of the insects they must constantly find in order to fuel themselves...for more ceaseless roving.

With its extensive coverage of conifer forests, Humboldt County provides a great deal of favored habitat for Golden-crowned Kinglets. Historically, this has been a routinely encountered common resident and breeding species across the portions of the county that are forested with short-needle conifers, and it was considered a common resident and breeder by Harris (1996). During the study period, Golden-crowned Kinglets were encountered in 241 (57%) of the 425 total blocks. These kinglets appear to be most numerous in Humboldt within the coast redwood zone; they also show an affinity for Sitka spruce (Hill 1977). In redwood forests, they reach their greatest numbers in older stands (Hazard and George 1999). They also occur along major river corridors, and at middle and higher elevations in the interior. In the Klamath Mountains they are associated with older stands of larger conifers, especially white fir, and with higher elevations (Sniado 1999), and show a strong negative association with hardwoods and a strong positive association with conifers (Ralph *et al.* 1991). Kinglets are present, but less numerous, throughout the middle-elevation mixed conifer–hardwood forests dominated by tanoak and Douglas-fir. These kinglets have probably decreased in population size or distribution in Humboldt as a result of clearcutting (Marcot 1984).

Stands of pines of any species are sparsely used, to the point of seeming avoidance. Forests dominated by hardwoods or dry-site conifers, or that have been intensively thinned or are otherwise of low canopy closure are also not preferred nesting habitat for Golden-crowned Kinglets. Within this drier, more hardwood-grown interior forest mosaic, kinglets appear to prefer clumps of Douglas-fir with high canopy closure and few hardwood trees.

The earliest report of probable breeding was of multiple males singing on territory 10 Apr 1999 (g33; *fide* RHw). The earliest confirmation was a bird mist-netted at the Wright Refuge (f16) with physiological evidence noted 25 May 1995 (L. George). There were no reports of the carrying of nest material. Only two nests were reported, both with young: 28 Jun 1998 (g15; DFx) and 26 Jul 1995 (L25; M. Morris). The 1998 nest in Bayside (g15) was placed in Douglas-fir foliage near the end of a limb in the tree's lowermost whorl of branches. Most confirmations were of fledglings (18 blocks), which spanned the dates 13 Jun 1997 (F09; GCH) and 28 Jul 1996 (m17; JCP). The latest breeding confirmation was made by the feeding of young seen 17 Aug 1996 (q10; JMH).

From September to April, Golden-crowned Kinglets spread out from their preferred breeding habitat and may appear in any manner of habitats with tree cover. Their appearance in ornamental trees and shrubbery in cities, as well as at the Fairhaven willow thickets and in Eel River riparia, typically takes place about the time "Myrtle" Yellow-rumped Warblers have begun to move through the area. Cottonwood and alder riparia, residential areas, and even oak woodlands may support many of these birds in fall and winter.

BLUE-GRAY GNATCATCHER *Polioptila caerulea*

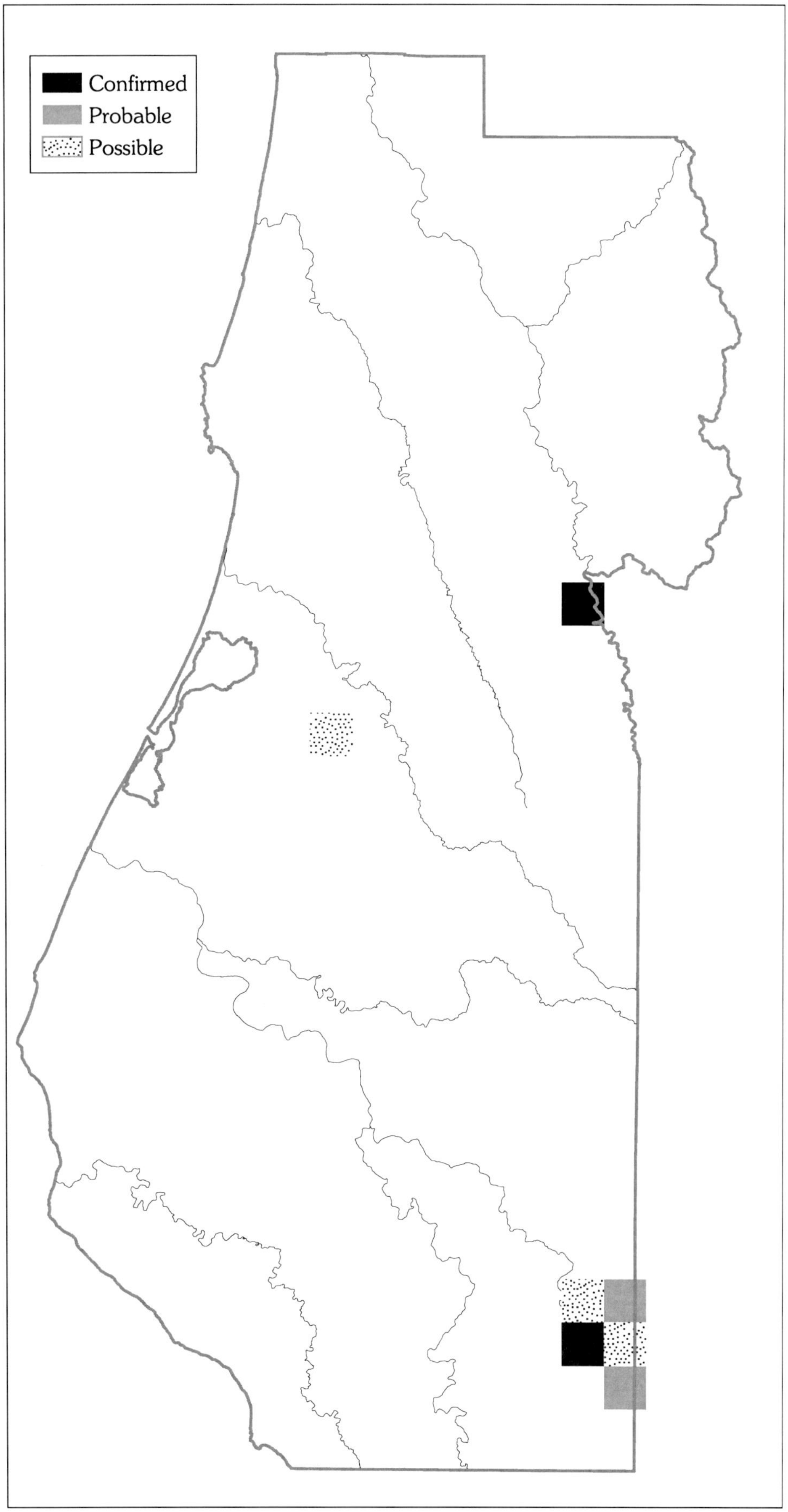

Sun-parched foothill slopes with chaparral, overshaded here and there by sprawling oaks, attract most of northern California's breeding Blue-gray Gnatcatchers. The semi-arid brush communities and dry-site oak woodlands occurring here and there across the southern one-third of Humboldt County would appear to offer habitat appealing to gnatcatchers. However, Atlas work revealed that they nest in only two areas of the county. Harris (1996) characterized the Blue-gray Gnatcatcher in northwestern California overall as a "rare summer resident and breeder in the drier southern and eastern portions of the region." It is apparent that even the choicest patches of "SoHum chaparral" must lack one or more biophysical elements defining the more extensive, more Mediterranean habitat extending southeastward just outside the county. This is a theme typifying a small yet significant suite of Humboldt birds more commonly represented in the state's Inner North Coast Range.

The places where breeding Blue-gray Gnatcatchers were encountered during Atlas surveys comprise a small northern site and a larger southern area, separated by about 50 miles. On lower Friday Ridge southeast of Willow Creek (n14), a nest with young was discovered 22 Jun 1998 (TWL). The nest was 12 feet up in a California black oak in a southeast-facing brushfield with oak, madrone, and other hardwoods at the junction of Forest Service roads 6N06 and county road 8L100 (intersection of Madden Creek and Friday Ridge roads). Humboldt's first and only pre-Atlas breeding record was from this site; a nest with at least three young about 10 feet up in an oak was discovered 20 May 1992 (COg). Three pairs were noted there 24 May 1992 (B. Allen, B. Rodstrom). Since that initial discovery of breeding gnatcatchers, the place has become known in the local birding community as "Gnatcatcher Junction". Seasonally, the earliest date they have been seen there has been 8 Mar 1994

(TWL). Of interest at this site is the considerable stand of buck brush, a shrub favoring hot southern exposures in the mountains of northwestern California. At several locations in southwestern Oregon, gnatcatchers prefer brush communities dominated by this shrub. Few such expanses of buck brush occur in Humboldt County. Other Atlas-period records of gnatcatchers belonging to this somewhat disjunct population include at least one bird seen there 23 Apr–30 Jun 1995 (GCH, TWL, S. Kallager) and two males and a female chasing each other 28 Apr 1996 (JEH).

In far southeastern Humboldt County, a bird was watched carrying food just above the upper Eel River near Cain Rock, southeast of Alderpoint, 30 May 1998 (N31; JEH) and another bird was observed there 5 Jul 1998 (COg). Farther up the Eel (o32), a pair was observed 30 May 1998 (JEH). These reports suggest that a small population of Blue-gray Gnatcatchers probably summers regularly in this area. A rich mosaic of vegetation occurs along the western side of the canyon, dominated by interior and canyon live oaks and California black oak, beneath much of which exists a continuous brush layer. The opposing east side of the canyon bottom is stonier and more sparsely grown to drier-site hardwoods, discontinuous clumps of brush, and grassy openings. This distinctive interior foothill habitat extends along the Eel River northwest to the vicinity of Fort Seward and southeast to the Trinity County line and beyond. Access to this reach of the Eel was made by walking south of Alderpoint along several miles of disused Northwestern Pacific Railroad trackbed. Other southeastern Humboldt reports of gnatcatchers during the Atlas period included a pair (no details) near Zenia Bluffs (o30; S. McAllister). Precipitous slopes densely grown to oaks, conifers, and brush may afford habitat for a few gnatcatchers at this location. A single outlier observation of a bird "acting territorial" was reported 1 mile below Kneeland School (H17) 5 Jul 1997 (S. Fitton, A. Desch). Given no history of gnatcatchers there, local observers do not suspect breeding in that area, but the record was retained as a suggestion for further searching there.

Single Blue-gray Gnatcatchers occasionally appear in the coastal lowlands in early fall, in mid-spring, and, exceptionally, in winter. Overshoots or vagrants remain essentially unknown at any season away from the coast. The comparatively small size of possible contributory populations within the region to the north and east of Humboldt County may make chance encounters with any inland straggler unlikely.

WESTERN BLUEBIRD *Sialia mexicana*

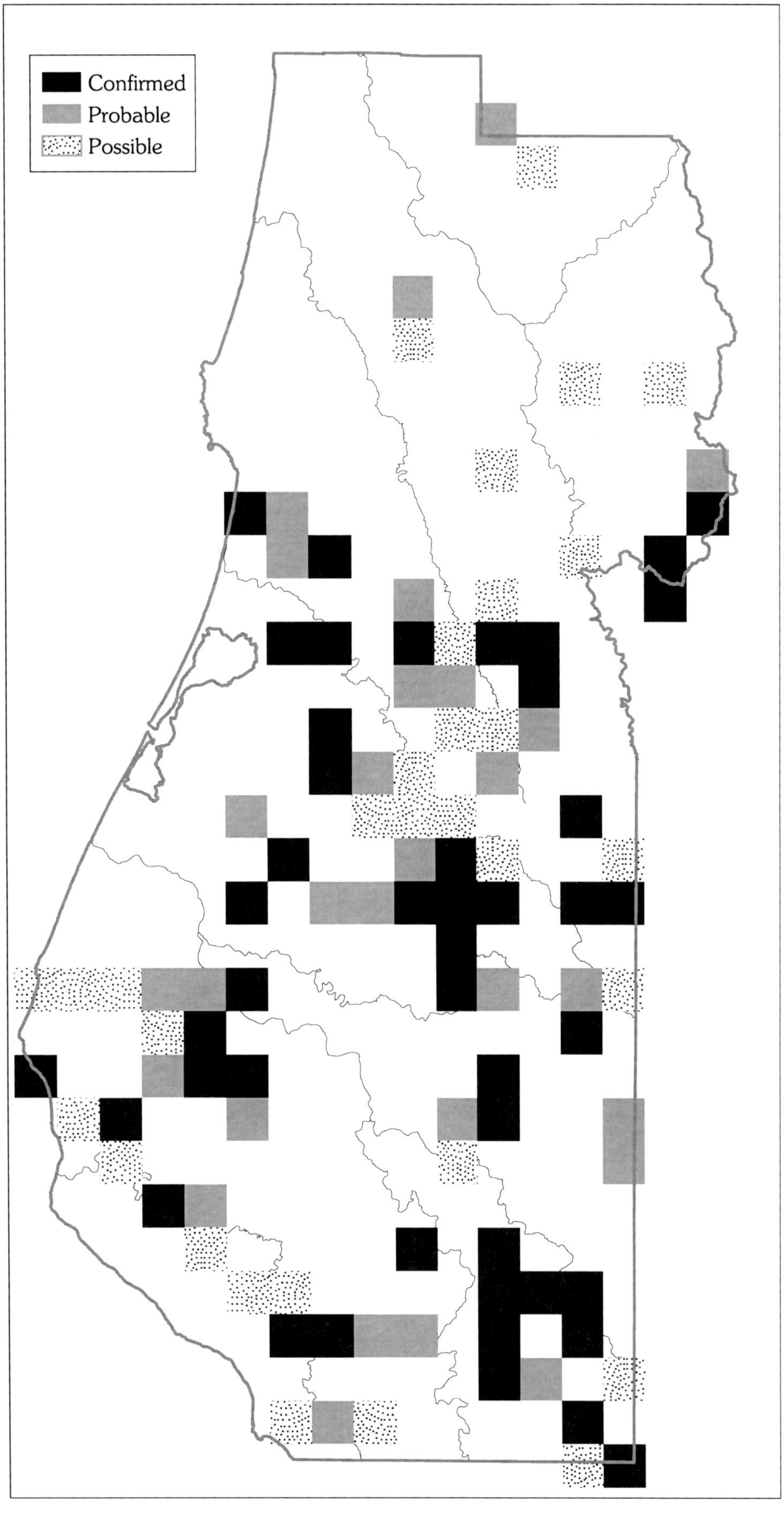

Throughout nearly all of their extensive range, Western Bluebirds occupy habitats characterized by a distinctly rural feel. During summer they favor oak woodlands, chaparral, forest edge, burns, logged areas, pastoral glades below the heads of meandering foothill drainages, and the more open or broken conifer forests of the mountains. Nearly all of this countryside typically is but lightly peopled. While living in some places around human settlement, most Western Bluebirds would appear to spend their lives in the hinterlands. It would seem likely that a far smaller proportion of the Western Bluebird population is enjoyed by human observers than is true for the Eastern Bluebird (*S. sialis*).

In northwestern California, the Western Bluebird is an uncommon resident (Harris 1996) outside the coastal fog zone. During the Atlas period, this species was detected in 106 (25%) of the 425 total blocks and was confirmed breeding in 46. These birds are fairly common in the oak woodlands and savannahs that clothe much of interior Humboldt at lower and middle elevations. They follow this band of vegetation a short way westward into the zone of summer overcast. Elevation is no barrier within the county, as broken montane forest and meadow edge settings were found inhabited locally to the higher ridgelines. Western Bluebirds can also show up at unlikely times and places in the coastal lowlands, especially in late fall and winter, when larger numbers drift widely throughout the county. Breeding confirmations in two outer-coastal blocks (a25, f12) were interesting but not surprising, as both blocks offer sizable expanses of suitable habitat for nesting and foraging.

Western Bluebirds are also common in recently cutover timberlands and in clearcuts that have failed to grow back with trees. Fisher (1902b) termed the species "not uncommon in cleared land." These occupied cut areas typically have a good complement of snags for nesting (Marcot 1983), grass and low brush, little or no tree over-story, and are quite often situated on ridges. It is unknown if an association with ridges is due to some selection for topography, or if ridges are where the roads—and atlasers—were, or if these denuded thin-soiled sites simply fail to regrow trees and persist as suitable habitat for as long as the snags do. Historically, these bluebirds have been rather numerous along the upper slopes of the South Fork Mountain ridge at the edges of forests with white fir, marking them as another among a large suite of songbirds nesting in Humboldt that occur in true fir forests of the mountains.

One of the more appealing aspects of Western Bluebirds is that they rarely occur alone, with reports generally of several birds or more. In northwestern California, as in the rest of their range, bluebirds are a sociable lot. Newly fledged family groups forage and roost together throughout the spring and summer months. This species was confirmed as breeding in nearly one-half of the blocks in which it was found, probably because of its conspicuous nature and the openness of favored habitat. The early breeding season was represented by records of an adult carrying food 28 May 1999 (o34; JEH) and one building a nest 31 May 1998 (H13; GAS). The earliest fledgling was reported 6 Jun 1996 (N31; M. Bailey). A nest with young was found 1 Aug 1995 (h18; JCP), and the feeding of young was recorded as late as 18 Aug 1995 (d28; C. Keil). A pair at the Bayside Golf Course (g15) pulled off two broods; the second one fledged 22 Jul 1995 (PFS).

Nowhere in Humboldt can Western Bluebirds be described as abundant, although they are easy enough to find in appropriate breeding habitat during the summer months. Breeding numbers have declined in the lowlands since the early 20th century, possibly a result of competition from European Starlings for nest cavities (Harris 1996), and possibly because of a reduction in snags and large oaks containing cavities. These birds persist in the hills and mountains distant from starling foraging locales. Timber harvest may have increased the amount of breeding habitat in some areas (Marcot 1984).

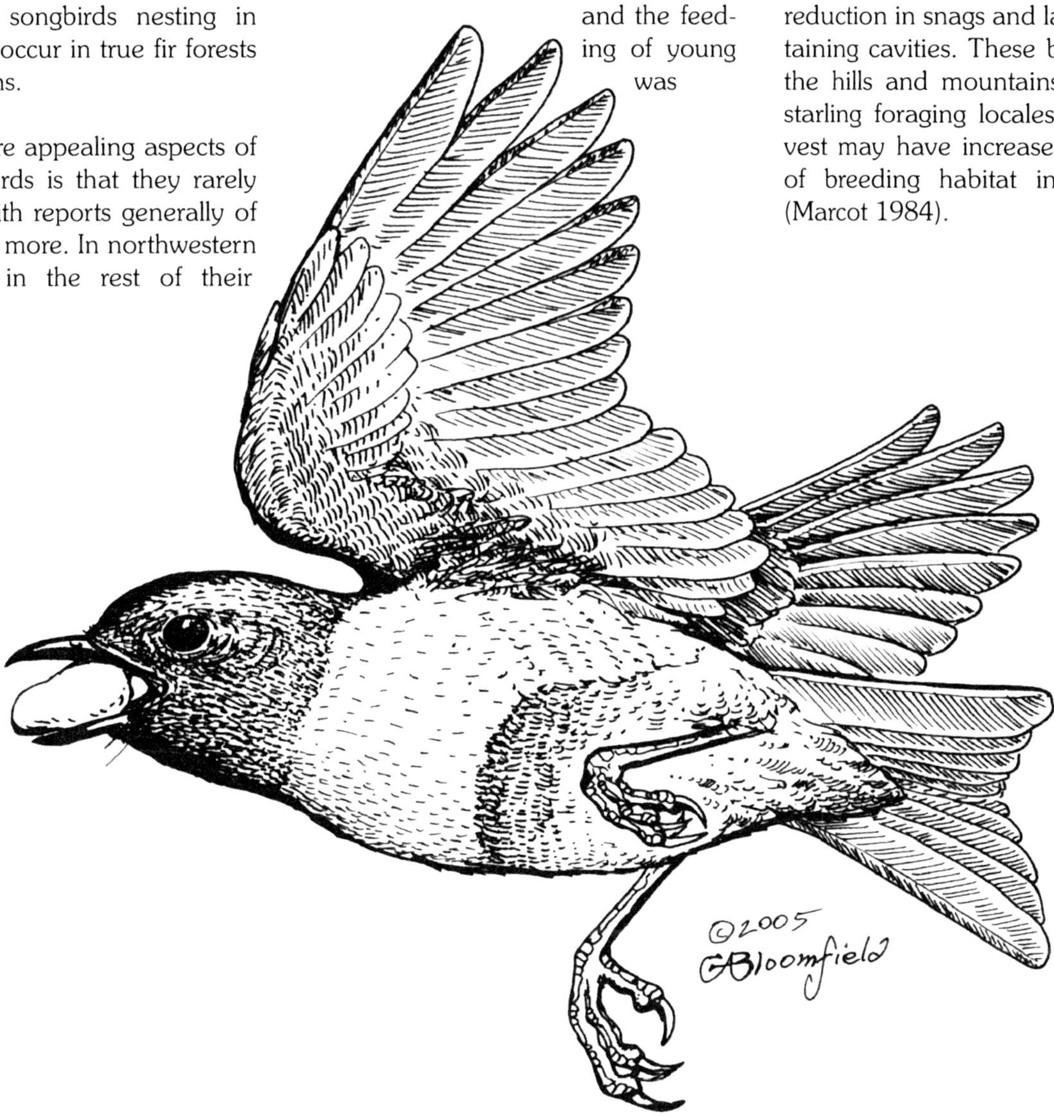

MOUNTAIN BLUEBIRD *Sialia currucoides*

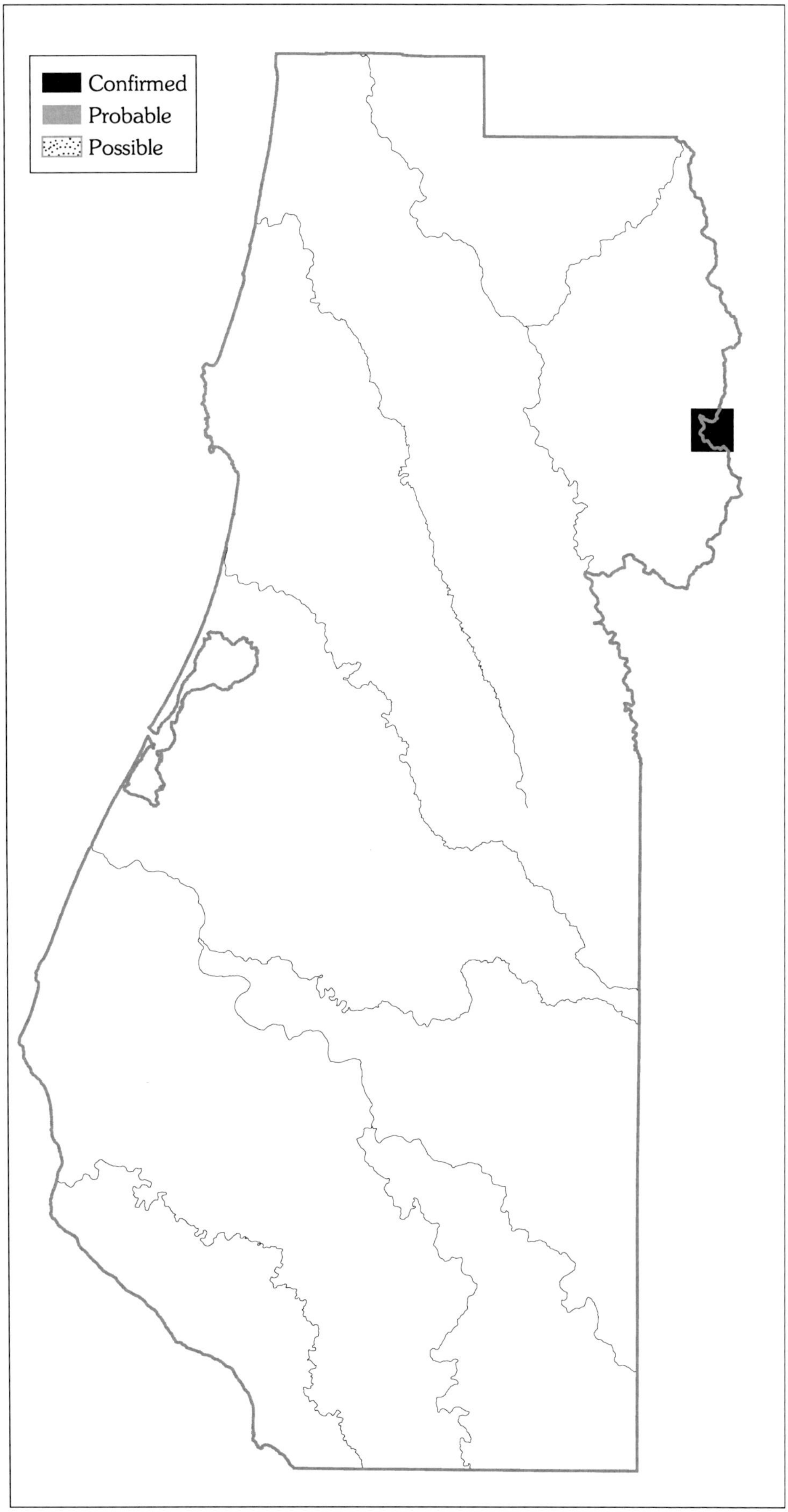

As dramatically blue a bird as any, the male Mountain Bluebird delights upon first acquaintance. Deep azure upperparts and paler blue underparts cause this bird of western mountains and plains to draw comment wherever it is seen. Females, clothed in mousy tones of gray scarcely tinged with blue, are far from "show-stopping" but lend their measure of grace to the places they inhabit. In summer, Mountain Bluebirds live in subalpine meadows, open pine and fir forests, aspen stands, juniper woodlands, sagebrush flats, and cattle range of the high country. They also summer above timberline wherever sufficient protected rock crannies can be found in which to nest.

This species has an extensive range in western North America. In California, the "Blue Mountainbird" is represented in appropriate habitat in the Cascades, the Warner Mountains, the Sierra Nevada, the San Bernardino Mountains, in Great Basin sagebrush and juniper stands, and locally on ridges and peaks at the highest elevations in the Klamath Mountains. It is at best a peripheral or rare breeding bird in Humboldt County. During the Atlas period, Mountain Bluebirds were encountered in only one block at the far eastern edge of the county. On 17 Aug 1996 a family group of five birds was observed flying from Trinity County into Humboldt County at the Trinity Summit Guard Station (q10; JMH). Similar records, such as fledged young near Salmon Mountain at the "triple junction" of Humboldt, Siskiyou, and Trinity counties 8 Jul 1989 (R. Erickson, G. Lester), also suggest breeding in Humboldt County proper. But since records such as these do not rule out nest sites in adjacent Siskiyou and Trinity counties where more extensive habitat exists, this species is considered a probable but unconfirmed breeder in Humboldt County proper.

The Mountain Bluebird has been known chiefly as a casual winter visitor in Humboldt County. Harris (1996) mentioned apparent high elevation inland breeding locations in northwestern California, including the Humboldt County locations of North Trinity Mountain, Salmon Mountain, and the Trinity Mountains. These areas are remote, and road and trail access typically is limited by snowpack until well into late spring or early summer. As a result, the token surveys accomplished in these areas by atlasers failed to shed much light on the actual breeding distribution of this bird.

TOWNSEND'S SOLITAIRE *Myadestes townsendi*

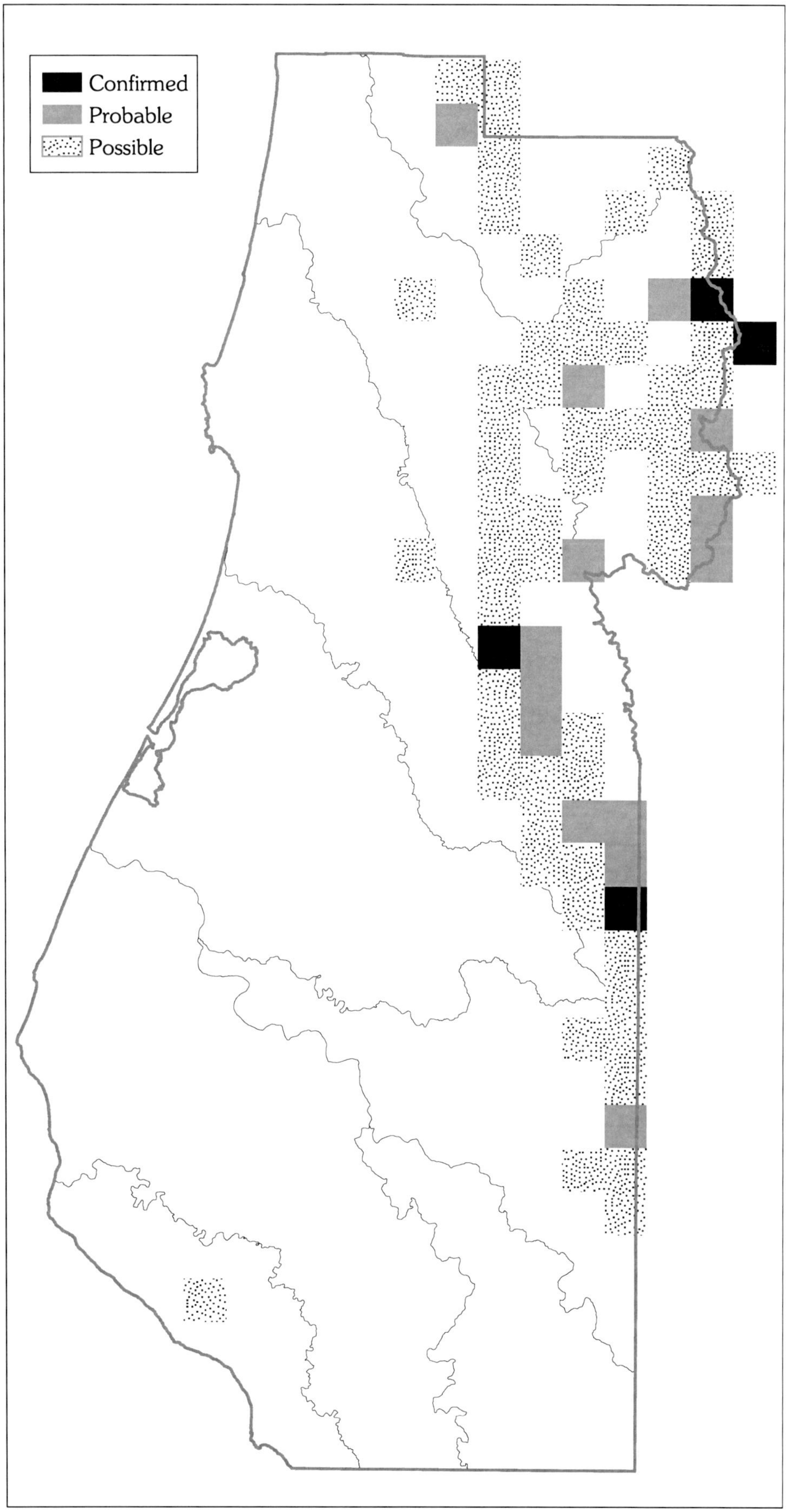

Townsend's Solitaire is a characteristic breeding bird of mountain forests throughout western North America. It is typically found in greatest abundance in broken forest or in natural openings within a heavily forested expanse. But the solitaire has also followed closely behind the chainsaw in northwestern California, where it is considered a rare to uncommon summer resident and breeder (Harris 1996). These birds pioneer clearcuts and partial-cuts not only within "ancestral" haunts at higher elevations, but also at middle elevations well downslope within the temperate Douglas-fir zone. Fresh openings created by industrial forestry at, and below, the lower edge of the montane conifer zone would appear to have essentially teased these birds into accepting what constitutes a significant amount of additional, human-made habitat. Most of these openings eventually grow back to conifer regeneration or brushfields, ultimately excluding the open-ground loving solitaires. Thus, the effect is ephemeral, the intensity of its impact on the birds' distribution largely tied to forestry practices and intensity of harvest.

It is apparent that burns, high meadow edge, rocky ridge-tops, lower-subalpine openings and meadows, rock-slides, and similar microsites at the higher elevations in eastern Humboldt County supported modest numbers of Townsend's Solitaires before the advent of logging in the region. This is suggested by the regular occurrence of the birds in such places today. A patchy scattering of such sites occurs throughout the higher elevations in the mountains, and solitaires were noted in them during the Atlas period.

During Atlas surveys, Townsend's Solitaires were detected almost exclusively in the mountains of eastern and northeastern Humboldt County. The range encompasses ridge-tops and upper mountain slopes from above the lower Klamath River south-southeastward across the higher portions of the Hoopa Valley Indian Reservation, the Salmon River/Trinity River divide country, the South Fork Mountain ridge, the Snow Camp Mountain ridge, and about the Lassics uplift shortly west of the southern Humboldt–Trinity county line.

Marking the western edge of this region were several detections worthy of discussion. A pair of birds, the male singing, 24 Jul 1999 at 3300 feet in upper Indian Creek of the lower Klamath River (k03; DFx) occupied a fresh clearcut that had burned too hot during a prescription burn, resulting in nearly complete loss of down woody material smaller than logs and expanses of exposed inorganic soil. This location was within the coastal fog zone and lacked any high-elevation conifers. One solitaire north of the main Bald Hills ridge (J07) 3 Jul 1998 was also in a burned clearcut at about 2400 feet (JCP)—with tanoak already vigorously stump-sprouting! A solitaire immediately above Lord-Ellis Summit along Highway 299 (J13) 22 Jul 1995 (JCP) was at another site well below the normal elevational threshold for this species, at 2300 feet. Although these reports probably pertained to breeding, no direct evidence was obtained.

The only Townsend's Solitaire detected in Humboldt's coastward mountains was a bird on North Slide Peak in the King Range (e30) 25 Jul 1995 (R. Sutherland). The King Range crests at 4087 feet at the summit of King Peak, only 3 miles from the ocean. Marginal breeding habitat appears to exist there. Dry-site Douglas-fir, tanoak, canyon live oak, knobcone pine, and *Ceanothus* dominate much of the upper slopes. In the ages preceding aggressive wildfire control, lightning burns created scorched openings along the ridgelines. One might conjecture that such forest opening mosaics once may have provided more suitable breeding habitat.

Key elements of Townsend's Solitaire habitat in the county are frequent commanding perches such as isolated saplings, snags, or residual green trees; a sparse, snow-compacted shrub layer; and exposed earth or only minimal duff and litter. White fir or (at higher elevations) Shasta red fir is usually present, although, as noted above, they may be absent. Huckleberry oak, Brewer's oak, or manzanita dominates the shrub layer in many locations.

Solitaires generally were encountered individually, commonly as singing or calling birds heard at a distance. Pairs were noted in five blocks. Multiple males singing on territory within a single block was reported only twice, on Last Chance Ridge (N19) 23 Jun 1995 (GAS) and on nearby Blake Mountain (o20) 24 Jun 1995 (GAS). This area is well within the montane forest zone, and may support some of the highest densities of solitaires within Humboldt County. The nesting season is represented by agitated behavior suggesting a nearby nest 16 Jun 1999 (m17; DFx) and the feeding of young 30 Jul 1998 (r08; GCH). The only actual nest reported was one found 16 Jul 1997 on the side of a bank, with a nearby adult performing distraction displays, along the Salmon Summit Trail near Whiteys Peak (q07; JEH, GCH). The earliest fledgling was seen 29 Jul 1995 (r08; TWL).

SWAINSON'S THRUSH *Catharus ustulatus*

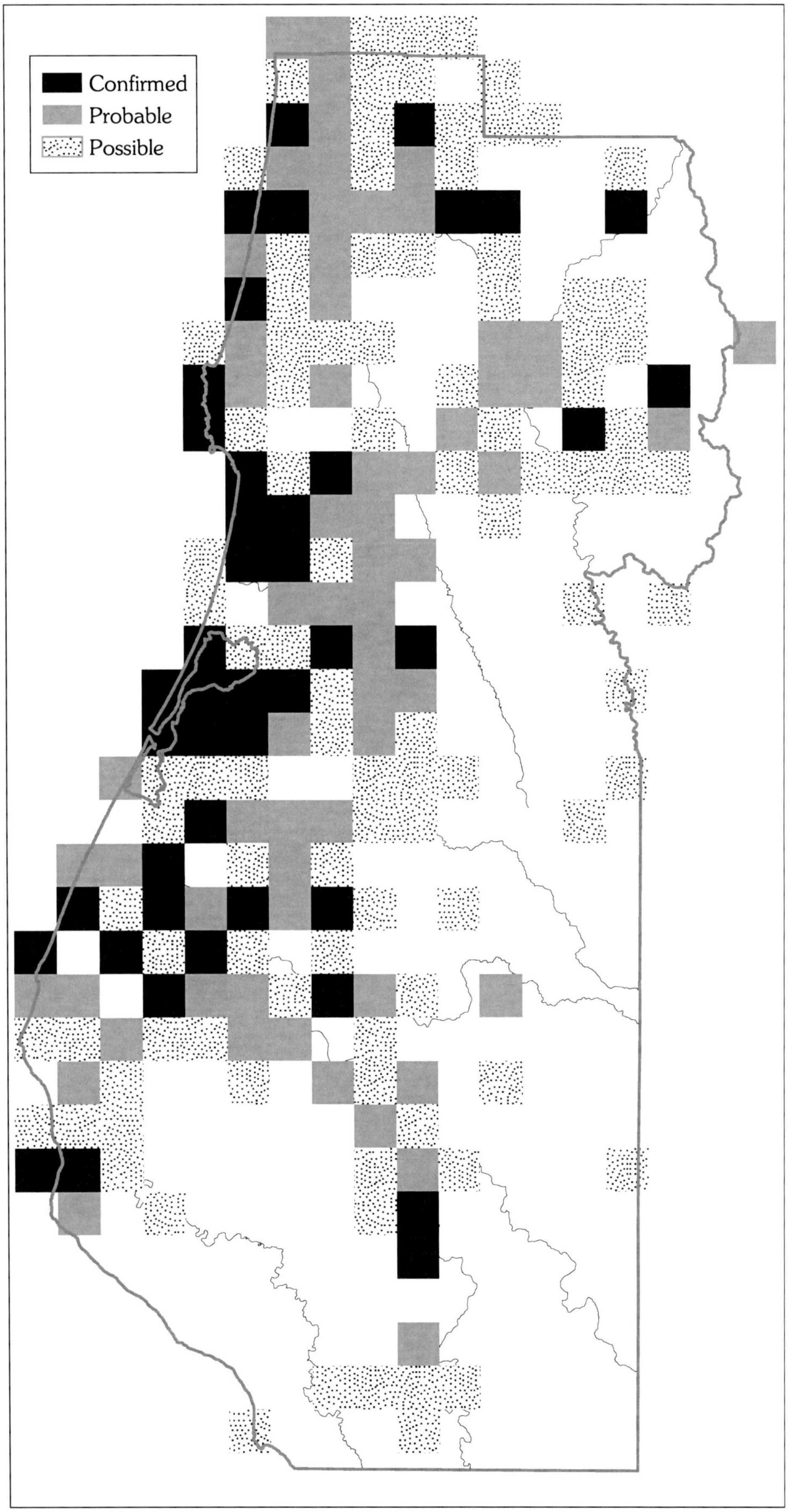

The pleasing songs and interrogative call-notes of Swainson's Thrushes are familiar to many persons attuned to the sounds of the outdoors. From early May through July the dawn chorus of songbirds across much of western and northern Humboldt County is dominated by the voices of these shy forest birds. Swainson's Thrushes steadfastly remain in dappled shadow, and are typically heard before they are seen. Seldom are these thrushes observed foraging along roadsides, in suburban gardens, or in forest clearings, as are Hermit Thrushes. To truly get to know the behaviors of this shy species requires patience, attentiveness, an ability to glimpse movement in low light, and a tolerance of mosquitoes.

Prior to Atlas fieldwork, Swainson's Thrushes were known to be more common and widespread within the coastal fog zone and general coast redwood belt than in the drier forests in the eastern part of the county. A sparser population, localized in pockets of suitable habitat, was recognized to occur in northeastern Humboldt. Harris (1996) characterized Swainson's Thrush in northwestern California generally as a "common migrant, summer resident, and breeder in coastal lowlands and some moist mountain riparian areas; accidental in winter." The results of the Atlas generally confirmed these perceptions, although the inland component of the population turned out to be more apparent than many atlasers might have predicted.

Within the fog belt, Swainson's Thrushes are routine nesting birds, found in both old-growth and regrowth forest (especially redwood), being most common where canopy openings have allowed luxuriant growth of seral red alder and associated moist-site shrubs and forbs. Shrub-trees such as red elderberry, red-osier dogwood, and Indian plum, and broadleaf shrubs like salmonberry and thimbleberry are usually present in the thickets to which these birds are partial. Riparian areas are well used by these thrushes. Standing or flowing fresh water or, lacking that, persistently damp or saturated ground is often a feature of their preferred habitat. Kelly (1987) found them to be the third most abundant breeder in the cottonwood–alder forests he studied in the Eel River delta, favoring areas with dense and diverse shrub and ground layers and with comparatively

less alder overstory. In 1999, during point counts within riparian habitats along the lower Eel River, they were detected during 52% of all counts (R. Hewitt, unpubl. data). Wooded gulches and ravines in and around McKinleyville, Arcata, and Eureka also support their share of these thrushes in summer. Swainson's Thrushes have higher densities, smaller territories, and lower nest success along urban/redwood forest edges than in forest interiors (Allen 2002).

Birds detected in northeastern Humboldt were typically found along streamsides, at seeps, and on moist benches in the combined Klamath and lower Trinity River watershed. Some reports were from wet meadows at or above 4000 feet. At least four countersinging males were noted in the wet meadows just west of Salmon Summit, Trinity County (r08), 22 Jul 1998 (JEH, GCH). Easy access to riparian and valley-edge sites at lower elevations within the South Fork Eel River drainage allowed coverage of localized patches of appropriate habitat for Swainson's Thrushes, which resulted in documentation of these birds extending southward to the vicinity of Garberville.

Contrasting with the thorough distribution of these birds in the western part of the county was their near absence from much of the interior. This is directly related to scarcity of habitat, as those mapped areas with few or no detections lie within a region characterized by intense summer heat and, consequently, only localized moist-site vegetation. Swainson's Thrushes are seemingly nearly absent along the South Fork Mountain ridge. Slope uniformity (which minimizes incised and mesic drainages), a sparse broadleaf shrub layer, and dominance by montane conifers at the higher elevations would appear to be factors limiting the birds there. Virtually no detections were made in the drier country of interior southwestern Humboldt County, where hillside forests of tanoak, Douglas-fir, and Pacific madrone offer too open and too dry an expanse for Swainson's Thrushes. Although streamside habitat grown to white alder surely exists locally within this area, nearly all of it lies deeply enfolded within private land, to which atlasers had only minimal access. Absence of birds in the southeastern corner of the county probably reflects a truer scarcity. Nearly that entire region is a summer-hot, oak and grass dominated landscape in which riparia is not well represented.

Obtaining firm breeding evidence for Swainson's Thrush was not readily accomplished by the casual observer, as nesting adults are quite reticent, so much so that there was only a single record of the carrying of nest material, 27 May 1995 (F11; JEH). There also was only a single record of a nest during the entire Atlas period, that of a used nest (no details) 16 Jun 1997 (H11; GAS). Yet despite all the secrecy, the persistence of atlasers and the sheer abundance of these thrushes allowed a strong complement of confirmed and probable breeding blocks. Sixty-three percent of the 43 breeding

Continued on page 393

HERMIT THRUSH *Catharus guttatus*

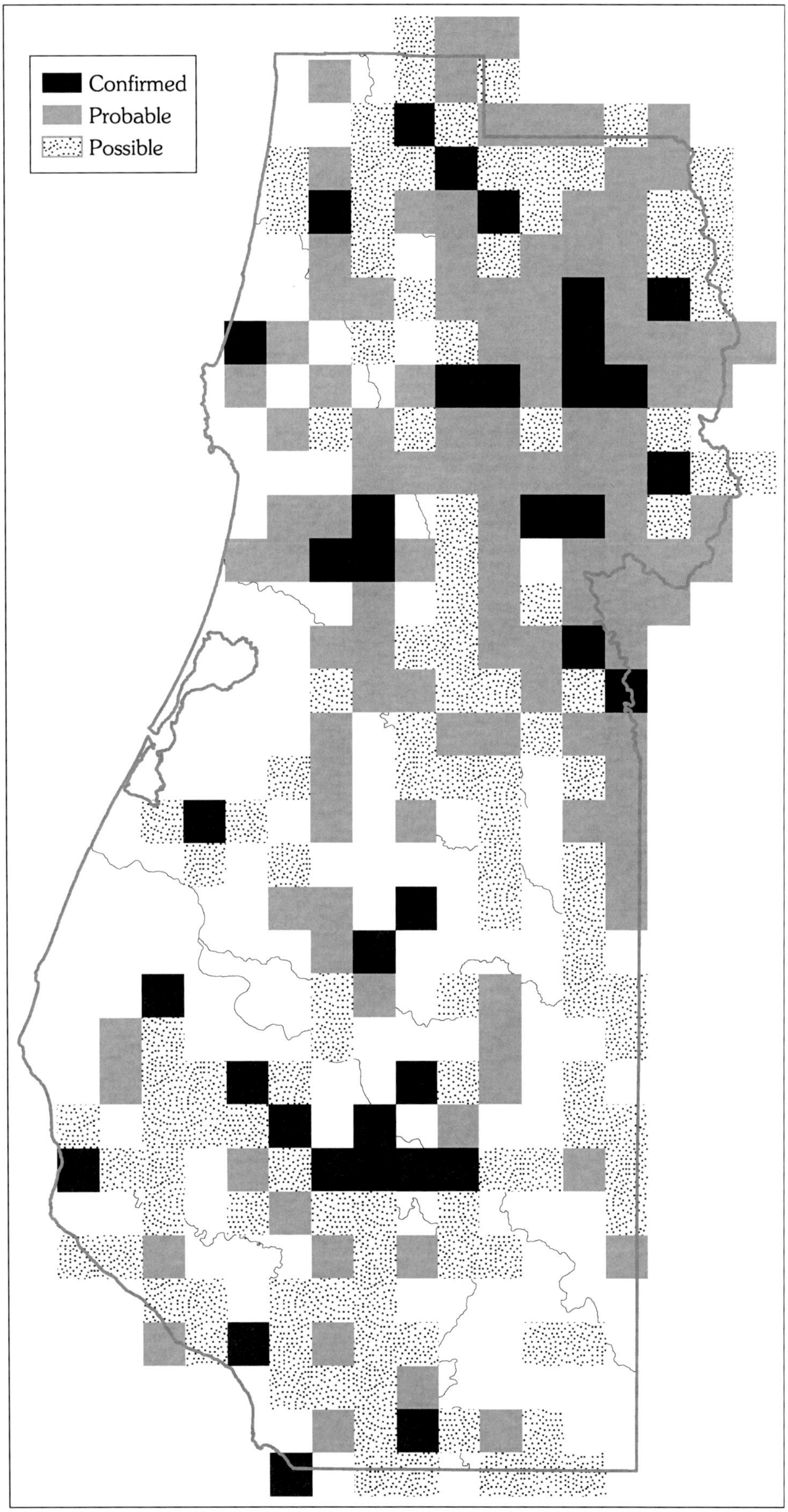

Were the song of the Hermit Thrush to be described in the language of wine criticism, it may be said to begin with a bouquet as natural as all outdoors, followed by hints of conifer resin and the intricate play of light and shadow. The finish is lingering and evocative, an exalted outcry celebrating existence!—yet with subtle overtones of stale bug dope and sweat.

Birders and naturalists who live up back roads, or whose work or play takes them into the woods of interior Humboldt, come to recognize that forested slopes across most of the county support Hermit Thrushes in abundance from April into September. Atlas results and the general experience of birders involved in the survey effort indicated that this species can be expected in well-drained forests of diverse description and age over most of the county east of the immediate coast, above the coastal lowlands, and outside old-growth coast redwood forest. Harris (1996) termed the species a common resident and breeder in northwestern California.

Hermit is by far the more common of the two spot-breasted *Catharus* thrushes in heavily-wooded northeastern and southwestern Humboldt County. The reason is apparent in a broad-brush examination of the habitat needs of each species with respect to the location of its habitat in the county. Swainson's Thrushes require moist sites, characterized by a lush and diverse broadleaf understory and a comparatively dense canopy cover. Although such places commonly occur well up on hillsides within the near-coastal fog belt, these sites are greatly restricted eastward, with Swainson's there occupying well-watered gulches, benches, and creek bottoms—situations comprising altogether but a fraction of the total landscape. Conversely, Hermit Thrushes avoid mesic sites, instead establishing their summer territories across the entire span of hillside forest from just above the riparian zone upslope to the ridge-tops, thus occupying a great sweep of interior Humboldt County in thorough abundance. They were found in 266 (63%) of the 425 total blocks, while Swainson's Thrushes were found in 206 (48%).

Hermit Thrushes summer in Humboldt County in a rather wide spectrum of forest types. Provided there is acceptable closure to the canopy and an open forest floor, the birds inhabit Douglas-fir stands, Douglas-fir/tanoak stands, and tanoak, madrone, and canyon live oak stands with a variable (often minimal) conifer component. While less common, they are also found in stands dominated by Jeffrey pine, with a shrub-form oak and manzanita understory, along ridge-tops such as at Horse Mountain, and sites where white or Shasta red fir is prevalent. This encompasses an elevational range of from 200 feet (in the Klamath River canyon and locally near the coast) to above 5000 feet. This broadly spans the ecotone between the warmer, evergreen hardwood-dominated forests of the foothills and canyons and the cooler, conifer-dominated stands well within the winter snowpack zone.

With local exceptions, coastal-slope woods grown to Sitka spruce, Douglas-fir, grand fir, or redwood and undergrown with red elderberry, salmonberry, and salal seem not to be used during the breeding season. However, Hermit Thrushes come into their own at the western limit of somewhat drier, summer-hot stands in which immature tanoak dominates in the understory. This boundary between coastal and interior forest types is irregular both in definition and in distance from the coast. In general, it occurs shortly within the eastern edge of the fog belt, and is indicated by a westward extension quite near the coast in northern Humboldt and almost to the immediate coast in the southwestern corner of the county.

In Humboldt County, Hermit Thrushes are not among the abundant colonizers of recent clearcuts. However, they are regular in Douglas-fir or tanoak regrowth of 20–50 years in which dense shade discourages a lush shrub layer. Logging would appear to have diminished the numbers of these birds (Raphael *et al.* 1988), since their abundance increases with increasing stand age (Raphael 1984). However, at sites that once were forested with old-growth redwood, the establishment of seral stands with little underbrush may benefit them. Hazard and George (1999) detected them in greatest numbers in 21- to 60-year-old redwood stands.

The earliest probable breeding record was of a male singing on territory for at least a week by 6 May 1996 in a known area of summer residency (q13; TWL). The earliest breeding confirmation was a record of adults carrying nest material 5 Jun 1995 (B27; S. Levy), and the latest was of young seen fed 3 Aug 1998 (L09; GCH). In winter, these birds become scarce in the mountains and appear in coastal lowlands, where they are absent during summer. As Hermit Thrushes are highly migratory, it is likely that this movement is not a down-mountain drift of locally breeding birds, but involves the appearance of members of subspecies originating to the north.

AMERICAN ROBIN *Turdus migratorius*

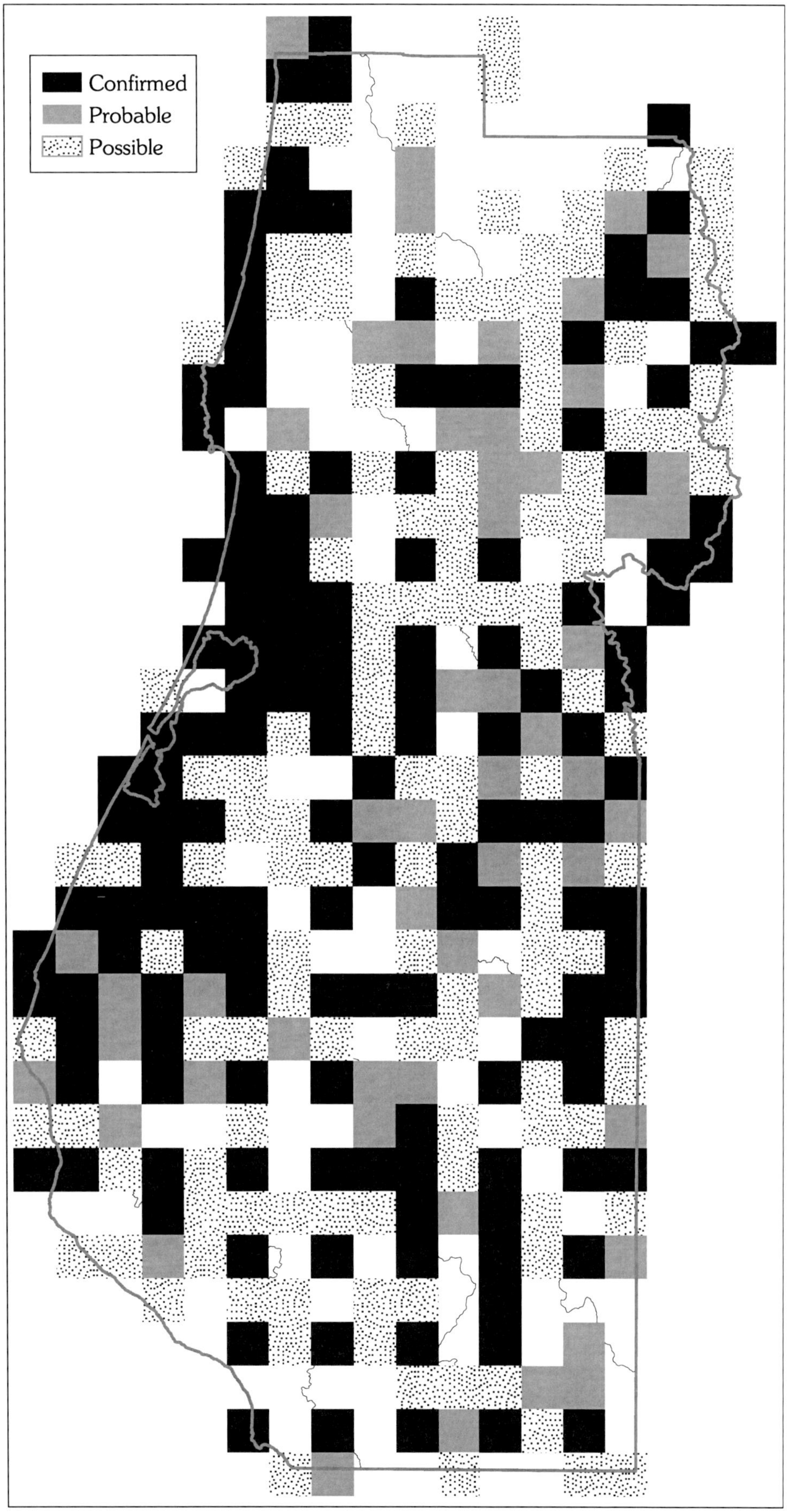

Familiar and abundant, American Robins may be seen nearly anywhere at any time in Humboldt County. This is among the most conspicuous nesting birds. Robins are common summer residents and breeders throughout northwestern California (Harris 1996). This species' confiding nature and readily observed breeding behaviors made it an easy bird for atlasers to confirm nesting. Confirmations were reported in 143 blocks, making the robin the third most confirmed species. The large summering population in residential neighborhoods, parks, and in the coastal dairylands was well surveyed by backyard atlasing, resulting in a nearly continuous strip of confirmation-blocks from Redwood National Park to Cape Mendocino.

American Robins may be uncommon or less thoroughly distributed within extensive tracts of old-growth forest, and in unmanaged second-growth densely undergrown with tanoak. If this is so, it begins to explain the comparative scarcity of observations of breeding within the private timberlands of extreme northern Humboldt County. No confirmed reports were generated in the course of surveys in priority blocks in that area, suggesting that robins may be somewhat spotty in distribution or less numerous there. Again, owing to the dense nature of the habitat there, thorough survey coverage occurred only along roads, and some off-roadside nesting was undoubtedly overlooked.

Suggestions of other variability in distribution within the county might also be noted. For example, minor concentrations of confirmed breeding reports appeared to fall within the middle Eel River corridor. Much typical-to-excellent breeding habitat for American Robins occurs throughout the riverine bottoms and sprawling lower foothill slopes in this area. Robins were also well confirmed in the country centered along the southwestern flanks of the South Fork Mountain uplift. This part of Humboldt is characterized by a mosaic of mixed stands of oak and Douglas-fir interspersed with hillside pastures and seasonally moist draws. Additionally, the extra hours and visits invested in completion of priority blocks in forested expanses of the southwestern part of the county paid off in confirmations that, judging from sparse evidence in the framing non-priority blocks, might otherwise have been missed.

Robins ask for little more than a forest edge or a small stand of trees, an expanse of short grass or open ground upon which to feed, a source of water for drinking and bathing, and mud for nest construction. During the long breeding season, two or three broods may be fledged (Ehrlich *et al.* 1988). This extended breeding season is represented by records of "birds carrying things": nest material, 14 Apr 1996 (b24; L. Bacon-Ogden) and food, 7 Aug 1999 (j26; JG). The earliest fledgling was seen 2 May 1996 (g14; RJA).

Much of the lowland breeding population appears to disperse gradually after nesting, thinning out noticeably by August. An annual influx of migrant and wintering flocks increases overall numbers greatly following the passage of mid- or late fall storms. More than 50,000 American Robins have been counted on some Willow Creek CBCs during years of high Pacific madrone berry production.

VARIED THRUSH *Ixoreus naevius*

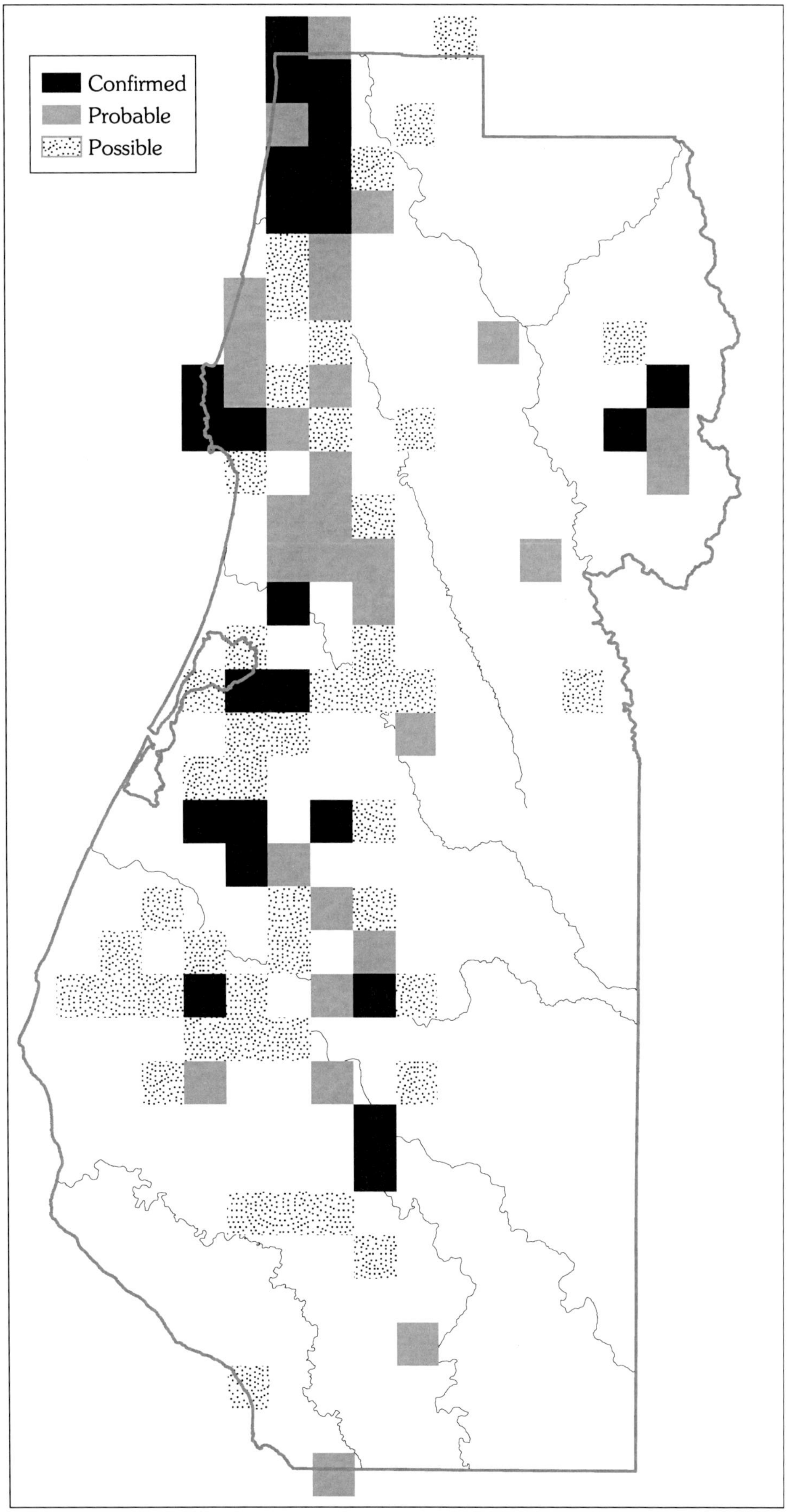

The eerie song of the Varied Thrush is the breath of dense coniferous forests in the spring. Like a high whistle superimposed on a low buzz, the song is unlike any other sound produced by a North American bird. It often remains a sound of mystery for many, but even those who fail to place a name to the voice may hold it in mind as a special sound heard in old-growth forests. Although a muted "tuning-up" version can be heard in winter in the wet, shadowed places these birds prefer, Varied Thrushes seldom sing persistently away from nesting sites. It is only in those forests lush, dark, and damp enough to support numerous breeding pairs that the full effect of multiple singers defending their territories can be savored.

The breeding range of the Varied Thrush takes in the zone of cool conifer-dominated forests along the coast and locally inland from southern Alaska to northern California. Habitat occupied by these birds from April through September is typically thick, shady forest characterized by cool to cold winters, cool to warm summers, much precipitation, and, along the coast, appreciable summer fog-drip as well. Such conditions exist southward to northwestern California, where Harris (1996) characterized Varied Thrush as a "common resident and breeder in moist, old-growth redwood and Douglas-fir forests. Common migrant and winter resident along inland valleys and ridges." Fisher (1901) provided an early published account of agitated behavior (*i. e.*, probable breeding) in a Humboldt coast redwood forest.

During the Atlas period, Varied Thrushes were detected during the nesting season in 99 (23%) of the 425 blocks, and were confirmed breeding in 24. The breeding range widely preconceived by the Humboldt birding community was reiterated during the course of the project. Varied Thrush is a member of a small suite of forest birds breeding in the county whose

distributions appeared to constitute two subregional groups. Both a coastal/near-coastal and a much smaller interior mountain population were noted, with but a weak suggestion of "bridging" blocks connecting the two segments.

The westerly, coastal or near-coastal portion of the range of Varied Thrush in Humboldt County in spring and summer extended from the Del Norte County line southward to the Mendocino County line near Whitethorn (h34). Judging from the reasonable contiguity of blocks with detections—including many reports of confirmed and probable breeding—this species was more at home in the far northwestern section of Humboldt than elsewhere. This area of comparative plenty reached south to the area northeast of Humboldt Bay, and extended inland to the east edge of Prairie Creek Redwoods State Park and Redwood National Park, thence taking in managed timberland of the fog belt south to the vicinity of Fieldbrook. South of that region, distribution changed to a heavy smattering of blocks with reports in west-central Humboldt. This included managed timberland above the lower Eel River and remnant stands of old trees in the area inland from Cape Mendocino. Old-growth redwood preserves along the South Fork Eel and Van Duzen rivers—Humboldt Redwoods and Grizzly Creek Redwoods state parks—supported Varied Thrushes as well.

Although redwoods occur commonly in forests used by Varied Thrushes in the near-coastal portion of their Humboldt breeding range, the absence of that tree is by no means a factor limiting the occurrence of the birds. Redwood does dominate the overstory in a big way within the old-growth parks, yet it is sparse or absent in other places frequented by these thrushes. Examples are Sitka spruce-dominated forests and regrowth in a narrow strip from the immediate coast inland variably some 3 to 10 miles, and stands heavy with Douglas-fir, grand fir, and western hemlock and with only a minority of spruce or redwood or neither present.

While many of the blocks in which these birds were found are intensively managed for timber production, this species clearly selects larger forest patches for nesting. In one local study, these thrushes were present during the breeding season in 16 of 17 forest fragments that were greater than about 40 acres in size, but were found in only one of 21 fragments that were smaller than 40 acres (Hurt 1996). Other landscape and vegetation characteristics were of less apparent importance, although Varied Thrush presence was related to the presence of larger diameter redwood trees (Hurt 1996). Brand (1998) found fewer of these thrushes near edges and characterized them as birds of the forest interior. While patch size and tree size are probably correlated, Varied Thrushes are known to use the tops of larger trees or snags for song posts (Beck and George 2000). Song posts are also often located on steep slopes, surrounded by a high density of trees, and somewhat near water (Beck and George 2000). Atlasers observed that previously managed stands used by these thrushes are usually dense, quite shady, and often with the summer fog-drip typical in older forests.

The interior-mountain population of Varied Thrushes encountered during the Atlas period was restricted in extent and involved, arguably, birds in as few as five blocks. Confirmed breeding was reported in only two of these five clustered blocks. One record was of an adult carrying food 1 Aug 1996 where Forest Service road 10N02 crosses a large alder wetland (P09; JEH). This area stretches from near Mill Creek Gap (o08) and southward along Packsaddle Ridge and Forest Service road 10N02 to near the headwaters of Tish Tang Creek (P11). It is a section of Humboldt County receiving high annual precipitation in normal years, much of which is in the form of snow. Although removed from the immediate coastal fog belt, the high ridges and windward slopes where these birds live nevertheless see some days of persistent drizzle and low cloud cover during late spring and summer. Mountain sites at which Varied Thrushes were reported during the Atlas period were forested chiefly with mature or old-growth Douglas-fir and white fir (and, at the higher elevations, Shasta red fir) with high canopy closure.

Continued on page 394

WRENTIT *Chamaea fasciata*

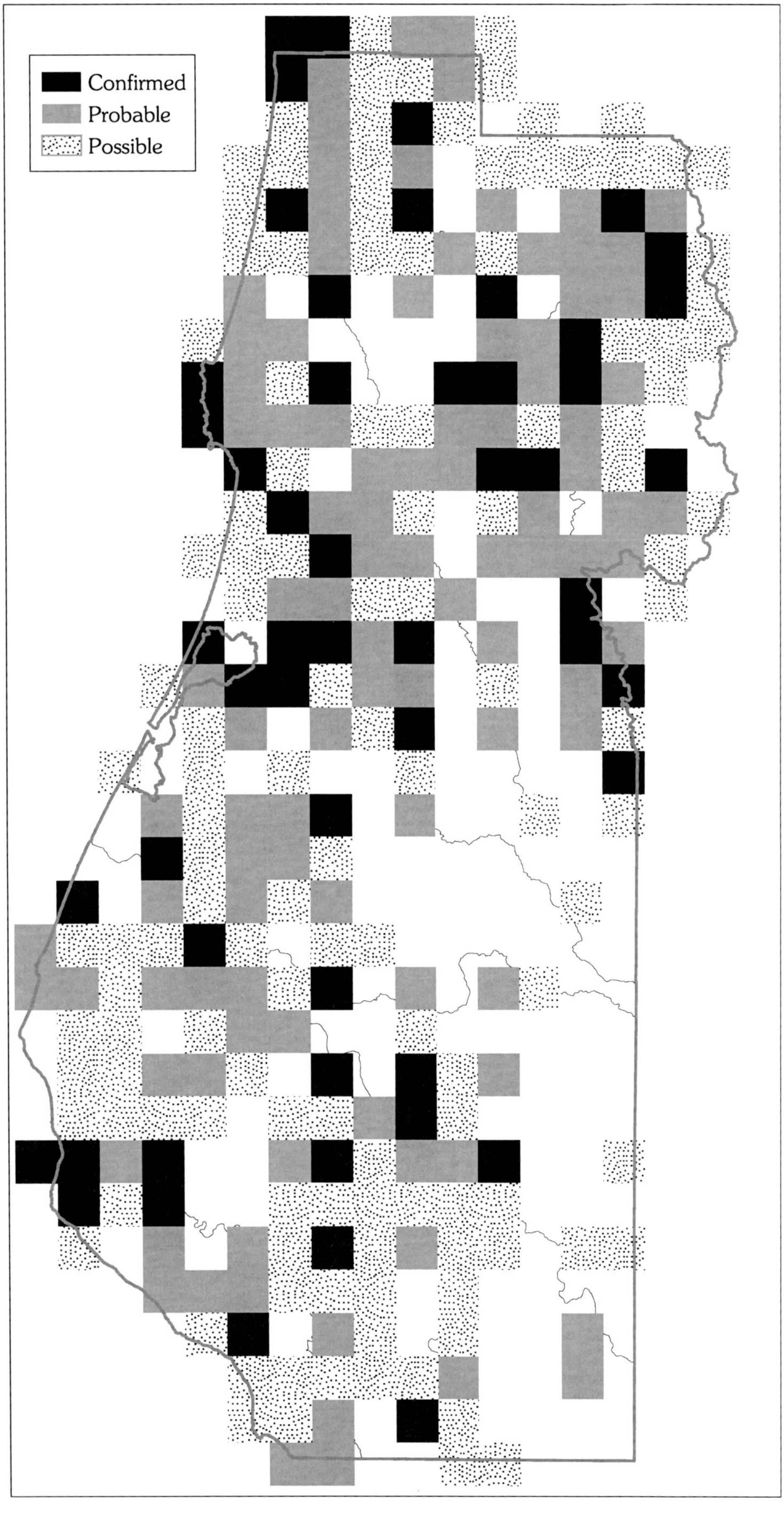

Wrentits are common residents across most of northwestern California (Harris 1996). Because they are birds of dense brush, much of Humboldt County's landscape appeals to them. The continuous mixed shrub layer so characteristic of forests in the region offers a great spectrum of habitat for Wrentits. Brushfields with few to many trees are thoroughly inhabited. The lightly managed seral growth so common in openings created by industrial forestry is widely used. Wrentits also occupy territories within closed-canopy conifer stands in which stringers or clumps of especially dense shade-tolerant brush persist. Rhododendron, evergreen huckleberry, salmonberry, and salal are among the shrubs typical at such sites. The birds' ability to remain common within the innermost bounds of the rural/suburban edge in towns is made possible by forest edge, shrubby backyards, wooded ravines, and streamside blackberry corridors. Not surprisingly, during 1995 surveys in redwood stands of different ages, Wrentits were detected more in shrub-stage stands than in other age classes (Hazard and George 1999). Wrentits were found in 282 (66%) of the 425 total blocks during the Atlas period, and were confirmed breeding in 53 of them.

Examination of the distribution map suggests that Wrentits were somewhat localized during the study over a considerable expanse of open, grassy or oak-dotted southerly exposures in the middle Mad River drainage. While this is plainly a sizable region in which Wrentit habitat is more or less wanting, access limitations may have colored survey results. Routes of public travel through the area fall chiefly along exposed slopes and ridge-tops; patches of habitat tucked near watercourses and in isolated brushfields could not be thoroughly sampled because of private ownership. Wrentits also were spotty or in otherwise undetectable numbers within portions of interior southern Humboldt extensively grown to dense, drier-site immature Douglas-fir/hardwood forest with little underbrush. Harris (1996) stated that they are more common at low elevations than in high mountain chaparral. Atlas results supported this, as fewer birds were detected in blocks within the snow zone.

The early breeding season in Humboldt is represented by a record of multiple males singing on territory in a single block 9 Apr 1995 (g12; GCH). The retiring habits of the Wrentit have been universally mentioned in field guides and other literature. It is not surprising, then, that only two nests, both with young, were located during the Atlas period: 31 May 1995 (e10; G. Lester) and 8 Jul 1996 (L09; JMH). Although the birds construct well-hidden nests, pairs respond promptly to close human intrusion, frequently popping into view with a freshly procured food item apparently intended for dependent young. Forty-seven percent of confirmed breeding reports were generated through observation of adults carrying food, which spanned the dates of 14 May 1998 (e15; GCH) to 18 Aug 1998 (d28; C. Keil). The latter record was the latest breeding confirmation. The earliest fledgling reported was seen 19 May 1996 (g01; M. Kenney).

NORTHERN MOCKINGBIRD *Mimus polyglottos*

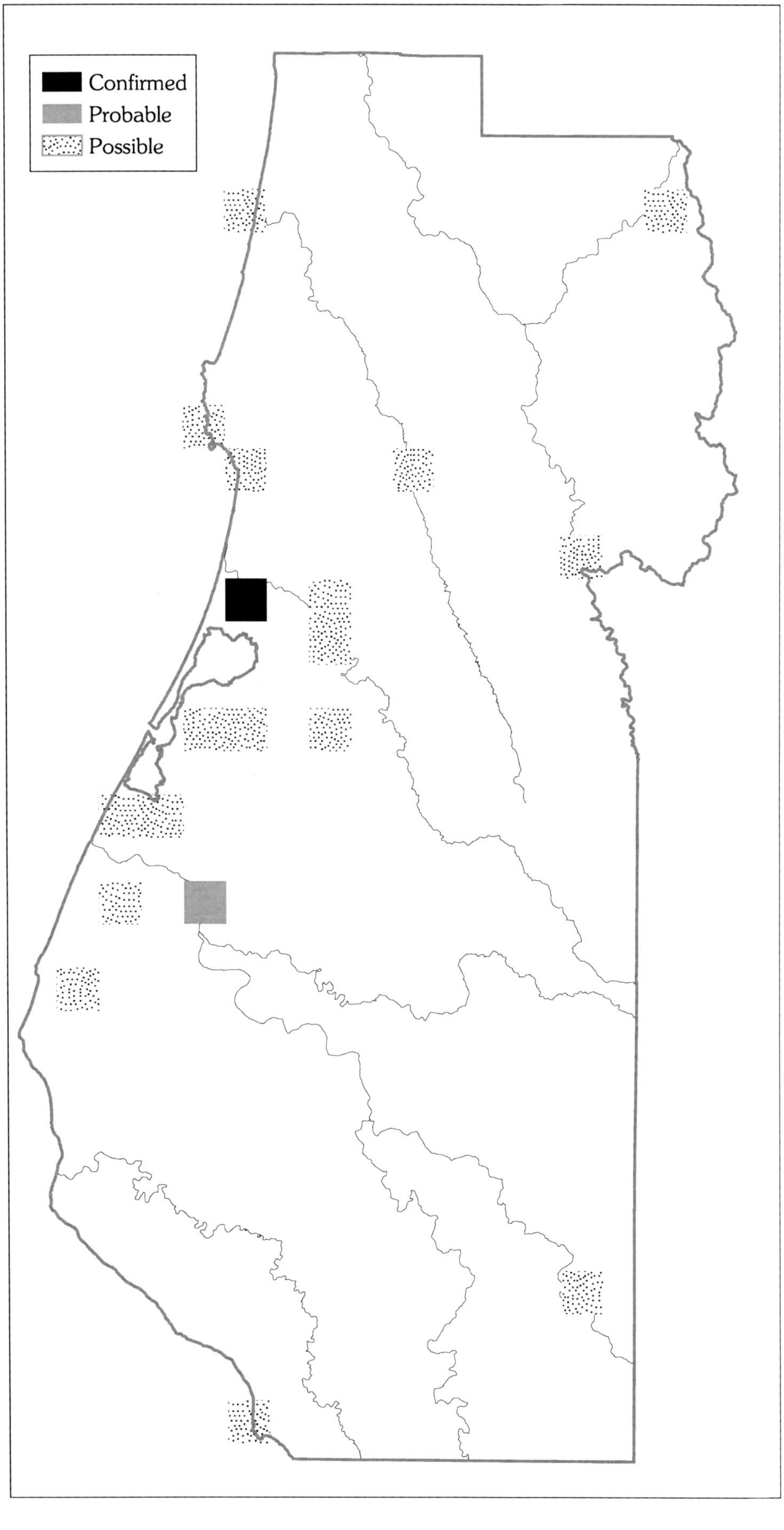

The virtuosity of song, jaunty postures, and territorial feistiness of Northern Mockingbirds are familiar to people living in regions where these birds are numerous. These behaviors are not often witnessed in Humboldt County. Mockingbirds are common across much of lowland California, yet their normal breeding range does not extend into the far northwestern corner of the state. However, the county lies sufficiently near the population frontier that individual mockingbirds are detected more or less regularly. Pairs are encountered less frequently. Some mockingbirds are known to have lingered near the point of initial discovery for months.

Northern Mockingbirds advanced northward as residents or breeding birds during the 20th century in California, with expansion taking place beyond the valleys and deserts of the southern part of the state (Arnold 1980). Having colonized the length of the Central Valley, during the 1980s mockingbirds then became established as breeding birds in the Rogue River Valley in interior southwestern Oregon, well northeast of Humboldt County. Although no great increase in sightings has occurred in Humboldt, it is conceivable that another pulse of pioneering spring overshoots, such as took place in 1995, may result in the tentative establishment of a resident population.

By and large, most of the historical reports for Humboldt County have fallen within the non-breeding season, during which time out-of-range stragglers may be most expected to pop up. Harris (1996) described the mockingbird in northwestern California generally as a rare to uncommon migrant and winter resident and a casual to rare summer resident and breeder. He noted that more than 70% of the 325+ reports in the region occurred in the 5-month period from October through February, whereas only 18% occurred during the period April–August. Humboldt's first confirmed breeding record was of fledglings seen in Eureka 24 Jul 1982 (PFS). Mockingbirds also bred in Bayside in 1994 (Harris 1996). Other early published records of this species in Humboldt were discussed by Wilder (1923), Talmadge (1953b), Legg (1959), and Harris (1996). In stark contrast to their status in Humboldt, mockingbirds are regular breeders and fairly common in some portions of Mendocino County (Carter 1998).

No evidence for any regular resident population emerged during the Atlas period. However, the confirmation of breeding by a pair that fledged several young at Moxon Dairy just west of Arcata (f14) in 1998 ranked among the marquee highlights of the project. One or two birds were first reported at this site 3 May (DFx), and were later seen carrying food 12 Jun (JEH), carrying nest material 23 Jun (RJA), and feeding young 8 Aug (M. Morris). A mockingbird singing from 29 May to 13 Jun 1997 in Fortuna (e21; R. LeValley) indicated probable breeding.

A remarkable springtime incursion of mockingbirds in 1995 resulted in a scattering of sightings from April into late summer, chiefly in the coastal lowlands. Reports during that time accounted for more than one-half of all Atlas detections. Birds seemingly swept up in this northward dynamic were detected at the California Department of Transportation yard at Willow Creek (N13) 13 May 1995 (TWL), at Alderpoint (n30) 9 May 1995 (M. Morris), and in Trinidad (e10) 2 Jul 1995 (G. Kenny); one was observed in Eureka (e17) 27 May–10 Jun 1995 (P. Spruance). A few mockingbirds probably also went undetected or unreported. It is intriguing that no breeding was confirmed in 1995. The documented successful nesting at Moxon Dairy occurred during 1998, a year in which—as is more typical—there was but one additional spring sighting in the county known to atlasers. In 1999, birds were seen 6 Jul at Eel River Wildlife Area (c19; DFx) and 3 Aug at Orleans Ranger Station (P05; COg).

Northern Mockingbirds were found during the Atlas period in 19 blocks, with most sightings in and about towns, residential areas, dairies, farms, and orchards, and in exotic plantings. Some were seen but once, while others lingered for days or even months in the vicinity of one or more favored feeding stations. Coincident with the pattern of human settlement, by far the greatest portion of sightings historically has been in the general coastal lowlands from Ferndale north to McKinleyville, with far fewer reports elsewhere. During the Atlas period, this long-term pattern continued, with mockingbirds encountered coastally from Trinidad to Shelter Cove and inland around human settlements.

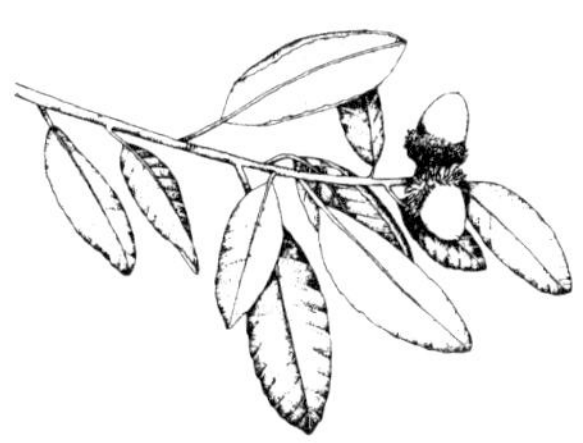

CALIFORNIA THRASHER *Toxostoma redivivum*

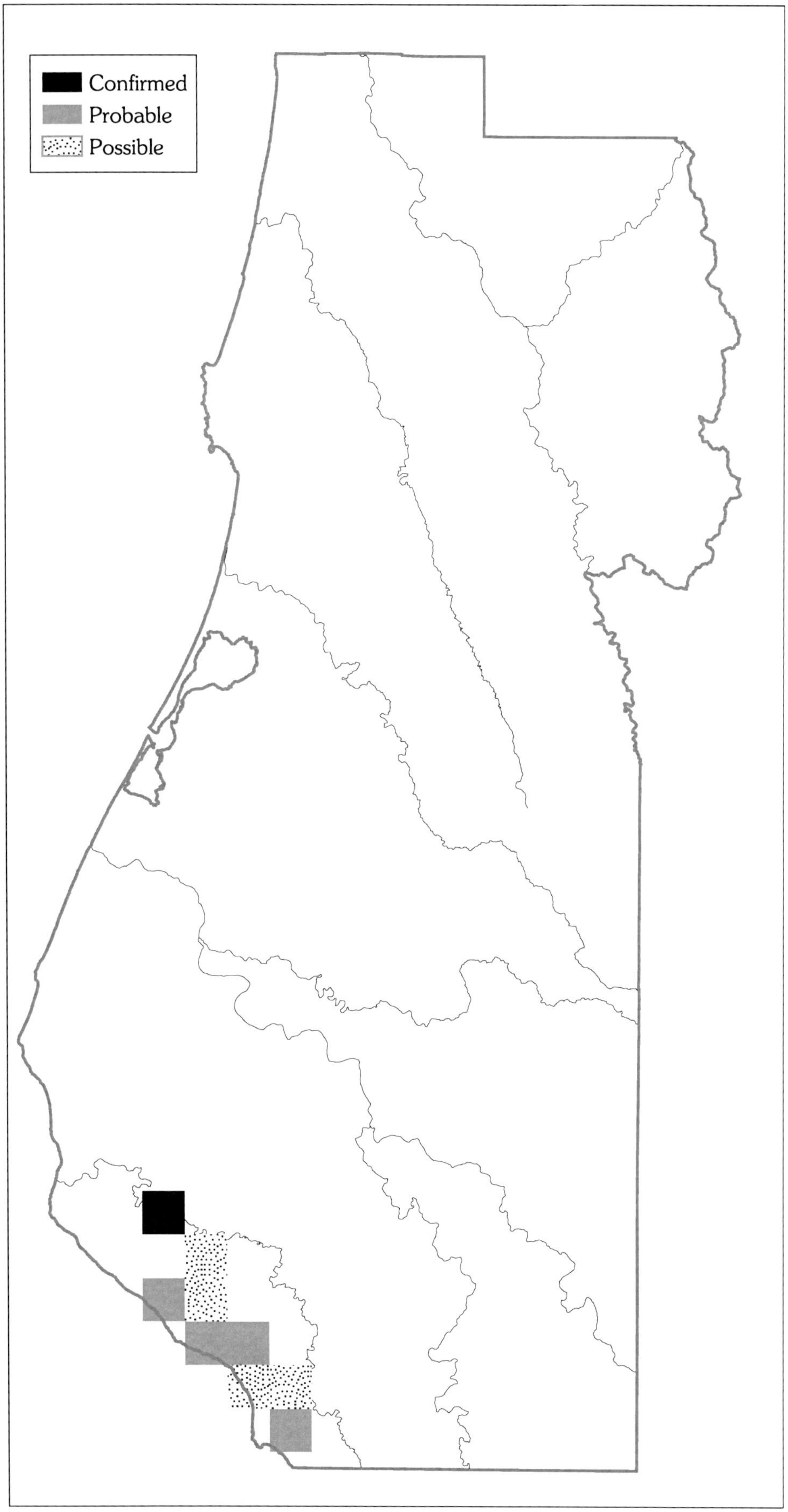

The large size, decurved bill, reclusive haunts, and restricted range of the California Thrasher lend a subtle charisma to the species. These birds are a California regional specialty, resident outside the state only in northern Baja California and recorded elsewhere only four or five times in interior southwestern Oregon (Marshall *et al.* 2003). Like its close relative the Northern Mockingbird, the thrasher is an uncommon sight in Humboldt County. Each species is encountered more routinely and in greater numbers at sites within the more classically Mediterranean climate zone of central and southern California valleys and foothills.

Unlike the case with the mockingbird, a substantial population of California Thrashers inhabits Humboldt County throughout the year. This population is restricted to the King Range of the southwestern corner of the county. That the thrasher is a bird of Humboldt County has been known by ornithologists since at least 1921, when three specimens were collected in whitethorn brush near Thorn in early October (Mailliard 1922; CAS 24024). Reports in the decades since have been infrequent, perhaps due to the remoteness of this species' range in the county.

In 1975, California Thrashers were reported as "locally numerous" in the vicinity of Whitethorn Junction (R. Sutherland). In 1979 they were observed to be abundant on North Slide Peak (R. Sutherland); this uplift stands very near the north end of the present known distribution. A pair of California Thrashers was found along the lower Mattole River at Petrolia in spring 1974 (AB 28:849), a few miles beyond the northernmost detection made during Atlas fieldwork. The first confirmed breeding record for Humboldt was of nest building seen at McKee Flat 4 Mar 1975 (R. Sutherland). During the 1980s and into the '90s, birders with a keen interest in finding thrashers in Humboldt found them on hillsides above Shelter Cove as well as along the upper slopes of the King Range proper. Many encounters were solicited by broadcasting recorded calls. Given the inconspicuous nature of this bird as well as the travails of accessing better habitat, the small yet contiguous spread of nine blocks with reports during the Atlas period suggests that fair numbers continue to occur locally in extreme southwestern Humboldt County.

Steep slopes thoroughly and extensively grown with mature dense brush provide habitat for California Thrashers. In the hills above Shelter Cove, long a customary spot to attempt to encounter a thrasher, mid-seral brush is dominated by tall blue blossom which occurs in sprawling patches, much of it competing with interspersed young Douglas-fir and hardwoods. Farther up into the King Range proper, thrashers are known to use brushfields in which not only *Ceanothus* spp. but manzanita, currant, silk tassel, and other components of a generally harsher-site shrub community are found. These brushy places offer nearly complete overarching canopy shading an open litter layer, a structure favored by the birds elsewhere within their range. As expanses of such habitat are not typical of the remainder of Humboldt County, it is not especially surprising that none were encountered elsewhere. A history of wildfire in the hills above Shelter Cove suggests the importance of fire as an agent of habitat creation, as it is at Red Mountain, another thrasher site about 20 miles east of the Humboldt line in southwestern Trinity County.

During the Atlas period, males singing on territory for a week were reported 11 Apr 1995 in the hills above Shelter Cove (g33; K. Dyer), and a pair was recorded in the same block 30 Jul 1999 (TWL). On 1 May 1999 countersinging males were recorded along Saddle Mountain Road (F31; TWL, JEH), and a pair was seen just northeast of Fire Hill (e31; JEH, TWL). The only breeding confirmation during the Atlas period was furnished by a bird with an obvious brood patch seen on Kinsey Ridge 28 May 1995 (d30; R. Sutherland). Countersinging males were also heard 21 Jun 1997 (D29; TWL).

Atlas results suggest that the birds in Humboldt occur as a disjunct population. That there were no outlying sightings suggests that preferred habitat for these birds does not presently occur in Humboldt County outside the King Range/Shelter Cove area. Should any additional colonies exist, they must be sufficiently small, or tucked into such remote haunts, that they have eluded detection. Southwestern Humboldt is exceedingly rural. Outside of the King Range National Conservation Area, most land is privately owned, and landowners are vigilant. This caused a great deal of the area to have remained essentially off limits to birders. It is conceivable that California Thrashers may occur spottily in appropriate habitat throughout this region. At the eastern edge of southern Humboldt very near the Trinity County line, an unstated number of birds was reported in January 1978 (R. Hill) 2 miles east of Alderpoint. No further sightings have come from southeastern Humboldt, nor have any thrashers been detected in the intervening country along the extreme southern edge of the county. The nearest known site in Trinity County where they have been located regularly is the Red Mountain area (Hunter and Hazard 1998).

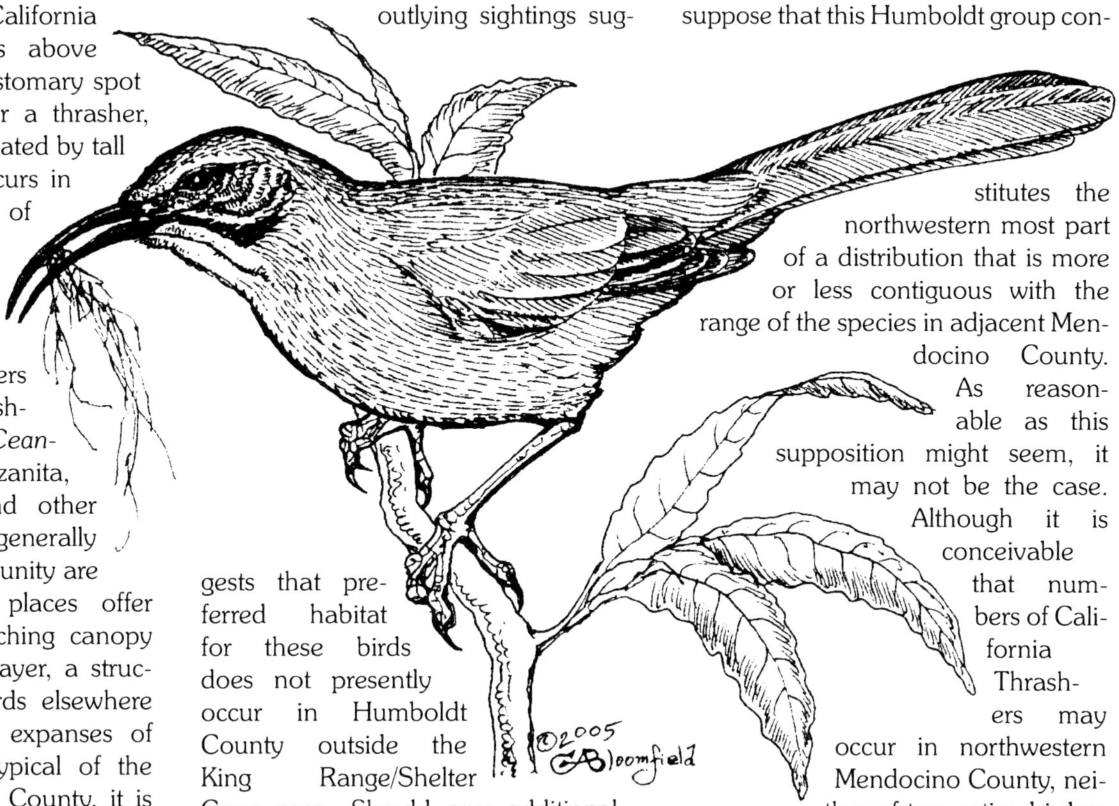

The population of California Thrashers in the King Range represents birds at the northwestern extreme of their range. It might strike one as intuitive to suppose that this Humboldt group constitutes the northwestern most part of a distribution that is more or less contiguous with the range of the species in adjacent Mendocino County. As reasonable as this supposition might seem, it may not be the case. Although it is conceivable that numbers of California Thrashers may occur in northwestern Mendocino County, neither of two active birders (D. Tobkin, R. Keiffer, pers. comm.) have found them north of the Navarro River, some 50 miles south of the Humboldt County line. The southernmost locality where this bird has been reported in Humboldt is Whale Gulch (Harris 1996) which is a few miles southeast of where atlasers found them. In Mendocino County, a thrasher was collected (CAS 24023) near Cummings (C. Lamb; Mailliard 1922), about 20 miles to the southeast of the Humboldt population.

EUROPEAN STARLING *Sturnus vulgaris*

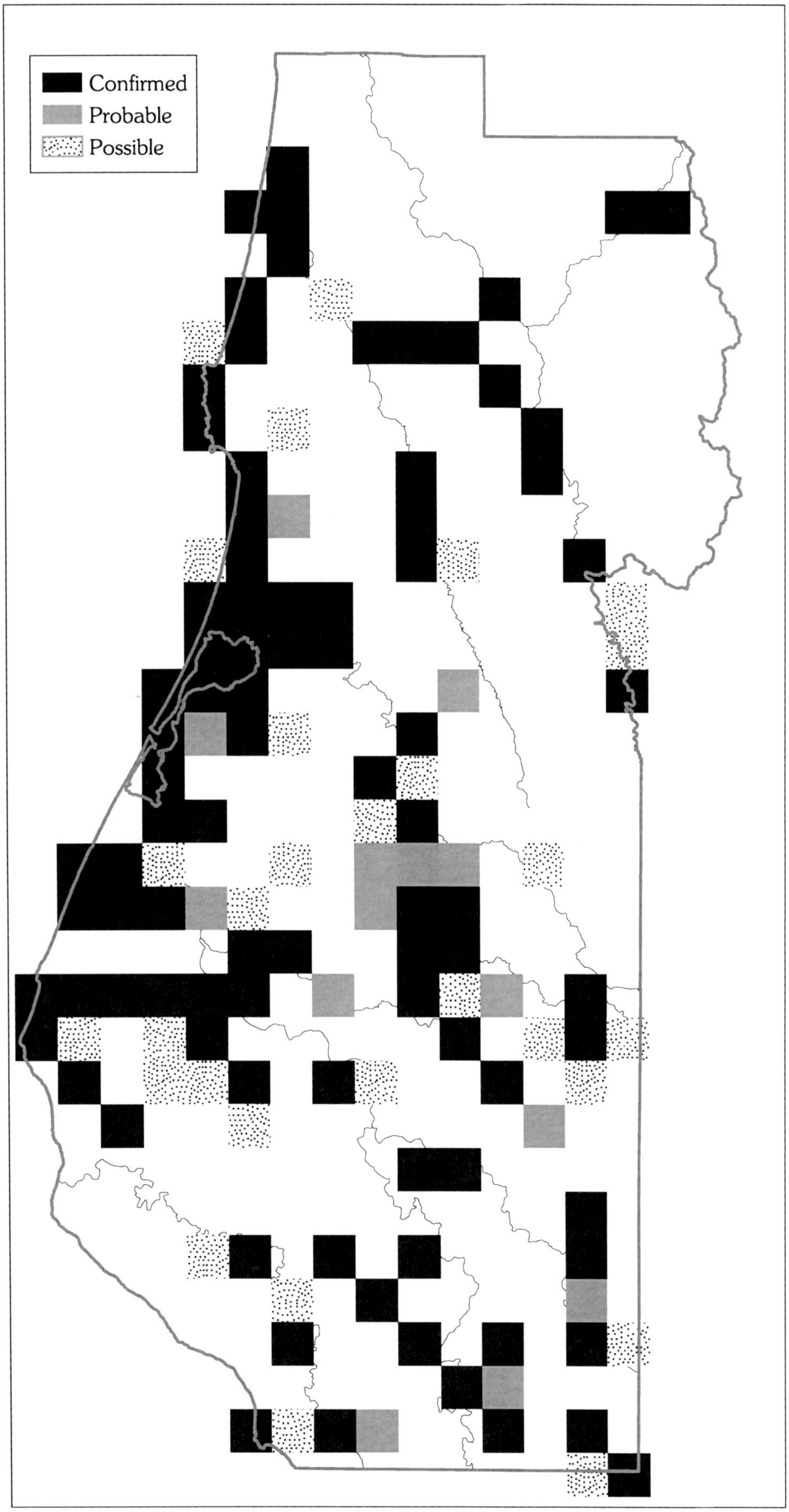

Introduced animals and plants tend to do best where the human footprint is broadest. European Starlings exemplify this concept. These birds occur widely in Humboldt County, but vary from uncommon in interior landscapes dominated by dense conifers to grossly abundant in the settled, cultivated, and grazed lowlands. European Starling is a common resident and breeder in northwestern California (Harris 1996).

In many places, starlings nearly saturate breeding habitat. The cities, towns, and pasturelands of Humboldt's central-coastal bottomlands support great numbers of nesting birds. While not as concentrated, they also breed and summer in river valleys and lower foothills within the major drainages and sparingly elsewhere. It is likely that no human settlement of any consequence in the county is free from European Starlings. While the greatest numbers of birds are found in the agricultural and human population centers around Humboldt Bay, starlings are also quite common in the Orick area, at Orleans, along the Trinity River from Weitchpec to Willow Creek, in the middle Mad River corridor, around Rio Del and Scotia, in Garberville–Redway, and spottily across the interior southern portion of the county. They were observed in 31% of all blocks during the Atlas period and were confirmed breeding in 90 of those.

Atlas results suggested that there are significant areas within Humboldt which appear to host few or no European Starlings. The northern part of the county is densely forested in intensively managed second-growth. With open ground in this region restricted to roadways, clearcut landings, and the like, it is possible that insufficient substrate for foraging exists. The extensively wooded expanses of rural Humboldt, as well as the higher mountains, similarly offer only localized and restricted habitat. This starling-poor area extends from the northern edge of Humboldt southward irregularly through the higher mountains of east county, the private timberland of the rolling country southeast of Humboldt Bay, the Humboldt Redwoods State Park area, and the King Range.

This species first appeared in the north coastal region of California in the 1950s. After the first two Humboldt County reports in 1952 (Glover 1953b), starlings had become common winter birds by the late 1950s (Yocom 1959). However, the first evidence of actual nesting was not recorded until 10 Apr 1968, when birds were observed entering a hole in a piling along Gannon Slough, near Jacoby Creek (S. Harris); by that year, starlings were very easy to find in the coastal plain April–June (S. Harris). Humboldt County was among the last regions in the lower 48 states in which starlings became established. It is clear that much appropriate habitat existed well before this time, and that the comparatively recent arrival of the species was due to geographical isolation behind the Redwood Curtain rather than the abrupt creation of breeding sites.

Although Harris (1996) stated that European Starlings commonly nest in suitable open or cutover habitats offering nesting cavities, it was apparent during the Atlas effort that most clearcuts and other logged areas in fact do not support these birds. Breeding in clearcuts does occur locally at lower and middle elevations. These secondary cavity nesters gladly take possession of old woodpecker holes and natural cavities. Nest building was observed 20 May 1995 in a cavity in an oak near Dinsmore (n24; M. Shreeve). Starlings also readily use human-provided structures; *e. g.*, young fledged 15 May 1995 from a nest in a tin roof near Blue Lake (H15; B. Hawes).

As is the case nearly everywhere else this species is found, starlings in Humboldt County routinely raise two, possibly three, broods during the long breeding season. Nearly synchronous fledging of first broods is evident each year during the last half of May, indicating that most nesting begins within a fairly confined temporal window in April. For some birds, breeding starts even earlier. Copulation was observed 10 Jan 1999 near Bayside (g15; JEH) and a nest containing young was found 15 Apr 1995 near Blue Lake (h14; C. Keil). Nests with eggs and young were reported in every month from April through July. The latest report of an occupied nest was 11 Jul 1998 (e24; TWL).

During late spring, the roadsides and open spaces across the lowlands teem with family groups of starlings. Larger concentrations consisting chiefly of fledged juveniles become conspicuous by early summer. By mid- to late summer much of the nesting population seems to disperse into favored feeding sites in farmland, about dairies, and in other open venues. Adults still feed some young as late as September, as evidenced by food-carrying seen 14 Sep 1996 (e15; JEH).

While the species is present throughout the year, much seasonal change in population is apparent. By late September, numbers seem to increase throughout the lowlands, with starlings sitting in long rows on power lines and foraging along roadsides. By this time they also assemble on power towers and at similar strategic gathering sites. During October and November, discrete flocks of European Starlings may be noticed flying southward across the dunelands and even over foothill ridges, evidently migrating by day. Other flocks settle in for the winter and become a significant part of the avian scene in the bottomlands. Pre-roost staging assemblages and nightly roosts commonly number in the thousands of birds by late fall. Upward of 10,000 starlings roosted in the two large old palms on the Arcata plaza each evening during the colder months of the Atlas period. These huge roosts make a tempting target for proponents of starling eradication, who are sadly aware that, come spring, many of these birds will disperse into nearby breeding habitats where they will aggressively usurp native cavity nesters.

CEDAR WAXWING *Bombycilla cedrorum*

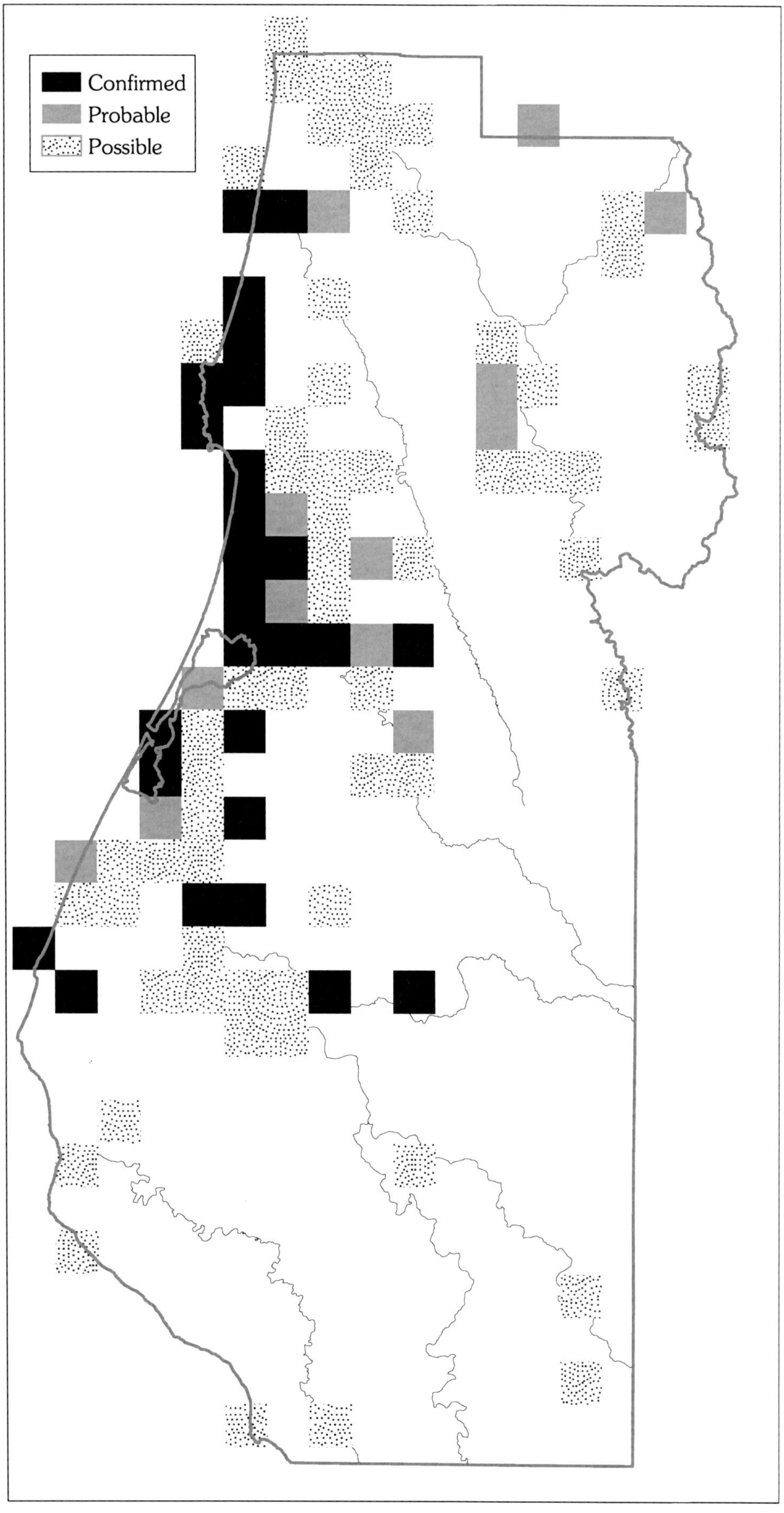

Satin-sleek plumage, an expressive crest, and waxy red feather-tips imbue Cedar Waxwings with a distinct appeal. Spilling in a hectic mass from fruiting trees and shrubs, wheezing their unvarying call-notes—and gobbling whatever red berries they can find—flocks of waxwings attract much attention in late summer and fall. Their fondness for the fruits of both native and exotic plants in residential settings cause them to incite wonder in the minds of both new and experienced observers.

Cedar Waxwings nest widely across much of temperate North America. In California, however, this species is best known as a migrant and winter visitor. It may surprise some bird students to learn that, until fairly recently, the breeding range within the state has been more or less restricted to the north coast. Only in recent years have scattered pairs and small groups lingered to nest in the San Francisco Bay area, in interior northern California, and occasionally southward well into southern California. The "stronghold" for breeding remains the moist near-coastal lowlands from the Oregon border south to coastal Humboldt County. In northwestern California, Harris (1996) considered Cedar Waxwing a "common migrant; uncommon summer resident and breeder; irregularly rare in winter." Humboldt's earliest historical breeding record was of nest building observed in Eureka 3 Aug 1913; the completed nest contained four eggs on 11 Aug (Davis 1914). Fraser (1926) reported another nest with four eggs and two additional nests with young in Eureka 23 Aug 1924.

During the Atlas period, Cedar Waxwings were found in 24% of all blocks. The distribution of these blocks revealed that waxwings may be expected annually in the nesting season within the coastal fog belt. Mirroring the known historical distribution, detections were rather strongly concentrated on the coastal slope, in the coastal lowlands, and along major river corridors. Cottonwood stands on alluvial flats or otherwise near streams were well inhabited. Waxwings were less commonly encountered in summer inland along the Klamath and Trinity rivers. Evidence of nesting was reported from shortly above the mouth of Redwood Creek south into the Eel River delta. Slightly outlying breeding confirmations were made northeast of False Cape, on lower Bear River Ridge, in two blocks along the lower Van Duzen River, and near Fernwood. This species appears to be nearly absent from extensive stretches of intensively managed forest and regrowth, as well as in the mountains generally. Closed-canopy settings of any description seem not to appeal to them.

Cedar Waxwings were only spottily detected in the interior of southern Humboldt County during the study period. That area is characterized by summer-hot foothills of oak, grass, and drier regrowth forest of Douglas-fir and hardwoods, lacking the mesic, bug-rich riparian woodland corridors favored by waxwings during the nesting season. Nevertheless, reports of possible breeding, which were somewhat discontinuous with the principal cluster of detections came from eight blocks in southern Humboldt. Waxwings were seen or heard in two blocks along the lower Mattole River; on the coast south of Punta Gorda and at Shelter Cove; at Whitethorn; along the South Fork Eel River at McCann; and in extreme southeast county at Alderpoint and near Harris. Curiously, the lower Mattole River, well-grown to white alder, seemed not to support Cedar Waxwings in detectable numbers upstream from Petrolia.

Perhaps suggesting that breeding Cedar Waxwings are more common with increasing summer moisture was that confirmed and probable breeding reports were most numerous nearer the coast, whereas detections without breeding evidence were more prevalent a bit farther inland. This effect may also have been due either to greater numbers of birds overall near the coast—providing more opportunities to confirm nesting—or to somewhat greater "backyard birding" effort in the better-settled lowlands through the 5 years of the project.

Cedar Waxwings increase in abundance in April and May each year. They typically are not early nesters; the earliest observation of nesting activity was of numerous birds carrying pampas grass up into the cypresses above Moonstone Beach (F11) 27 May 1995 (JEH). An interesting courtship display was observed 2 Jun 1996 when a pair was seen passing a large insect back and forth about four times (F17; H. Brown). The earliest fledgling was seen being fed 10 Jun 1995 (d18; P. Spruance). Nests are not especially difficult for patient observers to locate, as they typically are rather bulky, are poorly concealed, and are placed well out on a limb. A nest in a dogwood near Blue Lake (H15) had four downy young in it 26 Jun 1995, and three young fledged 6 Jul 1995 (C. Gutiérrez). Adults feeding young were reported as late as 29 Aug 1999 (F15; JCP).

After breeding, groups of dozens, sometimes hundreds, of waxwings concentrate along both slow-moving and higher-gradient streams, flycatching for the hatches of aquatic insects provided by clear waters. Streams with luxuriant, overarching foliage are particularly favored. Streak-breasted juveniles "hot off the press" often predominate in these assemblages, frequently remarked upon from mid-August through October.

Cedar Waxwings are uncommon in Humboldt County in the colder months, showing up erratically in suburban settings which offer late-fruiting shrubs and trees. In some years nearly none are reported. Inland, they seem to be rare in winter in years in which Pacific madrone fruit is sparse. In good madrone-fruit years, flocks of from a few individuals to dozens center their activity in and about lower foothill stands of regrowth forest and oak woodland containing abundant madrones. Such events often coincide with unusually large numbers of American Robins, with which the occasional bands of waxwings may loosely associate.

CEDAR WAXWING *Bombycilla cedrorum*

ORANGE-CROWNED WARBLER *Vermivora celata*

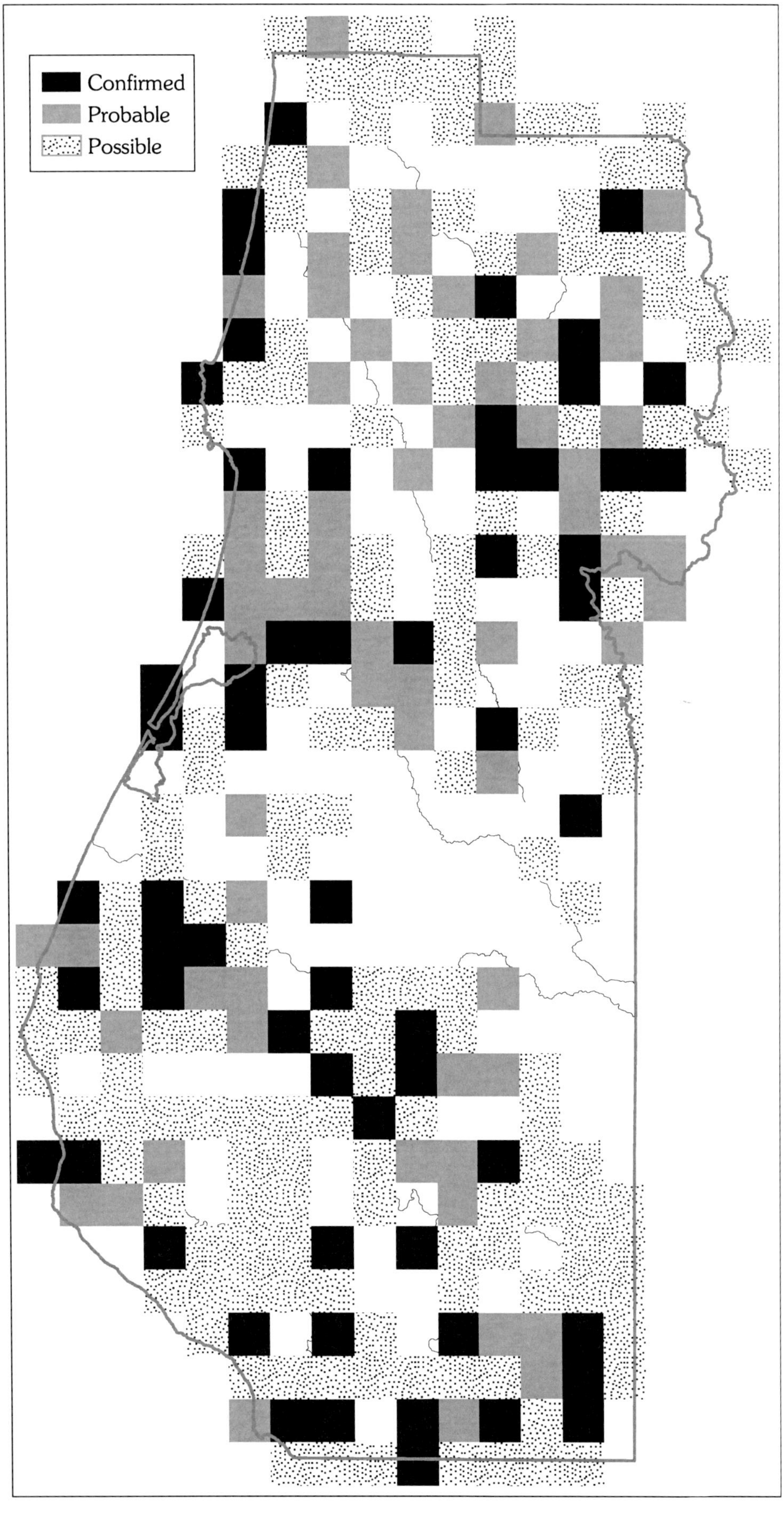

To watch birds for fun along the West Coast is to be well acquainted with Orange-crowned Warblers. They are abundant, and readily seen by those with an interest in raising binoculars at the various small birds of deciduous trees and shrubbery. This species is one of the most common breeding warblers in northwestern California, is numerous during migration, and may be encountered in small numbers in the coastal lowlands throughout the winter. Familiarity with this bird's simple good looks and appealing voice comes early in the experience of many observers.

Orange-crowned Warblers occurring in the breeding season in Humboldt County are of the subspecies *V. c. lutescens*, historically called the Lutescent Warbler. Birds of this subspecies are distinctively bright yellow below and have comparatively uniform olive-green upperparts. "Lutescent" Warblers are, with Common Yellowthroats, the earliest-arriving spring migrant warblers and are generally abundant for some weeks in April and early May. At times, flocks of from a dozen to 30 or more Orange-crowns may swarm in leafing willows, alders, and cottonwoods in stream corridors, and in flowering California black oaks across the foothills. At the peak of spring migration—with a great array of habitats occupied, and all elevations below the snow line inhabited—it is conceivable that tens of thousands of these birds may be present in the county. At the same time, differently-marked individuals, thought to belong to the northern, nominate *V. c. celata* or to *orestera* of the intermountain West, are occasionally seen. Orange-crowned Warblers that appear to be females predominate in the later part of the northward movement. The glut of passage birds abates by early to mid-May, leaving the numerous breeding birds and, by June, their progeny.

Orange-crowned Warblers proved to be the second most widely-distributed warbler species during the 5 years of Atlas surveys. They were detected in 278 (65%) of the blocks, compared to 73% of all blocks for Wilson's Warbler and 61% for Hermit Warbler. Breeding was confirmed in 60 blocks. Their range within Humboldt County is impressive. It includes all sections of the county but for the oak/grassland country of the central Mad River drainage and a contiguous cluster of no-detection blocks—comprising a spectrum of habitats—from near Showers Pass southeast into the Lassics. This zone of seeming absence was nearly continuous with scattered no-detection blocks reaching westward into the Eel River delta.

Habitats occupied by breeding Orange-crowned Warblers vary across the wide range of this species in the boreal North and in the West. Humboldt County offers these brush-loving birds a great many sites to choose from. This warbler is among the most routine summer residents of regrown, neglected slashings; riparian corridors; clearcuts, especially on northern exposures; in forest edge/brushfield situations; along power line rights-of-way; in naturally occurring brushfields; in chaparral on coastal slopes and bluffs, and—given sufficient shrub cover—in closed-canopy oak woodlands. Harris (1996) characterized Orange-crowned Warbler in northwestern California as a "common to abundant summer resident and breeder; uncommon to rare in winter." He described habitat as "brush/forest edges, riparian areas, and brushfields throughout the region." Atlas results supported these characterizations.

Multiple males singing in a block (o16) were reported 2 Mar 1997 (TWL), but some of these birds were probably migrants. The earliest report of a male singing on a breeding territory for a week was 20 Apr 1996 (f14; KMS). Blocks with breeding confirmations were numerous and were well distributed among all blocks with detections. The early fledgling reported 17 May 1997 (N31; M. & M. Bailey) was a temporal outlier. A nest with eggs on 20 May 1999 (H31; R. Sutherland) was the only one reported, as was the report of a bird carrying nest material 22 May 1997 (n08; JMH). Observations of adults carrying food and feeding young accounted for 91% of confirmations. Food-carrying was recorded as early as 7 May 1995 (k31; JEH), and the feeding of young was reported as late as 19 Aug 1998 (g13; J. Haggard).

Apparently lone adults and dispersed fledglings were observed from mid-July on and it appeared that post-breeding wandering or southward migration had begun, well within the breeding season of other brushland songbirds. During the latter half of July, atlasers exercised restraint in interpreting such observations as evidence of more than possible breeding status. Fall migration of Orange-crowns not only begins in midsummer, it is protracted, with apparent migrants expected outside of wintering habitat into October.

From November through March, most of Humboldt County appears to be vacated by these warblers. A barely annual presence on the Willow Creek CBC hints at only occasional birds remaining in the interior of the county; that count averaged less than one bird during 1995–1999. However, in the coastal lowlands, particularly around Humboldt Bay and in the Eel River delta, small numbers of Orange-crowns linger into or through the winter. During the colder months, they are detected most frequently in mixed native and exotic vegetation in towns, in riparian areas of any description, and in various other brushy situations on the coastal plain and in the lower river corridors. Himalayan blackberry was commonly present in places supporting wintering birds. Average numbers encountered from 1995–1999 on the Arcata and Centerville Beach to King Salmon CBCs were 28 and 36 birds, respectively.

NASHVILLE WARBLER *Vermivora ruficapilla*

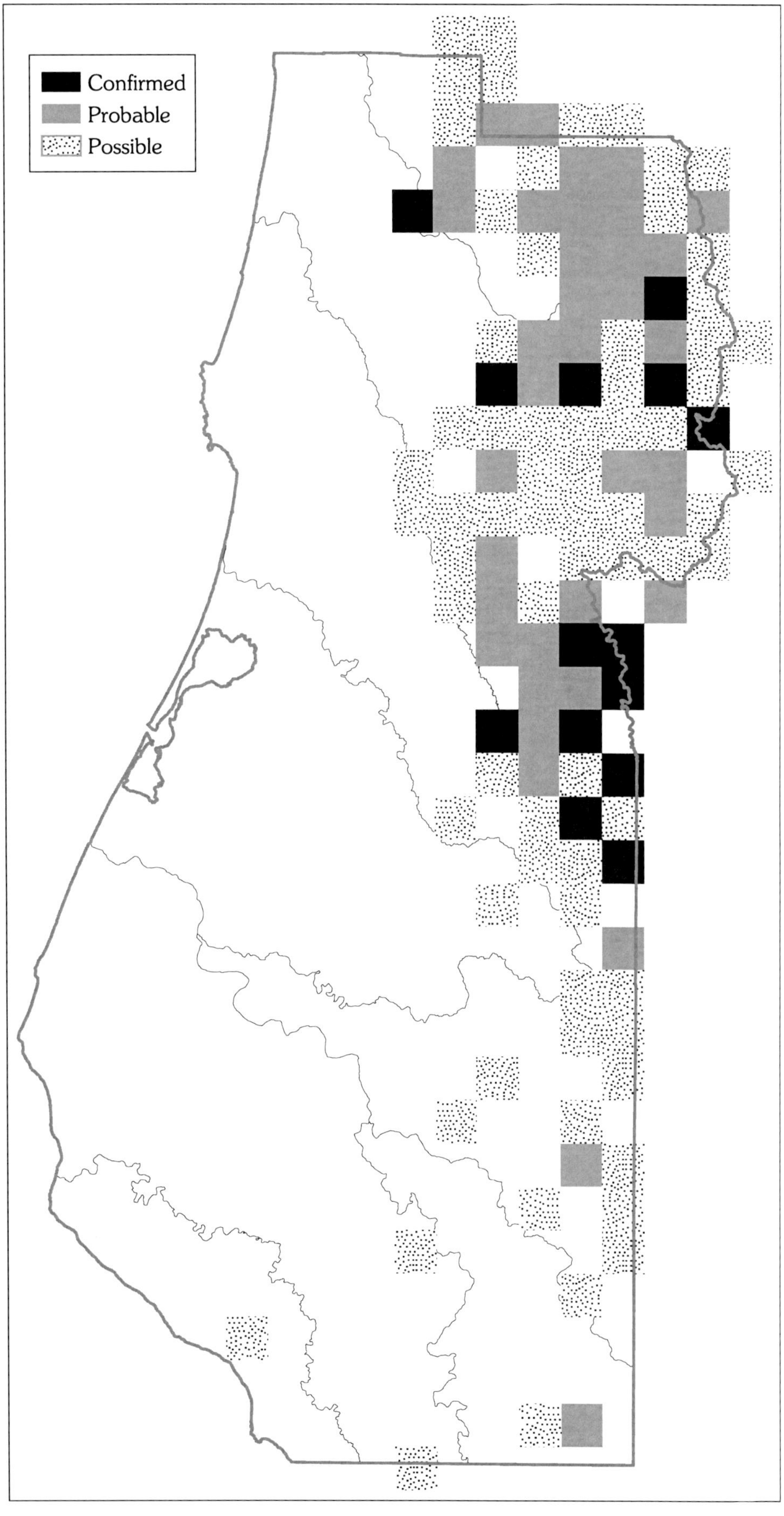

Active and attractive, Nashville Warblers can be coy with observers, often somehow managing to show but a part of their bright plumage at a single glimpse. These birds inhabit many different forest and shrub communities across their wide range during the breeding season. As if echoing this theme at the local level, their preferences in Humboldt County are not easily summarized. This is a spring and summer bird of canyon live oak and California black oak stands, clearcut brushfields, and old burns. Nashville Warbler has been considered a "common migrant, summer visitor and breeder" in northwestern California (Harris 1996). It was reported from 28% of all blocks. Confirmed breeding was established in 14 blocks.

Many, though not all, Nashville Warblers summering in Humboldt County occur at middle and higher elevations in the interior. Although that area overall constitutes but a modest portion of the county, a great deal of the suitable habitat therein is well populated with these warblers. This contention is supported by the noticeably thorough pattern of blocks with detections, as well as by the high proportion and telltale "richness" of blocks with confirmed and probable breeding there. This was a routinely encountered summer bird across much of northeastern Humboldt from along the lower Klamath River and along the Salmon/Trinity River divide southward along South Fork Mountain and neighboring slopes, through the higher country of the Lassics uplift, and continuing into the southeastern corner of the county.

The distribution outlined above is similar to that of many other mountain birds breeding in Humboldt County. However, Atlas results indicate that, unlike most of those species, Nashville Warblers also occur at lower elevations. A few were encountered during the breeding season at scattered locations some miles southwest of their primary range within the county. Such birds were found in suitable breeding habitat several miles east of Whitethorn (j34) 25 May 1998 (DFx) and in the King Range (F31) 27 May 1995 (KMS). A Nashville was seen in the Eel River corridor near Miranda (J29) 29 Jun 1996 (TWL).

A great many Nashville Warblers summering in Humboldt County are found in the seral brushfields of regenerating clearcuts. Owing to this predilection, they probably have increased in abundance because of logging (Raphael *et al.* 1988). Such sites typically are dense, mixed-species expanses of *Ceanothus* and *Arctostaphylos*, bitter cherry, currant, and similar shrubbery, typically dotted with residual trees or regrowth conifer saplings. Owing to variables of elevation, slope, aspect, and logging history, a wide range of brushfields was found to be occupied. Typically, Nashville Warblers appeared most numerous in clearcuts which were logged from 5–20 years previously, and which had not yet developed a closed canopy of young conifers. It would seem that wildfire suppression has probably diminished naturally occurring habitat for this species. In pre-settlement times, an irregular mosaic of uneven-aged forest cover must have existed over much of interior Humboldt County, which would have provided a more or less stable amount of appropriate habitat.

The rush to extract forest products from public and private timberlands during the several decades before the Atlas period created an unnaturally great amount of mountain brushfield. Given recent reductions in harvest levels on the Six Rivers National Forest, home to a large percentage of all Nashville Warblers summering within Humboldt County, it would appear that less human-made habitat is now being created. Habitat that is currently suitable will grow less so as the conifer component reasserts its dominance, resulting in loss of early-seral brush and in consequent abandonment of these places by the warblers, Dusky Flycatchers, and Fox Sparrows which find them attractive. A decrease in timber harvest during the 1990s may play a limited role in determining the abundance of such species. Perhaps an increasing proportion of their local populations may become confined either to naturally occurring forest openings or to those clearcuts in which conifer establishment is indefinitely retarded by biophysical conditions favoring persistent brushfields.

Of the 14 breeding confirmations, one was of a bird carrying a fecal sac 5 Jul 1997 near Oak Knob (N15; DFx). There were eight reports of adults carrying food, spanning the dates of 19 Jun 1995 (o18; JEH) to 17 Jul 1996 (L09; JMH). The five reports of fledglings ranged from 2 Jul 1997 (N09; JMH) to 17 Aug 1996 (q10; JMH).

Nashville Warblers arrive in Humboldt County in mid- and late April and are commonly seen into early August. The late summer exodus of breeding birds and progeny is seldom remarked upon and remains poorly perceived. Similarly, southbound migrants are infrequently observed. While nearly the entire population retires to tropical America for the colder months, Humboldt County is notable in supporting a significant complement of Nashvilles in late fall and early winter. Most of these reports originate from November through December. Historically, both the Arcata and Centerville Beach to King Salmon CBCs have recorded small numbers of such lingerers nearly annually. It seems likely that intensive birder coverage of favorable habitat microsites supporting these late fall birds prior to and during the CBC season, followed by minimal checking after the CBCs, has skewed "winter" records toward the earlier end of the season. Nevertheless, a smattering of reports during late January, February, and even early March suggests that at least a minority of late-lingering Nashville Warblers successfully overwinter in the Humboldt Bay and Eel River delta lowlands.

NORTHERN PARULA *Parula americana*

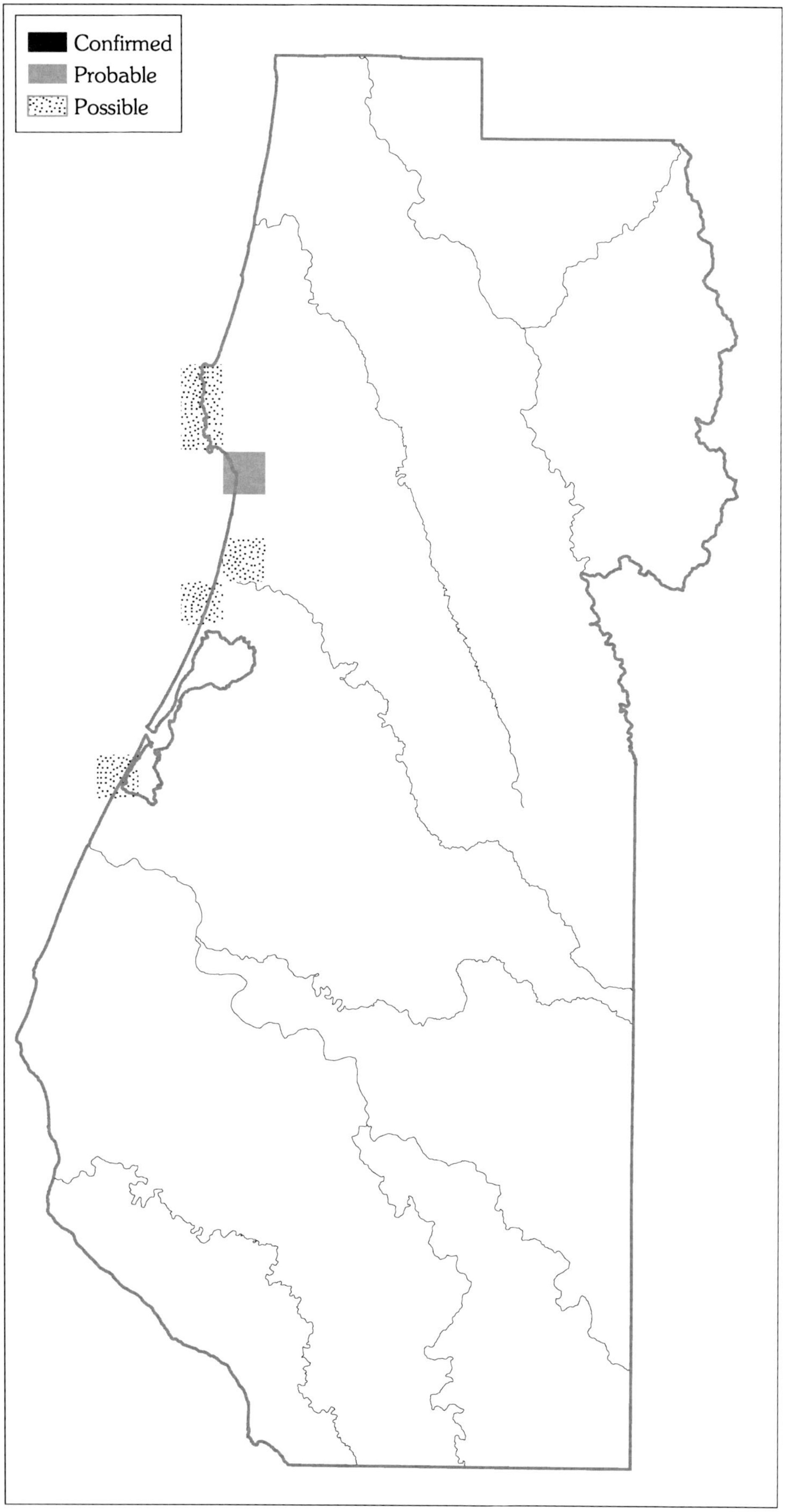

Of the 40 species of wood warblers that have occurred in Humboldt County, 10 breed regularly, one (American Redstart) does so irregularly or in tiny numbers, and another, Northern Parula, is known to have nested once. A pair, with the male present from 27 Apr, fledged three young from a nest along the main trail at Elk Head, Trinidad State Beach, in summer 1992 (G. Lester, DFx *et al.*). This surprise was made less shocking by two facts: the species had been one of the more regularly occurring vagrant warblers in spring, particularly from mid-May to early June; and breeding in California had been previously documented. California's first breeding was in 1952, when one male and two females attended two nests and fledged two young in Monterey County (Williams *et al.* 1958). Taking place as it did in an era preceding the mass search for vagrant landbirds at strategic coastal locations in California, the Monterey episode represented not only the first nesting but the first known occurrence of this species in California. See Shuford (1993) for information on subsequent nesting in Marin and San Mateo counties.

One would reasonably assume that occasional breeding by "eastern" warblers in California is correlated with unusually strong influxes of spring vagrants or overshoots. Indeed, Terrill *et al.* (1992) and Patten and Marantz (1996) provided discussions of the remarkable 1992 invasion, marked by scattered breeding records in the west by birds typical of the southeastern U.S. Although they were not found nesting in Monterey County during 1992, the 11 Northern Parulas reaching the Monterey coast that year were a record number for one season (Roberson and Tenney 1993). The seven records for northwestern California that year were likewise a record number (Harris 1996). Thus, the many parulas that reached northern California April through July 1992 undoubtedly helped set the stage for the pre-Atlas-period nesting at Elk Head.

The case of Northern Parula as a spring vagrant and as a breeding species in Humboldt County involves a well recognized history of both late spring appearances and apparent or proven over-summering. While this dynamic is not unlike that of other eastern warblers which have shown up in the county, it differs in its greater intensity. Through the spring of 1996, parulas had appeared 20 times in spring and early summer in northwestern California (Harris 1996), with most of those found along the outer coast of Humboldt. Many of the Humboldt breeding season records have been from Elk Head, with territorial males reported there in 1987, 1992, 1993 and 1996 (Harris 1996). Much of the habitat at the Elk Head nesting site was composed of large red alders mixed with Sitka spruce; both tree species hosted ample amounts of pendant, hair-like lichen. Northern Parulas habitually build nests in flags of lace lichens such as *Usnea* or *Ramalina*. This tendency is so pronounced that, should a pair find themselves without access to appropriate lichens, they may not attempt to nest.

The previously documented pattern of spring Northern Parulas persisted through the Atlas period, with birds recorded in six blocks in four different years. Although there were no instances of confirmed breeding during the Atlas period, probable breeding status was afforded the record of one or two males singing on territory 30 May–13 Jun 1999 along Crannell Road (F11; K. Irwin). In that same block, a male sang 9–11 Jun 1999 along Scenic Drive south of Trinidad, 10 Jun 1999 near Westhaven, and 20 Jun 1998 at the west end of Crannell Road (all K. Irwin). The remaining five records were as follows: at Lanphere Dunes 1 Jun 1995 (e14; C. J. Ralph); at Patricks Point State Park 4 Jun 1997 (e09; K. Irwin); a singing male at Table Bluff 29 Jun 1996 (c18; J. Danzenbaker); adjacent to a McKinleyville yard 2 Jul 1997 (F13; GAS); and at Elk Head 3–4 Jul 1996 (e10; S. Rovell, TWL). To complete the picture, it should be noted that single parulas at the Fairhaven willows (D17) 16 Jun 1999 (G. Lester) and 29 Jun 1995 (S. Rovell) were suspected to be merely extremely late, disoriented migrants without mates and with no real hope of nesting.

YELLOW WARBLER *Dendroica petechia*

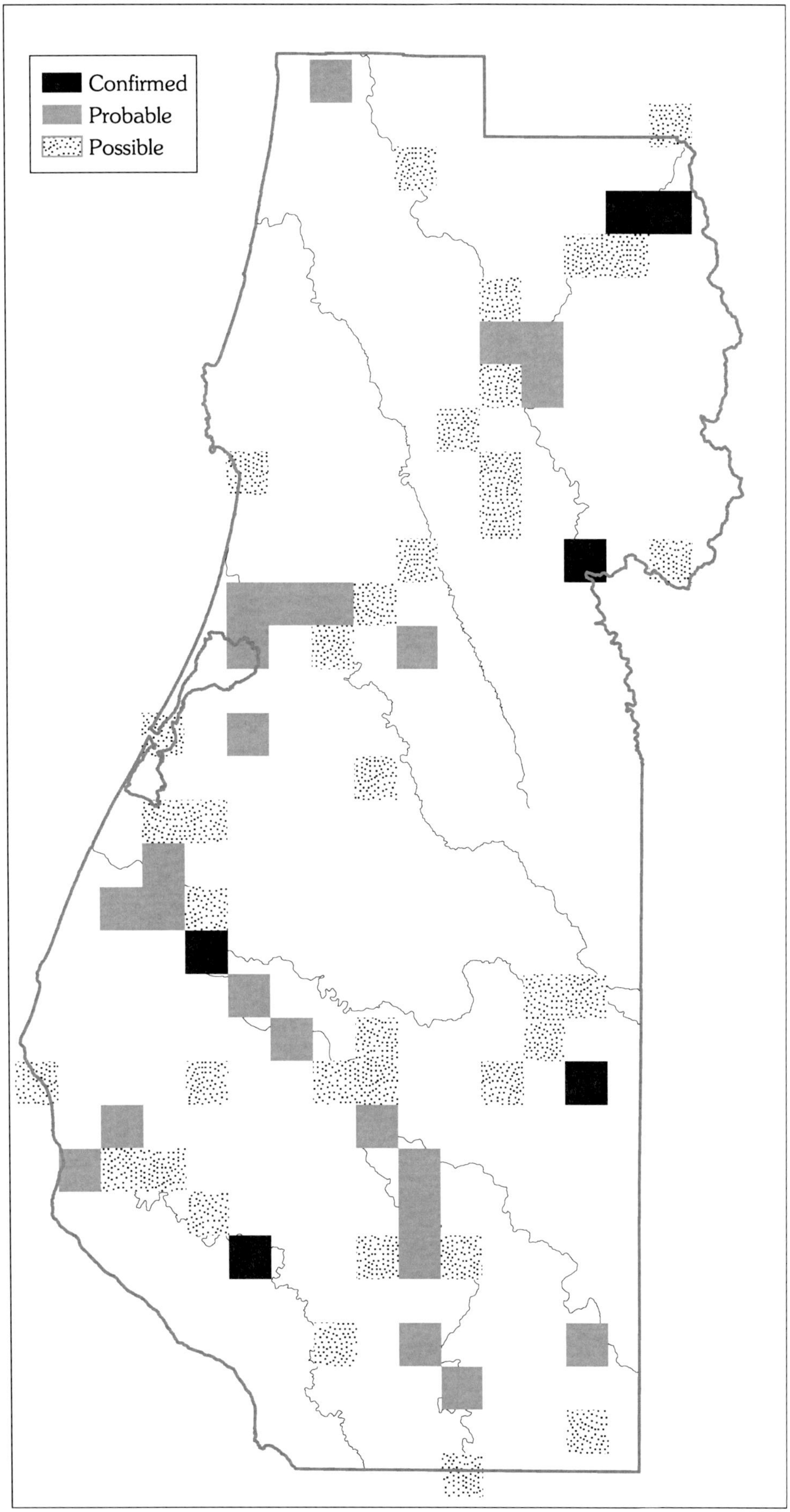

The Yellow Warbler is to be looked for in spring and summer in deciduous hardwoods along creeks and rivers, by lake shores, at the willow-grown edges of freshwater marshes, and in the shade trees of favored towns and cities. This warbler enjoys the most extensive breeding range of any North American parulid, occurring in appropriate habitat across most of the continent. Forty-three subspecies are recognized. The taxonomy of Yellow Warblers has been debated, with one school advancing the belief that these taxa can be sufficiently sorted into three groups of subspecies, each group together representing a separate species (Dunn and Garrett 1997). In California, members of the northern *D. p. aestiva* group of subspecies breed locally but widely, and migrate through the state in spring and fall. In recent decades, small numbers of these birds have become expected in winter at various sites, mostly along the coast and particularly in southern California.

Yellow Warblers have long been known to occur in Humboldt County. The first formal record was a specimen in the Fiebig collection taken in the vicinity of Eureka in 1889 (Harris 1996). Owing to habitat constraints clearly imposed by a corrugated and incised landscape, Yellow Warblers are not especially abundant over Humboldt County as a whole. However, they are locally numerous along the county's rivers and, to some extent, the lower reaches of their tributaries. Harris (1996) characterized the Yellow Warbler in northwestern California as a "locally common summer resident and breeder, common migrant; casual in winter. Breeders are associated closely with alder–cottonwood–willow stands in riparian cover." Yellow Warblers were reported during the Atlas period in 66 (16%) of the 425 total blocks, yet they were confirmed breeding in only six.

Results of Atlas surveys strongly suggest reliance by Yellow Warblers on vegetation along or near larger streams in Humboldt County. The map expresses a distribution similar to that of other aquatic or riparian bird species, as the stream courses of the Klamath, Trinity, lower Mad, Eel, Van Duzen, and Mattole rivers are suggested by the tracery of blocks with detections. Yellow Warblers were detected during 25% of all point counts within riparian habitats along the lower Eel River in 1999 (R. Hewitt, unpubl. data). During the Atlas period, they were found in an uninterrupted succession of blocks along the Eel/South Fork Eel corridor from Fernbridge (d20) to Miranda (J29). The lower Mad River bisects a cluster of probable breeding blocks (f14, F15, g15) and a possible breeding block (H15), yet no birds were detected above Blue Lake. It is likely that suitable riparian habitat occurs along some reaches of the Mad, as well as other streams, which atlasers could not access. Undetected, localized populations of Yellow Warblers presumably existed in at least a few blocks for which there were no reports.

Yellow Warblers were virtually undetected in the industrial timberlands of west-central and northern Humboldt County, regions under variably intensive management for coniferous timber. Creekside vegetation within the clearcut, brushfield, and regrowth mosaic is usually abundant, but as most small drainages are rather steep, such vegetation is characteristically restricted and simple. Narrow corridors of red or white alder undergrown to red elderberry appear to fall short of providing suitable breeding habitat for these warblers. In interior and southern Humboldt, rolling pastureland slopes dominated by grassland and oaks provide few opportunities for Yellow Warblers, nor do the extensive drier-site regrowth forests into which that zone broadly interdigitates. There were no detections at higher elevations. Unlike the pattern exhibited by many songbird species during the Atlas fieldwork, detections of Yellow Warblers were not increased within the rough crescent of blocks about Humboldt Bay. This would appear to reflect lack of suitable habitat in the area, which is either developed land or open, mostly treeless bottomland dotted with riparian fragments too small to support them.

Most Yellow Warblers summering in Humboldt County do so in riverside or floodplain stands of black cottonwood, often with other hardwood trees and shrub-trees. Tall and extensive willow or Oregon ash thickets occurring locally along each of the above-named rivers also are inhabited, as are riparian stringers on flats or on gently sloping ground. In Eel River delta riparian habitats, Yellow Warblers breed annually but only in low numbers, probably because they favor only those specific areas with low alder cover, high shrub cover, and high shrub and ground cover species richness (Kelly 1987). Yellow Warblers do not appear to summer in any numbers in residential areas or urban parks in Humboldt, but during the Atlas period they were noted around the edges of small towns in the interior of the county.

The proportion of detection-blocks with breeding confirmations for this species was conspicuously small. It is worth noting that, with the exception of two adjacent blocks with confirmations, these lay well scattered across the county. The two adjacent blocks represent a considerable population found in better-quality Klamath River riparian habitat at Orleans. In block o05, physiological evidence of breeding was noted during banding 15 Jun 1996 (J. Tietz), and in P05 adults feeding young were observed 26 Jun 1996 (KMS). The remaining four breeding confirmations were in four different watersheds. The carrying of nest material was recorded on the Trinity River at the confluence with Willow Creek (N13) 22 May 1996 (TWL), and on the Eel River at the confluence with the Van Duzen River (e22) 28 Jun 1998 (P. Herrera). Food-carrying was witnessed along Honeydew Creek at the confluence with Bear Trap Creek (F29) 1 Jun 1997 (JEH, L. Hunter). Finally, the only fledged young were seen on the Little Van Duzen River (N25) 11 Jul 1999 (DFx).

In the 24 blocks with probable breeding evidence, observations of pairs supported this status in nine. The most coastward of these sites with pairs were near the Mattole River mouth (B27) 3 Jun 1995 (GCH), on North Fork Ah Pah Creek of the lower Klamath River (h02) 13 Jun 1998 (M. Kenney), and near Ferndale (c21) 5 Jul 1997 (JEH). The earliest probable breeding record was of countersinging males at Tooby Park, Garberville (k32) 7 May 1995 (JG, JEH *et al.*).

Yellow Warblers are not among the neotropical migrants that arrive relatively early in Humboldt County; they typically begin to be noticed only in the last week of April. A general wave of northbound migrants moves through various leafy habitats—including the more luxuriant microsites in towns and cities—during early and mid-May. Many of these birds are conspicuous singing males, which seem to remain from one to several days before disappearing. After the nesting season, dispersing adults and increasing numbers of juveniles make use of nearly any broadleaf habitat available at lower elevations, particularly in late August and September. They are common inland as well as in the coastal bottomlands at this time; the total volume of fall passage through the county appears to greatly exceed that of local breeders. A few late migrants were reported from October into December during the period of the Atlas. This is believed to represent a normal pattern of occurrence. One or two birds are sometimes encountered at lowland sites near the coast in winter.

While declines in Yellow Warblers have been noted elsewhere in California because of loss of riparian habitat, any trends in Humboldt have been undocumented. Riparian areas used

Continued on page 394

YELLOW-RUMPED WARBLER *Dendroica coronata*

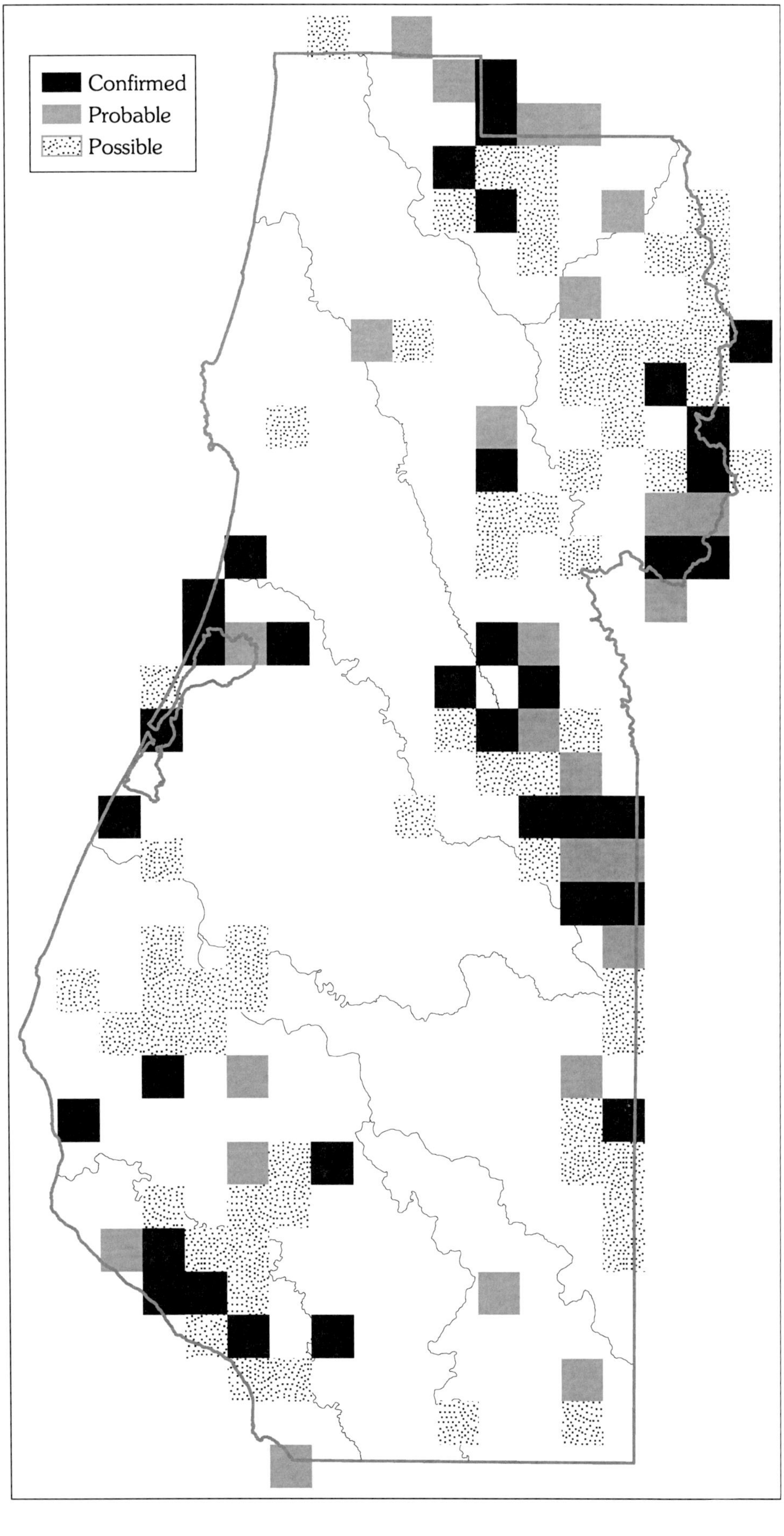

The bird called Yellow-rumped Warbler comprises several subspecies, two of which were historically recognized as species. They are comfortably referred to by their old names whenever a distinction is to be made. Each of these forms, the "Myrtle Warbler" (*D. c. coronata*) and "Audubon's Warbler" (*D. c. auduboni*), occurs commonly in California. Myrtle Warblers breed in Alaska, in the northeast U.S., and across boreal Canada, wintering along both coasts and in the southern U.S. and Mexico. They are abundant in migration and in winter along and near the coast of Humboldt County, and uncommon inland.

Audubon's Warblers occur widely throughout lowland and foothill Humboldt during migration and are common in winter in river valleys and in the coastal lowlands. While the Myrtle Warbler is unknown here in late spring and summer, Audubon's Warbler has been characterized as a "common summer resident and breeder in higher (above 4000 feet) conifer forests (pines, open-canopy Douglas-fir, and other firs)" of northwestern California (Harris 1996). Harris (1996) further noted that limited numbers of Audubon's also breed in shore pines on the north spit of Humboldt Bay. During the Atlas period, birds of this subspecies were found in 123 (29%) of the 425 total blocks. Breeding was confirmed in 35 blocks.

Atlas fieldwork indicated that the distribution of this species in Humboldt County in summer is somewhat patchy, with slightly different habitats occupied in different parts of the county. Audubon's Warbler is most common as a breeding bird in the higher montane forests, and specifically preferring stands of conifers whose crowns only infrequently touch. With respect to the latter trait, it is felt that this consistent preference is key to making sense of the otherwise somewhat puzzling countywide distribution of this species.

This is one among the characteristic breeding birds routinely inhabiting forests within the zone of winter snows in eastern and northeastern Humboldt County. Open or uneven-aged forests of pine, true fir, Douglas-fir, and incense-cedar are thoroughly occupied, with the clear air full of this species' song nearly everywhere from early April into mid-summer. This region extends from mountains above the Klamath River canyon south through the higher country near the Salmon–Trinity River divide, on Snow Camp and South Fork mountains, and southward into the Lassics. During extensive surveys in the Klamath Mountains, Sniado (1999) found this species strongly associated with white fir, and with an affinity for higher elevations (between about 500 and 5400 feet, but with a median elevation of 4200 feet). For this high-elevation subpopulation of "butter butts" in Humboldt, breeding confirmations spanned the dates of a 4 Jun 1997 nest with young (L11; JMH), to a 30 Jul 1998 report of adults feeding young (r08; GCH).

The penchant for open stands evidenced by these warblers manifests itself in a different fashion elsewhere in the county. Elevation would appear to be not especially important, provided that appropriate tree structure is present. Several kinds of tree associations different from those found in the mountains are inhabited during the summer. Scattered across the moister flanks of the ridges of southwestern Humboldt are old stands of Douglas-fir with widely spaced trees and limbs extending nearly to the ground. Such sites are used by Audubon's Warblers in the Bear River and Mattole River watersheds and in the King Range. Stands of tall Douglas-firs undergrown to a considerable degree with California black oak are occupied locally on ridges in the central Mad River drainage and in interior southwestern Humboldt.

Along the coast, adults were watched carrying nest material in shore pine forest near Manila (e15) 20 Apr 1996 (GCH) and feeding young in the same habitat at Lanphere Dunes (e14) 1 Jul 1995 (JEH). At the isolated Monterey cypress grove near the base of the north jetty of Humboldt Bay (D17), breeding was confirmed 28 May 1998 (carrying food), 2 Jun 1999 (feeding young), 7 Jun 1995 (carrying food), 7 Jun 1996 (carrying food), and 9 Jun 1998 (fledglings; all DFx). A few birds may summer in the Arcata Community Forest, and these warblers nested just above the eastern edge of the coastal pasturelands at Bayside (g15), as evidenced by a female feeding a recently fledged juvenile 7 Jun 1999 (DFx).

A broad zone of scarcity or absence is discernible on the map. This zone lacks the open forest structure that nesting Yellow-rumped Warblers favor. Instead, it is characterized by abundant tanoak or coast redwood. Reproducing in its own shade, tanoak commonly becomes the most abundant tree by stem count on many sites it occupies. This causes a dense canopy closure and minimal projection of individual crowns above and apart from each other. Much of interior Humboldt at middle elevations is densely clothed with such forests. These warblers are also absent from grassland/oak savannah habitats. As a result, there was a nearly continuous band of no-detection blocks from 15–25 miles wide. This extended from the northwestern corner of the county south through the redwood and lower Douglas-fir/tanoak zone, southeast through the southern extent of the redwood forests, and through the oak and grasslands of the southeastern corner of the county.

Early in the spring migration, Audubon's Warblers begin to withdraw from pastoral and other outlying habitats, and begin to occupy valley-edge and foothill stands of trees of any description. This form is abundant for some weeks in April as the northbound breeders move through the area. California black oaks support great numbers of them during the period of flowering and leaf-out. By early May, the pulse of the migration around Humboldt Bay lessens. Rare stragglers may be encountered as late as Memorial Day in places where summering would appear unlikely. Returning migrants in fall typically are first noted in early September and are abundant by late in the month. The large numbers of Audubon's Warblers on the move are augmented in October by droves of southbound and wintering Myrtle Warblers. Near-coastal woodlots and thickets teem with this combined glut for weeks, with numbers only gradually seeming to diminish.

In fall and winter, flocks of one or another or both forms of Yellow-rumped Warbler forage through the tangled and varied vegetation of lowlands and stream corridors. The degree to which the two subspecies intermingle varies greatly from site to site. On the whole, Audubon's Warblers exhibit an attraction to elements of human-altered landscapes. They use exotic trees and shrubs more frequently than do Myrtle Warblers. As a rule, Audubon's outnumber Myrtles in towns, with some exceptions, and can be among the more numerous species using Monterey pine and cypress, eucalyptus, and other plantings. Weedy lots, shrubby fencerows, suburban neighborhoods, old log ponds, and estuarine-edge environments are well populated with frequent flocks of Audubon's Warblers throughout the winter. Dairies, pasture edges, shrubby marshes, idle stacks of irrigation pipe, roof gutters, log decks, blackberry-covered outbuildings, rockworks, and even the outer-coastal driftwood receive the attention of flocks of various sizes. On occasion the warblers feed in the company of robins or bluebirds on lawns and golf courses.

BLACK-THROATED GRAY WARBLER *Dendroica nigrescens*

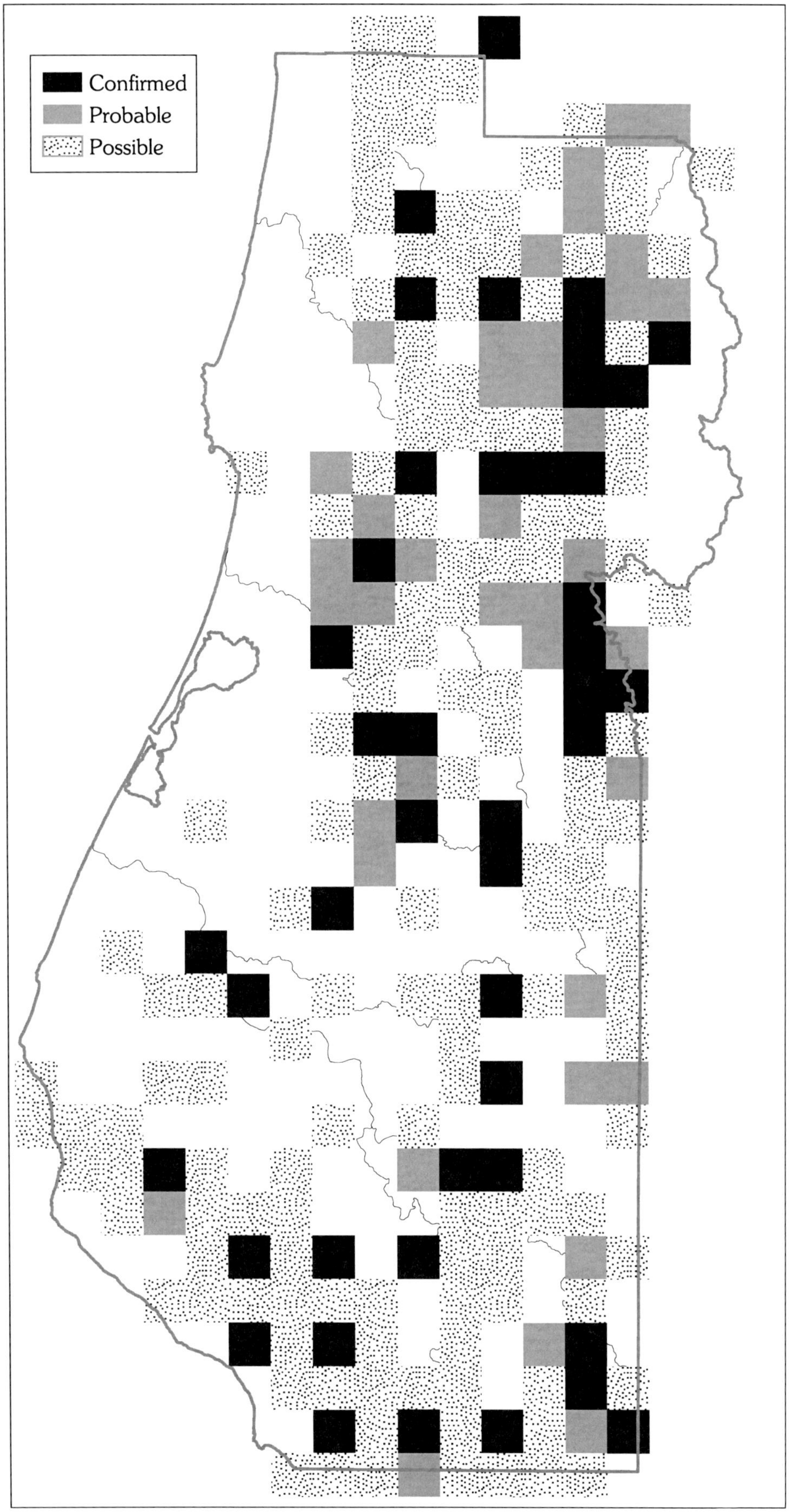

From early April into July, the shrill songs of male Black-throated Gray Warblers are a constant element of the birdsong chorus over most of Humboldt County. In June and July, the lisping calls of well-grown nestlings and dependent fledglings may be heard, accompanied by soft contact notes given by adults. These warblers are well known to birders as an inquisitive species, given to prompt and persistent investigation of "spishing" noises and pygmy-owl imitations made to attract small birds. It is unfortunate that these attractive birds seem destined to remain unrecognized by most people. The hillside woods they inhabit during the breeding season are for the most part rural places, well removed from the backyards and city parks where the uninitiated might otherwise marvel at the sight of one of these striking creatures. Fairly common spring and fall migrants through the settled lowlands, Black-throated Grays nevertheless escape attention by lingering within the interior shadows of leafy trees, progressing with incremental foraging movements, avoiding sparsely vegetated settings, and seldom dropping to the ground.

Across the species' extensive range, Black-throated Gray Warblers occupy a rather remarkable array of sites and vegetation communities. Spanning disparate endpoints of luxuriant regrowth timber near sea level in Washington to sun-scorched pine and juniper in the mountains of the Southwest are forests and woodlands in which an abundance of oaks is an oft-repeated theme. This is plainly the case in Humboldt County. Atlas data indicated that, by and large, the bulk of the county's summering Black-throated Gray Warblers inhabit Humboldt's expansive oak zone. They were found in 52% of all Atlas blocks. Breeding was confirmed in 44 blocks and was considered probable in 39 others.

Locally, these birds are not particularly common in simple California black or Oregon white oak woodland, instead favoring denser stands typically dominated by oaks but with other hardwoods and Douglas-firs present. As the map indicates, coastal and fog belt forests of coast redwood, Sitka spruce, and grand fir do not appeal to them. Like Hermit Thrushes, Black-throated Gray Warblers appear in numbers some miles inland, wherever coastal coniferous forest is replaced by drier stands containing much tanoak. This takes place much nearer the outer coast in the southwest portion of the county, where dry slopes of the King Range closely approach the ocean. The western edge of the general summer range includes areas within the fog belt, including the lower Klamath River canyon, lower Redwood Creek, Bear River Ridge, and the lower Mattole River corridor.

Spotty representation in the south-central section of Humboldt County may be related to the less frequent, less extensive patches of appropriate drier-site habitat in and around the large redwood parks. Seemingly confluent with this region of few reports is the rolling oak-dotted pastureland of the central Mad River country, in which access by atlasers to conifer-mixed hardwood stands was perhaps insufficient to detect birds tucked into draws on private lands. Certainly, the great expanses of open, grassy countryside so prevalent on southern exposures in interior southern Humboldt offers habitat for these birds only where stands of oaks and Douglas-fir occur about their edges.

Black-throated Gray Warblers are routine breeding birds of the nondescript Douglas-fir regrowth clothing much of interior Humboldt. Their songs are a familiar sound over Humboldt County's managed industrial timberland. This species pioneers regenerating clearcuts once young trees have formed a reasonable canopy and there is some semblance of a hardwood-dotted understory developing. In Douglas-fir forests, their abundance decreases as conifer cover increases (Ralph *et al.* 1991). The steep slopes of summer-hot canyons heavily grown to canyon live oak, Pacific madrone, *Ceanothus*, and a smattering of taller Douglas-firs are among their favorite situations. The Klamath, Trinity, and South Fork Trinity river canyons, in particular, offer much habitat of this description. Provided that forest cover is sufficient and conifers do not exclude hardwoods, Black-throated Grays occur widely at middle elevations, ranging upward to about the altitudinal limit of tree-size oaks. Maturing second-growth timber too scanty, broken or oak-grown for Hermit Warblers is widely inhabited. Black-throated Gray Warblers are replaced by Hermits on cooler northern exposures and everywhere within the mountain forests where winter snowpack suppresses hardwood trees. In the Klamath Mountains, Black-throated Grays are generally found at lower elevations than in areas with larger trees and lower amounts of alder and shrub cover, and they are least likely to be found in riparian habitat (Sniado 1999).

Black-throated Gray Warblers spend a great deal of time foraging not in the taller treetops but rather within the welter of lower limbs created by the mingling of taller shrubs, struggling understory trees, and the lower and middle canopy of whatever conifer trees may be present in a given territory. Their curiosity, lack of shyness, and commonness along wooded roadsides allowed atlasers many opportunities to confirm breeding. Extra survey effort in priority blocks is believed to have resulted in more confirmations in those blocks.

These warblers are among the earlier arboreal neotropical migrants to arrive in Humboldt County, typically first appearing in early April and becoming common in appropriate habitat by early May. Multiple males singing in suitable habitat were recorded as early as 31 May 1998 (H13; GAS and k33; DFx). The earliest confirmed breeding record was of an adult carrying food 9 Jun 1996 (o33, JEH). The earliest fledgling was one being fed 18 Jun 1995 (H33; K. Dyer). Only a single nest was reported, which contained young when discovered 2 Aug 1995 (H15; PFS). While the feeding of young by adults was observed as late as 15 Aug 1996 (o09; *fide* JMH), by August fall migration appears to be in full swing. Scattered migrants are detected throughout September, with far greater numbers believed to migrate inland than along the coast. Sightings after early October are unusual.

Harris (1996) summarized the status of Black-throated Gray Warbler in northwestern California as a common migrant, summer resident, and breeder. Results of Atlas fieldwork indicate no change to this characterization.

HERMIT WARBLER *Dendroica occidentalis*

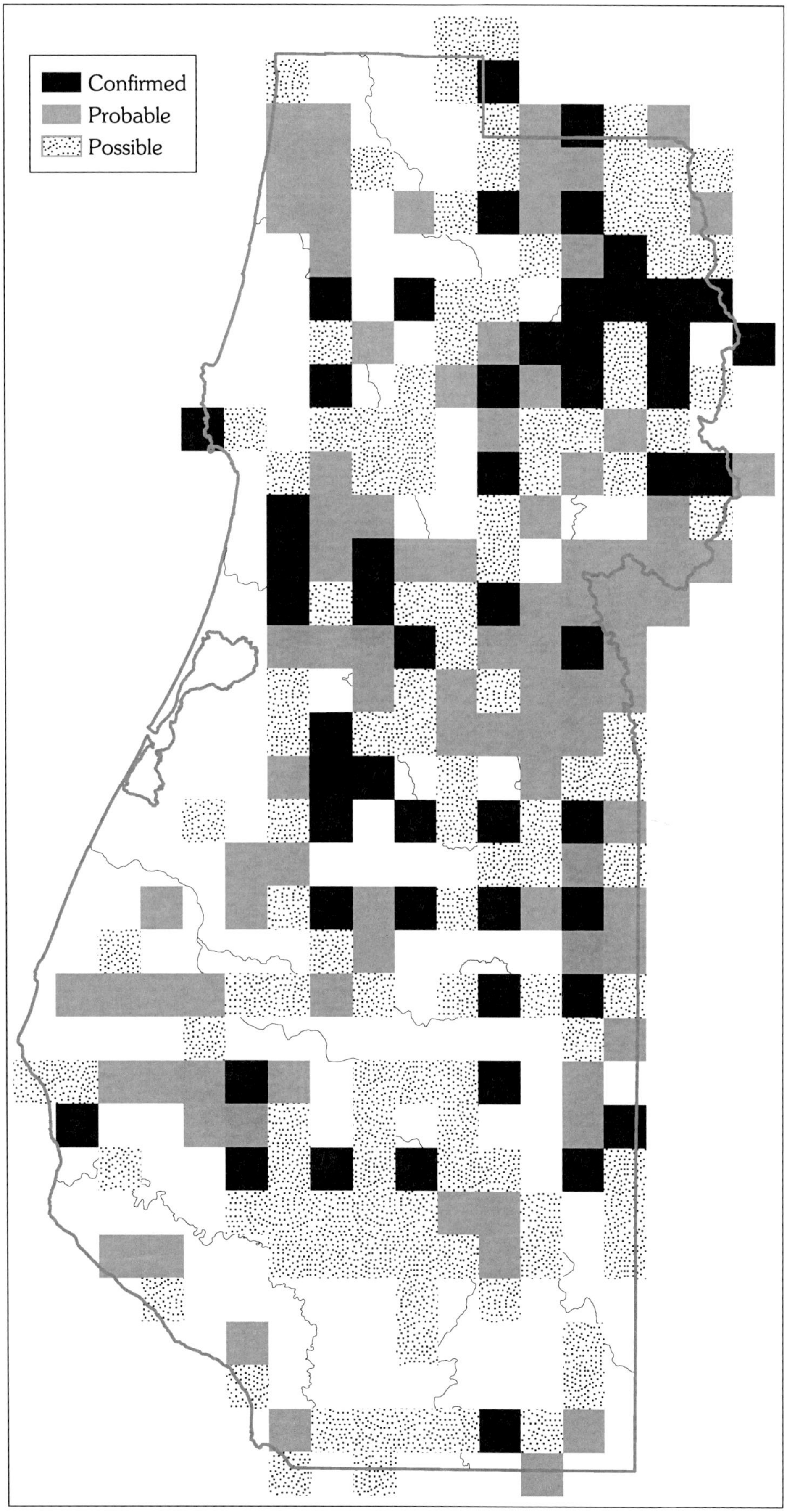

Hermit Warblers are poorly named if one judges by behavior. Although flocks of any size are not often seen, there is nothing in the manner of the birds to suggest any but the typical dynamics of territoriality, pair formation, breeding, and migration. If, however, one is to conjure an image of comparative seclusion from the human realm, then the name reflects the forested expanses inhabited by these birds in spring and summer. Hermit Warblers are most common in landscapes where coniferous forests are the rule and other habitats are exceptions. The range of this parulid encompasses the Pacific Slope from western Washington south across much of western Oregon and northern California, extending southward along the west side of the Sierra Nevada and in coastal mountains south to Marin County. Smaller, more southerly outlying populations occur in the Santa Cruz, San Bernardino, and San Gabriel Mountains (Dunn and Garrett 1997). East of this range Hermit Warblers are at best uncommon migrants, even in the adjacent Great Basin. They also are uncommon, rare, or nearly unknown at various well-birded lowland sites outside breeding habitat, but which otherwise lie within the range of the species.

Miller (1933) documented Humboldt's first historical breeding confirmation, of physiological evidence from a male collected 26 May 1932 north of Garberville (A. Miller; MVZ 81470). Yocom (1968) provided additional Humboldt records, including one of a female feeding a juvenile Brown-headed Cowbird in Sunny Brae. Harris (1996) termed the Hermit Warbler a "common summer resident and breeder; casual in winter." This condition has probably been the case for ages and it appears to remain true in spite of 150 years of logging. While Hermit Warblers favor older stands, that they widely accept stands of even the rattiest Douglas-fir or coast redwood regrowth belies any notion that this species is an old-growth obligate.

Hazard and George (1999) found that while Hermit Warblers occur more often in older redwood stands, they are not especially numerous in old-growth and can be common in regrowth. A capacity to live in old-growth, but to sustain reasonable abundance in the face of industrial logging, is a boon this species shares with few others.

A glance at the map shows that much suitable breeding habitat for Hermit Warblers existed in Humboldt County during the Atlas period. Hermit Warblers were encountered in 61% of the blocks, making this the third most widely distributed breeding warbler in Humboldt after Wilson's and Orange-crowned warblers. During point counts in interior Douglas-fir forests of northwestern California and southwestern Oregon, Hermit Warbler was the most abundant species recorded (Ralph *et al.* 1991). During the present study, breeding was confirmed in 53 Atlas blocks and probable breeding was reported in 91 others.

The range of this warbler in Humboldt takes in nearly the entire county except the coastal lowlands and several areas in the interior in which appropriate habitat is either sparse or, in some blocks, altogether lacking. Hermits are among the routine and abundant summer birds of conifer forests, including those dominated by coast redwood, Sitka spruce, Douglas-fir, white and Shasta red firs and, to a lesser extent, pine. Throughout nearly all of the Douglas-fir and tanoak forests of northern and central Humboldt, this warbler was one of the most frequently reported species. In the true fir zone above about 4000 feet, Hermit Warblers remain abundant, even though they share habitat with much greater numbers of Yellow-rumped Warblers. In the Klamath Mountains, Sniado (1999) found them at slightly more than one-half of the white fir points, and at slightly less than one-quarter of the Douglas-fir points, with occurrence more frequent in stands with larger trees and a closed canopy. They occur essentially to the coast in the redwood parks of extreme northwestern Humboldt and at Trinidad, follow regrowth timber stands well toward the coastal lowlands around Humboldt Bay and above the Eel River, and in the lower Mattole River country and in the King Range.

Although Hermit Warblers show a strong positive association with the presence of conifers (Ralph *et al.* 1991), during the Atlas period they were also found in many situations where oaks (particularly tanoak and California black oak), were abundant. Douglas-fir is typically dominant within the canopy in such situations, but the amount of oak by tree count can be significant. Sites with a minimal conifer overstory or which are truly heavy with oak—particularly canyon live oak—generally are more favored by Black-throated Gray Warblers. These two species sometimes were found to occur in the same place.

Absence of Hermit Warblers tended to be patchy in northern Humboldt County. Aside from the coastal lowlands, there was one sizable area where no-detection blocks formed a cluster: in extreme northern Humboldt and adjacent Del Norte County. As has been noted elsewhere in this work, in this region nearly all of the original forest cover present at the time of modern settlement has been logged. The resulting mosaic of various-aged conifer plantations, fresh and recent clearcuts, and tatty streamside buffers appears not to attract these birds in summer. Hermit Warblers were otherwise undetected in northern Humboldt in some blocks in the Bald Hills and in Redwood Valley. In southern Humboldt the situation was distinctly different, with corridors or small clusters of blocks without detections in the central Mad and Van Duzen river drainages, in the upper Mattole Valley area, west of Briceland, and in blocks dominated by grass-and-oak country near the southeastern corner of the county.

The carrying of nest material was reported only once, 3 Jul 1998 (J07; JCP). Only two nests were reported, both with young: 19 Jun 1998 (L33; JCP) and 20 Jun 1999 (H19; DFx). Adults carrying food was the most-reported best evidence in blocks where confirmed breeding was established. The span of dates ranged from 10 Jun 1995 (e10; TWL) to 31 Jul 1998 (N7; GCH). Three of the reports of food-carrying fell in the first half of June, eight in the second half, 12 in early July, and three in late July. Adults feeding young was ranked the highest behavior in eight blocks, from 20 Jun 1998 (N05) to 15 Jul 1995 (g14; both JEH). Fledged young provided the highest evidence in 15 blocks, from 21 Jun 1997 (P11; K. & E. Bentler) to 6 Aug 1996 (N07; *fide* JMH).

Hermit Warblers arrive from their wintering grounds, chiefly south of the U.S., during late April and early May. It is typical for first reports of this species to come during the last 10 days of April. By mid-May, the birds are present in numbers beyond estimation over most of the forested portion of Humboldt County. Generally, where one is heard singing, two or three others are also heard. By mid-June, many males have stopped singing their primary songs and have switched to softer, wheezier secondary song types. During early July the intensity of song greatly decreases. Throughout July the contact calls of adults and the lisps of nestlings or dependent fledglings are commonly heard. By early August, the show seems over, and only the occasional electric *chip* of a warbler somewhere overhead betrays the species' continued presence. Fall migration is seldom remarked upon. Hermit Warblers are rarely found in the coastal lowlands at any time; inland, spring migrants have been seen foraging in California black oaks in river corridors. Most Hermit Warblers have departed Humboldt County by mid-September, with later reports unusual.

AMERICAN REDSTART *Setophaga ruticilla*

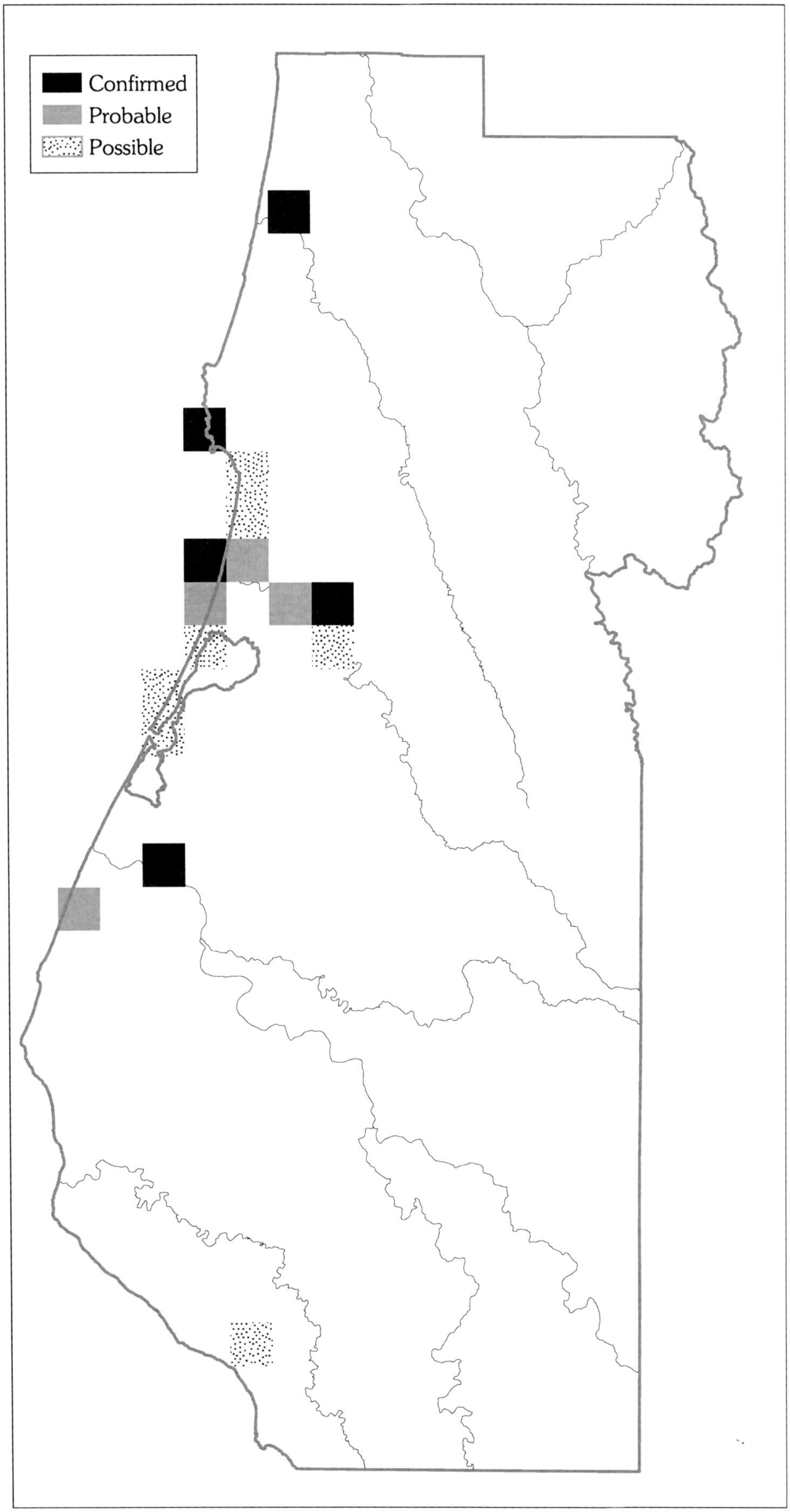

The American Redstart is widely distributed as a breeding bird in temperate North America, chiefly across Canada and in the U.S. east of the Rockies. It ranges locally westward into British Columbia, Washington, and central and northeastern Oregon. Redstarts were first detected breeding in California 22 Jul 1972, through discovery of a nest with young in an alder–willow thicket at the west end of Lanphere Road in the Arcata Bottoms (Binford and Stallcup 1972). Since that time, singing males, pairs, birds carrying food (Home Creek in Prairie Creek Redwoods State Park in 1973), and fledged young (Fairhaven in 1980 and Mad River County Park in 1994) have been reported at a number of coastal-lowland sites in Humboldt (Harris 1996). It has become apparent that, in some years, a very few American Redstarts may be expected in Humboldt County from late May into August. It seems likely that the size of the breeding season population varies greatly from year to year; in some years the birds may be not present. During the Atlas period their distribution and abundance seemed above average, with possible breeders recorded in all years, and birds found in 16 Atlas blocks.

Owing both to atlaser effort and to the serendipitous discoveries by birders working off the beaten path in early summer, an amazing five nesting attempts were documented, apparently providing Humboldt's fifth through ninth breeding confirmations. A pair attending a nest with two eggs was found in the "Boat Ramp Patch" at Mad River County Park (e13) on 3 Jul 1995 and, remarkably, the same observer (JEH) found a second nest with a female on two eggs the same day at Elk Head (e10). The following year, on 27 Jun at the Boat Ramp Patch, a male chased a female carrying nest material, while yet another male sang nearby (JEH). A female carrying a billful of cottonwood fluff along Redwood Creek just north of Orick (g05) 2 Jul 1996 provided evidence of nesting, while a second-year male "orangestart" sang nearby (DFx). Fledged young, the products of what may have been earlier nests, were found 10 Jul 1997 along the Mad River (h14; P. Herrera) and 14 Jul 1997 along Singley Road east of Loleta (d20; P. Herrera). During 1995–1997 there were six other reports of probable breeding including pairs, courtship or copulation, and males singing on territory for at least a week. By the end of the Atlas period things seemed to have cooled down for these warblers, since only two cases of possible breeding were reported in 1998 and 1999.

Most recent and historical breeding locations are within the immediate floodplains of rivers and streams. In Humboldt County, American Redstarts seem to show a preference for older and more highly developed riparian patches, woodlots, and dune hollows containing larger willows. A variable mixture of willow, red alder, red elderberry, salmonberry, and mixed underbrush dominated by blackberry tangles typifies such places. Many stretches of riverine woods and thickets, particularly along the Eel River, look good for redstarts but appear not to support them. However, general birding effort in Eel River riparia has not been intense enough to examine all potential habitat thoroughly for presence or absence of these birds.

Although this warbler occasionally has nested elsewhere in the state, the alder–willow thickets of northern Humboldt County and the neighboring Klamath River estuary in Del Norte County (Harris 1996) clearly support the closest semblance to a regular breeding population in California. If these important riparian habitats are protected, the token presence of American Redstarts in Humboldt may be ensured in coming years. However, the establishment of a regular breeding population in these patchy habitats may be unlikely given the persistent presence of Brown-headed Cowbirds associated with adjacent pasturelands. Binford and Stallcup (1972) observed the male redstart at their Lanphere Road nest feed a juvenile cowbird 29 Jul 1972. The eggs observed in the two nests found 3 Jul 1995 were suspected to belong to cowbirds (JEH), but this was not confirmed.

MacGILLIVRAY'S WARBLER *Oporornis tolmiei*

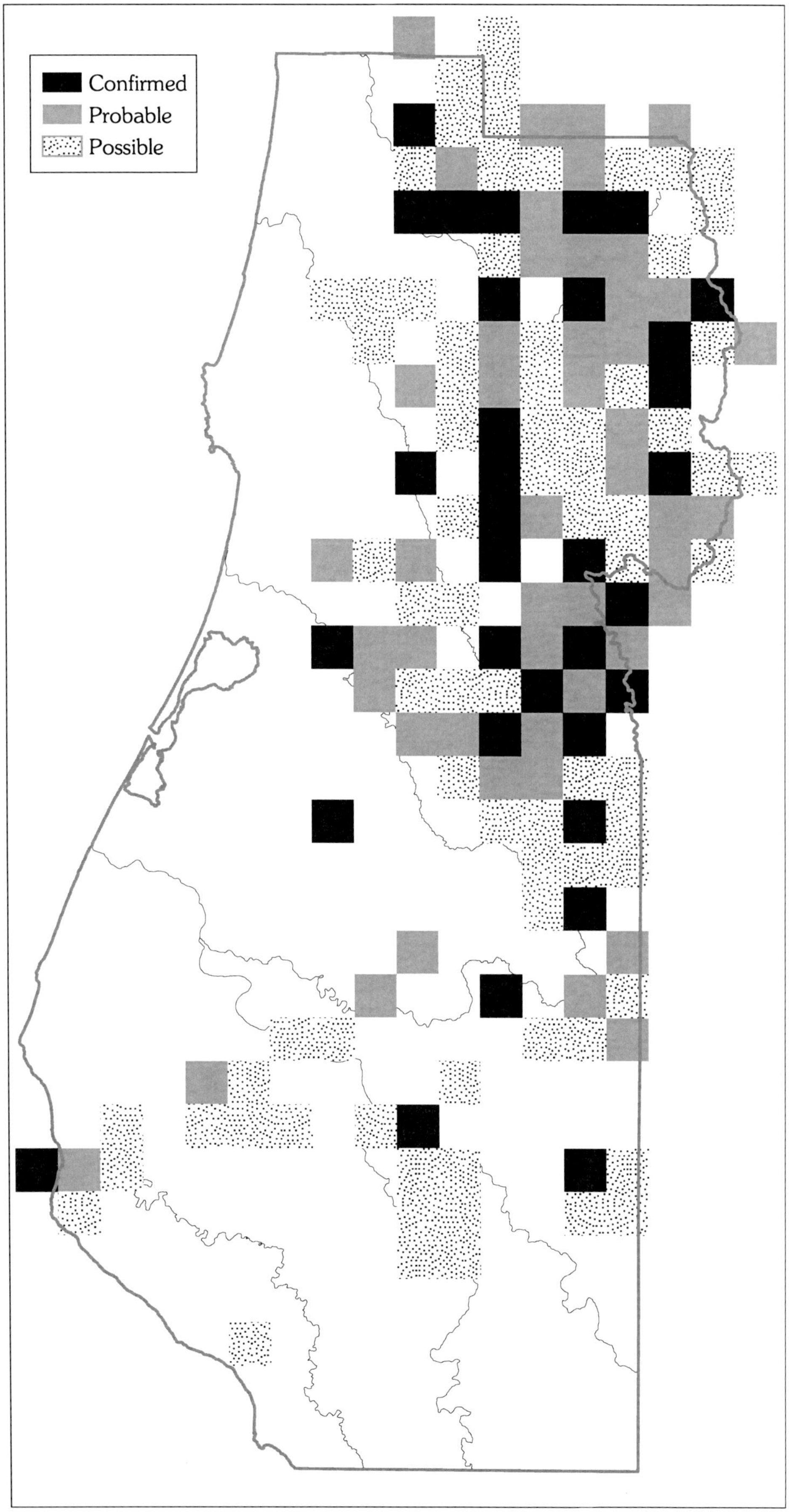

A tendency to forage low within dense tangles of brush and to keep to themselves often causes MacGillivray's Warblers to go unnoticed by those who have only recently begun birding. This is a pretty warbler, lively and loud-voiced. It is a fairly common breeding bird and regular migrant across much of the West. Yet its reclusive behaviors and backwoods hangouts in the nesting season make it known to many lowland- and city-based birders only through occasional encounters. With increasing experience, one learns that this species can be detected and teased into view during migration (particularly in fall) at almost any suitable brush-patch, especially at ponds or edges of wet meadows, or along tiny creeks. Prolonged "spishing," as well as close attention paid to each of the furtive greenish birds one glimpses among the shadows, will in time reward those for whom the "Mac" persists as a bird of mystery.

Because MacGillivray's Warblers are such skulkers, it might be imagined that evidence of their presence would generally be hard to acquire. However, owing chiefly to the loud and far-carrying songs delivered by males during the nesting months, MacGillivray's Warblers were detected in 158 (37%) of the 425 blocks. And, although members of a secretive species, adults involved in reproduction betrayed their activities in 33 blocks yielding reports of confirmed breeding. Probable breeding was established in 48 additional blocks.

The mapped distribution of MacGillivray's Warbler as a breeding bird in Humboldt County suggests an association with middle and higher elevations east of the coastal fog zone. That this is a species chiefly of northeastern Humboldt during spring and summer was made clear. Only slightly more than one-fifth of all blocks with detections were south of the Van Duzen River. Interesting was that, of those thirty-some scattered blocks, few were at higher elevations (notably, a small cluster in and about the Lassics), and some were quite near the coast.

During the Atlas period, this species was a regular inhabitant of brushfields, forest edges, and riparian corridors. While quite numerous, it seemed not as abundant as some other warblers. The apparent habitat needs of local MacGillivray's Warblers seem somewhat more limited than are those of Nashville Warblers, Dusky Flycatchers, and other brush-loving birds found across the same sections of the county. Without question, Macs favor a more mesic setting. This is plainly indicated by the extent to which northerly exposures attract them. Moisture-loving broad-leafed shrubs or shrub-trees such as thimbleberry, ninebark, and any of several wet-site currants typify vegetation in places where the birds are found. Seeps, springs, and other persistent water sources are often present. Further suggesting an affinity for dampness is that, at the scale of a typically corrugated mountainside, MacGillivray's Warblers tend to occur within ravines or draws rather than on ridges or convex slopes.

In Humboldt County this warbler is found most often in forested countryside, yet it is not really a creature of the forest. The heavy shade cast by mature trees prevents luxuriant growth of the brush species used by the birds. However, areas where varying amounts of tree overstory have been removed are commonly inhabited. As a result, these birds have probably increased in abundance in Humboldt because of timber harvest (Raphael *et al.* 1988). Habitat in regenerating clearcut units is often dense snowbrush mixed with greenleaf manzanita, bitter cherry, currant, and similar shrubs. With respect to the presence of young conifers, these logged areas span a complete spectrum: from seral brush with almost no conifer saplings (typically on southern exposures) to vigorous saplings at densities of thousands of young trees per acre (usually the northerly exposures). In the latter case, Douglas-fir generally predominates, although on some sites above 4000 feet, white fir seedlings and saplings furnish much of the vegetative cover. While a great many clearcuts in the hills and mountains of Humboldt host this species, others do not. Sun-parched southerly exposures with abundant Nashville Warblers often support no MacGillivray's. Where present in such places, the Macs can be anticipated in the taller shrubbery lining narrow draws bisecting the clearcut.

In south-central Humboldt, MacGillivray's Warblers were found in a variety of situations including clearcuts, naturally occurring brushy areas, shady forest edges, and along streams. Many sites at which they were detected in that section of the county were at low elevations, such as Humboldt Redwoods State Park, the Mattole River valley around Petrolia, and a cluster of blocks taking in the lower main stem Eel and South Fork Eel rivers. Within a stone's throw of the surf, a bird carried nest material in the willows at the mouth of the Mattole River (a27) 23 Jun 1996 (GCH). However, the Macs found along Rainbow Ridge and about the Lassics were certainly in hilly country. Throughout the species' breeding range in the county, riparian corridors were attractive if they provided extensive interconnected shrubbery, were not degraded by cattle, and did not suffer abuse at the hands (or the four wheels) of humans. None of these warblers was found anywhere along the lower Eel River or within its floodplain.

Six behavior codes indicating confirmed breeding by this species were reported during the Atlas period. By a wide margin, the most frequently reported among these were observations of fledged young (constituting 42% of all confirmations) and adults carrying food (32%). Fledged young were noted as early as 22 Jun 1997 (P11; K.& E. Bentler), but the 12 other such reports fell during July, evenly split between the first and last halves of that month; the latest were 31 Jul 1996 (p08; JEH) and 31 Jul 1998 (N07; GCH). Adults carrying food were witnessed from 7 Jun 1998 (o14; M. Wachs) to 13 Jul 1996 (H19; GAS). These 10 reports were also evenly distributed across the span of dates. Other confirmed breeding evidence included four instances of adults feeding young, reported as early as 23 Jun 1998 (q07; GCH) and as late as 27 Jul 1995 (H15; PFS); adults carrying fecal sacs 29 Jun 1999 (L17; JCP) and 20 Jul 1995 (P09; JEH); physiological evidence from a bird in hand 15 Jun 1996 (o05; J. Tietz); and the earliest breeding confirmation: an adult carrying nest material 9 May 1996 (N13; TWL). No nests were reported.

Continued on page 394

COMMON YELLOWTHROAT *Geothlypis trichas*

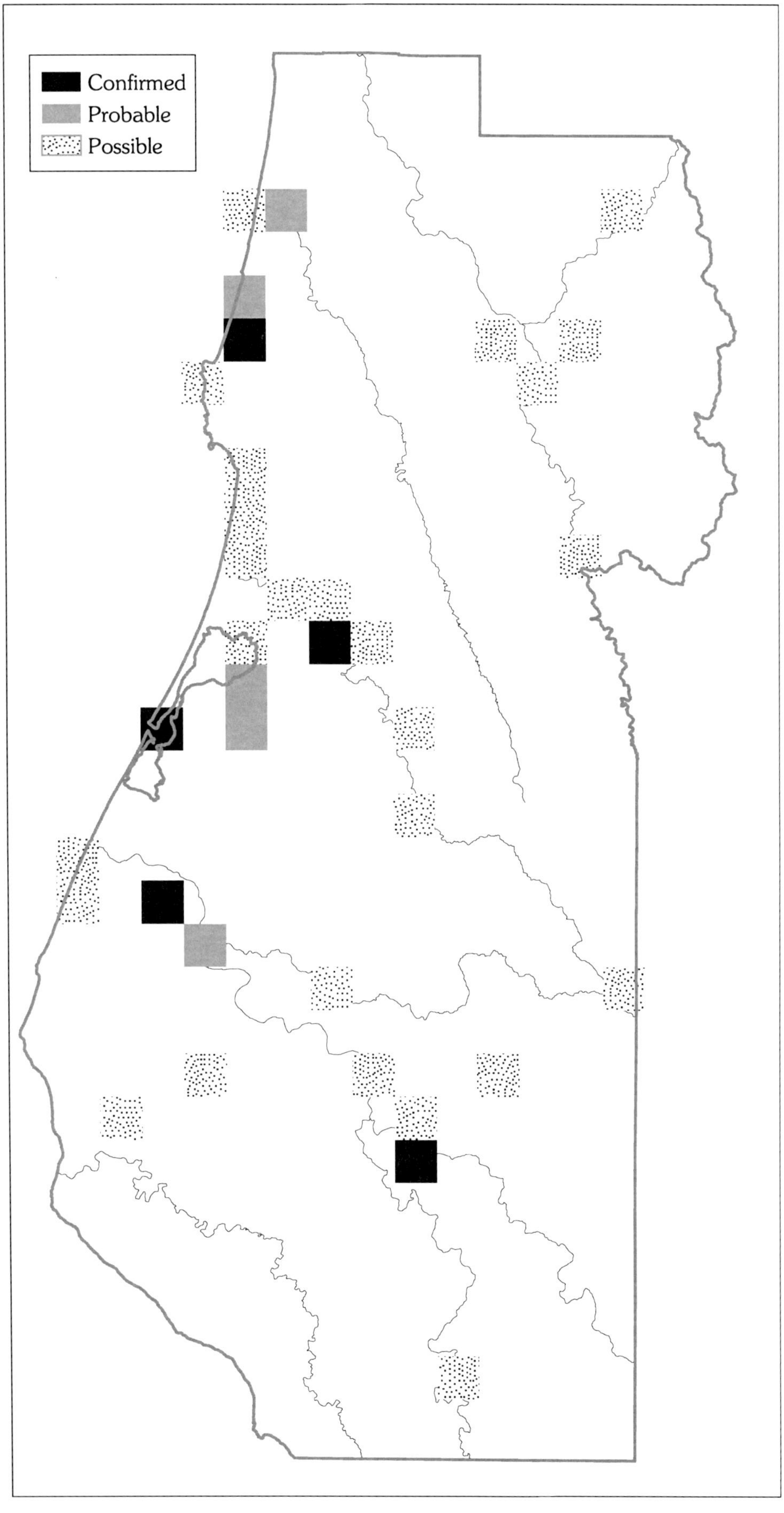

As the songs of a newly arrived male Common Yellowthroat carry across a marsh, so, too, is the listener carried—back to another April morning during the Beginner Days, in which one's first glimpse of a yellowthroat may have taken place in a similar setting. Few local warblers select such distinctive habitat as consistently as do members of this species. That the Common Yellowthroat breeds commonly across most of temperate North America is testimony both to its acceptance of varied wetlands and its ability to locate even highly isolated patches of suitable marsh.

During the course of Atlas fieldwork it was apparent that Common Yellowthroats occurred widely within Humboldt County, yet nowhere was it an abundant species. Yellowthroats were reported in 36 (8%) of the 425 total blocks and were confirmed as breeding in five of these. Harris (1996) summarized their status in northwestern California generally as a "locally uncommon to common migrant, summer resident, and breeder; casual to rare in winter." He stated further that "breeders appear in late Mar–early Apr around pond edges and in riparian cover along the coast and inland rivers and ponds."

Throughout much of the corrugated landscape of Humboldt County, habitat appropriate for Common Yellowthroats is limited, and it occurs mostly at lower elevations. Although bush-dotted seeps occurring in shallow draws or gullies may host a few, by and large this warbler is absent on hillsides, in contrast to most other warblers found here in summer. During Atlas surveys, most yellowthroats were encountered on nearly flat ground, plainly because of their preference for at least minimal (and sometimes quite ephemeral) standing or slowly flowing fresh water supporting emergent vegetation. In addition to the presence of grasses, cattails, and bulrush at a site may be an edge or intrusion of dense low shrubs, a neglected fenceline, or mounds of blackberry.

These requirements are met locally in rural streambottoms, in the lower floodplains of the larger rivers, and in and about well-vegetated ponds. Broken or disturbed cottonwood riparian habitat with wet spots, thoroughly undergrown with horsetail and containing clumps of willows, is commonly inhabited along the lower reaches of major rivers. Persistently wet corners of dairy pasture fronting willow thickets are occupied. Cattail beds, old millponds, and tangles of tall forbs near sewage ponds are also used. As a result of these habitat preferences, Atlas blocks with detections generally lay along principal streamcourses such as the Eel, Mad, and Trinity rivers, in the Eel delta, and about Humboldt Bay and at lagoons of the northern Humboldt coast.

Sizable stretches of the survey area were without detections, including the timberlands of north-central county; the highly incised Redwood Creek drainage; oak/grassland of east-central and southeastern Humboldt; the general South Fork Mountain Ridge; and the rugged, well-drained expanse of the southwestern corner of the county.

Observations of probable and confirmed breeding were sparse, with only five of each reported. The earliest confirmations were of adults feeding young 24 Jun 1996 (D17; GCH) and a nest with eggs at the lower edge of Eel River riparian habitat near McCann 24 Jun 1999 (J27; JCP). Although female Brown-headed Cowbirds have little difficulty in finding Common Yellowthroat nests—this species is the cowbird's third-most-frequently parasitized host (Ehrlich *et al.* 1988)—atlasers attempting the same feat with more benign intent met with but one success: the above-mentioned nest near McCann which, in addition to containing a full clutch, also held a cowbird egg! The three remaining breeding confirmations were adults carrying food 5 Jul 1997 (D21; GAS), a fledgling seen 9 Jul 1997 (f08; GCH), and the feeding of young 16 Jul 1995 (H15; PFS).

Common Yellowthroats arrive in Humboldt County each spring beginning in late March and early April. By mid-April they are regularly reported from sites at which they are expected. The fall migration appears to extend over several months, with stragglers lingering into November in some years. This movement is notable for a big pulse of passage birds which occurs annually in late August and the first half of September, during which time it seems that nearly any brushy or marshy place at lower elevations may support these wren-like warblers. As many as 40 birds have been counted within the Arcata Marsh and around the edges of the adjacent Arcata sewage ponds in a single outing during these migrant waves. Occasional individuals are detected on coastal CBCs and, rarely, later in winter.

WILSON'S WARBLER *Wilsonia pusilla*

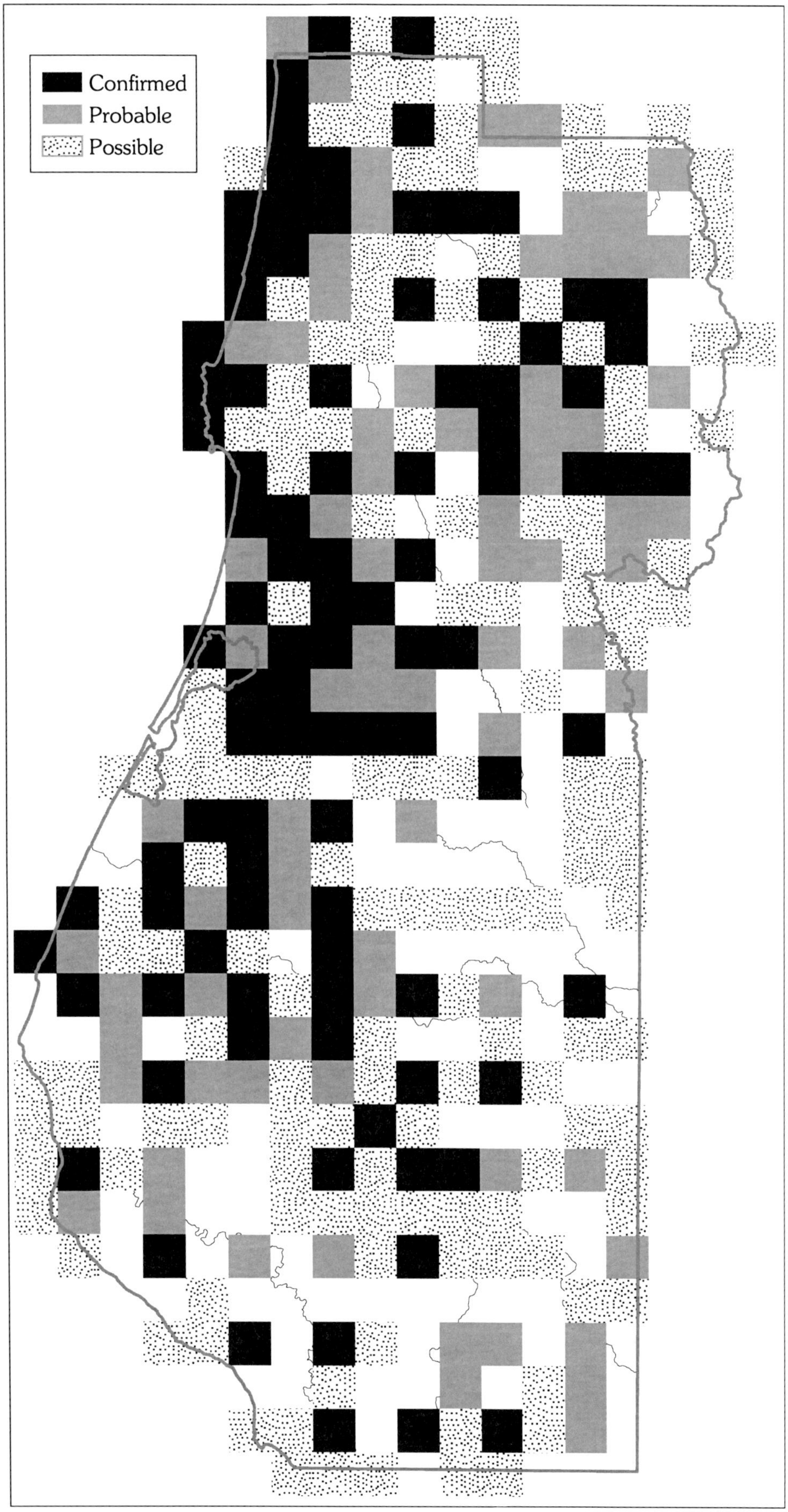

Wilson's Warblers are characterized by bright plumage, comparatively delicate legs and feet, well developed rictal bristles, and ceaseless energy. They are denizens of leafy shadows, constantly glancing upward and frequently making short hovering sallies to glean insects from the undersides of leaves. Living in the low light of the forest interior, they also have rather large eyes, no doubt aiding their search for morsels as small as aphids. In California, Wilson's Warblers occupy the southwesternmost extent of a huge summer range, stretching from the limit of willow scrub in Alaska across boreal Canada and into the extreme north-central and northeastern U.S. In northwestern California, the species is a common summer resident and breeder, and casual in winter (Harris 1996).

A mere glance at the map for Wilson's Warbler shows clearly that this was one of the most widely encountered species during the Atlas period. These warblers were reported in 312 (73%) of the 425 total blocks, making them the most widespread breeder among any of the 11 species of warblers known to nest in Humboldt County. Breeding was confirmed in 95 blocks and was judged probable in 76 others. Because suitable habitat was abundant, they probably nested also in the great majority of the blocks in which they were detected.

Although their breeding habitat in California includes coastal scrub, aspen groves, and willow thickets, Wilson's Warblers nesting in Humboldt County are typically associated with coniferous and mixed forests. Locally, a strong relationship with coast redwood stands is obvious. Very few blocks entirely within the redwood forest zone had no reports of these birds. Indeed, Wilson's Warbler is not only among the most abundant species of Humboldt's fog-zone forests, it is also common in Douglas-fir/tanoak forests and regrowth. Although not shown well by the map, they are definitely less common at higher elevations.

Worthy of mention is the ability of Wilson's Warblers to prosper in regenerating clearcuts. While a good many clearcuts fail to meet their needs because of insufficient moisture or simple lack of vegetation, scores of others do. These sites typically have northern exposures, are within the fog belt, and are grown to unthinned sapling conifers at densities of thousands of trees per acre. The presence of western hemlock, salmonberry, or chain fern in such situations is a reliable indicator of suitable habitat for Wilson's Warblers.

There were perceptibly fewer detections in the interior of the county, particularly in the oak-and-grassland country of the middle Mad River drainage and in parts of southern Humboldt. However, indicating that pockets of appropriate habitat occur widely even in sun-parched "SoHum" was the fact that only three priority blocks (F27, J31, N29) across that section of the county lacked detections. Wilson's Warblers also are common breeders in lowland riparian forests, such as those in the Eel River delta, where they tend to use areas with lower tree density, greater amounts of ground cover, and higher species richness of shrub and ground-level plants (Kelly 1987).

Eight different codes supported confirmed breeding in the 95 blocks reporting this status. Rather amazingly, 95% of all confirmations were made through observation of just three behaviors: adults carrying food (in 32 blocks) or feeding young (30), and fledged young (29). Food-carrying was seen as early as 6 Jun in 2 years, 1997 (k27; JG) and 1998 (j01; JCP). Indicating a minor peak occurring during a roughly 30-day period was the fact that six reports fell in early June, 10 in late June, 10 in early July, and six in late July. The latest report of adults carrying food was 28 Jul 1995 (B27; S. Levy). Juveniles were seen fed by adults as early as 9 Jun 1997 (H11; GAS). Again pointing to a peak shortly after the summer solstice was the fact that two of these reports were during early June, 13 in late June, seven in early July, three in late July, and four in early August. The latest was 10 Aug 1996 (e08; JEH).

The first fledgling was noted 30 May 1996 (g06; GCH), two were in early June, three in late June, eight in early July, and 14 were in late July. Somewhat surprising was that none was reported during August, with the latest on 31 Jul 1996 (k09; JMH). Complementing the "big three" behavioral codes were five others, each represented by one report. These were a nest with eggs 16 May 1999 (H31; R. Sutherland), a nest with young 27 May 1997 (f20; D. Garrison), physiological evidence from a bird in the hand 4 Jun 1995 (f16; L. George), an occupied nest 12 Jun 1997 (N09; JMH), and a distraction display 8 Jul 1996 (l18; D. Carpenter).

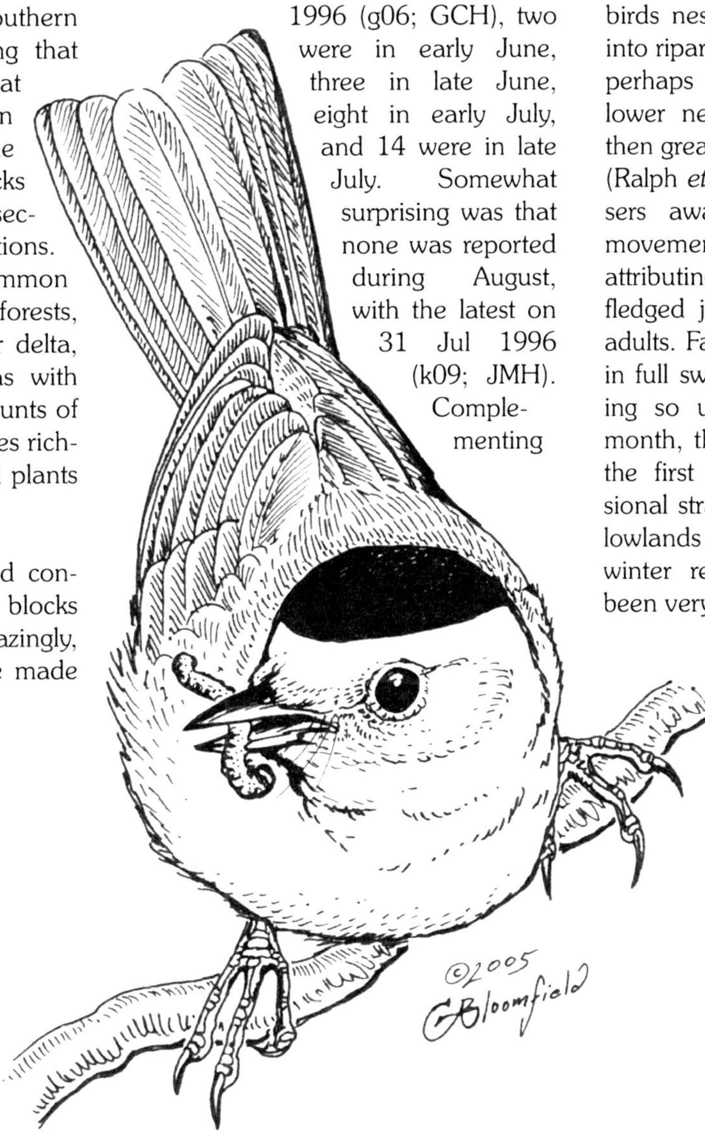

During spring migration, these warblers usually have been first reported in Humboldt County in the first week of April (Harris 1996). After nesting, Wilson's vies with Orange-crowned Warbler as the most conspicuous early-dispersing songbird. Adults out of breeding habitat were observed as early as the first week in July in three of the 5 Atlas years in Bayside (DFx, JCP). Banding studies suggest that birds nesting in coastal forests move into riparian areas in July and August, perhaps taking advantage of both lower nest predation in forests and then greater food availability in riparia (Ralph *et al.* 1996). By late July, atlasers aware of such post-breeding movements were cautious about attributing local reproduction to fledged juveniles unaccompanied by adults. Fall migration in this species is in full swing by mid-August, continuing so until about the end of the month, then diminishing rapidly after the first week in September. Occasional stragglers appear in the coastal lowlands as late as December, but true winter records in the county have been very few.

YELLOW-BREASTED CHAT *Icteria virens*

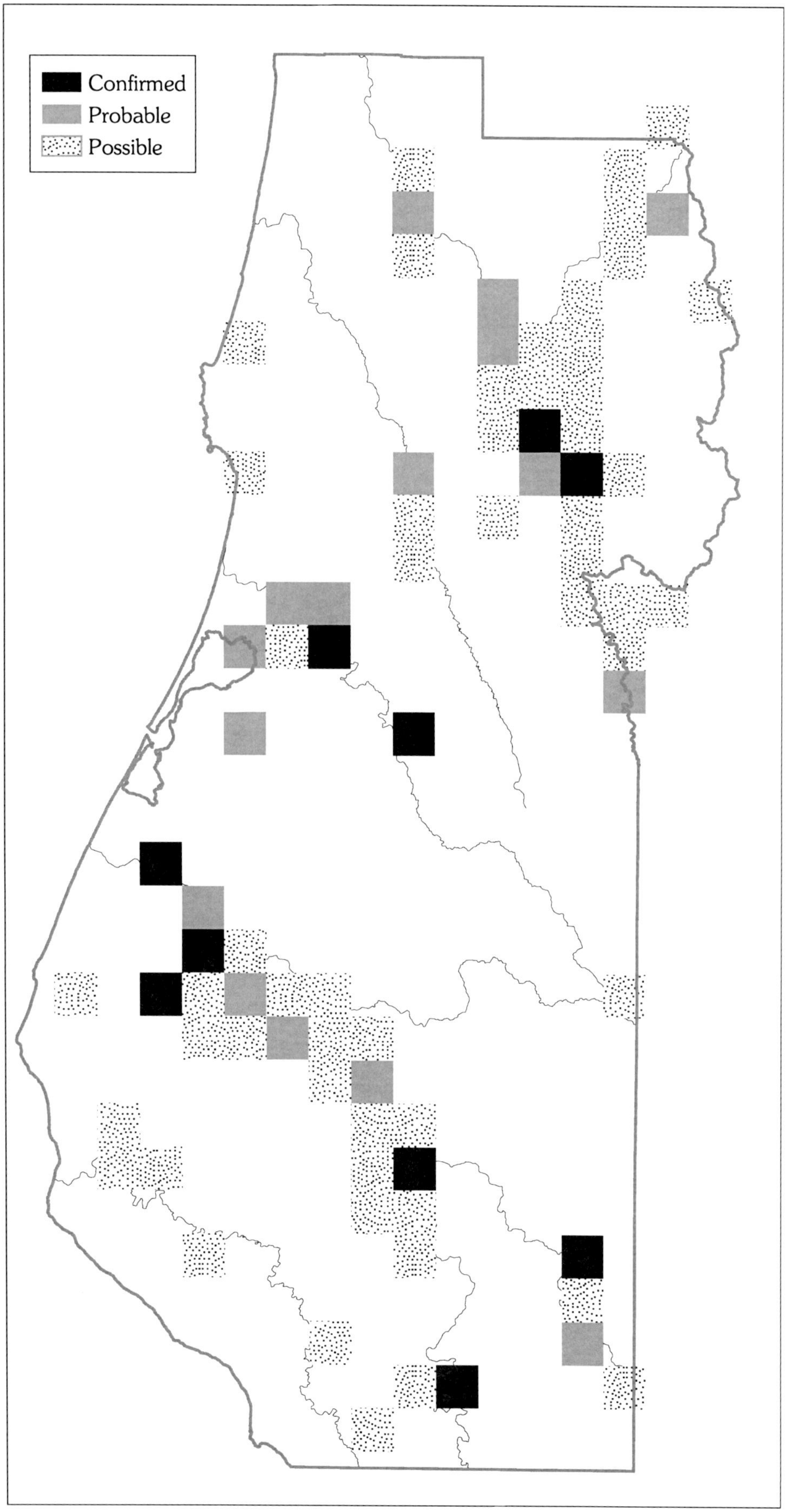

A late-rising third-quarter moon peeks now above Big Hill, bathing the Hoopa Valley in midnight light. Nothing would seem to stir and, aside from a single cricket near at hand and a ruckus raised by dogs far across the Trinity River, there is silence. Then, from somewhere neither near nor far, there comes a single upslurred report: *Queet!* Following a long moment, yet another odd call is heard: *Kook!* These separated calls precede a series of harsh notes suggesting some jay or...is it a mockingbird? Yet as the minutes pass no single phrase is ever closely repeated. The halting mixture of mellow and guttural voices betrays the identity of the unseen singer. It is a Yellow-breasted Chat, member of a most unusual warbler species, one well known for singing at night during the ardor of the spring season. Down along the riverbank in its bramble-strewn territory, the bird sits at attention just beneath the tip of a sapling oak, interspersing bouts of song with idle preening. The moon's lambent beaming is familiar to the chat, its rising face compelling the shy creature to give forth what music it can offer.

So is the Yellow-breasted Chat sometimes witnessed by those who know to listen for it. Although its voice can be heard day or night, getting a good long look at a chat requires talent and patience—or, sometimes, just dumb luck. These outsized warblers are reclusive, going about most of their business within extensive tangles of vines, shrubs, and small trees. Riparian strips, willow-choked alluvial meanders, and the thickets of blackberry, poison-oak, and hawthorn growing untidily across the slopes of neglected pastures attract these birds from April into late summer. Even with the relatively large size of Atlas sampling units (*i. e.*, blocks), Atlas results suggest an association with riparian vegetation. The spatial pattern of detections clearly indicates that most of the birds encountered were in blocks bisected by streams or rivers. Sniado (1999) considered this species to be a "riparian specialist" and found it at 17% of her riparian sample points in the Klamath Mountains. Within riparian habitats along the lower Eel River, these birds were detected during 10% of all point counts in 1999 (R. Hewitt, unpubl. data).

Harris (1996) termed the Yellow-breasted Chat in northwestern California a "locally uncommon to common summer resident and breeder; accidental in winter." Atlasers found them in 80 (19%) of the 425 total blocks. There were two distinct clusters of detection-blocks. The cluster discernible in northeastern Humboldt takes in most of the Klamath, Trinity, and lower South Fork Trinity rivers. Vividly apparent in the southwest is the corridor taking in the South Fork Eel River and the Eel below their confluence.

Birds were less apparent in the bottomlands of the Mattole River (c26 to i33); the upper Middle Fork Eel (N29 to o32); the lower Mad River (g14 to J17); that segment of Redwood Creek which is in Redwood Valley (J11 to J13); and several blocks along the Van Duzen River (f22, H23, o23). Yellow-breasted Chats were also found at the south end of Big Lagoon (f08), at the Little River estuary (F11), in Arcata (F15) and Bayside (g15), at Freshwater (F17), and at Capetown (B23).

Habitat used by these birds in Humboldt varies from site to site. Along the Klamath River, chats summer in the riverine growth sprouting from the rocky banks, and which is dominated by white alder, Oregon ash, and a great many shrubs and shrub-trees, many of which are deciduous, some of which are not. These birds are quite common along the Klamath, extending out of Humboldt into neighboring Del Norte County all the way to the river's estuary, even though the suitable habitat present in most places is severely confined by the steep, hot canyon slopes. The amount of appropriate habitat is augmented locally where small creeks tumble into the master stream. In the vicinity of Pecwan the birds were noted shortly above the river on west-facing shrubby slopes edged by oaks and Douglas-firs. The situation is somewhat different up the Trinity River, where undeveloped or neglected lands such as old mill flats in the narrow Hoopa Valley offer blackberry patches and black cottonwood thickets. Along the lower Mad River and in the coastal lowlands around Humboldt Bay, cottonwood- and willow-dominated sites are occupied. The lower Eel River system offers any number of sites, ranging from the edges of cottonwood stands to the straggling thickets of willows established on old gravel bars. The main (Middle Fork) Eel mirrors the situation along the Klamath, offering a narrow but well-used corridor of habitat adjacent to the river, extending upslope in places where dense coyote brush and blackberry have overrun untended land.

Because Yellow-breasted Chats are furtive birds that stick tenaciously to extensive cover, it is not surprising that confirmed breeding was noted in only 10 Atlas blocks, with probable breeding found in 16 others. The carrying of nest material was reported as the highest-ranking code in one block (m10; JMH) 28 May 1997. Adults carrying food were noted in five blocks, spanning 15 Jun 1997 (N29; RJA) to 26 Jul 1996 (k32; JEH). Of these five, three were in June and two were in July. Adults feeding young furnished the highest-ranking breeding evidence in two blocks, 28 Jun 1997 (D23; COg) and 13 Jul 1998 (d20; KMS). Two observations of fledged young rounded out the roster of confirmations, 11 Jul 1995 (e22; *fide* RHw) and 13 Jul 1997 (N11; *fide* JMH).

Continued on page 394

WESTERN TANAGER *Piranga ludoviciana*

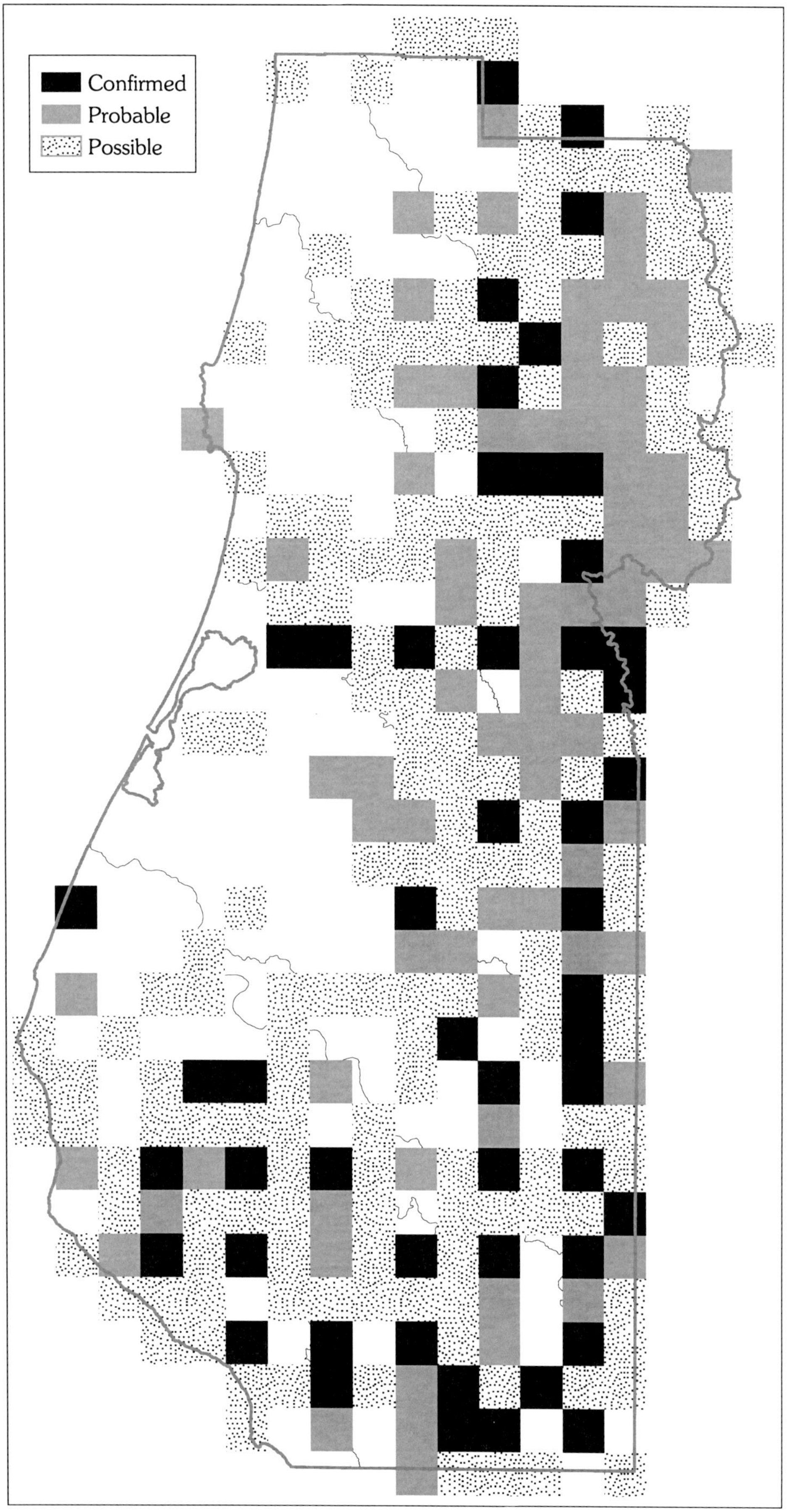

From early May through late summer, dry *kritik* notes in the treetops of forested Humboldt County betray the presence of Western Tanagers. The eye-catching plumage of male tanagers draws exclamation from the most casual observer during the species' spring migration, when numbers of these colorful birds may feed in loose flocks in lowland shade trees and planted conifers. During the course of nesting activities which follow promptly thereafter, they retire to the forest canopy, and for the balance of the year, they are heard far more often than they are seen. Experienced birders may "encounter" scores of tanagers in a given breeding season—yet most of these encounters will be of half-heard snatches of song coming from a stand of tall trees in the middle distance, glazed with the glimmer of high summer sun and abuzz with the droning of insects.

Atlas results indicate that tanagers were abundant and widespread across most of Humboldt County. They were found in 67% of the blocks and were confirmed in 51 blocks. Indeed, Harris (1996) characterized Western Tanager in northwestern California as a "common migrant, summer resident, and breeder; casual in winter." They are among the routine and expected nesting birds of the interior and southern sections of the county. Raphael (1984) found them in 93–100% of his plots in 50- to 350-year-old Douglas-fir stands. It is suspected that nesting probably takes place in nearly every one of the blocks in which this bird was reported. The extra effort expended in priority blocks resulted in a weak pattern of more confirmed and probable breeding reports in those blocks.

During the breeding season, Western Tanagers occupy many forest habitats across their range within Humboldt County. In a general sense, these birds would appear to increase in numbers and distribution with increasing cover of mature forest. Excellent habitat is offered by the drier, tanoak- and Douglas-fir-dominated forests at middle elevations in the interior of the county. Both older and younger stands of conifers mixed with oaks, bigleaf maples, and other hardwoods support these birds. Mixed-conifer stands within the zone of winter snows at higher elevations, in which hardwoods are scarce or absent, are also well populated; tanagers are common in white fir, Jeffrey pine, and even pure Shasta red fir stands at high elevations in extreme northeastern Humboldt and along the ridgeline of South Fork Mountain.

Where sufficient forest exists, breeding also occurs nearly to valley bottoms along the major streams, as along the Klamath, Trinity, Mad, Van Duzen, Eel, and Mattole rivers. In the hilly open country of the east-central and southeastern parts of the county, tanagers were encountered in fairly open, conifer-poor California black oak woodlands. They appear somewhat tolerant of interruptions in forest cover, and do much foraging within the forest canopy at or close to openings and edges. Western Tanagers are not among the early colonizers of regenerating clearcuts.

The distribution map illustrates a fairly conspicuous pattern of few or no reports from northwestern and west-central Humboldt. This apparent absence or scarcity may reasonably be attributed to inappropriate or insufficient forest habitat created by intensive logging, agriculture, and other human land use. It is worth noting that the region seemingly poor in tanagers conforms with the county's coastal summer fog belt. During 1995 counts in different-aged coast redwood stands, Hazard and George (1999) apparently did not record this species. Much of that area is occupied by old-growth redwood stands and younger regrowth largely dominated by redwoods. Might it be that forests within this zone somehow fail to offer this species the habitat it demands? Data gathered during the Atlas period can provide no clear explanation, they merely suggest the question.

The rather sluggish behavior and canopy hangouts of Western Tanagers caused Atlas surveyors to use patience and persistence to confirm breeding by this species. These efforts paid dividends, with breeding confirmed in 51 blocks. The earliest such report was of the birds carrying nest material 16 May 1995 (N33; C. Secor). Of the blocks with confirmed breeding, one-quarter each are represented by observations of food-carrying and observations of adults feeding young. Just under two-thirds of the food-carrying reports and just over two-thirds of the feeding young reports fell in July. This suggests that the peak period of egg-laying was in early June. Only six nests were reported and, of those, only two nests with young: 19 Jun 1998 (B21; COg) and 11 Jul 1999 (N25; DFx). No nests with eggs were reported. The paucity of nest reports is surely due to this species' habit of building nests upon foliated limbs high within the crowns of tall trees. The fledgling reported 31 May 1999 (o23; C. Cool) was the earliest. The latest breeding confirmation was of adults feeding young 21 Aug 1995 (H15; PFS).

Western Tanagers are not among the earlier-arriving spring migrant songbirds. First detection dates in Humboldt County typically fall in late April, and it is clear that the bulk of migration does not take place until well into May. By mid-September numbers of apparent migrants have peaked. A few birds are noted into October each year, but any thereafter are unusually tardy.

GREEN-TAILED TOWHEE *Pipilo chlorurus*

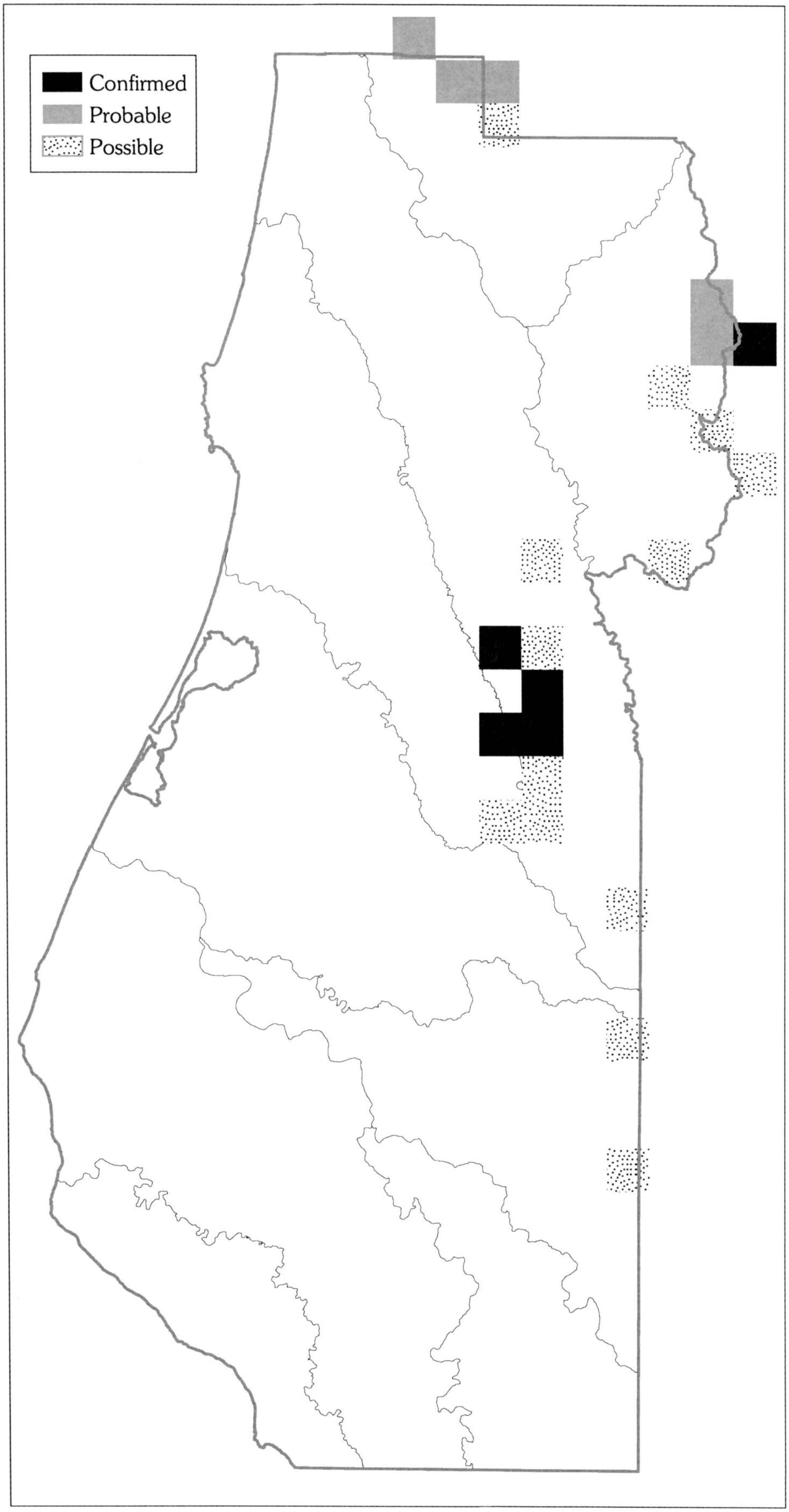

Living in backcountry haunts and little given to introduce itself to passing humans, the Green-tailed Towhee is a bird known to comparatively few people. This towhee is a locally common summer resident of mountain brushfields and higher sagebrush-steppe communities in the western U.S. from the Rockies to the crest of the Pacific Slope. In many areas it can be among the more abundant breeding birds. Yet the species remains scarcely known, other than to those who look for it. No sacks of birdseed are emblazoned with its appealing image. It is the emblem of no lodge or fraternity. Unique would be the needlework or ceramic creation featuring its colors. It is left to the bird-curious to go make acquaintance with it, suss out its wiles, and get to appreciate the subtle beauties of the towhee's chosen patch.

Green-tailed Towhees are, overall, rather uncommon in Humboldt County. They were found in 23 (5%) of the 425 total blocks during the Atlas period. Harris (1996) considered the species a "locally uncommon to rare summer resident and breeder in inland mountain brushfields, especially mountain chaparral" in northwestern California. This is one of the many mountain birds whose range in Humboldt is restricted to the eastern and northeastern parts of the county. Distribution in Humboldt is similar to that of Fox Sparrow, with which it often occurs, but appears somewhat less extensive. As is true for other birds of the forests and openings of the Humboldt County fir zone, Green-tailed Towhees occur in several areas separated by the unsuitable lowland habitat of intervening river corridors.

North of the Klamath River, the birds were detected in a cluster of four Atlas blocks stretching from the south end of the Siskiyou Wilderness southeast to about Onion Mountain. South of the Klamath, they were found in seven blocks taking in high country around the Trinity–Salmon River divide from Whiteys Peak south to Trinity Mountain and Groves Prairie. Between the Trinity and Mad rivers, towhees were found in 10 blocks, eight of which were adjacent. This cluster includes the slopes east of Three Creeks Summit, higher portions of Snow Camp Mountain, and several blocks along the South Fork Mountain ridge south to Last Chance Ridge and Blake Mountain. South of the Mad and Van Duzen rivers, Green-tailed Towhees were found in the Lassics. There were no observations anywhere outside the higher mountains of northern and eastern Humboldt during the 5-year Atlas period. Regionally, the species' range extends beyond the county eastward through the Klamath Mountains and locally southward in the Coast Range.

Old burns and brushfields occupying persistently harsh ground on sunny exposures attract Green-tailed Towhees. Many of the birds detected in Humboldt County during the Atlas period were encountered in or near clearcuts. As is the case for a number of other mountain songbirds found only in eastern Humboldt, the practices of industrial forestry certainly provide habitat for Green-tailed Towhees. However, this site alteration does not appear to benefit towhees to the extent that it does Fox Sparrows, since towhees seem to favor natural stands of brush and are pickier about where they live. It was apparent that stretches of mountainside or ridgeline offering an ideal shrub composition and structure may support some numbers of these towhees, whereas they may be absent at sites seeming to differ only slightly.

In Humboldt County, Green-tailed Towhees are fond of 5–20-year-old clearcuts with limited conifer reestablishment. Sites at which these birds are found undergo continual plant succession; as one site grows in too densely, another more recent clearcut may become suitable. Seral shrub communities in favored places are often dominated by any of several *Ribes* spp., *Ceanothus* spp., *Arctostaphylos* spp., silk tassel, and similar montane shrubs tolerant of summer heat and dryness. Clumps and individual seedling or sapling white fir, Douglas-fir, pine, or incense-cedar are usually present. Exposed stumps, varying amounts of large woody material, and the skid roads and landings of shallowly sloping ground are also features of human-made Green-tailed Towhee habitat. It was the impression of atlasers that these birds ask for somewhat less confined sites than do Fox Sparrows, seeming to require less overhead cover, less shrub-to-shrub continuity, and greater plant species diversity.

These birds sing loudly on summer territories. The song of the Green-tailed Towhee is remarkably similar to that of the Fox Sparrow, and nearly all birders find that considerable practice is required to consistently distinguish the two. Later in summer, a characteristic *mew* call is often the tip-off. Green-tailed Towhee nests are cleverly concealed, and breeding pairs are well recognized for furtiveness in the presence of humans. For these reasons, it is not surprising that none of the five confirmed breeding reports involved directly observed nests. Confirmed breeding was recorded when adults were seen carrying food on Horse Mountain (L15) 20 Jun 1998 (COg) and on Snow Camp Mountain (L17) 14 Jul 1999 (JCP). The additional confirmations were reports of fledged young near the head of Friday Ridge (m16) 5 Jul 1996 (GCH), on Trinity Mountain (r08) 16 Jul 1997 (JEH), and on Grouse Mountain (m17) 28 Jul 1996 (JCP). In the course of Atlas fieldwork, the earliest date reported was 6 May 1996 at Groves Prairie (P13; TWL), while the earliest date listed by Harris (1996) was 4 May 1980 (S. Harris). The earliest probable breeding evidence received during the Atlas period was of multiple males singing 11 Jun 1997 (j01; KMS). The latest Atlas-period report was of possible breeding 30 Jul 1998 (r08; GCH).

Ehrlich *et al.* (1988) listed information on courtship by Green-tailed Towhee with a question mark, indicating that little or nothing was known of the behavior at that time. An Atlas observation provided a glimpse of this. On 12 Jun 1999 at Snow Camp Mountain (L17), a male (known as such moments later at copulation) joined a female atop a 4-foot diameter stump, circled the female several times, hopped into the air repeatedly, and then walked slowly several inches from her, holding one wing outstretched vertically high above his body. Both birds then met breast-to-breast with their upper bodies extended vertically and tails out. Copulation followed. The entire display took about 15 seconds (DFx).

SPOTTED TOWHEE *Pipilo maculatus*

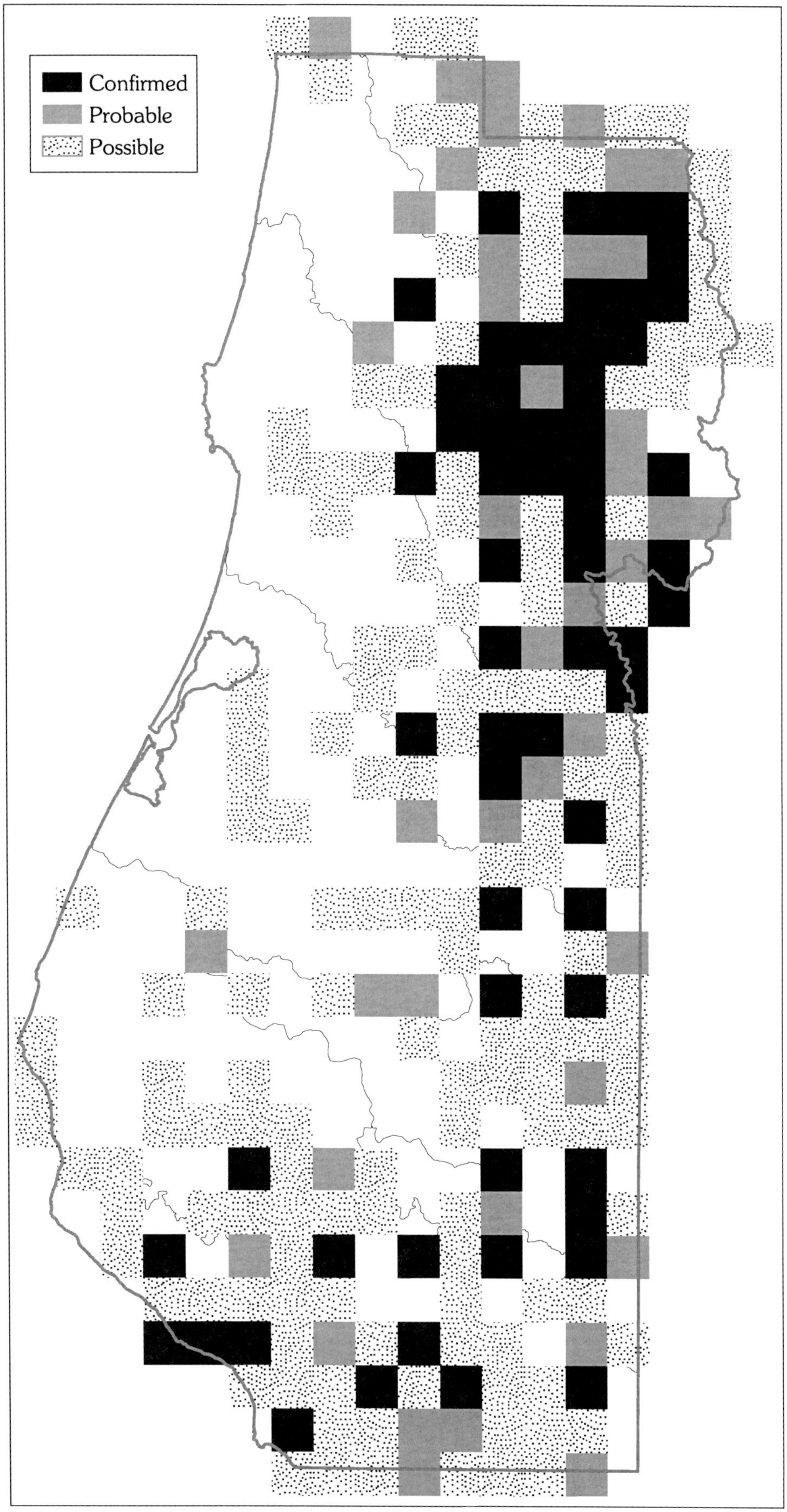

Nearly always within a short flight distance of shrubbery, the Spotted Towhee pursues life within tortuous grottos and scurry-ways of dappled shadow. Throughout its range this species is tied to expanses of well-established brush, which provides both shelter and an undisturbed layer of cast foliage supporting the invertebrates upon which the birds feed. Should a Spotted Towhee be neither vocalizing nor sitting atop an exposed twig, it may nevertheless be detected by the telltale rustling created as it kicks vigorously into dry leaf-litter.

Harris (1996) considered the Spotted Towhee in northwestern California to be a common resident and breeder. Atlasers also found Spotted Towhees to be common over much of Humboldt County. They were encountered in 248 (58%) of the 425 total blocks. The northeastern portion of the county, in particular, seemed to appeal greatly to these birds. A cluster of breeding confirmations centered in and around the lower Trinity River country is easily noted on the distribution map. It is worth noting, too, that this area also supports great numbers of Spotted Towhees in winter. Drier, semi-open, shrubby forests and woodlands of southern Humboldt also support large numbers of towhees. The distribution map shows breeding towhees to be either absent, present in low numbers, or patchily distributed throughout most of the coastal fog belt north of Cape Mendocino. The almost complete lack of coastal breeding in that area surprised some atlasers.

Spotted Towhees accept a wide range of brushy habitats within Humboldt County. Riparian corridors, foothill woodlands with clumps or stringers of low vegetation, early-seral regenerating clearcuts, mountain brushfields, Douglas-fir/tanoak forest edge, and varied "soft chaparral" patches are inhabited during the breeding season. Characteristic habitat elements include some expanse of well-established brush that produces significant leaf litter, some open ground beneath, and, often, an abrupt edge or small opening adjoining the vegetation. Higher-elevation clearcuts in which clumps of manzanita and currant are dotted with infrequent conifer saplings are examples of more sparsely vegetated sites, while the margins of a long-overgrown road bisecting an old, densely regrown clearcut unit with multi-layered vegetation represents a closed-canopy extreme.

In areas in which it is numerous, the Spotted Towhee is difficult to overlook. The distinctive song, whining call-note, and flash of white tail-corners as a bird flits across an opening render it conspicuous despite its retiring foraging habitats. Given to agitation at the presence of humans and expressing somewhat overt provisioning behaviors, this species provided no particular challenge to atlasers striving to detect breeding evidence; they were confirmed in 60 blocks. Spotted Towhees breeding in Humboldt County would appear to start nesting chiefly from early May to early June. Some of the earliest evidence of nesting activity was a bird carrying nest material 5 May 1997 (o08; JMH). Only 8% of confirmations involved the observation of a nest. Two nests with eggs were discovered, one 8 Jul 1996 (l18; D. Carpenter) and the other 22 Jun 1999 (P13; GCH). Two nests with young were observed, 3 Jul 1998 (J07; JCP) and 9 Jul 1996 (m11; JMH). Actual nest building was reported but once, 21 May 1995 (J11; D. Kriger). The low number of nests reported attests to the secrecy with which the early stages of the breeding cycle are carried out. The earliest fledglings were seen 15 Jun 1997 (n28, N29; both RJA). The latest confirmation was of the feeding of young 3 Aug 1998 (GCH; L09).

The extent to which Spotted Towhees linger in winter at higher elevations in the county is not well understood. Perhaps persistent snow cover prevents access to the forest floor, resulting in abandonment or low use of such areas. Although Spotted Towhees are permanent residents in Humboldt County, considerable migration is evidenced each fall and winter by the appearance of numbers of towhees in run-of-the-mill brushy places across coastal lowlands and on the near-coastal slope, where they may be absent or uncommon in summer.

CALIFORNIA TOWHEE *Pipilo crissalis*

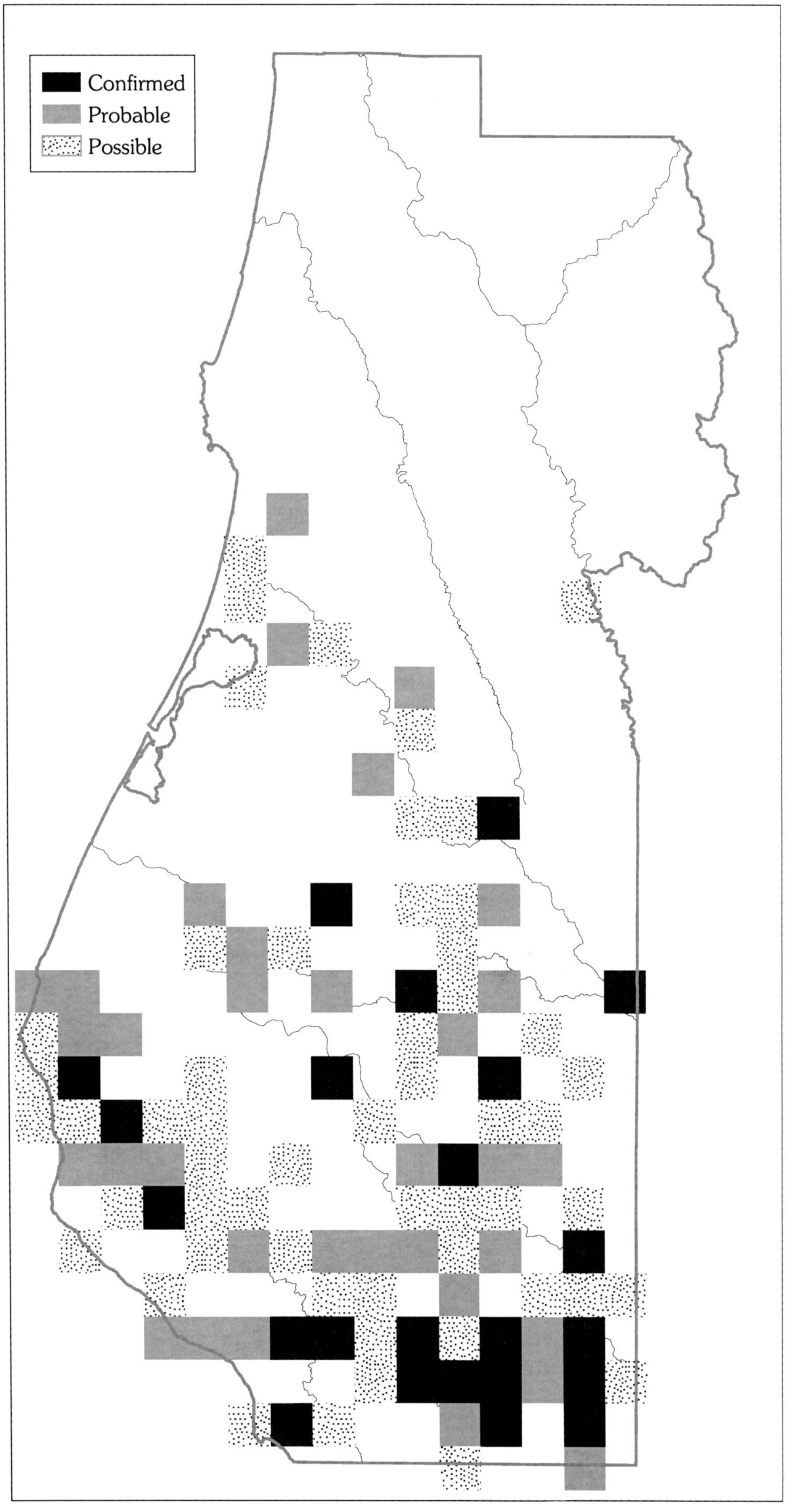

The penetrating metallic call-note of the California Towhee frequently breaks the hush of mid-afternoon in southern Humboldt County. While singers of most species have fallen silent for the duration of the daily siesta, California Towhees continue to announce their presence from a fallen oak, a tall shrub, or the top of a roadside power-pole. These towhees are common and widespread in appropriate habitat throughout most of northwestern California with the exception of Del Norte County and northern Humboldt County. They were found in 111 (26%) of the 425 total blocks and were confirmed breeding in 23.

Indicating a preference for drier and warmer sites, these towhees have been routinely encountered only in the southern one-third of the county. The species is widespread and conspicuous at lower elevations throughout southern Humboldt. However, a northward extension involving a considerable population occurred during the study in the Mad River corridor, with greatest numbers seeming to reside southeast of the Iaqua Buttes. California Towhees were present in appropriate habitat quite near the outer coast from the vicinity of Shelter Cove north to Cape Mendocino. They are known to occur north of the Eel River to Fortuna (occasionally in Loleta) and on lower foothill slopes and brushy flats within the Van Duzen River drainage. During the Atlas period, outlying single birds or pairs were found as far north as Bayside, Fieldbrook, Blue Lake, and Sandy Bar along the South Fork Trinity River. The northernmost Atlas record was of a pair near Fieldbrook (g12) 21 May 1998 (C. Lawrence).

Favored settings in the county are characterized by clumped, dense native brush adjacent to semi-open ground grown to sparse grass, dry-site forbs, and with perhaps a bit of surface rock or gravel. Shrub-dotted edges of grasslands and pastures on south and west exposures are commonly inhabited, as are ranches, riverside parks, and the drier-site brush communities rambling across the upper floodplains of the larger streams in "SoHum". Along and shortly above the lower Mattole River, California Towhees may also be found at the edges of scrubby young regrowth Douglas-fir woods, especially adjacent to small pastures or neglected disturbed ground near residences. California bay and California buckeye may also be present at these sites. Shallow draws bisecting pasturelands, grown to stringers of unmanaged native brush, are a common situation in the Mad River foothills. In some locations favored by the birds, vegetation other than grass may be limited to an irregular mosaic of poison-oak, coyote brush, and similar sunny-site shrubs tolerant of summer dryness and favoring full sunlight.

This species clearly does not climb the foothill slopes into the mountain brushfields, dropping from the bird scene well before the snow zone is reached. Much Atlas fieldwork in the mountains affirmed a preexisting perception that the numerous *Ceanothus*- and currant-dominated brushfields at higher elevations host very few of these birds.

Although the spectrum of breeding behaviors reported broadly overlapped in time, 87% of the confirmations were noted in June and July. The earliest courtship was reported 20 Apr 1995 (n32), and the earliest fledglings were reported in that same block 13 Jun 1995 (both M. Bailey). One report of the carrying of nest material was received, 31 May 1999 (o23; C. Cool). Only one nest with young was found, 10 Jun 1996 (N33; C. Secor). The latest confirmation was of young seen fed 18 Aug 1998 (d28; C. Keil).

Historically, California Towhees appeared to have occupied a somewhat greater range within Humboldt County than was indicated by Atlas surveys. Harris (1996) noted that the birds were known north to Redwood Valley and Hoopa. He cited Fisher (1904), who stated that at Hoopa it was "common . . . a characteristic bird of the valley" in 1899. During the Atlas period, no reports came from either of these areas. The only detection north of Highway 299 was of the aforementioned pair at Fieldbrook. Reasonably thorough coverage of Redwood Valley was impractical owing to limited public access. While it is possible that a small population occurs there, no recent evidence supports this. Much birding of the Hoopa Valley, including near-saturation coverage of likely California Towhee habitat during the Willow Creek CBCs, has resulted in only a few observations. The absence of the species from the Trinity River corridor in the vicinity of Hoopa seems puzzling. A great deal of nice-looking habitat exists at many spots on and near the valley floor, and, although these birds are considered largely sedentary, a possible "contributory" population exists to the south along the South Fork Trinity River. One bird was present "all summer" at Sandy Bar (n14) in 1997 (G. Lester).

Harris (1996), referring to northwestern California in general, stated that this species was an "uncommon resident and breeder in dry brushlands and edges in valleys of southern and eastern portion of region," adding that vagrants had been seen in the coastal lowlands throughout the year. This description fairly hits the mark, although the experience of atlasers within the time frame of the project was that California Towhees are, in a broad sense, rather common across much of lowland interior Humboldt. Certainly, atlasers helped to greatly detail the distribution of this species and to delineate its dynamic northern limit.

RUFOUS-CROWNED SPARROW *Aimophila ruficeps*

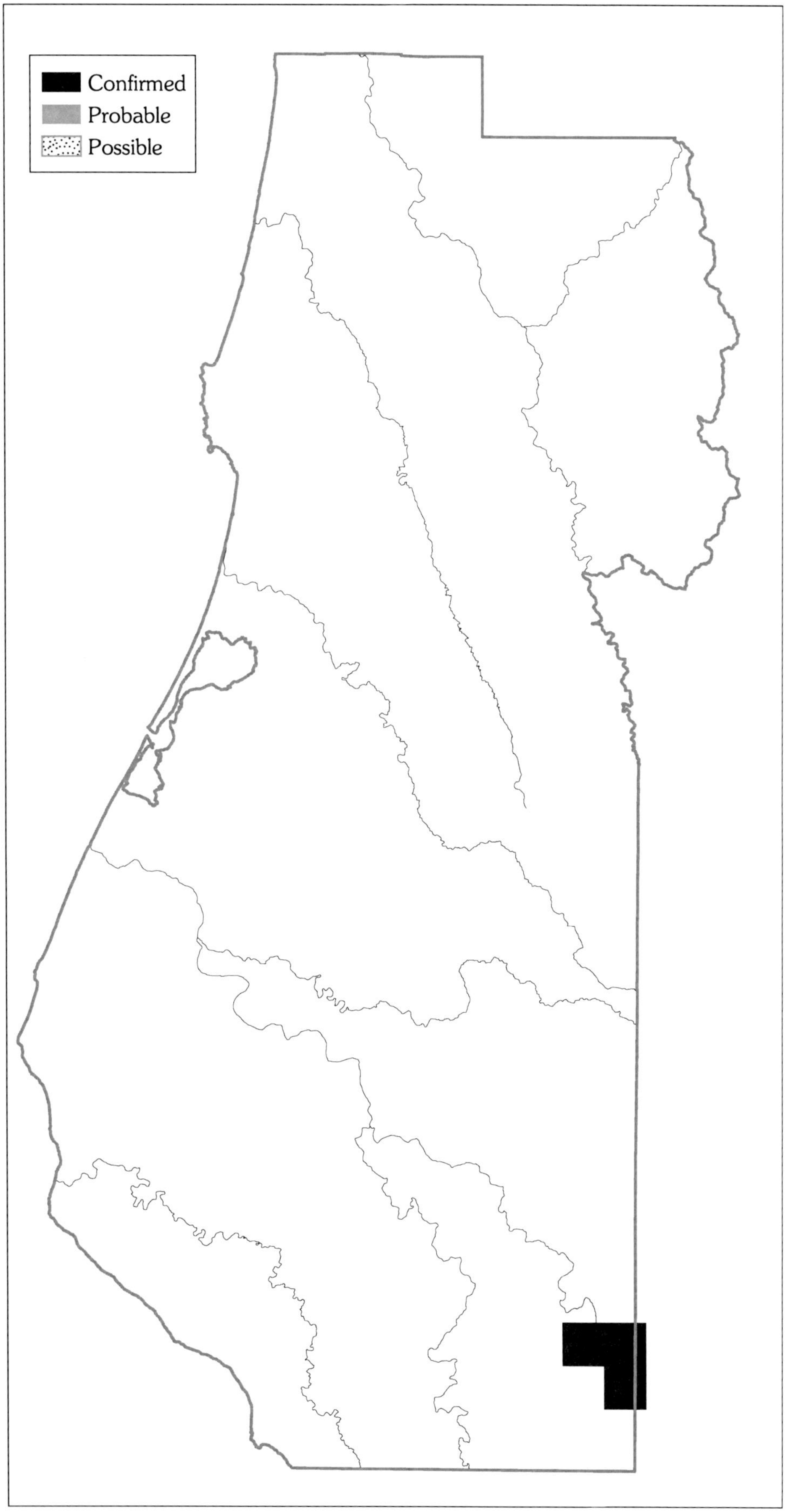

The history of Rufous-crowned Sparrow as a bird of Humboldt County is brief. On 24 May 1998 a small party of atlasers (JEH, TWL) diligently searched a remote portion of southeast county along the Northwestern Pacific Railroad tracks beside the Eel River (N31). Aware of the known presence of this species as a breeder in nearby Mendocino County in the vicinity of Dos Rios (Harris 1996), and the more recent (14 Jun 1996) discovery of a single bird along the same railroad tracks at Island Mountain in nearby Trinity County (RHw), Rufous-crowned Sparrow was among the litany of "fantasy" species being recited that day. And behold, up pops Humboldt County's first verified Rufous-crowned Sparrow—and carrying a food item, no less! The party went on to count nine adults that day (of which at least eight were in pairs, and including others carrying food), along the 2 miles of railroad tracks south of Cain Rock Crossing (N31, o31). Surely, this was one of the finest achievements of the Atlas project.

Since that discovery, 16 individuals (including two fledglings and numerous singing males) were counted 30 May 1998 along about 5 miles of track from Cain Rock Crossing to about one-half mile west of the Humboldt–Trinity county line (N31, o31, o32; JEH, TWL, GCH). Birds were also reported in the area the following year on 26 Jun 1999 (B. Deuel). That Rufous-crowned Sparrows were year-round residents here was made apparent by the observation of a pair near Cain Rock (N31) 10 Jan 1999 (JEH, TWL, DFx, JCP). While not all birders searching for them have met with success, most have, helped by persistent effort and a bit of luck. It should also be noted that a possible singing Rufous-crowned Sparrow was heard 25 Jul 1995 near the King Crest trailhead (d30; R. Sutherland); atlasers searched for that bird on the 27th but failed to detect it. Habitat there was mostly manzanita brushfields with Douglas-fir reproduction, a few rocky outcrops, and very little grassy cover (JEH, TWL).

It is unknown whether Rufous-crowned Sparrows had been present all along in southeastern Humboldt County, or if they are recent colonizers. G. Atwell collected eggs at all of the railroad stops along this section of track in the late 1920s and early 1930s and his reports do not mention Rufous-crowned Sparrows. Exploratory birding by a party (R. Willmarth, T. Schulenberg) who walked the 5 miles of track between Alderpoint and the Trinity County line 3 Jun 1973 did not reveal the presence of this species (Yocom and Harris 1975). Elsewhere in California, the northern extent of the range—or at least its perception by ornithologists—has undergone slight, incremental changes since the time of Grinnell and Miller (1944). Increasing investigation by curious birders pushed the historical edge of the range north during the 1980s on both sides of the northern Sacramento Valley (AB 38:956, AB 39:208, AB 42:318), and in the Coast Ranges north to Mendocino County (AB 40:521).

Throughout their range, these sparrows are widely distributed at lower and middle elevations in arid canyons and desert foothills of the southwestern U.S. and Mexico. In northern California, they are found along the drier portions of the coastal slope, on lower slopes of the Inner Coast Range, and along the western base of the Sierra Nevada. The range of Rufous-crowned Sparrow extends quite a bit farther north in interior California than it does elsewhere. The northern limit of the range in California is not an abrupt front; rather, the population appears to "finger" northward in segments or patches of habitat in river canyons and adjacent slopes. The distribution in Humboldt County probably represents the northern tip of one such finger of more or less continuous distribution that extends southward along the Eel River canyon to Dos Rios and beyond.

The places inhabited by Rufous-crowned Sparrows in the restricted Humboldt portion of their range are all within the immediate Eel River canyon, on the sunny western or southerly exposures. These are seasonally semi-arid sites, dry and very warm during much of the year. Recalling a more typically interior Mendocino County-like setting, interior live oak, Pacific madrone, California buckeye, toyon, birch-leaf mountain-mahogany, manzanita, and Diogenes' lantern are conspicuous plants. The sparrows have been found more or less randomly scattered along several miles of disused railroad trackbed, a sinuous corridor characterized by scattered shrubs, grasses, forbs, and much surface rock.

It should be mentioned that exposed rock occurring naturally along the immediate riverbank has been augmented by sidecast and blasted rock resulting from the construction of the rail line. In places this rubble, in combination with the native outcroppings and talus, has created boulder-strewn habitat of considerable extent. With growth of trees and shrubs limited within such microsites, grasses and forbs have become the dominant vegetal cover. Rufous-crowned Sparrows are well known for selecting sparsely vegetated slopes with scattered rocks and patches of shrubs and grass (Ehrlich *et al.* 1988). Thus, while naturally occurring habitat elements present within the Humboldt County portion of the Eel River canyon may suffice, it is possible that human activities have actually enhanced the fortunes of these sparrows locally. This species should also be looked for farther north in Humboldt along the Eel River in what appears to be additional suitable habitat.

CHIPPING SPARROW *Spizella passerina*

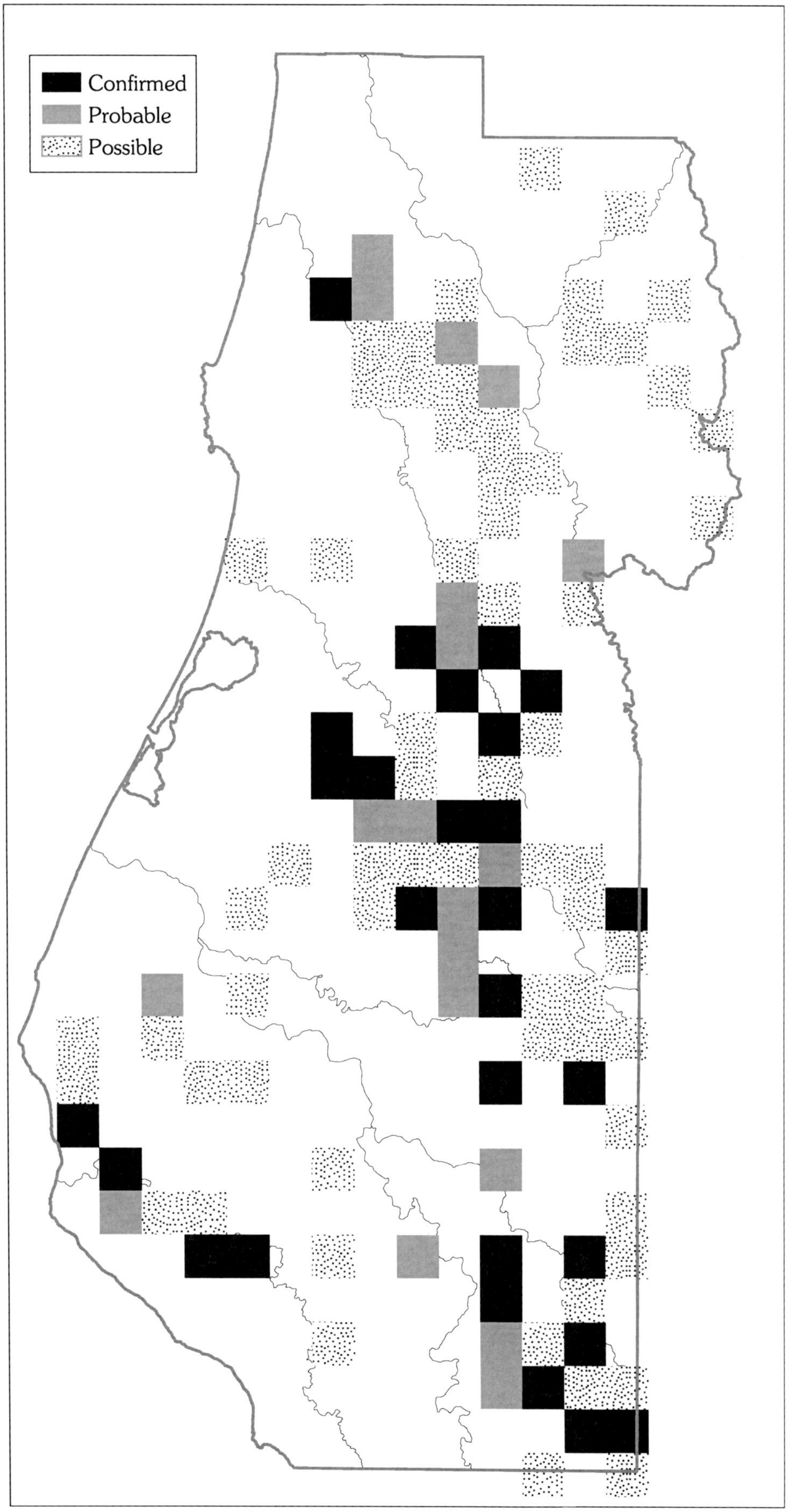

In an era now largely past, Chipping Sparrows first came to the attention of young naturalists through their skill in creating delicate nests from horsehair. Haunting stables and barnyards adjacent to their grassy hangouts, these sparrows gathered strand by strand a collection of materials that once warmed creatures immensely larger than the wee nestlings those same hairs would soon swaddle.

Among the six currently recognized species of *Spizella* sparrows occurring in North America north of Mexico, the Chipping Sparrow is the most familiar and most widely seen. This bird enjoys a great range during the nesting months, occurring from the Northwest Territories to the Maritime Provinces and southward across the U.S. Only in the southern Great Plains is it absent over a large area in summer. It is also locally scarce or absent in high mountains, in densely forested sections, and in the most severe deserts. In migration it is found widely, particularly in spring, when adults in natty plumage come to the notice of feeder-keepers unaccustomed to hosting them. Found most often in open or semi-open surroundings, Chipping Sparrows are widespread in California's grassy expanses and oak woodlands.

Chipping Sparrows were well observed during the course of Atlas fieldwork, and were found in 111 (26%) of the 425 total blocks. The Atlas map suggests an affinity with interior-county landscapes dominated by broken forests and oak woodlands. In so doing, it largely recapitulates a theme presented elsewhere many times in these pages. Yet, as is true for every species herein treated, it exhibits a singular twist here, an odd brow-raising there. No doubt attends either of two assertions: that semi-open country attracts these sparrows most consistently, and that oak woodland is important to them. However, the degree of landscape openness varies quite a bit, while oak woodland, though well inhabited, is far from the only cover type used by the birds in Humboldt.

The map suggests that some degree of "mountaineering" takes place in eastern Humboldt County, and this is indeed the case. Scattered locations at higher elevations north of the Klamath River, around the east side of the Hoopa Valley Indian Reservation, on the South Fork Mountain ridge, and in the Lassics were found to support Chippies in summer. In such places, oaks may be absent, and the ragged forest edge afforded by clearcuts, meadows, and burns provides an acceptable structural substitute. Chipping Sparrows are not unexpected well up into the white fir zone, where shrubby conifers dot the openings and the summer growth of grasses and forbs must spring each year from snow-compacted ground. Fisher (1904) found them common in gray pine and *Ceanothus* thickets at Hoopa in 1899. While of no great interest otherwise, this remark is one of exceedingly few historical references to the occurrence of any local bird species in association with gray pine, a tree with a low profile among many birders and a restricted inland range in Humboldt. Across the county toward the coast, the windswept mountain prairies and similar grassy sites shortly inland from Cape Mendocino also hosted their share of these birds. Truly, however, Chipping Sparrows come into their own in the summer-hot oak woodlands of central and southeastern Humboldt.

These sparrows were confirmed breeding in 28 blocks, and were judged a probable breeder in 19 others. Six behavioral codes were reported in the blocks with confirmed breeding. Birds carrying nest material were reported from three blocks, 21 Jun 1998 (o33; C. Secor), 28 Jun 1997 (H17; JCP), and 5 Jul 1998 (k19; TWL). Each of these observations seems to have fallen quite late in the breeding cycle. Three nests were discovered in the course of the project: an occupied nest 6 Jun 1997 (b26; DFx), and nests with young 8 Jun 1995 (e29; R. Sutherland) and 25 Jun 1995 (N33; D. Roberson). As was true for many songbirds encountered during the project, observations of fledged young or adults carrying food provided the largest share of breeding evidence. The carrying of food furnished the best code in nine blocks, spanning 6 Jun 1996 (N31; M. Bailey) to 27 Jul 1995 (F29; JEH). Two-thirds of these reports were in June. Fledged young were noted in 10 blocks, as early as 25 Jun 1995 (m32; D. Roberson) and as late as 28 Jul 1996 (m16; DFx). It was not surprising that these sightings of juveniles averaged later than reports of food-carrying; indeed, all but the earliest report fell in July. Adults feeding young were noted in three blocks, 15 Jun 1997 (N29; R. Adams), 20 Jul 1999 (J21; DFx), and 24 Jul 1996 (J15; JCP).

Referring in general to northwestern California, Harris (1996) termed the Chipping Sparrow a "common summer resident and breeder in open woodlands and brushlands. More common inland than along the coast." The earliest Atlas record of a bird in suitable breeding habitat was 14 Apr 1996 (d24; COg). Harris (1996) further indicated that most spring migrants arrive in April, with fall migration mostly in October. Chipping Sparrows are rarely seen in the coastal lowlands in fall, with the closely related Clay-colored Sparrow (*S. pallida*) having proven more frequent there from September into late winter.

LARK SPARROW *Chondestes grammacus*

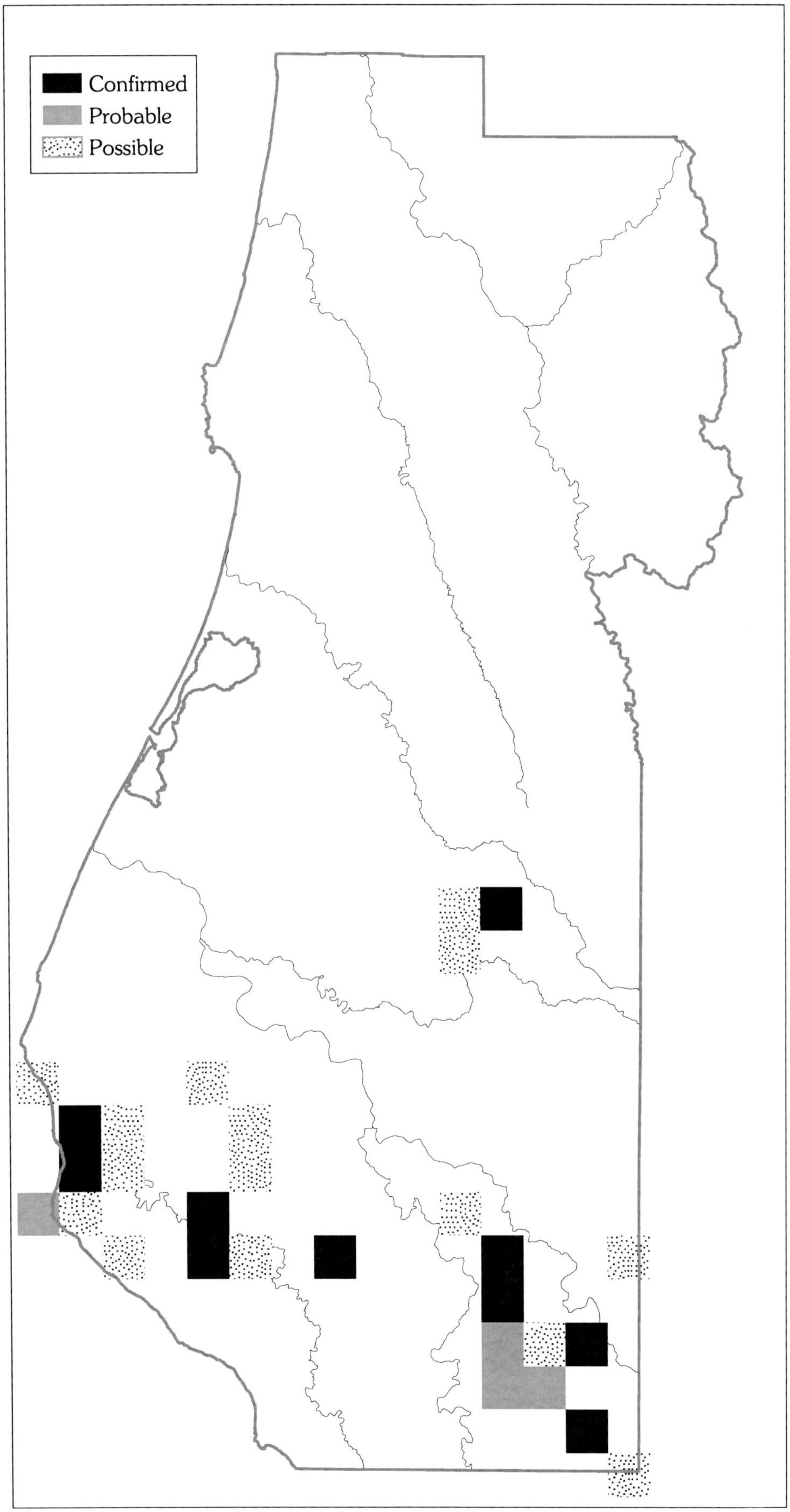

Lark Sparrows are uncommon summer residents of open country across the south-facing foothill slopes of southern interior Humboldt County. They are among the characteristic southern Humboldt species of tree-dotted pastureland which is grazed and seasonally hot. Expanses of grass interrupted by scattered trees or small groves, fence-lines, or rock outcrops offer preferred habitat locally. Although not especially abundant, breeding pairs are generally well scattered throughout much of their range within the county. The species is expected annually in most places it occurs. During the Atlas period, Lark Sparrows were reported in 29 (7%) of the 425 total blocks. Breeding was confirmed in 10 blocks.

Although the total number of reports was not great, and coverage of potential habitat was far from complete, it appeared that the population breeding in southern Humboldt County occupied several areas in greatest numbers, with a suggestion of sparser numbers in between these centers. Exposed, nearly treeless hillsides along the southern portion of the greater Mail Ridge uplift, separating the Eel and South Fork Eel rivers, provided much habitat along a northwest-southeast axis. In this pastoral roll of grass was found a smattering of pairs of Lark Sparrows, extending from around Garberville eastward into the Alderpoint area. Greatest numbers appeared to occupy the open, hardscrabble slopes of shallow draws throughout the vicinity of the Dyerville Loop, where territories seemed to abut in the best habitat. Between 1990–1999, an annual average of one, one, and five Lark Sparrows were recorded on the Honeydew, Holmes, and Alderpoint BBS routes, respectively.

To the west-northwest some 20 miles was another sizable population, occurring at the edges of oak woodland on lower foothill slopes within the upper Bear River-Mattole River complex of hillside grasslands and pastures. These birds were detected westward along the Mattole River corridor nearly to its entry into the ocean, nesting on the windswept hillsides from southeast of Cape Mendocino to Punta Gorda.

A third area of occurrence during the study period was in the rolling, highly rural countryside at middle elevations in the Mad River drainage southward from near Showers Pass. To what extent this smaller, northerly cluster of reports indicates a population disjunct from that on Mail Ridge is unclear. Habitat that seems appropriate occurs at least spottily in the intervening miles.

Lark Sparrows are known to arrive in Humboldt County during April. That there were only eight Atlas reports before 1 Jun may be a reflection of the timing of most surveys in southern Humboldt. The earliest confirmed breeding record involved the carrying of nest material 17 May 1997 (N31; M. & M. Bailey). Detections were generally spread throughout the nesting season. Two nests with eggs were found, 20 May 1995 (e29) and 27 Jun 1995 (e28; both R. Sutherland). Owing to the brilliant singing of the males, the selection of high song perches (where available), and other readily observed behaviors along roadsides, Lark Sparrows are not prone to being overlooked, and can be watched at length. Seven of the 10 confirmed breeding reports involved the conspicuous behaviors of carrying food or feeding young. One rather late young bird was seen being fed 16 Jul 1999 (I30; JG).

Harris (1996) characterized Lark Sparrow in northwestern California generally as a local, uncommon migrant, summer resident, and breeder. However, he noted that it bred in loose colonies on Cape Mendocino and near Kneeland; during the Atlas period, there were no reports from either area. Do these places still support Lark Sparrows, or have numbers fallen to levels below a threshold of ready detection? Further, Fisher (1904) reported that Lark Sparrows were "abundant in open areas" in the Hoopa Valley in 1899. As significant Atlas effort was invested at Hoopa without detections of these sparrows, clearly the species did not summer there during the study period. Interestingly, a single bird was found in a semblance of breeding habitat at the Hoopa airstrip on the 23 Dec 1995 Willow Creek CBC (DFx).

The Lark Sparrow is among a small suite of songbirds breeding in Humboldt County that tolerate, or prefer, treeless ground. Throughout the nearly bald, grassy, rock-dotted upper slopes of the Dyerville Loop area, pairs of Lark Sparrows occupy territories in which the only elevated song perch may be a low rock or hard-earth outcrop, a single sagging fenceline supporting no shrubs and only minimal forb cover, or the butt of a fallen oak. These Lark Sparrows constitute the northwesternmost extension of their distribution in California. This pattern is reflected in the ranges of a number of other passerines—species better represented in California to the southeast of Humboldt—whose drier habitats are found locally within the county, but which ultimately are replaced northward by wetter forests and shrub communities.

SAVANNAH SPARROW *Passerculus sandwichensis*

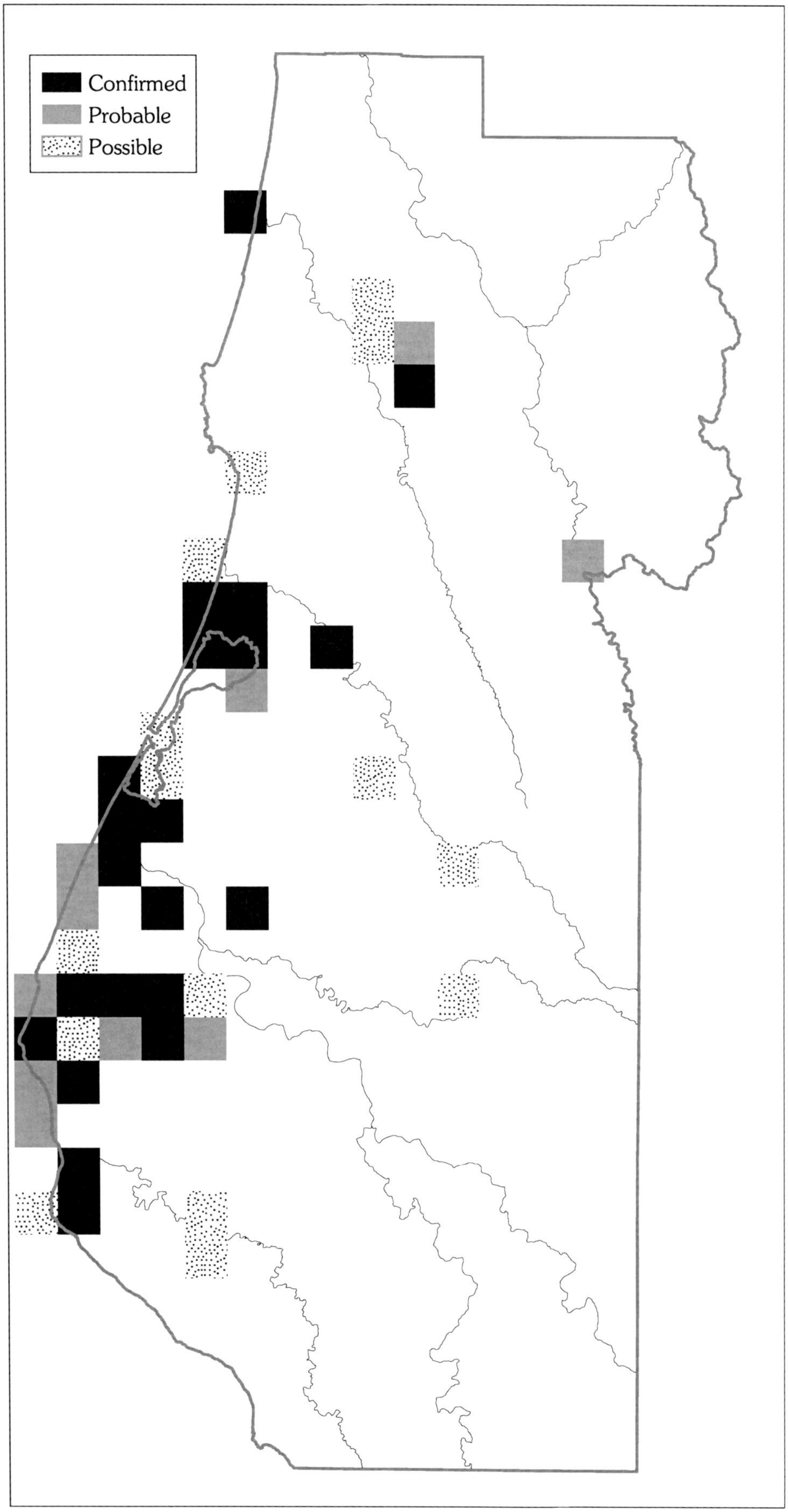

Unremarkable coloration and a streaky plumage pattern conceal Savannah Sparrows as they spend their lives within thickets of grass. These are small, plain birds, known well only to those who spend time in the birds' preferred haunts. Their retiring habits are balanced by their abundance during migration, especially in fall, when open-country expanses across North America offering even minimal habitat may crawl with them. As breeding birds, Savannah Sparrows are partial to grasslands of many descriptions. Ranging in summer from Arctic tundra southward, they nest in prairies, dunelands, marshes, alpine meadows, neglected fields, abandoned or undeveloped industrial footprint, crop fields, and similar country. This sparrow is widespread and common in many parts of California.

Harris (1996) described Savannah Sparrow as a "common resident and breeder in grasslands and saltmarshes of the coastal lowlands and along river valleys and well developed mountain prairies" in northwestern California. Given that so much of Humboldt County is forested, brushy, or otherwise unsuitable for Savannah Sparrows, it would follow that the summer distribution of the species might be restricted or patchy. Indeed this was the situation revealed by Atlas results. In the 5 years of the project, they were found during the nesting season in 46 (11%) of the 425 total blocks.

The range of Savannah Sparrow in Humboldt is chiefly coastal and near-coastal. The birds were found well within the interior of the county only at several locations. The bulk of the summering population occurs in sparsely vegetated dunes; around Humboldt Bay and in the lower Mad River coastal plain; in the Eel River delta; along and a short distance east of the immediate coast from near Cape Mendocino to near Punta Gorda (a28); and in the mountain prairies of the Bear River and Cape Mendocino ridge systems. Habitat along the northern coast of Humboldt County is limited, so it was not surprising that this species was encountered only in the small area of open country in the Redwood Creek estuary (f05). It was in the coastal portion of the range that nearly all of the breeding confirmations were made.

Inland, a somewhat isolated population was noted in four contiguous blocks in oak woodlands and grasslands of the Bald Hills (i07 to J09). Other detections included a pair at the Willow Creek airstrip (N13) 19 May 1995 (TWL); and singles near the Kneeland airstrip (i18) 14 Jul 1998 (RJA); shortly northwest of Showers Pass (k20) 14 Jun 1998 (KMS); near Bridgeville (k23) 9 Jun 1996 (*fide* RHw); and along the Mattole River west of Honeydew (e28, e29).

Breeding was confirmed in 21 blocks and judged probable in 10 others. Among those with confirmed breeding, the carrying of nest material was reported from a single block (F15) 28 Apr 1997 (GCH). A nest with young was found 10 May 1995 (D19; KMS), and a grass nest with eggs located on the ground was found 22 May 1996 at Humboldt Bay National Wildlife Refuge (D19; C. Goss). An adult was seen carrying a fecal sac 19 May 1996 (e15; GCH), which contrasts with a much later report of the same behavior 30 Jul 1998 (a24; DFx). There were three reports of adults feeding young, 8 Jun 1995 (B27; R. Sutherland), 27 Jun 1998 (D21; D. Anderson), and 2 Jul 1995 (c20; JEH). The highest-ranking breeding evidence in 12 blocks was the observation of fledged young. These reports spanned 1 Jun 1996 (c23; COg) to 6 Aug 1997 (J09; TWL). Eight of these fell during July. Adults carrying food were noted 3 Jun 1995 (b28; GCH), 1 Jul 1995 (f14; JEH), and 18 Aug 1995 (H15; PFS). The rather broad range of dates of the various breeding behaviors suggests that these birds were double-brooded during the study period.

During the 5 Atlas years, Savannah Sparrows were encountered chiefly in or around the edges of pastures stocked with dairy cows; in upland range frequented by small numbers of beef cattle; in fallow pastures; and in dunes grown to introduced European beachgrass. Kwasny (2000) studied the relationship between grazing and Savannah Sparrow reproduction at Humboldt Bay National Wildlife Refuge, and found that territory density and reproductive success were higher in ungrazed treatments compared to grazed treatments. That author indicated that delaying grazing until after 1 Jun would increase the density and nesting success of these sparrows.

It is interesting that, with the exceptions of sightings at Kneeland, near Showers Pass, and at Bridgeville, none of these birds was detected in the extensive grasslands of the interior portions of central and southern Humboldt County. What is it about those places that Savannah Sparrows find unattractive? Are hillsides not to their liking? Are these grasslands too hot and dry? Or were they somehow inexplicably overlooked? These questions will await the investigations of future workers, but the lack of breeders in hot, interior grasslands was consistent with some other atlas projects (*e. g.,* Roberson and Tenney 1993).

Interestingly, atlasers did not report any Savannah Sparrows nesting in saltmarsh, although countless numbers of these sparrows stop during migration, or spend the entire winter, in these habitats. Two observers recently queried on this subject reported that they have never found these sparrows nesting in coastal marsh habitats (J. Sterling, DFx). Jacobson (1986) indicated that Savannah Sparrows at her study site adjacent to Humboldt Bay were most common in grassy areas and actually declined after tidal intrusion was restored by dike breaching. Perhaps Humboldt Bay saltmarshes are no longer suitable due to the shift from pickleweed to cordgrass. Bryant's Marsh Sparrow (currently identified as *P. s. alaudinus*) was thought to be found principally in tidal marshes from about Humboldt Bay south to Morro Bay (Grinnell and Miller 1944), but they also have been recorded as possibly nesting at 2800 feet and 18 miles inland near Kneeland (Mailliard 1916). Fisher (1902b) indicated that this was "an abundant breeding race in the marshes and meadows about Humboldt Bay," and Mailliard (1916) reported that Bryant's was "common on the marshes around Humboldt Bay." Dwarf Savannah Sparrow (*P. s. brooksi*) was thought to breed as far south as Trinidad (Bishop 1915, Grinnell and Miller 1944). Unfortunately, Atlas work did not improve the understanding of either saltmarsh nesting or subspecific distribution in Humboldt County, and these remain subjects for future study.

In Humboldt County, Savannah Sparrows are present throughout the year, but their seasonal abundance varies considerably. Numbers are lowest in summer. The fall migration begins with the appearance of a telltale few birds on lowland dikes, at shorebird flats, in tidal wrack lines, and other nonbreeding settings in August. By mid-September, this trickle has become a torrent, with great numbers thoroughly inhabiting all manner of semi-open grassy or edge habitats from the outer coast into the mountain meadows and even the less-vegetated clearcuts. For the next month, Savannah Sparrows are almost underfoot at most of the well-birded open-country sites frequented by birders. Surely, many thousands of these sparrows might be present on a given day in early October in Humboldt's coastal and near-coastal lowlands. This glut diminishes by late fall, with birds retreating from the higher country and leaving the less-favored lowland habitats.

Winter concentrations assemble in weedy fields, vacant lots, farmland, ditches, sparse willow thickets, and estuarine edge habitats. During the Atlas years of 1995–1999 the Centerville Beach to King Salmon CBC recorded 125–233 birds and the Arcata CBC found 18–165.

Continued on page 395

GRASSHOPPER SPARROW *Ammodramus savannarum*

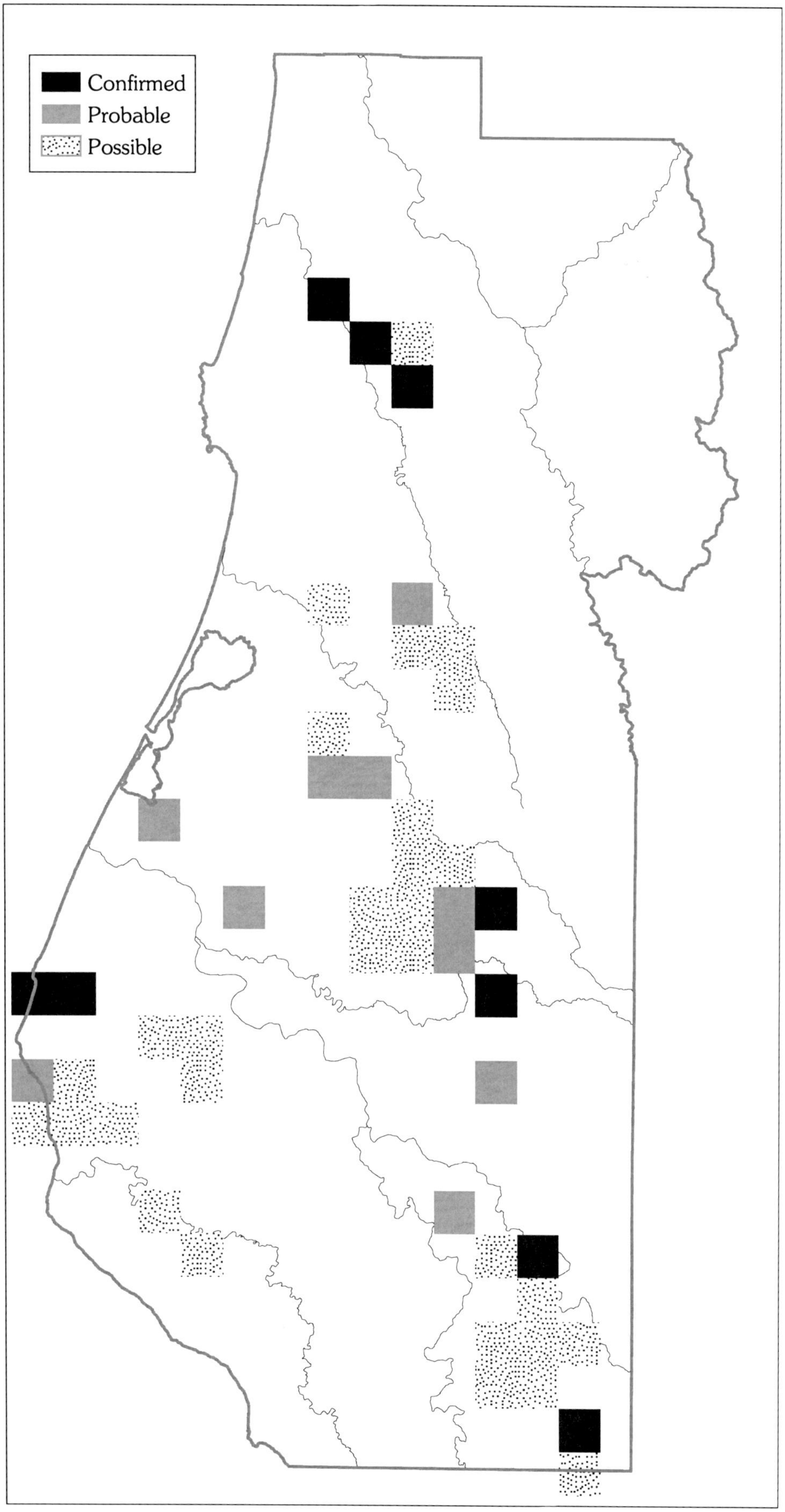

The genus *Ammodramus* comprises 10 species of New World sparrows which, as a group, are partial to low dense cover of grasses, sedges, and similar vegetation. Some prefer a moist or marshy environment, while others like it dry. Seven species occur in North America north of Mexico. Of these birds, the Grasshopper Sparrow is the most widespread, with a range encompassing temperate North America, Central America, northwestern South America, and some islands of the West Indies. Across much of their range these birds are fairly common, although a long-term decline has been noted outside of California. Not surprisingly, they are best represented on the Great Plains, with counts on BBS routes highest from the Dakotas south to the Texas panhandle. In regions deficient in habitat they are localized, rare, or absent.

Grasshopper Sparrows can be considered "stealth" birds in Humboldt County. They are inconspicuous creatures, known well by few people. Knowledge of Grasshopper Sparrow occurrence in the county is surely incomplete, because the birds are small, shy, plainly colored, soft-voiced, spend much time in dense cover, and live where the human population is sparse. Nevertheless, they were rather frequently reported during the Atlas period. It appears that they may be expected each year from May to July in several sections of the county. The earliest report in the course of the study was 5 May 1997 (m30; DFx) and the latest was 26 Jul 1995 (L25; M. Morris). Eight reports fell in May, 27 in June, and 14 in July, and these birds were reported in each of the 5 Atlas years. Grasshopper Sparrows were recorded in 12% of the 425 total blocks. Confirmed breeding was recorded in nine blocks and probable breeding in 11 others.

Areas within Humboldt County supporting Grasshopper Sparrows are characterized by their highly rural and bucolic nature. They are mostly quiet places, where human sounds are infrequently heard and are usually distant. Sites may be treeless. There may be a clump of oaks or California bay or an isolated California buckeye—perhaps even a timber edge of some description—yet everywhere round about there is much grass. The birds are often encountered on southerly slopes fully exposed to the summer sun, where the afternoon breezes blow in undulant waves across featureless hillsides of green or tan. Habitat used by Grasshopper Sparrows in Humboldt is generally rangeland, stocked very lightly, or optimally not at all, with beef cattle or horses. Some sites grow hay. The extent to which haying of occupied habitat in the summer months may reduce nest success has not been studied locally.

The distribution of Grasshopper Sparrow in Humboldt County is coincident with landscapes dominated by grass. Such places occupy portions of north-central, central, and southeastern Humboldt as well as the coastal and near-coastal grasslands inland from Cape Mendocino. More specifically, these areas can be defined as the Bald Hills (H07 to J09); interior Humboldt in the Mad River watershed (h14 to k16; H17 to L25) and on the divide between the Eel and South Fork Eel rivers (k28 to n34); in coastal and near-coastal grasslands from Cape Mendocino (a23) inland to Bear River Ridge; and in the Mattole River valley (d28, e29). Breeding confirmations were reported from each of these four general areas. Of interest as outliers were a pair on Table Bluff (D19) 31 May 1998 (TWL), and a male singing over a span of at least 1 week near Rohnerville (F21) 22 Jun 1997 (GCH). Proof that birds occupy sites along the immediate coast was the report of two birds singing 27 Jun 1998 along Mattole Road 1 mile north of McNutt Gulch (a26; G. Lester).

Large parts of the county lack appropriate habitat because they are heavily wooded. Ranch yards, small isolated pastures, vacant ground in river corridors, and similar sites which interrupt the closed nature of forested Humboldt County appear not to attract these birds. While Grasshopper Sparrows may have occurred in some blocks in which they were not reported, Atlas effort probably did not overlook any significant outlying populations in the county.

Two early reports of probable breeding were of a male singing on territory for a week 19 May 1996 (m32; M. Bailey) and multiple males singing on territory 22 May 1996 (L21; PFS). Both of these reports, however, were outside of "safe dates" for migrants and were not considered to be evidence of probable breeding. There were no reports of nests, but there was a single report of the carrying of nest material, 19 Jun 1996 (L23; M. Morris). There was a single report of the feeding of young: on 21 Jun 1996 a female was observed stuffing grasshoppers down the throats of two fledglings on a fence wire along Island Mountain Road while the male sang nearby (N33; C. Secor). There was also a single report of food-carrying, 14 Jul 1997 (L21; PFS). Each of the remaining six breeding confirmations were of fledglings, which spanned the dates of 18 Jun 1995 (a23, B23; COg, D. Ogden; both on Bear River Ridge) to 17 Jul 1997 (H07; JCP). The latter record was in the Bald Hills and was the northernmost observation of this species. An adult out of breeding habitat 21 Jul 1999 near Shelton Butte (N07; GCH) suggests timing of adult out-migration from the region.

Continued on page 395

FOX SPARROW *Passerella iliaca*

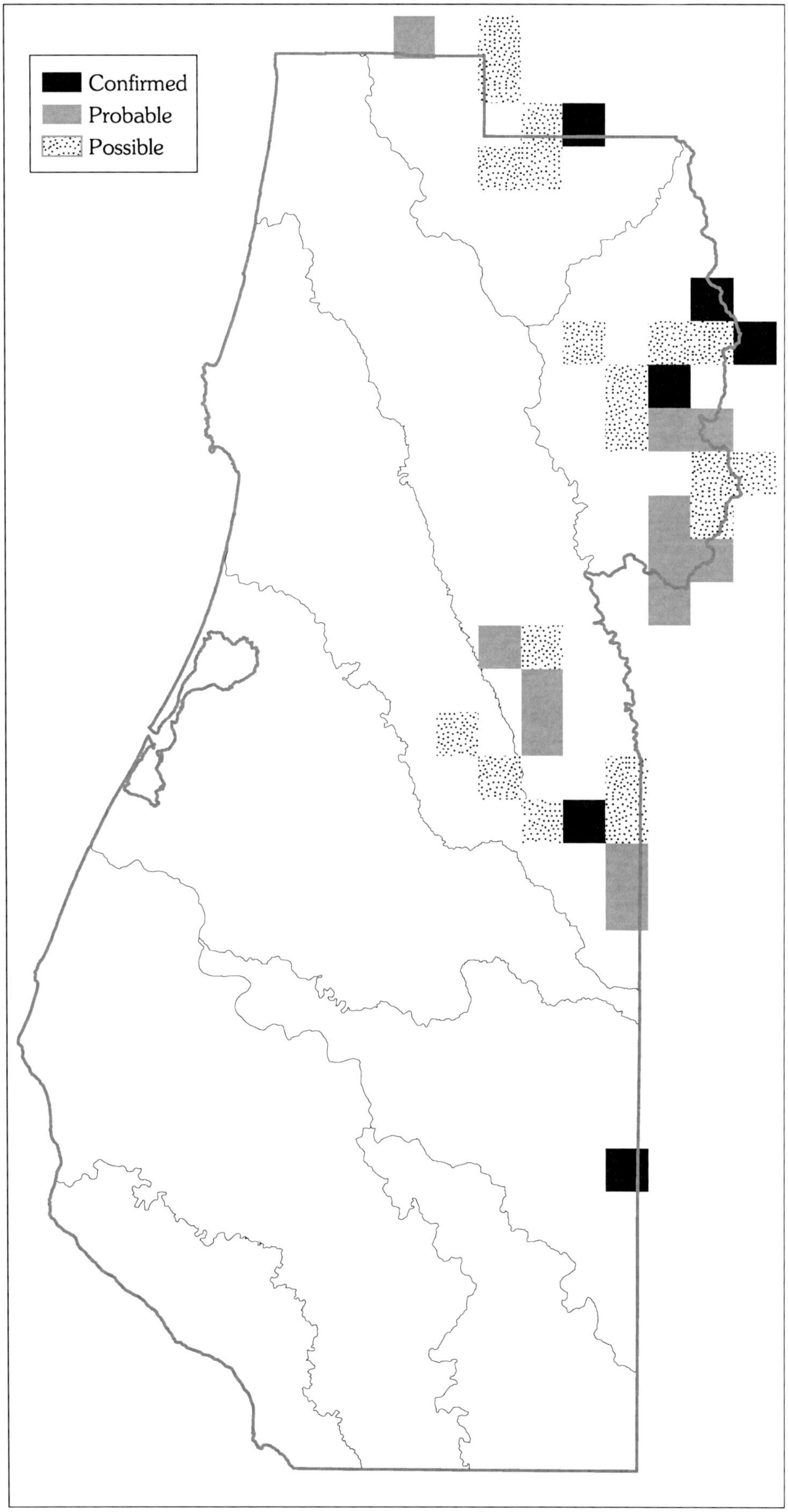

A Fox Sparrow lives its life in shadowed thickets, the glare of full sunlight its frontier. Kicking with vigor at the leaf-litter with both feet in the search for food, one of these birds can create a rustle that the unaccustomed observer might attribute to a larger animal.

Fox Sparrows of one or another well recognized form occur in Humboldt County throughout the year. From mid-September through April, great numbers of "Sooty" Fox Sparrows (primarily *P. i. fuliginosa*) migrate through or spend the winter in the area. Some "Slate-colored" Fox Sparrows, such as *P. i. schistacea* or *altivagans*, may also be present in winter. From April into early fall, breeding birds of the "Thick-billed" form (*P. i. brevicauda*, and possibly *megarhyncha*) inhabit mountain brushfields. Although Fox Sparrows are quite common at higher elevations in Humboldt during summer, birds of the breeding form are almost unknown in the county outside nesting areas, even as migrants in the intervening lowlands. Adding variety to this mix, an occasional "Red" Fox Sparrow (believed to represent *P. i. zaboria*) appears occasionally on the coastal plain, generally at backyard feeding stations or in with the throngs of other winter sparrows in suburban or rural settings.

During the Atlas period, Fox Sparrows were detected in 37 blocks in the mountains of eastern and northeastern Humboldt County, at elevations from about 2800 to 6000 feet. Four areas separated by the canyons of the Klamath, Trinity, and Van Duzen rivers constitute the range. Each of these areas represents a western extension of a greater regional distribution throughout the interior Klamath Mountains. The northernmost of these extensions into Humboldt occurs around the Del Norte County line. Blocks with detections form a small cluster reaching southeast from the vicinity of South Red Mountain (just into Del Norte) to about Slate Creek Butte, northwest of Orleans. A zone of unsuitable habitat at lower elevations within the Klamath River canyon separates the northernmost group of blocks from the next. This second group extends from upper Red Cap Creek across the east side of the Hoopa Valley Indian Reservation and south to Happy Camp Mountain. Three of the six confirmed breeding reports for Fox Sparrow came from this second cluster. Southwest of this group of blocks, another well-defined lowland corridor devoid of appropriate habitat, through which the Trinity River runs, dissects the range. Beyond this gap, the Snow Camp Mountain and South Fork Mountain ridges support more Fox Sparrows.

Evidence of breeding obtained by atlasers was diverse. Five different types of probable breeding behavior were recorded. Among the earliest such reports was multiple males singing in suitable habitat 7 Jun 1998 (p14; KMS). There were confirmations in only six blocks. The only occupied nest found during the Atlas period was one observed 13 Jun 1996 in an outlying block just southwest of the Lassics (o27; B. Damschroeder). In addition, observers reported adults carrying food 7 Jun 1995 (o27; JEH) and 8 Aug 1995 (o27; GCH) and a fecal sac 23 Jun 1995 (N19; GAS), and performing distraction displays 8 Jul 1995 (P09; JEH). Fledged young were noted 17 Jul 1997 (q07; JEH).

Fox Sparrows summering in Humboldt County are partial to expanses of tall shrubs on sloping ground. Although some natural montane brushfields occur as a result of wildfire or in association with serpentine soils, most are regenerating clearcuts or partial-cuts. Places preferred by the sparrows lie within the zone of winter snows, where broadleaf shrubs and forbs more characteristic of lower elevations are diminished in number and variety or are absent. Various species of *Ceanothus*, manzanita, and currant usually dominate such sites. Along with silk tassel, bitter cherry, Brewer's oak, and similar shrubs, these effect a more or less continuous shrub cover. On northern exposures, seedling or sapling Douglas-firs or white firs are often present and may be abundant. Indicating their preference for this type of site, Nashville Warblers and Dusky Flycatchers are also found at sites supporting Fox Sparrows. Indeed, the combined voice of these three species is a signature chorus of mountainsides logged within the past 15 years.

Because the Fox Sparrow owes its existence as a breeding bird in Humboldt in large part to the activities of the forest products industry, its persistence in numbers is dependent upon continued cutting of forests at middle and higher elevations. Raphael *et al.* (1988) estimated that this sparrow may have increased more than 240% as a result of timber harvest. After decades of intense logging within the mountains of interior county, a reduction in cutting took place in the 1990s. As seral succession ultimately causes brushfields to be replaced by young conifers, habitat suitable for Fox Sparrows may diminish.

SONG SPARROW *Melospiza melodia*

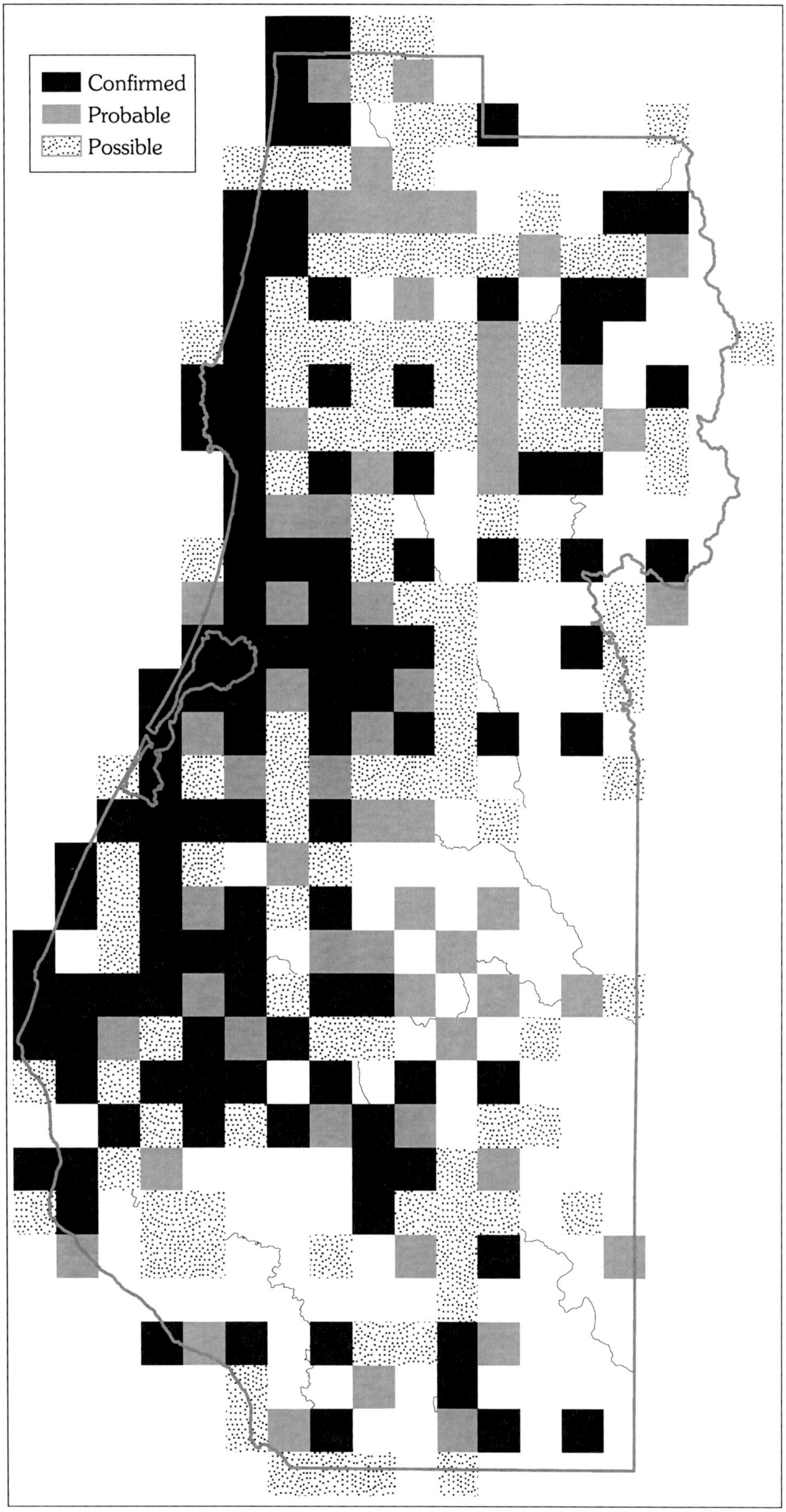

Song Sparrows are very common throughout the year in Humboldt County. They are found across most of the area except for the conifer-clad mountains and the open rangelands. Numbers in fall, winter, and early spring are greatly augmented by migrants and wintering birds originating to the north.

On a county-wide scale, this species exhibits a distribution similar to that of the Winter Wren. It is a common breeding bird in appropriate habitat nearly throughout the fog zone from the Del Norte County line south to Cape Mendocino. Fisher (1902b) stated that the species "swarms in some places, and is the commonest bird in deforested areas." Although habitat is restricted along the outer coast of southern Humboldt, stream outflows along the seaward base of the King Range are inhabited. The birds occur across much of the northern one-third of the county, throughout the near-coastal lowlands, and in all stream valleys where human settlement and consequent land disturbance has taken place. An extension to the east and southeast, somewhat out of the fog belt, was noted along the general courses of the larger rivers and their tributary streams.

Song Sparrows are routine breeding birds in cities and towns, asking for little more than a protected brush patch confluent with surrounding vegetation, and, almost always, some standing or running water in the vicinity. Harris (1996) stated that the bird was a "common resident and breeder in brushy edges of riparian areas, cutover forest, residential areas, roadsides, dikes, and marshes" in northwestern California. During the Atlas period, Song Sparrows were detected in 64% of all blocks and were confirmed breeding in 112 blocks.

Atlas fieldwork made clear that there are sections of Humboldt County in which Song Sparrows are sparse or lacking. As a rule, this sparrow is not found at high elevations, and this was supported by the surveys. There were only a few detections in blocks taking in the Trinity–Salmon River divide, the general South Fork Mountain uplift, and the Lassics area. Shrub communities in the mountains are generally summer-parched, and poor in the deciduous broadleaf shrubs commonly present in better Song Sparrow habitat.

Detections were nearly lacking over a sizable part of east-central and southeastern Humboldt County. Many blocks in the rolling grass-and-oak country of the interior Mad River drainage and in the upper Eel River offer little in the way of protected riparian corridors or ungrazed swales. Blocks in the vicinity of Humboldt Redwoods State Park and southward took in expansive areas of closed-canopy mid-slope forest. The familiar issue of restricted access to many blocks in private ownership in the county's hinterlands probably diminished detections of Song Sparrows, as it did efforts to detect other species whose habitat needs are fulfilled only locally in southern Humboldt. It is felt that these sparrows probably bred at one place or another within many of the blocks in which none were detected.

While Song Sparrows are found in a variety of both upland and riparian habitats that provide the brushy cover they need, a strong affinity for riparia has been well documented in Humboldt. Surveys in the Klamath Mountains indicated that this sparrow can be categorized as a "riparian specialist" there, where they occur more commonly at sites with small trees, and from lower elevations up to 2275 feet (Sniado 1999). On the coastal slope, the Song Sparrow is the most abundant species found in riparian habitat in the Eel River delta (Kelly 1987). Ralph *et al.* (1996) also characterized this species as a riparian specialist in their coastal redwood study areas. Its occurrence in these lowland riparian habitats is strongly influenced by low tree density and large amounts of ground cover. Burton (1972) also reported that Song Sparrows were the most abundant species in saltmarsh and other habitats on Indian Island.

These residents can start acting "nesty" remarkably early in the year. One member of an apparent pair was seen collecting nest material 20 Feb 1995 (F11; JEH, M. Schwarz), although actual nest building was not recorded until 13 Apr 1996 (F15; C. Meyers). While nesting can start early in the season, 62% of all reports of adults carrying nest material were from May. Exactly 50% of the 112 confirmations in the final map database were the result of observation of adults either carrying food or feeding young. These records spanned the dates of 14 May 1995 (a23; GCH) to 3 Aug 1997 (e09; GAS). The first nest with eggs and the first fledgling was reported 18 May 1996 (N33; C. Secor), and 30 May 1997 (f12; GCH). The latest nest with young was reported 28 Jul 1996 (e19; TWL).

LINCOLN'S SPARROW *Melospiza lincolnii*

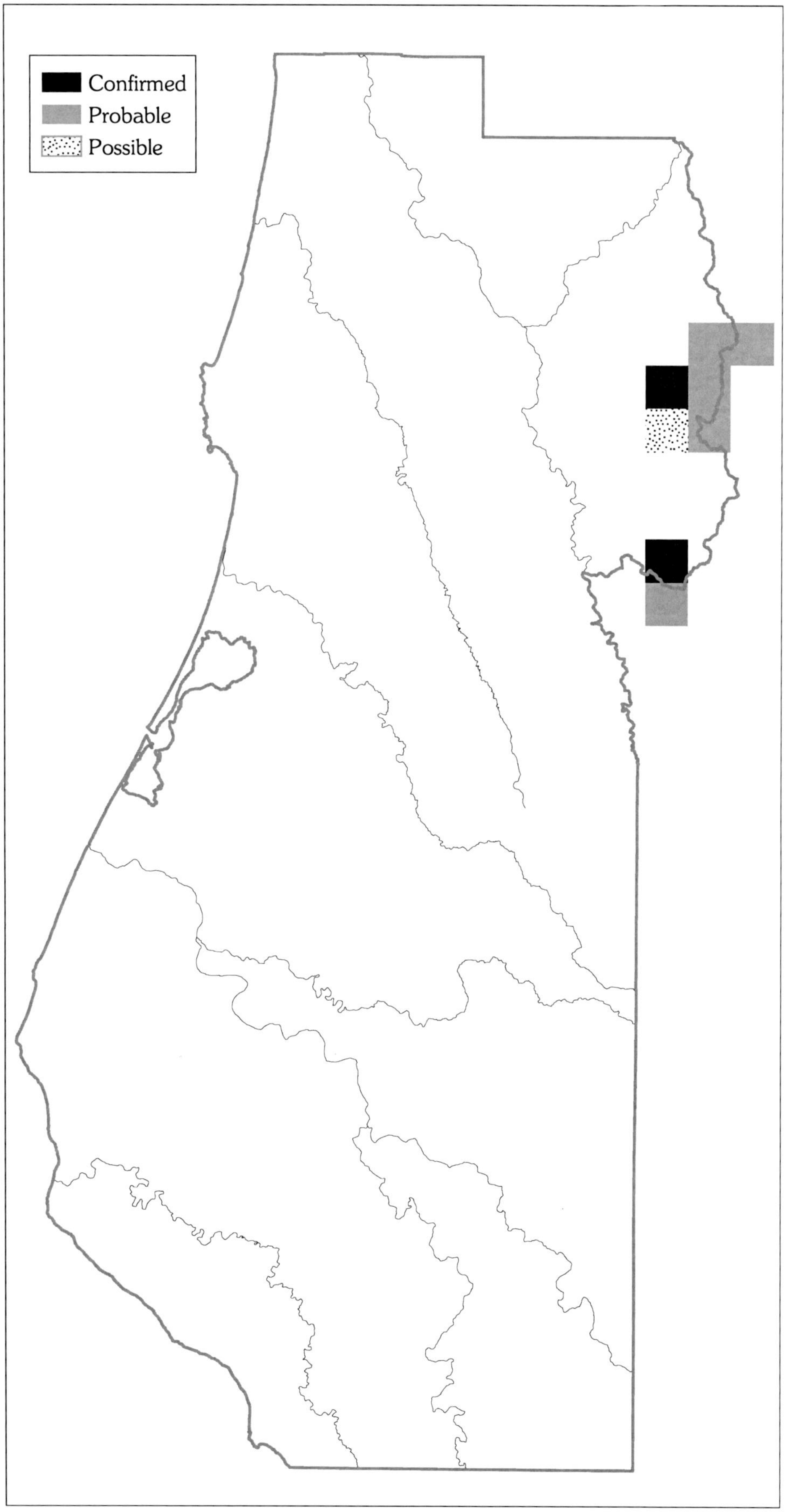

The shy beauty known as Lincoln's Sparrow is seen by most birders during the colder months. At that time, damp thickets and weed-patches of the Pacific Coast and the southern U.S. host these small sparrows, which associate loosely with the mixed assemblages of seed-eating birds partial to such sites. Northerners in summer, Lincoln's Sparrows inhabit the low growth of forbs and shrubs dotting the damp margins of forested lakes, stream meanders, and beaver ponds. Like many other transboreal Alaskan and Canadian breeding birds, they extend locally into the northeastern U.S. and well southward in mountains of the West. Compared to the abundance of such sparrows as Song, White-crowned, and Savannah, the number of Lincoln's Sparrows nesting in California is not great, yet the species maintains an annual summer presence at higher elevations. The Cascades, the Klamath Mountains, the inner northern Coast Range, the Sierra Nevada, and the San Bernardino and San Jacinto mountains support a scattered population centered about subalpine lakes and wet meadows ringed by *Spiraea* spp. and corn lily, shadowed by pines and firs.

Habitat in northwestern California suitable for Lincoln's Sparrows is limited. The boreal lake shores preferred by them exist in isolated and remote locations well up into the region's mountains. Tucked closely beneath the highest ridges and peaks in the region lies a handful of sites known to host these birds during the nesting months. The subalpine lakes and wet meadows of the Klamath Mountains and northern Coast Range are precious places, their beauty far transcending their status as tiny blue specks lost amid the scribble of ridgelines and streams on regional maps. It is at these high-country wetlands that Lincoln's Sparrows sing their songs and undertake nesting efforts in northwestern California.

The first published evidence of the presence of Lincoln's Sparrow during the breeding season in Humboldt County was the report of up to 12 birds at a wet meadow on North Trinity Mountain 4–6 Jul 1974 (D. Anderson; Yocom and Harris 1975). Humboldt's earliest historical report of confirmed nesting was of a nest with young at Groves Prairie 19 Jun 1980 (B. Marcot, L. Doerflinger). Seven were reported at Water Dog Lakes 15–16 Jul 1989 (L. Lester; Harris 1991). Thus, at the outset of the Atlas project it was apparent that this species was a peripheral breeding bird within Humboldt County, better represented in parts of adjoining counties. Visits to known or potential nesting sites by Atlas workers resulted in a fine-tuning of this perception. Lincoln's Sparrows were detected in eight blocks. Breeding was confirmed in two blocks and judged probable in five others.

As indicated by the map, these sparrows were encountered in only a small portion of the study area. This region is in the high country along the general Humboldt/Siskiyou/Trinity borderlands. Countersinging males were recorded 21 Jul 1998 west of Salmon Summit, Trinity County (r08), and the following day six countersinging males were recorded west of Red Cap Lake q08; (both JEH, GCH). On 7 Jul 1995, food-carrying was recorded at Water Dog Lakes west (P09), and agitated behavior was noted at Water Dog Lakes east (q09; both JEH, TWL). Distraction display was recorded 22 Jul 1999 at upper Mill Creek Lake (P09; GCH). The only other breeding confirmation was evidenced by food-carrying at Groves Prairie (P13) 12 Jul 1997 (JEH). A bird (or birds) singing on territory for at least a week was recorded 7 Jun 1998 along Forest Service road 6N10 near Happy Camp Campground (p14; KMS). Singing birds were also recorded 22 Jul 1999 (p10, q10; both GCH).

Habitat where these Lincoln's Sparrows were encountered varied somewhat from site to site, but corn lily was present in most, if not all, situations. *Spiraea* spp. was also present (especially at slightly drier sites), as was mountain alder and yellow coneflower (*fide* TWL). Swampy ground, in the form of either a wet meadow, riparian strip, or lake margin, seemed to be a common denominator at all sites.

Results of Atlas surveys helped to shed light on this elusive, marginal Humboldt County breeder. While a few very small lakes or wet meadows in northeastern Humboldt not visited by atlasers may possibly support pairs of Lincoln's Sparrows in the breeding season, unknown sites are not suspected to exist. All told, the 5-year Atlas snapshot probably generated a fairly accurate portrait of the whereabouts and status of these sparrows during the nesting season.

Lincoln's Sparrows are vastly more numerous during migration and winter in Humboldt than in summer. These birds are commonly seen from late September through April in wet or damp thickets, with greatest numbers found in the coastal and near-coastal lowlands. Unmanaged, ungrazed ground covered with several years' growth of grasses, wild teasel, fennel, thistle, and scattered mounds of Himalayan blackberry routinely supports considerable numbers of these birds, particularly during the peak of fall migration in October and early November. Many hundreds of them are believed present in the combined Humboldt Bay/Eel River lowlands at that time. In winter, the more exposed situations such as pasture fencerows and dunelands are less frequented, with modest concentrations and scattered individuals occurring in and about wetlands, along riparian corridors, and to a limited extent in city parks and urban gulches. During the 1995–1999 Atlas period, the Centerville Beach to King Salmon CBC recorded 29–60 Lincoln's Sparrows, with 8–94 during the Arcata CBC and only 5–26 inland during the Willow Creek CBC. Harris (1996) described the species in northwestern California as a "common fall migrant; uncommon winter resident and spring migrant; locally rare summer resident and breeder in high inner coast range." Except that these birds also breed in the Klamath Mountains, this characterization continued to be true through the course of the Atlas period.

WHITE-CROWNED SPARROW *Zonotrichia leucophrys*

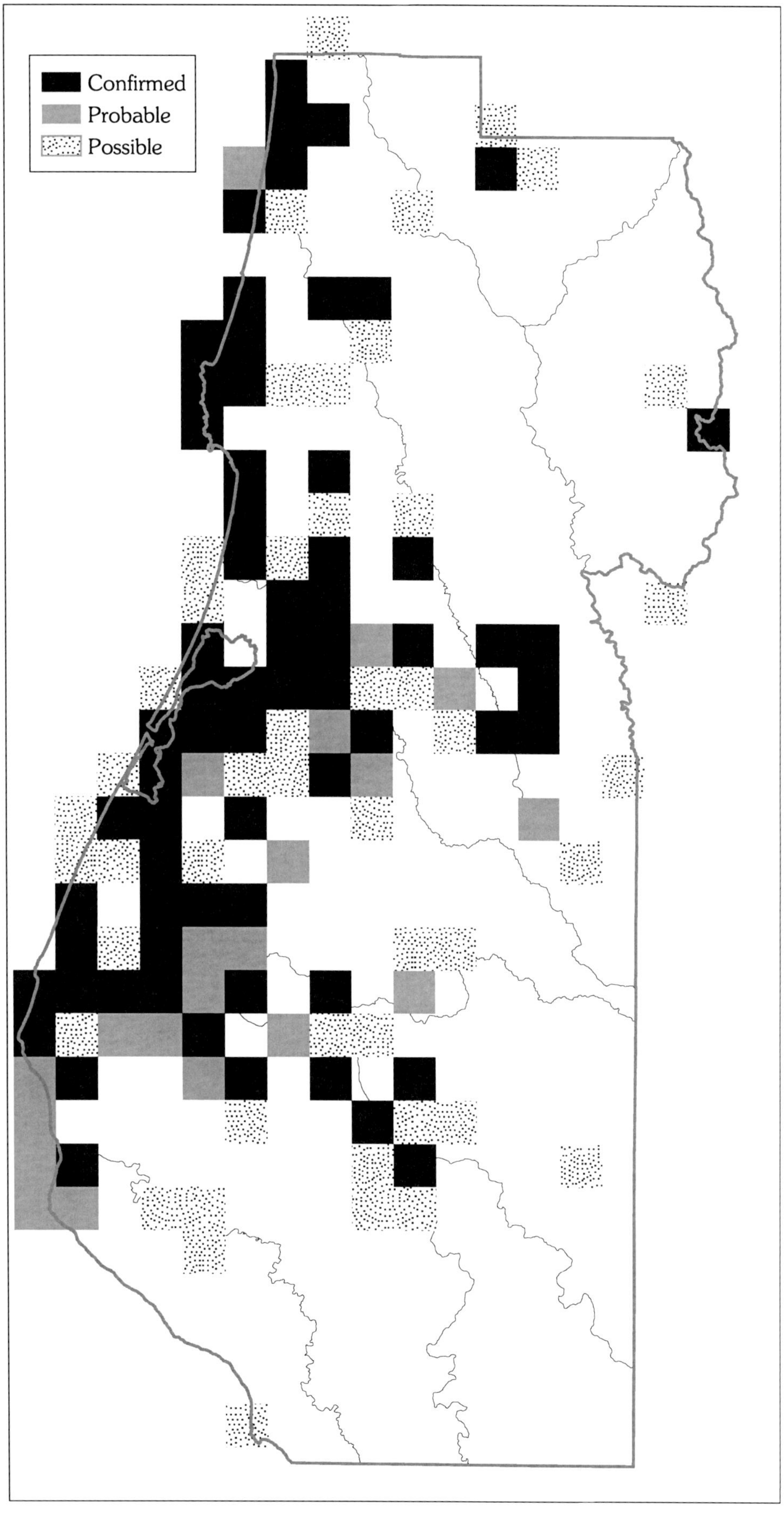

Anyone living in Humboldt County who looks at birds will come to know the White-crowned Sparrow. It is a common species, found in numbers at some time of year everywhere in the area. This holds most true during fall, winter, and early spring. During the nesting season, however, many sections of Humboldt appear not to support these birds. During the Atlas period, White-crowned Sparrows were confirmed or probable breeders in a somewhat restricted range encompassing west-central Humboldt County and northward near the coast. The influence of summer overcast may play a role in determining where these sparrows nest in greatest abundance in the county, as nearly all of their range lies either within the zone of coastal fog or in sections of the county in some way affected by it. This included the coast and near-coastal areas from the Del Norte County line south to near Punta Gorda (b28); Redwood Creek; the Mad, lower Eel, and lower Van Duzen river drainages, and sparingly elsewhere.

During the 5-year Atlas period, White-crowned Sparrows were detected in 135 (32%) of all blocks, with confirmed breeding established in 67. That nearly one-half of all blocks with detections ultimately carried confirmed breeding status suggests that this species' reproductive efforts can be readily witnessed, and indeed this was the case. A telling point is that fledged young were observed in 27 of the 67 blocks with confirmations. Streaky, geeky young White-crowns were often seen with adults in family groups along roadsides. These reports spanned the period from 1 Jun 1996 (c23; COg) to 11 Aug 1996 (f05; JEH). Five were in June, 19 in July, and three in August. Preceding the median date for fledged young were the 13 blocks with an adult feeding young, as two fell in early June, seven in late June, two in July, one in August, and one in September; the date span was from 6 Jun

1996 (g14; C. Ragatz) to 7 Sep 1998 (F13; GAS).

Active nests confirmed breeding in six blocks. Three nests with eggs were discovered, 11 May 1996 (D21; L. Karsteadt), 20 May 1995 (e16; GCH), and 4 Jun 1996 (D19; K. Foerster). Nests with young provided the highest-ranking code 16 Jun 1996 (H07; GAS) and 19 Jun 1999 (H03; JEH). The only instance in which an adult carrying nest material furnished the highest evidence obtained in a block came, oddly, 4 Jul 1995 near the end of Centerville Road (b22; JEH); this rather late initiation may have been a second nesting attempt.

During the breeding season, White-crowned Sparrows typically inhabit vegetated dunes and coastal scrub; brushy edges of gravel bars and willow thickets; clearcuts and very young conifer plantations within the fog belt; sapling-dotted pastureland; and farms. In summer they also occur in cities and towns. This is a routine nesting species in downtown Eureka and in residential neighborhoods of the several largest towns in the county. Neglected or vacant industrial-zoned land with sparse grass and shrub cover also attracts them.

Harris (1996) called White-crowned Sparrow a "common resident and breeder" in northwestern California that is "generally absent as a breeder from most of the interior." At least two subspecies of White-crowned Sparrow probably breed in Humboldt County. The widespread northwestern form *Z. l. pugetensis* occurs throughout the nesting range in the western section of the county, and occurs at inland sites such as on Coyote Peak at 3000 feet and 16 miles inland, and at Maple Creek some 18 miles inland (Grinnell and Miller 1944). Judging chiefly from differences in song, breeding birds from Cape Mendocino southward are probably *Z. l. nuttallii*.

The discovery of nesting White-crowns at some far-inland sites during Atlas fieldwork surprised local birders and prompted some to question whether "Mountain" White-crowned Sparrows (*Z. l. oriantha*) may also occur as a breeding form. Whatever their affiliation, these inland sparrows were encountered along semi-open streamsides and brushy wet meadows, and upslope in forest-edge situations and mesic brushy clearcuts, as at Horse Mountain (L15), Friday Ridge (m15), Grouse Mountain (m17), Snow Camp Mountain (L17), Board Camp Mountain (m19), and Pilot Ridge (n20). Especially intriguing were reports of White-crowns at some far-inland and high elevation sites, such as two birds singing at 4950 feet on Horse Mountain (L15) 11 Jun 1997 (TWL), and an occupied nest 22 Jul 1999 above 5000 feet (q10; GCH). Further work on the breeding distribution of White-crowned Sparrow subspecies in northwestern California is needed. Even if the far-inland birds are in fact *pugetensis* and not *oriantha*, the more extensive inland occurrence of this species as a breeder in Humboldt was nevertheless unexpected by many.

White-crowned Sparrows are migratory within Humboldt County, as they are nearly everywhere else within their range. The floodgates of fall migration open in mid-September. By mid-October, flocks of from several to several dozen *pugetensis* White-crowns may be encountered in nearly any semi-open setting providing a seed source and at least minimal escape cover. Particularly in the more exposed and less-vegetated places, such as dunelands and along roads through dairy pasture, they tend to occur in mixed-species flocks, in which from a few to many Golden-crowned Sparrows (*Z. atricapilla*) are found. White-crowns also constitute a variably significant component of the classic large mixed-sparrow flocks found widely at lower elevations in the colder months, and which typically include Song, White-throated (*Z. albicollis*), Lincoln's, and Fox sparrows, and Dark-eyed Juncos. Spring passage can involve the sudden appearance of flocks of nattily plumaged, high-spirited migrants in early and mid-April, but more often it is perceived simply as the disappearance of birds from the local winter sparrow-patch.

DARK-EYED JUNCO *Junco hyemalis*

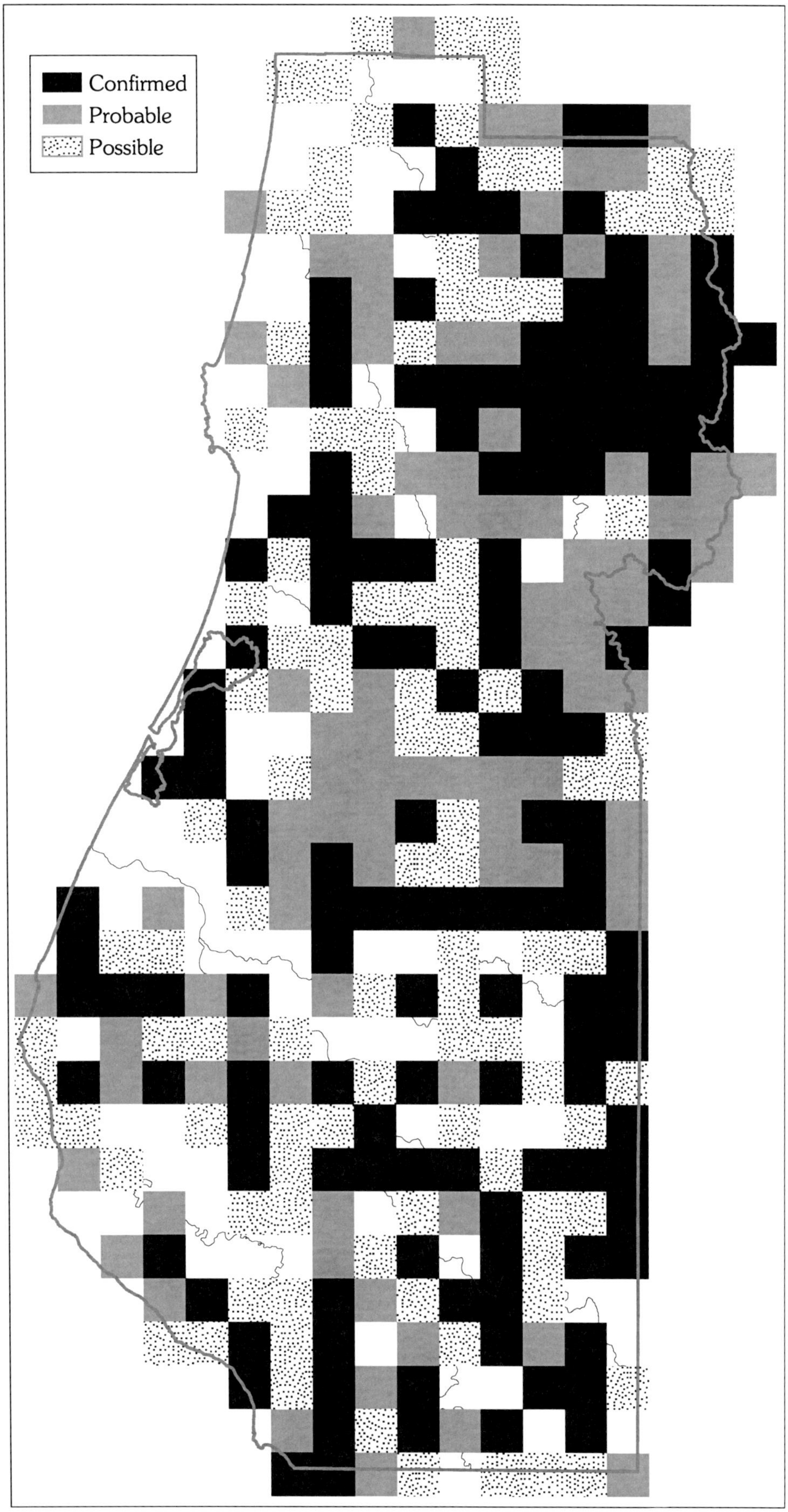

Juncos are pink-billed, dark-headed sparrows with white outer tail feathers. They are common, cold-hardy, and social; they eat seeds and tolerate human activity—all of these traits contributing to their status as among the best known backyard birds. Generally, juncos are found in northern and western montane forests from late spring through early autumn. They are migratory, leaving the boreal forest and higher elevations to the snows of winter. At that season, flocks of Dark-eyed Juncos are widely seen at bird feeders and along roadsides in the U.S. and southern Canada. Early in a naturalist's experience comes familiarity with the birds' insistent call-notes and simple song; the flash of white tail-sides; and a feather-loud scattering as the group retreats to shrubbery at the hint of a threat.

The taxonomy of the genus *Junco* is complex. Relationships between some of the many subspecies are controversial. The widespread *J. h. oregonus*, one of several forms comprised in the subspecies group known as "Oregon" Junco, occurs as a breeding bird in northwestern California. Elsewhere within its range, members of this race may hybridize and intergrade with those of other subspecies, but this is not the case locally.

The Dark-eyed Junco has long been recognized as a common to abundant bird in Humboldt County. The species is present throughout the year and is typically common wherever it is found. It is one of the most numerous and omnipresent birds of lowland and foothill openings and forest edge during the colder months. Unlike many other wintering sparrows, it remains widespread and common as a breeding bird in the county, found in spring and summer across nearly all of Humboldt except for parts of the immediate coast. During the Atlas period, the junco was found in an impressive 78% of all blocks, making it the fifth most widely distributed breeding species. There were no obvious regional gaps in distribution within the county. Blocks without detections of juncos were well scattered and conspicuous in the southern one-third of Humboldt, while comparatively few inland blocks in the northern two-thirds of the county were without reports.

During the breeding season, Dark-eyed Juncos in Humboldt County appear to require little more than stands of conifers with an open or minimal shrub layer. These settings may be as scanty as a thicket of young "volunteer" Douglas-firs along the edge of a suburban yard, or as full-blown as an old-growth coast redwood forest. In between lie any number of situations that appeal to the birds. Douglas-fir and tanoak regrowth at lower and middle elevations across the interior of Humboldt County is well populated. The montane forests of true fir, pine, and incense-cedar that clothe much of the higher elevations in the county host great numbers of these birds in the nesting season. Forests with much California black oak but with some conifers are also used by them. The clearcuts so common throughout the industrial timberlands are pioneered by juncos within a few years of logging, particularly on northern exposures and where clumps of residual trees persist. Dark-eyed Juncos are the most numerous birds in cutover Douglas-fir lands, and numbers reach a peak 3–6 years after a clearcut, as brush becomes dominant (Hagar 1960, Marcot 1984). On Kinsey Ridge they favor open herbaceous foraging areas, and often use forest-brushfield edges and open riparian habitat (Block 1985). Only on extensive southern exposures in the summer-hot oak-and-grassland country of interior southern Humboldt are juncos truly unexpected.

Dark-eyed Junco also was among the species most often confirmed. The 139 blocks with confirmed breeding ranked it fourth among all species detected during the project. Confirmations came from throughout the range of the species in the county. Although juncos are fewer nearer the coast than inland, breeding records nevertheless came from many blocks at the western edge of the range. Inland, extensive fieldwork on the Hoopa Valley Indian Reservation resulted in near-saturation of blocks with breeding confirmations. As was the case for many species, slightly higher levels of breeding evidence were obtained in priority blocks.

The carrying of nest material was reported as early as 4 May 1996 (m10; E. Elias). The latest nest with young was found 9 Jul 1996 (m11; *fide* JMH), and the latest observation of the carrying of food was 30 Jul 1998 (B25; JCP). Atlasers birding along country roadsides everywhere found the high visibility of family flocks to assist in establishing breeding. Small groups of juveniles, accompanied by their parents, were seen throughout the county during each June and July of the Atlas period. Fledged young accounted for a hefty 49% of all blocks with confirmations. A breakdown per 2-week period revealed that one report of fledged young was in April (29 Apr 1998; i27; JG), yet none was in May; three were in the first half of June, 15 were in the second half of June, 26 were in the first half of July, 19 were in the second half of July, and four were in the first half of August. The latest report of fledged young was 10 Aug 1997 (L05; P. Herrera).

Harris (1996) called the Dark-eyed Junco a "common resident and breeder" in northwestern California, adding that the population swells in winter with the arrival of migrants. He stated that they mostly breed in pine–fir forest above 4000 feet, but that some remain at lower elevations to nest. Atlas results are at some variance with this characterization, as the map indicates that lower- and middle elevation blocks were well populated during the breeding season. Block (1985) found this to be the most abundant species in his brushy study area on Kinsey Ridge.

Dark-eyed Juncos occur in flocks of various sizes during the fall, winter, and early spring. Most of the population wintering in Humboldt County is believed to reside at middle and lower elevations. However, access to higher elevations during mid-winter historically has been greatly restricted by snow, thus knowledge of the extent to which juncos may linger in the mountains is limited. In March and April, winter flocks disperse, migrate, or otherwise disappear, and breeding pairs settle into appropriate habitat throughout the area. While the open coastal lowlands and the Eel River delta appear to support few or no juncos after April, much of the rest of Humboldt hosts great numbers which remain to breed.

BLACK-HEADED GROSBEAK *Pheucticus melanocephalus*

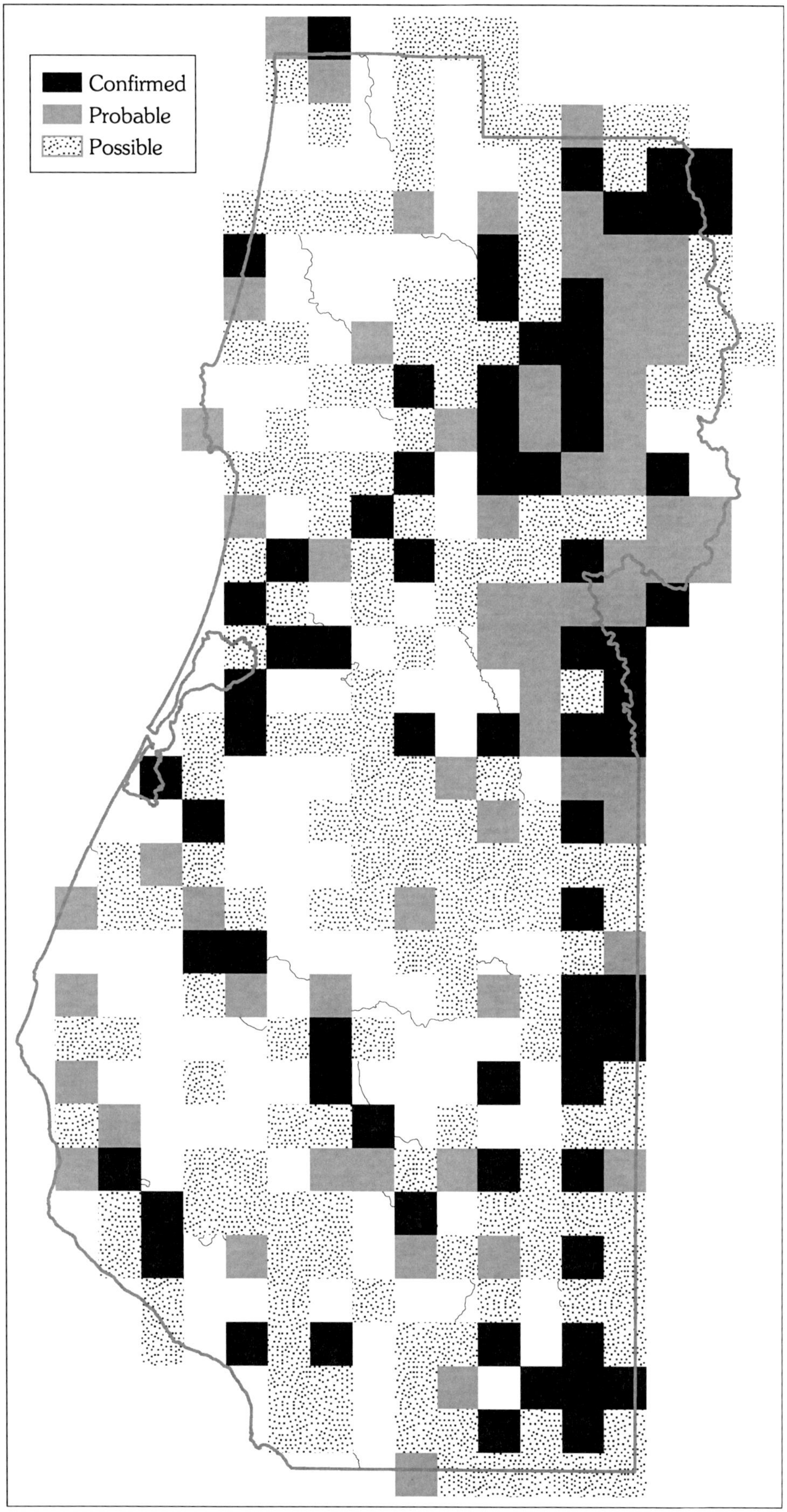

The Black-headed Grosbeak is among the most familiar neotropical migrant songbirds throughout the West. The loud, strikingly sweet singing of the male typically is delivered from a shaded perch in the upper crown of a tree. Its syncopated phrasing causes the song to stand out from the mixed chorus of forest bird songs heard from early May into July. Females are far less readily detected, though their call—a penetrating *eek!*—is distinctive once an observer learns it. Family groups are encountered frequently in mid-summer at forest's edge or in woodlands with tall, leafy shrubs.

One would choose a broad brush with which to paint a picture of the distribution and habitat preferences of this common summer resident (Harris 1996) in northwestern California. These grosbeaks are found nearly throughout Humboldt County during their summer residency and occupy an enormous sweep of landscape. They range from very near the coast and across the interior, occurring from riverbottom to ridge-top in many places and extending well into the middle and higher elevations of east county. Black-headed Grosbeaks were detected in 290 (68%) of the 425 blocks, making this species the tenth most widely distributed breeding bird noted in the course of the Atlas study.

Most stands of trees of reasonable size offering a fairly closed canopy and at least a modest hardwood midstory provide habitat for Black-headed Grosbeaks. Suggesting the importance of deciduous hardwoods is the commonness of the bird in conifer-dominated forest with a significant component of California black oak or bigleaf maple. Indeed, these birds are detected in greater numbers as the abundance of hardwoods increases and the abundance of conifers decreases (Ralph *et al.* 1991). Cottonwood-, alder-, and willow-dominated riparian areas are also well used. These birds clearly follow stands of California black oak and its leafy-shrub associates well into the middle elevations upslope to their uppermost extent, where the oak communities interdigitate with montane conifer stands and ultimately are replaced by them.

While no region within Humboldt plainly lacks Black-headed Grosbeaks, two small areas provided atlasers with comparatively few confirmed or probable breeding detections. One of these is in the northwestern corner of the county. It encompasses both old-growth coast redwoods and commercial timberland that is heavily logged and sprayed with herbicide. Although edges and openings in and around the redwood parks support small numbers of grosbeaks, by and large this species was uncommon in such places during the study period. It is possible that the high conifer canopy closure and scarcity of lush deciduous hardwood layers are limiting factors. Conversely, the commercial timberland to the east has been so intensively managed that large expanses presently exist without hardwoods or are in an early-seral stage, providing little habitat. The other minor "hole in the map" exists in extensive regrowth conifer timberlands in the upper Yager Creek drainage, where redwood reproduction is prompt and vigorous. However, limited survey effort may have also played a role in creating either or both of these areas of minimal confirmed and probable breeding observations.

Black-headed Grosbeaks arrive in Humboldt County from the south well along into the general spring migration of landbirds, with first reports usually not until late April. They rapidly become common thereafter. Multiple males singing in a block (m29) were reported as early as 2 May 1997 (JG) and adults carrying nest material were reported as early as 19 May 1996 (p04; RHw). A small minority of breeding confirmations consisted of observations of adults carrying nest material (two reports), nest building (one report), or discovery of a nest with eggs (one report), suggesting that these birds are secretive during the earlier portion of their reproductive cycle. The nest with eggs was found 11 Jul 1998 (n08; JMH). However, 20% of confirmations were based on adults carrying food. Later still, the reticence of adults seems to disappear, as observations of young being fed constituted 34% of the total confirmations. This nearly matched the percentage of confirmations involving the sighting of fledged young (36%), the most numerous "confirming" behavior code reported. The earliest fledged young was seen 21 May 1996 (g13; J. Haggard). The latter part of the breeding season was represented by the observation of an adult feeding young 12 Aug 1998 (N17, o17; TWL).

LAZULI BUNTING *Passerina amoena*

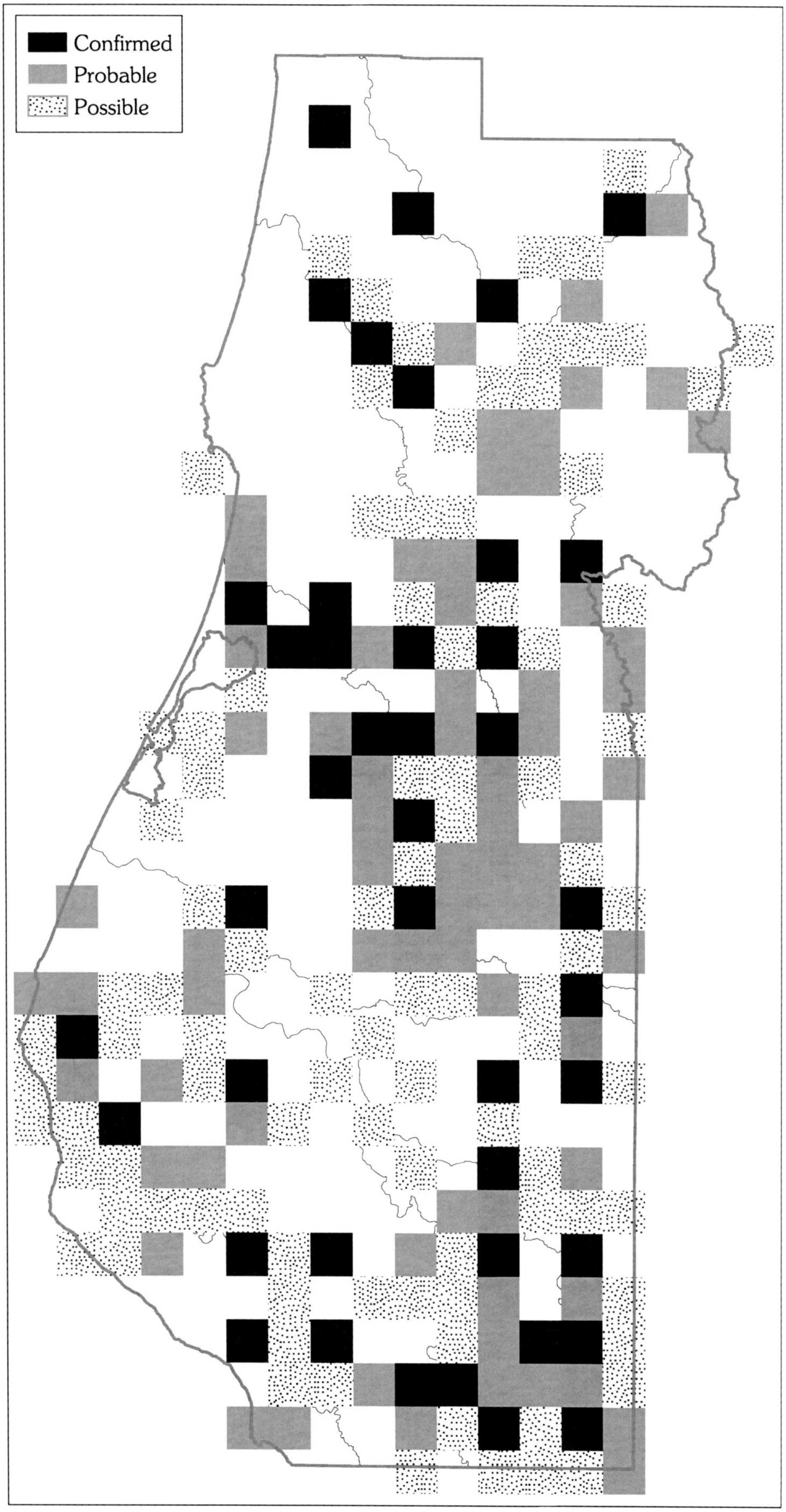

The brief finchy songs of male Lazuli Buntings may be heard in open or semi-open countryside nearly throughout Humboldt County from mid-April into July. Where there is one singer, there are ordinarily more in the neighborhood; a scattered chorus of these tiny turquoise birds engages in tireless counter-singing from dawn to dusk. So persistent are their territorial announcements that not even the shimmering heat of mid-afternoon in the county's sun-parched interior diminishes their vocal efforts. Keeping more to brushy edges and involved in no such advertisement, the females are far less frequently detected by human observers. Their soft but incisive *spit* call-notes betray their presence here and there among the patches of brambles, forbs, and grasses in which they place their nests. By mid-July, birds of both sexes become more difficult to see as singing diminishes and the task of feeding hidden nestlings takes the fore. As the breeding season becomes a memory, family groups disperse, the fall exodus happens, and the oak-dotted hillsides fall silent until the following spring.

The Lazuli Bunting is one of the most common summer songbirds in Humboldt County, occurring from sea level to some of the highest elevations. This species was encountered in 49% of all blocks. They are not especially partial to one or another plant community, demanding only that the open ground upon which they forage is dotted with prominent singing perches. Sniado (1999) found the abundance of this species to be negatively related to the amount of total tree cover. In the most open situations, territories may entirely lack trees or even brush. In such places, singing is done from the tops of boulders, abandoned fence lines, or even the tallest weed within a territory. Power lines are commonly used as song perches. Although they are generally birds of drier sites, Lazuli Buntings do intrude well westward into the zone of coastal low overcast, occurring regularly above Cape Mendocino, on Bear River Ridge, and on the choicer sites around the inland margin of the coastal lowlands around Humboldt Bay. They are annual in and about Arcata, but their abundance there varies from year to year.

Certainly the greatest numbers of Lazuli Buntings in Humboldt are found at mid-elevation in the rolling pastureland and oak country of interior and south county. A look at the distribution map reveals that atlasers found them in most blocks in the Mad, Mattole, and Trinity River drainages and in the southeastern portion of the county. In east-central and southern Humboldt this is truly among the most abundant landbirds in summer. Outside stands of dense forest, one might find it a tough chore indeed to find a spot from which one or more might not be heard or seen.

Lazuli Buntings readily colonize clearcuts throughout most of the private and federal timberlands, extending well up into the mountains. As a result of forest clearing, they may have increased more than 200% (Raphael *et al.* 1988). It is not uncommon to encounter them in meadow edges or along roadsides in the zone of winter snows, where they may spend the summer in the company of Fox Sparrows, Dusky Flycatchers, House Wrens, and even Townsend's Solitaires. When the open ground in logged areas ultimately disappears beneath a mantle of vigorous conifer or tanoak regrowth, buntings become less numerous. Most Lazuli Buntings are found on hillsides, but riparian corridors threading flat pastures or meadows are also frequented. Many were found in the open grassy hillsides in the Bald Hills above lower Redwood Creek. The outer coast north of Cape Mendocino was found to be uninhabited over the course of the study. Buntings also were seemingly absent from the exceptionally luxuriant coast redwood regrowth characterizing commercial timberland in the northwest portion of the county.

Probable breeding behaviors were first observed in early May, *e.g.*, territoriality seen 7 May 1995 (k32; JEH) and courtship 8 May 1997 (m11; *fide* JMH). Only three active nests were found, the earliest one with eggs 14 May 1995 (N33; C. Secor) and the latest one with nestlings 19 Jun 1996 (N23; GAS). The earliest fledgling observed was 26 Jun 1997 (h14; KMS). The feeding of young by adults was observed as late as 7 Aug 1995 (N13; TWL).

INDIGO BUNTING *Passerina cyanea*

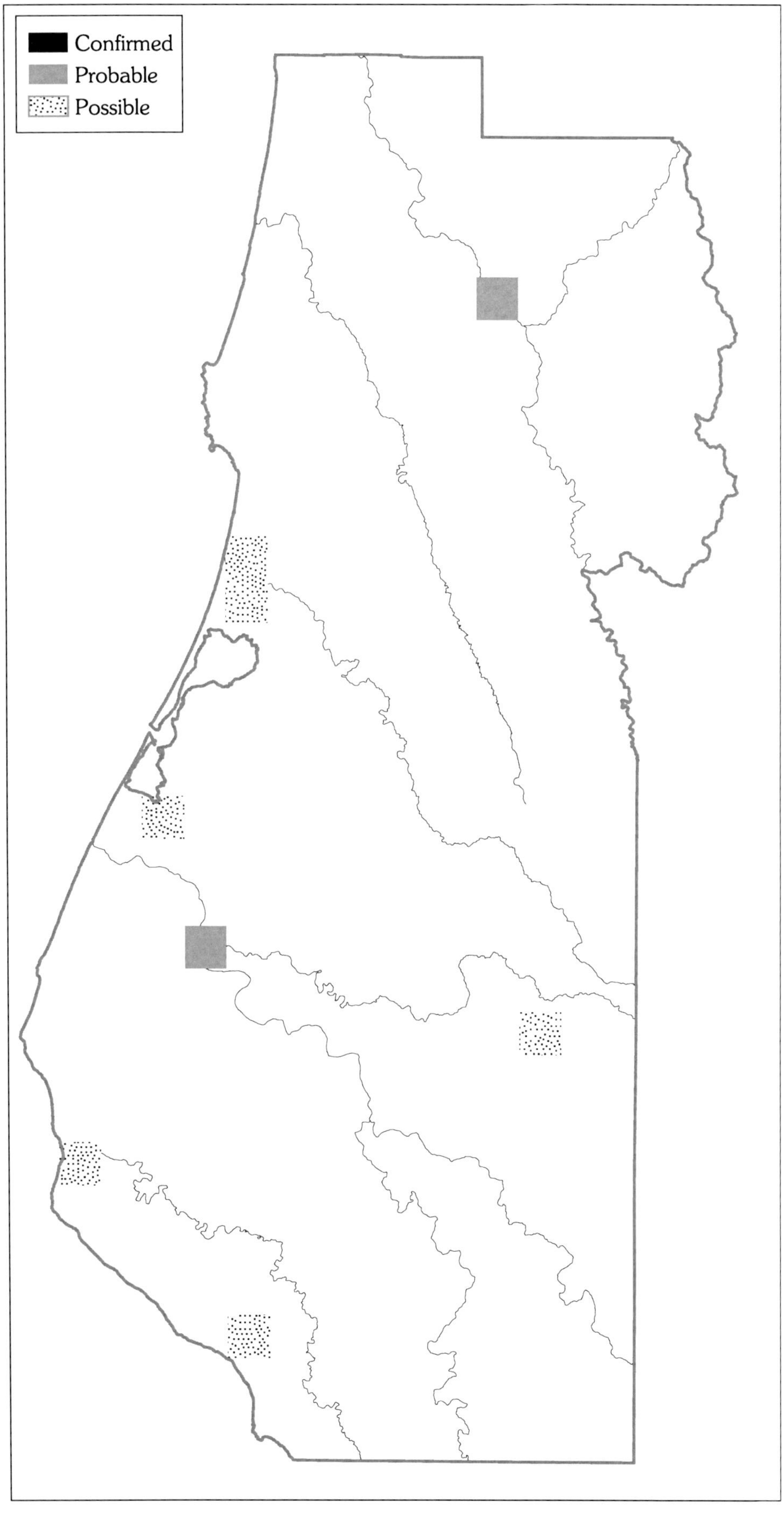

The eye-catching blue flash of a male Indigo Bunting flying up from a roadside or singing from a telephone line has become more familiar in northwestern California in recent decades. It is reasonable to associate this increase with the expansion of the breeding range into westernmost Texas, New Mexico, Arizona, southern Nevada, and (occasionally) California that has taken place over the past half-century. Cited as possibly the first nesting record by this species in northern California was a pair feeding young near Pescadero, San Mateo County 14-21 Aug 1994 (FN 49:99). Although pairings by male and female Indigos remain quite rare in the state, California has seen numerous instances of male Indigos paired with female Lazuli Buntings. A few of these partnerships have resulted in the production of offspring. Such a situation was reported from Cape Mendocino in Humboldt County 5 Aug 1979, when a male Indigo and female Lazuli were seen feeding young (B. Clow, R. LeValley; Harris 1996). Many other cases of mixed pairings of these closely related species in northern California have been reported (AB 33:895, NASFN 49:978, Bleitz 1958, Rowe and Cooper 1997).

During the Atlas period, Indigo Buntings were detected in eight (2%) of the 425 total blocks. All eight birds reported were males. A similar situation exists in Monterey County, where Roberson (2002) noted that the vast majority of spring vagrants recorded were males. Two of the birds encountered by Humboldt atlasers were known to have sung over a period of at least 1 week, signifying probable breeding status. One was in riparian edge at the confluence of the Van Duzen River with the Eel River 14-28 Jul 1998 (e22; W. Widdowson, P. Herrera). The other was in a patch of oak woodland a short distance upslope from the Klamath River near Martins Ferry (L07) 14 Jun-8 Jul 1998 (JCP, DFx, JEH, GCH). Records in the remaining six blocks were of single birds deemed to be in suitable breeding habitat. These sightings occurred 22 Jun 1995 (m24; S. McAllister); near Petrolia (B27) 22–23 Jun 1996 (GCH); at a bird feeder in McKinleyville (F13) 28 Jun 1996 (R. Ridenhour); at Table Bluff (D19) 29 Jun 1996 (J. Danzenbaker); along Mad River Road in the Arcata Bottoms (f14) 10 Jun 1997 (K. Irwin), and along Buck Creek trail in the King Range National Conservation Area (F31) 2 Jul 1999 (KMS). There were other reports of Indigo Bunting during the Atlas period that were deemed to be vagrants, and not possible breeders. For example, birds were reported at the "Entrance Patch" near Fairhaven (d16) 20 Jun 1996 (GCH), near the Mad River Slough 16 Jul 1999 (e15; C. J. Ralph), and at Lanphere Dunes (e14) 24 Jul 1995 (C. J. Ralph).

Rather than appearing almost entirely along the coast like most classic "eastern vagrant" landbirds, a disproportionate number of Indigo Buntings occurring in California in spring have also shown up well inland. A very few are now expected both coastally and inland each spring and summer in northwestern California. If one assumes that this phenomenon is part of their ongoing range expansion, then many of the Indigo Buntings appearing in California in spring may simply represent pioneering overshoots, scarcely to be described any longer as out of range.

Reversing the pattern of nearly all true eastern vagrant landbirds occurring in the state, Indigo Buntings show up more commonly in spring than in fall. Roberson (1980) noted that 65% of California records up to that year were in spring. Furthermore, many of the Indigo Buntings appearing in May or early June in California have remained for weeks. Although one might point to examples of many species of eastern vagrants spending time at one site, Indigo Buntings do this more often than most. This pattern was remarked upon by Roberson (1980), who noted that 5% of recent California records at that time had been of birds remaining to summer locally.

Humboldt's first record of Indigo Bunting was of a male present 9 Jan–7 Apr 1963 in Ferndale (Wilbur 1963, Yocom 1970). Humboldt's second record was of a male 30 Jun–21 Jul 1970 along Salmon Creek (Yocom 1970). The second bird observed was in breeding plumage and singing on territory, and was with some young buntings and an adult Lazuli on 13 Jul (Yocom 1970). Since then, there have been more than 30 May–June records of Indigo Buntings from northwestern California (Harris 1996). While none of the records establish confirmed nesting by Indigo Buntings, the report of a "small flock, both sexes" at Briceland 7 Jul 1980 (R. Sutherland) may have constituted a family group (Harris 1996). Roberson (2002) commented that "Occasional nesting in California is no longer a surprise." Thus, given the well established history of late spring and early summer birds in Humboldt, it seems likely that this probable breeder will eventually be confirmed in the county.

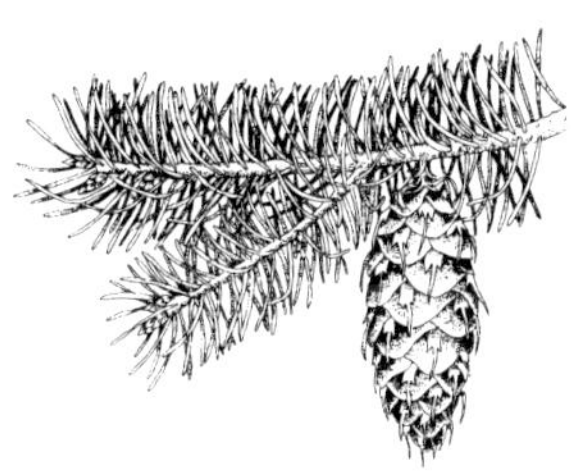

RED-WINGED BLACKBIRD *Agelaius phoeniceus*

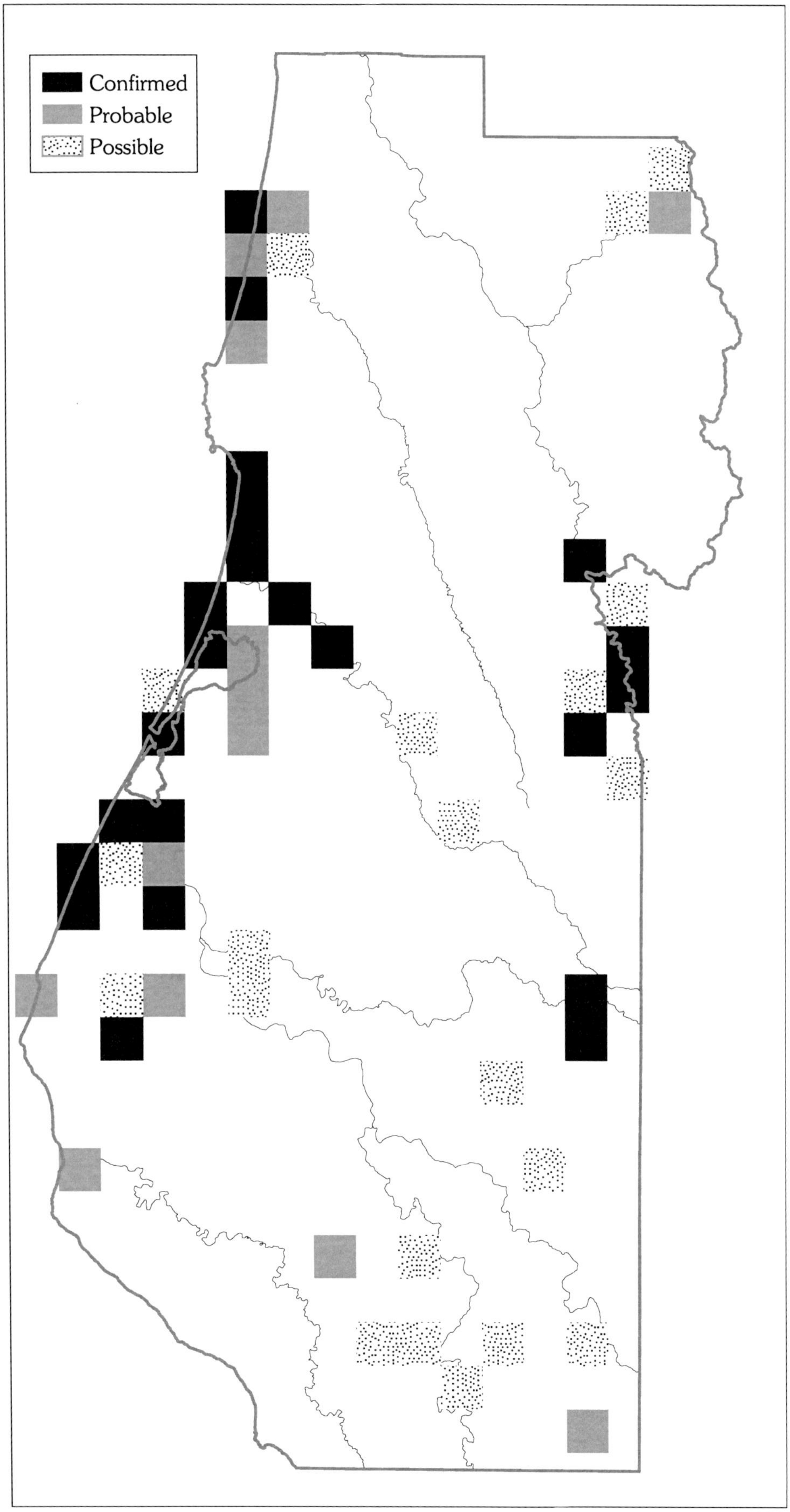

Red-winged Blackbirds are known by millions of people. It is scarcely necessary to own, let alone raise, binoculars to claim an easy familiarity with these abundant birds. In their own millions these blackbirds sway from cattail stalks, flock in crop fields, and stipple evening skies as they pass overhead toward night roosts. Highly social when not breeding, they may assemble in enormous concentrations in the southern states in winter. Then, with lengthening days causing the press of hormones to alter their behavior, the great swarms migrate northward and disperse. Soon the conspicuous red epaulets of adult males flash from the dried stalks of the previous year's growth of marsh plants as they advertise for mates. While males are first on the scene, females are not many days behind, and soon the reproductive cycle takes its turn.

Red-wings breed in nearly every marsh worthy of the description in the U.S. and temperate Canada. While in many regions they are well distributed and are commonly seen in summer, in others they are known only from precious few wetlands, or are absent because of lack of any habitat. They benefit from an uncanny ability to locate and make do with even the smallest wet spot providing tall emergent vegetation and near sufficient open-country foraging areas. The profile of the Red-wing in Humboldt County illustrates this. Marshes exist in only a fraction of the county's total landscape, but the birds know these places, and there they may be common. During Atlas fieldwork, Red-winged Blackbirds were found in 56 (13%) of the 425 total blocks, yet the total land area they occupied during the nesting months was a vanishingly small percentage of that 13%.

In the greater northwestern California area the Red-winged Blackbird is known as a common resident and breeder (Harris 1996). In Humboldt County, the spring and summer distribution of Red-winged Blackbirds is coincident with that of freshwater marshes. The coastal portion of the range recalls that of most other wetland-breeding birds detected during the project, but it was more extensive. It took in the Redwood Creek estuary and Humboldt Lagoons; the Clam Beach lagoons; the lower Mad River; various sites in the Humboldt Bay lowlands; the Eel River delta; Capetown, and the mouth of the Mattole River.

The inland distribution presented a somewhat different picture. With no need for expanses of open water nor for a food base sporting gills, fins, or nutritious stems, Red-winged Blackbirds were encountered in the course of the study in many places plainly unsuited for such marsh-dwellers as Pied-billed Grebes or American Coots. This tolerance of smaller, shallower, or only seasonally available wetlands probably explains the greater abundance of Red-wings in the interior of the county than was noted for either of those species. They were found in the Orleans area; around Willow Creek; at ponds between the South Fork Mountain ridge and the South Fork Trinity River; in the central Mad River corridor (J17, k19); up the Little Van Duzen River (N23, n24); along the Eel River between Bridgeville and Fort Seward (L25, m27); up Salmon Creek (H29); and in seven other loosely clustered blocks in southeastern Humboldt.

Red-winged Blackbirds nesting in Humboldt County rely largely on human-created wetlands. Important breeding sites for this species include the Arcata Marsh, the south Eureka "sewage marshes" north of the end of Pound Road, and the managed impoundments of the Eel River Wildlife Area. Stock ponds in the summer-dry interior of the county are important to them. They occupy naturally occurring habitats both near the coast and inland, such as the marshy edges of Freshwater and Stone lagoons, the extensive bulrush beds at Dry Lagoon and at the south end of Big Lagoon, and numerous swales and creek meanders elsewhere. It is unclear from Atlas data whether Redwings use non-wetland habitats for nesting less often in Humboldt than they do elsewhere in California, or if specific details on non-wetland habitat nesting were not reported by atlasers.

Confirmed breeding was reported from 22 blocks, and was judged probable in 13 others. An adult was seen carrying nest material 26 Apr 1997 (e15; P. Herrera), the earliest confirmation. Two occupied nests were reported as best evidence, 5 May 1999 (g14; GAS) and 23 May 1996 (D19; K. Foerster). In the 22 blocks with breeding confirmations, adult Red-wings feeding young proved the most commonly witnessed behavior supporting this status, reported in nine blocks. Dates on which this was observed spanned 10 Jun 1997 (c24; COg) and 10 Jun 1998 (N13; TWL) to 2 Aug 1995 (H15; PFS). Fledged young were seen in six blocks, ranging from 4 Jun 1995 (F11; JEH) to 10 Jul 1998 (b20; DFx); these records were split evenly among the two months in which they occurred. Adults carrying food furnished the highest code in three blocks, from 23 Jun 1999 (f12; GAS) to 5 Jul 1997 (D21; GAS). Rounding out the list of confirmations was an adult carrying a fecal sac 26 Jun 1998 (o15; *fide* RHw).

As with other species breeding in Humboldt, Atlas work did little to help elucidate the distribution of Red-winged Blackbird subspecies. The few published accounts of breeding Red-wing subspecies in Humboldt all refer to the Northwestern Red-winged Blackbird (*A. p. caurinus*; Mailliard 1916, Wilder 1916a, Talmadge 1953b). On 10 Apr 1996, two of seven male Red-wings observed on territory in a small marsh near Orick (g05) were identified as typical Bicolored Red-winged Blackbirds (*A. p. mailliardorum*), while the remainder were Northwesterns (A. Jaramillo). Jaramillo (pers. comm.) noted that he would have expected that most of the birds there would be intermediate between the two forms. Add this to the list of topics for future study in Humboldt County.

After nesting, Red-wings assemble into flocks that tend to be of one sex. Especially in autumn when the exotic cordgrass seed crop is heavy, Red-winged Blackbirds make greater use of Humboldt Bay saltmarsh than do starlings or any other icterids. Red-wing numbers swell in September and remain high through late winter, certainly the result of an influx of migrants from interior Humboldt and elsewhere. Much foraging takes place in highly rural locations, with seemingly less use of backyard bird feeders here than elsewhere in California. The largest winter concentrations occur around dairy operations in coastal lowlands, and these swarms typically involve numerous other "blackish-birds" including European Starlings, Brewer's Blackbirds, and Brown-headed Cowbirds. During the Atlas period 1995–1999, 389–912 Red-winged Blackbirds were counted on the Centerville Beach to King Salmon CBC, 115–618 on the Arcata CBC, and 45–83 on the Willow Creek CBC.

TRICOLORED BLACKBIRD *Agelaius tricolor*

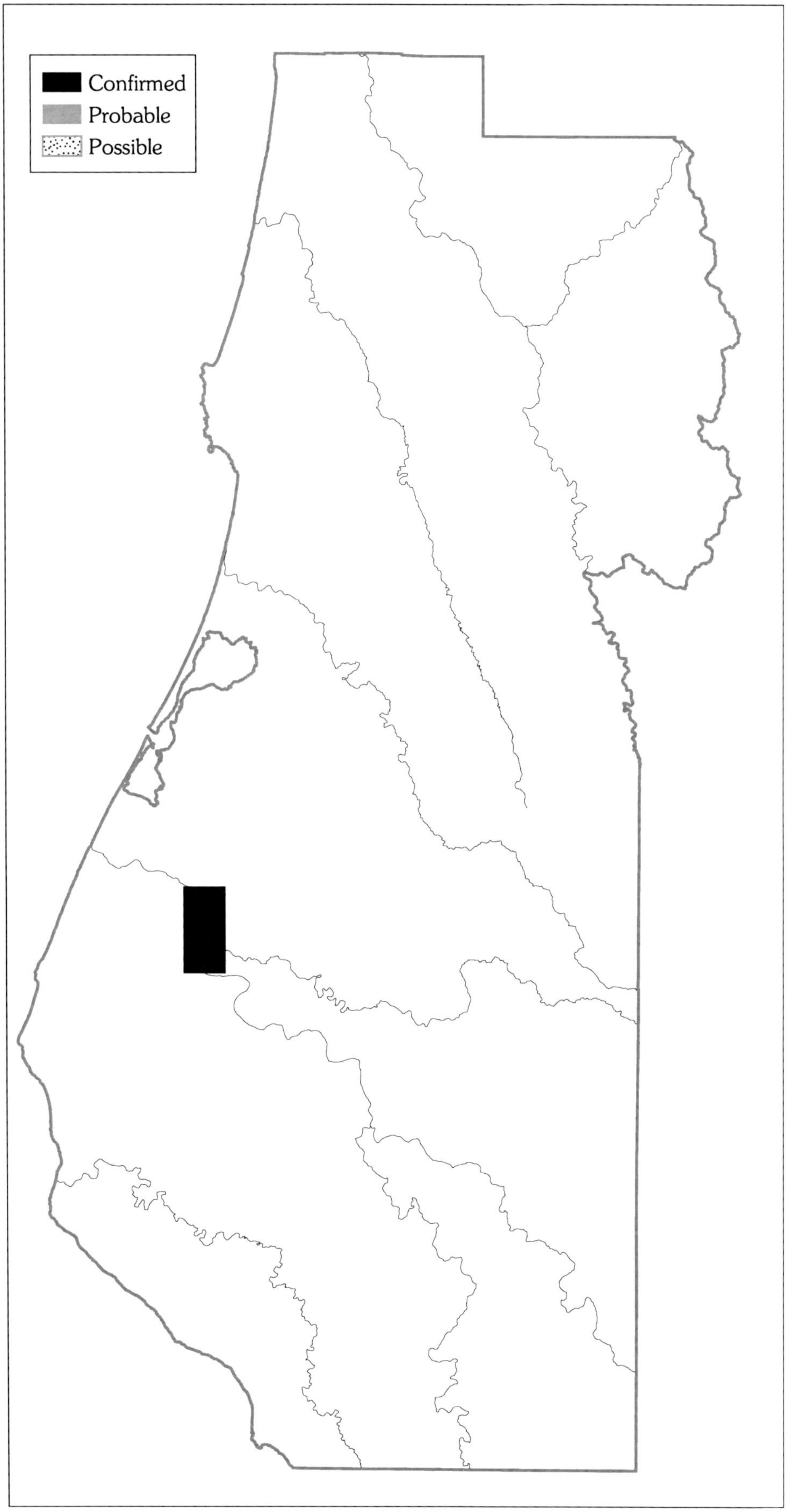

Often nesting in great colonial assemblages, the Tricolored Blackbird is an enigmatic and sensitive species. Until the last few decades, it was restricted as a breeding bird to interior California, the Klamath Basin, and the Rogue River Valley in southwestern Oregon. The range has expanded to include nesting sites farther northeast in Oregon, and in 1998 this blackbird was first reported from Washington, where it now breeds. It was historically known in Humboldt County only as an occasional visitor.

On 6 Jul 1992 a small breeding colony of Tricolored Blackbirds was discovered at the intersection of Drake Hill Road and Highway 101, near Alton (e21; R. LeValley, S. Harris). This colony was present each year subsequently and through the end of the Atlas period. While no more than five singing males were ever reported at any one time in any year, some numbers reported from the Drake Hill Road colony include 18 males and two females 8 Mar 1996 (DFx); 13 males 7 Mar 1997 (PFS); 35 birds on 31 May 1996 (S. Harris); and 70 birds 12 Jun 1997 (S. McAllister). This colony is the northernmost coastal breeding site known. Tricolored Blackbird has been considered a rare, local breeder in both Humboldt and Mendocino counties (Harris 1996).

When the Drake Hill Road birds were discovered in 1992, the colony was in a rank weed patch along railroad tracks between Highway 101 and a frontage road. Vegetation at the site was rather scant and nondescript, a rambling stringer of bushes and tall forbs typical of vegetation immediately beyond the reach of routine right-of-way management. Himalayan blackberry, fennel, mustard, and scattered saplings constituted the bulk of this. The Drake Hill Road colony was confirmed as active early in the Atlas period when birds carrying nest material were seen 7 May 1995 (e21; GCH, JEH).

By 16 Jul 1998, when an adult was seen popping a cutworm into the mouth of a begging juvenile, the colony had moved across the highway to Sandy Prairie Road (e22; JCP). This new site was about one-half mile south of Drake Hill Road. Habitat consisted of a blackberry tangle ingrown with tall fennel, nettles, and roses on an unmanaged corner of land adjacent to an old farmstead. The birds foraged somewhere in and across the expansive pastureland west of the colony sites. Occasional individuals or small groups of Tricolored Blackbirds have also been observed in this vicinity during winter. As had been the case prior to the beginning of the Atlas project, blackbirds identified as Tricoloreds continued to be occasionally reported in fall and winter among the blackbird flocks at dairies in the Eel River delta and in the Arcata Bottoms.

In contrast to the situation in Humboldt County, Tricolored Blackbirds elsewhere commonly situate colonies in large marshes. Many colonies in California's Central Valley historically contained tens of thousands of birds. Though exhibiting long-term abandonment and recolonization of some sites and irregular swings in numbers, the population overall has decreased in recent decades (Beedy and Hamilton 1997). Investigations made in coming years will indicate how long the isolated Humboldt colony may persist. Potential threats to Tricolored colonies include habitat loss to agriculture and urbanization, contaminants, and native predators such as Black-crowned Night-Herons and Common Ravens.

WESTERN MEADOWLARK *Sturnella neglecta*

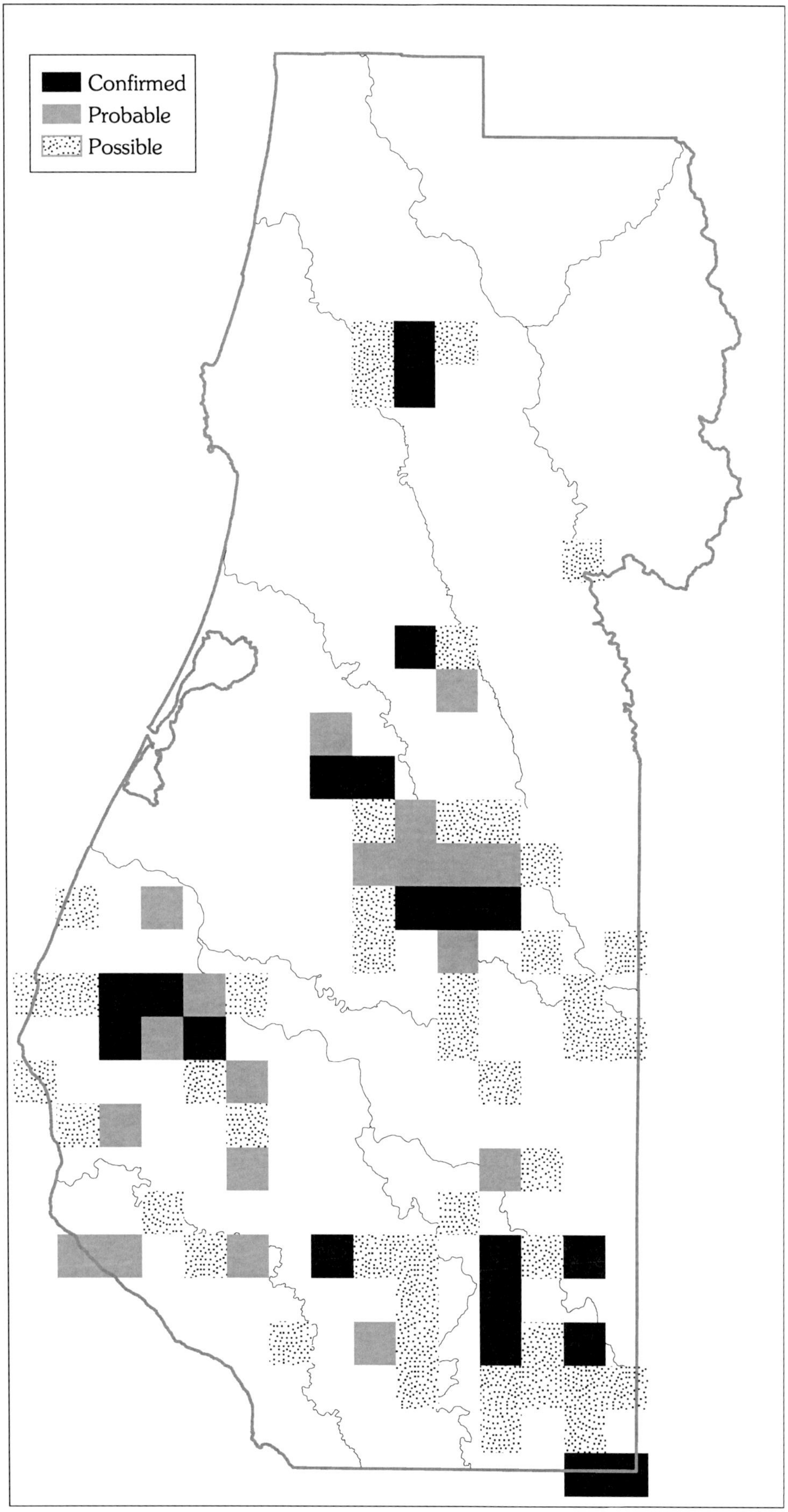

Meadowlarks are stout, short-tailed icterids with streaked backs and distinctive, pleasing songs. They live where there is much grass and where trees, if present, are discontinuous and are but a minor element of the landscape. Following the pattern of distribution elsewhere in their range, Western Meadowlarks are fairly common in California wherever suitable habitat occurs. As might be expected, these regions in a broad sense take in most of the unforested and non-mountainous sections of the state. Mirroring a theme exhibited by other open-country species in Humboldt, Western Meadowlarks were found to be locally common in portions of the southern one-half of the county. Atlas fieldwork found them in 84 (20%) of the 425 total blocks.

Within the county, Western Meadowlarks were well scattered across southern Humboldt, but were otherwise only sparingly reported during the study. Most conspicuous on the map generated by Atlas surveys are clusters of blocks centered in the central Mad River watershed (*e. g.*, k20) and on Bear River Ridge (d24). The clumping of blocks with probable and confirmed breeding within these clusters suggests that greater numbers of birds occur in these areas. Habitat in the Mad River country and on Bear River Ridge is relatively extensive. Near the Mad River, this habitat is edged and broken somewhat by dry oak groves and narrow upslope stringers of California bay and California buckeye. Along Bear River Ridge, the southerly exposures are grown to grass and little else near ridgelines, and are frequently dotted with coyote brush or poison-oak in the mid-slope settings.

Western Meadowlarks were commonly noted in highly rural, bucolic hilly country of interior southern Humboldt. Grasslands there are often abruptly fronted by Douglas-fir, tanoak or other hardwood stands or by well grazed pastures. Meadowlarks were particularly common on sunny exposures of the general Mail Ridge uplift, separating the Eel and South Fork Eel rivers. Elsewhere, a disjunct population of some size was apparent in the Bald Hills (j08 and vicinity), while other birds were found northwest of Lake Prairie on Snow Camp Mountain (k15 and vicinity), at Willow Creek (N13), and in the southwestern corner of the Eel River delta at Centerville Beach (B21) and outside Ferndale (D21).

Forested or otherwise thickly vegetated countryside offers little attraction to meadowlarks, an aversion well indicated by the Atlas map. None were reported from a large area of northern Humboldt, nor were any found along the coast north of Centerville Beach. The absence of these birds during the breeding season from the northern coast, as well as throughout the Humboldt Bay and lower Mad River lowlands, was known to the birding fraternity before the inception of the project and was not a surprising Atlas result. Clearly, Western Meadowlarks do not summer on the county's extensive sandspits, and show no attraction to saltmarsh, nor to those lands managed by the dairy industry—although each of these situations is well frequented in fall, winter, and early spring. Based on Fisher's (1902b) description of Western Meadowlarks as "common in open meadowland near the coast" and the large number of egg sets collected 1905–1930 around Humboldt Bay, Harris (1996) thought that this species may have declined as a breeder in the coastal lowlands over the past 40–50 years. Further evidence of local declines of this species is that Fisher (1904) considered this species "common" in the Hoopa Valley, yet there were no Atlas reports from that area.

Breeding by Western Meadowlarks was confirmed in 20 blocks and was judged probable in 19 others. The earliest evidence of probable breeding was of multiple males singing 22 Mar 1997 along Snow Camp Road (k16; JEH, L. Hunter). An occupied nest in the Bald Hills 25 May 1997 (j08; L. Bacon-Ogden), located through persistent observation of a furtive adult, was the only nest reported in the course of the study. Observations of adults carrying food supported confirmed breeding status in more than two-thirds of the 20 blocks with confirmations. The earliest such report was 28 May 1999 (o34; JEH). Five were in June and eight in July, the latest 20 Jul 1996 (D23; COg). Fledged young provided the best evidence in five blocks, ranging from 6 Jul 1997 (N31; DFx) to 3 Aug 1995 (L29; JG).

Harris (1996) described the Western Meadowlark in northwestern California as a "common resident and breeder. Winters in numbers in the lowlands; moves upslope to mountain prairies in summer. Now rare in lowlands in summer." This characterization remains quite applicable to Humboldt County. Spring migration is scarcely noticed in this species locally. Small cohesive flocks of birds, many singing, are occasionally encountered in the coastal lowlands in February and March, but to what extent these represent northbound migrants as opposed to wintering birds "tuning-up" is difficult to determine. A decrease in numbers and, eventually, disappearance from the near-coastal lowlands takes place at that time. Conversely, the southward movement is made apparent by numerous flocks of meadowlarks that appear from early September onward in places at which the species did not summer.

During the colder months, individuals and flocks of from a half-dozen to several dozen birds are widely noted in open or semi-open countryside along the coast and throughout the bottomlands. In addition to habitats listed above, they also resort to abandoned mill sites, airstrips, highway median corridors, and small farms. During the 1995–1999 Atlas period, 307–652 Western Meadowlarks were reported on the Centerville to King Salmon CBC, 48–564 on the Arcata CBC, and 1–30 on the Willow Creek CBC. Winter birding before and during the Atlas period in southern Humboldt, on Bear River Ridge, on Mail Ridge and elsewhere indicated that they are to be expected essentially throughout the breeding range during that season.

BREWER'S BLACKBIRD *Euphagus cyanocephalus*

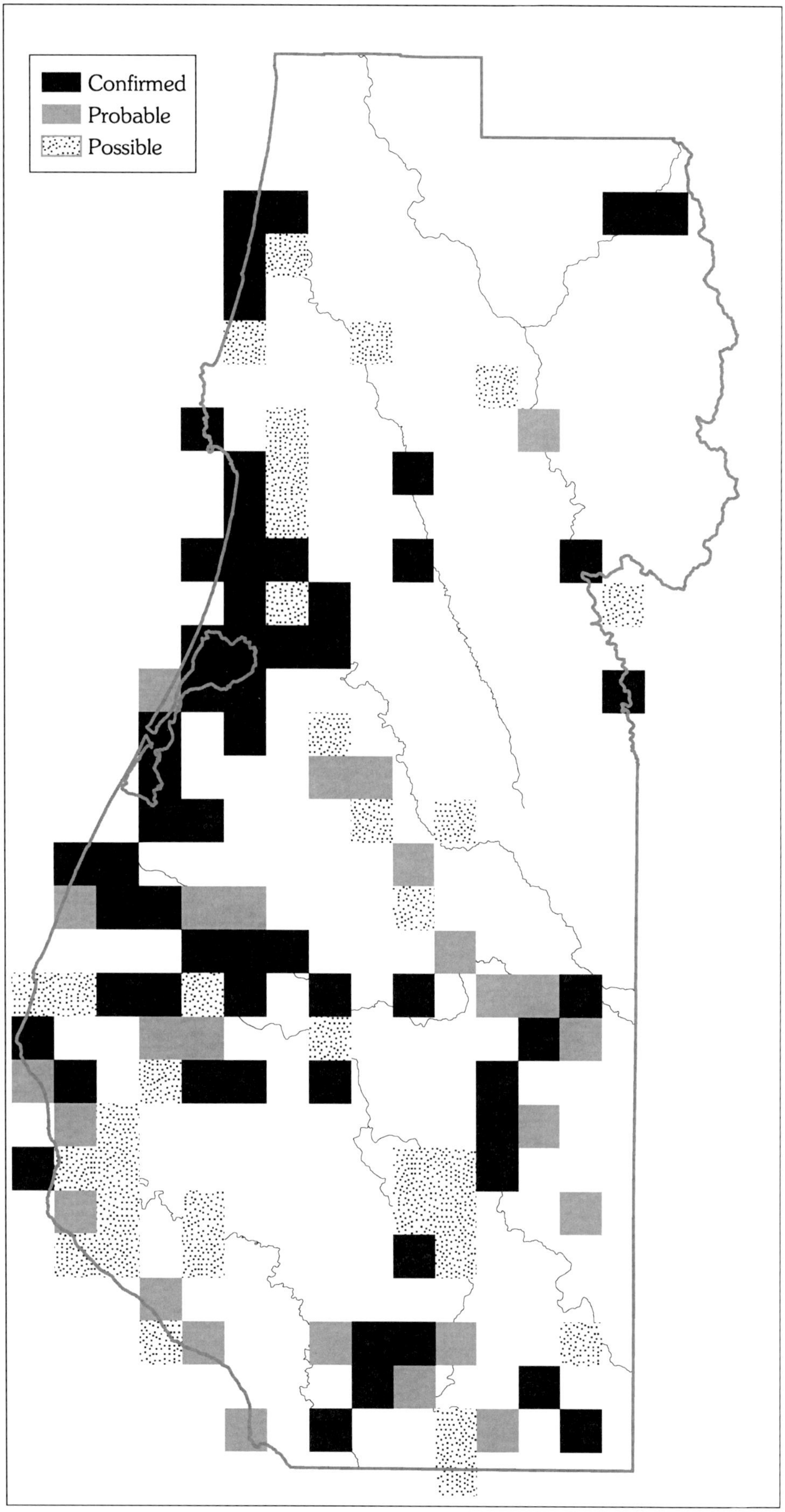

Brewer's Blackbirds are conspicuous throughout the valleys, open foothills, and rangeland of the West. Their preference for treeless or sparsely forested expanses has been well served by the land-clearing activities of humans. Today, they are as at home in cities and heavily settled residential areas as in the native grasslands they once may have favored. The planting of windbreaks, orchards, and shade trees in proximity to open ground may have provided enhanced nesting opportunities. California's immense system of rural and suburban roads clearly has created a steady source of food for adults to provide to nestlings—in the form of insects stunned or killed by vehicles. Parking lots at malls, stadiums, and shopping centers also attract Brewer's Blackbirds intent on procuring either insects or tidbits of human food. Farms, dairies, ranches, and varied semi-open country with exposed ground offer foraging sites for the mixed-species blackbird flocks into which this species aggregates in fall and winter. All in all, this species would appear to have fared well in the company of people.

Harris (1996) considered Brewer's Blackbird to be a common resident and breeder in northwestern California. Atlas surveys revealed that these blackbirds were indeed widespread in Humboldt County during the nesting season. They were detected in 28% of all blocks and confirmed as breeding in 59 of them. These birds are associated with settlements, floodplain dairy pastureland, river corridors, foothill beef-cattle pasturage, and the interior oak-grassland mosaic. Given their habitat preferences, it is not surprising that the greatest clustering of blocks with evidence of confirmed breeding lie in the Humboldt Bay lowlands and in the countryside along and above the lower Eel River. This species was present along most of the immediate coast south of Prairie Creek Redwoods State Park, in the Mattole River country, in interior southern Humboldt, and throughout a corridor broadly corresponding to oak-and-pasture rangeland of the upper Mad River and Van Duzen river foothills. In addition, confirmed breeding in Redwood Valley, at Orleans, and along the Trinity River indicated populations in those areas.

While habitat for Brewer's Blackbirds clearly is scattered throughout much of the county, large expanses of Humboldt plainly lack an environment suitable for these birds. Generally speaking, the large redwood parks, stretches of intensively managed commercial timberland, and openings within the higher conifer forests of northeastern and extreme east-central county seemed to host few or none of these blackbirds in the breeding season. Brewer's Blackbirds appear to require considerable open ground on which to forage. As so much of the freshly logged acreage in Humboldt County regenerates to ground-obscuring density within 1 or 2 years, clearcuts offer no attraction for them.

Brewer's Blackbirds nest semi-colonially. Several nests may be situated in a stand of one or a few dense trees. In urban and suburban areas, sidewalk or streetside trees that have been repeatedly topped below power lines, and consequently "butched" into dense masses of limbs and foliage, frequently are used for nesting. Adult Brewer's make no secret of their whereabouts whenever a person passes closely beneath a nest tree, clattering anxiously and sometimes diving upon hapless passers-by until some threshold of tolerance is again regained. This behavior marked the earliest breeding confirmation, a record of distraction display 14 Apr 1996 west of Rio Dell (e23; COg). The following day, nest building was recorded in Redwood Valley (J11; D. Kriger). The earliest fledglings were reported 17 May 1996 (f14; GAS).

While discoveries of actual nests are uncommon owing to stealthy approaches and obscuring foliage, adult blackbirds carrying food from dispersed foraging sites toward nests are quite conspicuous, and were well noted during the Atlas study. Forty-seven percent of breeding confirmations were based upon observations of provisioning adults. Such behavior noted 19 Jul 1998 (N33; C. Secor) represented the end of the breeding season fairly well, while the only confirmed breeding report for August, of a nest with young 15 Aug 1996 (H33; K. Dyer), was a temporal outlier.

BROWN-HEADED COWBIRD *Molothrus ater*

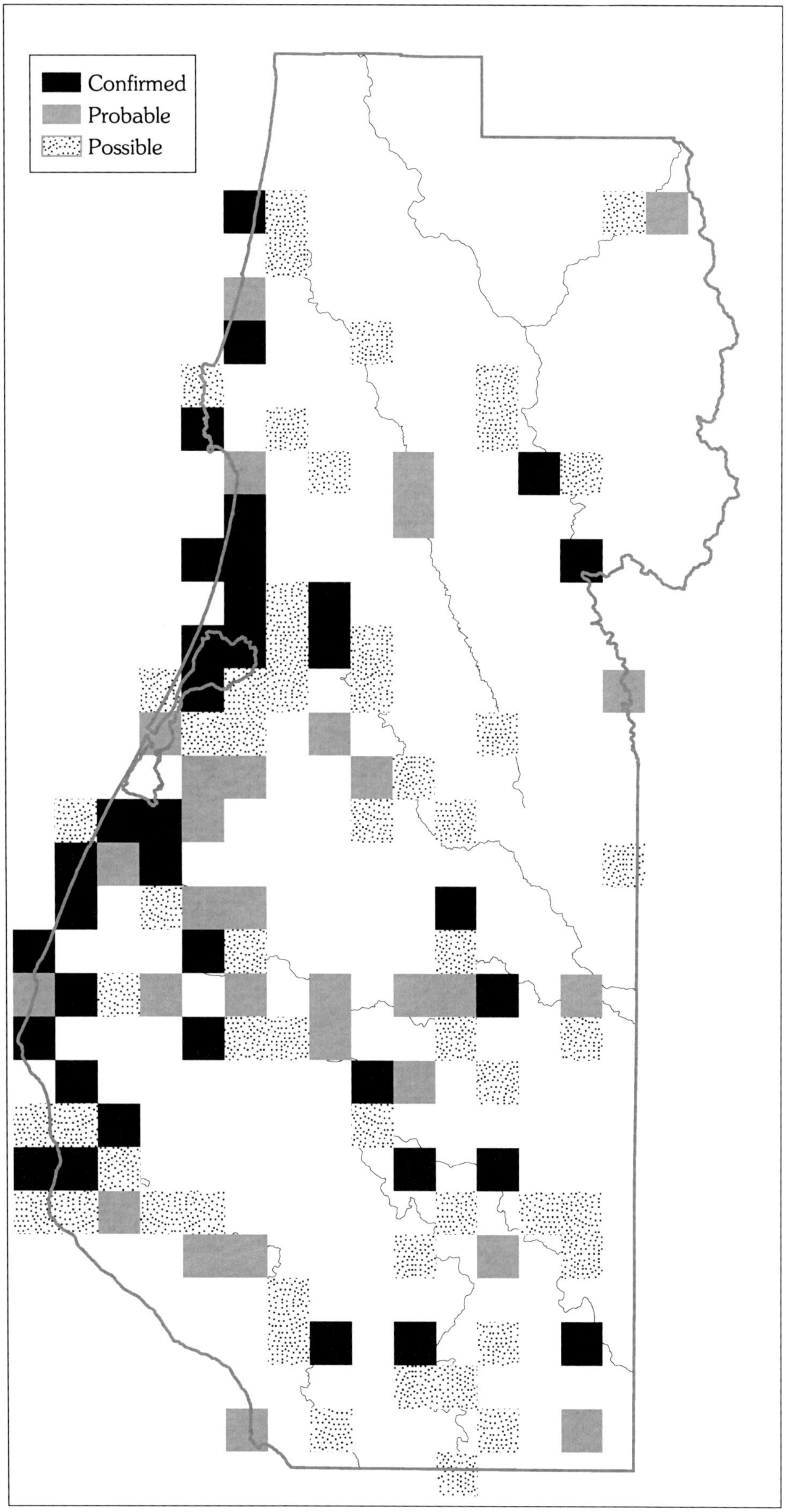

Cowbirds are known for nest parasitism, in which females lay eggs among clutches of other species' eggs, abandoning them to be incubated and raised by accepting parents. As juvenile cowbirds hatch in less time than other North American songbirds, they gain an early competitive advantage, reducing fledging success among foster siblings. Brown-headed Cowbird eggs have been found in the nests of more than 220 species of North American songbirds (Ehrlich *et al.* 1988), and more than 150 of these have fledged young cowbirds. In recent historical times, Brown-headed Cowbirds are believed to have been restricted to the Great Plains, where they benefited from an association with grazing bison (*Bison bison*), whose hooves stirred forth insects from hiding. With the clearing of eastern hardwood forests and introduction of cattle to western ranges, these birds were able to colonize many new settings and to prosper in most.

As recently as a century ago, Brown-headed Cowbirds were scarce or as yet unknown in some areas of California in which they are now common. This species has been stated as having "invaded the northcoast in the past 60 years" (Harris 1996). Harris (1996) also noted that no specimens or observations are known from the records of early Humboldt County ornithologists. The first breeding records in the county (cowbird eggs seen in an unknown warbler nest and in a Pacific-slope Flycatcher nest) were from near Fernbridge in June 1941. Yet, intriguingly, the first adult birds were not recorded in Humboldt until June 1948, when a pair was seen at Hoopa (Talmadge 1948). Talmadge (1948) also found a cowbird egg in a Chipping Sparrow nest at Hoopa and another in a Yellow Warbler nest near Miranda in June 1948.

Since that time, Brown-headed Cowbirds have become familiar over much of Humboldt County. They were widely reported by atlasers at lower and middle elevations, and were recorded in 28% of the blocks. Cowbirds appeared to be quite common nearly throughout the unforested portions of the coastal lowlands from the Redwood Creek estuary south to Punta Gorda, with a far-southern coastal detection of copulation at Shelter Cove 1 May 1999 (f33; JEH). This region encompasses the Humboldt Lagoons; all of the county's extensive dairyland; higher-quality as well as fragmented riparia; the valley-edge forest/opening mosaic, and light urban and residential areas.

The map indicates a strong presence locally inland as well, particularly in the southern one-third of the county. Farm clearings, pastures, some degree of dispersed human settlement and presence of livestock, and an abundance of deciduous and mixed forest edge provide habitat for Brown-headed Cowbirds nearly across the entire breadth of that area. Cowbirds were frequently reported and confirmed as breeders in riparian and lower foothill settings along Redwood Creek and along the general courses of the Trinity, Mad, Eel, South Fork Eel, Van Duzen, and Mattole rivers. They were reported from two outlier blocks along the upper Klamath River; one of these records was of copulation observed 25 May 1996 at Orleans (P05; KMS). It is likely that a population of Brown-headed Cowbirds is present around Orleans during the breeding season, but the birds seem confined to the small area of farmland within a couple miles of town. However, most of the immediate Klamath River canyon within Humboldt County is well clothed in dense coniferous forest and is deficient in extensive stream meanders offering suitable riparia. Representing the highest elevation record was a possible breeder near Blake Mountain (o20) 20 Jun 1995 (GCH).

That Brown-headed Cowbirds do not penetrate to any great extent into extensive areas of closed-canopy forest is indicated by their apparent absence from commercial timberlands in extreme northern and west-central Humboldt and from the greater portion of the redwood parks. They seem to occur in clearcuts to only a minor extent. During 1995 surveys in different-aged redwood stands, cowbirds were detected only in shrub-stage stands (Hazard and George 1999). They were also generally absent in the present study in the corrugated regrowth hills of the interior southwestern part of the county, in the King Range, and in the higher mountains.

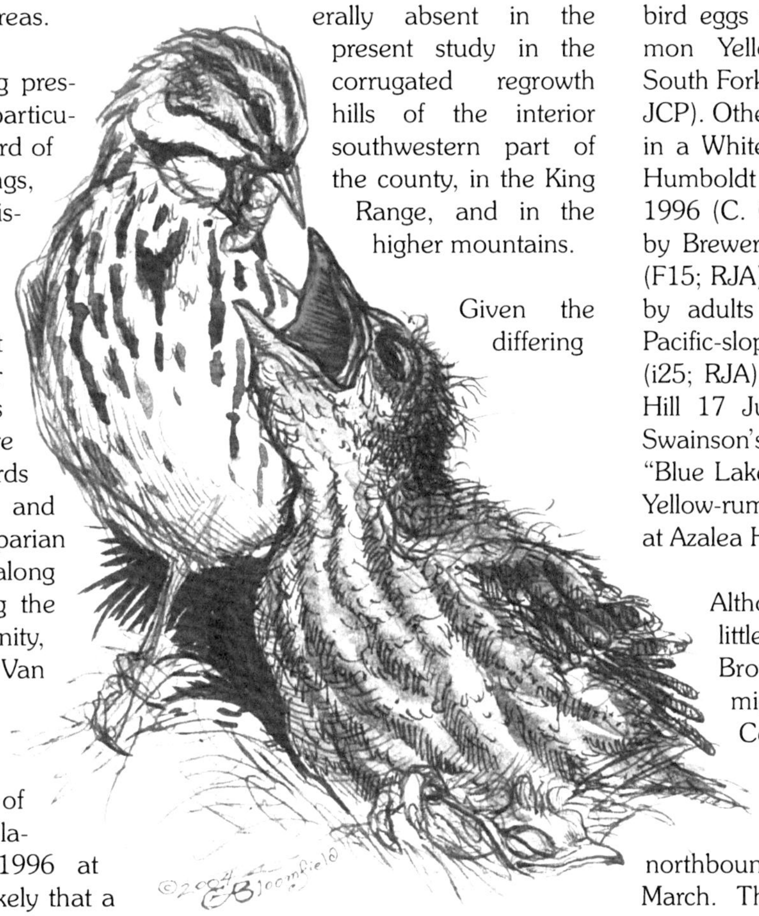

Given the differing breeding chronologies of potential host species, it is perhaps not surprising that confirmed breeding reports for Brown-headed Cowbirds fell during an extended period. Sixty-seven percent of the 36 Atlas confirmations were based on the observation of fledged young. A distinct peak in these (71% of such reports) fell during the 1-month period comprising the second half of June and the first half of July. Reports of adults feeding young comstituted only 28% of all confirmed reports, but spanned a three-and-a-half-month period from 1 May 1998 at Blue Lake (h14; KMS) to 15 Aug 1995 in Eureka (e16; L. Bacon-Ogden).

Host species were not consistently reported by atlasers, but nevertheless, nine species were identified during Atlas fieldwork. The most frequently reported host was Song Sparrow (4 reports). A measure of irony attends the records of adult House Sparrows feeding juvenile cowbirds at Redway (J31) 7 May 1996 (RJA) and in Eureka (e16) 18 Jul 1996 (RJA)! Cowbird eggs were also seen in a Common Yellowthroat nest along the South Fork Eel River at McCann (J27; JCP). Other host records include: eggs in a White-crowned Sparrow nest at Humboldt Bay Refuge (D19) 4 Jun 1996 (C. Goss); a fledgling attended by Brewer's Blackbirds 16 Jun 1997 (F15; RJA); and juvenile cowbirds fed by adults of the following species: Pacific-slope Flycatcher 29 Jun 1997 (i25; RJA), Hermit Warbler at Azalea Hill 17 Jul 1999 (F13; G. Lester), Swainson's Thrush 18 Jul 1999 at the "Blue Lake riparia" (H15; KMS), and Yellow-rumped Warbler 19 Jul 1999 at Azalea Hill (F13; G. Lester).

Although their movements are little remarked upon, Brown-headed Cowbirds are migratory in Humboldt County, and their abundance and distribution changes with the seasons. Each year sees an influx of northbound birds beginning in late March. This arrival seems to take place over a period of weeks, and eventually there are large numbers of cowbirds spread throughout most of the range in the county. The gurgling songs and display posturings of males become apparent in mid-April, and these behaviors are readily noted in May. Females are less conspicuous than males during the breeding season, but experienced observers know to listen for the telltale splutter they

Continued on page 395

HOODED ORIOLE *Icterus cucullatus*

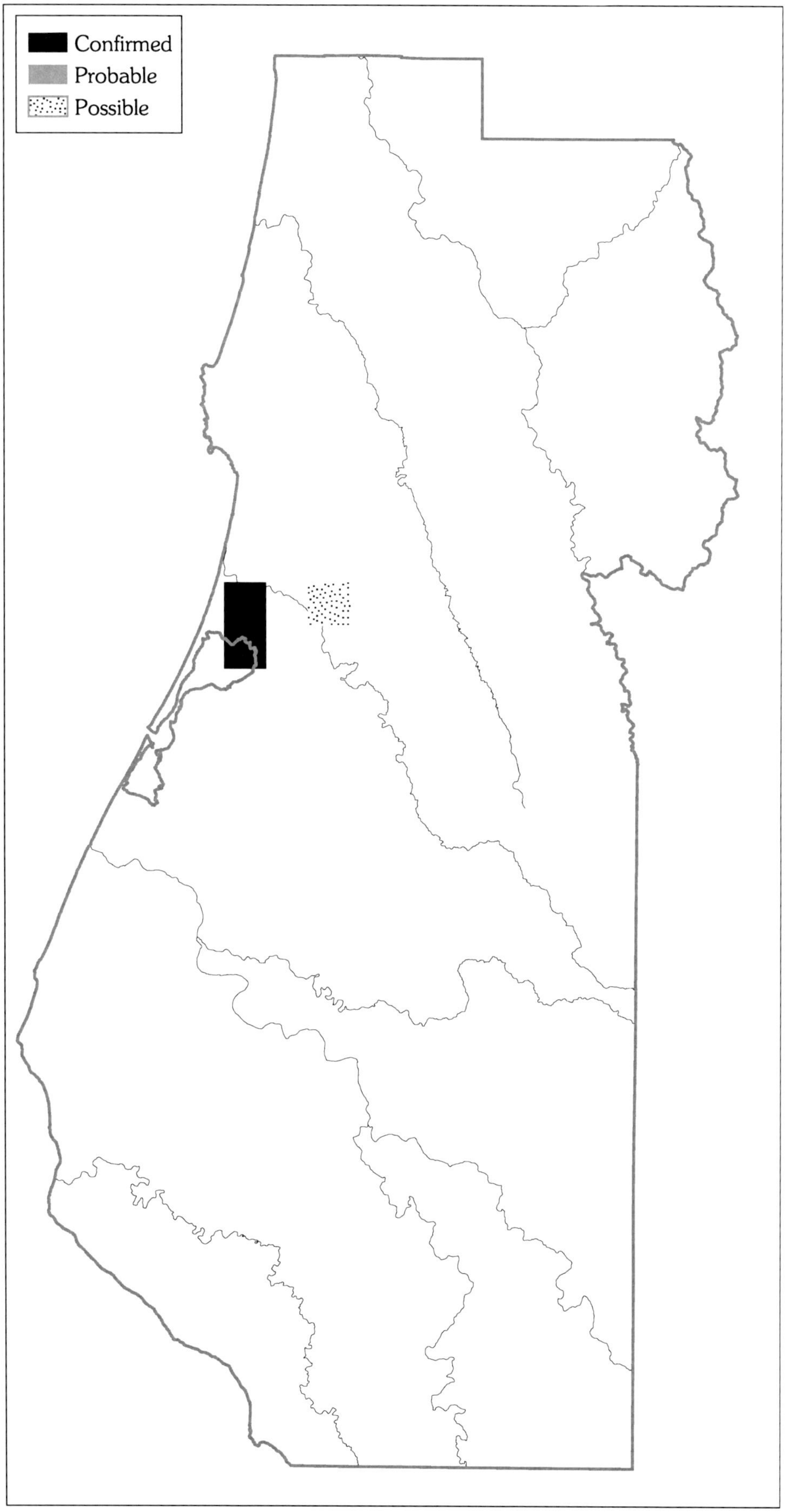

In the U.S., Hooded Orioles historically were restricted to the desert Southwest. Over the past century they have moved northward several hundred miles in California, and now breed in many places where they were once unknown. Shuford (1993) provided a detailed outline of the expansion of Hooded Oriole through interior central California and along the state's coast. The establishment of this oriole in northern California is the result of the planting of palms in and around urban centers. A very strong reliance by female Hooded Orioles upon palm fibers as nest material and, to a somewhat lesser extent, the trees themselves as nest sites, has limited this species' natural range and allowed it to increase as a result of human activities.

This species made its first appearance in Humboldt County when a female was seen in a residential yard in Ferndale 23 Apr 1972. It was later joined by a male, and the pair nested in a nearby palm and fledged two young (D. & S. Anderson; AB 26:807, 903). This pair or another nested at the same site for at least three successive years (S. Harris). A single Hooded Oriole in Eureka 12 Jul 1980 (W. Woodroof) may have been the first reported from the Humboldt Bay area. The species nested in Arcata in 1981 (Harris 1996). Beginning in 1984, intensive surveys of palms were initiated in Eureka and Arcata, and in that year two pairs were found in each city (A. Desch, S. Fitton). The following year, nine pairs were in Eureka and one was in Arcata. These birds produced a total of 15 offspring, the latest of which fledged 1 Aug 1985. A late summer total population of 40+ birds was estimated for that year (S. Fitton; AB 39:961). One of the Arcata nests was in a bigleaf maple (A. Desch). Despite comparatively few reports, it was nevertheless apparent that the species continued to occur in spring and summer in Eureka and Arcata in the late 1980s and early

1990s. No further observations were reported from Ferndale.

During the Atlas period, Hooded Orioles were encountered only in two adjacent blocks in Arcata and in another shortly to the east at Blue Lake. A pair brought off two broods of young in 1995 at the corner of A and Spring streets in Arcata (F15). They first appeared 11 Apr, the first fledglings were seen 7 Jun, the second brood was seen 21 Jul, adults were last observed 9 Aug, and the last fledgling was gone by 15 Aug 1995 (C. Walling). This pair or another was also present at the site in 1997, with the male first noticed 21 Mar (R. Novick, E. Lund). Also that year, a pair fledged three young on Spear Avenue in Arcata (f14), with the juveniles seen 17 Jun (RJA, R. Burt). A single oriole in the twin palms of the Arcata plaza 11 May 1997 (f14; R. Burt) was at a site used for nesting in 1987. In 1998 a pair was noted at 14th and I streets in Arcata (f14) 14-28 Apr, and two fledglings were subsequently reported near there (K. Nelson, S. Pagliughi, RJA, DFx). Nesting had also been confirmed at the latter site in 1993. Each of these places had stately palms and were among the better known Hooded Oriole sites in Arcata at the inception of the project. A single bird was also seen at 10th and G streets in Arcata (f14) 29 Jun 1999 (DFx). Rounding out the list of Atlas period records was a single oriole in the Blue Lake bottoms (h14) 25 Apr 1995 (M. Colwell).

The seeming absence of Hooded Orioles from Eureka during the Atlas period was puzzling, especially as that city had been an important center for them 10 years or more previously. No perceptible change in the number nor condition of Eureka's dozens of palms would seem to have taken place since oriole surveys were conducted there in the mid-1980s. At least one thorough check of the Eureka sites was conducted in 1997, with no evidence of Hooded Orioles found (A. Desch).

This species was unknown to early collectors and ornithologists working in Humboldt County; a century ago, the nearest populations lay far to the southeast in southern California. Those workers could scarcely know that, during the time of their own investigations, the future of the Hooded Oriole on the north coast was literally unfolding before their eyes. Palms of several species were planted on residential properties in Eureka, Arcata, and Ferndale during the late Victorian era (*e. g.*, the twin palms on the Arcata plaza date from about 1895). Though they grew more slowly than those enjoying warmer climates farther south, they prospered in a botanical zone friendly to plants prone to frost injury.

Surely helping to set the stage for the first appearance of Hooded Orioles here was a significant range expansion in the San Francisco Bay area, a phenomenon that was apparent from the 1930s onward. Originally confined to native fan palm groves and riparian woodlands with sycamores, these orioles began to exploit the numerous plantings of California and Mexican fan palms, Canary Island date palms, and other palms in cities of the Bay Area (Shuford 1993). At the time of the first Humboldt record in 1972, the nearest significant population of Hooded Orioles was in eastern Marin County. Today there are many birds in and around Redding, but the extent of that population was not understood until 1985 (AB 39:348). These orioles were considered rather rare as far north as Chico into the early 1970s (AB 26:903); it seems doubtful that birds initially prospecting in coastal Humboldt County would have originated from the Sacramento Valley. In any event, it was the fancy landscaping touch added to properties in the older residential sections of Eureka, Arcata, and Ferndale that played the most crucial role in the colonization of Humboldt County by Hooded Orioles.

Hooded Orioles are occasionally detected in the central Humboldt coastal lowlands some distance from suitable nesting sites. An example was a spring migrant that probably had just arrived at the Monterey cypress grove at the base of the north Humboldt Bay jetty 8 Jun 1999 (DFx). One at the Butcher Slough logpond at the Arcata Marsh 8 Aug 1997 (*fide* S. Harris) may have been a post-breeding dispersant.

BULLOCK'S ORIOLE *Icterus bullockii*

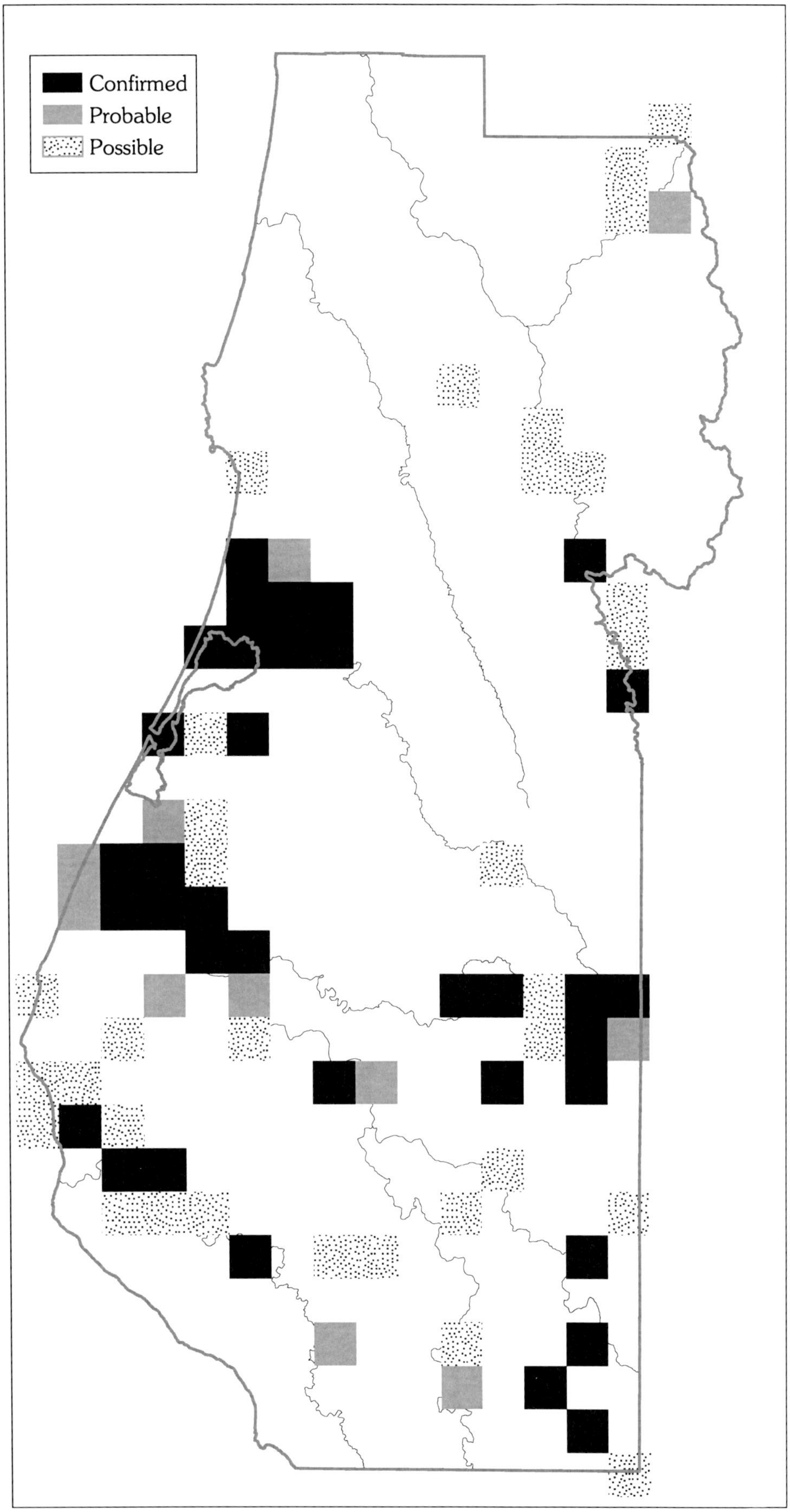

From late April to early August, stands of shade trees are given voice by the call-notes and piping songs of Bullock's Orioles. These birds of summer are more often heard than seen. Females and juveniles wear shades of muted gray and yellow above and whitish below. Deep leafy shadows of midday conceal the intense orange, black and white pattern of adult males. Even a frequent singer close at hand may require moments of patience and anticipation to pinpoint.

Throughout their range in temperate western North America, Bullock's Orioles are associated with stands of deciduous trees. Widely different settings are inhabited. Windbreaks of the Great Plains, gallery riparia in the arid Southwest, aspens in the Great Basin ranges, and coastal alders within easy reach of sea breezes support these orioles during the warmer months. Many smaller towns and cities with abundant shade trees also host them. They are routine migrants and breeding birds across much of California wherever their needs are met. Although nearly all of the population within the state departs for more southerly regions by November, small numbers remain in winter each year, chiefly along the coast from San Francisco Bay south, generally in and around plantings of exotic vegetation.

Harris (1996) characterized Bullock's Oriole in northwestern California as a "common migrant, summer resident, and breeder in residential areas, open woodland, and riparian areas throughout the region." Atlas results indicate that Bullock's Orioles occupy a somewhat restricted distribution in Humboldt County. They were detected in 19% of all blocks. While orioles are locally common during the breeding season, most of the actual land area of the county is unoccupied by them. This is underscored by the fact that atlasers generally found them to be concentrated along river corridors and floodplains and in lesser numbers elsewhere. This pattern of distribution was understood long before the project, and was supported by Atlas results. Nearly all detections of orioles were in blocks through which the principal streams flow, or in blocks comprising bottomland and towns around Humboldt Bay. Occurrence appeared quite heavily skewed to the southern one-third of the county, suggesting a greater abundance of suitable habitat there.

Corridors and patches of black cottonwoods along the banks of the Trinity, Mad, Eel, South Fork Eel, Van Duzen, and Mattole rivers attracted considerable numbers of these birds. In such centers of abundance, Bullock's Orioles can be quite numerous, with one or more frequently heard close by. Orioles were reported along the Klamath River only from near Orleans and Somes Bar. They were noted to extend into the foothills along shallow draws or at seeps and benches, particularly where Oregon ash or tall willows occupy sunny hollows. City parks, urban gulches, and older residential neighborhoods with appropriate exposed deciduous hardwoods in and about McKinleyville, Arcata, Eureka, Fortuna, and Ferndale supported these orioles. Red alders and ornamental birches have been noted as nest sites in such places. Pushing the habitat envelope were singing Bullock's Orioles in dry, open oak woodland with little more than grass and wildflowers beneath 28 May 1999 on the upper slopes of Island Mountain (o34; JEH) and 19 Jun 1998 above Salmon Creek (i29; DFx).

The extensive closed-canopy forests and incised streamcourses of most of central-interior and northern Humboldt County clearly do not offer Bullock's Orioles sufficient semi-open riparia. Clearly, these birds are out of place in the commercial timberland mosaic, in the redwood parks, in the oak-and-grassland country of the central Mad River, and anywhere at higher elevations.

Bullock's Orioles typically build a sac-like nest in the outer portions of boughs in the lower or middle canopy of hardwood trees. Its exact whereabouts may be guarded by the furtive approaches of adults, whose spirited chatter more often than not ceases at the nest. Eight active nests were reported; the earliest 19 May 1997 (c27; S. Petersen) and the latest (with young) 18 Jun 1996 (e22; RHw). As is the case elsewhere in the extensive range of this oriole, nests constructed along streamsides frequented by anglers are commonly built with much fishing line. Such nests are remarkably persistent. Their gaunt monofilament superstructures may remain in place for years, offering evidence of breeding attempts by birds that, in many cases, may no longer be alive. Old oriole nests (built with or without fishing line) were commonly observed in the expected range of the species in each of the 5 years of the Atlas; *e. g.*, numerous previously used nests in leafless trees 7 Jan 1996 served as the highest-attained breeding evidence at the Fortuna sewage ponds (e21; JEH).

Evidence other than completed nests was also frequently noted. The earliest confirmation involved nest building 25 Apr 1998 (g14; S. McVicor), and the first fledgling was seen 1 Jun 1996 (N33; C. Secor). Dependent fledglings appear to remain near the site with parents for some days after leaving the nest, allowing loose family groups to be readily noted. Not surprisingly, the latest breeding confirmation noted was fledged young 29 Jul 1998 (o16; TWL).

PURPLE FINCH *Carpodacus purpureus*

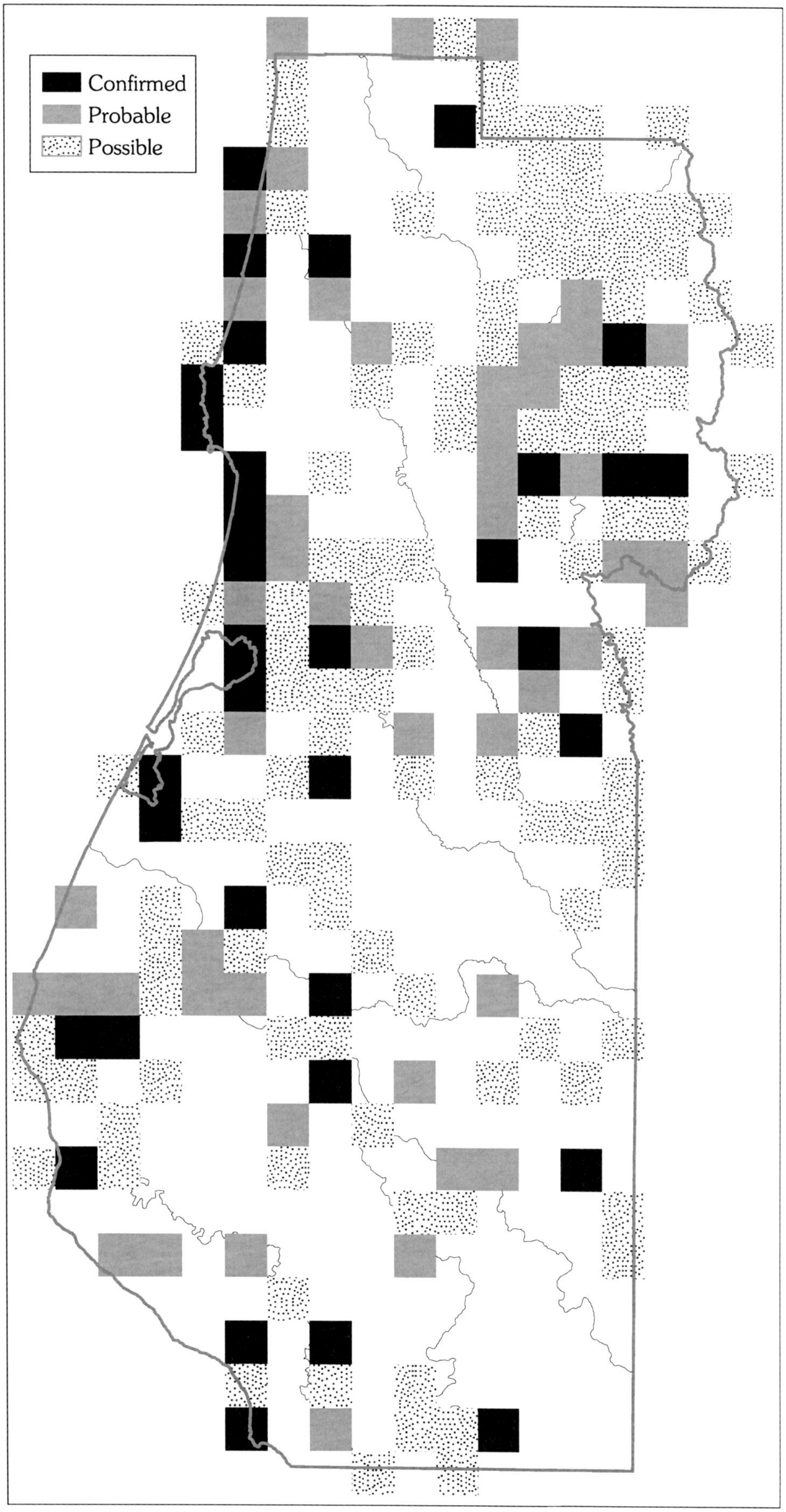

Seeming fond of fresh air and good scenery, male Purple Finches advertise their presence during the breeding season by singing from the tip-tops of tall conifers. At the extreme point of a towering coast redwood one may ultimately find the source of the exuberant outbursts: a diminutive darkish finch, remaining frozen to its perch for long minutes, delivering song after song as moments in the forest world unfold 300 feet below. When finally its energies must take it elsewhere, it launches quickly from the pulpit and undulates across intervening space, its tiny form nearly lost against the sky, yet betrayed from point to point by soft *pik* notes. Females are given to no similar displays, spending much time within tangles of midstory foliage or feeding silently on the ground. At all seasons, Purple Finches exhibit in many of their movements a vague hesitance, as if unsure what to do next. A measure of this vigilance is relaxed during trips paid to bird feeders, at which time they may shoulder right in among the goldfinches, House Finches, sparrows, and other seed-eating birds with which they compete for human handouts.

Purple Finches are widespread small songbirds, found chiefly in the North and West in summer but given to irregular southward incursions in fall; this is one of the classic "winter finches" of midcontinent. In the western U.S., they are largely restricted to the moist conifer forests and regrowth timberlands of the Pacific Slope. This species is found as a breeding bird in California in the Cascades, the western Sierra Nevada, the Klamath Mountains, the Coast Range south to Ventura County, and the mountains of southern California. In northwestern California, Harris (1996) described this species as a "common resident and breeder. Breeds in open and semi-open woodlands containing conifers. In non-breeding season also found in riparian and dune hollow deciduous cover." Over the 5 years of the Atlas project, Purple Finches were reported from 193 (45%) of the 425 total blocks.

This species was found to be quite broadly distributed in Humboldt during the breeding season, occurring nearly throughout the forested sections of the county. Recalling the maps of many of the county's forest birds, blocks with detections were most frequent in the northeastern mountains, along or near the northern coast, and scattered elsewhere in the interior. Stands inhabited by Purple Finches in spring and summer can scarcely be described without listing most of the forest types known in the county. As the map hints, no single tree community is especially favored. While the expansive cluster of blocks with detections in northeastern Humboldt may lead one to suspect that montane forests of true firs and pines may be preferred, such is not the case. This area takes in diverse forests, ranging from Douglas-fir/oak/madrone stands through the zone of mixed conifers, with comparatively few birds at the highest elevations. Redwood forest edge is commonly used, yet in the strip of blocks along the northern Humboldt coast these birds were more routinely encountered in cutover lands supporting red alder with scattered Sitka spruce, grand fir, and western hemlock. Northerly exposures shortly above incised streambottoms—dotted with bigleaf maples, often attractive to Purple Finches—were also found to support these birds.

Purple Finches were conspicuously absent from several large areas in the county during the 5 Atlas years. These places included land under intensive management for forest products in extreme northwestern Humboldt and inland from Trinidad, the rolling oak woodlands and grasslands of the central Mad River watershed, and many contiguous blocks of summer-parched territory in the interior southwestern and extreme southeastern parts of the county.

Breeding was confirmed in 34 blocks and was rated as probable in 49 others. Of the breeding confirmations, fledged young accounted for the greatest share of evidence, noted in 17 blocks on dates spanning 10 Jun 1997 (c24; COg) to 28 Jul 1996 (F21; GAS). Eleven of these reports fell during June. Second in importance were the nine observations of adults feeding young, ranging over nearly a 10-week period from 11 Jun 1996 (m11; JMH) to 22 Aug 1996 (m15; TWL). Purple Finches in the western portion of their range are known to be double-brooded (Ehrlich *et al.* 1988); the considerable span of dates among the sightings of young being fed suggests this took place. Reports pertaining to four other behaviors used to establish confirmed breeding were far fewer. Nest building was witnessed three times across a 5-day seasonal period: 24 May 1998 (o11; K. Bentler), 27 May 1995 (F31; KMS), and 28 May 1995 (f04; JEH). No other nests were reported. Greatly contrasting with most other passerines in this study was that only two instances of adults carrying food served as highest-ranking behavior, 18 Jun 1995 (b24; D. Ogden) and 13 Jul 1998 (H23; JEH). Adults carrying nest material were also seen but twice, 21 May 1996 (D19; KMS) and 4 Jun 1995 (F11; JEH). Finally, physiological evidence from a bird-in-hand supported one confirmation 6 Jul 1995 (f16; *fide* L. George).

Although these birds are generally thought to be resident within their breeding range in northwestern California, flocks of from several to a dozen Purple Finches typically visit riverbottom, agricultural edge, urban areas and other non-breeding habitats in the colder months. Stringers and thickets of red or white alder, swales of Oregon ash, and rambling mounds of Himalayan blackberry all attract a fair share of Purple Finches in winter. Thus, the appearance of small numbers of them outside breeding habitat from September to April suggests some sort of migration or dispersal at that time of year, possibly in the form of a downslope movement as winter conditions take hold in the higher mountains. Even if some birds shift to coastal areas in winter, numbers at lower elevations in the interior mountains seem to be slightly higher than on the coast, as evidenced by CBC numbers: during the 1995–1999 Atlas period, the Willow Creek CBC recorded 10–73 birds, while the Centerville Beach to King Salmon and Arcata CBCs recorded 1–37 and 2–14 birds, respectively.

CASSIN'S FINCH *Carpodacus cassinii*

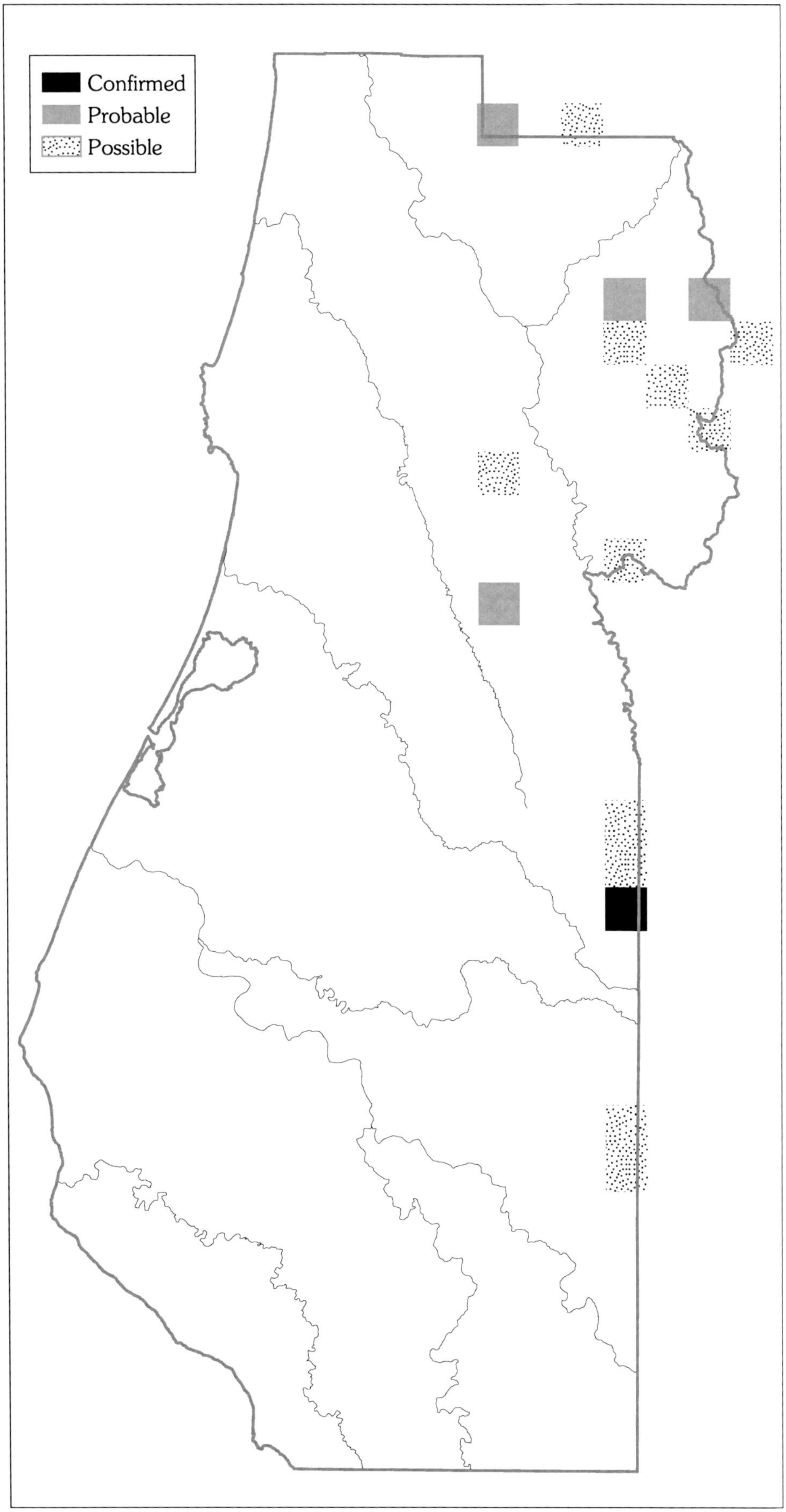

The genus *Carpodacus* comprises 21 species of small to medium-sized finches with variably reddish plumage. Most of these birds are Asiatic. Many have restricted ranges and are rare or otherwise obscure. Others are widespread and well known. The three North American species—Purple, Cassin's, and House finches—each enjoy a broad distribution. Of these three, only Cassin's is limited to the West. Branding it as characteristically Western as a Stetson hat, its range closely parallels that of ponderosa pine and, locally, Jeffrey pine.

Cassin's Finches occur widely in the higher mountains nearly throughout California. The range includes the higher forested ridges and slopes of the Klamath Mountains. The western frontier of their range laps into eastern Humboldt County. These finches also occur in the highest portions of the northernmost Coast Ranges. Cassin's Finches were detected during each of the 5 survey years in very small numbers, and were reported from 16 (4%) of the 425 blocks. Breeding was confirmed in one block on South Fork Mountain.

During Atlas fieldwork, Cassin's Finches were found in higher-elevation conifer forests on the upper slopes of the taller main dividing ridges such as Salmon Mountain and South Fork Mountain. Like other montane species in Humboldt County, the birds occupy several areas, each separated by intervening deep river courses and corridors of unsuitable lower-elevation habitat. Although it might seem that these represent island populations of sorts, the places where Cassin's Finches were reported in Humboldt are actually more or less adjacent to the population in southeastern Del Norte and western Siskiyou and Trinity counties.

North of the Klamath River, Cassin's Finches were detected in two blocks, with countersinging males noted at Onion Mountain (I03) 8 Jul 1999

(GCH). Among seven blocks with detections between the Klamath and Trinity rivers, probable breeding was reported in two, with pairs seen shortly northeast of the Hoopa Valley Indian Reservation (o07) 5 May 1997 (JMH) and near Whiteys Peak (q07) 16 Jul 1997 (JEH). Possible breeding was reported from near Sugar Pine Mountain (L11), between Salyer and Waterman Ridge (o13), and in two adjacent blocks in the Lassics.

On the South Fork Mountain ridge, a pair at Horse Mountain (l14) 28 Jun 1998 (M. Wachs) was complemented by birds in three adjacent blocks from Last Chance Ridge (o19) south to Blake Mountain (o21). Possible breeding was reported from two of those blocks, but in a third, an adult carrying food at the summit of Blake Mountain (o21) 6 Jul 1995 (GAS) furnished the only confirmed breeding record for the Atlas period. Confirming this known local breeder had been on the atlasers' "to-do" list from the start of the project. This target was accomplished in Atlas Year One, but as it happened, not again.

Habitat over much of the range of Cassin's Finch is usually mature stands of montane conifers, with ponderosa (or Jeffrey) pine or red fir the most common trees in much of the California range. The mix of tree species is variable. As many as seven or eight conifer species may provide suitable habitat at sites within the "Klamath knot"—the region wherein unique Klamath Mountain conifers and tree species of typically Cascadian and Sierran affiliations meet and mix. Unmanaged forests above 4000 feet within Humboldt County offer suitable tree species. Two general forest types in which Cassin's Finches are detected in Humboldt are Jeffrey pine, alone or with sugar pine and Shasta red fir, either by itself or with minor amounts of other species such as white fir and incense-cedar. Many of the forest communities in which Cassin's Finches live are found in parts of Humboldt County with ultramafic geologic parent material (T. Leskiw, pers. comm.). Douglas-fir is only a minor component of stands of trees where Cassin's Finches live in Humboldt, though it is almost universally present.

Mountain forests inhabited by Cassin's Finch in Humboldt County occupy only a small part of the county's total land area. Some of this habitat is protected, but much is not. It is possible that logging has reduced the available habitat; however, no real study of this species and its requirements has been undertaken locally. Almost nothing is known of the status of Cassin's Finches in Humboldt outside the breeding season. The places they live in summer are difficult or impossible to get to during winter, when nearly all higher roads lie beneath snow or are seasonally closed. The earliest and latest Atlas reports of birds in suitable breeding habitat were 28 Apr 1997 (o08; JMH) and 18 Aug 1996 (q10; JMH), respectively. Reports of these birds in the county anywhere outside the pattern of normal occurrence are most unusual.

Harris (1996) summarized Cassin's Finch in northwestern California as a "rare resident and breeder in coniferous forests of the higher inland mountains." He listed the species as having occurred at six locations in Humboldt County, all within the area in which Atlas detections were noted. The Atlas effort suggested that Cassin's Finches are either less abundant, or less readily detected, than are most of Humboldt's other "mountain specialties".

HOUSE FINCH *Carpodacus mexicanus*

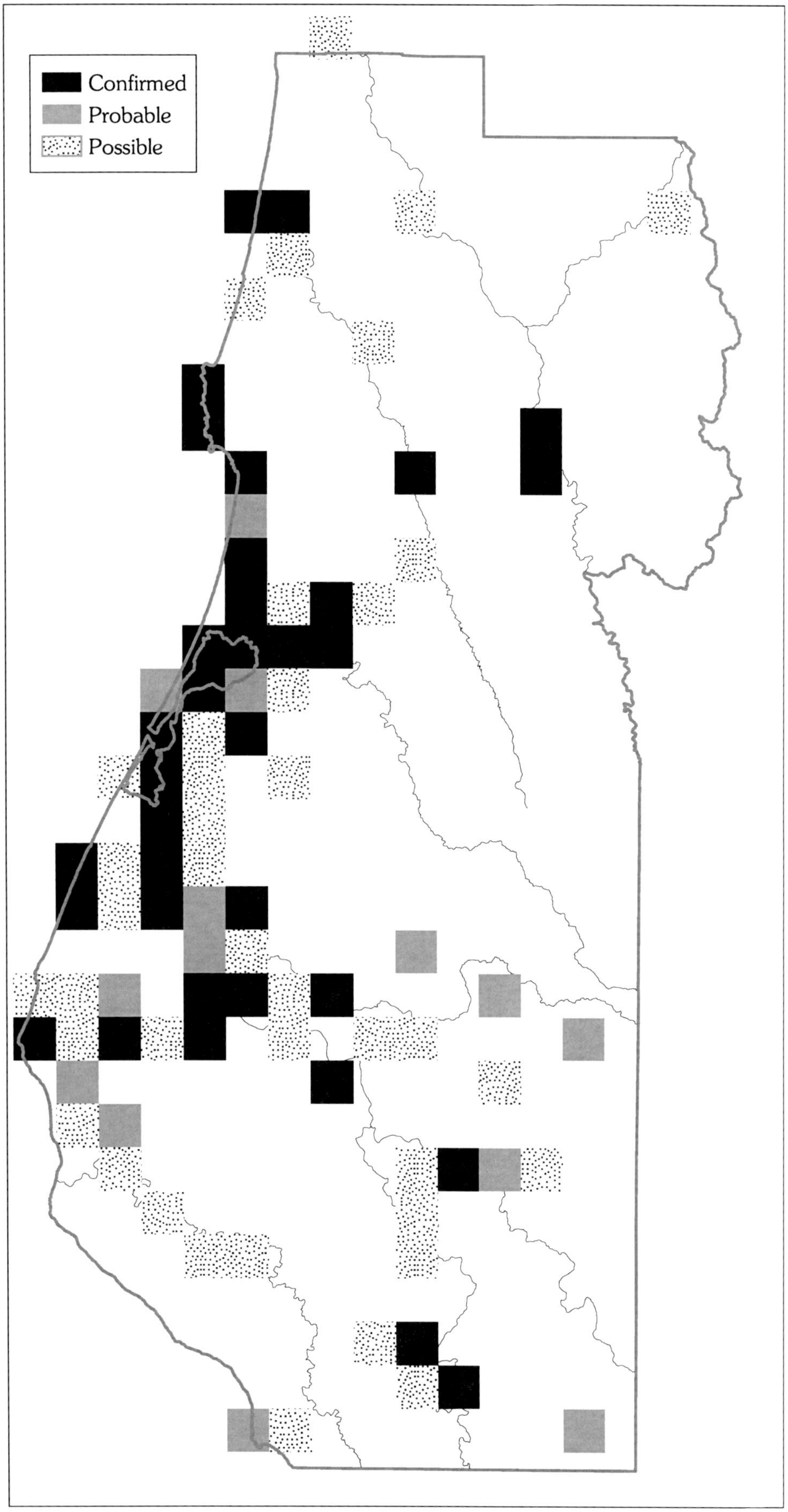

Valleys, canyons, and lower foothill slopes of the arid West probably were the ancestral haunts of House Finches. Today, however, these birds are found from the Pacific to Atlantic coasts and nearly everywhere in between at lower elevations. Ironically, they have become so abundant in suburbs of the Northeast that recent declines in House Sparrows have been attributed to them (Ehrlich *et al.* 1988). One might ask: what caused House Finches to spread eastward in the last half-century? That question cannot be answered, because House Finches actually invaded *from* the east. Owners of cage-bird shops on Long Island that had been selling these North American natives let their illegal charges fly off on the sly in 1940 to avoid prosecution. Finding the Empire State to their liking, the birds reproduced and so began an inexorable if unlikely march westward. Spreading decade by decade, House Finches claimed one eastern state and southern Canadian province after another, their incremental advance well recorded by backyard-feeder watchers. Within 50 years, the final gaps separating the indigenous western population from eastern birds on the move had become filled in.

Harris' (1996) description of the House Finch in northwestern California as a "common resident and breeder in open areas, mostly at lower elevations, especially in the vicinity of towns, farms, and ranches" captures the essence of this species' status in Humboldt County in far fewer words than are expended below. House Finches are present in Humboldt County throughout the year, but they are common only in a rather circumscribed region, outside of which their occurrence is spotty and the birds are considerably less abundant. Over much of the county the species is essentially unknown. During the Atlas period, these birds were found in 89 (21%) of the 425 total blocks.

The distribution of House Finches during the 5 years of the Atlas study revealed that an array of habitats was used. Falling short of describing a bimodal theme, nevertheless the two general endpoints along this spectrum seemed most important to the birds. One element of this was the species' presence in and around cities, towns, and even tiny settlements. House Finches are, with Violet-green Swallows and White-crowned Sparrows, on a brief list of native songbirds that penetrate as breeding species to the gritty urban core. A long-neglected ornamental juniper shrugging in defeat against the side of an abandoned building may provide a pair of House Finches just what they want in the way of nest cover. At home in any congested cityscape, the birds would seem to continually mistake power poles for trees, wires for greenery. In 1995 at the Blue Lake Fish Hatchery (H15), these birds nested in a metal pump tower and in a Barn Swallow nest on top of a light (C. Gutiérrez). With the single exception of Willow Creek, all of the county's incorporated towns and a number of smaller communities were known to support House Finches during the course of the project.

The other element of this species' general habitat in Humboldt during the nesting months is quite different. Open countryside with only scattered, or even no, human constructions in highly rural sections of Humboldt also hosted House Finches. Rangeland and other grassy, treeless, or tree-dotted settings in which only few people reside attracted them, as in Redwood Valley (J11) and up the Little Van Duzen River (n24). An isolated detection came from the vicinity of Pecwan (J05) along the lower Klamath River and another from shortly west of the Klamath some miles downstream (h01). They were also found in essentially unpeopled locations, such as along Monument Ridge (e23) and Bear River Ridge (c23); near the junction of Redwood House Road and Kneeland-Bridgeville Road (j22); near Cape Mendocino (a23, a24), and up lower Davis Creek (B25). In these places they seem to get by with scraps of riparian habitat or clumps of brush with occasional trees, set in expanses of grass. Typically, some source of fresh water is available.

Spanning these urban and pastoral extremes were situations in which people live at low density yet effect a definite human impact on the land. Examples were House Finches reported from southwest of Orick (g06), near Dry Lagoon (F07), at Patricks Point (e09), south of Blue Lake (H15), southwest of Kneeland (g18), near Harris (N33), and north of Fort Seward (m27). They were also reported from a number of contiguous blocks along the lower Mattole River. They also consistently used dairies from the lower Mad River to Fernbridge.

Large areas of Humboldt provided nothing but unsuitable habitat for these finches. Landscapes dominated by trees and with small openings clearly did not offer what the species needs. House Finches were reported in none of the truly mountainous or forested sections of the county. Indicating this well was the absence of reports from anywhere on the Hoopa Valley Indian Reservation other than at and around the town of Hoopa (m10, m11).

Confirmed breeding was reported from 35 blocks and probable breeding from 14 others. Probably because of their commonness around human dwellings, the proportion of detection-blocks with breeding confirmations was quite high: 39%, compared to 18% for the more reclusive Purple Finch. Among the 35 blocks in which breeding was confirmed, two behavioral codes—fledged young, and adults feeding young—were most prevalent, together constituting 80% of all confirmations for the species. Fledged young were the best evidence of nesting in 15 blocks, on dates ranging from 27 Apr 1995 (h14; C. Keil) to 26 Jul 1996 (k32; JEH). Six of these fell during June, eight in July, and none in May. Young being fed were seen in 13 blocks, from 15 May 1995 (d18; P. Spruance) to 6 Aug 1995 (F15; C. Meyers). Direct breeding evidence of five other types was reported, none more than twice. Most noteworthy among these records were a nest with eggs 27 Jun 1996 (D19; K. Foerster) and a nest with young 19 Jul 1995 (H15; PFS). Considered altogether, these data make it plain that House Finches made two, perhaps even three, nesting attempts per season during the 5-year run of the Atlas project.

After breeding, House Finches assemble in flocks of from a half-dozen to several dozen birds. "Swarms" of 100+ birds have been noted in late fall and winter at favored feeding sites. They concentrate noticeably in rank streambottom habitats thick with tall forbs such as fennel, radish, wild carrot, and poison hemlock. They appear impartial to the thistle seed so heavily eaten by American Goldfinches. Himalayan blackberry is a particularly important food during the colder months; the shrub itself continues to be used as escape cover even after every last hard, underripe berry has been sought out and consumed. A subtle increase in use of coastal locations takes place during the colder months as well. During the Atlas period, flocks were commonly seen in bush lupine-dominated portions of the Humboldt Bay sandspits and in weedy pastures throughout the Eel River delta.

RED CROSSBILL *Loxia curvirostra*

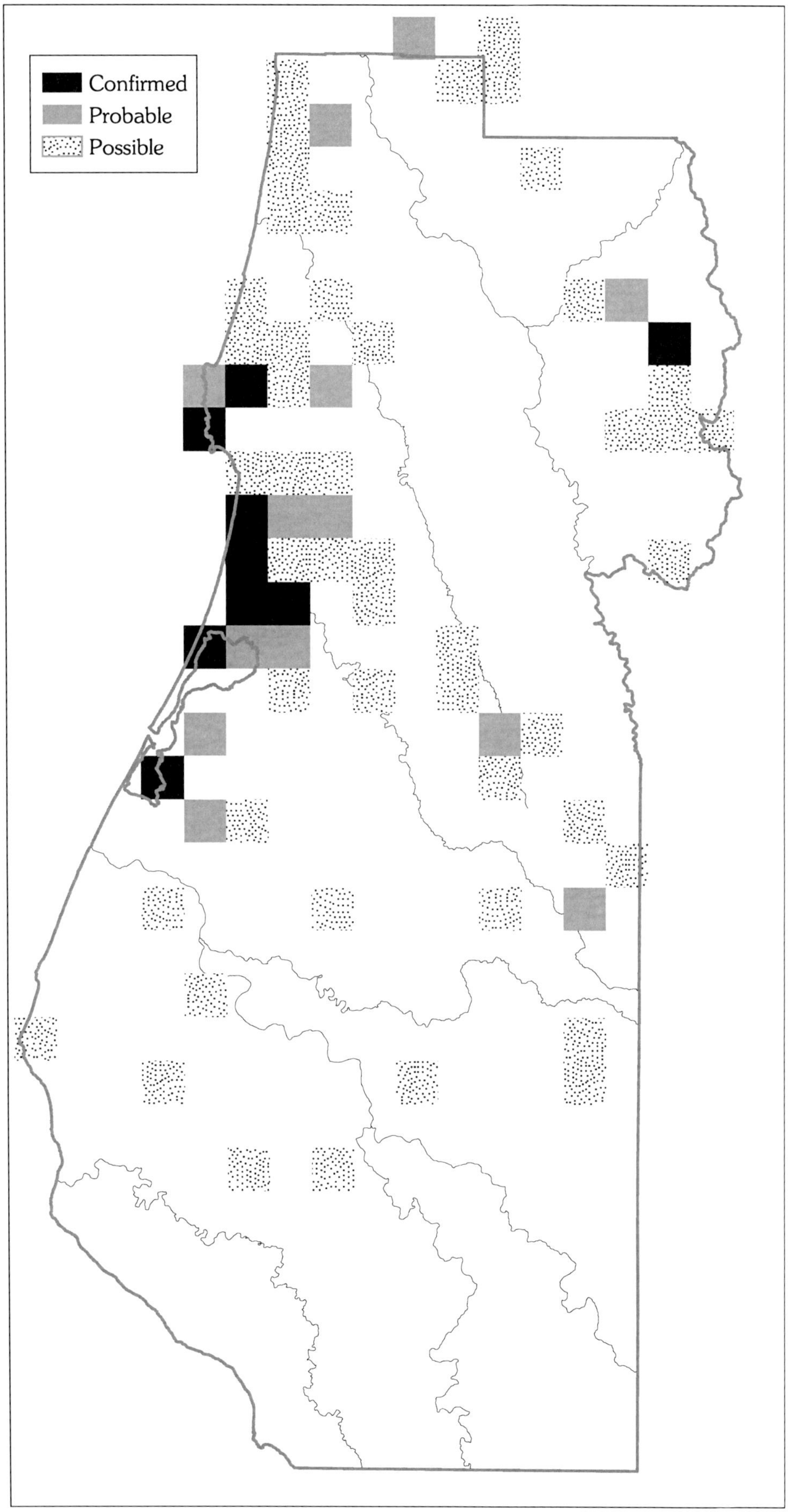

Red Crossbills are strange and enigmatic birds of conifer forests throughout the Northern Hemisphere. They possess mandibles with uniquely crossed tips, an adaptation enabling the birds to grasp and remove the ripening seeds of conifers from within the confines of their enclosing scales. These birds are nomadic, appearing in a given region in close accordance with the availability of maturing conifer seed crops. Although the Red Crossbill has been long recognized as a bird of Humboldt County, little beyond its' mere presence and absence is presently understood. This has remained so despite the fact that these birds are fairly often seen or heard. While they typically stay high in treetops and are not easily viewed, birders familiar with their telltale call-notes detect them with sufficient frequency to know the species is probably present at one place or another within the county throughout the year. It is worth noting that most of the reports of crossbills in the course of the Atlas project were made by the more experienced birders in the community.

The Red Crossbill is polytypic (Groth 1988, 1993). As currently perceived, the species in North America comprises eight closely related forms which may or may not occur together at the same time. The species of trees present and the maturity of seeds largely determine which type or types of crossbills are present in a forest. Bill morphology and call-notes differ among these forms, providing a challenge for birders bold enough to attempt to separate them. Commonly referred to as "types", and granted simple numerical designation, different forms of Red Crossbill are known to prefer seeds of particular conifer groups; *i. e.*, spruces, hemlocks, firs, pines, and others. Five types have been identified as occurring in California (DeBenedictis 1995). The separation, indeed the very definition, of Red Crossbill types is still under investigation. Because of this, the taxonomy of the superspecies will probably remain a matter of contention well into the future. It is possible that some forms, or groups of like forms, may prove to constitute distinct species. Certainly, Atlas surveys were poorly suited for assignment of Red Crossbills to type and, with one exception noted below, none of the detections reported during the Atlas period were assigned to one type or another. On one occasion, two unseen individuals uttered call-notes recognized by the observer as those commonly given by "large-billed" birds he had seen previously in the southern Oregon Cascades. This took place 29 Jun 1999 in forest containing six species of conifers at Snow Camp Mountain (L17; DFx). Davis (1933) observed and collected specimens of *L. c. sitkensis* and *L. c. bendirei* near Samoa during January–February 1923.

With 18 species of conifers native to Humboldt County and many of those widespread, it would seem on the surface that Red Crossbills of one type or another might be expected nearly anywhere. Results of Atlas fieldwork did not support such a sweeping supposition. Yet the reports received established that, during the 5 Atlas years, these birds were present in scattered locations across much of the county. This species was best represented in northern Humboldt, especially near the coast, and only a few blocks in the southern third of the county had detections. Further, extreme southern Humboldt had no detections. All told, Red Crossbills were encountered in 70 (16%) of 425 total blocks.

Slightly more than one-half of the blocks with reports of Red Crossbills were within forests or regrowth timber in the fog zone, in stands generally dominated by a variable mixture of coast redwood, Sitka spruce, grand fir, Douglas-fir, western hemlock, and token amounts of western red cedar and other conifers. These blocks were loosely clustered from Redwood National Park (g02) south to Ferndale (D21) and inland to lower Redwood Creek (i08, J11), Maple Creek (J17), Yager Creek (H21), and Grizzly Creek Redwoods State Park (J23). At the base of the Samoa Peninsula (e15) stands some of the small amount of shore pine forest extant in the county. Harris (1996) stated that small numbers of crossbills were resident and breeding in this location.

Away from that region, detections fell in two weakly defined clusters of detection-blocks roughly corresponding to mountainous country in northeastern Humboldt and along the Snow Camp Mountain and South Fork Mountain ridges, and in fewer than 10 scattered blocks extending from near Cape Mendocino randomly eastward into the Little Van Duzen River drainage. Habitat in the mountains contained several species of montane conifers as well as the nearly ubiquitous Douglas-fir. Above the North Fork Mattole River southeast of Pig Toe (D25) the birds were in mature Douglas-fir well mixed with oaks and other hardwoods.

Breeding was confirmed in nine blocks and was deemed probable in 13. Sightings of fledged young provided evidence in seven of the nine blocks in which breeding was confirmed. These reports were made 10 Jun 1999 (e10; TWL), 17 Jun 1995 on Humboldt Hill (d18; P. Spruance), 12 Jul 1998 (f12; GAS), 27 Jul 1997 on Azalea Hill (F13; G. Lester), 28 Jul 1997 at Glendale (g14; L. Parkinson), 3 Aug 1997 at Manila (e15; GCH), and on an unstated date in 1995 (f14; P. Stebbins). The carrying of nest material was also reported on Azalea Hill (F13) 21 Jun 1997 (G. Lester). Adults were seen feeding young 12 Jun 1999 (F09; GAS) and 31 Jul 1996 on Packsaddle Ridge (p08; JEH). Pairs of adults supported probable breeding in every one of the 13 blocks gaining that status.

Harris (1996) termed Red Crossbill an "uncommon resident and breeder; sporadically common migrant and winter resident" in northwestern California. Harris (1996) further noted that flocks of from a few to hundreds of birds invaded the region at unpredictable intervals. Yocom and Harris (1975) stated that the most recent large invasion was in the winter of 1969–1970, when 200–300 were reported from Humboldt and Del Norte counties 21–23 Feb 1970 (AFN 24:537). Serving as a very crude index to yearly abundance, the highest evidence obtained in blocks during the Atlas years 1995–1999 numbered 8, 18, 18, 8, and 17 reports, respectively, for a total of 69 (a record of possible breeding in F19 lacked a specific year). Thus, no great swing from near absence to "invasion" levels was apparent during the course of the project.

Not much more about crossbills can be learned from Atlas results. There are so many aspects of this species' life history in Humboldt County that remain mysterious that those interested in the species here are left with little more than a map painted with a broad brush. Efforts to learn more about crossbills in Humboldt will require a far more focused approach.

PINE SISKIN *Carduelis pinus*

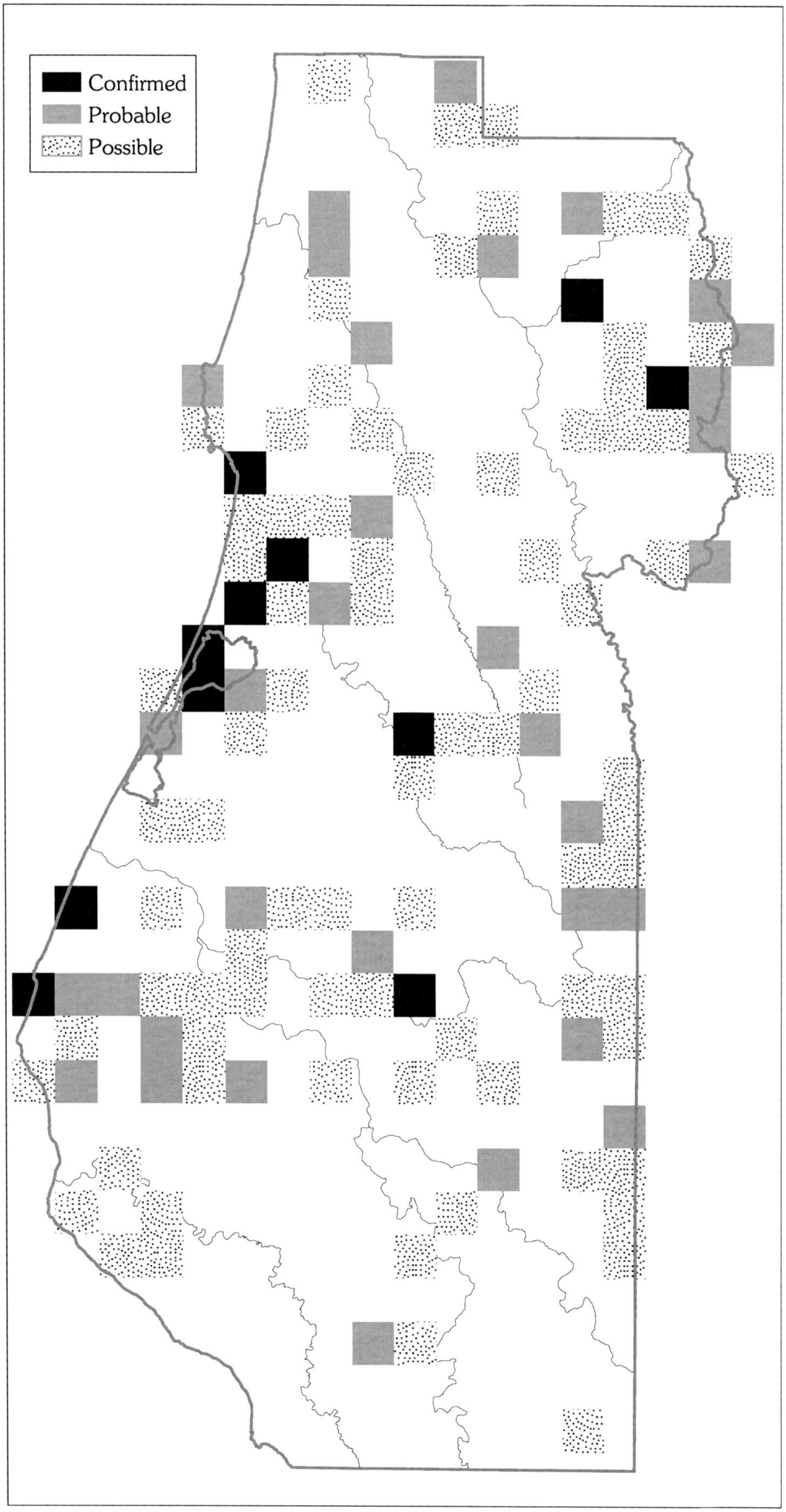

Dressed in drab and streaky plumage and usually holed up in the backwoods, the Pine Siskin is the hillbilly goldfinch. It shows its close relationship with others in the genus *Carduelis* through its notched, narrow-based tail; subtle yellowish patches; its voice; and by a strong tendency to occur in flocks when not nesting. Pine Siskins are known to many for their irregular appearance in regions where they do not breed. Particularly east of the Rockies, big invasions of siskins become the stuff of legend among people who maintain feeding stations stocked with sunflower or thistle seed. These events may involve the persistence for months of countless thousands of birds, with scores or hundreds at a time visiting yards offering goodies for the taking. Swarming over the top and sides of a feeder, they waste no time in adroitly shelling the seeds; a 10-pound sack may be emptied in a matter of days. No feeder can at one time support the droves of birds involved in the larger incursions, and so dozens await their turn gathered at perches surrounding the largesse. Continually engaged in close-contact bluffs and tussles while grabbing food—and stepping unavoidably in each other's feces—they are prone to contract and spread salmonellosis, also known as "slow siskin disease". Untended house cats, skilled at capturing healthy siskins, take an even greater share of sickened birds. These problems can be prevented by washing feeding stations weekly with a mild bleach solution and keeping domestic cats indoors.

The Pine Siskin as a breeding species of California is widespread, inhabiting much the same range as other forest birds. It is found widely in the mountains and wooded foothills of northwestern California during the nesting season, and is no stranger in the lowlands at other times. Harris (1996) termed the Pine Siskin in northwestern California a "common resident and breeder in coniferous forests, mixed woodlands, and riparian areas." Although siskins are present here throughout the year, as elsewhere, their abundance waxes and wanes on no apparent schedule. Because Pine Siskins are expected across much of forested Humboldt, local population fluctuations may be overlooked.

During the Atlas period they were detected in 123 (29%) of the 425 total blocks and were found in many sections of Humboldt County. The map shows no apparent clustering, but it does indicate that, in some portions of the county, these birds were scarce or absent in 1995–1999. Those areas within the county in which detections were most common were the timberlands northeast of Humboldt Bay; the mountainous country of northeastern Humboldt; varied lands along the eastern edge of the county taking in the South Fork Mountain ridge, the Van Duzen River corridor, and the Lassics uplift; and a broad, diffuse band of blocks extending inland several dozen miles from Cape Mendocino that escapes meaningful description.

In a vague sense, siskin occurrence tended to fall within the coast redwood and Douglas-fir/tanoak forest zones. Although many mid- and higher-elevation parts of Humboldt County supported these birds, many blocks with detections lay at low elevations. Seven of the 11 blocks with confirmed breeding were within a few hundred feet of sea level. This makes clear that Pine Siskins are not a "mountain specialty" in Humboldt. No association with streamcourses was apparent.

Portions of Humboldt County in which no birds were reported included the extreme north coast; the west side of the Hoopa Valley Indian Reservation and most of Redwood Valley; managed timberlands southwest of Kneeland; the extensive oak-grassland country of the central Mad River watershed; and most of the southern one-quarter of the county.

Local nesting habitat for Pine Siskins typically consists of broken forest, forest edge, or clumps of conifers adjacent to openings grown to preferred forage plants. Some degree of openness seems important to them; they cannot be described as birds of the forest interior. Clearcuts or overgrazed pastures containing much thistle attract them. While flowering California black oaks sometimes draw large numbers in April and May, it is unclear if this attraction in itself spurs the birds to breed in oak-rich woodland or forest edge. Ralph *et al.* (1991) found them to be negatively associated with the abundance of hardwood trees, and positively associated with the abundance of conifers. As shown by the map, they were found in summer from the coast to the highest elevations in the county. Wooded edges of McKinleyville, Arcata, and Eureka hosted at least a few breeding pairs.

Confirmed breeding was reported from 11 blocks, probable breeding from an additional 33. Siskins were confirmed breeding during the study in a comparatively low 9% of blocks with detections. Taking to extremes a theme expressed often in these pages, fledged young and observations of adults feeding young accounted for all but one of the confirmations. Fledged young were reported in six blocks on dates ranging from 18 Jun 1998 (J23; L. Wu) to 28 Jul 1996 (B21; COg). All but one of these occurred in July. Adults feeding young were noted in four blocks on dates describing a 2-week period from 21 Jun 1997 (a23; TWL) to 3 Jul 1995 (F11; JEH). The remaining confirmation was supported by the only nest found, one with young discovered 18 May 1998 near the Redwood Sciences Lab (f14; observer unknown). Harris (1991) stated that the Humboldt State University and Eureka High School collections contained 25 egg sets taken locally from April to August, thus extending the breeding season beyond what was made evident by Atlas survey results. Among 33 blocks with probable breeding status, 23 involved pairs of birds. This is not surprising, as it seemed as if two siskins flying over together were nearly as likely as single birds.

Although they are reasonably numerous in appropriate habitat in midsummer, Pine Siskins are most often remarked upon in fall, winter, and early spring. At those times they are assembled in flocks numbering from a few individuals to hundreds of birds. They may be encountered nearly anywhere during the colder months, even well out into the near-coastal lowlands, where they loosely accompany goldfinches and other seed-eating birds at food sources. In some winters they are abundant; in others, they are notable only for their scarcity. Status in the higher mountains in winter is poorly known because of travel limitations enforced by snow. The cutover slopes of the northern coastal foothills and riparian corridors inland support Pine Siskins when the seeds of alders are available. Few birds make such great use of alders as do siskins when the trees hang heavily with ripe catkins. So intent are they upon their silent feeding that a flock of several hundred siskins may remain unknown to the nearby observer until a disturbance sends them swinging skyward in a synchronized froth of tightly packed birds, the silence of the woods now broken by excited voices, then once again restored as the group settles quickly back into the leafless trees.

LESSER GOLDFINCH *Carduelis psaltria*

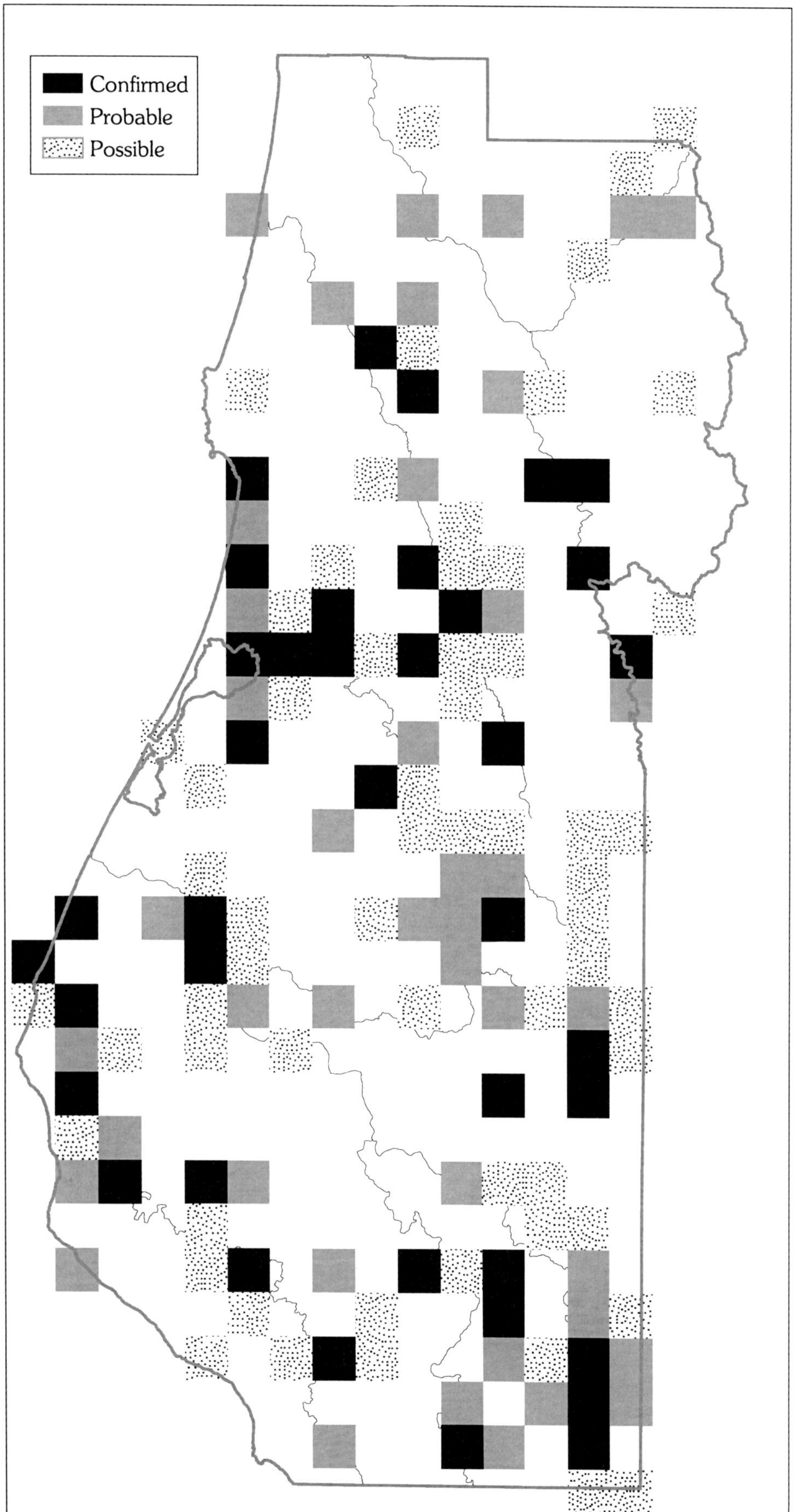

Effervescence scarce-contained within a wee package, the Lesser Goldfinch is among the most attractive of Humboldt County birds. They are seemingly at the center of the show wherever seed-eating birds flock at food sources. This is a species characteristic of rolling terrain at lower elevations in which tree cover, if present, is sparse or discontinuous; forbs such as wild teasel, fennel, thistle, and the like are abundant; and water for drinking and bathing is to be found. During the Atlas period, Lesser Goldfinches were encountered in one-third of all blocks.

In northwestern California, Lesser Goldfinches have been considered common summer residents and breeders, and uncommon in winter (Harris 1996). The Atlas map might suggest a more-or-less random breeding distribution, but the clusters of blocks with detections correspond closely to the actual "braided mosaic" in which this species appears to occur on the Humboldt County landscape. The special attraction exhibited by Lesser Goldfinches for small farms, weedy pastures, and fencerows in the bottomlands—as well as their consistent presence in towns and along roadways—contributed to the dispersed pattern of detections during the Atlas period. The courses of the Mattole, Mad, Trinity, and Eel rivers are crudely approximated. Abundant habitat exists in the hot, open countryside of Redwood Valley, the oak-dotted central Mad River foothills, and many blocks within the upper Van Duzen River drainage. Lesser Goldfinches were especially common during the study period in extreme southeastern Humboldt County, where excellent habitat is more or less continuous.

Lesser Goldfinches were regular in summer during the Atlas period in Arcata and Eureka, but not along the north coast (except one block), in the west Arcata Bottoms, in the lower Eel River delta, nor on the immediate coast south of Cape Mendocino. Other areas within the county in which these birds were encountered either infrequently or not at all included the mountains of northeastern Humboldt, much of South Fork Mountain, the Lassics, extensive parcels of commercial timberland, and Humboldt Redwoods State Park.

Somewhat tree-poor, summer-hot, and "wildly weedy" conditions typify preferred sites across the county. Because they prefer the sparser, sun-tolerant vegetation prevalent across southerly exposures, the birds are most numerous where this condition exists. Early seral brushfields in the regenerating clearcuts scattered throughout commercial timberland appeared to be lightly used by these goldfinches during the Atlas period. Lesser Goldfinches make good use of riparian corridors, "suburban woodland", parks, orchards, old mill sites, oak groves, and pasture edges. They appear to occur upslope to about the limit of oak-grassland vegetation; this is not a species to be expected in the high mountain brushfields.

A notable 57% of the 142 blocks with detections were of either confirmed or probable breeding. This species probably bred in nearly all of the blocks in which it was detected. Lesser Goldfinch nests are not difficult to locate, as they are typically placed near the tip of a lower bough of a hardwood or conifer, generally on the side of the tree facing the largest nearby opening in the vegetation. Five active nests were located. Breeding confirmations spanned the rather broad dates of 16 Apr 1997 (nest building; GCH) to 9 Sep 1998 (nest with young; PFS). Interestingly, both of these extreme dates were from the same block (H15). The earliest fledgling reported was seen 13 Jun 1999 (e27; DFx).

These tiny birds occur in pairs from spring into summer, with family groups becoming increasingly subsumed into little gangs of goldfinches that gather at concentrated food sources during the balance of the year. From two to six birds are most commonly met with. Ornamental birches and thistle-seed feeders may attract as many as several dozen, especially in winter. Lesser Goldfinches mix to some extent with Pine Siskins but considerably less with American Goldfinches.

LAWRENCE'S GOLDFINCH *Carduelis lawrencei*

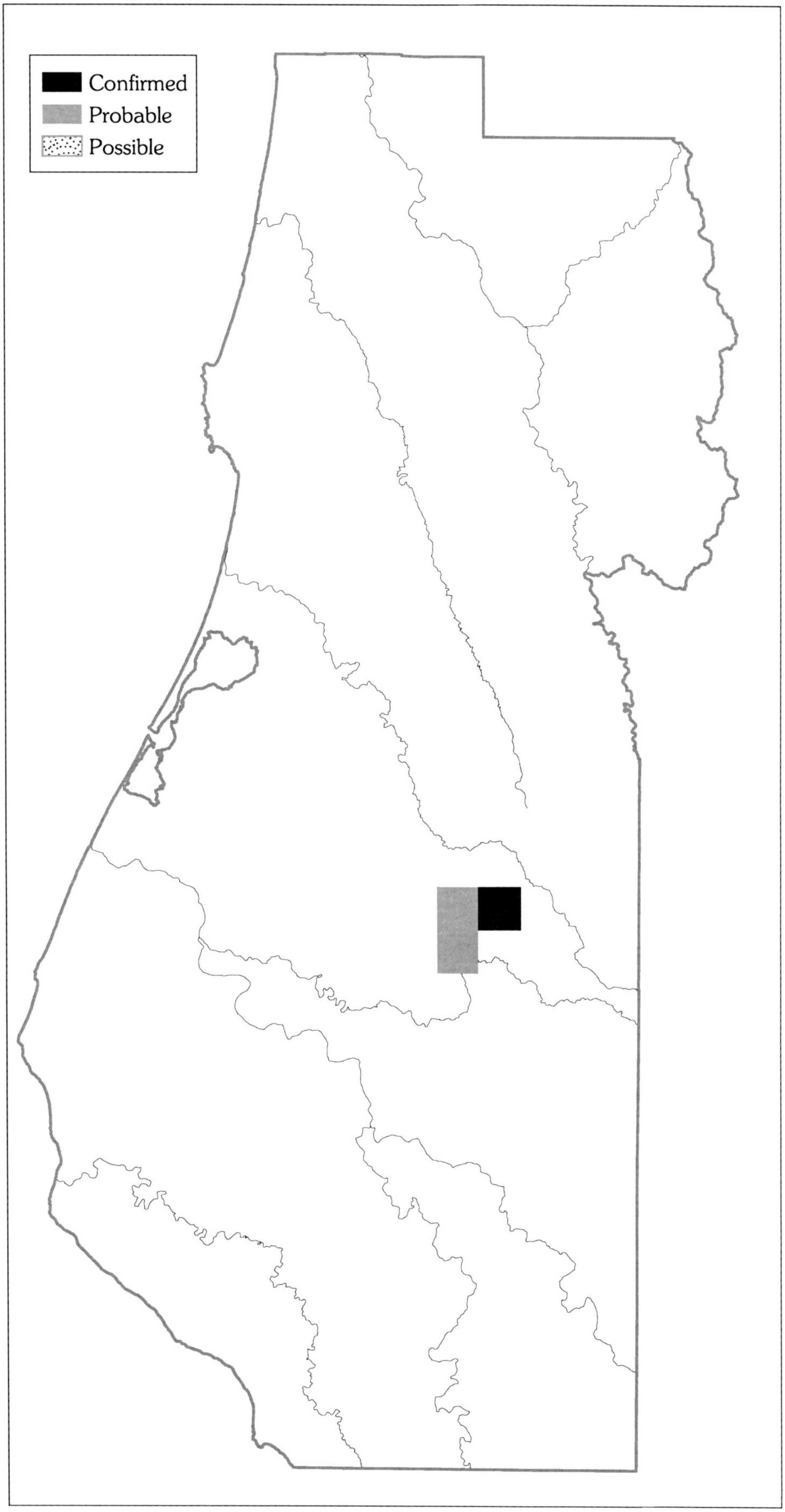

Lawrence's Goldfinch is a petite songbird of restricted range, lovely coloration, and remarkable vocal prowess. The breeding season finds nearly the entire world population confined to arid oak-grassland savannah and broken chaparral of California and northern Baja California. Irregular movements in fall and winter take variable numbers southward into the Republic of Mexico and eastward into Arizona, New Mexico, and, rarely, as far as west Texas. A narrow, vertically aligned black mask, yellow breast and wing markings, and gray upperparts mark the male; the female is known chiefly by her yellowish wingbars. Like the Lesser Goldfinch, Lawrence's routinely incorporates the voices of other species into its songs, but its efforts surpass those of the former species (Remsen *et al.* 1982). It may be the second-most-accomplished vocal mimic among North American birds, its virtuosity exceeded only by the Northern Mockingbird.

Given a state map of any size, a knowledgeable birder would not attempt to outline the range of Lawrence's Goldfinch with an indelible marker. The birds are well recognized for erratic shifts in presence and abundance, breeding commonly in some areas over several years, then abandoning them in favor of other places. Perhaps these shifts are in response to varying wet-season conditions, which either encourage or preclude growth of fiddleneck and other food plants upon which they may depend during the following spring and summer (Burridge 1995). The occurrence of these goldfinches in northern California west of the Sacramento Valley foothills is irregular. During the course of the 1986–1991 Sonoma County Breeding Bird Atlas project, they were detected in only five out of 195 blocks, with two breeding confirmations, in the drier eastern part of the county (Burridge 1995). In northwestern California, Lawrence's Goldfinch has been known to be present in many years in interior Mendocino and Trinity counties (with breeding confirmed in each) and, less frequently, in Siskiyou and Humboldt counties.

A northwesterly position, fuller exposure to the tracks of Pacific winter storms, and lack of true semi-arid oak woodland and chaparral may work together to contribute to the historical scarcity of Lawrence's Goldfinch in Humboldt. At the inception of the Atlas project, there had been 10 reports of the species from the county. These were a flock of 20 seen (but not described nor any collected) at Kneeland 29 Jun 1939 (C. Clay); one at Fairhaven 17 Jun 1979 (R. Erickson); one or two at McKee Flat near Shelter Cove 2 Nov 1979 (R. Sutherland); a pair found at Petrolia about 22 May 1984 and whose nest with young was discovered 1 Jul 1984 (M. Smith, L. McCullough, J. Sterling, K. Rosenberg; AB 38:1060); a pair on Cape Mendocino Ridge 24-29 May 1984 (J. Brack, L. McCullough *et al.*); two at Davison Dairy near Orick 3 Jun 1984 (K. Rosenberg, J. Sterling); one at Bigfoot Golf Course near Willow Creek 9 May 1987 (COg, J. Brack, T. McKay); one in the Horse Pasture willow patch at Fairhaven 4 Jul 1987 (G. Lester); two at Ammon Prairie southeast of Willow Creek 13 Sep 1989 (COg), and a pair at Garberville 27 Aug 1992 (G. Lester).

Although this list of observations is brief, it suggests that Lawrence's Goldfinch occurs with sufficient frequency that Humboldt County can be considered—by generous interpretation—to lie within the extreme bounds of the range of this species. That the 10 reports were made during only 6 years; that six of them involved birds at or near the coast; that a flock of 20 was noted; that pairs were seen four times, and that breeding was confirmed all suggest that when conditions are right, Lawrence's Goldfinches "invade" Humboldt County and may call it home during the nesting season. Adding weight to this school of thought is that reports from Trinity County have been semiregular over the years, especially near Hayfork. In 1996–1997 there were numerous records from southwestern Trinity County, including confirmed breeding (Hunter and Hazard 1998).

The listing of historical and regional records detailed above lends valuable context to the events that took place during the Atlas period in Humboldt. Working in the seldom-birded Showers Pass area (k21) 1 Jun 1997, atlasers (PFS, V. Springer) found one male Lawrence's Goldfinch feeding on the gravel road at the junction of Showers Pass Road and Fort Baker Ranch Road. Later that day they found three males and a female about one-half mile south of there on a fence along Showers Pass Road (k22). On 7 Jun 1997 additional birds, including at least one pair, were seen and photographed at a salt lick put out for livestock about 7 miles up Showers Pass Road (k21; G. Lester). Another trip into the general area (L21) in search of proof of breeding 3 Jun 1997 yielded only a male chasing a female (JEH, TWL), but a trip on 15 Jul 1997 resulted in success when fledged young were found (PFS, V. Springer). Thirteen years after the first nesting in the county had been reported, Lawrence's Goldfinches were once again breeding here, although they were not reported in subsequent Atlas years. The discovery of this incursion by Lawrence's Goldfinches ranked among the more exciting finds in the 5-year run of the Atlas study.

Countryside in the vicinity of Showers Pass is a wide-open sweep of grassy hillside that extends for miles across much of the central Mad River watershed. Much evidence of cattle was noted during the time the goldfinches were present. Along the upper slopes and ridges, trees are few. Far downhill at the lower edges of this grassland and marching upslope in the moister draws are oaks and California bay and occasional California buckeyes and Douglas-firs. All in all the land is very open; it is typically profoundly silent but for scattered bird and insect sounds; and that element of the Humboldt County experience having anything to do with redwoods, ferns, or banana slugs (*Ariolimax columbianus*) is not brought to mind. While this patch of oak/grassland mosaic falls short of the glories one encounters farther south in California, a handful of Lawrence's Goldfinches found it acceptable during the summer of 1997.

AMERICAN GOLDFINCH *Carduelis tristis*

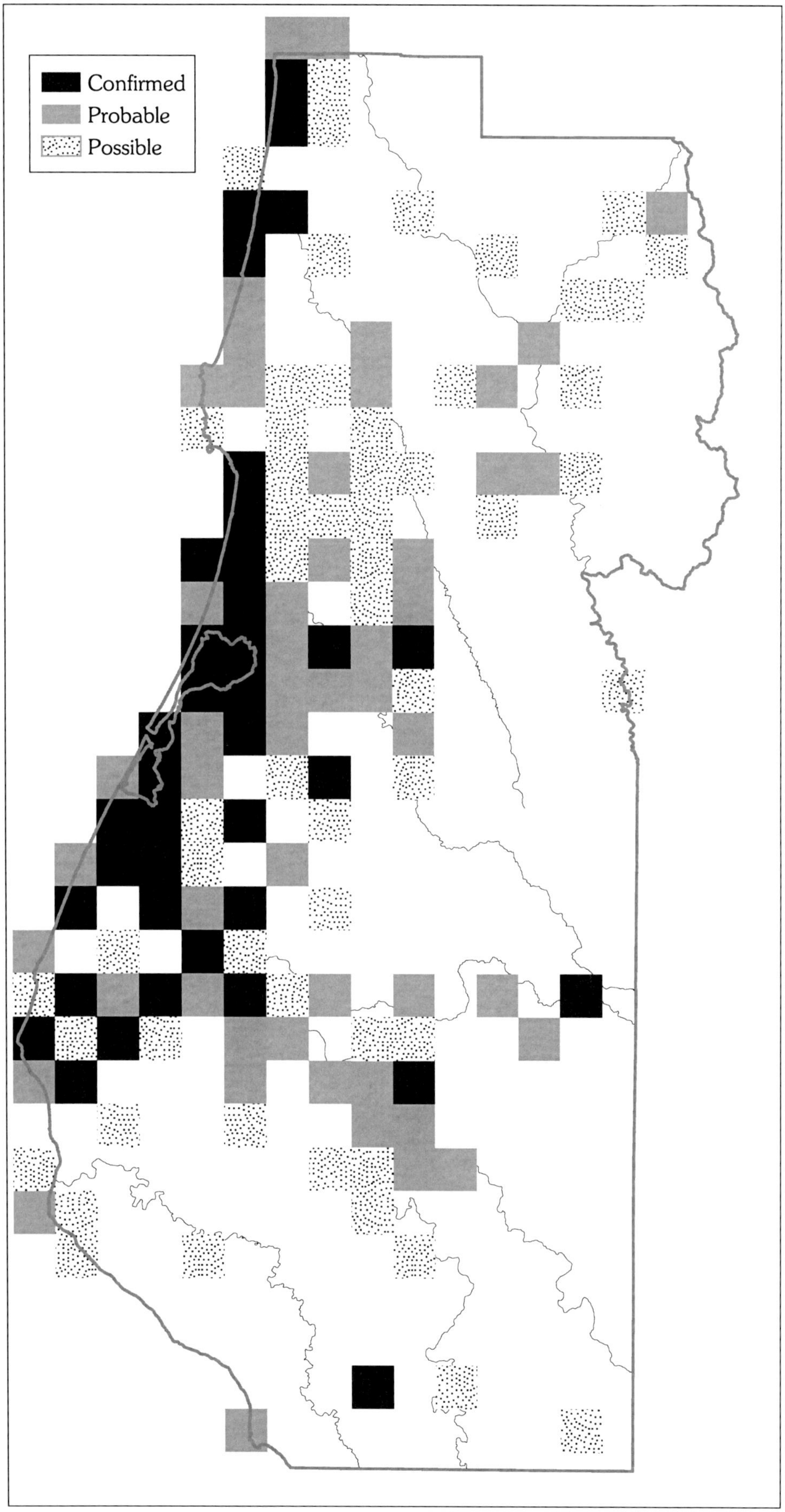

Displaying unceasing liveliness, American Goldfinches color summer thickets with a bold startlement of yellow. As autumn's onrush burnishes their weedy hide-outs with spent shades of brown, the birds acquire a plumage offering concealment while they yet clamber about clustered weed-stalks. Though they may remain inconspicuous for long moments, eventually a ragged band of dozens may erupt from a feeding site, giving this species' characteristic *po-ta-to-chip!* flight call.

Harris (1996) considered this species in northwestern California a common summer resident and breeder along the coast, less common inland, and uncommon in winter. In portions of Humboldt County in which they find plenty of preferred habitat, American Goldfinches are indeed common and conspicuous. They are a routine sight throughout the year, particularly in summer and fall in the coastal lowlands. Atlas surveys indicated that the distribution inland in Humboldt during the 5-year survey period was, with a few exceptions, at lower elevations and chiefly along stream corridors. The map shows an inland extension through blocks containing reaches or tributaries of the larger rivers. They were recorded in 34% of all Atlas blocks and were confirmed as breeding in 38 of them.

American Goldfinches are found locally inland along semi-open streamsides and in brushy riverbottom country with willows, alders, and an abundance of annual forbs. These goldfinches are among the more common birds in near-coastal pastureland and in landscapes dominated by dune or beach topography. Overgrazed pastures infested with thistle are heavily used around Humboldt Bay, in the Eel River delta, and elsewhere. Unmanaged bottomland grown to blackberry and wild teasel attracts goldfinches, perhaps accounting for their regular presence around the rural settlements scattered along so many of the smaller drainages in Humboldt County. American Goldfinch was a common species during 1995 surveys in different-aged coast redwood stands, and was detected most often in shrub-stage stands (Hazard and George 1999). During 1999 point counts in lower Eel River riparia, they were detected during 50% of all counts (R. Hewitt, unpubl. data).

American Goldfinches nest in deciduous shrubs and trees, with willow of nearly any species or structure affording a favored substrate. Pairs, or groups apparently consisting of pairs, spend much time at seed sources, including bird feeders. They are well scattered and at times abundant in late summer and fall in places well stocked with weeds. They seem not to occupy excessively open expanses nor very heat-blasted, oak-dominated sites as commonly as do Lesser Goldfinches. In interior valleys and lower foothills where both American and Lesser goldfinches are found, observers noted that Americans prefer luxuriant deciduous growth, while Lesser is the goldfinch of open oak woodlands and mixed-species forest edge.

American Goldfinches appear to have a long breeding season locally. Within a single block (F13), the carrying of nest material (spider silk) was observed as early as 7 May 1995 (JEH) and the feeding of young was seen as late as 22 Sep 1995 (GAS). American Goldfinches gathering spider webs were also reported 27 May 1996 (d18; GAS) and 31 May 1996 (D19; GAS). The earliest fledglings were reported 1 Jun 1996 (D23; COg).

EVENING GROSBEAK *Coccothraustes vespertinus*

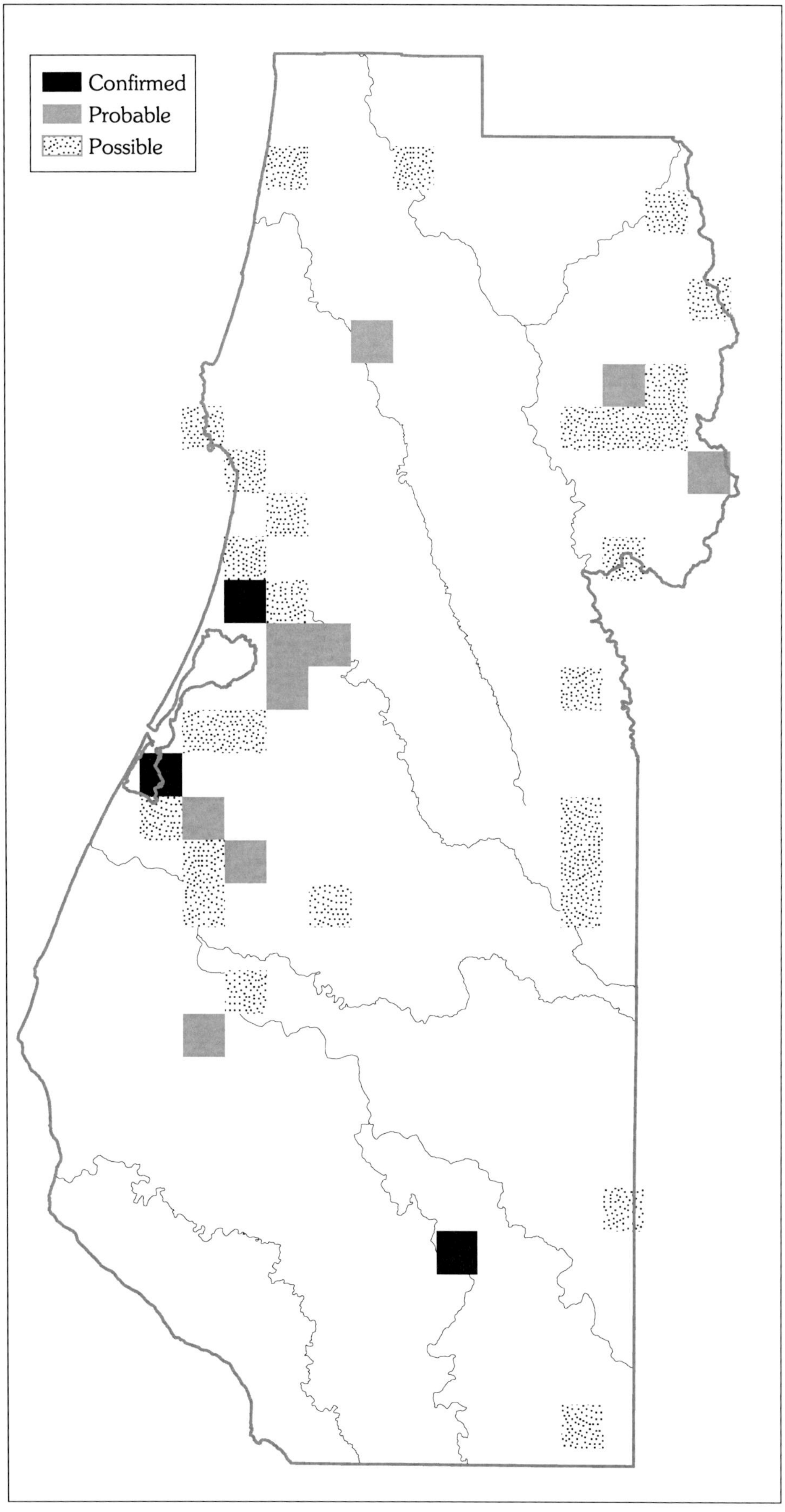

A combination of striking looks, social habits, and an appetite for sunflower seeds has lent the Evening Grosbeak a unique charisma. The species' unpredictable appearance in the lowlands—indeed, its erratic presence or absence across most of its range—adds a measure of enigma to the mix. It is well known as a "winter finch" in most of the central and eastern U.S. and southern Canada. In the coniferous forests of the West, it may be a more or less permanent resident. These birds have long been known to occur in Humboldt County, yet their status here remains unclear.

As a rule, Evening Grosbeaks are only infrequently observed in Humboldt County. Harris (1996) considered Evening Grosbeak in northwestern California to be an uncommon, sporadic visitor and breeder, most common in the coastal lowlands in spring and early summer. That they become so apparent in spring may be due to their frequent visitation at sunflower seed feeders in suburban yards. The birds also assemble during April and May in cottonwood riparian areas along rivers, and at middle elevations in bigleaf maples that are well-along into flowering and leaf-out. Pairs dispersing from mid- and late spring flocks may later breed in the same vicinity.

When not encountered in abundance, these grosbeaks are detected either in very small numbers or not at all. Distant, half-heard call-notes from one or two birds at long intervals may be the only indication of their presence. Evening Grosbeaks are stated in the literature to have a song, yet despite years of collective birding experience, atlasers remained unaware of it. Song was generally not used to estimate the breeding distribution of grosbeaks during Atlas fieldwork. Sufficient reports accrued to indicate that small numbers of Evening Grosbeaks are resident or otherwise to be anticipated in summer in the central coastal foothills and in the mountain forests of interior Humboldt. In inland Douglas-fir forests, Raphael (1982) detected them in 15–24% of his survey plots.

During the Atlas period, the number of detections of Evening Grosbeaks was small relative to the detections of other widespread and conspicuous songbirds. The species was recorded in only 39 blocks. A distinct concentration of blocks with detections occurred in the near-coastal lowlands and foothills from Trinidad south to Monument Ridge shortly southwest of Rio Dell. Three blocks with detections were scattered some miles apart in northwestern Humboldt County. A smaller region of clustered blocks with detections occurred along the east side of the Hoopa Valley Indian Reservation, with a few additional reports from South Fork Mountain, in the Lassics, on Island Mountain near Harris, and along the South Fork Eel River near Miranda.

Sightings of pairs accounted for 78% of the probable breeding reports. These sightings spanned the dates 5 May 1996 (g15; TWL) to 22 Jul 1995 (H15; PFS). Adults accompanying juveniles and feeding them were observed at a backyard feeder on Humboldt Hill, south of Eureka (d18) 3 Jul 1995 (P. Spruance). A female was observed feeding young in Miranda (k29) 11 Jul 1997 (JG). Davis (1922) found a nest with four eggs near Eureka 14 Jun 1914, and Wilder (1916b) noted young birds feeding on cascara berries during the last week of July 1916.

HOUSE SPARROW *Passer domesticus*

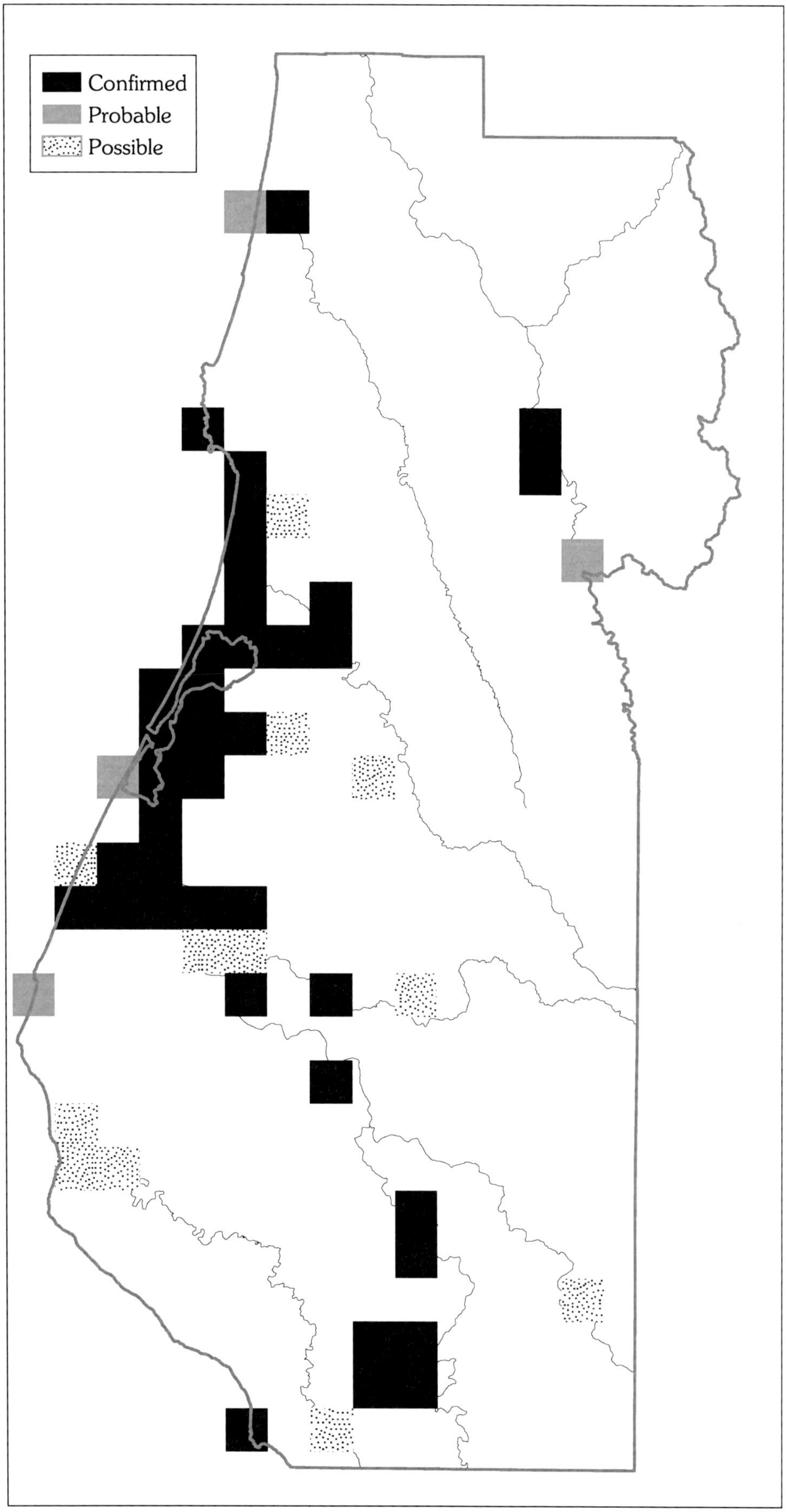

House Sparrows were first introduced to North America in 1850 in Brooklyn, New York, primarily to help control cankerworms. Many more introductions took place subsequently as far west as Oregon (Gabrielson and Jewett 1940), resulting in rapid spread across the continent. Ironically, the desired control of crop pests never took place, as these birds are chiefly vegetarian. Today, House Sparrows are widespread and abundant in North America, ranging from the Pacific to Atlantic coasts, northward well into Canada, and southward into Mexico. During the nineteenth century they prospered in cities because of the abundance of horses, in whose droppings they searched for edible matter. Now they fare well in urban settings offering waste or spilled human food, grain, and food put out for pigeons.

House Sparrows have flourished nearly anywhere people live or do business, occurring throughout the year in residential neighborhoods and industrial areas and at farms, ranches, grain elevators, and wharves. They place their nests under eaves and cornices, among exposed bundles of cables or pipes, in barn rafters, broken signs, old machinery, unscreened ventilation holes, and nest boxes erected for native species. Their usurpation and aggressive defense of nest boxes irritates many people. A bird was incubating eggs in a Cliff Swallow nest near Blue Lake (H15) 28 Jun 1995 (B. Hawes). It might be borne in mind that this behavior generally harms but a relatively small proportion of songbirds affected. At any distance from cities, towns, and farms, House Sparrows are distinctly out of place. The House Sparrow's status as a tolerated commensal of human daily life and commerce offers great opportunity, yet it imposes restrictions. Forests, mountains, grasslands, wetlands, and deserts in which people only visit would seem inhospitable to these birds. Yet they seem able to cross nearly any such barrier, and may be seen popping food into the mouths of begging juveniles at isolated hamlets.

This species was stated by Harris (1996) to have first appeared in Eureka in 1885. Whether Humboldt County's first House Sparrows arrived overland or made a human-assisted journey at sea is unknown. It is plain that the birds have accomplished in this region what they have elsewhere, becoming variably numerous residents of settled areas and outlying farming and dairy lands. Harris (1996) described the House Sparrow in northwestern California as "an introduced common resident and breeder around towns, farms, and ranch buildings. Closely associated with human habitations." During the Atlas period, House Sparrows were reported from 54 (13%) of the 425 total blocks.

A glance at the map makes it clear that House Sparrows were found most commonly in Humboldt's central-coastal lowlands. They were detected during the project in a band of blocks extending from Trinidad (e10) south through McKinleyville, Arcata, Manila, Eureka, Fernbridge, Ferndale, and Fortuna to Rio Dell (F23). Other clusters of detection-blocks correspond with the Redway–Garberville area in southern Humboldt (i31 to j32); the lower Mattole River community (b26 to c27, no breeding noted); Orick (f05, g05); Hoopa (m10, m11); and Myers Flat (j28) and Miranda (J29). Birds were also noted at Willow Creek (N13), at an unstated location along the Van Duzen River (H23), along the Eel River at Holmes (H25), and at Shelter Cove (f33). Possible breeding was reported from Fieldbrook (g12), Freshwater (g17), the south end of Kneeland (i18), the confluence of the Van Duzen and Eel rivers (e22), Hydesville (f22), along the Van Duzen at Swains Flat (J23), at Alderpoint (n30), and at Whitethorn (H33).

Given their nesting habits, it is not surprising that breeding was confirmed in 70% of all blocks with detections. Probable breeding was reported in four blocks. Six behaviors were observed in the course of confirming breeding in 38 blocks. Unlike the situation noted for many species detected during the Atlas study, there was comparatively equitable distribution among these, with no one or two confirmation codes dominating. Adults carrying nest material were observed seven times from 8 Mar 1996 (F13; L. Bacon-Ogden) to 28 May 1995 (h14; C. Keil), and were seen carrying food four times, spanning 25 May 1996 (F15; GAS) to 3 Aug 1996 (m10; JMH). Occupied nests were most frequently reported, the nine such records spanning 9 Apr 1995 (F11; JEH) to 28 Jul 1996 (F21; JEH). Although these records were subsequently superseded by higher records, some dates for nest building and a nest with eggs were 16 May 1999 (F13; JEH) and 9 May 1996 (f14; RJA), respectively. Seven nests with young ranged from 19 May 1996 (F15; C. Meyers) to 14 Jul 1996 (F23; RJA). Fledged young also served as the best-record-in-block seven times, on dates from 15 Jun 1996 (i31, i32, both RJA) to 28 Jul 1996 (B21; COg). Adults feeding young were reported as the highest-ranking code in four blocks, from 7 May 1996 (J31; RJA) to 18 Jul 1996 (e16; RJA).

House Sparrows in Humboldt County clearly are tied to human settlements. Flocks straggle from dairies in the open bottomlands out along adjacent fencerows, but as a rule the birds are rarely seen away from ground that is either paved with asphalt or well trodden with the hoofprints of livestock. Great stretches of Humboldt are uninhabited by House Sparrows and probably always will be. The distribution map may deceive somewhat, in that the actual land area occupied by the birds is considerably less extensive than the territory enclosed by the blocks.

WHITE-TAILED KITE
Continued from page 74

The species is considered to be a common resident and breeder throughout the coastal plain of northwestern California, with occasional birds or pairs at inland locations such as Snow Camp Road, Bald Hills, and Garberville (Harris 1996).

Kites were first reported in the Trinity River corridor in the mid-1990s. By the end of the Atlas period, they seemed present throughout the year there in small numbers. One was seen carrying nest material near Willow Creek (N13) 4 May 1995 (TWL).

Although not strongly migratory in the classic sense, White-tailed Kites undertake some sort of movement in fall, resulting in increased numbers throughout the coastal lowlands and in the semi-open near-coastal foothills. From late fall through late winter, they occur as singles, pairs, or small assemblages in appropriate habitat. Winter roosts have been well documented in Humboldt County. A roost on an island in the Eel River delta, attended each winter, supported as many as 120 kites in December and January during the mid-1990s (PFS *et al.*). As many as 47 kites were seen gathering at a pre-roost staging site in the south Arcata Bottoms concurrently. At each of these sites, Northern Harriers also were present and appeared to roost in close association with kites. It is possible that kite roosts exist elsewhere, such as in the Trinity River canyon between Willow Creek and Hoopa, in the Redwood Creek estuary, or near Garberville. Winter kite roosts are known to serve birds dispersing during the day over a radial distance of many miles from the site.

The retention and restoration of adequate, interconnected or adjacent parcels of lightly grazed or ungrazed, unplowed grassland in a semi-open landscape would appear to be of critical important in the conservation of White-tailed Kites in Humboldt County. Local observers note that in the past decade or so these birds have become much less common in the McKinleyville area, coincident with the rapid urbanization of grassy marine terraces in that area. Much habitat occurs on lands administered by the U.S. Fish and Wildlife Service and the California Department of Fish and Game. As periodic "treatments" of habitat in these places typically address the needs of waterfowl and those who hunt them, the requirements of such species as White-tailed Kites, Northern Harriers, and Short-eared Owls are sometimes given short shrift into the bargain. The trend in Humboldt is for kite habitat to be destroyed or degraded more often than it is created. It is to be hoped that the value of unmanaged grasslands can be raised in the eyes of the public as well as on the spreadsheets of professional wildlife managers. They are important and productive places, worthy of respect and protection.

PEREGRINE FALCON
Continued from page 94

probable breeding reports. In addition to coastal environments, such habitats as Douglas-fir/tanoak forest, oak woodland, and riparian habitat occur in the vicinity of Humboldt County nest sites.

Certainly the most interesting phenomenon pertaining to this species during the Atlas period was the discovery by Simpson Timber Company biologists of Peregrine Falcons nesting in residual old-growth redwood trees. As this story starts, one or two peregrines were seen in June and July around the northeastern corner of Humboldt Bay in most years up to and through the Atlas period. Fueling suspicions that nesting might be taking place nearby were an adult and two juveniles seen at Arcata Marsh 2 Aug 1988 (F. Broerman), an adult and begging juvenile there 23 Jun 1990 (S. Harris), and an adult seen to take an adult Black-necked Stilt at Fay Slough Wildlife Area shortly northeast of Eureka 6 Jul 1994 (M. Fisher). That peregrines were indeed breeding in the vicinity was revealed to be the case in spring 1999. First one pair 23 Mar 1999, then, on 20 Apr 1999, a second pair were discovered nesting in large (both about 135 inch diameter and 200 feet tall) residual old-growth redwoods on private timberland east of North Humboldt Bay (L. Diller, pers. comm.).

Tree-nesting by Peregrine Falcons had been documented previously in such far-flung locations as Kenya, Germany, and Sweden (Cade *et al.* 1988), Kansas (Goss 1878), Illinois (Ridgway 1895), British Columbia (Campbell *et al.* 1977), and Alaska (Van Horn *et al.* 1982). Yet it remains a spectacular and remarkable breeding strategy, and apparently one that was previously undocumented in the western conterminous U.S. That peregrines should honor Humboldt's redwoods with the status accorded the tall cliffs they typically nest upon seems somehow fitting. The use of residual old-growth redwoods by peregrines for nesting provides still more data regarding the great biological importance of these individual trees in intensively managed landscapes (Hunter and Bond 2001).

SNOWY PLOVER
Continued from page 112

but away from the beginning of the zone of annual forbs, such as exotic sweet-clover, which establishes itself each spring as flows in the river decrease. Scouring each year by winter floods normally erases each summer's growth of annual vegetation on the bars, cleaning the slate for the next season's nesting attempts.

It remains unclear whether post-breeding season commercial gravel extraction on river bars harms plovers, or if it actually helps maintain the vegetation-free areas they require for nesting. It is also unclear if bars on other Humboldt rivers, as well as additional bars farther up the Eel, are wide enough to provide the conditions these birds require. Frequent flooding events in the Eel River basin have created an exaggerated alluvial plain within the main stream corridor, through which the river courses between long, barren gravel bars. Much less of the water that normally would flow through the Eel River in summer actually does so, owing to diversion by Pacific Gas & Electric's Potter Valley Project. Because the river remains low during the breeding season, more plover habitat is exposed.

During the 5 years of the Atlas project, Snowy Plovers were confirmed breeding in 11 blocks, and possible breeding was reported from two more. Many of these confirmed blocks had multiple confirmations in any given year, and confirmations for multiple years. The pattern of confirmations reflects the localized distribution of this species in Humboldt County during the nesting season. Depending on how they are defined, two or three centers of occurrence existed during the Atlas period. These were Clam Beach (F11, f12) south to the Mad River mouth (e13); ocean beaches and dunes in portions of five blocks from below the west end of Table Bluff (c18) to Centerville Beach (B21); and gravel bars of the Eel River in four blocks, from shortly below Fernbridge (c20) upstream to near the confluence of the Van Duzen River (d20, e21, e22).

The distribution of confirmed blocks in the lower Eel suggests that the beach-nesting and gravel bar-nesting subpopulations are confluent. However, the map might mislead; this impression was generated by the distribution of plovers on the lower bars in blocks adjacent to those in which beach-nesting occurs. During the Atlas period, pairs breeding on the Eel River sandspits from Table Bluff to Centerville Beach were thought to be separated from the gravel bar population by several miles of unsuitable nesting habitat along shorelines within the lower Eel River estuary. It can also be noted that the Clam Beach group and the Eel River beach group were formerly more confluent during the time when Snowy Plovers nested commonly on the north and south spits of Humboldt Bay.

Some representative Atlas breeding records include nests with eggs 6 Apr 1995 (F11; M. Fisher), 13 May 1998 (e22; R. LeValley), 23 Jun 1999 (c18 and c19; S. McAllister), and 18 Jul 1999 (c18; R. LeValley). Precocial young were reported 28 May 1999 (b20; R. LeValley), 14 Jun 1999 (b19; S. McAllister), 9 Jul 1996 (e21; R. LeValley), and 13 Aug 1999 (F11; S. McAllister).

In winter it is likely that the local breeding population is augmented by migrants from interior North America and other Pacific Coast locations, and many of the local breeders themselves may winter elsewhere. It is currently unclear if there is any significant interchange of breeding birds between the interior and coastal populations. However, Humboldt's beach and gravel bar plovers do mix between seasons and years. Although many important questions remain, color and number banding is clearly an essential tool for understanding Snowy Plover movement patterns.

As mentioned above, Snowy Plovers face many threats in Humboldt County. Consideration by the U.S. Fish and Wildlife Service of which areas should be designated as "critical habitat" for the plover probably has been biased toward beaches humans visit least often (*e. g.*, Stone Lagoon), rather than which beaches the plovers seem to prefer (*e. g.*, Clam Beach). Augmented by exposed garbage, increases in populations of nest predators such as Western Gulls, American Crows, Common Ravens, striped skunks (*Mephitis mephitis*), and raccoons (*Procyon lotor*) surely add to the problem. Starting in 1999, wire-mesh exclosures designed to protect plover nests from both predators and humans have been erected. While labor-intensive, this method of increasing nest success has been used elsewhere on the Pacific Coast. The overall effect of the exclosure program at Clam Beach has been mixed, but it shows promise as a short-term conservation tool. Unfortunately, exclosures have drawn attention to some nests that were subsequently destroyed by humans or by other exclosure-foiling predators. Merlins (*Falco columbarius*) have been observed in coastal Oregon using Snowy Plover nest exclosures as hunting perches (D. Lauten and K. Castelein, pers. comm.). Restrictions on human use of beaches in order to protect nesting plovers has raised the ire, as well as the sympathy, of many citizens concerned about beaches. As it is with so many of our imperiled birds, the fate of the Snowy Plover will ultimately be determined by politicians.

MARBLED MURRELET
Continued from page 128

Smith Creek; west of Myers Flat in the headwaters of Canoe, Salmon, Kerr and Coon creeks; upper Lawrence and Booth's Run creeks, and possibly in the Bear and Mattole river drainages. The noteworthy exception is the known occupied site on the Bureau of Land Management parcel along Lacks Creek (k11), where a bird visiting a probable nest site was recorded 3 Aug 1995 (J. Price) and a bird was heard only 16 Jul 1998 (JEH). This is the only verified non-redwood site occupied in California. Slightly east of there, a murrelet detected 3 May 1995 east of Pine Ridge (l10; H. Brown) may have been associated with Lacks Creek, or perhaps represented yet another undiscovered site.

While Marbled Murrelets nesting in the state clearly favor redwood stands (Paton and Ralph 1990, Miller and Ralph 1995), of the 10 nest trees known in California by 1995, only one-half were redwoods, while four were Douglas-firs and one was a western hemlock (Hamer and Nelson 1995b). These murrelets do not build a nest, but rather lay their single egg on a large, well sheltered limb, often with a soft carpet of mosses. Old-growth redwoods certainly have large limbs and moss, but other tree species typically support more luxuriant moss cover. Suggesting that a cool and moist marine influence plays an important role in determining the inland extent of Marbled Murrelet in California is that the birds occur almost exclusively in redwood stands in California, and that they often nest on moss. Extensive surveys in the summer-hot inland Douglas-fir forests of northwestern California failed to find any "fog larks," probably since these areas are well beyond the influence of summertime marine air because of the intervening steep and rugged coastal topography (Hunter *et al.* 1998). Where cool moisture-laden marine air extends inland in some large river drainages, both murrelets and redwoods may still be found. One such location is Grizzly Creek Redwoods State Park (i23), an area of old-growth redwood located about 25 miles inland that receives almost daily incursions of fog and low overcast in the summer. This is the farthest-inland site in California with documented continuous occupancy by murrelets (O'Donnell 1993). The risk of predation on adults and the energetic demands of repeatedly carrying a single fish miles inland to a nestling probably also limit how far inland these seabirds can breed.

Marbled Murrelets were once thought to nest only in large, contiguous stands of old growth. To be sure, large stands provide more potential nest sites and fewer edge-associated nest predators such as corvids. However, murrelets are also known to nest in small patches of trees and even in single residual old-growth redwoods in California (Hunter and Bond 2001), although reproductive success in such situations is probably low. Miller and Ralph (1995) found that large stands were not more likely to have murrelets than smaller ones, and that more behaviors indicative of breeding were recorded at lower elevations in major drainages than on ridges. Sites lower on slopes typically have soil and other biophysical conditions favorable to large tree (and limb) growth, and are also more sheltered from wind and heat. These factors probably explain why many of the occupied sites in northwestern California are situated in the bottoms of rather steep gulches and valleys. It may also explain why murrelets are not found in Sitka spruce and other forests directly on the coast, where exposure to both high winds and abundant predators (*e. g.*, gulls) probably reduce suitability for nesting.

The Marbled Murrelet remains one of the most imperiled species in California and the Pacific Northwest. Based on observations of nest failures and exceedingly low numbers of hatch-year birds seen at sea, reproduction appears inadequate to support a self-sustaining population in California. While it is likely that they were once well distributed throughout the extensive old-growth redwood forests of Humboldt County, today less than 5% of those forests exist. Most of these remaining old-growth forests—and murrelets—can be found in two general areas of the county: in the Redwood National and State parks along the north coast, and in and around Humboldt Redwoods State Park and Pacific Lumber Company lands in south-central county. In 1999 the U.S. Fish and Wildlife Service and the California Department of Fish and Game authorized the cutting of about 15,000 acres of potential murrelet nesting habitat (about 2200 acres of which were occupied) as part of Pacific Lumber Company's Habitat Conservation Plan. Thus, ongoing habitat removal continues to be the single greatest problem for this species. Cutting of any old-growth redwood for any reason, on both private and public lands, should be halted immediately.

In addition, murrelets are greatly at risk from oil spills. After spills in Humboldt County in 1997 and 1999, a total of 34 dead murrelets were recovered. If, as estimated for the 1997 spill (Ford *et al.* 2002), direct mortality from these spills was actually as much as 15 times greater than the number of birds found, then a significant proportion of the local population was killed. Thus, the long-term future of this species in Humboldt County is highly questionable. One wonders if the few birds persisting in the redwood parks a century from now will consist of a remnant breeding population, or if they will be solely recent dispersants from a larger source population in Canada or Alaska?

SPOTTED OWL

Continued from page 152

also afford protection from potential competitors and predators such as large diurnal birds of prey (Hunter *et al.* 1993) and Great Horned Owls (Franklin *et al.* 2000). Structurally diverse stands also contain more potential nest sites, which include cavities in large broken top and dead trees, debris accumulations, old abandoned nests, and other platforms in both conifers and hardwoods (Folliard 1993, LaHaye and Gutiérrez 1999). Multiple canopy layers also enhance protection from summer heat and inclement weather. A combination of these factors may explain why owl sites in managed stands within the redwood zone that have greater amounts of residual conifers from previous harvests appear to have higher reproductive success (Thome *et al.* 1999).

Areas of Humboldt County in which Spotted Owls were not detected during the Atlas period coincided with sections deficient in suitable habitat. This includes nearly all of the outer coast and most of the coastal lowlands from Little River into the Eel River delta; the mountains of northeastern Humboldt; the oak-grassland country of the central Mad River drainage; and numerous blocks in interior southern Humboldt.

Humboldt's Spotted Owls are probably among the most-studied birds on Earth. A wide range of research has been conducted here, and even a brief mention of all of these scientific studies is far beyond the scope of this work. Habitat associations at a variety of spatial scales are well documented, including at nest trees (Folliard 1993,

LaHaye and Gutiérrez 1999); within nest and roost stands (Blakesley *et al.* 1992); within home ranges (Solis and Gutiérrez 1990); within landscapes (Hunter *et al.* 1995, Gutiérrez *et al.* 1998, Folliard *et al.* 2000), and province-wide (Zabel *et al.* 2003). Most habitat studies indicate a strong association with older, more structurally diverse stands. Mean summer home ranges in the Douglas-fir zone are about 1020 acres, and home ranges averaged about 45% mature and old-growth forest (Solis and Gutiérrez 1990). In addition, Spotted Owl studies in Humboldt have delved into genetics (Barrowclough and Gutiérrez 1990, Barrowclough *et al.* 1999); parasites (Gutiérrez 1989, Young *et al.* 1993, Hunter *et al.* 1994); age and sex characteristics (Blakesley *et al.* 1990, Moen *et al.* 1991); behavior (Carlson *et al.* 1998); research methodology (Paton *et al.* 1991, Franklin *et al.* 1996a); and density. Some estimates of the crude density of Spotted Owls in Humboldt's Douglas-fir and redwood zones are 0.61 owls per square mile and 0.54 owls per square mile, respectively (Franklin *et al.* 1990, Diller and Thome 1999). Extensive, long-term (since 1985) demographic studies in Douglas-fir forests indicate that cold and wet spring weather is correlated with lower reproductive success, and that greater amounts of interior forest is correlated with higher adult survivorship (Franklin *et al.* 2000). See Franklin *et al.* (1996b) and Thome *et al.* (2000) for additional information on population dynamics in Douglas-fir and redwood forests, respectively.

Spotted Owl biologists typically survey at night by mimicking territorial calls (*i. e.*, "hooting") and then the following morning seeking any owls located. Roosting owls are approached and provided with a prey item, which the owls readily take with apparently little fear of the humans standing nearby. Nesting owls typically take prey items to nests or young, while non-nesters usually either eat or cache the food for later consumption. Most Atlas records were gathered using this standard methodology (USFWS 1991), which, over the years, has provided a great deal of information on the breeding chronology of these owls. Considering this, it is unlikely that the relatively sparse Atlas data would shed much new light on the local breeding chronology.

These owls typically begin nesting in early April, although pair bonding probably starts as early as February. The earliest Atlas record of a pair calling was 13 Mar 1995 (h32; *fide* G. Gould), and the earliest occupied nest reported was 25 Mar 1995 (J15; C. Hibbard). Sixty-five of 67 probable Atlas breeding records were of pairs, most of which represented a male and female heard calling during a night survey. Spotted Owls usually fledge young by mid-June; the earliest Atlas report of a fledgling was of two young birds north of Friday Camp 29 May 1996 (m15; A. Franklin). After fledging, one or two (occasionally three) young continue to receive food from the parents until late August or early September; the feeding of young by adults was reported as late as 6 Sep 1995 (L17; C. Hibbard). By late October, most young owls have dispersed from the natal area (Gutiérrez *et al.* 1984).

If the current intensive management for Spotted Owls continues in Humboldt, this species may ultimately represent an example of success in recovery of an endangered species. However, the owls' future is not certain here. Constant, well financed efforts seek to reduce protections in order to increase timber yields. Current forestry regulation and management practices do not adequately protect existing structures such as residual old-growth trees and snags, nor do they provide for the future recruitment of these important habitat elements (Hunter and Bond 2001). The invasion of the range of the Spotted Owl by the Barred Owl in California (Dark *et al.* 1998) has raised questions regarding the capacity for the two species to coexist. Barred Owls may out-compete, usurp, or even kill (Leskiw and Gutiérrez 1998) Spotted Owls. Hybridization between the two species has proven less frequent than some biologists had feared, yet it has been documented in all three Pacific Coast states and in Humboldt County.

COMMON NIGHTHAWK
Continued from page 162

the business districts of Arcata and Eureka. Alarmingly, observers living in the Humboldt Bay area noted that, by the close of the Atlas period, these once familiar urban populations were gone or nearly so. Reasons for the decline of the rooftop-nesting population in Eureka and Arcata are unclear, although it has been suggested that predation by increasing numbers of American Crows may have played a role (R. LeValley, pers. comm.). Similarly, local observers have noticed that birds associated with beach areas are much less commonly encountered than they were just a few years ago. The near constant presence of humans (often with unleashed dogs) may result in a level of disturbance that precludes these ground nesters from using many beaches and dune areas.

VAUX'S SWIFT
Continued from page 166

was present 28 May and two were seen 4 Jun; three broken eggs were found below the nest 9 Jun; more copulations were observed 11 Jun; two new eggs were present 18 Jun, and four fledged young were seen returning to the nest 31 Jul (C. Gutiérrez). Some of the nest-with-young and occupied nest records probably involved fledged young returning to the nest site to roost, rather than nestlings *per se*. Two confirmations were of adults carrying nest material, 31 May 1998 (H31; GAS) and 5 Jun 1996 (H07; GCH).

Family groups of Vaux's Swifts are frequently noticed in late June through early August, the adults conspicuous by their wing molt. An adult was seen feeding young 15 Jun 1997 (J31; RJA), and fledged young were reported 29 Jun 1996 (J29; GAS), 29 Jun 1997 (h24; RJA), and 30 Jul 1999 (i23; GCH). By late August, breeding birds and their progeny appear to disperse, depart, or otherwise become less noticeable. Fall migration is evidenced by flocks of Vaux's Swifts thoroughly mixed with the throngs of swallows on the move. A loose association with Violet-green Swallows is sometimes remarked upon.

During fall migration, spectacular concentrations of these birds can be observed at dusk swirling whirlpool-like into a suitable chimney or hollow tree to roost, or erupting out come morning. Infrequent reports of 50–500 seen entering residential chimneys in Arcata, Eureka, and Ferndale have provided local examples of this phenomenon. Elsewhere, these roosts can number in the tens of thousands of birds, such as at the furnace chimney in Healdsburg, Sonoma County (Burridge 1995) and at sites in interior valleys of western Oregon (Marshall *et al.* 2003). As is the case with other species whose populations are greatly concentrated at favored staging sites, such observations may delude one into imagining that limitless numbers must exist. Most summering birds and migrants have left Humboldt by the end of September.

The loss of unique and long-lasting nest structures such as large hollow redwoods probably has had the single greatest impact on Humboldt's breeding population of Vaux's Swifts. Retention of these large old trees, and providing for the recruitment of such trees in the future, is unfortunately entirely voluntary for land managers. Older chimneys, smokestacks, and other human-made structures are increasingly being replaced by newer models that are seemingly unsuitable as nest sites. Mass mortality events are occasionally reported when birds become trapped inside structures or are killed when furnaces or fireplaces are stoked (*e. g.*, Michener 1933, Huey 1960, AB 31:1044). Educating the public about the roosting and nesting behavior of these birds might reduce the likelihood of these unfortunate incidents. Vaux's Swift should be considered for listing as a "threatened" species in California, but the extensive range of this bird outside of California might make such a designation politically difficult. It is unknown what, if any, effects on this species may occur on the winter range due to unregulated pesticide use and habitat loss.

GRAY JAY
Continued from page 216

breeding was established, fledged young supported this status in 16; the distinctive plumage worn by juveniles was the most important evidence indicating breeding during the Atlas period. The earliest such report was 9 Jun 1996 at Russ Park in Ferndale (c21; M. Varga), the latest 30 Jul 1998 in the Cutten section of Eureka (e17; G. Lester). Adults feeding young confirmed breeding in five blocks. Davis (1938) described some May 1904 nests with eggs he found near Arcata and Eureka.

Gray Jays are considered essentially resident across their extensive North American range, and this seems to hold true in Humboldt County. During some winters, occasional wandering individuals or small flocks are reported from suburban neighborhoods in McKinleyville, Arcata, and Eureka that are close to occupied summer habitat. These movements are small in scale and seem to involve but few birds. A notable local winter concentration developed in each Atlas year in the vicinity of a yard shortly east of Centerville Beach, where dozens awaited daily handouts. Through 1999, Gray Jays remained unknown or nearly so over most of the coastal plain from the Mad River estuary south through the Eel River delta.

Harris (1996) characterized Gray Jay in northwestern California as an uncommon resident and noted that his files contained reports from about 40 locations throughout the region covered by his work. He further reported that Gray Jays were usually associated with stands of large Sitka spruce/redwood, Sitka spruce/Douglas-fir, Douglas-fir, or redwood. He did not directly mention occurrence well inland, nor did he describe habitat at inland sites. Harris (1996) believed that this species had probably declined with the removal of old-growth forests and that, as of 1996, the population was fragmented and confined to the remaining spruce/redwood groves. This appeared to remain the case during the Atlas period although, as described above, Gray Jays were also detected in the interior mountains in somewhat different settings.

SWAINSON'S THRUSH
Continued from page 280

confirmations were of adults carrying food, these records spanning the dates of 25 May 1998 (g05; COg) to 4 Aug 1996 (F13; JEH). A noteworthy observation was of an adult carrying a Pacific treefrog (*Pseudacris regilla*) 1 Jul 1995 at the Horsepasture Patch near Fairhaven (d16; JEH). All records of fledglings, except the one being fed 28 Jun 1998 (a22; COg), were from July. The latest report of the feeding of young was 14 Jul 1998 (e22; P. Herrera). It was felt that Swainson's Thrushes probably breed in most of the blocks in which they were reported. No suggestion of increased breeding confirmation in priority blocks due to greater observer effort was apparent.

This species is a relatively late migrant in spring, ordinarily not reported until late April and not becoming common until early May. The earliest Atlas report of a migrant was 14 Apr 1996 (c23 and D23; COg), and the earliest report of birds possibly already on summer territories was of countersinging males 14 May 1998 (F21; JEH). Banding studies suggest that Swainson's Thrushes nesting in coastal forests move into riparian habitats in July and August, perhaps taking advantage of lower nest predation in forests and then greater food availability in riparian areas (Ralph *et al.* 1996). Night migrants are widely known for their nocturnal calls, heard from overhead in the hours of darkness and especially just before dawn. Calm evenings in May, September, and early October are enlivened by the frequent call-notes of these birds. The latest report of a probable fall migrant during the Atlas period was of one 8 Sep 1997 (o11; JMH).

VARIED THRUSH
Continued from page 286

Trees such as Jeffrey, sugar, and western white pines, Brewer spruce, and deciduous hardwoods were either represented very sparingly or were absent.

Intriguing were detections of Varied Thrushes in three blocks which fell roughly between the near-coastal and mountain population centers. A pair was noted 15 May 1995 near Willow Creek (m13; M. Steenblock-Smith), a single bird was detected 21 May 1996 in n16 (GAS), and a male was singing for a week or more near Weitchpec (l08) in May 1996 (R. LeValley).

The earliest reports of multiple males singing in a block were 14 Apr 1998 (g04; DFx) and 20 Apr 1998 (H03; *fide* RHw). Among the 22 blocks with confirmed breeding by Varied Thrushes, the highest-ranking evidence in 10 was adults carrying food, the earliest of which was 19 Apr 1996 (g16; M. Bailey). Only one report each was received for adults carrying nest material, 15 May 1997 (f10; M. Morrissette); for the feeding of young, 20 Jun 1995 (g04; M. Beck); and of a nest with eggs, found 14 Jul 1997 (f20; D. Garrison). The earliest fledgling reported was 13 May 1996 at Catfish Lake (e19; M. Wachs). After the young birds fledge, Varied Thrushes frequently forage on fruits, especially those of the red huckleberry (Beck and George 2000).

From about late September into early April, Varied Thrushes may be found in lowland and foothill areas in which they do not nest. Coniferous and mixed conifer–hardwood forests and thick, damp woodlands—as well as shaded ravines in urban and residential areas—support these thrushes. River canyons, rural roadcuts along the more shadowed northern exposures, and stands containing a strong madrone component attract numbers of Varied Thrushes. Widely varying CBC totals from the region suggest that the extent or magnitude of this movement appears to change from year to year.

YELLOW WARBLER
Continued from page 304

by Yellow Warblers are degraded by human activities such as off-road vehicle driving, grazing cattle, levee construction and, possibly, by gravel extraction. The size and quality of riparian parcels also fluctuate naturally in response to such events as the migration of river bars and winter flood scouring. While cottonwood riparian habitat has been greatly reduced since the onset of modern settlement (Kelly 1987), most of the places in Humboldt County currently hosting Yellow Warblers appear to remain reasonably stable.

MacGILLIVRAY'S WARBLER
Continued from page 314

Addressing its status across northwestern California generally, Harris (1996) termed MacGillivray's Warbler an "uncommon summer resident and breeder . . . Breeds mostly inland on cutover brushy slopes and along river valleys." Harris (1996) stated that the birds occur in migration in riparian, dune hollow, and coastal brushfields in the coastal lowlands. MacGillivray's Warblers typically arrive in early and mid-April and become numerous by early May. This species has the curious habit of sometimes appearing to remain at breeding sites until well into late summer, long after other locally nesting warblers have departed; singing may continue sporadically through August and even into the first days of September. Fall migrants are uncommonly met with in the lowlands as well as in mountain brushfields during late August and September. Only occasional stragglers are seen past the end of that month.

YELLOW-BREASTED CHAT
Continued from page 320

Chats arrive in Humboldt County from their tropical wintering range beginning in mid-April, but they are not commonly encountered until about the first week in May. The earliest Atlas report was of a bird 14 Apr 1996 (c23; COg), possibly a migrant still on the move. A phenomenon not often remarked upon, but apparent during the 5 Atlas years, was the appearance of singing birds at outer- or near-coastal locations where they were not seen for long. Such birds have been found in the Fairhaven willow patches, at the Arcata Marsh, and at old mill flats elsewhere in the coastal lowlands. Clearly, however, some such sites do support chats throughout the breeding season. After nesting, the population within the county seems to melt away. Alluding to this seeming lack of any perceptible fall movement, Harris (1996) stated that the species "disappears quietly from breeding areas in Aug–Sep." Extremely few fall migrants have been encountered at any of the well worked birding sites during late August and September, a period during which nearly all other West Coast-breeding warblers appear in at least small numbers. Secretive behavior, lack of singing, and the often impenetrable thickets favored by these birds may cause them to be overlooked.

Although the riverbanks of northwestern California probably still support relatively high densities of chats, in many regions habitat for Yellow-breasted Chats has been significantly altered or destroyed. The human hand has erased chat hangouts to some degree nearly everywhere. Asphalt fails to offer suitable hiding cover; channelized streamcourses safeguard towns and improve agricultural efficiency but produce few insects. Dunn and Garrett (1997) stated that these birds "have declined precipitously over much of their range, although they still occupy at least sparsely most of their historical range . . . Populations are generally stable in much of the West, although urbanization, flood control, and perhaps cowbirds have greatly reduced populations in California and some adjacent regions." While riparian habitats along the Klamath, Trinity, and Eel rivers would appear to be free from most overt threats, these places are now subject to water control projects that greatly reduce dry season flows. The Eel, Mattole, and Trinity rivers, for example, can be little more than a series of shallow, stagnant puddles festooned with blue-green algae by early fall. The long-term effects of these water regimes on riparian vegetation, and on the welfare of animals living therein, beg further scrutiny and study.

SAVANNAH SPARROW
Continued from page 336

Magnitudes fewer are believed to winter inland in the county; for example, the Willow Creek CBCs for that same period reported only single birds in 1995 and 1999. Spring migration is apparent with the arrival of small flocks and the first singing territorial birds in March. By mid-April, breeding territories are occupied, and the cycle repeats itself.

GRASSHOPPER SPARROW
Continued from page 338

Harris (1996) called this species a "locally rare summer resident and breeder; casual in winter." This statement may be amended with specific reference to Humboldt County. Typically, when atlasers found Grasshopper Sparrows, numerous birds were present in the general area. Further, a great deal of habitat lay in remote areas, and most of it is unbisected by public roads. Before the Atlas project, Grasshopper Sparrows were regularly reported at previously known sites, or at occasional points in between. Coverage demanded by the Atlas project led the surveyors to a great many out-of-the-way places. This unique effort over 5 years resulted in detections of Grasshopper Sparrows at many sites for which there had been few or no reports. It appears likely that the population present in Humboldt County consists of at least many hundreds of birds. For these reasons, the species may be termed an uncommon but locally numerous summer resident and breeder.

Data supporting a meaningful trend analysis in Humboldt are limited. However, as is the case for many obligate grassland bird species, Grasshopper Sparrows elsewhere have experienced declines. Humboldt's grassland habitats are vulnerable to overgrazing and degradation by cattle, and replacement by asphalt and buildings.

BROWN-HEADED COWBIRD
Continued from page 364

give upon taking flight. Midsummer sees callow juvenile cowbirds begging in blackberry thickets and similar exposed shrubbery.

In August and September, independent juveniles begin to accrete into small flocks, at first only of their own kind. It seems typical for occasional individuals or flocks to appear in odd settings, such as along ocean beach wrack lines, on river cobble bars, in parking lots, and even some miles out at sea—they occasionally land on fishing boats. An example of this species' late summer dispersal power was observed by atlasers 7 Aug 1999 when a juvenile cowbird landed on the *M/V Blitzen* about 45 nautical miles off the southern Humboldt coast (JEH, DFx, GCH *et al.*), remaining aboard until the vessel was tied up inside San Francisco Bay. As fall progresses, young cowbirds increasingly associate with larger assemblages of blackbirds, starlings, and adult cowbirds. The winter population in Humboldt County is not especially large, with most of the breeding range abandoned and nearly all of the birds associated with large flocks of blackbirds and starlings at dairies in the coastal lowlands.

The ability of Brown-headed Cowbirds to exploit a great spectrum of wooded edge habitats has contributed to long-term decreases in populations of a number of threatened or endangered North American songbirds. While carefully implemented control programs have been effective in enhancing nesting success for some species on a local scale, some controversy exists about the efficacy of these programs, especially when they are applied at a large scale or during the non-breeding season. Cowbird control has not been attempted in Humboldt, probably since no apparent situation where a declining host species would benefit from such an effort has yet been identified. Unfortunately, it would be expected that any such control project proposed for Humboldt County, regardless of how well considered, would likely be met with emotionally charged opposition from animal rights advocates.

GREEN-WINGED TEAL

Anas crecca

Harris (1996) termed Green-winged Teal a "common migrant and winter visitor; casual to rare in summer" in northwestern California, although he did not consider it to be a breeder. Grinnell and Miller (1944) reported a "definite record" of nesting in Humboldt 3 miles south of Eureka. On 19 May 1935 B. Betterley collected a nest and eggshell fragments (MVZ 4710) thought to be Blue-winged Teal, and sent them to Joseph Grinnell at MVZ. Grinnell concluded that he was "almost 100%" sure the down feathers in the nest belonged to a Green-winged Teal, and that the eggshells "seem to fit this species closest" (J. Grinnell, 20 Jun 1935, unpubl. correspondence on file at MVZ). This record is cautiously accepted as Humboldt's first and only confirmed nesting of Green-winged Teal.

Green-wings typically depart the area by mid-May, with very few birds detected in June and July. Atlas-period Green-winged Teal considered to be non-breeders were recorded in or near the Arcata Marsh 1 Jun 1998, 24 Jun 1998, and 28 Jun 1996 (S. Harris). Fall dispersants or migrants usually appear by August, and these teal are common to abundant in the colder months. Nearly all of the great numbers which migrate through California or spend the winter retire to breeding grounds well to the north of the state.

NORTHERN PINTAIL

Anas acuta

Northern Pintails nesting in California do so primarily east of the Cascade–Sierra Nevada axis, although some nest in coastal central California (Roberson and Tenney 1993). A single confirmed nesting record is known from Del Norte County, but no conclusive evidence exists for Humboldt County (Harris 1996). Yocom (1957) published a report he received of a female flushed from nine eggs near Table Bluff on 21 Apr 1956 (E. Gibbs). Lacking any independent verification of that observation, this species is considered as only a possible breeder in Humboldt.

There were no records suggesting breeding during the Atlas period; a pair at Ocean Ranch (c19) on 2 Jul 1995 (S. McAllister, JEH) may have summered, but were probably very early post-breeding dispersants since they were not subsequently reported. Like Green-winged Teal, Northern Pintail is among the earliest-arriving waterfowl in autumn migration. In northwestern California, Northern Pintail is a common migrant, locally common in winter, and casual to rare in summer (Harris 1996).

BLUE-WINGED TEAL

Anas discors

In northwestern California, Blue-winged Teal is a rare migrant and casual in winter (Harris 1996). They are the latest-migrating ducks in spring along the Pacific Coast, typically showing up in numbers only in May and early June. The majority of Blue-wings encountered in northwestern California in late spring occur as pairs, groups of pairs, or threesomes. These facts, combined with the species' tendency to use shallow freshwater marshes with pools resembling its breeding habitat elsewhere, makes their appearance in May and June confusing. While this duck is a rare breeder in Del Norte County (Harris 1996), conclusive evidence of nesting in Humboldt County is lacking. Clay (1939) discussed a pair seen near Arcata 21 Jun 1938. Wheeler (1965) provided few details on "mated pairs" seen in the Humboldt Bay and Petrolia areas 1959–1964. Harris and Wheeler (1965) reported a hybrid male Blue-winged × Cinnamon Teal attending a female teal sp. and brood near Arcata 10 May 1964. Blue-winged Teal is considered a possible breeder in Humboldt County. It is possible that females with broods may be overlooked or misidentified as Cinnamon Teal.

Eighteen records of this species in May through early July were submitted during the Atlas period, but all were thought to be of non-breeders. Some Atlas-era records were sufficiently late in the season to merit mention. For example, three males were attempting to court a female Cinnamon Teal attending a brood at the Arcata Marsh (F15) 5 Jun 1998 (S. Harris) and 14 birds were seen 2 days later at that same location (M. Morrissette). Five males were there 1 Jul 1998 (S. Harris). Blue-winged Teal pairs also occasionally made appearances at the Arcata Marsh (F15); for example, they were reported there 14 May 1995 (E. Elias) and 21 Jun 1997 (JCP). A group of four (a pair, with two other males) was seen 29 Jun 1998 at Redwood Creek estuary (f05) and a male was at Freshwater Lagoon the same day (f06; B. Freeman). Three were reported at an unspecified location in the Arcata Bottoms 4 Jul 1995 (M. Varga).

RUDDY DUCK

Oxyura jamaicensis

In northwestern California, Harris (1996) considered this species to be a "common migrant and winter visitor; rare to uncommon summer visitor; rare to casual breeder." This species is known as a casual breeder in Del Norte and Humboldt counties, with small numbers known to oversummer at the Arcata Marsh and Arcata sewage ponds in some years (Harris 1996). The Humboldt County breeding confirmations are of three broods at the sewage ponds September and October 1968 (R. Gerstenberg; Harris 1996) and a female with two young there 30 Aug and 8 Sep 1974 (S. Harris; Harris 1996). In 1968, the first of the three broods was observed 5 Sep (R. Gerstenberg). No records suggesting breeding were submitted during the Atlas period.

RING-NECKED DUCK

Aythya collaris

Harris (1996) categorized this duck as a "locally common migrant and winter visitor; casual to rare summer resident and breeder" in northwestern California. While there are a handful of confirmed breeding records from Del Norte County (Harris 1996), there are no such records from Humboldt. The report of a brood at Essex Pond 6 May 1984 (P. & V. Springer) first published in Harris (1991) was an error; this group of Ring-necked Ducks was actually composed of three adult males and an adult female (S. Harris, pers. comm.). This duck is considered a possible, yet unconfirmed, breeder in Humboldt County.

There were no records suggesting breeding during the Atlas period, but a few spring–summer migrants or lingerers were reported. Two females at Ocean Ranch (c19) 2 Jul 1995 (JEH) were of passing interest. These attractive ducks favor quiet log ponds and other still backwaters with heavily vegetated margins that provide seclusion. Breeding by this species in Humboldt should be watched for at freshwater impoundments in the coastal lowlands.

LESSER SCAUP

Aythya affinis

In northwestern California, Lesser Scaup is a "common migrant and winter visitor; rare in summer; accidental breeder" (Harris 1996). The only breeding confirmation in northwestern California was a Humboldt County record of a female with nine newly hatched young at the Arcata sewage ponds 5 Aug 1994 (S. Harris, B. Cariss; Harris 1996).

The only record suggesting breeding for this species during the Atlas period was that of an adult scaup (not identified to species) with a brood of five near Teal Island, Hookton Slough, Humboldt Bay National Wildlife Refuge (D19) 26 Jul 1997 (M. Kuehner). As Lesser Scaup is a regular nesting species in California and has recently increased somewhat as a breeding bird in the San Francisco Bay area, and because Greater Scaup (*Aythya marila*) is not known to nest in the U.S. outside of Alaska, it is very likely that this record referred to nesting by Lesser Scaup. One of the few late spring reports of this species received during the Atlas period was of a bird at Humboldt Bay National Wildlife Refuge 22 May 1996 (D19; K. Foerster), but this observation was not considered evidence of possible breeding.

CLAPPER RAIL

Rallus longirostris

The California Clapper Rail (*R. l. obsoletus*) is one of about 24 subspecies of Clapper Rails that occur from the northeastern U.S. to Brazil and Peru. The endemic California subspecies once occurred in the tidal saltmarshes of central coastal and northern coastal California (Humboldt to Morro Bay). Widespread loss of habitat, overhunting, reductions in its range and corresponding population declines led to the listing of this subspecies as "endangered" by the U.S. Fish and Wildlife Service in 1973, and by the State of California in 1976 (Gill 1979). These rails are now almost entirely restricted to the marshes of the San Francisco Bay estuary.

Harris (1996) discussed some unverified Clapper Rail nest records from Indian (Gunther) Island and Mad River Slough from 1932 (Burton 1972), and some other questionable reports through 1966. There have been no subsequent reports, although thorough and systematic surveys have not been conducted around the bay. However, older reports from Humboldt Bay (Cooper and Suckley 1859, Storer 1915; HSU 2228), along with a specimen taken at Herrick Marsh near Eureka 18 Jan 1917 (HSU 5144; Harris 1996) appear to confirm that this species indeed occurred here in an earlier era. Harris (1996) suggested that the shift in local saltmarshes from pickleweed to cordgrass may be to blame for the loss of this species, although the original population may have been very small or consisted entirely of dispersants from the south. While the species possibly once nested around Humboldt Bay, it is not considered a confirmed breeder in Humboldt County. No Clapper Rails were reported in Humboldt during the Atlas period.

BLACK-NECKED STILT

Himantopus mexicanus

Historically, Black-necked Stilts were considered rare spring migrants and winter visitors in Humboldt County. After the first county record of three near Ferndale 18 Apr 1964 (C. Crane), only four more reports of one or two birds accrued until 1980 (Harris 1996). Thereafter the picture began to change. Stilts were detected in 16 of the years from 1980–1999. Most of these were apparent migrants in April and May, but some remained to nest. The upswing in sightings in the 1980s may have been related to high levels of spring floodwater in the northern Great Basin, an event which persisted over several years and was thought responsible for record numbers of spring migrants in western Oregon and Washington (AB 38:949).

Confirmed nesting in Humboldt occurred at the Arcata Marsh in 1985, 1986, and 1987, at the Vigo Street Marsh in 1985, and at the Fay Slough Wildlife Area in 1994 (Harris 1996). The earliest historical breeding record was of an active nest 29 Jun 1985 at Vigo Street (Eureka) Marsh. This was preceded by the discovery of two pairs there 13 Jun 1985 (G. Lester), with both eventually nesting and young reported 11 Jul 1985 (R. Erickson, G. Lester). Six juveniles from these attempts fledged and, along with the adults, dispersed north to Arcata in August (AB 39:958). One pair nested at the Arcata Marsh in 1986, with two young flying by 14 Aug (Higley 1989). In 1987 two pairs nested there (AB 41:484); and a nest with eggs was reported 2 Jun 1987 (F. Broerman). At Fay Slough there were six or seven adults and three nests 26 Apr–24 May 1994 (M. Fisher).

During the Atlas period there were no observations of stilts suggesting breeding. Five mid- to late April reports were considered spring overshoots; two birds seen along the lower Eel River (e21) 11 Jun 1997 (R. LeValley) were not in suitable nesting habitat and were not reported again.

FORSTER'S TERN

Sterna forsteri

In northwestern California, Forster's Tern is a "common fall migrant; uncommon local winter visitor and spring migrant; casual in summer" (Harris 1996). Townsend (1887) considered them "plentiful at Humboldt Bay in the fall." Harris (1996) reported that a nest with two small tern eggs was photographed on an unnamed island in North Humboldt Bay on 18 Jun 1969 (S. Harris). While adult Forster's Terns were nearby, they were not actually observed on the nest, which was located in an active Caspian Tern colony. This observation is considered confirmation of a breeding attempt in Humboldt County. The nearest coastal breeding sites are in San Francisco Bay.

During the Atlas period, no observations of Forster's Tern suggesting breeding were made. The only late May through July observation of this species during the Atlas period was of a single bird near Manila (e15) on 4 Jun 1996 (GCH). About 20 of these terns began to winter around North Humboldt Bay in 1982, and wintering birds numbered 45 in 1997 (S. Harris, pers. comm.).

CASPIAN TERN

Sterna caspia

Like those of many other colonial bird species, the breeding sites of Caspian Terns shift irregularly in response to environmental conditions. Appropriate nesting areas must have a high degree of freedom from disturbance by people and predators. Until recent decades, Caspian Terns were unknown as breeding birds along most of the Pacific Coast, and they were unknown in northwestern California until 22 Jul 1954, when an adult with a juvenile was seen at the Eel River estuary (R. Mall), and eight were reported at the Mad River Estuary (W. Stumpf; Harris 1996). In Humboldt County, Caspian Terns are common migrants, fairly common summer visitors, and are rare in winter.

Harris (1996) discussed confirmed breeding on Sand Island and another small island in North Humboldt Bay from 1962 to 1969. Numerous adults and young were seen on the unnamed island 14 May 1962 (S. Harris). More than 100 adults flying over Sand Island, along with 10 nests with eggs, were recorded 28 Apr 1963 (S. Harris). On 16 Jun 1966, 144 young were banded on Sand Island (S. Harris). Sand Island was not used for nesting in 1969, but 27 nests were found on the unnamed island (Harris 1996). It is unknown why these islands, originally created by dredge spoils, were subsequently abandoned as nesting sites. S. Harris (pers. comm.) wondered whether increased use of the nesting islands by loafing gulls associated with the nearby Arcata landfill (capped in 1980) may have diminished attendance and nest success by these terns. The colonization—albeit temporary—of North Humboldt Bay corresponded to colony expansion seen in Washington state, on San Francisco Bay, and at Moss Landing (Gill and Mewaldt 1983).

The small islands used by Caspian Terns for nesting in the 1960s are situated at a point in the North Bay that precludes anything other than distant scope views from the shoreline. Approach can be made by boat, but this was never undertaken by atlasers. While regular checks of the islands in North Humboldt Bay were not made during the Atlas period, no nesting activity was noted and it is unlikely that breeding went undetected. While the following breeding behaviors were observed during the Atlas period, they were probably related to known breeding sites to the north of Humboldt in Oregon or Washington: carrying food 21 Apr 1996 near Trinidad (e10; TWL); copulation observed at the mouth of Mad River (f12; TWL) 30 May 1997; food transfer between adults (*i. e.*, courtship) along the lower Mad River (F13) 24 Jun 1997 (S. Levy); carrying food 15–17 Jul 1995 at Manila (e15; GCH); feeding young capable of flight 21 Jul 1996 at Big Lagoon (f08; TWL); and feeding young 2 Aug 1995 at Manila (e15; GCH). Sand Island, as well as Teal Island in South Humboldt Bay, should be monitored in the future for nesting activity.

RING-NECKED PHEASANT

Phasianus colchicus

Although male Ring-necked Pheasants can be regularly heard crowing in some areas of inland Mendocino County, this introduced species no longer breeds in Humboldt. Harris (1996) discussed an apparently self-sustaining population in the Humboldt Bay area in the 1920s through the 1950s, noting that about 750 birds per year were taken by hunters 1938–1940. Pheasants were probably first introduced in the 1910s (S. Harris, pers. comm.). Some confirmed Humboldt breeding records include nests at Arcata 23 May 1927 (HSU 1751) and Rohnerville 6 Apr 1931 (both G. Atwell; Harris 1991), and a brood of one-third-grown young at Petrolia 7 Jun 1976 (S. Harris; Harris 1991). The Humboldt population probably died out as a result of the change from grain farming to dairy pastures (S. Harris, pers. comm.). Similarly, populations in other regions of the United States have declined as a result of modern, "clean" agricultural practices which decrease cover for rearing chicks (Leopold *et al.* 1981).

No records suggestive of breeding were received during the Atlas period. At the Eel River Wildlife Area and other locations, some birds are still released for sport hunting. Escapees are sometimes noted in odd locations, such as a male observed in pasturelands in the Redwood Creek estuary (f05) 29 Aug 1995 (DFx).

YELLOW-BILLED CUCKOO

Coccyzus americanus

The Yellow-billed Cuckoo has not been known to nest anywhere in northwestern California, although seemingly suitable habitat exists along the lower reaches of the Eel River. These riparian stands of black cottonwood and willow were much more extensive in the past (Kelly 1987). Historically, breeding was known from near Sebastopol in northern Sonoma County (Shelton 1911). Although there were no reports of this species in Humboldt County during the Atlas period, Harris (1996) provided four records from before the Atlas period that suggested the possibility of breeding: one bird was reportedly heard calling on the north spit of Humboldt Bay 13 Jul 1947 (B. Anderson); one bird was at Ferndale 24 May 1958 (C. Crane; AFN 12:383); a bird was mist-netted at Lanphere Dunes 4 Jun 1992 which showed a brood patch (C. J. Ralph; AB 46:1175); and a bird was reported at Rohnerville 27 May 1993 (W. & I. Crane). Since it is likely that these records involved lone migrants rather than breeders, Yellow-billed Cuckoo is classified here as a possible breeder in Humboldt County.

While the potential source populations for this species' recruitment to Humboldt County have drastically declined because of habitat loss (Hughes 1999), it is conceivable that this casual visitor may someday be found nesting here. They were formerly more common in riverside cottonwoods and willows throughout the Central Valley and locally elsewhere in California. Continued efforts to restore bottomland riparian habitat along the lower Eel at sites such as Cock Robin Island should be encouraged and could improve the chances of future nesting.

RUFOUS HUMMINGBIRD

Selasphorus rufus

For birders in both northwestern California and southwestern Oregon, the breeding status of Rufous Hummingbird in those areas has been mysterious. Allen's Hummingbird has long been recognized as the breeding *Selasphorus* from coastal Curry County, Oregon, southward. The southernmost limits of Rufous as a nesting bird, however, remain unclear. Do some extend into the northwestern corner of California?

Harris (1996) described this species as an "uncommon to locally common spring migrant; casual in early summer. Rare in late spring." Yet, due to similarities with Allen's Hummingbird, the true status of Rufous Hummingbird as a fall migrant in Humboldt County is unclear. Harris (1996) also indicated that there was no positive evidence of breeding in northwestern California. A nest collected by C. Clay 19 Apr 1930 near Table Bluff (HSU 1267) is of uncertain identity (Harris 1991) and almost certainly Allen's Hummingbird at that coastal location. However, Talmadge (1953a) reported that he collected a Rufous Hummingbird nest with two eggs and the incubating female 4 May 1952 at Mud Springs, Humboldt County. The nest was located 6 feet above the ground along the edge of a clump of white fir in a wet meadow at 4500 feet. This female specimen collected by Talmadge (CAS 61311) was recently rechecked and was indeed found to have measurements typical of Rufous Hummingbirds (D. Mitchell, pers. comm.). Thus, Talmadge's (1953a) nesting record is considered to be Humboldt's first and only confirmed breeding record for this species. In years preceding this collection, Talmadge (1953a) also observed other high elevation nests possibly belonging to this species; the only other location given was Trinity Summit.

During the Atlas period, *Selasphorus* thought to be this species were observed at suspicious inland locations such as Happy Camp Campground (p14) 7 Jun 1998 (KMS), Mill Creek (n08) 13 Jun 1996 (JMH), north of Twin Lakes (m04) 16 Jun 1996 (KMS), and Mud Springs (o10) 7 Jul 1998 (two were banded; P. Herrera). However, because of the late dates and lack of conclusive identification for some, none of the observations were considered evidence of breeding. In Oregon, following nesting in May (Gabrielson and Jewett 1940),

Rufous Hummingbirds move to higher elevations in preparation for southward migration. Sightings in July, June, and possibly even late May in northwestern California are more likely to be migrants rather than breeding birds.

BANK SWALLOW

Riparia riparia

Bank Swallows are small swallows known for nesting colonially in burrows they excavate in soil, sand, or loose rock bluffs and, rarely, other substrates. The species is widely distributed in the Northern Hemisphere. As it is elsewhere, distribution in California is patchy, with occurrence of the birds during the breeding season dependent on appropriate bluffs and banks. Because of extensive loss of suitable nesting habitat, this species is currently listed by the State of California as "threatened" (CDFG 1992). The Bank Swallow was considered a rare spring and fall migrant in northwestern California at the beginning of the Atlas period, and was known to nest at a single site along the lower Smith River in Del Norte County (Harris 1996).

The historical status of Bank Swallow as a breeding bird in Humboldt County is somewhat puzzling. It is known as a confirmed breeder based solely on an account of Talmadge (1947). In the summer of 1946 five isolated nesting pairs, each of which eventually fledged young, were observed nesting at five separate locations. Breeding sites listed were a highway cutbank south of Orick, a sandy cliff at the mouth of Luffenholtz Creek, a clay-and-sand bluff at Essex Rock on the Mad River, sandy bluffs at Table Bluff, and a sandy face above the Van Duzen River. Talmadge (1947) made a distinction between these birds and the sympatric and similar Northern Rough-winged Swallow. These uncorroborated sightings remain Humboldt's only breeding records, and, if correct, they constitute a remarkable phenomenon standing in contrast to the typical behavior of this species.

Efforts to verify two other purported Humboldt breeding occurrences listed in CDFG (1992) were unsuccessful: a 1904 set of six eggs (WFVZ 134659) was determined to be too large for Bank Swallow (R. Corbado, WFVZ, pers. comm.), and a "published" 1956 breeding record at Prairie Creek Redwoods State Park cited as "AFN 11:408" apparently does not exist.

During the Atlas period, there were a few records of single Bank Swallows in late May through July in Humboldt, but all were considered migrants. For example, 18 May 1995 at the Arcata sewage ponds (F15; DFx), 17 and 22 Jun 1995 at Lanphere Bridge over the Mad River Slough (e14; JEH, GCH), and 20 Jul 1999 near Fernbridge (D21; S. McAllister, S. Tucker). Seemingly suitable nesting habitat exists, especially along the Eel River, and this species may well reappear as a Humboldt breeder in the future. This species will readily colonize new sites, as they did during 1995 when a new nesting colony was discovered in Del Norte County at the Crescent City landfill 9 Aug (R. LeValley).

YELLOW-THROATED WARBLER

Dendroica dominica

Harris (1996) considered Yellow-throated Warbler to be a casual spring migrant and accidental in fall in northwestern California. Harris (1996) also reported two birds together 11 Jun–1 Jul 1982 in Eureka, one of which was observed carrying nest material (PFS). While carrying nest material can be considered as confirmation of breeding under Atlas methodology, given how unusual it would be for this species to nest in Humboldt County, this evidence was considered inadequate to establish breeding. Thus, this species is categorized as only a possible breeder in Humboldt County. There were no breeding season reports of Yellow-throated Warblers in Humboldt during the Atlas period, but there was a bird at Loleta 25 Jan–2 Mar 1996 (R. LeValley *et al.*; Harris 1996).

There have been no other reports of nesting by this species in California. Of interest, however, is that this species' seasonal occurrence in California is different from that of most vagrant warblers, with spring records outnumbering those in autumn (Roberson 1980, Harris 1991). This pattern is similar to that of Northern Parula, an "eastern" warbler that has bred several times in California, and in Humboldt County.

ROSE-BREASTED GROSBEAK

Pheucticus ludovicianus

Rose-breasted Grosbeak is a rare migrant in northwestern California (Harris 1996), mostly along the coast. This is among the more frequently detected "eastern" vagrant songbirds. Harris (1996) listed 66 records for the region covered by his work, with 64 of those since 1972. As no remarkable increase in range or numbers elsewhere has been noted, it seems an inescapable conclusion that the increase in coverage of coastal willow thickets and greater attention paid to birds using feeders is responsible for this phenomenon.

Gilbert (1916) related that both males and females of this "abundant" species were observed in an orchard at the Myer's Ranch 1 Jul 1897, and speculated that breeding was likely. McLain (1898) indicated that several heads of this species that had been earlier shot as pests were collected at Myer's Ranch on that day. Efforts to locate these specimens were unsuccessful. Considering how unlikely it would be for Rose-breasted Grosbeaks to be "abundant" anywhere in the western U.S., these records are questionable and possibly refer to Black-headed Grosbeaks. Accordingly, this species is not considered a confirmed Humboldt breeder.

There were nine records of single birds in the county between late May and mid-July during the Atlas period; *e. g.*, 28 May 1999 (H15; S. Colvig), 6 Jun 1997 (e10; K. Irwin), 27 Jun 1997 (e10; TWL), 1 Jul 1995 (D17; JEH), 14 Jul 1996 (F13; K. Miller), and 17 Jul 1998 (e17; L. Warmuth). However, based on factors such as age and length of stay, all birds appeared to be transients and none were considered possible breeders. Since Rose-breasteds have nested in California and Oregon in the past (often pairing with Black-headeds; *e. g.*, AFN 23:693, AB 38:1060, AB 46:1173), this species is considered a possible future breeder in Humboldt.

APPENDIX B: BLOCK STATISTICS

Priority block numbers are capitalized. PO = Possible Breeding; PR = Probable Breeding; CO = Confirmed Breeding. Total area includes only the terrestrial portion of Humboldt County for each block and excludes ocean waters. The %Water column includes bodies of freshwater and brackish estuaries.

BLOCK NUMBER	BLOCK NAME	BLOCK OWNER	TOTAL AREA (KM^2)	%AG/ URBAN	%RED-WOOD	%OAK/ GRASS	%DOU-GLAS-FIR	%HIGH ELEVATION	%WATER	SPECIES PER BLOCK			
										PO	PR	CO	TOTAL
a22	False Cape	Ogan, C.	6.2	0	56	14	16	0	14	5	9	21	35
a23	**Capetown**	**Bacon-Ogden, L.&D.**	**12.7**	**0**	**2**	**52**	**41**	**0**	**5**	**18**	**16**	**14**	**48**
a24	Cape Mendocino	Bacon-Ogden, L.&D.	19.7	0	0	43	55	0	2	24	2	15	41
a25	**Devils Gate**	**Fix, D./Adams, R. J.**	**10.1**	**0**	**0**	**73**	**25**	**0**	**2**	**36**	**6**	**7**	**49**
a26	Mussel Rock	Lester, G.	2.8	0	0	44	28	0	27	28	2	1	31
a27	**Mattole Beach**	**Hazard, G.**	**0.4**	**0**	**0**	**36**	**12**	**0**	**52**	**5**	**1**	**11**	**17**
a28	Punta Gorda	Hazard, G.	2.4	0	0	53	37	0	10	7	3	0	10
b19	**Crab Park North**	**Fix, D.**	**0.6**	**100**	**0**	**0**	**0**	**0**	**0**	**4**	**0**	**1**	**5**
b20	Salt River	Fix, D.	9.9	100	0	0	0	0	0	16	7	10	33
B21	**Centerville**	**Ogan, C.**	**20.7**	**42**	**40**	**15**	**0**	**0**	**3**	**17**	**17**	**36**	**70**
b22	Guthrie Creek	Ogan, C.	25.0	0	76	14	10	0	0	7	7	5	19
B23	**Bear River Ridge West**	**Bacon-Ogden, L.&D.**	**25.0**	**0**	**1**	**55**	**44**	**0**	**0**	**24**	**19**	**29**	**72**
b24	Mendocino Ridge West	Bacon-Ogden, L.&D.	25.0	0	0	27	71	0	2	26	3	8	37
B25	**Branstetter Ridge**	**Power, J.**	**25.0**	**0**	**0**	**41**	**59**	**0**	**0**	**21**	**16**	**19**	**56**
b26	McNutt Gulch	Lester, G.	24.9	1	0	61	39	0	0	39	3	6	48
B27	**Mouth of the Mattole**	**Hazard, G.**	**23.9**	**2**	**0**	**34**	**57**	**0**	**7**	**15**	**22**	**29**	**66**
b28	Uncle Tommy	Hazard, G.	24.0	0	0	19	81	0	0	14	7	6	27
b29	**Cooskie Creek**	**Taylor, H.**	**6.4**	**0**	**0**	**64**	**32**	**0**	**4**	**37**	**4**	**5**	**46**
c18	Table Bluff	Fix, D.	3.8	100	0	0	0	0	0	12	3	9	24
c19	**Ocean Ranch**	**Kidd, J.**	**21.9**	**100**	**0**	**0**	**0**	**0**	**0**	**9**	**3**	**21**	**33**
c20	Cock Robin Island	Hunter, J.	25.0	100	0	0	0	0	0	15	4	18	37
c21	**Ferndale**	**Hunter, J.**	**25.0**	**52**	**41**	**6**	**1**	**0**	**0**	**11**	**3**	**20**	**34**
c22	Mattole Road	Slauson, K.	25.0	0	35	6	59	0	0	24	1	4	29
c23	**Malfunction Junction**	**Ogan, C.**	**25.0**	**0**	**0**	**48**	**51**	**0**	**0**	**14**	**14**	**15**	**43**
c24	Middle Bear River	Ogan, C.	25.0	0	0	28	70	0	2	19	18	7	44
c25	**Hollister Creek**	**Power, J.**	**25.0**	**0**	**0**	**30**	**70**	**0**	**0**	**19**	**7**	**1**	**27**
c26	Apple Tree Ridge	Fix, D./Power, J.	25.0	1	0	35	62	0	2	24	10	10	44
c27	**Petrolia**	**Leskiw, T.**	**25.0**	**5**	**0**	**25**	**66**	**0**	**4**	**44**	**6**	**10**	**60**

BLOCK NUMBER	BLOCK NAME	BLOCK OWNER	TOTAL AREA (KM^2)	%AG/URBAN	%RED-WOOD	%OAK/GRASS	%DOUGLAS-FIR	%HIGH ELEVATION	%WATER	SPECIES PER BLOCK			
										PO	PR	CO	TOTAL
c28	Cooskie Mountain	Leskiw, T.	25.0	2	0	14	85	0	0	20	7	2	29
c29	Spanish Ridge	Leskiw, T.	23.9	0	0	41	58	0	1	19	8	1	28
c30	Spanish Creek	Leskiw, T.	4.4	0	0	41	53	0	6	3	0	0	3
d16	Fairhaven	Hunter, J.	5.0	100	0	0	0	0	0	14	4	13	31
D17	North Jetty	Schmidt, G./Hunter, J.	10.5	98	1	1	0	0	0	18	4	42	64
d18	Humboldt Hill	Spruance, P.	14.2	63	31	6	1	0	0	18	1	24	43
D19	Hookton Slough	Foerster, K.	24.5	90	8	1	1	0	0	21	15	38	74
d20	Loleta	Hewitt, R.	25.0	86	11	2	0	0	0	20	15	29	64
D21	Waddington	Karsteadt, L.	25.0	88	8	2	1	0	0	29	15	36	80
d22	Crosby Road	Leskiw, T.	25.0	1	50	21	28	0	0	25	2	6	33
D23	Bear River Ridge East	Ogan, C.	25.0	0	9	36	55	0	0	23	19	27	69
d24	Kinman Pond	Ogan, C.	25.0	0	1	39	58	0	1	29	8	3	40
D25	North Fork Mattole	Power, J./Fix, D.	25.0	0	0	14	86	0	0	21	10	18	49
d26	Buckeye Mountain	Leskiw, T.	25.0	0	0	23	77	0	0	21	0	1	22
D27	Burgess Ridge	Leskiw, T.	25.0	0	0	18	81	0	1	17	12	15	44
d28	Arthur Way County Park	Leskiw, T.	25.0	4	0	20	74	0	2	39	9	15	63
D29	Moorehead Ridge	Leskiw, T.	25.0	0	0	11	89	0	0	17	14	14	45
d30	Kinsey Ridge	Leskiw, T.	21.6	0	0	4	95	0	0	16	4	3	23
d31	Big Flat	Slauson, K.	4.3	0	0	1	98	0	1	15	5	2	22
e04	Redding Rock	Hunter, J.				OCEAN BLOCK				3	0	1	4
e08	Agate Beach	Hunter, J.	1.2	8	63	0	4	0	25	21	0	2	23
e09	Patricks Point	Schmidt, G.	11.7	4	92	1	3	0	1	21	10	23	54
e10	Trinidad	Schmidt, G.	9.1	6	87	4	1	0	2	18	8	43	69
e11	Pilot Rock	Hunter, J.				OCEAN BLOCK				2	1	3	6
e13	Mad River County Park	Fix, D.	1.5	100	0	0	0	0	0	14	1	10	25
e14	Lanphere Dunes	Hunter, J.	8.7	100	0	0	0	0	0	8	5	13	26
e15	Manila	Fix, D.	7.3	100	0	0	0	0	0	6	2	30	38
e16	Woodley Island	Fix, D.	14.7	93	6	1	0	0	0	9	9	33	51
e17	Southeast Eureka	Springer, P.&V.	25.0	38	55	7	0	0	0	23	7	18	48
e18	Elk River Road	Springer, P.	25.0	14	82	3	1	0	0	22	6	7	35
e19	Catfish Lake	Wachs, M.	25.0	0	100	0	0	0	0	26	13	27	66
e20	Palmer Creek	Wachs, M.	25.0	5	89	6	0	0	0	38	0	3	41

BLOCK NUMBER	BLOCK NAME	BLOCK OWNER	TOTAL AREA (KM^2)	%AG/ URBAN	%RED- WOOD	%OAK/ GRASS	%DOU- GLAS-FIR	%HIGH ELEVATION	%WATER	SPECIES PER BLOCK			
										PO	PR	CO	TOTAL
e21	Fortuna	Chinnici, S.	25.0	49	30	17	3	0	0	26	20	19	65
e22	Metropolitan	Hewitt, R.	25.0	39	42	18	0	0	1	18	21	28	67
e23	Howe Creek	Ogan, C.	25.0	2	67	14	17	0	0	30	22	9	61
e24	Monument Road	Bacon-Ogden, L.&D.	25.0	0	17	30	53	0	0	35	6	14	55
e25	Windy Nip	Hiss, J.	25.0	0	0	17	81	0	1	37	16	6	59
e26	Rainbow Ridge South	Morris, M.	25.0	0	0	19	80	0	0	22	4	2	28
e27	Van Choick Ridge	Fix, D.	25.0	0	0	23	76	0	1	23	5	2	30
e28	Honeydew	Leskiw, T.	25.0	0	0	46	53	0	1	34	3	4	41
e29	Bear Trap	Hunter, J.	25.0	1	0	14	84	0	1	41	3	4	48
e30	North Slide Peak	Hunter, J.	25.0	0	0	0	99	1	0	19	1	3	23
e31	Fire Hill	Slauson, K.	19.1	0	0	1	98	0	0	24	5	1	30
f04	Gold Bluffs Beach	Hunter, J.	1.3	0	100	0	0	0	0	21	1	1	23
f05	Redwood Creek Estuary	Hunter, J.	5.8	41	59	0	0	0	0	24	11	27	62
f06	Freshwater Lagoon	Power, J./Fix, D.	11.0	2	66	0	0	0	32	9	7	17	33
F07	Dry Lagoon	Hunter, J.	16.7	5	80	0	0	0	14	22	18	22	62
f08	Big Lagoon	Hunter, J.	24.0	1	71	3	0	0	25	20	16	24	60
F09	Another Maple Creek	Morrissette, M.	25.0	0	100	0	0	0	0	16	16	14	46
f10	Fox Farm Road	Morrissette, M.	25.0	0	99	1	0	0	0	17	5	4	26
F11	Crannell	Hunter, J.	18.8	2	86	8	0	0	3	29	17	51	97
f12	Dows Prairie	Hunter, J.	20.3	47	43	8	0	0	1	20	10	34	64
F13	McKinleyville	Schmidt, G.	24.7	77	21	2	0	0	0	28	8	44	80
f14	Arcata North	Stebbins, P.	25.0	91	8	1	0	0	1	11	11	46	68
F15	North Arcata Bay	Meyers, C.	13.8	92	6	1	1	0	0	18	18	47	83
f16	Fay Slough	George, L.	20.3	63	30	6	0	0	1	17	14	31	62
F17	Freshwater West	Springer, P.&V.	25.0	5	88	5	1	0	0	22	11	43	76
f18	Elk River Scout Camp	Springer, P.	25.0	0	100	0	0	0	0	20	3	4	27
F19	Falk	Wachs, M.	25.0	0	100	0	0	0	0	17	9	18	44
f20	Felt Springs	Garrison, D.	24.9	0	100	0	0	0	0	9	9	11	29
F21	Rohnerville East	Hunter, J.	25.0	9	80	11	0	0	0	23	12	27	62
f22	Hydesville	Hunter, J.	25.0	46	38	15	0	0	1	33	6	11	50
F23	Scotia	Bacon-Ogden, L.&D.	25.0	13	79	6	1	0	0	30	20	31	81
f24	Giacomini Prairie	Hewitt, R.	25.0	1	83	4	12	0	1	24	12	7	43

BLOCK NUMBER	BLOCK NAME	BLOCK OWNER	TOTAL AREA (KM^2)	%AG/ URBAN	%RED-WOOD	%OAK/ GRASS	%DOU-GLAS-FIR	%HIGH ELEVATION	%WATER	SPECIES PER BLOCK			
										PO	PR	CO	TOTAL
F25	Mount Pierce	Morris, M.	25.0	0	3	5	91	0	1	19	16	21	56
f26	Harmonica Creek	Gaffin, J.	25.0	0	7	2	91	0	0	25	3	3	31
F27	Fox Camp Spring	Leskiw, T.	25.0	0	0	16	83	0	1	22	13	13	48
f28	Windy Nip Gap	Leskiw, T.	25.0	0	0	17	83	0	1	24	2	3	29
F29	Honeydew Southeast	Hunter, J.	25.0	0	0	12	86	0	2	18	17	23	58
f30	Wilder Ridge	Hunter, J.	25.0	0	0	0	100	0	0	18	3	0	21
F31	Saddle Mountain	Slauson, K.	25.0	0	0	0	100	0	0	17	9	28	54
f32	Tolkan Campground	Hunter, J./Leskiw, T.	16.1	0	0	0	100	0	0	18	0	4	22
f33	Shelter Cove	Dyer, K.	8.4	3	0	19	77	0	1	34	9	11	54
g01	Alder Conservation Camp	Kenney, M.	2.6	0	100	0	0	0	0	11	24	7	42
g02	Upper Prairie Creek	Kenney, M.	19.3	0	100	0	0	0	0	21	5	19	45
g03	Fern Canyon	Anderson, D.	21.1	0	100	0	0	0	0	14	10	15	39
g04	Lower Prairie Creek	Maxwell, M.	24.6	0	97	3	0	0	0	25	14	12	51
g05	Orick	Devlin-Craig, B.&P.	25.0	16	80	3	0	0	0	27	14	30	71
g06	McArthur Creek	Devlin-Craig, B.&P.	25.0	1	99	0	0	0	0	21	6	10	37
g07	Fortyfour Creek	Power, J.	25.0	0	100	0	0	0	0	20	3	0	23
g08	Rodgers Peak	Schmidt, G.	25.0	0	100	0	0	0	0	16	6	1	23
g09	North Fork Maple Creek	Schmidt, G.	25.0	0	100	0	0	0	0	28	1	0	29
g10	The Gap	Kenney, M.	25.0	0	100	0	0	0	0	29	11	3	43
g11	Railroad Creek	Schmidt, G.	25.0	0	100	0	0	0	0	29	3	0	32
g12	Mather Creek	Schmidt, G.	25.0	2	96	2	0	0	0	16	16	20	52
g13	Fieldbrook South	Schmidt, G.	25.0	1	88	10	0	0	0	18	14	24	56
g14	Essex Pond	Hunter, J.	25.0	10	71	14	2	0	3	29	20	29	78
g15	Lower Fickle Hill	Springer, P.&V.	25.0	12	82	6	1	0	0	14	9	45	68
g16	Greenwood Heights	Springer, P.	25.0	0	100	0	0	0	0	29	8	19	56
g17	Freshwater East	Hewitt, R.	25.0	1	97	2	0	0	0	28	3	12	43
g18	Freshwater Creek	Sundeen, K.	25.0	0	100	0	0	0	0	22	3	5	30
g19	South Branch North Fork Elk River	Hiss, J.	25.0	0	100	0	0	0	0	20	9	1	30
g20	Corner Creek	Hiss, J.	25.0	0	100	0	0	0	0	21	15	6	42
g21	Yager Camp	Chinnici, S.	25.0	1	99	0	0	0	0	18	10	4	32
g22	Cuddeback	Marsh, P.	25.0	9	82	7	0	0	1	12	3	6	21
g23	Stitz Creek	Hunter, J.	25.0	1	98	0	0	0	1	29	3	3	35

BLOCK NUMBER	BLOCK NAME	BLOCK OWNER	TOTAL AREA (KM^2)	%AG/ URBAN	%RED-WOOD	%OAK/ GRASS	%DOU-GLAS-FIR	%HIGH ELEVATION	%WATER	SPECIES PER BLOCK			
										PO	PR	CO	TOTAL
g24	Stafford	Hewitt, R.	25.0	1	96	2	1	0	0	46	12	3	61
g25	Harrow Prairie	Power, J.	25.0	0	53	3	44	0	0	20	5	1	26
g26	Albee Creek	Garrison, D.	25.0	0	59	5	36	0	1	23	7	9	39
g27	Bull Creek	Garrison, D.	25.0	0	17	3	79	0	0	27	6	1	34
g28	Panther Gap Road	Leskiw, T.	25.0	0	0	4	96	0	0	27	2	0	29
g29	Duncan Creek	Power, J./Fix, D.	25.0	0	0	9	89	0	1	19	2	1	22
g30	Sholes Creek	Fix, D./Power, J.	25.0	0	0	6	91	0	2	25	0	0	25
g31	Ettersburg Town	Carratello, R./Roberson, D.	25.0	1	0	11	88	0	0	23	2	6	31
g32	Queens Peak	Carratello, R./Roberson, D.	25.0	0	0	3	97	0	0	18	1	1	20
g33	Shelter Cove Highlands	Dyer, K.	24.5	0	0	3	97	0	0	20	7	14	41
g34	Chemise Mountain	Leskiw, T.	5.4	0	0	2	98	0	0	18	2	4	24
h01	Tarup Creek	Kenney, M.	3.4	0	100	0	0	0	0	6	18	9	33
h02	North Fork Ah Pah Creek	Kenney, M.	25.0	0	100	0	0	0	0	17	17	7	41
H03	Big Tree North	Hunter, J.	25.0	0	93	0	7	0	0	13	15	15	43
h04	Boyes Creek	Maxwell, M.	25.0	0	99	0	1	0	0	7	12	8	27
H05	Lost Man Creek	Devlin-Craig, B.&P.	25.0	0	99	0	1	0	0	14	13	12	39
h06	Bald Hills	Devlin-Craig, B.&P.	25.0	0	99	0	1	0	0	18	18	4	40
H07	Immortal Trees	Power, J.	25.0	0	92	4	4	0	0	23	19	22	64
h08	Bridge Creek	Schmidt, G.	25.0	0	100	0	0	0	0	17	4	1	22
H09	Devils Pass	Schmidt, G.	25.0	0	100	0	0	0	0	19	11	14	44
h10	Squaw Tit	Schmidt, G.	25.0	0	99	0	1	0	0	23	5	0	28
H11	Little River	Schmidt, G.	25.0	0	99	0	0	0	0	16	16	15	47
h12	The Basin	Schmidt, G.	25.0	0	100	0	0	0	0	17	19	8	44
H13	Tip Top Ridge	Schmidt, G.	25.0	0	99	1	0	0	0	19	15	23	57
h14	Blue Lake	Brown, H.	25.0	17	68	14	1	0	0	23	15	44	82
H15	Mad River Hatchery	Springer, P.	25.0	16	75	1	4	0	3	15	10	68	93
h16	Boynton Prairie	Springer, P.&V.	25.0	0	66	1	32	0	1	19	9	6	34
H17	Kneeland	Levy, S.	25.0	0	56	4	39	0	0	28	14	19	61
h18	Kneeland Prairie	Desch, A.	25.0	0	37	27	36	0	0	6	20	11	37
H19	Lawrence Creek	Power, J./Fix, D.	25.0	0	84	2	13	0	0	20	10	17	47
h20	Shaw Creek	Hewitt, R.	25.0	0	91	4	5	0	0	20	4	2	26
H21	Yager Creek Swamp	Hunter, J./Leskiw, T.	25.0	0	96	1	1	0	2	24	14	21	59

BLOCK NUMBER	BLOCK NAME	BLOCK OWNER	TOTAL AREA (KM^2)	%AG/URBAN	%RED-WOOD	%OAK/GRASS	%DOU-GLAS-FIR	%HIGH ELEVATION	%WATER	SPECIES PER BLOCK			
										PO	PR	CO	TOTAL
h22	Cummings Creek Road	Hunter, J.	25.0	0	98	0	2	0	0	15	6	3	24
H23	Van Duzen County Park	Powell, C.	25.0	1	98	1	0	0	0	31	18	37	86
h24	Pepperwood	Hunter, J.	25.0	6	93	0	0	0	0	20	3	8	31
H25	Holmes Flat	Adams, R. J./Anderson, D.	25.0	5	91	1	4	0	0	24	10	31	65
h26	Big Tree South	Garrison, D.	24.9	0	97	2	1	0	0	15	4	5	24
H27	Grasshopper Peak	Leskiw, T.	25.0	0	75	2	23	0	0	13	9	16	38
h28	Upper Bull Creek	Leskiw, T.	25.0	0	19	3	78	0	0	18	3	0	21
H29	Gilham Butte	Fix, D./Power, J.	25.0	0	0	22	78	0	0	19	9	22	50
h30	Crooked Creek	Power, J./Fix, D.	25.0	0	0	11	88	0	1	17	1	2	20
H31	Ettersburg Bridge	Roberson, D./Carratello, R.	25.0	0	0	11	88	0	1	17	11	29	57
h32	Ettersburg Road East	Carratello, R./Roberson, D.	25.0	0	0	3	97	0	0	21	1	3	25
H33	Thorn South	Dyer, K.	25.0	0	6	5	89	0	0	21	14	17	52
h34	Whale Gulch	Leskiw, T./Hunter, J.	9.2	0	1	0	98	0	0	16	4	2	22
i01	Starwein	Fix, D./Power, J.	3.6	0	30	1	59	0	10	18	2	1	21
i02	Blue Creek	Power, J./Fix, D.	25.0	0	48	2	42	0	8	17	2	5	24
i03	Bear Creek	Fix, D./Power, J.	25.0	0	25	0	69	0	6	19	0	4	23
i04	Tectah Creek	Power, J./Fix, D.	25.0	0	27	0	73	0	1	16	7	1	24
i05	Johnson Road North	Fix, D./Power, J.	25.0	0	26	1	74	0	0	11	10	1	22
i06	Johnson Road South	Power, J./Fix, D.	25.0	0	25	1	74	0	0	23	3	1	27
i07	Bear Prairie	Fix, D./Power, J.	25.0	0	6	7	86	0	0	17	7	2	26
i08	Childs Hill Prairie	Ogan, C.	25.0	0	47	18	34	0	0	37	30	11	78
i09	Lyons Ranch	Leskiw, T.	25.0	0	54	10	36	0	0	26	4	4	34
i10	Panther Creek	Leskiw, T.	25.0	0	32	0	68	0	0	22	3	1	26
i11	Dolly Varden	Kriger, D.	25.0	0	44	0	55	0	1	24	7	2	33
i12	Wiregrass Ridge	Kriger, D.	25.0	0	63	1	35	0	0	21	9	4	34
i13	Mule Creek	Power, J.	25.0	0	95	1	4	0	0	26	13	13	52
i14	Bald Mountain Creek	Power, J.	25.0	0	98	0	2	0	0	26	11	11	48
i15	Jiggs Creek	Hazard, G.	25.0	0	66	15	18	0	1	22	21	12	55
i16	Angel Ranch	Hazard, G.	25.0	0	54	2	41	0	3	24	17	8	49
i17	Butler Valley	Keil, C.	25.0	0	1	6	93	0	1	21	5	8	34
i18	Ashfield Ridge	Keil, C.	25.0	0	3	22	75	0	0	29	15	11	55
i19	Gordon RIdge	Slauson, K.	25.0	0	14	18	68	0	0	18	11	2	31

BLOCK NUMBER	BLOCK NAME	BLOCK OWNER	TOTAL AREA (KM^2)	%AG/ URBAN	%RED-WOOD	%OAK/ GRASS	%DOU-GLAS-FIR	%HIGH ELEVATION	%WATER	SPECIES PER BLOCK			
										PO	PR	CO	TOTAL
i20	Shaws Ranch	Slauson, K.	25.0	0	5	52	43	0	0	12	10	1	23
i21	Bootjack Prairie	Fix, D./Power, J.	25.0	0	43	20	37	0	0	30	5	2	37
i22	Redwood House	Hazard, G.	25.0	0	51	20	29	0	0	12	12	5	29
i23	Grizzly Creek State Park	Gilchrist, E./Sing Wu, L.	25.0	0	88	0	11	0	0	28	14	14	56
i24	Bridge Creek	Hunter, J.	25.0	0	96	0	3	0	0	29	6	1	36
i25	Larabee Town	Adams, R. J.	25.0	7	88	4	0	0	1	28	7	6	41
i26	Dyerville	Hewitt, R.	25.0	2	90	1	7	0	0	31	14	12	57
i27	Weott	Gaffin, J.	24.9	1	98	0	1	0	0	28	6	12	46
i28	Kerr Peak	Gaffin, J.	25.0	0	68	3	28	0	0	22	1	3	26
i29	Middle Salmon Creek	Gaffin, J.	25.0	0	1	47	52	0	0	24	2	4	30
i30	Goat Rock	Gaffin, J.	25.0	0	0	37	63	0	0	20	3	1	24
i31	Briceland North	Adams, R. J.	25.0	0	0	13	87	0	0	32	5	2	39
i32	Briceland South	Adams, R. J.	25.0	0	0	7	93	0	0	20	8	7	35
i33	Whitethorn School	Fix, D./Power, J.	25.0	0	3	0	97	0	0	20	2	1	23
i34	Our Lady of the Redwoods Abbey	Power, J./Fix, D.	9.4	0	25	2	73	0	0	19	3	1	23
j01	South Red Mountain	Fix, D./Power, J.	2.4	0	2	0	98	0	0	21	11	3	35
j02	Coyote Creek	Power, J./Fix, D.	25.0	0	6	0	93	0	1	15	5	1	21
J03	Bear Creek Headwaters	Power, J./Fix, D.	25.0	0	3	0	96	0	0	17	1	16	34
j04	Pecwan	LeValley, R.	25.0	0	5	0	92	0	2	27	7	2	36
J05	Mettah	Power, J.	25.0	0	0	1	99	0	0	26	16	21	63
j06	Roach Creek	Fix, D./Power, J.	25.0	0	0	2	98	0	0	22	1	1	24
J07	Shelton Prairie	Power, J./Fix, D.	25.0	0	0	3	97	0	0	16	14	15	45
j08	Schoolhouse Peak	Schmidt, G.	25.0	0	0	21	78	0	0	22	8	4	34
J09	Coyote Peak	Leskiw, T.	25.0	0	0	30	70	0	0	19	11	17	47
j10	Eagle Rock	Leskiw, T.	25.0	0	3	28	68	0	0	19	2	0	21
J11	Redwood Valley	Kriger, D.	25.0	0	4	6	90	0	0	21	18	26	65
j12	Camp Anderson	Kriger, D.	25.0	0	3	5	91	0	1	29	3	3	35
J13	Lord-Ellis	Power, J.	25.0	0	28	4	68	0	0	33	13	20	66
j14	Bald Mountain	Power, J.	25.0	0	25	4	71	0	0	17	4	4	25
J15	Canon Creek	Power, J.	25.0	0	19	9	73	0	0	20	8	29	57
j16	Maple Creek North	Power, J.	25.0	1	11	10	78	0	0	23	11	4	38
J17	Maple Creek Middle	Keil, C.	25.0	0	0	8	90	0	2	27	14	25	66

BLOCK NUMBER	BLOCK NAME	BLOCK OWNER	TOTAL AREA (KM^2)	%AG/ URBAN	%RED-WOOD	%OAK/ GRASS	%DOU-GLAS-FIR	%HIGH ELEVATION	%WATER	SPECIES PER BLOCK			
										PO	PR	CO	TOTAL
j18	Maple Creek South	Keil, C.	25.0	0	0	15	83	0	2	40	6	5	51
J19	Iaqua Buttes	Bunnell, S./Fix, D.	25.0	0	1	14	85	0	1	27	20	22	69
j20	Lone Star Junction	Slauson, K.	25.0	0	0	56	44	0	0	20	12	3	35
J21	Yager Creek Middle Fork	Fix, D.	25.0	0	0	30	70	0	0	20	14	15	49
j22	Bald Jessie	Fix, D.	25.0	0	0	60	40	0	0	16	6	1	23
J23	Bear River Ridge West	Gilchrist, E./Sing Wu, L.	25.0	1	6	33	60	0	1	20	14	22	56
j24	Pepperwood Falls	Gilchrist, E./Sing Wu, L.	25.0	1	17	3	78	0	1	22	3	7	32
J25	Arnold Gulch	Fix, D.	25.0	0	60	2	37	0	1	19	15	19	53
j26	Kapple Creek	Hewitt, R.	25.0	0	47	7	46	0	1	22	5	7	34
J27	McCann Creek	Gaffin, J.	25.0	0	88	1	10	0	0	24	23	22	69
j28	Myers Flat	Gaffin, J.	25.0	1	94	2	2	0	1	34	4	12	50
J29	Lower Salmon Creek	Gaffin, J.	25.0	0	33	6	60	0	1	29	18	29	76
j30	Bear Buttes	Gaffin, J.	25.0	0	0	11	89	0	0	18	2	3	23
J31	Redway West	Adams, R. J./Anderson, D.	25.0	3	19	12	65	0	1	18	19	18	55
j32	Little Sprowel Creek	Adams, R. J.	25.0	0	0	9	91	0	0	25	12	7	44
J33	Sprowel Creek	Power, J./Fix, D.	25.0	0	17	3	81	0	0	14	11	17	42
j34	Jones Creek	Fix, D./Power, J.	9.6	0	42	6	52	0	0	15	4	2	21
k01	Blue Creek	Hunter, J.	1.9	0	0	0	100	0	0	17	3	4	24
k02	Slide Creek	Hazard, G.	25.0	0	1	0	98	0	0	19	7	1	27
k03	West Fork Pecwan Creek	Fix, D./Power, J.	25.0	0	0	1	80	19	0	16	3	2	21
k04	East Fork Pecwan Creek	Hazard, G.	25.0	0	0	2	98	0	0	11	5	4	20
k05	Cappell Creek	Hazard, G.	25.0	0	0	5	95	0	0	10	5	7	22
k06	Cappell Flat	LeValley, R.	25.0	0	0	2	98	0	0	24	3	2	29
k07	Wiregrass Prairie	LeValley, R.	25.0	0	0	4	96	0	0	21	3	1	25
k08	French Camp Ridge	Fix, D./Power, J.	25.0	0	0	1	99	0	0	27	7	2	36
k09	Little Pine Creek	Higley, M.	25.0	0	0	0	100	0	0	33	7	16	56
k10	Hupa Mountain	Higley, M.	25.0	0	0	4	95	1	0	24	16	8	48
k11	Pine Ridge	Hunter, J.	25.0	0	0	2	96	2	0	16	6	1	23
k12	Upper Lacks Creek	Hunter, J.	25.0	0	0	8	92	0	0	21	3	1	25
k13	Circle Point	Leskiw, T.	25.0	0	0	11	89	0	0	24	5	0	29
k14	Green Point	Leskiw, T.	25.0	0	1	20	79	0	0	24	8	6	38
k15	Christmas Prairie	Fix, D.	25.0	0	2	7	91	0	0	22	3	2	27

BLOCK NUMBER	BLOCK NAME	BLOCK OWNER	TOTAL AREA (KM^2)	%AG/ URBAN	%RED-WOOD	%OAK/ GRASS	%DOU-GLAS-FIR	%HIGH ELEVATION	%WATER	SPECIES PER BLOCK			
										PO	PR	CO	TOTAL
k16	High Prairie	Fix, D.	25.0	0	0	14	85	0	1	33	12	7	52
k17	Snow Camp Road	Ragatz, C.	25.0	0	0	7	70	22	0	29	8	1	38
k18	Graham Ridge	Ragatz, C.	25.0	0	0	9	61	30	0	25	4	0	29
k19	Wilson Creek	Schmidt, K.	25.0	0	0	33	50	16	1	28	2	2	32
k20	McAlvey Ridge	Schmidt, K.	25.0	0	0	46	53	0	1	19	8	3	30
k21	Showers Pass Road North	Springer, P.	25.0	0	0	63	37	0	0	23	7	8	38
k22	Showers Pass Road South	Springer, P.	25.0	0	0	62	37	0	1	18	16	4	38
k23	Brown Creek	Hewitt, R.	25.0	0	0	49	50	0	2	30	4	3	37
k24	Hoagland Creek	Hunter, J.	25.0	0	1	24	75	0	0	24	3	4	31
k25	Maxwell Creek	Fix, D./Power, J.	25.0	0	4	14	81	0	0	25	3	0	28
k26	Brushy Mountain	Hewitt, R.	25.0	0	8	10	82	0	0	22	4	0	26
k27	Whitlow	Diehl, D.	25.0	0	51	9	38	0	2	29	12	15	56
k28	Mail Ridge	Diehl, D.	25.0	0	45	11	44	0	0	40	10	3	53
k29	Phillipsville/Miranda	Gaffin, J.	25.0	1	63	5	30	0	1	26	2	4	32
k30	Sylvandale	Gaffin, J.	25.0	0	10	39	50	0	1	26	5	2	33
k31	Redway East	Hunter, J.	25.0	4	15	36	42	0	3	27	5	4	36
k32	Garberville	Hunter, J.	25.0	7	14	36	43	0	1	33	11	22	66
k33	Reed Mountain Road	Power, J./Fix, D.	25.0	0	23	9	68	0	0	35	8	9	52
k34	Richardson Grove	Fix, D./Power, J.	9.8	0	39	6	55	0	0	27	2	2	31
l01	Nickowitz	Hunter, J.	0.3	0	0	0	100	0	0	22	4	2	28
l02	Soapstone Gulch	Hunter, J.	3.5	0	0	0	87	13	0	31	4	4	39
l03	Onion Mountain North	Hewitt, R.	8.5	0	0	0	34	66	0	20	21	4	45
l04	Onion Mountain South	Hunter, J.	25.0	0	0	0	80	20	0	21	1	1	23
L05	Bee Mountain	Herrera, P.	25.0	0	0	0	88	12	0	16	15	14	45
l06	Rube Creek	Herrera, P.	25.0	0	0	14	80	6	0	24	6	2	32
L07	Martins Ferry	Power, J./Fix, D.	25.0	0	0	8	91	1	0	20	16	22	58
l08	Bloody Camp	Higley, M.	25.0	0	0	2	98	0	0	27	27	4	58
L09	Hog Ranch Ridge	Higley, M.	25.0	0	0	8	92	0	0	25	23	23	71
l10	Lone Pine Mountain	Godfrey, T.	25.0	0	0	1	99	0	0	27	20	16	63
L11	Sugar Pine Mountain	Higley, M.	25.0	0	1	0	99	0	0	17	24	25	66
l12	Supply Creek	Higley, M.	25.0	0	0	2	79	19	0	35	16	8	59
L13	Three Creeks	Leskiw, T.	25.0	0	0	3	92	5	0	23	16	19	58

BLOCK NUMBER	BLOCK NAME	BLOCK OWNER	TOTAL AREA (KM^2)	%AG/ URBAN	%RED-WOOD	%OAK/ GRASS	%DOU-GLAS-FIR	%HIGH ELEVATION	%WATER	SPECIES PER BLOCK			
										PO	PR	CO	TOTAL
l14	Ruby Creek	Leskiw, T.	25.0	0	0	6	76	18	0	19	12	3	34
L15	Horse Mountain Southwest	Fix, D.	25.0	0	0	28	24	48	0	14	24	27	65
l16	Minon Creek	Fix, D.	25.0	0	0	33	64	3	0	25	1	1	27
L17	Snow Camp Mountain	Ragatz, C.	25.0	0	0	10	63	27	0	19	18	20	57
l18	Board Camp Mountain	Ragatz, C.	25.0	0	0	0	18	82	0	30	8	3	41
L19	Bug Creek	Schmidt, K.	25.0	0	0	9	54	37	0	19	10	17	46
l20	Gorden Road	Schmidt, K.	25.0	0	0	18	79	2	1	16	11	3	30
L21	Stapp Road	Springer, P.	25.0	0	0	59	34	6	0	11	15	22	48
l22	Bloody Run		25.0	0	0	57	41	0	1	0	0	0	0
L23	McClellan Mountain	Morris, M.	25.0	0	0	24	76	0	0	10	31	16	57
l24	Larabee Buttes	Hunter, J.	25.0	0	0	8	90	2	0	15	4	1	20
L25	Brightman Ranch	Morris, M.	25.0	0	0	24	74	1	0	25	20	31	76
l26	Frost Creek	Hewitt, R.	25.0	0	0	15	85	0	0	3	4	3	10
L27	Billy Goat Peak	Diehl, D.	25.0	0	0	20	77	0	3	16	14	30	60
l28	Eel Rock	Diehl, D.	25.0	0	0	19	80	0	1	29	10	8	47
L29	Hubbard Station	Gaffin, J.	25.0	0	0	24	76	0	0	20	14	20	54
l30	Ohman Creek	Gaffin, J.	25.0	0	0	45	55	0	0	13	5	8	26
L31	Dean Creek	Rovell, S.	25.0	0	0	76	24	0	0	12	15	15	42
l32	Little Buck Mountain	Rovell, S.	25.0	0	0	72	28	0	0	14	4	2	20
L33	Reed Mountain	Power, J./Fix, D.	25.0	0	0	17	82	0	0	16	7	22	45
l34	South County Line	Fix, D./Power, J.	9.9	0	1	15	84	0	0	23	2	0	25
m03	Deer Lick Lake	Hewitt, R.	5.7	0	0	0	42	58	0	24	12	0	36
m04	Rock Prairie	Slauson, K.	25.0	0	0	0	67	33	0	34	3	0	37
m05	Twin Lakes	Slauson, K.	25.0	0	0	0	99	1	0	20	13	0	33
m06	Fish Lake	Roberson, D./Carratello, R.	25.0	0	0	0	97	3	1	23	11	13	47
m07	Cavanaugh Creek	LeValley, R.	25.0	0	0	0	98	2	0	17	6	3	26
m08	Weitchpec	Higley, M.	25.0	0	0	0	100	0	0	27	22	15	64
m09	Mill Creek Ridge	Higley, M.	25.0	0	0	14	84	1	2	34	24	10	68
m10	Hoopa Valley	Higley, M.	25.0	15	0	12	68	0	4	33	15	15	63
m11	Hoopa Airport	Higley, M.	25.0	6	0	8	83	0	3	31	15	28	74
m12	Brannan Mountain	Pagan, D.	25.0	0	0	0	94	6	0	29	8	8	45
m13	Johnson Prairie	Leskiw, T.	25.0	0	0	18	82	0	0	26	6	4	36

BLOCK NUMBER	BLOCK NAME	BLOCK OWNER	TOTAL AREA (KM^2)	%AG/ URBAN	%RED-WOOD	%OAK/ GRASS	%DOU-GLAS-FIR	%HIGH ELEVATION	%WATER	SPECIES PER BLOCK			
										PO	PR	CO	TOTAL
m14	Brush Mountain Lookout	Leskiw, T.	25.0	0	0	0	96	3	0	16	9	4	29
m15	Friday Ridge	Leskiw, T.	25.0	0	0	1	74	25	0	20	16	7	43
m16	White Rock	Leskiw, T.	25.0	0	0	4	16	79	0	18	17	14	49
m17	Grouse Mountain	Schmidt, G.	25.0	0	0	8	18	74	0	20	21	9	50
m18	Cow Creek Ridge	Schmidt, G.	25.0	0	0	6	60	34	0	15	11	1	27
m19	Board Camp Butte	Schmidt, G.	25.0	0	0	8	29	62	0	24	1	7	32
m20	Deer Creek	Schmidt, G.	25.0	0	0	41	43	16	0	33	2	2	37
m21	Showers Mountain	Fix, D.	25.0	0	0	7	73	19	0	21	7	2	30
m22	Eaton Roughs	Herrera, P.	25.0	0	0	34	54	10	2	17	1	0	18
m23	McClellan Rock	Hazard, G.	25.0	0	0	38	59	0	3	31	2	0	33
m24	Larabee Valley	Shreeve, M.&J.	25.0	0	0	35	64	0	1	22	4	6	32
m25	Butte Creek	Hunter, J.	25.0	0	0	10	69	21	0	21	1	1	23
m26	Payton Ranch	Hunter, J.	25.0	0	0	50	43	6	1	25	2	0	27
m27	Upper Coleman Creek	Hunter, J.	25.0	0	0	23	76	0	1	32	3	2	37
m28	Blue Jacket Butte	Hunter, J.	25.0	0	0	38	62	0	0	20	0	0	20
m29	Boehne Butte	Gaffin, J.	25.0	0	0	39	61	0	0	16	4	1	21
m30	Hole in the Ground	Sooter, J./Taylor, M.	25.0	0	0	52	47	0	0	20	1	1	22
m31	Pratt Mountain	Bailey, M.&M.	25.0	0	0	38	62	0	0	16	5	4	25
m32	Mail Ridge Road	Bailey, M.&M.	25.0	0	0	77	23	0	0	22	12	11	45
m33	Panther Peak	Leskiw, T.	25.0	0	0	38	62	0	0	20	0	1	21
m34	Chervimov Opening	Leskiw, T.	10.1	0	0	21	79	0	0	18	2	0	20
n03	Catsup Creek	Hewitt, R.	5.6	0	0	0	96	4	0	15	8	9	32
n04	Cedar Camp Creek	Hewitt, R.	25.0	0	0	0	97	3	0	22	8	2	32
N05	Slate Creek	Hunter, J.	25.0	0	0	0	100	0	0	10	18	14	42
n06	Shelton Butte Lookout	Slauson, K.	25.0	0	0	0	100	0	0	29	16	2	47
N07	Hopkins Butte	Hazard, G.	25.0	0	0	0	99	1	0	19	14	30	63
n08	Bunch Grass Ridge	Higley, M.	25.0	0	0	0	100	0	0	25	15	21	61
N09	Marshall Crossing Creek	Brown, H.	25.0	0	0	0	100	0	0	26	11	29	66
n10	Big Hill	Higley, M.	25.0	0	0	1	99	0	0	37	12	12	61
N11	Tish Tang a Tang Creek	Power, J.	25.0	0	0	13	85	0	2	30	20	24	74
n12	Horse Linto	Power, J.	25.0	1	0	7	92	0	0	14	5	8	27
N13	Willow Creek	Leskiw, T.	23.0	4	0	15	81	0	0	32	16	42	90

BLOCK NUMBER	BLOCK NAME	BLOCK OWNER	TOTAL AREA (KM^2)	%AG/ URBAN	%RED-WOOD	%OAK/ GRASS	%DOU-GLAS-FIR	%HIGH ELEVATION	%WATER	SPECIES PER BLOCK			
										PO	PR	CO	TOTAL
n14	Friday Flat	Leskiw, T.	20.1	1	0	13	84	0	3	27	17	14	58
N15	Oak Knob	Leskiw, T.	24.9	0	0	2	98	0	0	21	11	24	56
n16	Hogback Ridge	Fix, D./Power, J.	25.0	0	0	2	80	18	0	26	14	7	47
N17	Big Lake	Schmidt, G.	25.0	0	0	1	75	24	0	24	10	23	57
n18	Mosquito Creek	Schmidt, G.	25.0	0	0	11	86	3	0	33	4	3	40
N19	Last Chance Ridge	Schmidt, G.	25.0	0	0	0	21	79	0	23	10	25	58
n20	Whiting Ridge	Schmidt, G.	25.0	0	0	8	37	55	0	37	8	7	52
N21	Holm Ridge	Schmidt, G.	25.0	0	0	27	47	25	0	23	9	21	53
n22	Anderson Ford	Schmidt, G.	25.0	0	0	36	62	1	1	26	2	0	28
N23	Dinsmore	Shreeve, M.&J.	25.0	0	0	29	69	0	2	24	14	37	75
n24	Burr Valley	Shreeve, M.&J.	25.0	0	0	24	44	33	0	27	10	18	55
N25	Little Van Duzen River	Fix, D.	25.0	0	0	25	71	3	1	20	11	21	52
n26	Charles Mountain	Hunter, J.	25.0	0	0	22	49	29	0	32	2	0	34
N27	Mountain Creek	Hunter, J.	25.0	0	0	57	34	9	0	19	8	20	47
n28	Blocksburg	Hunter, J.	25.0	0	0	49	51	0	0	23	2	3	28
N29	Fort Seward	Adams, R. J.	25.0	0	0	41	57	0	1	19	8	21	48
n30	Alderpoint	Hewitt, R.	25.0	1	0	50	48	0	1	33	7	1	41
N31	Cain Rock	Bailey, M.&M.	25.0	0	0	42	58	0	0	18	20	31	69
n32	Jewett Valley	Bailey, M.&M.	25.0	0	0	56	44	0	0	20	18	19	57
N33	Island Mountain Road	Secor, C.&E.	25.0	0	0	58	42	0	0	22	21	33	76
n34	Bell Springs Road	Secor, C.&E.	10.2	0	0	33	67	0	0	28	3	3	34
o03	G-O Road	Slauson, K.	5.7	0	0	0	100	0	0	19	3	4	26
o04	Black Mountain	Hewitt, R.	25.0	0	0	0	100	0	0	35	10	0	45
o05	Orleans West	Hewitt, R.	25.0	4	0	0	96	0	0	43	16	20	79
o06	Sarvorum Mountain	Slauson, K.	25.0	0	0	0	100	0	0	19	21	10	50
o07	Red Cap Creek	Brown, H.	25.0	0	0	0	100	0	0	19	18	17	54
o08	Mill Creek Gap	Higley, M.	25.0	0	0	0	84	16	0	38	7	9	54
o09	Long Ridge	Hunter, J.	25.0	0	0	0	65	35	0	25	10	6	41
o10	Box Camp	Hunter, J.	25.0	0	0	0	43	57	0	33	14	6	53
o11	Sign Board Gap	Higley, M.	25.0	0	0	0	100	0	0	19	17	9	45
o12	Waterman Ridge North	Leskiw, T.	25.0	0	0	0	100	0	0	24	12	1	37
o13	Waterman Ridge South	Leskiw, T.	15.5	0	0	0	100	0	0	19	22	4	45

BLOCK NUMBER	BLOCK NAME	BLOCK OWNER	TOTAL AREA (KM^2)	%AG/ URBAN	%RED-WOOD	%OAK/ GRASS	%DOU-GLAS-FIR	%HIGH ELEVATION	%WATER	SPECIES PER BLOCK			
										PO	PR	CO	TOTAL
o14	Salyer	Leskiw, T.								31	10	7	48
o15	Hennessey Ridge	Hunter, J.	7.2	0	0	6	85	0	9	25	20	19	64
o16	Todd Ranch	Hunter, J.	13.8	0	0	12	85	0	4	25	19	33	77
o17	Gainor Peak	Hunter, J.	17.5	0	0	11	84	3	2	28	3	2	33
o18	Devastation Slide	Hunter, J.	20.7	0	0	4	94	1	1	40	16	7	63
o19	Lake Mountain	Schmidt, G.	20.5	0	0	1	57	42	0	36	13	4	53
o20	Blake Mountain	Schmidt, G.	20.4	0	0	0	4	96	0	32	9	8	49
o21	South Fork Mountain	Schmidt, G.	20.1	0	0	12	41	48	0	26	10	12	48
o22	Eightmile Ridge	Schmidt, G.	20.0	0	0	22	51	27	0	23	10	4	37
o23	Mad River Ridge	Hunter, J.	20.0	0	0	47	45	5	2	41	4	11	56
o24	Buck Mountain	Hunter, J.	19.8	0	0	9	31	58	2	46	7	7	60
o25	Browns Canyon	Hunter, J.	19.5	0	0	17	16	67	0	18	4	1	23
o26	Black Lassic	Hunter, J.	19.1	0	0	0	30	70	0	30	9	7	46
o27	Red Lassic	Hunter, J.	18.8	0	0	1	12	87	0	35	8	5	48
o28	Pine Mountain	Hunter, J.	18.6	0	0	23	42	35	0	42	2	4	48
o29	Dobbyn Creek	Hunter, J.	18.6	0	0	55	45	0	0	33	9	2	44
o30	Zenia Bluffs	Hunter, J.	18.6	0	0	57	42	0	0	16	4	0	20
o31	Tichnor Creek	Hunter, J.	18.4	0	0	80	20	0	0	20	2	3	25
o32	Lauffer Ranch	Hunter, J.	18.2	0	0	80	20	0	0	27	6	5	38
o33	Chamise Creek	Secor, C.&E.	18.1	0	0	85	15	0	0	12	6	5	23
o34	Island Mountain	Hunter, J.	7.4	0	0	74	26	0	0	21	4	4	29
p03	Somes Bar	Hewitt, R.	4.5	0	0	0	100	0	0	28	8	11	47
p04	Ishi Pishi Road	Hewitt, R.	24.7	0	0	0	98	2	0	27	5	4	36
P05	Orleans East	Slauson, K.	25.0	0	0	0	78	22	0	29	19	23	71
p06	LePerron Flat	Slauson, K.	25.0	0	0	0	83	17	0	31	8	8	47
P07	LePerron Peak	Leskiw, T.	25.0	0	0	0	96	4	0	16	12	16	44
p08	Red Cap Mine	Widdowson, W.&M.	25.0	0	0	0	48	52	0	22	11	7	40
P09	North Trinity Mountain	Hunter, J.	25.0	0	0	0	0	100	0	25	10	27	62
p10	Bear Hole	Hunter, J.	25.0	0	0	0	3	97	0	29	3	3	35
P11	Tish Tang Trail	Bentler, K.&E.	25.0	0	0	0	45	55	0	16	10	19	45
p12	Lone Pine Ridge	Leskiw, T.	25.0	0	0	0	57	42	0	15	27	4	46
P13	Groves Prairie	Leskiw, T.	23.1	0	0	0	52	48	0	17	18	23	58

BLOCK NUMBER	BLOCK NAME	BLOCK OWNER	TOTAL AREA (KM^2)	%AG/ URBAN	%RED-WOOD	%OAK/ GRASS	%DOU-GLAS-FIR	%HIGH ELEVATION	%WATER	SPECIES PER BLOCK			
										PO	PR	CO	TOTAL
p14	Hawkins Creek	Leskiw, T.	1.3	0	0	0	16	84	0	19	25	11	55
q04	Monte Creek	Hunter, J.	0.3	0	0	0	92	8	0	27	2	1	30
q05	Somes Mountain	Slauson, K.	6.9	0	0	0	11	89	0	30	5	2	37
q06	Orleans Mountain	Slauson, K.	13.2	0	0	0	20	80	0	33	0	1	34
q07	Whiteys Peak	Hunter, J.	18.1	0	0	0	15	85	0	32	11	7	50
q08	Devils Backbone North	Hunter, J./Hazard, G.	22.9	0	0	0	25	75	0	26	6	4	36
q09	Devils Backbone South	Hunter, J.	19.2	0	0	0	18	82	0	14	7	2	23
q10	Trinity Summit	Hazard, G.	11.1	0	0	1	0	99	0	23	7	8	38
q11	Grizzly Camp	Hunter, J.	23.7	0	0	0	1	99	0	20	6	4	30
q12	East Fork Horse Linto	Wallace, T.	19.5	0	0	1	18	82	0	29	13	4	46
q13	Onion Camp	Leskiw, T.	7.5	0	0	0	14	86	0	16	16	5	37
r08	Salmon Summit	Hazard, G./Hunter, J.	1.5	0	0	0	0	100	0	25	10	14	49
r11	Trinity Mountain	Hunter, J.	1.5	0	0	0	0	100	0	19	4	1	24
									TOTALS:	9449	3884	4624	17957

APPENDIX C: SCIENTIFIC NAMES OF PLANTS

Scientific names follow Hickman (1993).

COMMON NAME	SCIENTIFIC NAME
TREES	
Alder, mountain	*Alnus incana* ssp. *tenuifolia*
Alder, red	*Alnus rubra*
Alder, white	*Alnus rhombifolia*
Ash, Oregon	*Fraxinus latifolia*
Aspen	*Populus* spp.
Aspen, quaking	*Populus tremuloides*
Bay, California	*Umbellularia californica*
Birch	*Betula* spp.
Buckeye, California	*Aesculus californica*
Cherry, bitter	*Prunus emarginata*
Chinquapin, giant	*Chrysolepis chrysophylla* var. *chrysophylla*
Cottonwood, black	*Populus balsamifera* ssp. *trichocarpa*
Cottonwood, Fremont	*Populus fremontii*
Cypress, Monterey	*Cupressus macrocarpa*
Dogwood, Pacific	*Cornus nuttallii*
Douglas-fir	*Pseudotsuga menziesii* var. *menziesii*
Eucalyptus	*Eucalyptus globulus*
Fir	*Abies* spp.
Fir, grand	*Abies grandis*
Fir, Shasta red	*Abies magnifica* var. *shastensis*
Fir, white	*Abies concolor*
Hawthorn	*Crataegus douglasii*
Hemlock	*Tsuga* spp.
Hemlock, western	*Tsuga heterophylla*
Incense-cedar	*Calocedrus decurrens*
Juniper	*Juniperus* spp.
Larch	*Larix* spp.
Larch, western	*Larix occidentalis*
Madrone, Pacific	*Arbutus menziesii*
Maple, bigleaf	*Acer macrophyllum*
Mountain-mahogany, birch-leaf	*Cercocarpus betuloides* var. *betuloides*
Oak, California black	*Quercus kelloggii*
Oak, canyon live	*Quercus chrysolepis*
Oak, interior live	*Quercus wislizenii*
Oak, Oregon white	*Quercus garryana*
Oak, valley	*Quercus lobata*
Palm, Canary Island date	*Phoenix canariensis*
Palm, fan	*Washingtonia* spp.
Pine	*Pinus* spp.
Pine, gray	*Pinus sabiniana*
Pine, Jeffrey	*Pinus jeffreyi*
Pine, knobcone	*Pinus attenuata*

COMMON NAME	SCIENTIFIC NAME
Pine, Monterey	*Pinus radiata*
Pine, ponderosa	*Pinus ponderosa*
Pine, shore	*Pinus contorta* ssp. *contorta*
Pine, sugar	*Pinus lambertiana*
Pine, western white	*Pinus monticola*
Port Orford cedar	*Chamaecyparis lawsoniana*
Red cedar, western	*Thuja plicata*
Redwood, coast	*Sequoia sempervirens*
Spruce	*Picea* spp.
Spruce, Brewer	*Picea breweriana*
Spruce, Sitka	*Picea sitchensis*
Sycamore	*Platanus* spp.
Tanoak	*Lithocarpus densiflorus* var. *densiflorus*
Willow	*Salix* spp.
SHRUBS	
Blackberry, Himalayan	*Rubus discolor*
Blue blossom	*Ceanothus thyrsiflorus*
Buck brush	*Ceanothus cuneatus* var. *cuneatus*
Cascara	*Rhamnus purshiana*
Chinquapin, bush	*Chrysolepis sempervirens*
Coyote brush	*Baccharis pilularis*
Currant	*Ribes* spp.
Currant, red-flowering	*Ribes sanguinuem* var. *glutinosum*
Deer brush	*Ceanothus integerrimus*
Diogenes' lantern	*Calochortus amabilis*
Dogwood, red-osier	*Cornus sericea*
Elderberry, red	*Sambucus racemosa* var. *racemosa*
Gooseberry, canyon	*Ribes menziesii*
Gooseberry, Sierra	*Ribes roezlii* var. *cruentum*
Huckleberry, evergreen	*Vaccinium ovatum*
Huckleberry, red	*Vaccinium parvifolium*
Lupine	*Lupinus* spp.
Lupine, yellow bush	*Lupinus arboreus*
Manzanita	*Arctostaphylos* spp.
Manzanita, green-leaf	*Arctostaphylos patula*
Marijuana	*Cannabis sativa*
Ninebark	*Physocarpus capitatus*
Oak, Brewer's	*Quercus garryana* var. *breweri*
Oak, huckleberry	*Quercus vaccinifolium*
Oceanspray	*Holodiscus discolor*
Oregon-grape	*Berberis aquifolium*
Plum, Indian	*Oemleria cerasiformis*
Poison-oak	*Toxicodendron diversilobum*
Redbud	*Cercis occidentalis*
Rhododendron, Pacific	*Rhododendron macrophyllum*
Rose	*Rosa* spp.

COMMON NAME	SCIENTIFIC NAME
Sagebrush	*Artemisia* spp.
Salal	*Gaultheria shallon*
Salmonberry	*Rubus spectabilis*
Silk tassel	*Garrya* spp.
Snowberry	*Symphoricarpus* spp.
Snowbrush	*Ceanothus velutinus*
Spiraea	*Spiraea* spp.
Thimbleberry	*Rubus parviflorus*
Toyon	*Heteromeles arbutifolia*
Wax myrtle	*Myrica californica*
Whitethorn, coast	*Ceanothus incanus*
FORBS AND GRASSES	
Beachgrass, European	*Ammophila arenaria*
Bulrush	*Scirpus acutus*
Carrot, wild	*Daucus carota*
Cattail, broad-leaved	*Typha latifolia*
Clover	*Trifolium* spp.
Cone-flower	*Rudbeckia* spp.
Cordgrass, dense-flowered	*Spartina densiflora*
Dogtail, hedgehog	*Cynosurus echinatus*
Eelgrass	*Zostera marina*
Fennel	*Foeniculum vulgare*
Fern, chain	*Woodwardia fimbriata*
Fern, sword	*Polystichum munitum*
Fiddleneck	*Amsinckia* spp.
Grass, pampas	*Cortaderia jubata*
Grass, sweet vernal	*Anthoxanthum oderatum*
Hemlock, poison	*Conium maculatum*
Horsetail	*Equisetum* spp.
Ice plant	*Carpobrotus* spp.
Lily, Corn	*Veratrum* spp.
Mustard	*Brassica* spp.
Nettle	*Urtica* spp.
Oatgrass, California	*Danthonia californica*
Pennywort, marsh	*Hydrocotyle ranunculoides*
Penstemon	*Penstemon* spp.
Pickleweed	*Salicornia virginica*
Pond-lily, yellow	*Nuphar luteum*
Radish	*Raphanus sativus*
Rush	*Juncus* spp.
Sedge	*Carex* spp.
Sunflower (seeds)	*Helianthus* spp.
Sweet-clover	*Melilotus alba*
Teasel, wild	*Dipsacus fullonum*
Thistle	*Cirsium* spp.
Thistle (seeds)	*Guizotia abyssinica*

APPENDIX D: NUMBER OF BLOCKS FOR EACH SPECIES, BY BREEDING CODE

NOTE: The possible breeding (PO) category has only a single code, "X". Therefore, the number of blocks for the X code is the same as the number presented in the Total PO column. PR = Probable Breeding; CO = Confirmed Breeding.

SPECIES	NUMBER OF BLOCKS	% OF TOTAL BLOCKS	CONFIRMED BREEDING CODES													PROBABLE BREEDING CODES								TOTAL	TOTAL	TOTAL
			NY	NE	FS	ON	FY	CF	PY	FL	UN	PE	NB	DD	CN	B	N	C	A	T	M	S	P	PO	PR	CO
Pied-billed Grebe	8	2							4									1					1	2	2	4
Fork-tailed Storm-Petrel	1	<1																						1		
Leach's Storm-Petrel	2	<1															1							1	1	
Brandt's Cormorant	8	2				4																		4		4
Double-crested Cormorant	7	2		1		3												1						2	1	4
Pelagic Cormorant	9	2	1			4																		4		5
American Bittern	8	2							1													2		5	2	1
Great Blue Heron	71	17	4			5		1		2							1						5	53	6	12
Great Egret	2	<1		1		1																				2
Snowy Egret	2	<1	1																					1		1
Cattle Egret	1	<1	1																							1
Green Heron	28	7				1				2					2		1		3				2	17	6	5
Black-crowned Night-Heron	14	3				2			1	3			1		1									6		8
Turkey Vulture	275	65	1	1		2				2							1						21	247	22	6
Canada Goose	20	5		1		2	1		7														3	6	3	11
Wood Duck	52	12		2		21			9		1								1				9	9	10	33
Gadwall	5	1		1					2														1	1	1	3
Mallard	59	14		1		1			23														18	16	18	25
Cinnamon Teal	9	2							4														4	1	4	4
Northern Shoveler	1	2							1																	1
Hooded Merganser	6	1		2																				4		2
Common Merganser	77	18				1			20								2						18	36	20	21
Osprey	120	28	7			41		10		1	3		1		3	1	1	2	1				6	43	11	66
White-tailed Kite	35	8				4		3		4					2		1	1	1				5	14	8	13
Bald Eagle	15	4	3			3																	1	8	1	6
Northern Harrier	31	7				5		1										1	1					23	2	6
Sharp-shinned Hawk	61	14	2			1		2		4							1		3				2	46	6	9
Cooper's Hawk	70	16	1			12	1	2		5								2	1	1			4	41	8	21
Northern Goshawk	20	5	1					1		1									4					13	4	3
Red-shouldered Hawk	103	24	5	2		39		2		3					1			1					6	44	7	52
Red-tailed Hawk	246	58	3			63	2	8		26								4	5				29	106	38	102
Golden Eagle	45	11				1	1	1		3	2							1					8	28	9	8
American Kestrel	77	18				2	1	5		2							1	4	3				12	47	20	10
Peregrine Falcon	20	5	3			5		1		1							4		1					5	5	10
Ruffed Grouse	62	15				1			21	1	1												1	37	1	24
Blue Grouse	106	25		1					13					1						4	1	9	2	75	16	15
Wild Turkey	29	7		1					9									1		1			3	14	5	10
Mountain Quail	178	42		2					55					2				2	1	4	3	7	17	85	34	59
California Quail	266	63		2					87								2		3		1	13	68	90	87	89
SPECIES	NUMBER OF BLOCKS	% OF TOTAL BLOCKS	NY	NE	FS	ON	FY	CF	PY	FL	UN	PE	NB	DD	CN	B	N	C	A	T	M	S	P	TOTAL PO	TOTAL PR	TOTAL CO
			CONFIRMED BREEDING CODES													PROBABLE BREEDING CODES										

SPECIES	NUMBER OF BLOCKS	% OF TOTAL BLOCKS	CONFIRMED BREEDING CODES													PROBABLE BREEDING CODES								TOTAL	TOTAL	TOTAL
			NY	NE	FS	ON	FY	CF	PY	FL	UN	PE	NB	DD	CN	B	N	C	A	T	M	S	P	PO	PR	CO
Virginia Rail	20	5							5													1	1	13	2	5
Sora	7	2																				2		5	2	
American Coot	8	2				1	1		1															5		3
Snowy Plover	13	3		6		1			4															2		11
Killdeer	65	15	1			1			8					13					6				4	32	10	23
Black Oystercatcher	6	1							2									1					1	2	2	2
American Avocet	1	<1		1																						1
Spotted Sandpiper	82	19		2					13					1			1	5	2				20	38	28	16
Western Gull	13	3	2	1		3				2							1							4	1	8
Common Murre	7	2	1			2											1							3	1	3
Pigeon Guillemot	8	2				3		1									2							2	2	4
Marbled Murrelet	41	10						2			3						26							10	26	5
Cassin's Auklet	1	<1																						1		
Rhinoceros Auklet	3	1								1													1	1	1	1
Tufted Puffin	1	<1				1																				1
Rock Dove	18	4	2			1									2		2						2	9	4	5
Band-tailed Pigeon	191	45		1		1				15					2		3	10		4		7	39	109	63	19
Mourning Dove	165	39		1						10								6		1	2	7	49	89	65	11
Barn Owl	41	10	1			6				2	1						1							30	1	10
Flammulated Owl	21	5	1							1								1						18	1	2
Western Screech-Owl	141	33	1	2		1				2								1		1	1	10	11	111	24	6
Great Horned Owl	134	32				5	1			3								1		1		4	13	106	19	9
Northern Pygmy-Owl	176	41	1							3									3	14		7	10	138	34	4
Spotted Owl	272	64	11			135	5			16							1					1	64	39	66	167
Barred Owl	38	9								2									1	4		6	7	18	18	2
Long-eared Owl	11	3																				1	1	9	2	
Short-eared Owl	2	<1																						2		
Northern Saw-whet Owl	111	26	1				1			1												4	4	100	8	3
Common Nighthawk	45	11	1											1			1	9	1			3	2	27	16	2
Common Poorwill	1	<1																				1			1	
Vaux's Swift	129	30	6			6	1			3					2		4	2	2			1	23	79	32	18
White-throated Swift	1	<1	1																							1
Anna's Hummingbird	148	35	2	2		1	3			7			1		4			9				1	4	114	14	20
Calliope Hummingbird	3	1																						3		
Allen's Hummingbird	240	56	2	2			4	1		10	1	1	3		10		2	9	2				6	187	19	34
Belted Kingfisher	66	16				2		12		4							1			1		1	10	35	13	18
Acorn Woodpecker	156	37	6			9	2	2		8							9		1	2		4	12	101	28	27
Red-breasted Sapsucker	137	32	12			5	8	3		11		1				1	1	2	5	1		2	10	75	22	40
Downy Woodpecker	122	29	5			2	5	4		15		1				1	2	1		3		1	7	75	15	32
Hairy Woodpecker	229	54	21			5	12	8		24							2	2		2		3	18	131	27	71
White-headed Woodpecker	19	4	1			2	2	1		1							1							11	1	7
Northern Flicker	325	76	10	1		15	8	3		17	1					2	6	3	4	15	1	7	16	216	54	55
Pileated Woodpecker	206	48	4			2	2			5	1					1		1	3	18		6	22	141	51	14
Olive-sided Flycatcher	184	43						1		8			1		2			2	4	13	3	17	3	130	42	12
Western Wood-Pewee	202	48	4	4		4	4	3		2		1			5		1		3	7	13	13	4	134	41	27
SPECIES	NUMBER OF BLOCKS	% OF TOTAL BLOCKS	NY	NE	FS	ON	FY	CF	PY	FL	UN	PE	NB	DD	CN	B	N	C	A	T	M	S	P	TOTAL PO	TOTAL PR	TOTAL CO
			CONFIRMED BREEDING CODES													PROBABLE BREEDING CODES										

SPECIES	NUMBER OF BLOCKS	% OF TOTAL BLOCKS	CONFIRMED BREEDING CODES													PROBABLE BREEDING CODES								TOTAL	TOTAL	TOTAL
			NY	NE	FS	ON	FY	CF	PY	FL	UN	PE	NB	DD	CN	B	N	C	A	T	M	S	P	PO	PR	CO
Willow Flycatcher	7	2					1															2		4	2	1
Hammond's Flycatcher	48	11					1	3											5	3	4	2		30	14	4
Dusky Flycatcher	68	16						2									2		1	13	4	2	1	43	23	2
Pacific-slope Flycatcher	351	83	5	5		6	13	8		21		2	3		6			4	8	31	35	29	9	166	116	69
Black Phoebe	127	30	9	1		9	10	5		20					2		1		1	2		2	10	55	16	56
Ash-throated Flycatcher	43	10				2		2					1		1		1			2		6	3	25	12	6
Western Kingbird	35	8	3			1	2	4		2			1						1	2		1	6	12	10	13
Cassin's Vireo	256	60				2	6	11		13			2		4			1	6	16	11	15	6	163	55	38
Hutton's Vireo	295	69	1	1		1	10	6		35	1		1		4				8	12	11	27	11	166	69	60
Warbling Vireo	287	68	3			3	9	8		12			1		2		1		7	28	16	28	8	161	88	38
Gray Jay	42	10					5			16			1										9	11	9	22
Steller's Jay	380	89	8	1	3	3	26	32		48	2		3		6			2	6	3	1		31	205	43	132
Western Scrub-Jay	184	43	2				10	15		20			1		3		1		3				18	111	22	51
American Crow	113	27	1	1		5	6	10		7	1		1		9				2	2			7	61	11	41
Common Raven	352	83	7			5	6	27		37		1			9		1	8	8	1			46	196	64	92
Horned Lark	6	1					2			1													1	2	1	3
Purple Martin	76	18	2			12				1							5		1	1	1	3	12	38	23	15
Tree Swallow	115	27	12	1	2	11	4			7			2				8	1	1	1	1		11	53	23	39
Violet-green Swallow	166	39	8		1	28	6			20			1		5		10	3		1	1		16	66	31	69
Northern Rough-winged Swallow	112	26	1		1	23	2	3		4			2		2		9	2	2	2			12	47	27	38
Cliff Swallow	128	30	8			42	1	1		9	2		8		3		4						5	45	9	74
Barn Swallow	163	38	24	4		37	5	1		14	2		2		4		5						7	58	12	93
Black-capped Chickadee	21	5				2	3	3		1			2										2	8	2	11
Mountain Chickadee	44	10	4				1	3		1					2				1	1		1	1	29	4	11
Chestnut-backed Chickadee	335	79	12			15	39	13		68			3		4			1		5	1	9	21	144	37	154
Oak Titmouse	7	2								1														6		1
Bushtit	150	35	3			3	10	5		26	1		1		6					3	3		22	67	28	55
Red-breasted Nuthatch	247	58	1			4	4	8		13			1						12	9	1	20	13	161	55	31
White-breasted Nuthatch	37	9	1			3	3	2															2	26	2	9
Brown Creeper	225	53	2	1		5	5	9		19			2		1		1	1	1	13	6	21	13	125	56	44
Rock Wren	8	2			1			2														1		4	1	3
Canyon Wren	3	1																				3			3	
Bewick's Wren	99	23				1	3	2		6			1				1		2	2		11	5	65	21	13
House Wren	74	17	4			7	3	5		2					1		2	1	1	2		4	2	40	12	22
Winter Wren	248	58	1		1	2	9	7		25		2				1			13	20	11	31	6	119	82	47
Marsh Wren	24	6	2	1			1	1		2						1			1		4	3		8	9	7
American Dipper	45	11	2			3		1		4	2				1		1		1	1			4	25	7	13
Golden-crowned Kinglet	241	57	2				4	4		18		1							2	7	17	23	11	152	60	29
Blue-gray Gnatcatcher	7	2	1					1															2	3	2	2
Western Bluebird	106	25	4	1	1	14	5	5		15			1				4	1		2		1	21	31	29	46
Mountain Bluebird	1	<1								1																1
Townsend's Solitaire	74	17				1	1			2							1		1	3	2	2	5	56	14	4
Swainson's Thrush	206	48					3	27		9	1	2			1		1		5	17	20	18	2	100	63	43
Hermit Thrush	266	63	1	1			4	11		15	1				3		1	1	7	31	44	27	2	117	113	36
American Robin	324	76	10	4		4	17	58		39	2	2	2		5				9	5	9	16	15	127	54	143
SPECIES	NUMBER OF BLOCKS	% OF TOTAL BLOCKS	NY	NE	FS	ON	FY	CF	PY	FL	UN	PE	NB	DD	CN	B	N	C	A	T	M	S	P	TOTAL PO	TOTAL PR	TOTAL CO
			CONFIRMED BREEDING CODES													PROBABLE BREEDING CODES										

SPECIES	NUMBER OF BLOCKS	% OF TOTAL BLOCKS	CONFIRMED BREEDING CODES													PROBABLE BREEDING CODES								TOTAL PO	TOTAL PR	TOTAL CO
			NY	NE	FS	ON	FY	CF	PY	FL	UN	PE	NB	DD	CN	B	N	C	A	T	M	S	P			
Varied Thrush	99	23		1		2	1	10		7	1	1			1				2	11	4	10	3	45	30	24
Wrentit	282	66	2		1		6	26		14		3			1			1	6	23	21	26	25	127	102	53
Northern Mockingbird	19	4					1															1		17	1	1
California Thrasher	9	2										1								2		1	1	4	4	1
European Starling	133	31	17	3	1	15	4	16		27			2		5		3					1	10	29	14	90
Cedar Waxwing	100	24	1			3	8	2		7			1		4			2					11	61	13	26
Orange-crowned Warbler	278	65		1			13	23		19		3			1		1	1	5	8	12	27	10	154	64	60
Nashville Warbler	121	28			1			8		5									3	16	9	7	1	71	36	14
Northern Parula	6	1																				1		5	1	
Yellow Warbler	66	16					2	1		1		1			1		1		1	4	3	6	9	36	24	6
Yellow-rumped Warbler	123	29	1			1	9	10		10			1		3			3	3	2	6	4	7	63	25	35
Black-throated Gray Warbler	222	52	1				14	12		14		1			2				2	8	5	19	5	139	39	44
Hermit Warbler	259	61	2				8	27		15					1		2		7	28	26	23	5	115	91	53
American Redstart	16	4		2						2					1							3	1	7	4	5
MacGillivray's Warbler	158	37			2		4	10		15		1			1		1	1	6	11	6	11	12	77	48	33
Common Yellowthroat	36	8		1			2	1		1										1		3	1	26	5	5
Wilson's Warbler	312	73	1	1		1	29	32		29		1		1				3	5	25	15	19	9	141	76	95
Yellow-breasted Chat	80	19					2	5		2					1			2		2	3	8	1	54	16	10
Western Tanager	284	67	2		1	3	13	14		7		1	1		9			1	5	16	22	11	19	159	74	51
Green-tailed Towhee	23	5						2		3									1	2	1		1	13	5	5
Spotted Towhee	248	58	2	2			11	14		26		1	1		3				2	7	4	12	17	146	42	60
California Towhee	111	26	1			2	4	6		9					1		1	1	5	1	3	4	19	54	34	23
Rufous-crowned Sparrow	3	1						2		1																3
Chipping Sparrow	111	26	2			1	3	9		10					3			1	1		7	2	8	64	19	28
Lark Sparrow	29	7		2			2	5		1										1	2	1		15	4	10
Savannah Sparrow	46	11	1		2		3	3		12										1	5	1	3	15	10	21
Grasshopper Sparrow	49	12					1	1		6					1					2	3	2	3	30	10	9
Fox Sparrow	37	9			1	1		2		1				1			1		2	4	3	2		19	12	6
Song Sparrow	273	64	5	4	2	2	22	34		38		1		1	3		1	1	8	5	15	14	14	103	58	112
Lincoln's Sparrow	8	2						2									1		1	2		1		1	5	2
White-crowned Sparrow	135	32	2	3		1	14	19		27					1		1	1	4	2	5	4	4	47	21	67
Dark-eyed Junco	331	78	10	4		8	23	22		68			1		3		1	4	12	11	5	20	37	102	90	139
Black-headed Grosbeak	290	68		1			23	14		27		2	1		2			3	2	20	7	18	15	155	65	70
Lazuli Bunting	208	49	2	1			10	21		6		1			1		3	4	2	6	23	19	15	94	72	42
Indigo Bunting	8	2																				2		6	2	
Red-winged Blackbird	56	13			1	2	9	3		6					1		2		3		2	1	5	21	13	22
Tricolored Blackbird	2	<1						1							1											2
Western Meadowlark	84	20				1		14		5									1	3	9	2	4	45	19	20
Brewer's Blackbird	120	28	4		5	1	9	25		8			1	1	5			2	4	1		1	18	35	26	59
Brown-headed Cowbird	121	28		2			10			24							1	8					21	55	30	36
Hooded Oriole	3	1								2														1		2
Bullock's Oriole	79	19	3		1	4	9	6		8	1		1		2					1		2	8	33	11	35
Purple Finch	193	45					9	2		17		1	3		2			1	1	10	2	17	18	110	49	34
Cassin's Finch	16	4						1												1			3	11	4	1
House Finch	89	21	1	1			13	2		15			1		2			1			1	1	11	40	14	35

SPECIES | NUMBER OF BLOCKS | % OF TOTAL BLOCKS | NY NE FS ON FY CF PY FL UN PE NB DD CN | CONFIRMED BREEDING CODES | B N C A T M S P | PROBABLE BREEDING CODES | TOTAL PO | TOTAL PR | TOTAL CO

SPECIES	NUMBER OF BLOCKS	% OF TOTAL BLOCKS	CONFIRMED BREEDING CODES													PROBABLE BREEDING CODES								TOTAL PO	TOTAL PR	TOTAL CO
			NY	NE	FS	ON	FY	CF	PY	FL	UN	PE	NB	DD	CN	B	N	C	A	T	M	S	P			
Red Crossbill	70	16					2			7													13	48	13	9
Pine Siskin	123	29	1				4			6							1	2		3	1	3	23	79	33	11
Lesser Goldfinch	142	33	2	1		2	13	5		12			1		3			1	1		1	1	38	61	42	39
Lawrence's Goldfinch	3	1								1													2		2	1
American Goldfinch	145	34	2	1	1	4	18	1		8		1			2			2	1	1	4	9	34	56	51	38
Evening Grosbeak	39	9					2			1										2			7	27	9	3
House Sparrow	54	13	7			9	4	4		7					7		2						2	12	4	38
TOTALS	425	-	346	94	29	755	639	800	288	1327	31	34	65	22	194	8	167	158	277	581	463	769	1461	9449	3884	4624
SPECIES	NUMBER OF BLOCKS	% OF TOTAL BLOCKS	NY	NE	FS	ON	FY	CF	PY	FL	UN	PE	NB	DD	CN	B	N	C	A	T	M	S	P	TOTAL PO	TOTAL PR	TOTAL CO
			CONFIRMED BREEDING CODES													PROBABLE BREEDING CODES										

LITERATURE CITED

AINLEY, D. G., AND M. C. WHITT. 1973. Numbers of marine birds breeding in northern California. Western Birds 4:65–70.

AINLEY, D. G., AND R. J. BOEKELHEIDE [EDS.]. 1990. Seabirds of the Farallon Islands: Ecology, dynamics, and structure of an upwelling-system community. Stanford University Press, Stanford, CA.

ALLEN, J. R. K. 2002. Edge effects on birds nesting in redwood forests. M.S. Thesis, Humboldt State University, Arcata, CA.

AMERICAN ORNITHOLOGISTS' UNION [AOU]. 1998. Check-list of North American Birds, seventh edition. American Ornithologists' Union, Washington, DC.

ANDRES, B. A., AND G. A. FALXA. 1995. Black Oystercatcher (*Haematopus bachmani*). *In* The Birds of North America. No. 155, A. Poole and F. Gill [EDS.], The Academy of Natural Sciences, Philadelphia, and The American Ornithologists' Union, Washington, DC.

ANONYMOUS. 1931. Turkeys do well. California Fish and Game 17:454–455.

ARNOLD, J. R. 1980. Distribution of the Mockingbird in California. Western Birds 11:97–102.

AYERS, D. 1975. Reproductive performance of the Double-crested Cormorant in Humboldt Bay, California. M.S. Thesis, Humboldt State University, Arcata, CA.

BARDOLF, J. M. 2001. Relative abundance and richness of forest birds in riparian and upland mixed-conifer forests of the Pilot Creek watershed. M.S. Thesis, Humboldt State University, Arcata, CA.

BARNHART, R. A., M. J. BOYD, AND J. E. PEQUEGNAT. 1992. The Ecology of Humboldt Bay, California: An Estuarine Profile. Biological Report 1. U.S. Fish and Wildlife Service, Washington, DC.

BARROWCLOUGH, G. F., AND R. J. GUTIÉRREZ. 1990. Genetic variation and differentiation in the Spotted Owl (*Strix occidentalis*). Auk 107:737–744.

BARROWCLOUGH, G. F., R. J. GUTIÉRREZ, AND J. G. GROTH. 1999. Phylogeography of Spotted Owls (*Strix occidentalis*) based on mitochondrial DNA sequences: Gene flow, genetic structure, and a novel biogeographic pattern. Evolution 53:919–931.

BASE, D. L. 1982. Composition, status, and changes of birds and other wildlife on a tidal marsh restoration site at Humboldt Bay. M.S. Thesis, Humboldt State University, Arcata, CA.

BEALS, J. W. 1994. Nest site characteristics of four species of Ardeids breeding at the Indian Island Rookery during the 1991 and 1992 breeding seasons. M.S. Thesis, Humboldt State University, Arcata, CA.

BECK, M. J., AND T. L. GEORGE. 2000. Song post and foraging site characteristics of breeding Varied Thrushes in northwestern California. Condor 102:93–103.

BEEDY, E. C., AND W. J. HAMILTON III. 1997. Tricolored Blackbird Status Update and Management Guidelines. Unpublished report on file, U.S. Fish and Wildlife Service, Sacramento, CA.

BEHLE, W. H. 1942. Distribution and variation of the Horned Larks (*Otocoris alpestris*) of western North America. University of California Publications in Zoology 46:205–316.

BENT, A. C. 1940. Life histories of North American cuckoos, goatsuckers, hummingbirds, and their allies. U.S. National Museum Bulletin 176.

BENTIVOGLIO, N., J. BALDWIN, P. G. R. JODICE, D. E. MACK, T. MAX, S. MILLER, S. K. NELSON, K. OSTROM, C. J. RALPH, M. RAPHAEL, C. STRONG, C. THOMPSON, AND R. WILK. 2002. Northwest Forest Plan Marbled Murrelet Effectiveness Monitoring: 2000 Annual Report. Unpublished report on file, U.S. Fish and Wildlife Service, Portland, OR.

BINFORD, L. C., AND R. W. STALLCUP. 1972. American Redstart breeding in California. California Birds 3:87–90.

BINFORD, L. C., B. G. ELLIOTT, AND S. W. SINGER. 1975. Discovery of a nest and the downy young of the Marbled Murrelet. Wilson Bulletin 87:303–319.

BISHOP, L. B. 1915. Description of a new race of Savannah Sparrow and suggestions on some California birds. Condor 17:185–189.

BLAKESLEY, J. A., A. B. FRANKLIN, AND R. J. GUTIÉRREZ. 1990. Sexual dimorphism in Northern Spotted Owls from northwest California. Journal of Field Ornithology 61:320–327.

BLAKESLEY, J. A., A. B. FRANKLIN, AND R. J. GUTIÉRREZ. 1992. Spotted Owl roost and nest site selection in northwestern California. Journal of Wildlife Management 56:388–392.

BLEITZ, D. 1958. Indigo Bunting breeding in Los Angeles County, California, Condor 60:408.

BLOCK, W. M. 1985. Habitat and morphological relationships of a guild of ground-foraging birds in northern California. M.S. Thesis, Humboldt State University, Arcata, CA.

BOLEYN, P. C. 1997. Pileated Woodpecker (*Dryocopus pileatus*) habitat use study on the Six Rivers National Forest, California. M.S. Thesis, Humboldt State University, Arcata, CA.

BRANDY, P. M. 2001. A hierarchical analysis of Olive-sided Flycatcher habitat in a managed landscape. M.S. Thesis, Humboldt State University, Arcata, CA.

BRAND, L. A. 1998. Edge effects in coast redwood forest fragments. M.S. Thesis, Humboldt State University, Arcata, CA.

BRENNAN, L. A. 1984. Summer habitat ecology of Mountain Quail in northern California. M.S. Thesis, Humboldt State University, Arcata, CA.

BRENNAN, L. A., W. M. BLOCK, AND R. J. GUTIÉRREZ. 1987. Habitat use by Mountain Quail in northern California. Condor 89:66–74.

BRENNAN, L. A., AND M. L. MORRISON. 1991. Long-term trends of chickadee populations in western North America. Condor 93:130–137.

BURKETT, E. E. 1995. Marbled Murrelet food habits and prey ecology, p. 223–246. *In* C. J. Ralph, G. L. Hunt, M. G. Raphael, and J. F. Piatt [EDS.], Ecology and conservation of the Marbled Murrelet. USDA Forest Service General Technical Report PSW-152.

BURRIDGE, B. [ED.]. 1995. Sonoma County Breeding Bird Atlas. Madrone Audubon Society, Santa Rosa, CA.

BURTON, T. S. 1972. Avian composition and utilization of Gunther Island, California. M.S. Thesis, Humboldt State University, Arcata, CA.

CADE, T. J., J. H. ENDERSON, C. G. THELANDER, AND C. M. WHITE. 1988. Peregrine Falcon Populations: Their Management and Recovery. The Peregrine Fund, Boise, ID.

CALIFORNIA DEPARTMENT OF FISH AND GAME [CDFG]. 1992. Recovery Plan: Bank Swallow (*Riparia riparia*). California Department of Fish and Game, Nongame Bird and Mammal Section Report 93.02. Nongame Bird and Mammal Section, Wildlife Management Division, Sacramento, CA.

CALIFORNIA DEPARTMENT OF FISH AND GAME [CDFG]. 1995. Report of the 1995 Game Take Hunter Survey. Upland Game/Waterfowl Program, Sacramento, CA.

CALIFORNIA DEPARTMENT OF FISH AND GAME [CDFG]. 1996. Report of the 1996 Game Take Hunter Survey. Upland Game/Waterfowl Program, Sacramento, CA.

CALIFORNIA DEPARTMENT OF FISH AND GAME [CDFG]. 1997. Report of the 1997 Game Take Hunter Survey. Upland Game/Waterfowl Program, Sacramento, CA.

CALIFORNIA DEPARTMENT OF FISH AND GAME [CDFG]. 1998. Report of the 1998 Game Take Hunter Survey. Wildlife Programs Branch, Sacramento, CA.

CALIFORNIA DEPARTMENT OF FISH AND GAME [CDFG]. 1999. Report of the 1999 Game Take Hunter Survey. Wildlife Programs Branch, Sacramento, CA.

CALIFORNIA DEPARTMENT OF WATER RESOURCES [CDWR]. 1965. Flood! December 1964–January 1965. Bulletin Number 161. California Department of Water Resources, Sacramento, CA.

CAMPBELL, R. W., M. A. PAUL, M. S. RODWAY, AND H. R. CARTER. 1977. Tree-nesting Peregrine Falcons in British Columbia. Condor 79:500–501.

CARLSON, P. C., W. S. LAHAYE, AND A. B. FRANKLIN. 1998. Incestuous behavior in Spotted Owls. Willson Bulletin 110:562–564.

CARTER, H. R., AND S. G. SEALY. 1987. Inland records of downy young and fledgling Marbled Murrelets in North America. Murrelet 68:58–63.

CARTER, K. 1998. A Guide to Finding Birds in Mendocino County. K. Carter, Albion, CA.

CHRISNEY, A. C. 1988. The abundance of avian predators and small mammals in three grassland habitats. M.S. Thesis, Humboldt State University, Arcata, CA.

CLAY, C. I. 1911. Some diving notes on cormorants. Condor 13:138.

CLAY, C. I. 1913. Artificial hatching of a Cassin Auklet. Condor 15:93.

CLAY, C. I. 1916. A new breeding record for California. Condor 18:205.

CLAY, C. I. 1917. Notes on the California Jay in Humboldt County. Condor 19:25.

CLAY, C. I. 1926. Unusual visitors in Humboldt County, California. Condor 28:98.

CLAY, C. I. 1939. White-throated Sparrow coincidence and other notes. Condor 41:121.

COLWELL, M. A., T. DANUFSKY, R. L. MATHIS, AND S. W. HARRIS. 2001. Historical changes in the abundance and distribution of the American Avocet at the northern limit of its winter range. Western Birds 32:1–12.

COMPTON, L. V. 1931. An unusual date for the occurrence of the young of the California Quail. Condor 33:249.

COOPER, J. G., AND G. SUCKLEY. 1859. The natural history of Washington Territory, with much relating to Minnesota, Nebraska, Kansas, Oregon, and California. Bailliere Brothers, New York, NY.

DANUFSKY, T., AND M. A. COLWELL. 2003. Winter shorebird communities and tidal flat characteristics at Humboldt Bay, California. Condor 105:117–119.

DARK, S. J., R. J. GUTIÉRREZ, AND G. I. GOULD, JR. 1998. The Barred Owl (*Strix varia*) invasion of California. Auk 115:50–56.

DAVIS, J. M. 1914. Cedar Waxwing nesting in Humboldt County, California. Condor 16:182.

DAVIS, J. M. 1918. Long waits for sets of Winter Wren. Condor 20:190.

DAVIS, J. M. 1922. Nesting of the California Evening Grosbeak. Condor 24:136.

DAVIS, J. M. 1924. Notes from Eureka, California. Condor 26:105.

DAVIS, J. M. 1933. New bird records for California. Condor 35:119.

DAVIS, J. M. 1938. Nesting dates from the Humboldt Bay region. Condor 40:182–183.

DAVIS, J. M. 1939. More shorebirds from the Humboldt Bay region. Condor 41:124.

DAVIS, J. M. 1940. Random notes from the Humboldt Bay area. Condor 42:222.

DAWSON, W. L. 1923. The Birds of California. Volumes 1–4. South Moulton Company, San Francisco, CA.

DEBENEDICTIS, P. A. 1995. Red Crossbills, one through eight. Birding 27:494–501.

DEVLIN-CRAIG, B. 1992. Species composition, distribution and seasonal abundance of waterfowl and raptors on the Humboldt Lagoon system, Humboldt County, California, 8 May 1990 to 8 May 1991. M.S. Thesis, Humboldt State University, Arcata, CA.

DESANTE, D. F., AND T. L. GEORGE. 1994. Population trends in the landbirds of western North America. Studies in Avian Biology 15:173–190.

DILLER, L. V., AND D. M. THOME. 1999. Population density of Northern Spotted Owls in managed young-growth forests in coastal northern California. Journal of Raptor Research 33:275–286.

DOLLAHITE, N., AND M. P. ANDERSON, JR. 1952. A new nesting record of the Cinnamon Teal in Humboldt County, California. Condor 54:320.

DREWIEN, R. C., R. J. VERNIMEN, S. W. HARRIS, AND C. F. YOCOM. 1966. Spring weights of Band-tailed Pigeons. Journal of Wildlife Management 30:190–192.

DUNK, J. R., AND R. J. COOPER. 1994. Territory-size regulation in Black-shouldered Kites. Auk 111:588–595.

DUNN, J., AND K. GARRETT. 1997. Warblers. Houghton Mifflin Company, Boston, MA.

EHRLICH, P. R., D. S. DOBKIN, AND D. WHEYE. 1988. The Birder's Handbook. Simon and Schuster/Fireside, New York, NY.

ELFORD, C. R., AND M. R. MCDONOUGH. 1964. The Climate of Humboldt and Del Norte Counties. Humboldt & Del Norte Counties Agricultural Extension Service, University of California, Eureka, CA.

EMLEN, J. T., JR. 1940. The midwinter distribution of the Crow in California. Condor 42:287–294.

EVANS, T. J. 1988. Habitat use and behavioral ecology of American Avocets (*Recurvirostra americana*) wintering at Humboldt Bay, California. M.S. Thesis, Humboldt State University, Arcata, CA.

EVANS, T. J., AND S. W. HARRIS. 1994. Status and habitat use by American Avocets wintering at Humboldt Bay, California. Condor 96:178–189.

FERRY, J. F. 1908. Notes from the diary of a naturalist in northern California. Condor 10:30–44.

FISHER, W. K. 1901. Breeding of *Hesperocichla naevia* in California. Condor 3:91.

FISHER, W. K. 1902a. The redwood belt of northwestern California: I. Faunal peculiarities of the region. Condor 4:111–114.

FISHER, W. K. 1902b. The redwood belt of northwestern California: II. Land birds. Condor 4:131–135.

FISHER, W. K. 1904. Notes on the birds of Hoopa Valley, California. Condor 6:50–51.

FOLLIARD, L. B. 1993. Nest site characteristics of Northern Spotted Owls in managed forests of northwest California. M.S. Thesis, University of Idaho, Moscow, ID.

FOLLIARD, L. B., K. P. REESE, AND L. V. DILLER. 2000. Landscape characteristics of Northern Spotted Owl nest sites in managed forests of northwestern California. Journal of Raptor Research 34:75–84.

FORD, R. G., G. K. H. BOOR, B. E. SHARP, AND J. L. CASEY. 2002. Estimates of bird impacts resulting from the *M/V Kure*/Humboldt Bay oil spill of November 5, 1997. Unpublished report on file, California Department of Fish and Game, Office of Spill Prevention and Response, Sacramento, CA.

FOREMAN, L. D. 1975. Status, habitat use and behavior of the Common Merganser in northwestern California. M.S. Thesis, Humboldt State University, Arcata, CA.

FOREMAN, L. D. 1976a. Nest site and activity of an incubating Common Merganser in northwestern California. California Fish and Game 62:87–88.

FOREMAN, L. D. 1976b. Observations of Common Merganser broods in northwestern California. California Fish and Game 62:207–212.

FRANKLIN, A. B., J. P. WARD, R. J. GUTIÉRREZ, AND G. I. GOULD, JR. 1990. Density of Northern Spotted Owls in northwest California. Journal of Wildlife Management 54:1–10.

FRANKLIN, A. B., D. R. ANDERSON, E. D. FORSMAN, K. P. BURNHAM, AND F. W. WAGNER. 1996a. Methods for collecting and analyzing demographic data on the Northern Spotted Owl. Studies in Avian Biology 17:12–20.

FRANKLIN, A. B., R. J. GUTIÉRREZ, B. R. NOON, AND J. P. WARD, JR. 1996b. Demographic characteristics and trends of Northern Spotted Owl populations in northwestern California. Studies in Avian Biology 17:83–91.

FRANKLIN, A. B., D. R. ANDERSON, R. J. GUTIÉRREZ, AND K. P. BURNHAM. 2000. Climate, habitat quality, and fitness in Northern Spotted Owl populations in northwestern California. Ecological Monographs 70:539–590.

FRASER, J. T. 1926. Late nesting of Cedar Waxwing. Condor 28:173.

FRASER, J. T. 1931. Late nesting of the Black Phoebe. Condor 33:34.

FRENCH, J. M. 1972. Distribution, abundance and breeding status of Ospreys in northwestern California. M.S. Thesis, Humboldt State Universtiy, Arcata, CA.

FRENCH, J. M., AND J. R. KOPLIN. 1977. Distribution, abundance, and breeding status of Ospreys in northwestern California, p. 223–240. *In* Ogden, J. C. [ED.]. Transactions of the North American Osprey Research Conference, Williamsburg, VI. National Park Service Transactions and Proceedings Series Number Two.

GABRIELSON, I. N., AND S. G. JEWETT. 1940. Birds of Oregon. Oregon State College, Corvallis, OR.

GALLAGHER, S. R. [ED.]. 1997. Atlas of Breeding Birds: Orange County, California. Sea and Sage Audubon Press, Irvine, CA.

GERSTENBERG, R. H. 1972. A study of shorebirds (Charadrii) in Humboldt Bay, California, 1968 to 1969. M.S. Thesis, Humboldt State University, Arcata, CA.

GILBERT, C. H. 1916. Nature of the occurrence of the Rose-breasted Grosbeak in Humboldt County, California. Condor 18:81.

GILL, R., JR. 1979. Status and distribution of the California Clapper Rail (*Rallus longirostris obsoletus*). California Fish and Game 65:36–49.

GILL, R., JR., AND L. R. MEWALDT. 1983. Pacific Coast Caspian Terns: Dynamics of an expanding population. Auk 100:369–381.

GILLIGAN, J., M. SMITH, D. ROGERS, AND A. CONTRERAS. [EDS.]. 1994. Birds of Oregon: Status and Distribution. Cinclus Publications, McMinnville, OR.

GLOVER, F. A. 1953a. A nesting study of the Band-tailed Pigeon in northwestern California. California Fish and Game 39:397–407.

GLOVER, F. A. 1953b. Record of Starling in Humboldt County, California. Condor 55:219.

GOSS, N. S. 1878. Breeding of the Duck Hawk in trees. Bulletin of the Nuttall Ornithology Club 3:32–34.

GRINNELL, J. 1915. A distributional list of the birds of California. Pacific Coast Avifauna 11:1–217.

GRINNELL, J. 1926. The evidence as to the former breeding of the Rhinoceros Auklet in California. Condor 28:37–40.

GRINNELL, J. 1931. The Streaked Horned Lark breeds in northwestern California. Condor 33:74–75.

GRINNELL, J. 1933. The vulture's fair-way. Condor 35:164–166.

GRINNELL, J., AND A. H. MILLER. 1944. The distribution of the birds of California. Pacific Coast Avifauna 27:1–615.

GROSZ, T. L., AND M. DOLE. 1967. A record of the Cattle Egret in Humboldt County, California. Condor 69:608–609.

GROSZ, R. 1992. Ecology of small marshes using three different water sources. M.S. Thesis, Humboldt State University, Arcata, CA.

GROTH, J. G. 1988. Resolution of cryptic species in Appalachian Red Crossbills. Condor 90:745–760.

GROTH, J. G. 1993. Evolutionary differentiation in morphology, vocalizations, and allozymes among nomadic sibling species in the North American Red Crossbill (*Loxia curvirostra*) complex. University of California Publications in Zoology 127:1–143.

GUTIÉRREZ, R. J. 1989. Hematozoa from the Spotted Owl. Journal of Wildlife Diseases 25:614–618.

GUTIÉRREZ, R. J., A. B. FRANKLIN, W. LAHAYE, V. J. MERETSKY, AND J. P. WARD. 1985. Juvenile spotted owl dispersal in northwestern California: Preliminary results, p. 60–65. *In* R. J. Gutiérrez and A. B. Carey. [EDS.], Ecology and Management of the Spotted Owl in the Pacific Northwest. USDA Forest Service General Technical Report PNW-185.

GUTIÉRREZ, R. J., J. E. HUNTER, G. CHÁVEZ-LÉON, AND J. PRICE. 1998. Characteristics of Spotted Owl habitat in landscapes disturbed by timber harvest in northwestern California. Journal of Raptor Research 32:104–110.

HAGAR, D. C. 1960. The interrelationships of logging, birds and timber regeneration in the Douglas-fir region of northwestern California. Ecology 41:116–125.

HALL, P. A. 1984. Characterization of nesting habitat of Goshawks (*Accipiter gentilis*) in northwestern California. M.S. Thesis, Humboldt State University, Arcata, CA.

HAMER, T. E., AND S. K. NELSON. 1995a. Nesting chronology of the Marbled Murrelet, p. 46–56. *In* C. J. Ralph, G. L. Hunt, M. G. Raphael, and J. F. Piatt [EDS.], Ecology and Conservation of the Marbled Murrelet. USDA Forest Service General Technical Report PSW-152.

HAMER, T. E., AND S. K. NELSON. 1995b. Characteristics of Marbled Murrelet nest trees and nesting stands, p. 69–82. *In* C. J. Ralph, G. L. Hunt, M. G. Raphael, and J. F. Piatt [EDS.], Ecology and Conservation of the Marbled Murrelet. USDA Forest Service General Technical Report PSW-152.

HAMM, K. A. 1995. Abundance of dusky-footed woodrats in managed forests of north coastal California. M.S. Thesis, Humboldt State University, Arcata, CA.

HARRIS, S. W. 1974. Status, chronology, and ecology of nesting storm petrels in northwestern California. Condor 76:249–261.

HARRIS, S. W. 1991. Northwestern California Birds, first edition. Humboldt State University Press, Arcata, CA.

HARRIS, S. W. 1996. Northwestern California Birds, second edition. Humboldt State University Press, Arcata, CA.

HARRIS, S. W., AND R. J. WHEELER. 1965. Hybrid of Blue-winged Teal × Cinnamon Teal in northwestern California. Condor 67:539–540.

HARRIS, S. W., AND C. F. YOCOM. 1967. A record of the Cattle Egret in Humboldt County, California. Condor 69:608.

HAZARD, G. C., AND T. L. GEORGE. 1999. Landbird abundance and diversity in different-aged stands of coast redwood forests in northwestern California. Northwestern Naturalist 80:99–109.

HAZARD, G. C. 2002. Landbird abundance in second-growth and old-growth coast redwood forests in northwestern California. M.S. Thesis, Humboldt State University, Arcata, CA.

HEAD, L. 1983. Canadian Geese for California's Northcoast: A Propagation Plan. Unpublished report on file, Northcoast Waterfowlers Association, Eureka, CA.

HELBING, G. L. 1977. Maintenance activities of the Black Oystercatcher, *Haematopus bachmanii* Audubon, in northwestern California. M.S. Thesis, Humboldt State University, Arcata, CA.

HERMAN, S. G. 1971. The Peregrine Falcon decline in California: 2. The breeding status in 1970. American Birds 25:818–820.

HICKMAN, J. C. [ED.]. 1993. The Jepson Manual: Higher Plants of California. University of California Press, Berkeley, CA.

HIGLEY, J. M. 1989. Waterbird and raptor use of the Arcata Marsh and Wildlife Sanctuary, Humboldt County, California, 1984–1986. M.S. Thesis, Humboldt State University, Arcata, CA.

HILL, R. L. 1977. Avian composition of coastal dune habitat on the Samoa Peninsula, California. M.S. Thesis, Humboldt State University. Arcata, CA.

HOFFMANN, R. 1927. Birds of the Pacific States. Houghton Mifflin Company, Boston, MA.

HOUSTON, D. B. 1963. A contribution to the ecology of the Band-tailed Pigeon *Columba fasciata* Say. M.A. Thesis, University of Wyoming, Laramie, WY.

HUEY, L. H. 1960. Notes on Vaux and Chimney Swifts. Condor 62:483.

HUGHES, J. M. 1999. Yellow-billed Cuckoo (*Coccyzus americanus*). *In* The Birds of North America, No. 418. A. Poole and F. Gill [EDS.], The Academy of Natural Sciences, Philadelphia, and The American Ornithologists' Union, Washington, DC.

HUNTER, J. E., C. JEWETT, AND A. PADILLA. 1993. Kleptoparasitism of Northern Spotted Owls. Northwestern Naturalist 74:28–29.

HUNTER, J. E., R. J. GUTIÉRREZ, A. B. FRANKLIN, AND D. OLSON. 1994. Ectoparasites of the Spotted Owl. Journal of Raptor Research 28:232–235.

HUNTER, J. E., R. J GUTIÉRREZ, AND A. B. FRANKLIN. 1995. Habitat configuration around Spotted Owl sites in northwestern California. Condor 97:684–693.

HUNTER, J. E. AND R. LEVALLEY. 1996. Improving the reliability of Marbled Murrelet surveys in low abundance areas. Pacific Seabirds 23(1):3–4.

HUNTER, J. E., AND G. C. HAZARD. 1998. Birds of the Mad River Ranger District, Six Rivers National Forest, California. Western Birds 29:65–82.

HUNTER, J. E., K. N. SCHMIDT, H. B. STAUFFER, S. L. MILLER, C. J. RALPH, AND L. ROBERTS. 1998. Status of the Marbled Murrelet in the Inner North Coast Ranges of California. Northwestern Naturalist 79:92–103.

HUNTER, J. E., AND M. L. BOND. 2001. Residual trees: Wildlife associations and recommendations. Wildlife Society Bulletin 29:995–999.

HUNTER, J. E., AND M. J. MAZUREK. 2003. Characteristics of trees used by nesting and roosting Vaux's Swifts in northwestern California. Western Birds 34:225–229.

HURT, M. C. 1996. The influence of forest fragmentation and vegetation structure on the breeding distribution of Varied Thrushes in redwood forests. M.S. Thesis, Humboldt State University, Arcata, CA.

IRWIN, K. 1985. Foraging ecology and reproduction of the Black Phoebe in Humboldt County. M.S. Thesis, Humboldt State University, Arcata, CA.

IVES, J. H. 1973a. Common Egret and Great Blue Heron nest study Indian Island, Humboldt County, California 1971–1972. Wildlife Management Branch Administrative Report Number 72–9. California Department of Fish and Game, Sacramento, CA.

IVES, J. H. 1973b. The breeding biology of the Common Egret on Humboldt Bay, California. M.S. Thesis, Humboldt State University, Arcata, CA.

JACOBSON, S. L. 1986. Vertebrate response to a tidal marsh restoration in Humboldt Bay, California. M.S. Thesis. Humboldt State University, Arcata, CA.

JIMERSON, T. M., E. A. MCGEE, D. W. JONES, R. J. SVILICH, E. HOTALEN, G. DENITTO, T. LAURENT, J. D. TENPAS, M. E. SMITH, K. HEFNER-MCCLELLAND, AND J. MATTISON. 1996. A Field Guide to the Tanoak and Douglas-fir Plant Associations in Northwestern California. R5-ECOL-TP-009. U.S. Forest Service, San Francisco, CA.

JUREK, R. M. 1989. Five-year Status Report: American Peregrine Falcon. Nongame Bird and Mammal Section Report. California Department of Fish and Game, Sacramento, CA.

JUREK, R. M. 1990. California Bald Eagle Breeding Population Surveys and Trend, 1970–90. Nongame Bird and Mammal Section Report. California Department of Fish and Game, Sacramento, CA.

KELLY, J. P. 1987. Habitat relationships among riparian forest birds in the Eel River Delta, California. M.S. Thesis, Humboldt State University, Arcata, CA.

KIMSEY, J. B. 1955. The White-tailed Kite in Humboldt County, California. Condor 57:190.

KROEBER, A. L. 1925. Handbook of the Indians of California. Bureau of American Ethnology Bulletin 78.

KWASNY, D. C. 2000. Reproductive response of Savannah Sparrows (*Passerculus sandwichensis*) to livestock grazing at Humboldt Bay National Wildlife Refuge. M.S. Thesis, Humboldt State University, Arcata, CA.

LAHAYE, W. S., AND R. J. GUTIÉRREZ. 1999. Nest sites and nesting habitat of the Northern Spotted Owl in northwestern California. Condor 101:324–330.

LAUGHLIN, S. D., D. P. KIBBE, AND P. F. J. EAGLES. 1982. Atlasing the distribution of the breeding birds of North America. American Birds 35:6–19.

LEGG, K. 1959. Occurrences of the Mockingbird at the northwestern margin of its range. Condor 61:438.

LEOPOLD, A. S., R. J. GUTIÉRREZ, AND M. T. BRONSON. 1981. North American Game Birds and Mammals. Charles Scribner's Sons, New York, NY.

LESKIW, T., AND R. J. GUTIÉRREZ. 1998. Possible predation of a Spotted Owl by a Barred Owl. Western Birds 29:225–226.

LEVENSON, H. 1976. Behavior and energetics of Ospreys nesting in northern California. M.S. Thesis, Humboldt State University, Arcata, CA.

LEVENSON, H., AND J. R. KOPLIN. 1984. Effects of human activity on productivity of nesting Ospreys. Journal of Wildlife Management 48:1374–1377.

LIEBEZEIT, J. R., AND T. L. GEORGE. 2002. A Summary of Predation by Corvids on Threatened and Endangered Species in California and Management Recommendations to Reduce Corvid Predation. Species Conservation and Recovery Report 2002-02. California Department of Fish and Game, Sacramento, CA.

MAILLIARD, J. 1916. Some bird notes from Humboldt Bay. Condor 18:198–200.

MAILLIARD, J. 1922. Sonoma Thrasher in Humboldt County, California. Condor 24:62–63.

MANUWAL, D. A., H. R. CARTER, T. S. ZIMMERMAN, AND D. L. ORTHMEYER. [EDS]. 2001. Biology and Conservation of the Common Murre in California, Oregon, Washington, and British Columbia. Volume I: Natural History and Population Trends. Information and Technology Report 2000-0012. U.S. Geological Survey, Washington, DC.

MARCOT, B. G. 1978. Flora and fauna of existing and potential slump pond sites in Six Rivers National Forest with recommendations for management. M.S. Thesis, Humboldt State University, Arcata, CA.

MARCOT, B. G. 1983. Snag use by birds in Douglas-fir clearcuts, p. 134–139. *In* J. W. Davis, G. A. Godwin, and R. A. Ockenfels [EDS.], Snag Habitat Management: Proceedings of the Symposium. USDA Forest Service General Technical Report RM-99.

MARCOT, B. G. 1984. Habitat relationships of birds and young-growth Douglas-fir in northwestern California. Ph.D. dissertation, Oregon State University, Corvallis, OR.

MARCOT, B. G., AND R. HILL. 1980. Flammulated Owls in northwestern California. Western Birds 11:141–149.

MARSHALL, D. B., M. G. HUNTER, AND A. L. CONTRERAS [EDS.]. 2003. Birds of Oregon: A General Reference. Oregon State University Press, Corvallis, OR.

MCGINNIS, W. J., R. H. PHILLIPS, AND K. P. CONNAUGHTON. 1996. County Portraits of Oregon and Northern California. USDA Forest Service General Technical Report PNW-377.

MCLAIN, R. B. 1898. The Rose-breasted Grosbeak in California. Auk 15:190.

MICHENER, H. 1933. Minutes of Cooper Club Meetings: Southern Division. Condor 35:211–212.

MILLER, S. L., AND C. J. RALPH. 1995. Relationships of Marbled Murrelets with habitat characteristics at inland sites in California, p. 205–215. *In* C. J. Ralph, G. L. Hunt, M. G. Raphael, and J. F. Piatt [EDS.], Ecology and conservation of the Marbled Murrelet. USDA Forest Service General Technical Report PSW-152.

MILLER, S. L., C. B. MEYER, AND C. J. RALPH. 2002. Land and seascape patterns associated with Marbled Murrelet abundance offshore. Waterbirds 25:100–108.

MILLER, A. M. 1933. Distributional notes from the Northwest Coast District of California. Condor 35:36–37.

MOEN, C. A., A. B. FRANKLIN, AND R. J GUTIÉRREZ. 1991. Age determination of subadult Northern Spotted Owls in northwest California. Wildlife Society Bulletin 19:489–493.

NOSS, R. F. [ED.]. 2000. The Redwood Forest. Island Press, Washington, DC.

O'DELL, T. E. 1996. Silviculture in the redwood region: An historical perspective, p. 15–17. *In* J. LeBlanc [ED.]. Proceedings of the Conference on Coast Redwood Forest Ecology and Management. University of California Cooperative Extension, Berkeley, CA.

O'DONNELL, B. P. 1993. Patterns of activity and vocalizations of the Marbled Murrelet, *Brachyramphus marmoratus*, within old-growth redwood stands in northwestern California. M.S. Thesis, Humboldt State University, Arcata, CA.

OSBORNE, T. O. 1972. Ecology and avian use of the coastal rocks of northern California. M.A. Thesis, Humboldt State University, Arcata, CA.

PACIFIC SEABIRD GROUP. 1994. Methods for Surveying for Marbled Murrelets in Forests: A Protocol for Land Management and Research. Unpublished report on file, Pacific Seabird Group, Seattle, WA.

PAGE, G. W., AND L. E. STENZEL. 1981. The breeding status of the Snowy Plover in California. Western Birds 12:1–40.

PAGEL, J. 1990. Status Report on Peregrine Falcons from Northwest California to Central Oregon. Unpublished report on file, Rogue River National Forest, Medford, OR.

PATON, P. W. C., AND C. J. RALPH. 1990. Distribution of the Marbled Murrelet at inland sites in California. Northwestern Naturalist 71:72–84.

PATON, P. W. C., C. J. ZABEL, D. L NEAL, G. N. STEGER, N. G. TILGHMAN, AND B. R. NOON. 1991. Effects of radio tags on Spotted Owls. Journal of Wildlife Management 55:617–622.

PATTEN, M. A., AND C. A. MARANTZ. 1996. Implications of vagrant southeastern vireos and warblers in California. Auk 113:911–923.

POSTERNAK, L. D. 1987. Foraging behavior of the Snowy Egret in Humboldt Bay, California. M.S. Thesis, Humboldt State University, Arcata, CA.

RALPH, C. J., P. W. C. PATON, AND C. A. TAYLOR. 1991. Habitat association patterns of breeding birds and small mammals in Douglas-fir/hardwood stands in northwestern California and southwestern Oregon, p. 379–393. *In* L. F. Ruggiero, K. B. Aubry, A. B. Carey, and M. H. Huff [EDS.], Wildlife and Vegetation of Unmanaged Douglas-fir Forests. USDA Forest Service General Technical Report PNW-285.

RALPH, C. J., AND S. L. MILLER. 1995. Offshore population estimates of Marbled Murrelets in California, p. 353–360. *In* C. J. Ralph, G. L. Hunt, M. G. Raphael, and J. F. Piatt [EDS.], Ecology and Conservation of the Marbled Murrelet. USDA Forest Service General Technical Report PSW-152.

RALPH, C. J., H. F. SAKAI, AND K. HOLLINGER. 1996. Monitoring populations and demography of landbirds in riparian corridors of the coastal redwood zone, p. 124–126. *In* J. LeBlanc [ED.]. Proceedings of the Conference on Coast Redwood Forest Ecology and Management. University of California Cooperative Extension, Berkeley, CA.

RALPH, C. J., L. L. LONG, S. L. MILLER, AND T. SHAW. 2000. Abundance, Distribution, and Productivity of Marbled Murrelets along the Northern California Coast and Southern Oregon in 1999. Unpublished report on file, U.S. Forest Service, Redwood Sciences Lab, Arcata, CA.

RAPHAEL, M. G. 1984. Wildlife populations in relation to stand age and area in Douglas-fir forests of northwestern California, p. 259–274. *In* W. R. Meehan, T. R. Merrell, Jr., and T. A. Hanley [EDS.], Fish and Wildlife Relationships in Old-growth Forests. The American Institute of Fishery Research Biologists, Morehead City, NC.

RAPHAEL, M. G. 1987. Wildlife-tanoak associations in Douglas-fir forests in northwestern California, p. 183–189. *In* T. R. Plumb and N. H. Pillsbury [EDS.], Multiple Use Management of California's Hardwood Resources. USDA Forest Service General Technical Report PSW-100.

RAPHAEL, M. G., K. V. ROSENBERG, AND B. G. MARCOT. 1988. Large scale changes in bird populations of Douglas-fir forests, northwestern California, p. 63–83. *In* J. Jackson [ED.]. Bird Conservation 3. University of Wisconsin Press, Madison, WI.

REMSEN, J. V., JR., K. GARRETT, AND R. A. ERICKSON. 1982. Vocal copying in Lawrence's and Lesser Goldfinches. Western Birds 13:29–33.

RIDGWAY, R. 1895. Nesting of the Duck Hawk in trees. Nidologist 3:42–44.

ROBERSON, D. 1980. Rare Birds of the West Coast of North America. Woodcock Publications, Pacific Grove, CA.

ROBERSON, D., AND C. TENNEY. [EDS.]. 1993. Atlas of the Breeding Birds of Monterey County, California. Monterey Peninsula Audubon Society, Carmel, CA.

ROBERSON, D. 2002. Monterey Birds, second edition. Monterey Peninsula Audubon Society, Carmel, CA.

ROSENBERG, K. V., AND M. G. RAPHAEL. 1986. Effects of forest fragmentation on vertebrates in Douglas-fir forests, p. 263–272. *In* J. Verner, M. L. Morrison, C. J. Ralph [EDS.], Wildlife 2000: Modeling Habitat Relationships of Terrestrial Vertebrates. University of Wisconsin Press, Madison, WI.

ROWE, S. P., AND D. S. COOPER. 1997. Confirmed nesting of an Indigo Bunting with a Lazuli Bunting in Kern County, California. Western Birds 28:225–227.

SAKAI, H. F. 1987. Response of Hammond's and Western Flycatchers to different aged Douglas-fir stands in northwestern California. M.S. Thesis, Humboldt State University, Arcata, CA.

SAKAI, H. F. 1988. Breeding biology and behavior of Hammond's and Western Flycatchers in northwestern California. Western Birds 19:49–60.

SAKAI, H. F., AND B. R. NOON. 1991. Nest-site characteristics of Hammond's and Pacific-slope Flycatchers in northwestern California. Condor 93:563–574.

SAKAI, H. F., AND B. R. NOON. 1993. Dusky-footed woodrat abundance in different-aged forests in northwestern California. Journal of Wildlife Management 57:373–382.

SCHLAFMANN, D. L. 1992. Use of five habitats by waterbird groups at the Arcata Marsh and Wildlife Sanctuary, Humboldt County, California, 1 July 1988–30 June 1989. M.S. Thesis, Humboldt State University, Arcata, CA.

SCHLORFF, R. W. 1978. Predatory ecology of the Great Egret at Humboldt Bay, California. M.S. Thesis, Humboldt State University, Arcata, CA.

SCHMIDT, K. N., J. E. HUNTER, H. B. STAUFFER, S. L. MILLER, C. J. RALPH, AND L. ROBERTS. 2000. Status and Distribution of the Marbled Murrelet in Interior Northwestern California: Final Report. Unpublished report on file, Six Rivers National Forest, Eureka, CA.

SFERRA, S. J. 1988. Bird use of the interior and edge of an isolated coniferous forest in Humboldt County, California. M.S. Thesis, Humboldt State University, Arcata, CA.

SHELTON, A. C. 1911. Nesting of the California Cuckoo. Condor 13:19–22.

SHUFORD, D. W. 1993. The Marin County Breeding Bird Atlas: A Distributional and Natural History of Coastal California Birds. Bushtit Books, Bolinas, CA.

SINGER, S. W., N. L. NASLUND, S. A. SINGER, AND C. J. RALPH. 1991. Discovery and observations of two tree nests of the Marbled Murrelet. Condor 93:330–339.

SMITH, F. J. 1916. Occurrence of the Condor in Humboldt County. Condor 18:205.

SNIADO, S. M. 1999. Bird habitat relationships in the Klamath Mountains of northwestern California. M.S. Thesis, Humboldt State University, Arcata, CA.

SOLIS, D. M., JR., AND R. J. GUTIÉRREZ. 1990. Summer habitat ecology of Northern Spotted Owls in northwestern California. Condor 92:739–748.

SOWLS, A. L., A. R. DEGRANGE, J. W. NELSON, AND G. S. LESTER. 1980. Catalog of California Seabird Colonies. FWS/OBS-80/37. U.S. Fish and Wildlife Service, Washington, DC.

SPITLER, J. 1985. Avian use of newly created wetlands, Arcata, California: April 1980 to April 1981. M.S. Thesis, Humboldt State University, Arcata, CA.

STERLING, J., AND P. W. C. PATON. 1996. Breeding distribution of Vaux's Swift in California. Western Birds 27:30–40.

STORER, T. I. 1915. Additional records of the California Clapper Rail and Red Phalarope in California. Condor 17:98.

STRONG, C. S., J. JACOBSEN, D. M. FIX, M. R. FISHER, R. LEVALLEY, C. STRIPLEN, W. R. MCIVER, AND I. GAFFNEY. 1997. Distribution, Abundance, and Reproductive Performance of Marbled Murrelets Along the Northern California Coast During the Summers of 1994 and 1995. Unpublished report on file, Crescent Coastal Research, Crescent City, CA.

TALMADGE, R. R. 1946. Visitants to Humboldt Bay, California. Condor 48:41.

TALMADGE, R. R. 1947. The Bank Swallow breeding in Humboldt County, California. Condor 49:38.

TALMADGE, R. R. 1948. The Cowbird moves northward in California. Condor 50:273–274.

TALMADGE, R. R. 1953a. Rufous Hummingbird nesting in California. Condor 55:155–156.

TALMADGE, R. R. 1953b. Range extensions in northern California. Condor 55:315–316.

TALMADGE, R. R. 1954. Turkey Vulture wintering in northern California. Condor 56:227–228.

TALMADGE, R. R. 1957. Nesting of the Ruffed Grouse in California. Condor 59:67–68.

TAYLOR, H. R. 1905. The nest and eggs of the Vaux Swift. Condor 7:177–179.

TERRILL, S., K. P. ABLE, AND M. A. PATTEN. 1992. The changing seasons: Summer 1992. American Birds 46:1109–1111, 1182.

THOME, D. M., C. J. ZABEL, AND L. V. DILLER. 1999. Forest stand characteristics and reproduction of Northern Spotted Owls in managed north-coastal California forests. Journal of Wildlife Management 63:44–59.

THOME, D. M., C. J. ZABEL, AND L. V. DILLER. 2000. Spotted Owl turnover and reproduction in managed forests of north-coastal California. Journal of Field Ornithology 71:140–146.

TING, T. 1998. The thermal environment of Northern Spotted Owls in northwestern California: possible explanations for use of interior old growth and coastal early successional stage forest. M.S. Thesis, Humboldt State University, Arcata, CA.

TOUSLEY, S. R. 1982. Diurnal water bird use of a large sewage pond. M.S. Thesis, Humboldt State University, Arcata, CA.

TOWNSEND, C. H. 1886. Four rare birds in northern California: Yellow Rail, Emperor Goose, European Widgeon, and Sabine's Ruffed Grouse. Auk 3:491.

TOWNSEND, C. H. 1887. Notes on California animals. Annual Proceedings of the U.S. National Museum 10:190–237.

TUTTLE, D. C., R. STEIN, AND G. LESTER. 1997. Snowy Plovers nesting on Eel River gravel bars, Humboldt County. Western Birds 28:174–176.

UDVARDY, M. D. F. 1981. An overview of grid-based atlas works in ornithology. Studies in Avian Biology 6:103–109.

UEOKA, M. L. 1974. Feeding behavior of Ospreys at Humboldt Bay, California. M.S. Thesis. Humboldt State University, Arcata, CA.

U.S. DEPARTMENT OF AGRICULTURE [USDA]. 1995. Land and Resource Management Plan: Six Rivers National Forest. Six Rivers National Forest, 1330 Bayshore Way, Eureka, CA.

U.S. DEPARTMENT OF AGRICULTURE [USDA]. 1997. CALVEG: A Classification of California Vegetation. U. S. Forest Service, Pacific Southwest Region, Regional Ecology Group, San Francisco, CA.

U.S. FISH AND WILDLIFE SERVICE [USFWS]. 1990. Endangered and threatened wildlife and plants; Determination of threatened status for the Northern Spotted Owl; Final rule. Federal Register 55:26114–26194.

U.S. FISH AND WILDLIFE SERVICE [USFWS]. 1991. Protocol for surveying proposed management activities that may impact Northern Spotted Owls. Revised March 1992. Unpublished report on file, U.S. Fish and Wildlife Service, Portland, OR.

U.S. FISH AND WILDLIFE SERVICE [USFWS]. 1992. Endangered and threatened wildlife and plants; Determination of threatened status for the Washington, Oregon, and California population of the Marbled Murrelet; Final rule. Federal Register 57:45328–45337.

U.S. FISH AND WILDLIFE SERVICE [USFWS]. 1993. Endangered and threatened wildlife and plants; Determination of threatened status for the Pacific Coast population of the Western Snowy Plover; Final rule. Federal Register 58:12864–12874.

U.S. FISH AND WILDLIFE SERVICE [USFWS]. 1995. Endangered and threatened wildlife and plants; Final rule determining endangered status for the Southwestern Willow Flycatcher. Federal Register 60:10694–10715.

U.S. FISH AND WILDLIFE SERVICE [USFWS]. 1999a. Endangered and threatened wildlife and plants; Proposed rule to remove the Bald Eagle in the lower 48 states from the list of endangered and threatened wildlife; Proposed rule. Federal Register 64:36454–36464.

U.S. FISH AND WILDLIFE SERVICE [USFWS]. 1999b. Endangered and threatened wildlife and plants; Final rule to remove the American Peregrine Falcon from the federal list of endangered and threatened wildlife, and to remove the similarity of appearance provision for free-flying peregrines in the conterminous United States; Final rule. Federal Register 64:46542–46558.

VAN HORN, D., G. MCDONALD, AND G. RAVENSFEATHER. 1982. Breeding populations of the Peregrine Falcon in southeast Alaska, p. 154. *In* W. N. Ladd and P. F. Schempf [EDS.], Raptor Biology and Management in Alaska and Western Canada. FWS/AK/PROC-82. U.S. Fish and Wildlife Service, Anchorage, AK.

VEIRS, S. D., JR. 1982. Coast Redwood Forest: Stand Dynamics, Successional Status and the Role of Fire, p. 119–141. *In* Proceedings of the Symposium on Forest Succession and Stand Development Research in the Northwest. Oregon State University, Corvallis, OR.

WAHL, T. R. 2001. Brandt's Cormorant sinks at sea. Western Birds 32:177–178.

WARD, J. P., JR., R. J. GUTIÉRREZ, AND B. R. NOON. 1998. Habitat selection by Northern Spotted Owls: The consequences of prey selection and distribution. Condor 100:79–92.

WHEELER, R. J. 1965. Pioneering of Blue-winged Teal in California, Oregon, Washington and British Columbia. Murrelet 46:40–42.

WHEELER, R. J., AND S. W. HARRIS. 1970. Duck nesting and production in the Humboldt Bay area of California. California Fish and Game 56:180–187.

WILBUR, S. R. 1963. A record of the Indigo Bunting in northwestern California. Condor 65:533–534.

WILDER, C. M. 1923. Mockingbird in Humboldt County, California. Condor 25:70.

WILDER, H. E. 1916a. Some distributional notes on California birds. Condor 18:127–128.

WILDER, H. E. 1916b. Humboldt County bird notes. Condor 18:204–205.

WILLIAMS, B. D. C. 1998. Distribution, habitat associations, and conservation of Purple Martins breeding in California. M.S. Thesis, California State University, Sacramento, CA.

WILLIAMS, L., K. LEGG, AND F. S. L. WILLIAMSON. 1958. Breeding of the Parula Warbler at Point Lobos, California. Condor 60:345–354.

WYCKOFF, S. S. 1983. The nesting of the Common Raven. M.A. Thesis, Humboldt State University, Arcata, CA.

YOCOM, C. F. 1957. Breeding record of Pintail in Humboldt County, California. Condor 59:340–341.

YOCOM, C. F. 1959. Migration of Starling into the Pacific coastal redwood region of northwestern California. Condor 61:301–302.

YOCOM, C. F. 1961. Shovelers nesting in Humboldt County, California. California Fish and Game 47:21–22.

YOCOM, C. F. 1968. Status of the Hermit Warbler in north western California. Murrelet 49:27–28.

YOCOM, C. F. 1970. Status of the Indigo Bunting in the Pacific states. Murrelet 51:40.

YOCOM, C. F. 1978. Status of the Oregon Ruffed Grouse in northwestern California. California Fish and Game 64:125–127.

YOCOM, C. F., AND S. W. HARRIS. 1975. Status, habitats and distribution of birds of northwestern California. Humboldt State University Bookstore, Arcata, CA.

YOUNG, K. E., A. B. FRANKLIN, AND J. P. WARD. 1993. Infestation of Northern Spotted Owls by hippoboscid (Diptera) flies in northwestern California. Journal of Wildlife Diseases 29:278–283.

YULL, P. F. 1972. Ecology of the Common Egret (*Casmerodes albus*) at Humboldt Bay, California. M.S. Thesis, Humboldt State University, Arcata, CA.

ZABEL, C. J., J. R. DUNK, H. B. STAUFFER, L. M. ROBERTS, B. S. MULDER, AND A. WRIGHT. 2003. Northern Spotted Owl habitat models for research and management applications in California (USA). Ecological Applications 13:1027–1040.

INDEX OF BIRD NAMES

ABOUT THE AUTHORS

John E. Hunter started working as a seasonal firefighter in 1978, and moved to Humboldt County in 1985 to accept a permanent position with the U.S. Forest Service. He received his B.S. in Wildlife Management in 1989, and his M.S. in Natural Resource Management in 1994, both from Humboldt State University. As a graduate student he studied Northern Spotted Owl landscape-level habitat associations, and has published a dozen bird-related technical or scientific papers. In 1998 John was hired by the U.S. Fish and Wildlife Service in Arcata, and currently works on endangered species issues. When not spending time with his wife Xiaoqing and daughter Libbie, John is an unabashed bird lister.

David Fix is a naturalist whose most enduring interest has been the relationship of birds to plant communities and landscapes. He gained experience with forest and mountain birds while employed by the U.S. Forest Service as a timber cruiser in Washington in the 1970s and as a silvicultural and biological technician in Oregon in the 1980s. Each spring for more than a decade, David has trained Marbled Murrelet forest surveyors at Prairie Creek Redwoods State Park through Mad River Biologists. He writes the "Bird of the Month" column for the *Sandpiper*, newsletter of Redwood Region Audubon Society, and is senior author of the regional field guide *Birds of Northern California*. A native of Oregon, David has lived in Humboldt County since 1992.

Gregory A. Schmidt was raised in a small town in northern Minnesota where he spent most of his time roaming the local woods and getting wet and cold in its many lakes and rivers. He received a B.A.S. in Life Science in 1986 and an M.S. in Biology in 1990, both from the University of Minnesota. Greg's focus was on mammals and he only became serious about bird watching when he moved to Humboldt County in 1990. He worked as a wildlife biologist for the Six Rivers National Forest for nine years where he studied Pacific fisher (*Martes pennanti pacifica*) ecology and developed his data management and GIS skills. Greg currently works as a wildlife biologist for the non-profit Institute for Wildlife Studies, which is involved in the conservation and recovery of threatened and endangered species. He currently resides in McKinleyville with his wife Kristin and their two sons, Sam and Carlo.

Jude Claire Power is the sole non-biologist among the authors. Nevertheless, she has enjoyed a lifelong interest in the natural world and is a leading woman birder in northwestern California. Jude is a subregional editor for *North American Birds*, past president of Redwood Region Audubon Society, and leads frequent birding field trips in Humboldt County. Her skill at interfacing with the community and providing background support to working groups, developed over the past 28 years as a school psychologist, served the Atlas effort well.